Applied Thermodynamics

for Engineering Technologists

By the same authors

T D Eastop and D R Croft *Energy efficiency*
T D Eastop and W E Watson *Mechanical services for buildings*

Applied Thermodynamics

For Engineering Technologists

Fifth Edition

T. D. EASTOP

B.Sc., Ph.D., C.Eng.,
F.I.Mech.E., F.C.I.B.S.E.,

*Formerly Head of the School
of Engineering at
Wolverhampton Polytechnic*

The late
A. McCONKEY

B.Sc., Ph.D., C.Eng.,
F.I.Mech.E.

*Formerly Head of the
Department of Mechanical and
Industrial Engineering at
Dundee College of Technology*

PEARSON
Prentice
Hall

Harlow, England • London • New York • Boston • San Francisco • Toronto
Sydney • Tokyo • Singapore • Hong Kong • Seoul • Taipei • New Delhi
Cape Town • Madrid • Mexico City • Amsterdam • Munich • Paris • Milan

Pearson Education Limited
Edinburgh Gate
Harlow
Essex CM20 2JE
England

and Associated Companies throughout the world

Visit us on the World Wide Web at:
http://www.pearsoned.co.uk

First published 1963
Second Edition 1969
Third Edition 1978
Fourth Edition 1986
Fifth Edition 1993

ISBN 0-582-09193-4

British Library Cataloguing in Publication Data

A CIP record for this book is available from the British Library

Library of Congress Cataloging-in-Publication Data

Eastop, T.D. (Thomas D.)
    Applied thermodynamics for engingeering technologists/T.D.
  Eastop, A. McConkey. -- 5th ed.
      p   cm.
    Includes bibliographical references and index.
    I. Thermodynamics.   I. McConkey, A. (Allan), 1927–
II. Title.
TJ265.E23   1993
621.402'1--dc20   92-38042
                                      CIP

Set by 6 in 10/12pt Monotype Lasercomp Times 569
Produced by Pearson Education Asia Pte Ltd
Printed in Singapore (COS)

15   14   13   12
07   06   05   04

Contents

Preface   xi
Acknowledgements   xiii
Nomenclature   xv

1   Introduction and the First Law of Thermodynamics   **1**

    1.1   Heat, work, and the system   2
    1.2   Units   6
    1.3   The state of the working fluid   9
    1.4   Reversibility   10
    1.5   Reversible work   11
    1.6   Conservation of energy and the First Law of Thermodynamics   15
    1.7   The non-flow equation   17
    1.8   The steady-flow equation   19
            Problems   23

2   The Working Fluid   **27**

    2.1   Liquid, vapour, and gas   27
    2.2   The use of vapour tables   30
    2.3   The perfect gas   39
            Problems   47

3   Reversible and Irreversible Processes   **51**

    3.1   Reversible non-flow processes   51
    3.2   Reversible adiabatic non-flow processes   59
    3.3   Polytropic processes   66
    3.4   Reversible flow processes   72
    3.5   Irreversible processes   73
    3.6   Nonsteady-flow processes   78
            Problems   84

4   The Second Law   **88**

    4.1   The heat engine   88

4.2	Entropy	90
4.3	The T–s diagram	93
4.4	Reversible processes on the T–s diagram	99
4.5	Entropy and irreversibility	109
4.6	Exergy	115
	Problems	121

5 The Heat Engine Cycle — **125**

5.1	The Carnot cycle	125
5.2	Absolute temperature scale	127
5.3	The Carnot cycle for a perfect gas	128
5.4	The constant pressure cycle	130
5.5	The air standard cycle	133
5.6	The Otto cycle	135
5.7	The diesel cycle	136
5.8	The dual-combustion cycle	138
5.9	Mean effective pressure	141
5.10	The Stirling and Ericsson cycles	143
	Problems	145

6 Mixtures — **147**

6.1	Dalton's law and the Gibbs–Dalton law	147
6.2	Volumetric analysis of a gas mixture	150
6.3	The molar mass and specific gas constant	151
6.4	Specific heat capacities of a gas mixture	157
6.5	Adiabatic mixing of perfect gases	162
6.6	Gas and vapour mixtures	166
6.7	The steam condenser	170
	Problems	173

7 Combustion — **176**

7.1	Basic chemistry	177
7.2	Fuels	178
7.3	Combustion equations	180
7.4	Stoichiometric air–fuel ratio	182
7.5	Exhaust and flue gas analysis	183
7.6	Practical analysis of combustion products	192
7.7	Dissociation	200
7.8	Internal energy and enthalpy of reaction	208
7.9	Enthalpy of formation	219
7.10	Calorific value of fuels	221
7.11	Power plant thermal efficiency	221
7.12	Practical determination of calorific values	223
7.13	Air and fuel-vapour mixtures	228
	Problems	230

8   Steam Cycles   **234**

8.1   The Rankine cycle   235
8.2   Rankine cycle with superheat   243
8.3   The enthalpy–entropy chart   245
8.4   The reheat cycle   246
8.5   The regenerative cycle   248
8.6   Further considerations of plant efficiency   253
8.7   Steam for heating and process use   255
   Problems   257

9   Gas Turbine Cycles   **260**

9.1   The practical gas turbine cycle   260
9.2   Modifications to the basic cycle   269
9.3   Combustion   281
9.4   Additional factors   283
   Problems   283

10   Nozzles and Jet Propulsion   **287**

10.1   Nozzle shape   287
10.2   Critical pressure ratio   289
10.3   Maximum mass flow   295
10.4   Nozzles off the design pressure ratio   298
10.5   Nozzle efficiency   300
10.6   The steam nozzle   304
10.7   Stagnation conditions   309
10.8   Jet propulsion   311
10.9   The turbojet   314
10.10   The turboprop   322
   Problems   325

11   Rotodynamic Machinery   **328**

11.1   Rotodynamic machines for steam and gas turbine plant   328
11.2   The impulse steam turbine   332
11.3   Pressure and velocity compounded impulse steam turbines   338
11.4   Axial-flow reaction turbines   346
11.5   Losses in turbines   358
11.6   Axial-flow compressors   360
11.7   Overall efficiency, stage efficiency, and reheat factor   363
11.8   Polytropic efficiency   368
11.9   Centrifugal compressors   372
11.10   Radial-flow turbines   375
   Problems   376

12   Positive Displacement Machines   **381**

12.1   Reciprocating compressors   382
12.2   Reciprocating compressors including clearance   388

12.3	Multi-stage compression	396
12.4	Steady-flow analysis	405
12.5	Rotary machines	406
12.6	Vacuum pumps	411
12.7	Air motors	412
	Problems	416

13   Reciprocating Internal-combustion Engines   419

13.1	Four-stroke cycle	421
13.2	Two-stroke cycle	424
13.3	Other types of engine	426
13.4	Criteria of performance	427
13.5	Engine output and efficiency	434
13.6	Performance characteristics	437
13.7	Factors influencing performance	442
13.8	Real cycles and the air standard cycle	447
13.9	Properties of fuels for IC engines	450
13.10	Fuel systems	452
13.11	Measurement of air and fuel flow rates	460
13.12	Supercharging	463
13.13	Engine emissions and legal requirements	470
13.14	Alternative forms of IC engines	475
13.15	Developments in IC engines	479
	Problems	481

14   Refrigeration and Heat Pumps   485

14.1	Reversed heat engine cycles	486
14.2	Vapour-compression cycles	491
14.3	Refrigerating load	499
14.4	The pressure–enthalpy diagram	501
14.5	Compressor type	503
14.6	The use of the flash chamber	507
14.7	Vapour-absorption cycles	511
14.8	Gas cycles	517
14.9	Liquefaction of gases	520
14.10	Steam-jet refrigeration	522
14.11	Refrigerants	522
14.12	Control of refrigerating capacity	528
	Problems	529

15   Psychrometry and Air-conditioning   533

15.1	Psychrometric mixtures	533
15.2	Specific humidity, relative humidity, and percentage saturation	534
15.3	Specific enthalpy, specific heat capacity, and specific volume of moist air	540
15.4	Air-conditioning systems	542

15.5   Cooling towers   553
        Problems   556

16   Heat Transfer   561

16.1   Fourier's law of conduction   562
16.2   Newton's law of cooling   565
16.3   The composite wall and the electrical analogy   568
16.4   Heat flow through a cylinder and a sphere   572
16.5   General conduction equation   577
16.6   Numerical methods for conduction   584
16.7   Two-dimensional steady conduction   587
16.8   One-dimensional transient conduction by finite difference   593
16.9   Forced convection   599
16.10   Natural convection   610
16.11   Heat exchangers   613
16.12   Heat exchanger effectiveness   623
16.13   Extended surfaces   627
16.14   Black-body radiation   633
16.15   The grey body   634
16.16   The Stefan–Boltzmann law637   637
16.17   Lambert's law and the geometric factor   639
16.18   Radiant interchange between grey bodies   643
16.19   Heat transfer coefficient for radiation   649
16.20   Gas radiation   650
16.21   Further study   651
        Problems   652

17   The Sources, Use, and Management of Energy   663

17.1   Sources of energy supply, and energy demands   664
17.2   Combined cycles   670
17.3   Combined heat and power (co-generation)   673
17.4   Energy management and energy audits   680
17.5   The technology of energy saving   688
17.6   Alternative energy sources   696
17.7   Nuclear power plant   699
        Problems   701

        Index   708

Preface to the Fifth Edition

This book aims to give students of engineering a thorough grounding in the subject of thermodynamics and the design of thermal plant. The book is comprehensive in its coverage without sacrificing the necessary theoretical rigour; the emphasis throughout is on the applications of the theory to real processes and plant. The objectives have remained unaltered through four previous editions and continuing interest in the book not only in the UK but also in most other countries in the English-speaking world has confirmed these objectives as suitable for students on a wide range of courses.

The book is designed as a complete course text for degree courses in mechanical, aeronautical, chemical, environmental, and energy engineering, engineering science, and combined studies courses in which thermodynamics and related topics are an important part of the curriculum. Students on technician diploma and certificate courses in engineering will also find the book suitable although the coverage is more extensive than they might require.

A number of lecturers in universities and polytechnics in the UK were asked for comments on the book before the fifth edition was prepared; the consensus was that the balance of the book was broadly correct with only minor changes needed, but a more modern format was thought to be desirable.

The fifth edition has therefore been completely recast in a new style which will make it more attractive, and easier to use. The opportunity has also been taken to rearrange the chapters in what seems to be a more logical order. Throughout the book the emphasis is now on the effective use of energy resources and the need to protect the environment. The chapter on energy sources, use and management (Ch. 17), has been improved and extended; it now includes a more extensive coverage of combined heat and power and a new section on energy recovery, including a brief mention of pinch technology. The material on gas turbines, steam turbines, nozzles, and propulsion (Chs 8–10) has been rewritten in a more logical format giving a more general treatment of blade design while still stressing the differences in design procedures for steam and gas turbines. In the chapter on refrigeration (Ch. 14) more emphasis is given to the heat pump and to vapour-absorption plant. A new section on refrigerants discusses the vitally important question of the thinning of the ozone layer due to CFCs; examples and problems in this chapter now use refrigerant 134a instead of refrigerant 12, and tables and a reduced scale chart for R134a are included by permission of ICI. Analysis of exhaust gases, emission control for IC engines, and the greenhouse effect are also included.

A new sign convention for energy transfer across a system boundary has come into general use in recent years and has therefore been introduced in this book. The convention is to treat both work and heat crossing a boundary as positive when it is transferred from the surroundings to the system. Also, there has been an international agreement to standardize symbols used for heat and mass transfer and the symbols in this text have been chosen accordingly. For example, the symbol for heat transfer coefficient is α, that for thermal conductivity λ, that for dynamic viscosity η, and that for thermal diffusivity κ. Molar quantities are now distinguished by the overscript, $\tilde{\ }$. Thanks are due to Dr Y.R. Mayhew for many helpful discussions on the use of physical quantities, units, and nomenclature. In the chapter on combustion (Ch. 7) the section on dissociation has been rewritten to conform with the use of a standard thermal equilibrium constant as tabulated in the latest edition of Rogers and Mayhew's *Thermodynamic and Transport Properties of Fluids*.

While preparing this new edition I have been ever conscious of the loss of my co-author and colleague for so many years, Allan McConkey, who died just after the publication of the previous edition in 1986. I would like to dedicate this edition to Allan with deep affection and gratitude for a long and fruitful collaboration.

TDE   1992

Acknowledgements

We are grateful to Blackwell Publishers for permission to include extracts from the Rogers and Mayhew *Thermodynamic and Transport Properties of Fluids* (*SI Units*) (4th ed.), 1988. Figure 15.4 is reproduced by permission of the Chartered Institution of Building Services Engineers, copies of the chart (size A3) for record purposes may be obtained from CIBSE, 222 Balham High Road, London SW12 9BS. Figure 14.14 is reproduced by permission of ICI; Table 14.1 is an extract with some interpolated values of thermodynamic properties of HFA 134a by permission of ICI, Runcorn, Cheshire.

The following sources have been drawn on for information: Figures 13.21 and 13.22 are adapted from *The Internal Combustion Engine in Theory and Practice* by C.F. Taylor, MIT Press. Section 13.13 includes material adapted from *Exhaust Emissions Handbook* published by Cussons Ltd. The data for Fig. 12.30 was provided by J.S. Milne of the Department of Mechanical and Industrial Engineering, Dundee College of Technology, from an original test carried out by him.

Nomenclature

A	air–fuel ratio; area
a	velocity of sound; acceleration; non-flow specific exergy
BDC	bottom dead centre
BS	British Standard
Bi	Biot number
b	steady-flow specific exergy
bmep	brake mean effective pressure
bp	brake power
C	velocity; constant; thermal capacity
CHP	combined heat and power
CI	compression ignition
COP	coefficient of performance
CV	calorific value
C_d	discharge coefficient
c	specific heat capacity
$\tilde{c}$	molar heat capacity
c_p	specific heat capacity at constant pressure
c_{pma}	specific heat capacity of air per unit mass of dry air
c_v	specific heat capacity at constant volume
$\tilde{c}_p$	molar heat capacity at constant pressure
$\tilde{c}_v$	molar heat capacity at constant volume
D, d	bore; diameter
E	emissive power; energy
e	eccentricity
F	force; geometric factor
FI	fuel injection
Fo	Fourier number
f	friction factor; frequency
fp	friction power
$\dot{G}$	irradiation
GCV	gross calorific value
Gr	Grashof number
g	gravitational acceleration

H	enthalpy; fundamental dimension of heat
HC	hydrocarbons
ΔH	enthalpy of reaction
h	specific enthalpy
Δh	specific enthalpy of reaction
$\tilde{h}$	molar enthalpy
$\Delta\tilde{h}$	molar enthalpy of reaction
h_f	specific enthalpy of a saturated liquid
h_fg	specific enthalpy of vaporization
h_g	specific enthalpy of a saturated vapour
I	electric current
IC	internal combustion
i	intensity of radiation
imep	indicated mean effective pressure
ip	indicated power
J	current density
$\dot{j}$	radiosity
j	Colburn factor for heat transfer
K	equilibrium constant
$K^{\ominus}$	standardized equilibrium constant
k	isentropic index for steam; blade velocity coefficient
L	stroke; fundamental dimension of length
l	length; characteristic linear dimension
M	fundamental dimension of mass
Ma	Mach number
m	mass
$\tilde{m}$	molar mass
$\dot{m}$	mass flow rate
N	rotational speed
NCV	net calorific value
NDIR	non-dispersive infra-red
Nu	Nusselt number
N_tu	number of transfer units
n	polytropic index; amount of substance; number of cylinders; nozzle arc length
ON	octane number
P	perimeter
PN	performance number
Pr	Prandtl number
p	absolute pressure; blade pitch
p_m	mean effective pressure
p_b	brake mean effective pressure
p_i	indicated mean effective pressure
Δp	pressure loss
Q	heat
$\dot{Q}$	rate of heat transfer
$\dot{q}$	rate of heat transfer per unit area
$\dot{q}_\mathrm{g}$	rate of heat transfer per unit volume

R	specific gas constant; thermal resistance; radius; ratio of thermal capacities
$\tilde{R}$	molar gas constant
RF	reheat factor
Re	Reynolds number
r	radius; expansion ratio
r_p	pressure ratio
r_v	compression ratio
S	entropy; steam consumption
SI	spark ignition
St	Stanton number
s	specific entropy
sfc	specific fuel consumption
T	absolute temperature; torque; fundamental dimension of time
TDC	top dead centre
t	temperature; fundamental dimension of temperature; blade thickness
Δt	temperature difference
$\Delta \bar{t}$	true mean temperature difference
$\Delta \bar{t}_a$	arithmetic mean temperature difference
$\Delta \bar{t}_{ln}$	logarithmic mean temperature difference
U	internal energy; overall heat transfer coefficient
ΔU	internal energy of reaction
u	specific internal energy
Δu	specific internal energy of reaction
$\tilde{u}$	molar internal energy
$\Delta \tilde{u}$	molar internal energy of reaction
V	volume
$\dot{V}$	rate of volume flow
v	specific volume
W	work; brake load
$\dot{W}$	rate of work transfer, power
X	temperature on any arbitrary scale
x	dryness fraction; nozzle pressure ratio; length
Z	height above a datum level
z	number of stages

Greek symbols

α	angle of absolute velocity; heat transfer coefficient; absorptivity for radiation
β	blade angle; coefficient of cubical expansion
γ	ratio of specific heats, c_p/c_v
δ	film thickness
Λ	degree of reaction
ε	emissivity; effectiveness of a heat exchanger

η	efficiency; dynamic viscosity
κ	thermal diffusivity
λ	thermal conductivity; wavelength
ν	kinematic viscosity
ρ	density; reflectivity
σ	Stefan–Boltzmann constant
τ	time; shear stress in a fluid; transmissivity
ϕ	relative humidity; angle
ω	specific humidity; solid angle
ψ	percentage saturation

Subscripts

AS	air standard
a	dry air; atmospheric; aircraft; absolute velocity
ai	absolute velocity at inlet
ae	absolute velocity at exit
B	black body
BT	brake thermal
b	blade velocity
C	cold fluid; compressor
c	condensate; convective; critical value; clearance
d	dew point; diagram
DB	dry bulb
e	exit; exhaust
F	fin; fluid
f	saturated liquid; fuel; film; flow velocity
fg	change of phase at constant pressure
g	saturated vapour; gases
gr	gross
H	hot fluid; high-pressure stage
hp	heat pump
I	intercooler
IT	indicated thermal
i	inlet; a constituent in a mixture; inside surface; intermediate; indicated; mesh point; injector
j	mesh point; jet
L	low-pressure stage
M	mechanical
m	mean
max	maximum
min	minimum
N	normal
net	net
0	stagnation condition; overall; outside; zero or reference condition

P	product of combustion
p	constant pressure
R	reactant
r	radiation; relative velocity
ref	refrigeration
re	relative velocity at exit
ri	relative velocity at inlet
s	isentropic
s	vapour; swept volume; stage; steam
T	throat; turbine; total
V	volumetric
v	constant volume
WB	wet bulb
w	water; wall; whirl
λ	monochromatic value at wavelength, λ
ϕ	radiation at angle, ϕ
∞_c	polytropic compression
∞_e	polytropic expansion

1

Introduction and the First Law of Thermodynamics

All living things depend on energy for survival, and modern civilizations will continue to thrive only if existing sources of energy can be developed to meet the growing demands. Energy exists in many forms, from the energy locked in the atoms of matter itself to the intense radiant energy emitted by the sun. Many sources of energy exist; many are known, some perhaps unknown; but when an energy source exists means must first be found to transform the energy into a form convenient to our purpose.

The chemical energy of combustion of fossil fuels (oil, coal, gas), and waste (agricultural, industrial, domestic), is used to produce heat which in turn is used to provide mechanical energy in turbines or reciprocating engines; uranium atoms are bombarded asunder and the nuclear energy released is used as heat; the potential energy of large masses of water is converted into electrical energy as it passes through water turbines on its way from the mountains to the sea; the kinetic energy of the wind is harnessed by windmills to produce electricity; the energy of the waves of the sea is converted into electrical power in floating turbines; the tides produced by the rotation of the moon produce electrical energy by flowing through turbines in large river estuaries; hot rocks and trapped liquids in the depths of the earth are made to release their energy to be converted to electricity; the immense radiant energy of the sun is tapped to heat water or by suitable device is converted directly into electricity. Figure 1.1 shows the various energy sources and the possible conversion paths with the more important transfers shown as bold lines; more information can be found in Chapter 4 of ref. 1.1 and the bibliography therein.

Applied thermodynamics is the science of the relationship between heat, work, and the properties of systems. It is concerned with the means necessary to convert heat energy from available sources such as fossil fuels into mechanical work. A *heat engine* is the name given to a system which by operating in a cyclic manner produces net work from a supply of heat. The laws of thermodynamics are natural hypotheses based on observations of the world in which we live. It is observed that heat and work are two mutually convertible forms of energy, and this is the basis of the First Law of Thermodynamics. It is also observed that heat never flows unaided from an object at a low temperature to one at a high temperature, in the same way that a river never

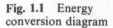

Fig. 1.1 Energy conversion diagram

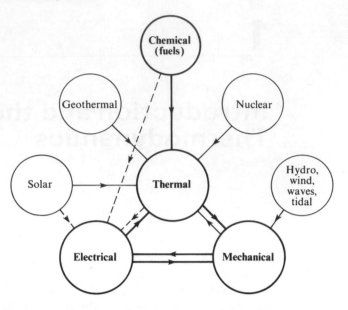

flows unaided uphill. This observation is the basis of the Second Law of Thermodynamics, which can be used to show that a heat engine cannot convert all the heat supplied to it into mechanical work but must always reject some heat at a lower temperature. These ideas will be discussed and developed in due course, but first some fundamental definitions must be made.

1.1 Heat, work, and the system

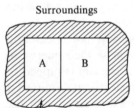

Fig. 1.2 Two isolated bodies in contact

In order to deal with the subject of applied thermodynamics rigorously it is necessary to define the concepts used.

Heat is a form of energy which is transferred from one body to another body at a lower temperature, by virtue of the temperature difference between the bodies.

For example, when a body A at a certain temperature, say 20 °C, is brought into contact with a body B at a higher temperature, say 21 °C, then there will be a transfer of heat from B to A until the temperatures of A and B are equal (Fig. 1.2). When the temperature of A is the same as the temperature of B no heat transfer takes place between the bodies, and they are said to be in *thermal equilibrium*. Heat is apparent during the process only and is therefore transitory energy. Since heat energy flows from B to A there is a reduction in the intrinsic energy possessed by B and an increase in the intrinsic energy possessed by A. This intrinsic energy of a body, which is a function of temperature at least, must not be confused with heat. Heat can never be contained in a body or possessed by a body.

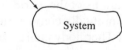

Fig. 1.3 Definition of a system

A *system* may be defined as a collection of matter within prescribed and identifiable boundaries (Fig. 1.3). The boundaries are not necessarily inflexible; for instance the fluid in the cylinder of a reciprocating engine during the expansion stroke may be defined as a system whose boundaries are the cylinder walls and the piston crown. As the piston moves so do the boundaries move (Fig. 1.4). This type of system is known as a *closed* system.

An *open* system is one in which there is a transfer of mass across the boundaries; for instance, the fluid in a turbine at any instant may be defined as an open system whose boundaries are as shown in Fig. 1.5.

Fig. 1.4 Fluid in a cylinder as a closed system

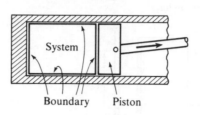

Fig. 1.5 Fluid in a turbine as an open system

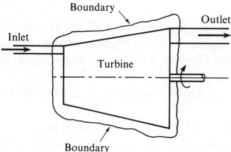

The *pressure* of a system is the force exerted by the system on unit area of its boundaries. Units of pressure are, for example, pascal, Pa (where $1 \text{ Pa} = 1 \text{ N/m}^2$), or bar; the symbol p will be used for pressure. Pressure as defined here is called *absolute pressure*. A gauge for measuring pressure (e.g. as shown in Fig. 1.6(a) and 1.6(b)), records the pressure above atmospheric. This is called *gauge pressure*, i.e. absolute pressure equals gauge pressure plus atmospheric pressure.

The gauge shown in Fig. 1.6(b) is called a Bourdon gauge. The absolute pressure of the system in a closed elliptical section tube forces the tube out of

Fig. 1.6 Two different pressure gauges

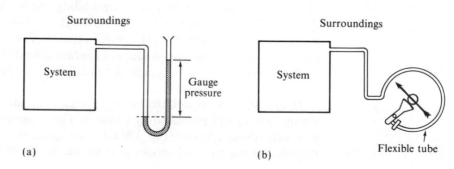

(a)

(b)

position against the pressure of the atmosphere. The tube's displacement is recorded by a pointer on a circular scale, which can be calibrated directly in bars.

When the pressure of a system is below atmospheric, it is called *vacuum pressure* (Fig. 1.7(a)).

When one side of a U-tube is completely evacuated and then sealed, the gauge will act as a *barometer* and the atmospheric pressure can be measured (Fig. 1.7(b)).

Fig. 1.7 Vacuum pressure and barometric pressure

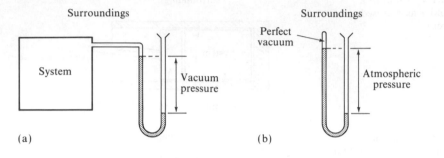

(a)   (b)

The gauges shown in Figs 1.6(a) and 1.7(a) measure gauge pressure in mm of a liquid of known relative density, and are called manometers.

For example, when water is the liquid, then

$$1 \text{ mm of water} = \frac{1}{10^3} \times 9806.65 \text{ N/m}^2 = 9.81 \text{ N/m}^2 = 9.81 \text{ Pa}$$

where 1 m³ of water weighs 9810 N, say.

Mercury (Hg) is very often used in gauges. Taking the relative density of mercury as 13.6, then

$$1 \text{ mm Hg} = \frac{1}{10^3} \times 13.6 \times 9810 \text{ N/m}^2 = 133.4 \text{ N/m}^2 = 133.4 \text{ Pa}$$

For a simple introduction to manometers and pressure measurement, see ref. 1.2.

The *specific volume* of a system is the volume occupied by unit mass of the system. The symbol used is v and the units are, for example, m³/kg. The symbol V will be used for volume. (Note that the specific volume is the reciprocal of density.)

Work is defined as the product of a force and the distance moved in the direction of the force. When a boundary of a closed system moves in the direction of the force acting on it, then the surroundings do work on the system. When the boundary is moved outwards the work is done by the system on its surroundings. The units of work are, for example, N m. If work is done on unit mass of a fluid, then the work done per kilogram of fluid has units of N m/kg.

Work is observed to be energy in transition. It is never contained in a body or possessed by a body.

Heat and work are both transitory energies and must not be confused with the intrinsic energy possessed by a system. For example, when a gas contained in a well-lagged cylinder (Fig. 1.8(a)) is compressed by moving the piston to the left, the pressure and temperature of the gas are observed to increase, and

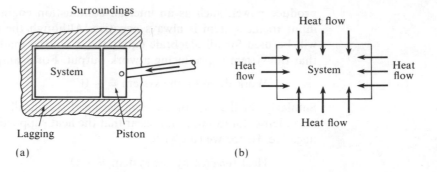

Fig. 1.8 Intrinsic energy increase by work or heat input

Surroundings

System

Lagging   Piston

(a)

Heat flow

Heat flow   System   Heat flow

Heat flow

(b)

hence the intrinsic energy of the gas increases. Since the cylinder is well lagged, no heat can flow into or out of the gas. The increase in intrinsic energy of the gas has therefore been caused by the work done by the piston on the gas.

As another example, consider a gas contained in a rigid container and heated (Fig. 1.8(b)). Since the boundaries of the system are rigidly fixed then no work is done on or by the system. The pressure and temperature of the gas are observed to increase, and hence the intrinsic energy of the gas will increase. The increase in intrinsic energy has been caused by the heat flow to the system.

In the example of Fig. 1.8(a) the work done on the system is energy which is apparent only during the actual process of compression. There is an intrinsic energy of the system initially and an intrinsic energy finally, but the work done appears only in transition from the initial to the final condition. Similarly, in the example of Fig. 1.8(b), the heat supplied appears only in transition from one state of the gas to another.

Another way in which work may be transferred to a system is illustrated in Fig. 1.9. The paddle wheel imparts a change of momentum to the fluid and a work input is required to turn the shaft. The kinetic energy attained by the fluid is dissipated by internal fluid friction, and friction between the fluid and the container. When the container is well lagged, all the work input goes to increasing the intrinsic energy of the system.

Paddle wheel

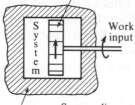

Work input

Lagging   Surroundings

Fig. 1.9 Paddle wheel work input

Convention

The sign convention used in this book assumes that all external inputs to a system are positive. That is

> Heat supplied to a system, Q, is positive.
> Work input to a system, W, is positive.

When a system boundary is drawn to define the system then it follows that heat supplied, Q, and work input, W, will always be shown by arrows pointing into the system. In algebraic equations it will be quite clear when numbers are substituted whether the value of Q and/or W is positive or negative; a negative value for Q will indicate that heat is rejected from the system; a negative value for W will indicate that work is done by the system on its surroundings.

In many cases it would cause unnecessary confusion by referring throughout to negative quantities; for example, it is clear that for a device designed to produce power, such as an internal combustion engine or turbine, the work input to the system is always negative. Although the above sign convention will be used for all algebraic equations it will be made clear in the wording that the system is producing a work output. For example

Work done *by* the system $= -W$

Similarly, for the case of a system designed specifically to cool a fluid, such as a condenser for example, it is clear that the heat supplied to the system is always negative. Hence we can write

Heat *rejected by* the system $= -Q$

1.2 Units

Throughout this book SI units will be used. The International System of Units (Système International d'Unités, abbreviation SI) was adopted by the General Conference of Weights and Measures in 1960 and subsequently endorsed by the International Organization for Standardization. It is a *coherent* system. In a coherent system all derived unit quantities are formed by the product or quotient of other unit quantities. In SI units six physical quantities are arbitrarily assigned unit value and hence all other physical quantities are derived from these. The six quantities chosen and their units are as follows: length (metre, m); mass (kilogram, kg); time (second, s); electric current (ampere, A); thermodynamic temperature (degree kelvin, K); luminous intensity (candela, cd).

Thus, for example, velocity = length/time has units of m/s; acceleration = velocity/time has units of m/s^2; volume = length $\times$ length $\times$ length has units of m^3; specific volume = volume/mass has units of m^3/kg.

Force, energy, and power

Newton's second law may be written as force $\propto$ mass $\times$ acceleration for a body of constant mass, i.e.

$$F = kma \tag{1.1}$$

where m is the mass of a body accelerated with an acceleration a, by a force F; k is a constant.

In a coherent system of units such as SI, $k = 1$, hence

$$F = ma$$

The SI unit of force is therefore $kg\,m/s^2$. This composite unit is called the *newton*, N, i.e. 1 N is the force required to give a mass of 1 kg an acceleration of $1\,m/s^2$.

It follows that the SI unit of work ($=$ force $\times$ distance) is the newton metre, N m. As stated earlier heat and work are both forms of energy, and hence both can have the units of $kg\,m^2/s^2$ or N m. A general unit for energy is introduced by giving the newton metre the name *joule*, J, i.e.   1 joule, J = 1 newton × 1 metre

or   1 J = 1 N m

The use of additional names for composite units is extended further by introducing the *watt*, W, as the unit of power,

i.e.   1 watt, W = 1 J/s = 1 N m/s

Pressure

The unit of pressure (force per unit area) is N/m^2 and this unit is sometimes called the *pascal*, Pa. For most cases occurring in thermodynamics the pressure expressed in pascals would be a very small number; a new unit is defined as follows:

$$1 \text{ bar} = 10^5 \text{ N/m}^2 = 10^5 \text{ Pa}$$

The advantage of using a unit such as the bar is that it is approximately equal to atmospheric pressure. In fact the standard atmospheric pressure is exactly 1.013 25 bar.

As indicated in section 1.1, it is often convenient to express a pressure as a head of a liquid. We have:

Standard atmospheric pressure = 1.013 25 bar = 0.76 m Hg

Temperature

The variation of an easily measurable property of a substance with temperature can be used to provide a temperature-measuring instrument. For example, the length of a column of mercury will vary with temperature due to the expansion and contraction of the mercury. The instrument can be calibrated by marking the length of the column when it is brought into thermal equilibrium with the vapour of boiling water at atmospheric pressure and again when it is in thermal equilibrium with ice at atmospheric pressure. On the Celsius (or Centigrade) scale 100 divisions are made between the two fixed points and the zero is taken at the ice point.

The change in volume at constant pressure, or the change in pressure at constant volume, of a fixed mass of gas which is not easily liquefied (e.g. oxygen, nitrogen, helium, etc.) can be used as a measure of temperature. Such an instrument is called a *gas thermometer*. It is found for all gases used in such thermometers that if the graph of temperature against volume in the constant pressure gas thermometer is extrapolated beyond the ice point to the point at which the volume of the gas would become zero, then the temperature at this point is $-273\,°C$ approximately (Fig. 1.10). Similarly if the graph of temperature against pressure in the constant volume gas thermometer is extrapolated to zero pressure, then the same zero of temperature is found. An absolute zero of temperature has therefore been fixed, and an absolute scale of temperature can be defined. Temperature on the absolute Celsius scale can be

Fig. 1.10 Graph of temperature against volume for a gas

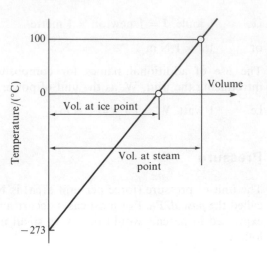

obtained by adding 273 to all temperatures on the Celsius scale; this scale is called the *Kelvin* scale. The unit of temperature is the degree kelvin and is given the symbol K, but since the Celsius scale which is used in practice has a different zero the temperature in degrees Celsius is given the symbol C (e.g. $20\,°C = 293$ K approximately; also, $30\,°C - 20\,°C = 10$ K). In this text capital T is used for absolute temperature and small t for other temperatures.

In Chapter 5 an absolute scale of temperature will be introduced as a direct consequence of the Second Law of Thermodynamics. It is found that the gas thermometer absolute scales approach the ideal scale as a limit. Also, with regard to the practical absolute temperature scale, there is an internationally agreed working scale which gives temperatures in terms of more practicable and more accurate instruments than the gas thermometer (see ref. 1.3).

Multiples and sub-multiples

Multiples and sub-multiples of the basic units are formed by means of prefixes, and the ones most commonly used are shown in the following table:

Multiplying factor	Prefix	Symbol
One million million	tera	T
One thousand million	giga	G
One million, 10^6	mega	M
One thousand, 10^3	kilo	k
One thousandth, 10^{-3}	milli	m
One millionth, 10^{-6}	micro	μ
One thousand millionth	nano	n
One million millionth	pico	p

For most purposes the multiplying factors shown in the above table are sufficient. For example, power can be expressed in either megawatts, MW, or kilowatts, kW, or watts, W. In the measurement of length the millimetre, mm, the metre, m, and the kilometre, km, are usually adequate. For areas, the difference in size between the square millimetre, mm^2, and the square metre, m^2, is large (a factor of 10^6), and an intermediate size is useful; the square centimetre, cm^2, is recommended for limited use only. For volumes, the difference between the cubic millimetre, mm^3, and the cubic metre, m^3, is much too great (a factor of 10^9), and the most commonly used intermediate unit is the cubic decimetre, dm^3, which is equal to one-thousandth of a cubic metre (i.e. 1 dm^3 = 10^{-3} m^3). The cubic decimetre can also be called the *litre*, l,

i.e.   1 litre, l = 1 dm^3 = 10^{-3} m^3

(Note, for very precise measurements, 1 litre = 1.000 028 dm^3.)

Certain exceptions to the general rule of multiplying factors are inevitable. The most obvious example is in the case of the unit of time. Instead of the centisecond, kilosecond, or megasecond, for instance, the minute, hour, day, etc. are used. Similarly, a mass flow rate may be expressed in kilograms per hour, kg/h, if this gives a more convenient number than when expressed in kilograms per second, kg/s. Also the speed of road vehicles is expressed in kilometres per hour, km/h, since this is more convenient than the normal unit of velocity which is metres per second, m/s.

1.3 The state of the working fluid

In all problems in applied thermodynamics we are concerned with energy transfers to or from a system. In practice the matter contained within the boundaries of the system can be liquid, vapour, or gas, and is known as the *working fluid*. At any instant the *state* of the working fluid may be defined by certain characteristics called its *properties*. Many properties have no significance in thermodynamics (e.g. electrical resistance), and will not be considered. The thermodynamic properties introduced in this book are pressure, temperature, specific volume, specific internal energy, specific enthalpy, and specific entropy. It has been found that, for any pure working fluid, only two independent properties are necessary to define completely the state of the fluid. Since any two independent properties suffice to define the state of a system, it is possible to represent the state of a system by a point situated on a diagram of properties. For example, a cylinder containing a certain fluid at pressure p_1 and specific volume v_1 is at state 1, defined by point 1 on a diagram of p against v (Fig. 1.11(a)). Since the state is defined, then the temperature of the fluid, T, is

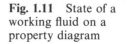

Fig. 1.11 State of a working fluid on a property diagram

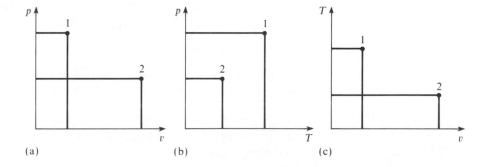

(a)   (b)   (c)

fixed and the state point can be located on a diagram of p against T and T against v (Figs 1.11(b) and 1.11(c)). At any other instant the piston may be moved in the cylinder such that the pressure and specific volume are changed to p_2 and v_2. State 2 can then be marked on the diagrams. Diagrams of properties are used continually in applied thermodynamics to plot state changes. The most important are the pressure–volume and temperature–entropy diagrams, but enthalpy–entropy and pressure–enthalpy diagrams are also used frequently.

1.4 Reversibility

In section 1.3 it was shown that the state of a fluid can be represented by a point located on a diagram using two properties as coordinates. When a system changes state in such a way that at any instant during the process the state point can be located on the diagram, then the process is said to be *reversible*. The fluid undergoing the process passes through a continuous series of equilibrium states. A reversible process between two states can therefore be drawn as a line on any diagram of properties (Fig. 1.12(a)). In practice, the fluid undergoing a process cannot be kept in equilibrium in its intermediate states and a continuous path cannot be traced on a diagram of properties. Such real processes are called *irreversible processes*. An irreversible process is usually represented by a dotted line joining the end states to indicate that the intermediate states are indeterminate (Fig. 1.12(b)).

Fig. 1.12 Reversible and irreversible processes

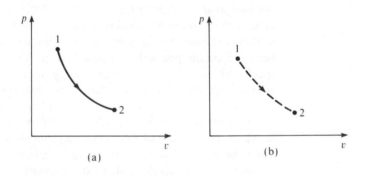

(a)   (b)

A more rigorous definition of reversibility is as follows:

When a fluid undergoes a reversible process, both the fluid and its surroundings can always be restored to their original state.

The criteria of reversibility are as follows:

(a) The process must be frictionless. The fluid itself must have no internal friction and there must be no mechanical friction (e.g. between cylinder and piston).
(b) The difference in pressure between the fluid and its surroundings during the process must be infinitely small. This means that the process must take place infinitely slowly, since the force to accelerate the boundaries of the system is infinitely small.

(c) The difference in temperature between the fluid and its surroundings during the process must be infinitely small. This means that the heat supplied or rejected to or from the fluid must be transferred infinitely slowly.

It is obvious from the above criteria that no process in practice is truly reversible. However, in many practical processes a very close approximation to an *internal reversibility* may be obtained. In an internally reversible process, although the surroundings can never be restored to their original state, the fluid itself is at all times in an equilibrium state and the path of the process can be exactly retraced to the initial state. In general, processes in cylinders with a reciprocating piston are assumed to be internally reversible as a reasonable approximation, but processes in rotary machinery (e.g. turbines) are known to be irreversible due to the high degree of turbulence and scrubbing of the fluid.

1.5 Reversible work

Consider an ideal frictionless fluid contained in a cylinder behind a piston. Assume that the pressure and temperature of the fluid are uniform and that there is no friction between the piston and the cylinder walls. Let the cross-sectional area of the piston be A, let the pressure of the fluid be p, let the pressure of the surroundings be $(p + dp)$ (Fig. 1.13). The force exerted by the piston on the fluid is pA. Let the piston move under the action of the force exerted a distance dl to the left. Then work done on the fluid by the piston is given by force times the distance moved,

i.e.   Work done, $dW = -(pA) \times dl = -p\,dV$

Fig. 1.13 Fluid in a cylinder undergoing a compression

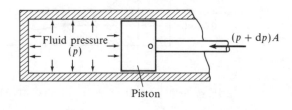

Piston

where dV is a small increase in volume. The negative sign is necessary because the volume is decreasing.

Or for a mass, m,

$$dW = -mp\,dv$$

where v is the specific volume. This is only true when criteria (a) and (b) hold as stated in section 1.4.

When a fluid undergoes a reversible process a series of state points can be joined up to form a line on a diagram of properties. The work done on the fluid during any reversible process, W, is therefore given by the area under the line of the process plotted on a p–v diagram (Fig. 1.14),

i.e. $$W = -m \int_1^2 p\,dv = m \int_2^1 p\,dv = m \text{ (shaded area on Fig. 1.14)} \qquad (1.2)$$

11

Fig. 1.14 Work done in a compression process

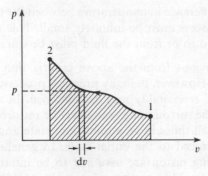

When p can be expressed in terms of v then the integral, $m \int_1^2 p \, dv$, can be evaluated.

Example 1.1   Unit mass of a fluid at a pressure of 3 bar, and with a specific volume of 0.18 m³/kg, contained in a cylinder behind a piston expands reversibly to a pressure of 0.6 bar according to a law $p = c/v^2$, where c is a constant. Calculate the work done during the process.

Solution   Referring to Fig. 1.15

Fig. 1.15 Pressure-specific volume diagram for Example 1.1

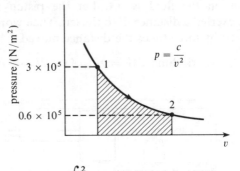

$$W = -m \int_1^2 p \, dv = -m \text{ (shaded area)}$$

i.e. $$W = -cm \int_{v_1}^{v_2} \frac{dv}{v^2} = -cm \left[-\frac{1}{v} \right]_{v_1}^{v_2}$$

also $c = pv^2 = 3 \times 0.18^2 = 0.0972$ bar (m³/kg)²

and $v_2 = \sqrt{\dfrac{c}{p_2}} = \sqrt{\dfrac{0.0972}{0.6}} = 0.402$ m³/kg

therefore

$$W = -0.0972 \times 10^5 \left(\frac{1}{0.18} - \frac{1}{0.402} \right) \text{N m/kg}$$

$$= -29\,840 \text{ N m/kg}$$

i.e.   Work done *by* the fluid $= +29\,840$ N m/kg

Fig. 1.16 Reversible expansion process on a *p–v* diagram

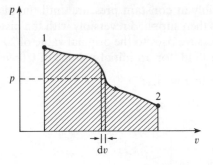

When an expansion process takes place reversibly (see Fig. 1.16), the integral, $\int_1^2 p \, \mathrm{d}v$, is positive, i.e.

$$W = -m \int_1^2 p \, \mathrm{d}v = -m \text{ (shaded area on Fig. 1.16)}$$

A process from right to left on the *p–v* diagram is one in which there is a work input to the fluid (i.e. *W* is positive). Conversely, a process from left to right is one in which there is a work output from the fluid (i.e. *W* is negative).

When a fluid undergoes a series of process and finally returns to its initial state, then it is said to have undergone a thermodynamic *cycle*. A cycle which consists only of reversible processes is a reversible cycle. A cycle plotted on a diagram of properties forms a closed figure, and a reversible cycle plotted on a *p–v* diagram forms a closed figure the area of which represents the net work of the cycle. For example, a reversible cycle consisting of four reversible processes 1 to 2, 2 to 3, 3 to 4, and 4 to 1 is shown in Fig. 1.17. The net work input is equal to the shaded area. If the cycle were described in the reverse direction (i.e. 1 to 4, 4 to 3, 3 to 2, and 2 to 1), then the shaded area would represent net work output from the system. The rule is that the enclosed area of a reversible cycle represents net work input (i.e. net work done on the system) when the cycle is described in an anticlockwise manner, and the enclosed area represents work output (i.e. work done by the system) when the cycle is described in a clockwise manner.

Fig. 1.17 Reversible cycle on a *p–v* diagram

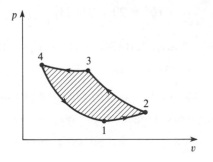

Example 1.2   Unit mass of a certain fluid is contained in a cylinder at an initial pressure of 20 bar. The fluid is allowed to expand reversibly behind a piston according to a law $pV^2 = \text{constant}$ until the volume is doubled. The fluid is then cooled

reversibly at constant pressure until the piston regains its original position; heat is then supplied reversibly with the piston firmly locked in position until the pressure rises to the original value of 20 bar. Calculate the net work done by the fluid, for an initial volume of 0.05 m³.

Solution   Referring to Fig. 1.18

$$p_1 V_1^2 = p_2 V_2^2$$

Fig. 1.18 Figure for Example 1.2

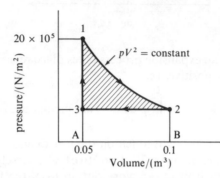

therefore

$$p_2 = p_1 \left(\frac{V_1}{V_2}\right)^2 = \frac{20}{2^2} = 5 \text{ bar}$$

$$W_{12} = -\int_1^2 p \, dV \text{ from equation (1.2)} = -\text{area 12BA1}$$

i.e. $$W_{12} = -\int_{V_1}^{V_2} \frac{c}{V^2} \, dV \quad \text{where} \quad c = p_1 V_1^2 = 20 \times 0.05^2 \text{ bar m}^6$$

therefore

$$W_{12} = -10^5 \times 20 \times 0.0025 \left[-\frac{1}{V}\right]_{0.05}^{0.1}$$

$$= -10^5 \times 20 \times 0.0025 \left(\frac{1}{0.05} - \frac{1}{0.1}\right) = -50\,000 \text{ N m}$$

$$W_{23} = \text{area 32BA3} = p_2(V_2 - V_3) = 10^5 \times 5 \times (0.1 - 0.05)$$

$$= 25\,000 \text{ N m}$$

Work done from 3 to 1 is zero since the piston is locked in position. Therefore

Net work done $= W_{12} + W_{23} = -(\text{enclosed area } 1231)$

$$= -50\,000 + 25\,000 = -25\,000 \text{ N m}$$

Hence the net work done by the fluid is $+25\,000$ N m.

It has been stated above that work is given by $-\int p \, dv$ for a reversible process only. It can easily be shown that $-\int p \, dv$ is not equal to the work done if a

Fig. 1.19
Compartments with
sliding partitions

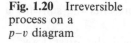

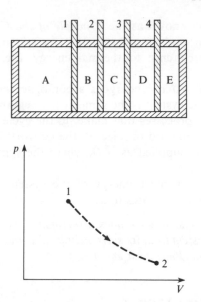

Fig. 1.20 Irreversible
process on a
$p-v$ diagram

process is irreversible. For example, consider a cylinder, divided into a number of compartments by sliding partitions (Fig. 1.19). Initially, compartment A is filled with a mass of fluid at pressure p_1. When the sliding partition 1 is removed quickly, then the fluid expands to fill compartments A and B. When the system settles down to a new equilibrium state the pressure and volume are fixed and the state can be marked on the $p-V$ diagram (Fig. 1.20). Sliding partition 2 is now removed and the fluid expands to occupy compartments A, B, and C. Again the equilibrium state can be marked on the diagram. The same procedure can be adopted with partitions 3 and 4 until finally the fluid is at p_2 and occupies a volume V_2 when filling compartments A, B, C, D, and E. The area under the curve 1–2 on Fig. 1.20 is given by $\int_1^2 p \, dV$, but no work has been done (apart from the negligible work required to move the partitions). No piston has been moved, no turbine wheel has been revolved; in other words, no external force has been moved through a distance. This is the extreme case of an irreversible process in which $\int p \, dV$ has a value and yet the work done is zero. When a fluid expands without a restraining force being exerted by the surroundings, as in the example above, the process is known as *free expansion*. Free expansion is highly irreversible by criterion (b), section 1.4. In many practical expansion processes some work is done by the fluid which is less than $\int p \, dv$ and in many practical compression processes work is done which is greater than $\int p \, dv$. It is important to represent all irreversible processes by dotted lines on a $p-v$ diagram as a reminder that the area under the dotted line does not represent work.

1.6 Conservation of energy and the First Law of Thermodynamics

The concept of energy and the hypothesis that it can neither be created nor destroyed were developed by scientists in the early part of the nineteenth century, and became known as the *Principle of the Conservation of Energy.* The First Law of Thermodynamics is merely one statement of this general principle with particular reference to thermal energy, (i.e. heat), and mechanical energy, (i.e. work).

When a system undergoes a complete thermodynamic cycle the intrinsic energy of the system is the same at the beginning and end of the cycle. During the various processes that make up the cycle work is done on or by the fluid and heat is supplied or rejected; the network input can be defined as $\sum W$, and the net heat supplied as $\sum Q$, where the symbol $\sum$ represents the sum for a complete cycle.

Since the intrinsic energy of the system is unchanged the First Law of Thermodynamics states that:

When a system undergoes a thermodynamic cycle then the net heat supplied to the system from its surroundings plus the net work input to the system from its surroundings must equal zero.

That is

$$\sum Q + \sum W = 0 \tag{1.3}$$

Example 1.3   In a certain steam plant the turbine develops 1000 kW. The heat supplied to the steam in the boiler is 2800 kJ/kg, the heat rejected by the steam to the cooling water in the condenser is 2100 kJ/kg and the feed-pump work required to pump the condensate back into the boiler is 5 kW. Calculate the steam flow rate.

Solution   The cycle is shown diagrammatically in Fig. 1.21. A boundary is shown which encompasses the entire plant. Strictly, this boundary should be thought of as encompassing the working fluid only. For unit mass flow rate

$$\sum dQ = 2800 - 2100 = 700 \text{ kJ/kg}$$

Let the steam flow be $\dot{m}$ kg/s. Therefore

$$\sum dQ = 700m \text{ kW}$$

and $$\sum dW = 5 - 1000 = -995 \text{ kW}$$

Fig. 1.21 Steam plant for Example 1.3

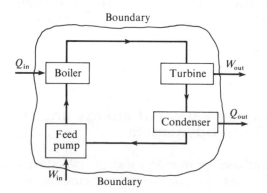

Then in equation (1.3)

$$\sum dQ + \sum dW = 0$$

i.e. $700 \, \dot{m} - 995 = 0$

therefore

$$\dot{m} = 995/700 = 1.421 \text{ kg/s}$$

i.e.   Steam mass flow rate required $= 1.421$ kg/s

1.7 The non-flow equation

In section 1.6 it is stated that when a system possessing a certain intrinsic energy is made to undergo a cycle by heat and work transfer, then the net heat supplied plus the net work input is zero.

This is true for a complete cycle when the final intrinsic energy of the system is equal to its initial value. Consider now a process in which the intrinsic energy of the system is finally greater than the initial intrinsic energy. The sum of the net heat supplied and the net work input has increased the intrinsic energy of the system, i.e.

Gain in intrinsic energy = Net heat supplied + net work input

When the net effect is to transfer energy from the system, then there will be a loss in the intrinsic energy of the system.

When a fluid is not in motion then its intrinsic energy per unit mass is known as the *specific internal energy* of the fluid and is given the symbol u. The specific internal energy of a fluid depends on its pressure and temperature, and is itself a property. The simple proof that specific internal energy is a property is given in ref. 1.4. The internal energy of mass, m, of a fluid is written as U, i.e. $mu = U$. The units of internal energy, U, are usually written as kJ.

Since internal energy is a property, then gain in internal energy in changing from state 1 to state 2 can be written $U_2 - U_1$.

Also, gain in internal energy = net heat supplied + net work input,

i.e. $$U_2 - U_1 = \sum_1^2 dQ + \sum_1^2 dW$$

This equation is true for a process or series of processes between state 1 and state 2 provided there is no flow of fluid into or out of the system. In any one non-flow process there will be either heat supplied or heat rejected, but not both; similarly there will be either work input or work output, but not both. Hence,

$$U_2 - U_1 = Q + W \quad \text{for a non-flow process}$$

or, for unit mass

$$Q + W = u_2 - u_1 \tag{1.4}$$

17

This equation is known as the *non-flow energy equation*. Equation (1.4) is very often written in differential form. For a small amount of heat supplied dQ, a small amount of work done on the fluid dW, and a small gain in specific internal energy du, then

$$dQ + dW = du \qquad (1.5)$$

Example 1.4   In the compression stroke of an internal-combustion engine the heat rejected to the cooling water is 45 kJ/kg and the work input is 90 kJ/kg. Calculate the change in specific internal energy of the working fluid stating whether it is a gain or a loss.

Solution $Q = -45 \, \text{kJ/kg}$

($-$ve sign since heat is rejected).

$W = 90 \, \text{kJ/kg}$
Using equation (1.4)

$$Q + W = u_2 - u_1$$
$$-45 + 90 = u_2 - u_1$$

therefore

$$u_2 - u_1 = 45 \, \text{kJ/kg}$$

i.e.   Gain in internal energy $= 45 \, \text{kJ/kg}$

Example 1.5   In the cylinder of an air motor the compressed air has a specific internal energy of 420 kJ/kg at the beginning of the expansion and a specific internal energy of 200 kJ/kg after expansion. Calculate the heat flow to or from the cylinder when the work done by the air during the expansion is 100 kJ/kg.

Solution   From equation (1.4)

$$Q + W = u_2 - u_1$$

i.e. $Q - 100 = 200 - 420$

therefore

$$Q = -120 \, \text{kJ/kg}$$

i.e.   Heat rejected by the air $= +120 \, \text{kJ/kg}$

It is important to note that equations (1.3), (1.4), and (1.5) are true whether or not the process is reversible. These are energy equations.

For reversible non-flow processes we have, from equation (1.2)

$$W = -m \int_1^2 p \, dv$$

or in differential form

$$dW = -m \, p \, dv$$

Hence for any reversible non-flow process for unit mass, substituting in equation (1.5)

$$dQ = du + p\,dv \tag{1.6}$$

or substituting in equation (1.4)

$$Q = (u_2 - u_1) + \int_1^2 p\,dv \tag{1.7}$$

Equations (1.6) and (1.7) can only be used for ideal reversible non-flow processes.

1.8 The steady-flow equation

In section 1.7, the specific internal energy of a fluid was said to be the intrinsic energy of the fluid due to its thermodynamic properties. When unit mass of a fluid with specific internal energy, u, is moving with velocity C and is a height Z above a datum level, then it possesses a total energy of $u + (C^2/2) + Zg$, where $C^2/2$ is the kinetic energy of unit mass of the fluid and Zg is the potential energy of unit mass of the fluid.

In most practical problems the rate at which the fluid flows through a machine or piece of apparatus is constant. This type of flow is called *steady flow*.

Consider a fluid flowing in steady flow with a mass flow rate, $\dot{m}$, through a piece of apparatus (Fig. 1.22). This constitutes an open system as defined in section 1.2. The boundary is shown cutting the inlet pipe at section 1 and the outlet pipe at section 2. This boundary is sometimes called a *control surface*, and the system encompassed, a *control volume*.

Fig. 1.22 Steady-flow open system

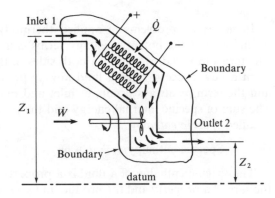

Let it be assumed that a steady rate of flow of heat $\dot{Q}$ units is supplied, and that $\dot{W}$ is the rate of work input on the fluid as it passes through the apparatus. Now in order to introduce the fluid across the boundary an expenditure of energy is required; similarly in order to push the fluid across the boundary at exit, an expenditure of energy is required. The inlet section is shown enlarged in Fig. 1.23. Consider an element of fluid, length l, and let the cross-sectional area of the inlet pipe be A_1. Then we have

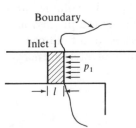

Fig. 1.23 Section at inlet to the system

Energy required to push element across boundary
$$= (p_1 A_1) \times l = p_1 \times \text{(volume of fluid element)}$$

therefore

Energy required for unit mass flow rate of fluid $= p_1 v_1$

where v_1 is the specific volume of the fluid at section 1.

Similarly it can be shown that

Energy required at exit to push unit mass flow rate of fluid across the boundary
$= p_2 v_2$

Consider now the energy entering and leaving the system. The energy entering the system consists of the energy of the flowing fluid at inlet

$$\dot{m}\left(u_1 + \frac{C_1^2}{2} + Z_1 g \right)$$

the energy term $\dot{m} p_1 v_1$, the heat supplied $\dot{Q}$, and the rate of work input, W. The energy leaving the system consists of the energy of the flowing fluid at the outlet section

$$\dot{m}\left(u_2 + \frac{C_2^2}{2} + Z_2 g \right)$$

and the energy term $\dot{m} p_2 v_2$. Since there is steady flow of fluid into and out of the system, and there are steady flows of heat and work, then the energy entering must exactly equal the energy leaving.

$$\dot{m}\left(u_1 + \frac{C_1^2}{2} + Z_1 g + p_1 v_1 \right) + \dot{Q} + \dot{W} = \dot{m}\left(u_2 + \frac{C_2^2}{2} + Z_2 g + p_2 v_2 \right)$$

(1.8)

In nearly all problems in applied thermodynamics, changes in height are negligible and the potential energy terms can be omitted from the equation. The terms in u and pv occur on both sides of the equation and always will do so in a flow process, since a fluid always possesses a certain internal energy, and the term pv always occurs at inlet and outlet as seen in the above proof. The sum of specific internal energy and the pv term is given the symbol h, and is called *specific enthalpy*,

i.e.   Specific enthalpy, $h = u + pv$ (1.9)

The specific enthalpy of a fluid is a property of the fluid, since it consists of the sum of a property and the product of two properties. Since specific enthalpy is a property like specific internal energy, pressure, specific volume, and temperature, it can be introduced into any problem whether the process is a flow process or a non-flow process. The enthalpy of mass, m, of a fluid can be written as H (i.e. $mh = H$). The units of h are the same as those of internal energy. Substituting equation (1.9) in equation (1.8)

$$\dot{m}\left(h_1 + \frac{C_1^2}{2} + Z_1 g \right) + \dot{Q} + \dot{W} = \dot{m}\left(h_2 + \frac{C_2^2}{2} + Z_2 g \right)$$

(1.10)

Equation (1.10) is known as the *steady-flow energy equation*. In steady flow the rate of mass flow of fluid at any section is the same as at any other section.

Consider any section of cross-sectional area A, where the fluid velocity is C, then the rate of volume flow past the section is CA. Also, since mass flow is volume flow divided by specific volume

$$\text{Mass flow rate, } \dot{m} = \frac{CA}{v} = \rho CA \qquad (1.11)$$

where v is the specific volume at the section and ρ the density at the section. This equation is known as the *continuity of mass equation*.

With reference to Fig. 1.22

$$\dot{m} = \frac{C_1 A_1}{v_1} = \frac{C_2 A_2}{v_2}$$

Example 1.6

In the turbine of a gas turbine unit the gases flow through the turbine at 17 kg/s and the power developed by the turbine is 14 000 kW. The specific enthalpies of the gases at inlet and outlet are 1200 kJ/kg and 360 kJ/kg respectively, and the velocities of the gases at inlet and outlet are 60 m/s and 150 m/s respectively. Calculate the rate at which heat is rejected from the turbine. Find also the area of the inlet pipe given that the specific volume of the gases at inlet is 0.5 m³/kg.

Solution

A diagrammatic representation of the turbine is shown in Fig. 1.24. From equation (1.10), neglecting changes in height

$$\dot{m}\left(h_1 + \frac{C_1^2}{2} \right) + \dot{Q} + \dot{W} = \dot{m}\left(h_2 + \frac{C_2^2}{2} \right)$$

Fig. 1.24 Gas turbine for Example 1.6

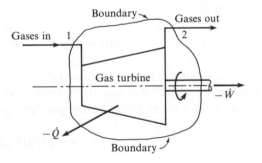

For unit mass flow rate:

$$\text{Kinetic energy at inlet} = \frac{C_1^2}{2} = \frac{60^2}{2} \text{ m}^2/\text{s}^2 = \frac{60^2}{2} \frac{\text{kg m}^2}{\text{s}^2 \text{ kg}}$$

$$= 1800 \text{ N m/kg} = 1.8 \text{ kJ/kg}$$

$$\text{Kinetic energy at outlet} = \frac{C_2^2}{2} = 2.5^2 \times (\text{kinetic energy at inlet})$$

$$= 11.25 \text{ kJ/kg (since } C_2 = 2.5C_1)$$

21

Also $\quad \dot{W} = -14\,000\ \text{kW}$

Substituting in equation (1.10)

$$17(1200 + 1.8) + \dot{Q} - 14\,000 = 17(360 + 11.25)$$

therefore

$$Q = -119.3\ \text{kW}$$

i.e. $\quad$ Heat rejected $= +119.3\ \text{kW}$

To find the inlet area, use equation (1.11), i.e.

$$\dot{m} = \frac{CA}{v} \quad \text{or} \quad A = \frac{\dot{m}v}{C}$$

therefore

$$\text{Inlet area, } A_1 = \frac{17 \times 0.5}{60} = 0.142\ \text{m}^2$$

Example 1.7

Air flows steadily at the rate of 0.4 kg/s through an air compressor, entering at 6 m/s with a pressure of 1 bar and a specific volume of 0.85 m^3/kg, and leaving at 4.5 m/s with a pressure of 6.9 bar and a specific volume of 0.16 m^3/kg. The specific internal energy of the air leaving is 88 kJ/kg greater than that of the air entering. Cooling water in a jacket surrounding the cylinder absorbs heat from the air at the rate of 59 kW. Calculate the power required to drive the compressor and the inlet and outlet pipe cross-sectional areas.

Solution

In this problem it is more convenient to write the flow equation as in equation 1.8, omitting the Z terms,

i.e. $\quad \dot{m}\left(u_1 + \dfrac{C_1^2}{2} + p_1 v_1 \right) + \dot{Q} + \dot{W} = \dot{m}\left(u_2 + \dfrac{C_2^2}{2} + p_2 v_2 \right)$

A diagrammatic representation of the compressor is shown in Fig. 1.25. Note that the heat rejected across the boundary is equivalent to the heat removed by the cooling water from the compressor. For unit mass flow rate:

$$\frac{C_1^2}{2} = \frac{6 \times 6}{2}\ \text{J/kg} = 18\ \text{J/kg} = 0.018\ \text{kJ/kg}$$

Fig. 1.25 Air compressor for Example 1.7

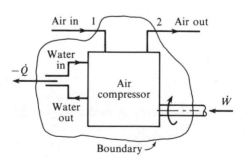

$$\frac{C_2^2}{2} = \frac{4.5 \times 4.5}{2} \, \text{J/kg} = 10.1 \, \text{J/kg} = 0.0101 \, \text{kJ/kg}$$

$$p_1 v_1 = 1 \times 10^5 \times 0.85 = 85\,000 \, \text{J/kg} = 85 \, \text{kJ/kg}$$

$$p_2 v_2 = 6.9 \times 10^5 \times 0.16 = 110\,400 \, \text{J/kg} = 110.4 \, \text{kJ/kg}$$

$$u_2 - u_1 = 88 \, \text{kJ/kg}$$

Also $\dot{Q} = -59 \, \text{kW}$

Now $\dot{Q} + \dot{W} = \dot{m} \left\{ (u_2 - u_1) + (p_2 v_2 - p_1 v_1) + \left(\frac{C_2^2}{2} - \frac{C_1^2}{2} \right) \right\}$

i.e. $-59 + \dot{W} = 0.4(88 + 110.4 - 85 + 0.0101 - 0.018)$

therefore

$$\dot{W} = 104.4 \, \text{kW}$$

(Note that the change in kinetic energy is negligibly small in comparison with the other terms.)

i.e. Power input required $= 104.4 \, \text{kW}$

From equation (1.11)

$$\dot{m} = \frac{CA}{v}$$

$$A_1 = \frac{0.4 \times 0.85}{6} = 0.057 \, \text{m}^2$$

i.e. Inlet pipe cross-sectional area $= 0.057 \, \text{m}^2$

Similarly

$$A_2 = \frac{0.4 \times 0.16}{4.5} = 0.014 \, \text{m}^2$$

i.e. Outlet pipe cross-sectional area $= 0.014 \, \text{m}^2$

In Example 1.7 the steady-flow energy equation has been used, despite the fact that the compression consists of suction of air, compression in a closed cylinder, and discharge of air. The steady-flow equation can be used because the cycle of processes takes place many times in a minute, and therefore the average effect is steady flow of air through the machine.

Problems

1.1 A certain fluid at 10 bar is contained in a cylinder behind a piston, the initial volume being 0.05 m³. Calculate the work done by the fluid when it expands reversibly:
 (i) at constant pressure to a final volume of 0.2 m³;
 (ii) according to a linear law to a final volume of 0.2 m³ and a final pressure of 2 bar;

(iii) according to a law $pV = $ constant to a final volume of 0.1 m³;
(iv) according to a law $pv^3 = $ constant to a final volume of 0.06 m^3;
(v) according to a law, $p = (A/V^2) - (B/V)$, to a final volume of 0.1 m³ and a final pressure of 1 bar, where A and B are constants.
Sketch all processes on a $p-v$ diagram.

(150 000 N m; 90 000 N m; 34 700 N m; 7640 N m; 19 200 N m)

1.2 1 kg of a fluid is compressed reversibly according to a law $pv = 0.25$, where p is in bar and v is in m³/kg. The final volume is $\frac{1}{4}$ of the initial volume. Calculate the work done on the fluid and sketch the process on a $p-v$ diagram.

(34 660 N m)

1.3 0.05 m³ of a gas at 6.9 bar expands reversibly in a cylinder behind a piston according to the law $pv^{1.2} = $ constant, until the volume is 0.08 m³. Calculate the work done by the gas and sketch the process on a $p-V$ diagram.

(15 480 N m)

1.4 1 kg of a fluid expands reversibly according to a linear law from 4.2 bar to 1.4 bar; the initial and final volumes are 0.004 m³ and 0.02 m³. The fluid is then cooled reversibly at constant pressure, and finally compressed reversibly according to a law $pv = $ constant back to the initial conditions of 4.2 bar and 0.004 m³. Calculate the work done in each process and the net work of the cycle. Sketch the cycle on a $p-v$ diagram.

(-4480 N m; $+1120$ N m; $+1845$ N m; -1515 N m)

1.5 A fluid at 0.7 bar occupying 0.09 m³ is compressed reversibly to a pressure of 3.5 bar according to a law $pv^n = $ constant. The fluid is then heated reversibly at constant volume until the pressure is 4 bar; the specific volume is then 0.5 m³/kg. A reversible expansion according to a law $pv^2 = $ constant restores the fluid to its initial state. Sketch the cycle on a $p-v$ diagram and calculate:
 (i) the mass of fluid present;
 (ii) the value of n in the first process;
 (iii) the net work of the cycle.

(0.0753 kg; 1.847; -640 N m)

1.6 A fluid is heated reversibly at a constant pressure of 1.05 bar until it has a specific volume of 0.1 m³/kg. It is then compressed reversibly according to a law $pv = $ constant to a pressure of 4.2 bar, then allowed to expand reversibly according to a law $pv^{1.7} = $ constant, and is finally heated at constant volume back to the initial conditions. The work done in the constant pressure process is -515 N m, and the mass of fluid present is 0.2 kg. Calculate the net work of the cycle and sketch the cycle on a $p-v$ diagram.

($+781$ N m)

1.7 In an air compressor the compression takes place at a constant internal energy and 50 kJ of heat are rejected to the cooling water for every kilogram of air. Calculate the work input for the compression stroke per kilogram of air.

(50 kJ/kg)

1.8 In the compression stroke of a gas engine the work done on the gas by the piston is 70 kJ/kg and the heat rejected to the cooling water is 42 kJ/kg. Calculate the change of specific internal energy stating whether it is a gain or a loss.

(28 kJ/kg gain)

1.9 A mass of gas at an initial pressure of 28 bar, and with an internal energy of 1500 kJ, is contained in a well-insulated cylinder of volume 0.06 m³. The gas is allowed to expand behind a piston until its internal energy is 1400 kJ; the law of expansion is $pv^2 =$ constant. Calculate:

(i) the work done;

(ii) the final volume;

(iii) the final pressure.

$(-100\,\text{kJ};\ 0.148\,\text{m}^3;\ 4.59\,\text{bar})$

1.10 The gases in the cylinder of an internal combustion engine have a specific internal energy of 800 kJ/kg and a specific volume of 0.06 m^3/kg at the beginning of expansion. The expansion of the gases may be assumed to take place according to a reversible law, $pv^{1.5} =$ constant, from 55 bar to 1.4 bar. The specific internal energy after expansion is 230 kJ/kg. Calculate the heat rejected to the cylinder cooling water per kilogram of gases during the expansion stroke.

$(104\,\text{kJ/kg})$

1.11 A steam turbine receives a steam flow of 1.35 kg/s and the power output is 500 kW. The heat loss from the casing is negligible. Calculate:

(i) the change of specific enthalpy across the turbine when the velocities at entrance and exit and the difference in elevation are negligible;

(ii) the change of specific enthalpy across the turbine when the velocity at entrance is 60 m/s, the velocity at exit is 360 m/s, and the inlet pipe is 3 m above the exhaust pipe.

$(370\,\text{kJ/kg};\ 433\,\text{kJ/kg})$

1.12 A steady flow of steam enters a condenser with a specific enthalpy of 2300 kJ/kg and a velocity of 350 m/s. The condensate leaves the condenser with a specific enthalpy of 160 kJ/kg and a velocity of 70 m/s. Calculate the heat transfer to the cooling fluid per kilogram of steam condensed.

$(-2199\,\text{kJ/kg})$

1.13 A turbine operating under steady-flow conditions receives steam at the following state: pressure, 13.8 bar; specific volume 0.143 m^3/kg, specific internal energy 2590 kJ/kg, velocity 30 m/s. The state of the steam leaving the turbine is as follows: pressure 0.35 bar, specific volume 4.37 m^3/kg, specific internal energy 2360 kJ/kg, velocity 90 m/s. Heat is rejected to the surroundings at the rate of 0.25 kW and the rate of steam flow through the turbine is 0.38 kg/s. Calculate the power developed by the turbine.

$(102.7\,\text{kW})$

1.14 A nozzle is a device for increasing the velocity of a steadily flowing fluid. At the inlet to a certain nozzle the specific enthalpy of the fluid is 3025 kJ/kg and the velocity is 60 m/s. At the exit from the nozzle the specific enthalpy is 2790 kJ/kg. The nozzle is horizontal and there is a negligible heat loss from it. Calculate:

(i) the velocity of the fluid at exit;

(ii) the rate of flow of fluid when the inlet area is 0.1 m^2 and the specific volume at inlet is 0.19 m^3/kg;

(iii) the exit area of the nozzle when the specific volume at the nozzle exit is 0.5 m^3/kg.

$(688\,\text{m/s};\ 31.6\,\text{kg/s};\ 0.0229\,\text{m}^2)$

References

1.1 EASTOP T D and CROFT D R 1990 *Energy Efficiency* Longman

1.2 DOUGLAS J F, GASIOREK J M and SWAFFIELD J A 1986 *Fluid Mechanics* 2nd edn Longman

1.3 BS 1041 *Temperature Measurement* HMSO; Section 2.1 1985 *Guide to Selection and Use of Liquid-in-glass Thermometers*; Part 3 1989 *Guide to Selection and Use of Industrial Resistance Thermometers*; Part 4 1966 *Thermocouples*; Part 5 1972 *Radiation Pyrometers*; Part 7 1988 *Guide to Selection and Use of Temperature-time Records*

1.4 ROGERS G F C and MAYHEW Y R 1992 *Engineering Thermodynamics, Work and Heat Transfer* 4th edn Longman

2

The Working Fluid

In section 1.3 the matter contained within the boundaries of a system is defined as the working fluid, and it is stated that when two independent properties of the fluid are known then the thermodynamic state of the fluid is defined. In thermodynamic systems the working fluid can be in the liquid, vapour, or gaseous phase. All substances can exist in any one of these phases, but we tend to identify all substances with the phase in which they are in equilibrium at atmospheric pressure and temperature. For instance, substances such as oxygen and nitrogen are thought of as gases; H_2O is thought of as liquid or vapour (i.e. water or steam); mercury is thought of as a liquid. All these substances can exist in different phases: oxygen and nitrogen can be liquefied; H_2O can become a gas at very high temperatures; mercury can be vaporized and will act as a gas if the temperature is raised high enough.

2.1 Liquid, vapour, and gas

Consider a $p-v$ diagram for any substance. The solid phase is not important in engineering thermodynamics, being more the province of the metallurgist or physicist. When a liquid is heated at any one constant pressure there is one fixed temperature at which bubbles of vapour form in the liquid; this phenomenon is known as boiling. The higher the pressure of the liquid then the higher the temperature at which boiling occurs. It is also found that the volume occupied by 1 kg of a boiling liquid at a higher pressure is slightly larger than the volume occupied by 1 kg of the same liquid when it is boiling at a low pressure. A series of boiling-points plotted on a $p-v$ diagram will appear as a sloping line, as shown in Fig. 2.1. The points P, Q, and R represent the boiling-points of a liquid at pressure p_P, p_Q, and p_R respectively.

When a liquid at boiling-point is heated further at constant pressure the additional heat supplied changes the phase of the substance from liquid to vapour; during this change of phase the pressure and temperature remain constant. The heat supplied is called the *specific enthalpy of vaporization*. It is found that the higher the pressure then the smaller is the amount of heat required. There is a definite value of specific volume of the vapour at any one

Fig. 2.1 Boiling-points plotted on a p–v diagram

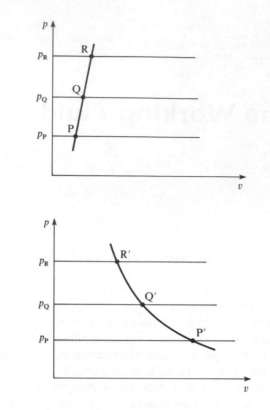

Fig. 2.2 Points of complete vaporization plotted on a p–v diagram

pressure, at the point at which vaporization is complete; hence a series of points such as P′, Q′, R′ can be plotted and joined to form a line as shown in Fig. 2.2.

When the two curves already drawn are extended to higher pressures they form a continuous curve, thus forming a loop (see Fig. 2.3). The pressure at which the turning point occurs is called the *critical pressure* and the turning point itself is called the *critical point* (point C on Fig. 2.3). It can be seen that at the critical point the specific enthalpy of vaporization is zero. The substance existing at a state point inside the loop consists of a mixture of liquid and dry vapour and is known as a *wet vapour*. A *saturation state* is defined as a state at which a change of phase may occur without change of pressure or temperature. Hence the boiling-points P, Q, and R are saturation states, and a series of such boiling-points joined up is called the *saturated liquid line*. Similarly the points P′, Q′, and R′, at which the liquid is completely changed into vapour, are saturation states, and a series of such points joined up is called the *saturated vapour line*. The word 'saturation' as used here refers to energy saturation. For example, a slight addition of heat to a boiling liquid changes some of it into a vapour, and it is no longer a liquid but is now a wet vapour. Similarly when a substance just on the saturated vapour line is cooled slightly, droplets of liquid will begin to form, and the saturated vapour becomes a wet vapour. A saturated vapour is usually called *dry saturated* to emphasize the fact that no liquid is present in the vapour in this state.

Lines of constant temperature, called isothermals, can be plotted on a p–v diagram as shown in Fig. 2.4. The temperature lines become horizontal between

Fig. 2.3 Wet loop plotted on a p–v diagram

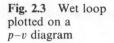

Fig. 2.4 Isothermals for a vapour plotted on a p–v diagram

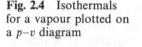

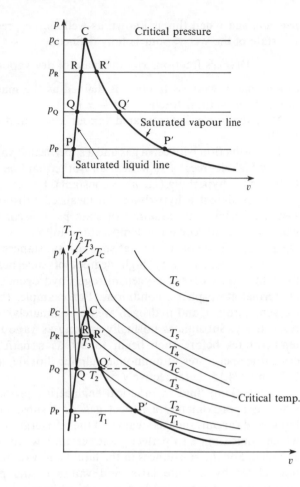

the saturated liquid line and the saturated vapour line (e.g. between P and P′, Q and Q′, R and R′). Thus there is a corresponding *saturation temperature* for each *saturation pressure*. At pressure p_P the saturation temperature is T_1, at pressure p_Q the saturation temperature is T_2, and at pressure p_R the saturation temperature is T_3. The critical temperature line T_C just touches the top of the loop at the critical point C.

When a dry saturated vapour is heated at constant pressure its temperature rises and it becomes *superheated*. The difference between the actual temperature of the superheated vapour and the saturation temperature at the pressure of the vapour is called the *degree of superheat*. For example, the vapour at point S (Fig. 2.4) is superheated at p_Q and T_3, and the degree of superheat is $T_3 - T_2$.

In section 1.5 it is stated that two independent properties are sufficient to define the state of a substance. Now between P and P′, Q and Q′, R and R′ the temperature and pressure are not independent since they remain constant for a range of values of v. For example, a substance at p_Q and T_2 (Fig. 2.4) could be a saturated liquid, a wet vapour, or a dry saturated vapour. The state cannot be defined until one other property (e.g. specific volume) is given. The condition or quality of a wet vapour is most frequently defined by its *dryness*

29

fraction, and when this is known as well as the pressure or temperature then the state of the wet vapour is fully defined.

Dryness fraction, x = the mass of dry vapour in 1 kg of the mixture

(Sometimes a wetness fraction is defined as the mass of liquid in 1 kg of the mixture, i.e. wetness fraction = $1 - x$.)

Note that for a dry saturated vapour $x = 1$, and that for a saturated liquid $x = 0$.

The distinction between a gas and a superheated vapour is not rigid. However, at very high degrees of superheat an isothermal line on the p–v diagram tends to become a hyperbola (i.e. pv = constant). For example the isothermal T_6 on Fig. 2.4 is almost a hyperbola. An idealized substance called a *perfect gas* is assumed to have an equation of state pv/T = constant. It can be seen that when a line of constant temperature follows a hyperbolic law then the equation pv/T = constant is satisfied. All substances tend to obey the equation pv/T = constant at very high degrees of superheat. Substances which are thought of as gases (e.g. oxygen, nitrogen, hydrogen, etc.) are highly superheated at normal atmospheric conditions. For example, the critical temperatures of oxygen, nitrogen, and hydrogen are approximately -119, -147, and $-240\,°C$ respectively. Substances normally existing as vapours must be raised to high temperatures before they begin to act as a perfect gas. For example, the critical temperatures of ammonia, sulphur dioxide, and water vapour are 130, 157, and 374.15 °C respectively.

The working fluid in practical engineering problems is either a substance which is approximately a perfect gas, or a substance which exists mainly as liquid and vapour, such as steam and the refrigerant vapours. For the substances which approximate to perfect gases certain laws relating the properties can be assumed. For the substances in the liquid and vapour phases the properties are not related by definite laws, and values of the properties are determined empirically and tabulated in a convenient form.

2.2 The use of vapour tables

Tables are available for a wide variety of substances which normally exist in the vapour phase. The tables which will be used in this book are those arranged by Rogers and Mayhew (ref. 2.1), which are suitable for student use. For more comprehensive tables for steam, ref. 2.2 should be consulted. The tables of Rogers and Mayhew are mainly concerned with steam, but some properties of refrigerants are also given.

Saturation state properties

The saturation pressures and corresponding saturation temperatures of steam are tabulated in parallel columns in the first table, for pressures ranging from 0.006 112 bar to the critical pressure of 221.2 bar. The specific volume, internal energy, enthalpy, and entropy are also tabulated for the dry saturated vapour

Table 2.1 Extract from tables of properties of wet steam

p	t	v_g	u_f	u_g	h_f	h_{fg}	h_g	s_f	s_{fg}	s_g
(bar)	(°C)	(m³/kg)	(kJ/kg)		(kJ/kg)			(kJ/kg K)		
0.34	72	4.649	302	2472	302	2328	2630	0.980	6.745	7.725

at each pressure and corresponding saturation temperature. The suffix g is used to denote the dry saturated stage. A specimen row from the tables is shown in Table 2.1. For example at 0.34 bar the saturation temperature is 72 °C, the specific volume of dry saturated vapour, v_g, at this pressure is 4.649 m³/kg, the internal energy of dry saturated vapour, u_g, is 2472 kJ/kg, and the enthalpy of dry saturated vapour, h_g, is 2630 kJ/kg. The steam is in the state represented by point A on Fig. 2.5. At point B dry saturated steam at a pressure of 100 bar and saturation temperature 311 °C has a specific volume, v_g, of 0.01802 m³/kg, internal energy, u_g, of 2545 kJ/kg and enthalpy, h_g, of 2725 kJ/kg.

Fig. 2.5 Points identified on a p–v diagram for steam

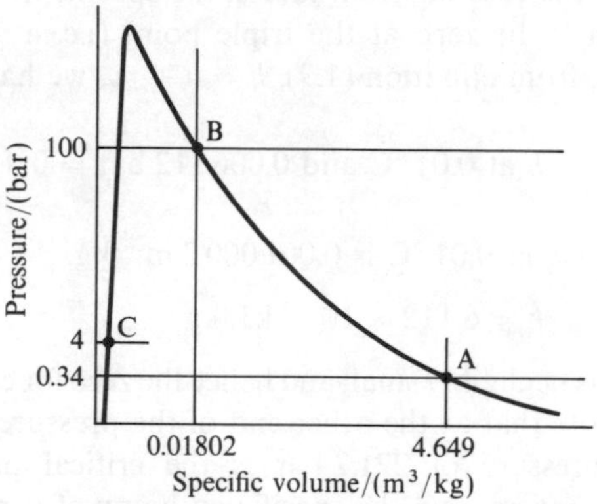

The specific internal energy, specific enthalpy, and specific entropy of saturated liquid are also tabulated, the suffix f being used for this state. For example at 4 bar and the corresponding saturation temperature 143.6 °C, saturated water has a specific internal energy, u_f, of 605 kJ/kg, and a specific enthalpy, h_f, of 605 kJ/kg. This state corresponds to point C on Fig. 2.5. The specific volume of saturated water, v_f, is tabulated in a separate table, but it is usually negligibly small in comparison with the specific volume of the dry saturated vapour, and its variation with temperature is very small; the saturated liquid line on a p–v diagram is very nearly coincident with the pressure axis in comparison with the width of the wet loop (see Fig. 2.5). As seen from the table, values of v_f vary from about 0.001 m³/kg at 0.01 °C to about 0.0011 m³/kg at 160 °C; as the pressure approaches the critical value, the increase of v_f is more marked, and at the critical temperature of 374.15 °C the value of v_f is 0.00317 m³/kg.

The change in specific enthalpy from h_f to h_g is given the symbol h_{fg}. When saturated water is changed to dry saturated vapour, from equation (1.4),

$$Q + W = u_2 - u_1 = u_g - u_f$$

Also $-W$ is represented by the area under the horizontal line on the $p-v$ diagram,

i.e. $\qquad W = -(v_g - v_f)p$

therefore

$$Q = (u_g - u_f) + p(v_g - v_g)$$
$$= (u_g + pv_g) - (u_f + pv_f)$$

From equation (1.9)

$$h = u + pv$$

therefore

$$Q = h_g - h_f = h_{fg}$$

The heat required to change a saturated liquid to a dry saturated vapour is called the specific enthalpy of vaporization, h_{fg}.

In the case of steam tables, the specific internal energy of saturated liquid is taken to be zero at the triple point (i.e. at $0.01\,°C$ and $0.006\,112$ bar). Then since, from equation (1.9), $h = u + pv$, we have

$$h \text{ at } 0.01\,°C \text{ and } 0.006\,112 \text{ bar} = 0 + \frac{0.006\,112 \times 10^5 \times 0.001\,000\,2}{10^3}$$

where v_f at $0.01\,°C$ is $0.001\,000\,2 \text{ m}^3/\text{kg}$,

i.e. $\qquad h = 6.112 \times 10^{-4} \text{ kJ/kg}$

This is negligibly small and hence the zero for enthalpy may be taken at $0.01\,°C$.

Note that at the other end of the pressure range tabulated in the first table the pressure of 221.2 bar is the critical pressure, $374.15\,°C$ is the critical temperature, and the specific enthalpy of vaporization, h_{fg}, is zero.

Properties of wet vapour

For a wet vapour the total volume of the mixture is given by the volume of liquid present plus the volume of dry vapour present. Therefore the specific volume is given by

$$v = \frac{\text{volume of liquid} + \text{volume of dry vapour}}{\text{total mass of wet vapour}}$$

Now for 1 kg of wet vapour there are x kg of dry vapour and $(1 - x)$ kg of liquid, where x is the dryness fraction as defined earlier. Hence,

$$v = v_f(1 - x) + v_g x$$

The volume of the liquid is usually negligibly small compared to the volume of dry saturated vapour, hence for most practical problems

$$v = xv_g \qquad\qquad (2.1)$$

The enthalpy of a wet vapour is given by the sum of the enthalpy of the liquid plus the enthalpy of the dry vapour,

i.e. $h = (1 - x)h_f + xh_g$

therefore

$$h = h_f + x(h_g - h_f)$$

i.e. $h = h_f + xh_{fg}$ (2.2)

Similarly, the internal energy of a wet vapour is given by the internal energy of the liquid plus the internal energy of the dry vapour,

i.e. $u = (1 - x)u_f + xu_g$ (2.3)

or $u = u_f + x(u_g - u_f)$ (2.4)

Equation (2.4) can be expressed in a form similar to equation (2.2), but equations (2.3) and (2.4) are more convenient since u_g and u_f are tabulated and the difference, $u_g - u_f$, is not tabulated in ref. 2.1.

Example 2.1   Calculate the specific volume, specific enthalpy, and specific internal energy of wet steam at 18 bar, dryness fraction 0.9.

Solution   From equation (2.1)

$$v = xv_g$$

therefore

$$v = 0.9 \times 0.1104 = 0.0994 \text{ m}^3/\text{kg}$$

From equation (2.2)

$$h = h_f + xh_{fg}$$

therefore

$$h = 885 + (0.9 \times 1912) = 2605.8 \text{ kJ/kg}$$

From equation (2.3)

$$u = (1 - x)u_f + xu_g$$

therefore

$$u = (1 - 0.9)883 + (0.9 \times 2598) = 2426.5 \text{ kJ/kg}$$

Example 2.2   Calculate the dryness fraction, specific volume and specific internal energy of steam at 7 bar and specific enthalpy 2600 kJ/kg.

Solution   At 7 bar, $h_g = 2764$ kJ/kg, hence since the actual enthalpy is given as 2600 kJ/kg, the steam must be in the wet vapour state. From equation (2.2), $h = h_f + xh_{fg}$,

i.e. $2600 = 697 + x2067$

33

therefore

$$x = \frac{2600 - 697}{2067} = 0.921$$

Then from equation (2.1)

$$v = xv_g = 0.921 \times 0.2728 = 0.2515 \text{ m}^3/\text{kg}$$

From equation (2.3)

$$u = (1 - x)u_f + xu_g$$

therefore

$$u = (1 - 0.921)696 + (0.921 \times 2573) = 55 + 2365$$

i.e. $\quad u = 2420 \text{ kJ/kg}$

Properties of superheated vapour

For steam in the superheat region, temperature and pressure are independent properties. When the temperature and pressure are given for superheated steam then the state is defined and all the other properties can be found. For example, steam at 2 bar and 200 °C is superheated since the saturation temperature at 2 bar is 120.2 °C, which is less than the actual temperature. The steam in this state has a degree of superheat of $200 - 120.2 = 79.8$ K. The tables of properties of superheated steam (ref. 2.1) range in pressure from 0.006 112 bar to the critical pressure of 221.2 bar, and there is an additional table of supercritical pressures up to 1000 bar. At each pressure there is a range of temperatures up to high degrees of superheat, and the values of specific volume, internal energy, enthalpy, and entropy are tabulated at each pressure and temperature for pressures up to and including 70 bar; above this pressure the specific internal energy is not tabulated. For reference the saturation temperature is inserted in brackets under each pressure in the superheat tables and values of v_g, u_g, h_g and s_g are also given. A specimen row of values is shown in Table 2.2. For example, from superheat tables at 20 bar and 400 °C the specific volume is 0.1511 m³/kg and the enthalpy is 3248 kJ/kg.

For pressures above 70 bar the internal energy can be found when required using equation (1.9). For example, steam at 80 bar, 400 °C has an enthalpy, h,

Table 2.2 Extract from tables of properties of superheated steam at 20 bar (saturation temperature 212.4°C)

	Temperature/(°C)						
	250	300	350	400	450	500	600
$v/(\text{m}^3/\text{kg})$	0.1115	0.1255	0.1386	0.1511	0.1634	0.1756	0.1995
$u/(\text{kJ/kg})$	2681	2774	2861	2946	3030	3116	3291
$h/(\text{kJ/kg})$	2904	3025	3138	3248	3357	3467	3690
$s/(\text{kJ/kg K})$	6.547	6.768	6.957	7.126	7.283	7.431	7.701

of 3139 kJ/kg and a specific volume, v, of 3.428×10^{-2} m^3/kg, therefore

$$u = h - pv = 3139 - \frac{80 \times 10^5 \times 0.034\,28}{10^3}$$

i.e. $u = 3139 - 274.2 = 2864.8$ kJ/kg

Example 2.3   Steam at 110 bar has a specific volume of 0.0196 m^3/kg, calculate the temperature, the specific enthalpy, and the specific internal energy.

Solution   First it is necessary to decide whether the steam is wet, dry saturated, or superheated. At 110 bar, $v_g = 0.015\,98$ m^3/kg, which is less than the actual specific volume of 0.0196 m^3/kg, and hence the steam is superheated. The state of the steam is shown as point A of Fig. 2.6.

Fig. 2.6
Pressure-specific volume
diagram for
Example 2.3

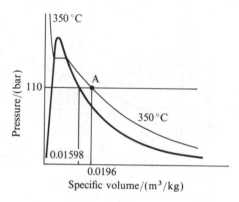

From the superheat tables at 110 bar, the specific volume is 0.0196 m^3/kg at a temperature of 350 °C. Hence this is the isothermal which passes through point A as shown. The degree of superheat in this case is $350 - 318 = 32$ K. From tables the enthalpy, h, is 2889 kJ/kg. Then using equation (1.9), we have

$$u = h - pv = 2889 - \frac{110 \times 10^5 \times 0.0196}{10^3}$$

i.e. $u = 2889 - 215.6 = 2673.4$ kJ/kg

Example 2.4   Steam at 150 bar has a specific enthalpy of 3309 kJ/kg. Calculate the temperature, the specific volume, and the specific internal energy.

Solution   At 150 bar, $h_g = 2611$ kJ/kg, which is less than the actual enthalpy of 3309 kJ/kg, and hence the steam is superheated. From superheat tables at 150 bar, $h = 3309$ kJ/kg at a temperature of 500 °C. The specific volume is $v = 0.020\,78$ m^3/kg. Using equation (1.9)

$$u = h - pv = 3309 - \frac{150 \times 10^5 \times 0.02078}{10^3} = 2997.3 \text{ kJ/kg}$$

Interpolation

For properties which are not tabulated exactly in the tables it is necessary to interpolate between the values tabulated. For example, to find the temperature, specific volume, internal energy, and enthalpy of dry saturated steam at 9.8 bar, it is necessary to interpolate between the values given in the tables.

$$t_g \text{ at } 9.8 \text{ bar} = (t_g \text{ at } 9 \text{ bar}) + \left(\frac{9.8 - 9}{10 - 9}\right) \times \{(t_g \text{ at } 10 \text{ bar}) - (t_g \text{ at } 9 \text{ bar})\}$$

Note that this assumes a linear variation between the two values (see Fig. 2.7),

i.e. $\quad t_g = 175.4 + \left(\dfrac{9.8 - 9}{10 - 9}\right) \times (179.9 - 175.4)$

Fig. 2.7 Interpolation for Example 2.4

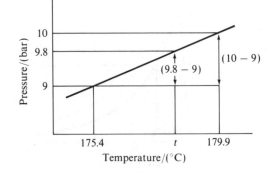

therefore

$$t_g = 175.4 + 0.8 \times 4.5 = 179\,°\text{C}$$

Similarly,

$$h_g \text{ at } 9.8 \text{ bar} = (h_g \text{ at } 9 \text{ bar}) + 0.8 \times (h_g \text{ at } 10 \text{ bar} - h_g \text{ at } 9 \text{ bar})$$

i.e. $\quad h_g \text{ at } 9.8 \text{ bar} = 2774 + 0.8 \times (2778 - 2774)$

$$= 2774 + 0.8 \times 4 = 2777.2 \text{ kJ/kg}$$

Also, $\quad u_g \text{ at } 9.8 \text{ bar} = 2581 + 0.8(2584 - 2581)$

$$= 2581 + (0.8 \times 3) = 2583.4 \text{ kJ/kg}$$

As another example consider steam at 5 bar and 320 °C. The steam is superheated since the saturation temperature at 5 bar is 151.8 °C, but to find the specific volume and enthalpy an interpolation is necessary,

$$v = (v \text{ at } 5 \text{ bar and } 300\,°\text{C})$$

$$+ \frac{20}{50}(v \text{ at } 5 \text{ bar and } 350\,°\text{C} - v \text{ at } 5 \text{ bar and } 300\,°\text{C})$$

therefore

$$v = 0.5226 + 0.4(0.5701 - 0.5226)$$

$$= 0.5226 + 0.019 = 0.5416 \text{ m}^3/\text{kg}$$

Similarly,

$$h = 3065 + 0.4(3168 - 3065) = 3065 + 41.2$$

i.e. $h = 3106.2 \text{ kJ/kg}$

In some cases a double interpolation is necessary. For example, to find the enthalpy of superheated steam at 18.5 bar and 432 °C an interpolation between 15 bar and 20 bar is necessary, and interpolation between 400 °C and 450 °C is also necessary. A tabular presentation is usually better in such cases (Table 2.3). First find the enthalpy at 15 bar and 432 °C,

$$h = 3256 + \frac{32}{50}(3364 - 3256) = 3256 + 0.64 \times 108$$

i.e. $h = 3325.1 \text{ kJ/kg}$

Table 2.3 Table showing a double interpolation

Pressure		Temperature/(°C)		
(bar)		400	432	450
15.0	h/(kJ/kg)	3256	?	3364
18.5	h/(kJ/kg)		?	
20.0	h/(kJ/kg)	3248	?	3357

Now find the enthalpy at 20 bar, 432 °C,

$$h = 3248 + 0.64(3357 - 3248) = 3248 + 0.64 \times 109$$

i.e. $h = 3317.8 \text{ kJ/kg}$

Now interpolate between h at 15 bar, 432 °C, and h at 20 bar, 432 °C in order to find h at 18.5 bar, 432 °C,

i.e. $h = 3325.1 - \dfrac{3.5}{5}(3325.1 - 3317.8)$

(Note the negative sign in this case since h at 15 bar, 432 °C is larger than h at 20 bar, 432 °C.) Then

$$h \text{ at } 18.5 \text{ bar, } 432\,°\text{C} = 3325.1 - (0.7 \times 7.3) = 3320 \text{ kJ/kg}$$

Example 2.5   Sketch a pressure–volume diagram for steam and mark on it the following points, labelling clearly the pressure, specific volume and temperature of each point.

(a) $p = 20$ bar, $t = 250\,°$C

(b) $t = 212.4\,°C$, $v = 0.09957\,\text{m}^3/\text{kg}$
(c) $p = 10\,\text{bar}$, $h = 2650\,\text{kJ/kg}$
(d) $p = 6\,\text{bar}$, $h = 3166\,\text{kJ/kg}$

Solution   Point (a): At 20 bar the saturation temperature is 212.4 °C, hence the steam is superheated at 250 °C. Then from tables, $v = 0.1115\,\text{m}^3/\text{kg}$.
Point (b): At 212.4 °C the saturation pressure is 20 bar and v_g is 0.09957 m³/kg. Therefore the steam is just dry saturated since $v = v_g$.
Point (c): At 10 bar, h_g is 2778 kJ/kg, therefore the steam is wet since $h = 2650\,\text{kJ/kg}$. Since the steam is wet, the temperature is the saturation temperature at 10 bar, i.e. $t = 179.9\,°C$. The dryness fraction can be found from equation (2.2),

$$h = h_f + x h_{fg}$$

therefore

$$x = \frac{2650 - 763}{2015} = \frac{1887}{2015} = 0.937$$

Then from equation (2.1)

$$v = x v_g$$

$$v = 0.937 \times 0.1944 = 0.182\,\text{m}^3/\text{kg}$$

Point (d): At 6 bar, h_g is 2757 kJ/kg, therefore the steam is superheated, since it is given that $h = 3166\,\text{kJ/kg}$. Hence from tables at 6 bar and $h = 3166\,\text{kJ/kg}$ the temperature is 350 °C, and the specific volume is 0.4743 m³/kg.

The points (a), (b), (c), and (d) can now be marked on a p–v diagram as shown in Fig. 2.8.

Fig. 2.8 Solution for Example 2.5

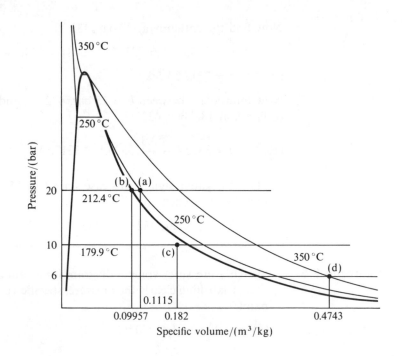

Example 2.6   Calculate the internal energy for each of the four states given in Example 2.5.

Solution   (a) The steam is superheated at 20 bar, 250 °C,

i.e. $u = 2681 \text{ kJ/kg}$

(b) The steam is dry saturated at 20 bar,

i.e. $u = u_g = 2600 \text{ kJ/kg}$

(c) The steam is wet at 10 bar with $x = 0.937$. Therefore

$$u = (1 - x)u_f + xu_g \quad \text{from equation (2.3)}$$

i.e. $u = (1 - 0.937)762 + (0.937 \times 2584) = 2470 \text{ kJ/kg}$

(d) The steam is superheated at 6 bar, 350 °C,

i.e. $u = 2881 \text{ kJ/kg}$

Example 2.7   Using the properties of ammonia given in ref. 2.1, calculate:

(i) the enthalpy at 1.902 bar, dryness fraction 0.95;
(ii) the enthalpy at 8.57 bar, 60 °C.

Solution   (i) From equation (2.2)

$$h = h_f + xh_{fg}$$

Therefore, at 1.902 bar,

$$h = 89.8 + 0.95(1420.0 - 89.8)$$

$$= 1353.5 \text{ kJ/kg}$$

(ii) At 8.570 bar the saturation temperature is 20 °C so the ammonia at 60 °C is superheated by $(60 - 20) = 40 \text{ K}$. It is therefore necessary to interpolate to find the enthalpy,

i.e. $h = 1462.6 + \dfrac{40}{50} \times (1597.2 - 1462.6)$

$$= 1570.3 \text{ kJ/kg}$$

2.3 The perfect gas

The characteristic equation of state

At temperatures that are considerably in excess of the critical temperature of a fluid, and also at very low pressures, the vapour of the fluid tends to obey the equation

$$\frac{pv}{T} = \text{constant} = R$$

No gases in practice obey this law rigidly, but many gases tend towards it. An imaginary ideal gas which obeys the law is called a *perfect gas*, and the equation, $pv/T = R$, is called the characteristic equation of state of a perfect gas. The constant, R, is called the *specific gas constant*. The units of R are N m/kg K or kJ/kg K. Each perfect gas has a different specific gas constant.

The characteristic equation is usually written

$$pv = RT \qquad (2.5)$$

or for a mass, m, occupying a volume, V,

$$pV = mRT \qquad (2.6)$$

Another form of the characteristic equation can be derived using the *amount of substance* (sometimes called the mole). The amount of substance is defined by the 1971 General Conference of Weights and Measures (CGPM) as follows:

The amount of substance of a system is that quantity which contains as many elementary entities as there are atoms in 0.012 kg of carbon-12; the elementary entities must be specified and may be atoms, molecules, ions, electrons, or other particles, or specific groups of such particles.

The normal unit symbol used for the amount of substance is 'mol'. In SI it is convenient to use 'kmol'.

The mass of any substance per amount of substance is known as the *molar mass*, $\tilde{m}$, i.e.

$$\tilde{m} = \frac{m}{n} \qquad (2.7)$$

where m is the mass and n is the amount of substance. The normal units used for m and n are kg and kmol, therefore the normal unit for $\tilde{m}$ is kg/kmol.

Relative masses of the various elements are commonly used, and physicists and chemists agreed in 1960 to give the value of 12 to the isotope 12 of carbon (this led to the definition of the amount of substance as above). A scale is thus obtained of *relative atomic mass* or *relative molecular mass* (e.g. the relative atomic mass of the element oxygen is approximately 16; the relative molecular mass of oxygen gas, O_2, is approximately 32).

The relative molecular mass is numerically equal to the molar mass, $\tilde{m}$, but is dimensionless.

Substituting for m from equation (2.7) in equation (2.6) gives

$$pV = n\tilde{m}RT \quad \text{or} \quad \tilde{m}R = \frac{pV}{nT}$$

Now *Avogadro's hypothesis* states that the volume of 1 mol of any gas is the same as the volume of 1 mol of any other gas, when the gases are at the same temperature and pressure. Therefore V/n is the same for all gases at the same value of p and T. That is, the quantity pV/nT is a constant for all gases. This constant is called the *molar gas constant*, and is given the symbol, $\tilde{R}$,

i.e. $$\tilde{m}R = \tilde{R} = \frac{pV}{nT} \quad \text{or} \quad pV = n\tilde{R}T \qquad (2.8)$$

or since $\tilde{m}R = \tilde{R}$ then

$$R = \frac{\tilde{R}}{\tilde{m}} \qquad (2.9)$$

The value of $\tilde{R}$ has been shown to be 8314.5 N m/kmol K.

From equation (2.9) the specific gas constant for any gas can be found when the molar mass is known, e.g. for oxygen of molar mass 32 kg/kmol, the specific gas constant

$$R = \frac{\tilde{R}}{\tilde{m}} = \frac{8314.5}{32} = 259.83 \text{ N m/kg K}$$

Example 2.8   A vessel of volume 0.2 m^3 contains nitrogen at 1.013 bar and 15 °C. If 0.2 kg of nitrogen is now pumped into the vessel, calculate the new pressure when the vessel has returned to its initial temperature. The molar mass of nitrogen is 28 kg/kmol, and it may be assumed to be a perfect gas.

Solution   From equation (2.9)

$$\text{Specific gas constant, } R = \frac{\tilde{R}}{\tilde{m}} = \frac{8314.5}{28} = 296.95 \text{ N m/kg K}$$

From equation (2.6), for the initial conditions

$$p_1 V_1 = m_1 R T_1$$

therefore

$$m_1 = \frac{p_1 V_1}{R T_1} = \frac{1.013 \times 10^5 \times 0.2}{296.95 \times 288} = 0.237 \text{ kg}$$

where $T_1 = 15 + 273 = 288$ K

The mass of nitrogen added is 0.2 kg, hence $m_2 = 0.2 + 0.237 = 0.437$ kg. Then from equation (2.6), for the final conditions

$$p_2 V_2 = m_2 R T_2$$

but $V_2 = V_1$ and $T_2 = T_1$, therefore

$$p_2 = \frac{m_2 R T_2}{V_2} = \frac{0.437 \times 296.95 \times 288}{10^5 \times 0.2}$$

i.e. $p_2 = 1.87$ bar

Example 2.9   A certain perfect gas of mass 0.01 kg occupies a volume of 0.003 m^3 at a pressure of 7 bar and a temperature of 131 °C. The gas is allowed to expand until the pressure is 1 bar and the final volume is 0.02 m^3. Calculate:

(i) the molar mass of the gas;
(ii) the final temperature.

Solution   (i) From equation (2.6)

$$p_1 V_1 = mRT_1$$

therefore

$$R = \frac{p_1 V_1}{mT_1} = \frac{7 \times 10^5 \times 0.003}{0.01 \times 404} = 520 \text{ N m/kg K}$$

where $T_1 = 131 + 273 = 404$ K.
  Then from equation (2.9)

$$R = \frac{\tilde{R}}{\tilde{m}}$$

therefore

$$\tilde{m} = \frac{\tilde{R}}{R} = \frac{8314.5}{520} = 16 \text{ kg/kmol}$$

i.e.   Molar mass = 16 kg/kmol
(ii) From equation (2.6)

$$p_2 V_2 = mRT_2$$

therefore

$$T_2 = \frac{p_2 V_2}{mR} = \frac{1 \times 10^5 \times 0.02}{0.01 \times 520} = 384.5 \text{ K}$$

i.e.   Final temperature = 384.5 − 273 = 111.5 °C

Specific heat capacity

The specific heat capacity of a solid or liquid is usually defined as the heat required to raise unit mass through one degree temperature rise. We have $dQ = mc\,dT$, where m is the mass, dT is the increase in temperature, and c is the specific heat capacity. For a gas there are an infinite number of ways in which heat may be added between two temperatures, and hence a gas could have an infinite number of specific heat capacities. However, only two specific heat capacities for gases are defined; the specific heat capacity at constant volume, c_v, and the specific heat capacity at constant pressure, c_p.

The definition must be restricted to reversible non-flow processes, since irreversibilities can cause temperature changes which are indistinguishable from those due to reversible heat and work quantities. Specific heat capacities can be introduced more rigorously as properties of a fluid. We have in the limit

$$c_v = \left(\frac{\partial u}{\partial T} \right)_v \quad \text{and} \quad c_p = \left(\frac{\partial h}{\partial T} \right)_p$$

A more rigorous treatment is given in ref. 2.3.

We can write:

$$dQ = mc_p \, dT \quad \text{for a reversible non-flow process at constant pressure} \tag{2.10}$$

and

$$dQ = mc_v \, dT \quad \text{for a reversible non-flow process at constant volume} \tag{2.11}$$

For a perfect gas the values of c_p and c_v are constant for any one gas at all pressures and temperatures. Hence integrating equations (2.10) and (2.11) we have for a reversible constant pressure process

$$Q = mc_p(T_2 - T_1) \tag{2.12}$$

for a reversible constant volume process

$$Q = mc_v(T_2 - T_1) \tag{2.13}$$

For real gases, c_p and c_v vary with temperature, but for most practical purposes a suitable average value may be used.

Joule's law

Joule's law states that the internal energy of a perfect gas is a function of the absolute temperature only, i.e. $u = f(T)$. To evaluate this function let unit mass of a perfect gas be heated at constant volume. From the non-flow energy equation, (1.5),

$$dQ + dW = du$$

Since the volume remains constant then no work is done, i.e. $dW = 0$, therefore

$$dQ = du$$

At constant volume for a perfect gas, from equation (2.11), for unit mass,

$$dQ = c_v \, dT$$

Therefore, $dQ = du = c_v \, dT$, and integrating

$$u = c_v T + K$$

where K is a constant.

Joule's law states that $u = f(T)$, hence it follows that the internal energy varies linearly with absolute temperature. Internal energy can be made zero at any arbitrary reference temperature. For a perfect gas it can be assumed that $u = 0$ when $T = 0$, hence the constant K is zero,

i.e.   Specific internal energy, $u = c_v T$ for a perfect gas   (2.14)

or for mass, m, of a perfect gas,

Internal energy, $U = mc_v T$ (2.15)

In any process for a perfect gas, between states 1 and 2, we have from equation (2.15),

$$\text{Gain in internal energy, } U_2 - U_1 = mc_v(T_2 - T_1) \tag{2.16}$$

The gain of internal energy for a perfect gas between two states is always given by equation (2.16), for any process, reversible or irreversible.

Relationship between the specific heat capacities

Let a perfect gas be heated at constant pressure from T_1 to T_2. From the non-flow equation (1.4), $Q + W = (U_2 - U_1)$. Also, for a perfect gas, from equation (2.16), $U_2 - U_1 = mc_v(T_2 - T_1)$. Hence,

$$Q + W = mc_v(T_2 - T_1)$$

In a constant pressure process the work done is given by the pressure times the change in volume, i.e. $W = -p(V_2 - V_1)$. Then using equation (2.6), $pV_2 = mRT_2$ and $pV_1 = mRT_1$, we have

$$W = -mR(T_2 - T_1)$$

Therefore substituting

$$Q - mR(T_2 - T_1) = mc_v(T_2 - T_1)$$

therefore

$$Q = m(c_v + R)(T_2 - T_1)$$

But for a constant pressure process from equation (2.12)

$$Q = mc_p(T_2 - T_1)$$

Hence by equating the two expressions for the heat flow, Q, we have

$$m(c_v + R)(T_2 - T_1) = mc_p(T_2 - T_1)$$

therefore

$$c_v + R = c_p$$

or $\qquad c_p - c_v = R \tag{2.17}$

Specific enthalpy of a perfect gas

From equation (1.9), specific enthalpy, $h = u + pv$.

For a perfect gas, from equation (2.5), $pv = RT$. Also for a perfect gas, from Joule's law, equation (2.14), $u = c_v T$. Hence, substituting

$$h = c_v T + RT = (c_v + R)T$$

But from equation (2.17)

$$c_p - c_v = R \quad \text{or} \quad c_v + R = c_p$$

Therefore, specific enthalpy, h, for a perfect gas is given by

$$h = c_p T \tag{2.18}$$

For mass, m, of a perfect gas

$$H = mc_p T \tag{2.19}$$

(Note that, since it has been assumed that $u = 0$ at $T = 0$, then $h = 0$ at $T = 0$.)

Ratio of specific heat capacities

The ratio of the specific heat capacity at constant pressure to the specific heat capacity at constant volume is given the symbol γ (gamma),

i.e. $$\gamma = \frac{c_p}{c_v} \tag{2.20}$$

Note that since $c_p - c_v = R$, from equation (2.17), it is clear that c_p must be greater than c_v for any perfect gas. It follows therefore that the ratio, $c_p/c_v = \gamma$, is always greater than unity. In general, γ is about 1.4 for diatomic gases such as carbon monoxide (CO), hydrogen (H_2), nitrogen (N_2), and oxygen (O_2). For monoatomic gases such as argon (A), and helium (He), γ is about 1.6, and for triatomic gases such as carbon dioxide (CO_2), and sulphur dioxide (SO_2), γ is about 1.3. For some hydrocarbons the value of γ is quite low (e.g. for ethane (C_2H_6), $\gamma = 1.22$, and for isobutane (C_4H_{10}), $\gamma = 1.11$).

Some useful relationships between c_p, c_v, R, and γ can be derived. From equation (2.17)

$$c_p - c_v = R$$

Dividing through by c_v

$$\frac{c_p}{c_v} - 1 = \frac{R}{c_v}$$

Therefore using equation (2.17), $\gamma = c_p/c_v$, then,

$$\gamma - 1 = \frac{R}{c_v}$$

therefore

$$c_v = \frac{R}{(\gamma - 1)} \tag{2.21}$$

Also from equation (2.20), $c_p = \gamma c_v$, hence, substituting in equation (2.21),

$$c_p = \gamma c_v = \frac{\gamma R}{(\gamma - 1)} \tag{2.22}$$

45

Example 2.10   A certain perfect gas has specific heat capacities as follows:

$$c_p = 0.846 \text{ kJ/kg K} \quad \text{and} \quad c_v = 0.657 \text{ kJ/kg K}$$

Calculate the gas constant and the molar mass of the gas.

Solution   From equation (2.17)

$$c_p - c_v = R$$

i.e. $R = 0.846 - 0.657 = 0.189 \text{ kJ/kg K} = 189 \text{ N m/kg K}$

From equation (2.9)

$$R = \frac{\tilde{R}}{\tilde{m}}$$

i.e. $\tilde{m} = \dfrac{8314.5}{189} = 44 \text{ kg/kmol}$

Example 2.11   A perfect gas has a molar mass of 26 kg/kmol and a value of $\gamma = 1.26$. Calculate the heat rejected:

(i) when unit mass of the gas is contained in a rigid vessel at 3 bar and 315 °C, and is then cooled until the pressure falls to 1.5 bar;

(ii) when unit mass flow rate of the gas enters a pipeline at 280 °C, and flows steadily to the end of the pipe where the temperature is 20 °C. Neglect changes in velocity of the gas in the pipeline.

Solution   From equation (2.9)

$$R = \frac{\tilde{R}}{\tilde{m}} = \frac{8314.5}{26} = 319.8 \text{ N m/kg K}$$

From equation (2.21)

$$c_v = \frac{R}{(\gamma - 1)} = \frac{319.8}{10^3 (1.26 - 1)} = 1.229 \text{ kJ/kg K}$$

Also from equation (2.20)

$$\frac{c_p}{c_v} = \gamma$$

therefore

$$c_p = \gamma c_v = 1.26 \times 1.229 = 1.548 \text{ kJ/kg K}$$

(i) The volume remains constant for the mass of gas present, and hence the specific volume remains constant. From equation (2.5),

$$p_1 v_1 = R T_1 \quad \text{and} \quad p_2 v_2 = R T_2$$

Therefore since $v_1 = v_2$ we have

$$T_2 = T_1 \frac{p_2}{p_1} = 588 \times \frac{1.5}{3} = 294 \text{ K}$$

where $T_1 = 315 + 273 = 588$ K.

Then from equation (2.13)

$$\text{Heat supplied per kg of gas} = c_v(T_2 - T_1) = 1.229(294 - 588)$$
$$= -1.229 \times 294 = -361 \text{ kJ/kg}$$

i.e.   Heat rejected per kilogram of gas $= +361$ kJ/kg

(ii) From the steady-flow energy equation, (1.10),

$$\dot{m}\left(h_1 + \frac{C_1^2}{2}\right) + \dot{Q} + \dot{W} = \dot{m}\left(h_2 + \frac{C_2^2}{2}\right)$$

In this case we are told that changes in velocity are negligible; also there is no work done.

Therefore we have

$$\dot{m}h_1 + \dot{Q} = \dot{m}h_2$$

For a perfect gas, from equation (2.18)

$$h = c_p T$$

therefore

$$\dot{Q} = \dot{m}c_p(T_2 - T_1) = 1 \times 1.548(20 - 280) = -403 \text{ kW}$$

i.e.   Heat rejected per kilogram per second $= +403$ kW

Note that it is not necessary to convert $t_1 = 280\,°\text{C}$ and $t_2 = 20\,°\text{C}$ into degrees Kelvin, since the temperature difference $(t_1 - t_2)$ is numerically the same as the temperature difference $(T_1 - T_2)$.

Problems

[Note: the answers to these problems have been evaluated using the tables of Rogers and Mayhew (ref. 2.1). The values of R, c_p, c_v, and γ for air may be assumed to be as given in the tables (i.e. $R = 0.287$ kJ/kg K; $c_p = 1.005$ kJ/kg K; $c_v = 0.718$ kJ/kg K; and $\gamma = 1.4$). For any other perfect gas the values of R, c_p, c_v, and γ, if required, must be calculated from the information given in the problem; the value of $\tilde{R}$ is given in the tables (ref. 2.1).]

2.1 Complete Table 2.4 (p. 48) using steam tables. Insert a dash for irrelevant items, and interpolate where necessary.

(see Table 2.6, p. 50)

2.2 A vessel of volume 0.03 m³ contains dry saturated steam at 17 bar. Calculate the mass of steam in the vessel and the enthalpy of this mass.

(0.257 kg; 718 kJ)

Table 2.4 Data for Problem 2.1

p (bar)	t (°C)	v (m³/kg)	x	Degree of superheat	h (kJ/kg)	u (kJ/kg)
	90	2.361				
20						2799
5		0.3565				
	188					2400
34			0.9			
	81.3		0.85			
3	200					
15		0.152				
130					3335	
	250	1.601				
38.2			0.8			
	297		0.95			
2.3	300					
44	420					

The completed table is given on p. 50 as Table 2.6.

2.3 Steam at 7 bar and 250 °C enters a pipeline and flows along it at constant pressure. If the steam rejects heat steadily to the surroundings, at what temperature will droplets of water begin to form in the vapour? Using the steady-flow energy equation, and neglecting changes in velocity of the steam, calculate the heat rejected per kilogram of steam flowing.

(165 °C; 191 kJ/kg)

2.4 0.05 kg of steam at 15 bar is contained in a rigid vessel of volume 0.0076 m³. What is the temperature of the steam? If the vessel is cooled, at what temperature will the steam be just dry saturated? Cooling is continued until the pressure in the vessel is 11 bar; calculate the final dryness fraction of the steam, and the heat rejected between the initial and the final states.

(250 °C; 191.4 °C; 0.857; 18.5 kJ)

2.5 Using the tables for ammonia given in ref. 2.1, calculate:
(i) the specific enthalpy and specific volume of ammonia at 0.7177 bar, dryness fraction 0.9;
(ii) the specific enthalpy and specific volume of ammonia at 13 °C saturated;
(iii) the specific enthalpy of ammonia at 7.529 bar, 30 °C.

(1251 kJ/kg, 1.397 m³/kg; 1457 kJ/kg, 0.1866 m³/kg; 1496.5 kJ/kg)

2.6 Using the property values for refrigerant HFA 134a given in Table 2.5, calculate:
(i) the specific enthalpy and specific volume of HFA 134a at −8 °C, dryness fraction 0.85;
(ii) the specific enthalpy of HFA 134a at 5.7024 bar, 35 °C.

(259.96 kJ/kg, 0.0775 m³/kg; 323.25 kJ/kg)

Table 2.5 Data for
Problem 2.6

	Saturation values				Superheat values degree of superheat 20 K
t_g	p_g	v_g	h_f	h_g	h
(°C)	(bar)	(m³/kg)	(kJ/kg)		(kJ/kg)
−10	2.0051	0.098	86.98	288.86	308.64
−5	2.4371	0.081	93.46	291.77	312.05
20	5.7024	0.036	126.92	306.22	328.93

2.7 The relative molecular mass of carbon dioxide, CO_2, is 44. In an experiment the value of γ for CO_2 was found to be 1.3. Assuming that CO_2 is a perfect gas, calculate the specific gas constant, R, and the specific heat capacities at constant pressure and constant volume, c_p and c_v.

(0.189 kJ/kg K; 0.819 kJ/kg K; 0.63 kJ/kg K)

2.8 Calculate the internal energy and enthalpy of 1 kg of air occupying 0.05 m³ at 20 bar. If the internal energy is increased by 120 kJ as the air is compressed to 50 bar, calculate the new volume occupied by 1 kg of the air.

(250.1 kJ/kg; 350.1 kJ/kg; 0.0296 m³)

2.9 Oxygen, O_2, at 200 bar is to be stored in a steel vessel at 20 °C. The capacity of the vessel is 0.04 m³. Assuming that O_2 is a perfect gas, calculate the mass of oxygen that can be stored in the vessel. The vessel is protected against excessive pressure by a fusible plug which will melt if the temperature rises too high. At what temperature must the plug melt to limit the pressure in the vessel to 240 bar? The molar mass of oxygen is 32 kg/kmol.

(10.5 kg; 78.6 °C)

2.10 When a certain perfect gas is heated at constant pressure from 15°C to 95 °C, the heat required is 1136 kJ/kg. When the same gas is heated at constant volume between the same temperatures the heat required is 808 kJ/kg. Calculate c_p, c_v, γ, R and the molar mass of the gas.

(14.2 kJ/kg K; 10.1 kJ/kg K; 1.405; 4.1 kJ/kg K; 2.028 kg/kmol)

2.11 In an air compressor the pressures at inlet and outlet are 1 bar and 5 bar respectively. The temperature of the air at inlet is 15°C and the volume at the beginning of compression is three times that at the end of compression. Calculate the temperature of the air at outlet and the increase of internal energy per kg of air.

(207 °C; 138 kJ/kg)

2.12 A quantity of a certain perfect gas is compressed from an initial state of 0.085 m³, 1 bar to a final state of 0.034 m³, 3.9 bar. The specific heat at constant volume is 0.724 kJ/kg K, and the specific heat at constant pressure is 1.020 kJ/kg K. The observed temperature rise is 146 K. Calculate the specific gas constant, R, the mass of gas present, and the increase of internal energy of the gas.

(0.296 kJ/kg K; 0.11 kg; 11.63 kJ)

Table 2.6 Solution to Problem 2.1 on p. 47

p (bar)	t (°C)	v (m³/kg)	x	Degree of superheat	h (kJ/kg)	u (kJ/kg)
0.70	90	2.361	1	0	2660	2494
20	212.4	0.09957	1	0	2799	2600
5	151.8	0.3565	0.951	—	2646	2471
12	188	0.1461	0.895	—	2576	2400
34	240.9	0.0529	0.9	—	2627	2447
0.5	81.3	2.75	0.85	—	2300	2165
3	200	0.7166	—	66.5	2866	2651
15	250	0.152	—	51.7	2925	2697
130	500	0.02447	—	169.2	3335	3017
1.5	250	1.601	—	138.6	2973	2733
38.2	247.6	0.04175	0.8	—	2456	2296
82.38	297	0.0216	0.95	—	2683	2505
2.3	300	1.184	—	175.8	3071	2808
44	420	0.0696	—	164.3	3254	2952

References

2.1 ROGERS G F C and MAYHEW Y R 1987 *Thermodynamic and Transport Properties of Fluids* 4th edn Blackwell

2.2 NATIONAL ENGINEERING LABORATORY *Steam Tables 1964* HMSO

2.3 ROGERS G F C and MAYHEW Y R 1992 *Engineering Thermodynamics, Work and Heat Transfer* 4th edn Longman

3

Reversible and Irreversible Processes

In the previous chapters the energy equations for non-flow and flow processes are derived, the concepts of reversibility and irreversibility introduced, and the properties of vapours and perfect gases discussed. It is the purpose of this chapter to consider processes in practice, and to combine this with the work of the previous chapters.

3.1 Reversible non-flow processes

Constant volume process

In a constant volume process the working substance is contained in a rigid vessel, hence the boundaries of the system are immovable and no work can be done on or by the system, other than paddle-wheel work input. It will be assumed that 'constant volume' implies zero work unless stated otherwise.

From the non-flow energy equation, (1.4), for unit mass,

$$Q + W = u_2 - u_1$$

Since no work is done, we therefore have

$$Q = u_2 - u_1 \tag{3.1}$$

or for mass, m, of the working substance

$$Q = U_2 - U_1 \tag{3.2}$$

All the heat supplied in a constant volume process goes to increasing the internal energy.

A constant volume process for a vapour is shown on a p–v diagram in Fig. 3.1(a). The initial and final states have been chosen to be in the wet region and superheat region respectively. In Fig. 3.1(b) a constant volume process is shown on a p–v diagram for a perfect gas. For a perfect gas we have from equation (2.13)

$$Q = mc_v(T_2 - T_1)$$

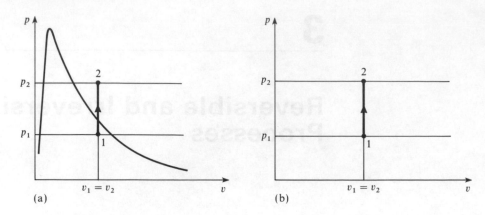

Fig. 3.1 Constant volume process for a vapour and a perfect gas

Constant pressure process

It can be seen from Figs 3.1(a) and 3.1(b) that when the boundary of the system is inflexible as in a constant volume process, then the pressure rises when heat is supplied. Hence for a constant pressure process the boundary must move against an external resistance as heat is supplied; for instance a fluid in a cylinder behind a piston can be made to undergo a constant pressure process. Since the piston is pushed through a certain distance by the force exerted by the fluid, then work is done by the fluid on its surroundings.

From equation (1.2) for unit mass

$$W = -\int_{v_1}^{v_2} p \, dv \quad \text{for any reversible process}$$

Therefore, since p is constant,

$$W = -p \int_{v_1}^{v_2} dv = -p(v_2 - v_1)$$

From the non-flow energy equation, (1.4),

$$Q + W = u_2 - u_1$$

Hence for a reversible constant pressure process

$$Q = (u_2 - u_1) + p(v_2 - v_1) = (u_2 + pv_2) - (u_1 + pv_1)$$

Now from equation (1.9), enthalpy, $h = u + pv$, hence,

$$Q = h_2 - h_1 \tag{3.3}$$

or for mass, m, of a fluid,

$$Q = H_2 - H_1 \tag{3.4}$$

A constant pressure process for a vapour is shown on a p–v diagram in Fig. 3.2(a). The initial and final states have been chosen to be in the wet region and the superheat region respectively. In Fig. 3.2(b) a constant pressure process for a perfect gas is shown on a p–v diagram. For a perfect gas we have from

Fig. 3.2 Constant pressure process for a vapour and a perfect gas

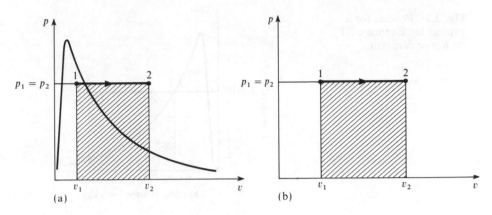

equation (2.12),

$$Q = mc_p(T_2 - T_1)$$

Note that in Figs 3.2(a) and 3.2(b) the shaded areas represent the work done *by* the fluid, $p(v_2 - v_1)$.

Example 3.1   A mass of 0.05 kg of a fluid is heated at a constant pressure of 2 bar until the volume occupied is 0.0658 m³. Calculate the heat supplied and the work done:

   (i) when the fluid is steam, initially dry saturated;
   (ii) when the fluid is air, initially at 130°C.

Solution   (i) Initially the steam is dry saturated at 2 bar, hence,

$$h_1 = h_g \text{ at } 2 \text{ bar} = 2707 \text{ kJ/kg}$$

Finally the steam is at 2 bar and the specific volume is given by

$$v_2 = \frac{0.0658}{0.05} = 1.316 \text{ m}^3/\text{kg}$$

Hence the steam is superheated finally. From superheat tables at 2 bar and 1.316 m³/kg the temperature of the steam is 300°C, and the enthalpy is $h_2 = 3072$ kJ/kg.

Then from equation (3.4)

$$Q = H_2 - H_1 = m(h_2 - h_1) = 0.05(3072 - 2707)$$

i.e.   Heat supplied $= 0.05 \times 365 = 18.25$ kJ

The process is shown on a *p–v* diagram in Fig. 3.3,

$$-W = p(v_2 - v_1) = \text{shaded area}$$

Now $v_1 = v_g$ at 2 bar $= 0.8856$ m³/kg, and $v_2 = 1.316$ m³/kg. Therefore

$$W = -2 \times 10^5(1.316 - 0.8856) = -86\,080 \text{ N m/kg}$$

Fig. 3.3 Process for a
vapour for Example 3.1
on a p–v diagram

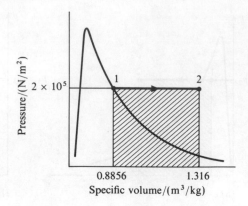

i.e.   Work done *by* the total mass present = $0.05 \times 86\,080$

$$= 4304\text{ N m} = 4.304\text{ kJ}$$

(ii) Using equation (2.6),

$$T_2 = \frac{p_2 V_2}{mR} = \frac{2 \times 10^5 \times 0.0658}{0.05 \times 0.287 \times 10^3} = 917\text{ K}$$

For a perfect gas undergoing a constant pressure process we have, from equation (2.12),

$$Q = mc_p(T_2 - T_1)$$

i.e.   Heat supplied = $0.05 \times 1.005(917 - 403)$

where $T_1 = 130 + 273 = 403$ K,

i.e.   Heat supplied = $0.05 \times 1.005 \times 514 = 25.83$ kJ

The process is shown on a p–v diagram in Fig. 3.4, i.e.

$$-W = p(v_2 - v_1) = \text{shaded area}$$

From equation (2.5), $pv = RT$, therefore

$$\text{Work done} = -R(T_2 - T_1) = -0.287(917 - 403)\text{ kJ/kg}$$

i.e.   Work done *by* the mass of gas present = $0.05 \times 0.287 \times 514$

$$= 7.38\text{ kJ}$$

Fig. 3.4 Process for a
perfect gas for
Example 3.1 on a
p–v diagram

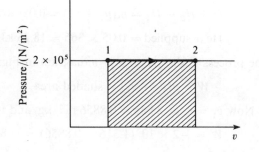

Constant temperature or isothermal process

A process at constant temperature is called an isothermal process. When a fluid in a cylinder behind a piston expands from a high pressure to a low pressure there is a tendency for the temperature to fall. In an isothermal expansion heat must be added continuously in order to keep the temperature at the initial value. Similarly in an isothermal compression heat must be removed from the fluid continuously during the process. An isothermal process for a vapour is shown on a $p–v$ diagram in Fig. 3.5. The initial and final states have been chosen in the wet region and superheat region respectively. From state 1 to state A the pressure remains at p_1, since in the wet region the pressure and temperature are the corresponding saturation values. It can be seen therefore that an isothermal process for wet steam is also at constant pressure and equations (3.3) and (3.4) can be used (e.g. heat supplied from state 1 to state A per kilogram of steam $= h_A - h_1$). In the superheat region the pressure falls to p_2 as shown in Fig. 3.5, and the procedure is not so simple. When states 1 and 2 are fixed then the internal energies u_1 and u_2 may be obtained from tables. When the property entropy, s, is introduced in Chapter 4, a convenient way of evaluating the heat supplied will be shown. When the heat flow is calculated the work done can then be obtained using the non-flow energy equation, (1.4), for unit mass

$$Q + W = u_2 - u_1$$

Fig. 3.5 Isothermal process for a vapour on a $p–v$ diagram

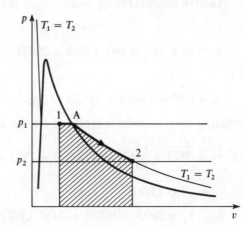

Example 3.2 Steam at 7 bar and dryness fraction 0.9 expands in a cylinder behind a piston isothermally and reversibly to a pressure of 1.5 bar. Calculate the change of internal energy and the change of enthalpy per kg of steam. The heat supplied during the process is found to be 547 kJ/kg, by the method of Chapter 4. Calculate the work done per kilogram of steam.

Solution The process is shown in Fig. 3.6. The saturation temperature corresponding to 7 bar is 165 °C. Therefore the steam is superheated at state 2. The internal energy at state 1 is found by using equation 2.3,

i.e. $u_1 = (1 - x)u_f + xu_g = (1 - 0.9) \times 696 + (0.9 \times 2573)$

Fig. 3.6 Isothermal process on a $p-v$ diagram for Example 3.2

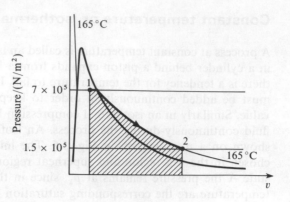

therefore

$$u_1 = 69.6 + 2315.7 = 2385.3 \text{ kJ/kg}$$

Interpolating from superheat tables at 1.5 bar and 165 °C, we have

$$u_2 = 2580 + \frac{15}{50}(2656 - 2580) = 2580 + 22.8$$

i.e. $\quad u_2 = 2602.8 \text{ kJ/kg}$

Therefore

$$\text{Gain in internal energy} = u_2 - u_1 = 2602.8 - 2385.3$$

$$= 217.5 \text{ kJ/kg}$$

$$h_1 = h_f + x h_{fg} = 697 + (0.9 \times 2067)$$

therefore

$$h_1 = 697 + 1860.3 = 2557.3 \text{ kJ/kg}$$

Interpolating from superheat tables at 1.5 bar and 165 °C, we have

$$h_2 = 2773 + \frac{15}{50}(2873 - 2773) = 2773 + 30$$

$$= 2803 \text{ kJ/kg}$$

i.e. $\quad h_2 - h_1 = 2803 - 2557.3 = 245.7 \text{ kJ/kg}$

From the non-flow energy equation, (1.4),

$$Q + W = u_2 - u_1$$

therefore

$$W = (u_2 - u_1) - Q = 217.5 - 547 = -329.5 \text{ kJ/kg}$$

i.e. $\quad$ Work done *by* the system $= 329.5 \text{ kJ/kg}$

(The work output is also given by the area on Fig. 3.6, $\int_{v_1}^{v_2} p \, dv$; this could only be evaluated graphically in this case.)

An isothermal process for a perfect gas is more easily dealt with than an isothermal process for a vapour, since there are definite laws for a perfect gas relating p, v, and T, and the internal energy u. We have, from equation (2.5),

$$pv = RT$$

Now when the temperature is constant as in an isothermal process then we have

$$pv = RT = \text{constant}$$

Therefore for an isothermal process for a perfect gas

$$pv = \text{constant} \tag{3.5}$$

i.e. $\quad p_1 v_1 = p_2 v_2$

In Fig. 3.7 an isothermal compression process for a perfect gas is shown on a p–v diagram. The equation of the process is $pv = \text{constant}$, which is the equation of a hyperbola. It must be stressed that an isothermal process is only of the form $pv = \text{constant}$ for a perfect gas, because it is only for a perfect gas that an equation of state, $pv = RT$, can be applied.

From equation (1.2) we have for unit mass

$$W = -\int_1^2 p \, dv = (\text{shaded area in Fig. 3.7})$$

Fig. 3.7 Isothermal process for a perfect gas on a p–v diagram

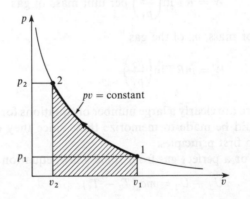

In this case, $pv = \text{constant}$, or $p = c/v$, where $c = \text{constant}$. Therefore

$$W = -\int_{v_1}^{v_2} c \frac{dv}{v} = -c[\ln v]_{v_1}^{v_2} = c \ln\left(\frac{v_1}{v_2}\right)$$

The constant c can either be written as $p_1 v_1$ or as $p_2 v_2$, since $p_1 v_1 = p_2 v_2 = \text{constant}$, c,

i.e. $\quad W = p_1 v_1 \ln\left(\dfrac{v_1}{v_2}\right)$ per unit mass of gas $\tag{3.6}$

or $\quad W = p_2 v_2 \ln\left(\dfrac{v_1}{v_2}\right)$ per unit mass of gas

For mass, m, of the gas

$$W = p_1 V_1 \ln\left(\frac{v_1}{v_2}\right) \quad (3.7)$$

Also, since $p_1 v_1 = p_2 v_2$, then

$$\frac{v_1}{v_2} = \frac{p_2}{p_1}$$

Hence, substituting in equation (3.6)

$$W = p_1 v_1 \ln\frac{p_2}{p_1} \text{ per unit mass of gas} \quad (3.8)$$

or for mass, m, of the gas

$$W = p_1 V_1 \ln\left(\frac{p_2}{p_1}\right) \quad (3.9)$$

Using equation (2.5)

$$p_1 v_1 = RT$$

Hence, substituting in equation (3.8)

$$W = RT \ln\left(\frac{p_2}{p_1}\right) \text{ per unit mass of gas} \quad (3.10)$$

or for mass, m, of the gas

$$W = mRT \ln\left(\frac{p_2}{p_1}\right) \quad (3.11)$$

There are clearly a large number of equations for the work done, and no attempt should be made to memorize these since they can all be derived very simply from first principles.

For a perfect gas from Joule's law, equation (2.14), we have

$$U_2 - U_1 = mc_v(T_2 - T_1)$$

Hence for an isothermal process for a perfect gas, since $T_2 = T_1$, then

$$U_2 - U_1 = 0$$

i.e. the internal energy remains constant in an isothermal process for a perfect gas.

From the non-flow energy equation (1.4) for unit mass

$$Q + W = u_2 - u_1$$

Therefore, since $u_2 = u_1$, then

$$Q + W = 0 \quad (3.12)$$

for an isothermal process for a perfect gas.

Note that the heat flow plus the work input is zero in an isothermal process *for a perfect gas only*. From Example 3.2 for steam it is seen that, although the process is isothermal, the change in internal energy is 217.5 kJ/kg, and therefore the heat supplied plus the work done is not zero.

Example 3.3   1 kg of nitrogen (molar mass 28 kg/kmol) is compressed reversibly and isothermally from 1.01 bar, 20 °C to 4.2 bar. Calculate the work done and the heat flow during the process. Assume nitrogen to be a perfect gas.

Solution   From equation (2.9), for nitrogen,

$$R = \frac{\tilde{R}}{\tilde{m}} = \frac{8.3145}{28} = 0.297 \text{ kJ/kg K}$$

The process is shown on a $p-v$ diagram in Fig. 3.8.

Fig. 3.8 Isothermal process on a $p-v$ diagram for Example 3.3

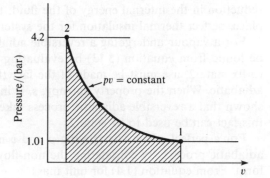

From equation (3.10)

$$W = RT\ln\left(\frac{p_2}{p_1}\right) = 0.297 \times 293 \times \ln\left(\frac{4.2}{1.01}\right) = 124 \text{ kJ/kg}$$

where $T = 20 + 273 = 293$ K.

i.e.   Work input = 124 kJ/kg

From equation (3.12), for an isothermal process for a perfect gas,

$$Q + W = 0$$

therefore

$$Q = -124 \text{ kJ/kg}$$

i.e.   Heat rejected = 124 kJ/kg

3.2 Reversible adiabatic non-flow processes

An *adiabatic* process is one in which no heat is transferred to or from the fluid during the process. Such a process can be reversible or irreversible. The reversible adiabatic non-flow process will be considered in this section.

From the non-flow equation (1.4),

$$Q + W = u_2 - u_1$$

and for an adiabatic process

$$Q = 0$$

Therefore we have

$$W = u_2 - u_1 \quad \text{for any adiabatic non-flow process} \tag{3.13}$$

Equation (3.13) is true for an adiabatic non-flow process whether or not the process is reversible. In an adiabatic compression process all the work done on the fluid goes to increasing the internal energy of the fluid. Similarly in an adiabatic expansion process, the work done by the fluid is at the expense of a reduction in the internal energy of the fluid. For an adiabatic process to take place, perfect thermal insulation for the system must be available.

For a vapour undergoing a reversible adiabatic process the work done can be found from equation (3.13) by evaluating u_1 and u_2 from tables. In order to fix state 2, use must be made of the fact that the process is reversible and adiabatic. When the property entropy, s, is introduced in Chapter 4 it will be shown that a reversible adiabatic process takes place at constant entropy, and this fact can be used to fix state 2.

For a perfect gas, a law relating p and v may be obtained for a reversible adiabatic process, by considering the non-flow energy equation in differential form. From equation (1.4) for unit mass

$$dQ + dW = du$$

Also for a reversible process $dW = -p\,dv$, hence for a reversible adiabatic process

$$dQ = du + p\,dv = 0 \tag{3.14}$$

Since $h = u + pv$

then $dh = du + p\,dv + v\,dp$

i.e. $du + p\,dv = dh - v\,dp$

and hence, using equation (3.14),

$$dh - v\,dp = 0$$

i.e. $dh = v\,dp$ \hfill (3.15)

Also, using equations (2.5) and (3.14), we have

$$du + \frac{RT\,dv}{v} = 0$$

From equation (2.14)

$$u = c_v T \quad \text{or} \quad du = c_v\,dT$$

therefore

$$c_v \, dT + \frac{RT \, dv}{v} = 0$$

Dividing through by T to give a form that can be integrated, i.e.

$$c_v \frac{dT}{T} + \frac{R \, dv}{v} = 0$$

Integrating

$$c_v \ln T + R \ln v = \text{constant}$$

Using equation (2.5) we have $T = (pv)/R$, therefore substituting

$$c_v \ln\left(\frac{pv}{R}\right) + R \ln v = \text{constant}$$

Dividing through by c_v

$$\ln\left(\frac{pv}{R}\right) + \frac{R}{c_v} \ln v = \text{constant}$$

Also, from equation (2.21),

$$c_v = \frac{R}{(\gamma - 1)} \quad \text{or} \quad \frac{R}{c_v} = \gamma - 1$$

Hence substituting

$$\ln\left(\frac{pv}{R}\right) + (\gamma - 1) \ln v = \text{constant}$$

or

$$\ln\left(\frac{pv}{R}\right) + \ln(v^{\gamma - 1}) = \text{constant}$$

therefore

$$\ln\left(\frac{pvv^{\gamma - 1}}{R}\right) = \text{constant}$$

i.e.

$$\ln\left(\frac{pv^{\gamma}}{R}\right) = \text{constant}$$

therefore

$$\frac{pv^{\gamma}}{R} = e^{(\text{constant})} = \text{constant}$$

or

$$pv^{\gamma} = \text{constant} \tag{3.16}$$

We therefore have a simple relationship between p and v for any perfect gas undergoing a reversible adiabatic process, each perfect gas having its own value of γ.

Using equation (2.5), $pv = RT$, relationships between T and v, and T and p, may be derived,

i.e. $pv = RT$

therefore

$$p = \frac{RT}{v}$$

Substituting in equation (3.16)

$$\frac{RT}{v}v^{\gamma} = \text{constant}$$

i.e. $Tv^{\gamma-1} = \text{constant}$ (3.17)

Also, $v = (RT)/p$; hence substituting in equation (3.16)

$$p\left(\frac{RT}{p}\right)^{\gamma} = \text{constant}$$

therefore

$$\frac{T^{\gamma}}{p^{\gamma-1}} = \text{constant}$$

or $\dfrac{T}{p^{(\gamma-1)/\gamma}} = \text{constant}$ (3.18)

Therefore for a reversible adiabatic process for a perfect gas between states 1 and 2 we can write as follows. From equation (3.16)

$$p_1 v_1^{\gamma} = p_2 v_2^{\gamma} \quad \text{or} \quad \frac{p_1}{p_2} = \left(\frac{v_2}{v_1}\right)^{\gamma}$$ (3.19)

From equation (3.17)

$$T_1 v_1^{\gamma-1} = T_2 v_2^{\gamma-1} \quad \text{or} \quad \frac{T_1}{T_2} = \left(\frac{v_2}{v_1}\right)^{\gamma-1}$$ (3.20)

From equation (3.18)

$$\frac{T_1}{p_1^{(\gamma-1)/\gamma}} = \frac{T_2}{p_2^{(\gamma-1)/\gamma}} \quad \text{or} \quad \frac{T_1}{T_2} = \left(\frac{p_1}{p_2}\right)^{(\gamma-1)/\gamma}$$ (3.21)

From equation (3.13) the work done in an adiabatic process per unit mass of gas is given by $W = (u_2 - u_1)$. The gain in internal energy of a perfect gas is given by equation (2.16),

i.e.   for unit mass $u_2 - u_1 = c_v(T_2 - T_1)$

therefore

$$W = c_v(T_2 - T_1)$$

Also, from equation (2.21),

$$c_v = \frac{R}{(\gamma - 1)}$$

Hence substituting

$$W = \frac{R(T_2 - T_1)}{(\gamma - 1)} \tag{3.22}$$

Using equation (2.5), $pv = RT$,

$$W = \frac{p_2 v_2 - p_1 v_1}{\gamma - 1} \tag{3.23}$$

A reversible adiabatic process for a perfect gas is shown on a p–v diagram in Fig. 3.9. We have

$$-W = \int_{v_1}^{v_2} p \, dv = \text{shaded area}$$

Fig. 3.9 Reversible adiabatic process for a perfect gas on a p–v diagram

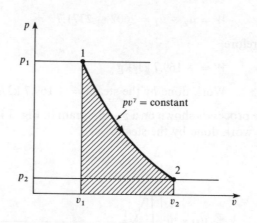

Therefore, since $pv^\gamma = \text{constant}, c$, then

$$W = -\int_{v_1}^{v_2} \frac{c \, dv}{v^\gamma}$$

i.e.

$$W = -c \int_{v_1}^{v_2} \frac{dv}{v^\gamma} = -c \left[\frac{v^{-\gamma + 1}}{-\gamma + 1} \right]_{v_1}^{v_2}$$

$$= -c \left(\frac{v_2^{-\gamma + 1} - v_1^{-\gamma + 1}}{1 - \gamma} \right) = -c \left(\frac{v_1^{-\gamma + 1} - v_2^{-\gamma + 1}}{\gamma - 1} \right)$$

The constant in this equation can be written as $p_1 v_1^\gamma$ or as $p_2 v_2^\gamma$. Hence

$$W = \frac{p_2 v_2^\gamma v_2^{1-\gamma} - p_1 v_1^\gamma v_1^{1-\gamma}}{\gamma - 1} = \frac{p_2 v_2 - p_1 v_1}{\gamma - 1}$$

This is the same expression obtained before as equation (3.23).

Example 3.4   1 kg of steam at 100 bar and 375 °C expands reversibly in a perfectly thermally insulated cylinder behind a piston until the pressure is 38 bar and the steam is then dry saturated. Calculate the work done.

Solution   From superheat tables at 100 bar and 375 °C,

$$h_1 = 3017 \text{ kJ/kg} \quad \text{and} \quad v_1 = 0.02453 \text{ m}^3/\text{kg}$$

Using equation (1.9)

$$u = h - pv$$

therefore

$$u_1 = 3017 - \frac{100 \times 10^5 \times 0.02453}{10^3} = 2771.7 \text{ kJ/kg}$$

Also, $u_2 = u_g$ at 38 bar $= 2602 \text{ kJ/kg}$

Since the cylinder is perfectly thermally insulated then no heat flows to or from the steam during the expansion; the process is therefore adiabatic. Using equation (3.13),

$$W = u_2 - u_1 = 2602 - 2771.7$$

therefore

$$W = -169.7 \text{ kJ/kg}$$

i.e.   Work done *by* the steam $= +169.7 \text{ kJ/kg}$

The process is shown on a p–v diagram in Fig. 3.10, the shaded area representing the work done by the steam.

Fig. 3.10 Reversible adiabatic process for steam on a p–v diagram for Example 3.4

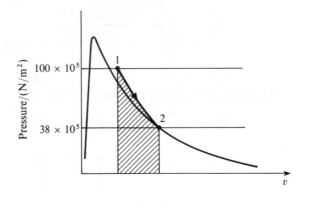

Example 3.5   Air at 1.02 bar, 22 °C, initially occupying a cylinder volume of 0.015 m³, is compressed reversibly and adiabatically by a piston to a pressure of 6.8 bar. Calculate the final temperature, the final volume, and the work done on the mass of air in the cylinder.

Solution   From equation (3.21)

$$\frac{T_1}{T_2} = \left(\frac{p_1}{p_2}\right)^{(\gamma-1)/\gamma} \quad \text{or} \quad T_2 = T_1 \times \left(\frac{p_2}{p_1}\right)^{(\gamma-1)/\gamma}$$

i.e. $T_2 = 295 \times \left(\dfrac{6.8}{1.02}\right)^{(1.4-1)/1.4}$

$$= 295 \times 1.7195 = 507.3 \text{ K}$$

where $T_1 = 22 + 273 = 295$ K; γ for air $= 1.4$,

i.e.   Final temperature $= 507.3 - 273 = 234.3 \,°C$

From equation (3.19)

$$\frac{p_1}{p_2} = \left(\frac{V_2}{V_1}\right)^{\gamma} \quad \text{or} \quad \frac{V_1}{V_2} = \left(\frac{p_2}{p_1}\right)^{1/\gamma}$$

therefore

$$\frac{0.015}{V_2} = \left(\frac{6.8}{1.02}\right)^{1/1.4} = 3.877$$

therefore

$$V_2 = \frac{0.015}{3.877} = 0.003\,87 \text{ m}^3$$

i.e.   Final volume $= 0.003\,87$ m³

From equation (3.13), for an adiabatic process

$$W = u_2 - u_1$$

and for a perfect gas, from equation (2.14), $u = c_v T$ per kg of gas, therefore

$$W = c_v(T_2 - T_1) = 0.718(507.3 - 295)$$

$$= 152.4 \text{ kJ/kg}$$

i.e.   Work input $= 152.4$ kJ/kg

The mass of air can be found using equation (2.6), $pV = mRT$. Therefore

$$m = \frac{p_1 v_1}{RT_1} = \frac{1.02 \times 10^5 \times 0.015}{0.287 \times 10^3 \times 295} = 0.0181 \text{ kg}$$

i.e.   Total work done $= 0.0181 \times 152.4 = 2.76$ kJ

The process is shown on a $p-v$ diagram in Fig. 3.11, the shaded area representing the work input per unit mass of air.

Fig. 3.11 Reversible
adiabatic process for air
on a p–v diagram for
Example 3.5

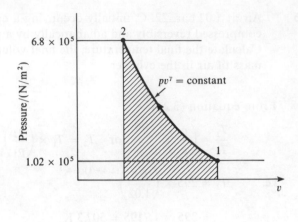

3.3 Polytropic processes

It is found that many processes in practice approximate to a reversible law of
the form $pv^n =$ constant, where n is a constant. Both vapours and perfect gases
obey this type of law closely in many non-flow processes. Such processes are
internally reversible.

From equation (1.2) for any reversible process,

$$W = -\int p\, dv$$

For a process in which $pv^n =$ constant, we have $p = c/v^n$, where c is a constant.
Therefore

$$W = -c \int_{v_1}^{v_2} \frac{dv}{v^n} = -c\left[\frac{v^{-n+1}}{-n+1}\right] = -c\left(\frac{v_2^{-n+1} - v_1^{-n+1}}{-n+1}\right)$$

i.e.
$$W = c\left(\frac{v_2^{1-n} - v_1^{1-n}}{n-1}\right) = \frac{p_2 v_2^n v_2^{1-n} - p_1 v_1^n v_1^{1-n}}{n-1}$$

since the constant, c, can be written as $p_1 v_1^n$ or as $p_2 v_2^n$,

i.e.
$$\text{Work input} = \frac{p_2 v_2 - p_1 v_1}{n-1} \tag{3.24}$$

Equation (3.24) is true for any working substance undergoing a reversible
polytropic process. It follows also that for any polytropic process we can write

$$\frac{p_1}{p_2} = \left(\frac{v_2}{v_1}\right)^n \tag{3.25}$$

Example 3.6 At the commencement of compression in the reciprocating compressor of a
refrigeration plant the refrigerant is dry saturated at 1 bar. The compression
process follows the law $pv^{1.1} =$ constant until the pressure is 10 bar. Using

Table 3.1 Properties of refrigerant for Example 3.6

Saturation values					Superheat values at 10 bar	
p_g	t_g	v_g	h_f	h_g	v	h
(bar)	(°C)	(m³/kg)	(kJ/kg)		(m³/kg)	(kJ/kg)
1	−30	0.160	8.9	174.2		
10	42	0.018	76.3	203.8	0.020	224.0

the properties of refrigerant given in Table 3.1, interpolating where necessary, calculate:

(i) the work done on the refrigerant during the process;

(ii) the heat transferred to or from the cylinder walls during the process.

Solution   (i) From Table 3.1, $v_1 = v_{g1} = 0.16 \text{ m}^3/\text{kg}$. We then have

$$\frac{v_2}{v_1} = \left(\frac{p_1}{p_2}\right)^{1/1.1} = \left(\frac{1}{10}\right)^{0.909} = 0.1233$$

therefore

$$v_2 = 0.1233 \times 0.16 = 0.01973 \text{ m}^3/\text{kg}$$

From equation (3.24)

$$W = \frac{p_2 v_2 - p_1 v_1}{n - 1}$$

$$= \left\{\frac{(10 \times 0.01973) - (1 \times 0.16)}{1.1 - 1}\right\} \times 10^5$$

$$= 37\,300 \text{ N m} = 37.3 \text{ kJ}$$

i.e.   Work done on the refrigerant = 37.3 kJ

(ii) To find the heat transferred it is first necessary to evaluate the internal energies at the end states. Using equation (1.9), $h = u + pv$, we have,

$$u_1 = u_{g1} = h_{g1} - p_{g1} v_{g1}$$

$$= 174.2 - \left(\frac{1 \times 10^5 \times 0.16}{10^3}\right) = 158.2 \text{ kJ/kg}$$

Interpolating from superheat tables at 10 bar, $v_2 = 0.01973 \text{ m}^3/\text{kg}$ we have

$$h_2 = 203.8 + \frac{(0.01973 - 0.018)}{(0.02 - 0.018)} \times (224 - 203.8)$$

$$= 221.3 \text{ kJ/kg}$$

67

Then using equation (1.9)

$$u_2 = 221.3 - \left(\frac{10 \times 10^5 \times 0.01973}{10^3} \right) = 201.6 \text{ kJ/kg}$$

From equation (1.4)

$$Q + W = (u_2 - u_1)$$

therefore

$$Q = -37.3 + (201.6 - 158.2) = 6.1 \text{ kJ/kg}$$

i.e. the heat transferred from the cylinder walls to the refrigerant during the compression process is 6.1 kJ/kg.

Consider now the polytropic process for a perfect gas. From equation (2.5)

$$pv = RT \quad \text{or} \quad p = \frac{RT}{v}$$

Hence, substituting in the equation $pv^n = $ constant, we have

$$\frac{RT}{v} v^n = \text{constant} \quad \text{or} \quad Tv^{n-1} = \text{constant} \tag{3.26}$$

Also, writing $v = RT/p$, we have

$$p \left(\frac{RT}{p} \right)^n = \text{constant} \quad \text{or} \quad \frac{T}{p^{(n-1)/n}} = \text{constant} \tag{3.27}$$

It can be seen that these equations are similar to the equations (3.17) and (3.18) for a reversible adiabatic process for a perfect gas. In fact the reversible adiabatic process for a perfect gas is a particular case of a polytropic process with the index, n, equal to γ.

Equations (3.26) and (3.27) can be written as

$$\frac{T_1}{T_2} = \left(\frac{v_2}{v_1} \right)^{n-1} \tag{3.28}$$

and

$$\frac{T_1}{T_2} = \left(\frac{p_1}{p_2} \right)^{(n-1)/n} \tag{3.29}$$

Note that equations (3.26), (3.27), (3.28) and (3.29) do not apply to a vapour undergoing a polytropic process, since the characteristic equation of state, $pv = RT$, which was used in the derivation of the equations, applies only to a perfect gas.

For a perfect gas expanding polytropically it is sometimes more convenient to express the work input in terms of the temperatures at the end states. From equation (3.24)

$$W = (p_2 v_2 - p_1 v_1)/(n - 1)$$

then, from equation (2.5), $p_1 v_1 = RT_1$ and $p_2 v_2 = RT_2$. Hence,

$$W = \frac{R(T_2 - T_1)}{n - 1} \tag{3.30}$$

or for mass, m,

$$W = \frac{mR(T_2 - T_1)}{n - 1} \tag{3.31}$$

Using the non-flow energy equation, (1.4), the heat flow during the process can be found,

i.e. $Q + W = u_2 - u_1 = c_v(T_2 - T_1)$

i.e. $Q + \dfrac{R(T_2 - T_1)}{(n - 1)} = c_v(T_2 - T_1)$

From equation (2.21)

$$c_v = \frac{R}{(\gamma - 1)}$$

Hence substituting

$$Q = \frac{R}{(\gamma - 1)}(T_2 - T_1) - \frac{R}{(n - 1)}(T_2 - T_1)$$

i.e. $Q = R(T_2 - T_1)\left(\dfrac{1}{\gamma - 1} - \dfrac{1}{n - 1}\right) = \dfrac{R(T_2 - T_1)(n - 1 - \gamma + 1)}{(\gamma - 1)(n - 1)}$

therefore

$$Q = \left(\frac{n - \gamma}{\gamma - 1}\right)\frac{R(T_2 - T_1)}{(n - 1)}$$

Now from equation (3.30), $W = R(T_2 - T_1)/(n - 1)$ per unit mass of gas, therefore

$$Q = \left(\frac{n - \gamma}{\gamma - 1}\right)W \tag{3.32}$$

Equation (3.32) is a convenient and concise expression relating the heat supplied and the work input in a polytropic process. In a compression process work is done on the gas, and hence the term W is positive. Therefore it can be seen from equation (3.32) that when the polytropic index n is greater than γ, in a compression process, then the right-hand side of the equation is positive (i.e. heat is supplied during the process). Conversely, when n is less than γ in a compression process, then heat is rejected by the gas. Similarly, the work input in an expansion process is negative, therefore when n is greater than γ, in an expansion process, heat is rejected; and when n is less than γ, in an expansion process, heat must be supplied to the gas during the process. It was shown in section 2.3 that γ for all perfect gases has a value greater than unity.

Example 3.7   1 kg of a perfect gas is compressed from 1.1 bar, 27 °C according to a law $pv^{1.3}$ = constant, until the pressure is 6.6 bar. Calculate the heat flow to or from the cylinder walls:

(i) When the gas is ethane (molar mass 30 kg/kmol), which has $c_p = 2.10$ kJ/kg K.

(ii) When the gas is argon (molar mass 40 kg/kmol), which has $c_p = 0.520$ kJ/kg K.

Solution   From equation (3.29), for both ethane and argon,

$$\frac{T_1}{T_2} = \left(\frac{p_1}{p_2}\right)^{(n-1)/n} \quad \text{or} \quad T_2 = T_1\left(\frac{p_2}{p_1}\right)^{(n-1)/n}$$

i.e. $T_2 = 300\left(\frac{6.6}{1.1}\right)^{1.3-1/1.3} = 300 \times 6^{0.231} = 300 \times 1.512 = 453.6$ K

where $T_1 = 27 + 273 = 300$ K.

(i) From equation (2.9), $R = \tilde{R}/\tilde{m}$, therefore, for ethane

$$R = \frac{8.3145}{30} = 0.277 \text{ kJ/kg K}$$

Then from equation (2.17), $c_p - c_v = R$, therefore

$$c_v = 2.10 - 0.277 = 1.823 \text{ kJ/kg K}$$

where $c_p = 1.75$ kJ/kg K for ethane. Then from equation (2.20)

$$\gamma = \frac{c_p}{c_v} = \frac{2.10}{1.823} = 1.152$$

From equation (3.30)

$$W = \frac{R(T_2 - T_1)}{n-1} = \frac{0.277 \times (453.6 - 300)}{1.3 - 1} = 141.8 \text{ kJ/kg}$$

Then from equation (3.32)

$$Q = \left(\frac{n-\gamma}{\gamma-1}\right)W = \left(\frac{1.3-1.152}{1.152-1}\right) \times 141.8 = 138.1 \text{ kJ/kg}$$

i.e.   Heat supplied = 138.1 kJ/kg

(ii) Using the same method for argon we have

$$R = \frac{8.3145}{40} = 0.208 \text{ kJ/kg K}$$

Also $c_v = 0.520 - 0.208 = 0.312$ kJ/kg K

therefore

$$\gamma = \frac{0.520}{0.312} = 1.667$$

Then the work input is given by

$$W = \frac{R(T_1 - T_2)}{n - 1} = \frac{0.208 \times (453.6 - 300)}{1.3 - 1} = 106.5 \text{ kJ/kg}$$

Then, $\quad Q = \left(\frac{n - \gamma}{\gamma - 1}\right) W = \left(\frac{1.3 - 1.667}{1.667 - 1}\right) \times 106.5 = -58.6 \text{ kJ/kg}$

i.e. $\quad$ Heat rejected = 58.6 kJ/kg

In a polytropic process the index n depends only on the heat and work quantities during the process. The various processes considered in sections 3.1 and 3.2 are special cases of the polytropic process for a perfect gas. When $n = 0$

$$pv^0 = \text{constant, i.e. } p = \text{constant}$$

When $n = \infty$,

$$pv^\infty = \text{constant} \quad \text{or} \quad p^{1/\infty} v = \text{constant, i.e. } v = \text{constant}$$

When $n = 1$

$$pv = \text{constant, i.e. } T = \text{constant}$$

since $pv/T = \text{constant}$ for a perfect gas.
When $n = \gamma$

$$pv^\gamma = \text{constant, i.e. reversible adiabatic}$$

This is illustrated on a $p-v$ diagram in Fig. 3.12. Thus,

state 1 to state A is constant pressure cooling ($n = 0$)

state 1 to state B is isothermal compression ($n = 1$)

state 1 to state C is reversible adiabatic compression ($n = \gamma$)

state 1 to state D is constant volume heating ($n = \infty$)

Fig. 3.12 General polytropic processes plotted on a $p-v$ diagram

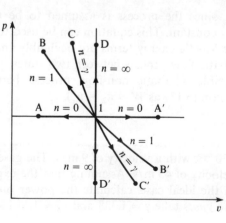

Similarly, 1 to A' is constant pressure heating; 1 to B' is isothermal expansion; 1 to C' is reversible adiabatic expansion; 1 to D' is constant volume cooling. Note that, since γ is always greater than unity, then process 1 to C must lie between processes 1 to B and 1 to D; similarly, process 1 to C' must lie between processes 1 to B' and 1 to D'.

For a vapour a generalization such as the above is not possible. A vapour may undergo a process according to a law pv = constant. In this case, since the characteristic equation of state, $pv = RT$, does not apply to a vapour, then the process is not isothermal. Tables must be used to find the properties at the end states, making use of the fact that $p_1 v_1 = p_2 v_2$. Expansion of steam in a reciprocating engine is found to approximate to a hyperbolic expansion (pv = constant); such engines are rarely used nowadays.

3.4 Reversible flow processes

Although flow processes in practice are usually highly irreversible, it is sometimes convenient to assume that a flow process is reversible in order to provide an ideal comparison. An observer travelling with the flowing fluid would appear to see a change in thermodynamic properties as in a non-flow process. For example, in a reversible adiabatic process for a perfect gas, an observer travelling with the gas would appear to see a process pv^γ = constant taking place, but the work input would not be given by $-\int p \, dv$, or by the change in internal energy as given by equation (3.13). Some work is done by virtue of the forces acting between the moving gas and its surroundings. For example, for a reversible adiabatic flow process for a perfect gas, from the flow equation (1.10), for unit mass flow rate

$$\left(h_1 + \frac{C_1^2}{2} \right) + Q + W = \left(h_2 + \frac{C_2^2}{2} \right)$$

Then since $Q = 0$

$$W = (h_2 - h_1) + \left(\frac{C_2^2}{2} - \frac{C_1^2}{2} \right)$$

Also, since the process is assumed to be reversible, then for a perfect gas, pv^γ = constant. This equation can be used to fix the end states. Note that even if the kinetic energy terms are negligibly small the work input in a reversible adiabatic flow process between two states is not equal to the work input in a reversible adiabatic non-flow process between the same states (given by equation (3.13) as $W = u_2 - u_1$).

Example 3.8

A gas turbine receives gases from the combustion chamber at 7 bar and 650 °C, with a velocity of 9 m/s. The gases leave the turbine at 1 bar with a velocity of 45 m/s. Assuming that the expansion is adiabatic and reversible in the ideal case, calculate the power output per unit mass flow rate. For the gases take $\gamma = 1.333$ and $c_p = 1.11$ kJ/kg K.

Solution Using the flow equation for an adiabatic process

$$W = \dot{m}\left\{(h_2 - h_1) + \left(\frac{C_2^2 - C_1^2}{2}\right)\right\}$$

For a perfect gas from equation (2.18), $h = c_p T$, therefore,

$$W = \dot{m}\left\{c_p(T_2 - T_1) + \left(\frac{C_2^2 - C_2^2}{2}\right)\right\}$$

To find T_2 use equation (3.21),

$$\frac{T_1}{T_2} = \left(\frac{p_1}{p_2}\right)^{(\gamma - 1)/\gamma}$$

i.e. $\dfrac{T_1}{T_2} = \left(\dfrac{7}{1}\right)^{(1.333 - 1)/1.333} = 1.626$

therefore

$$T_2 = \frac{T_1}{1.626} = \frac{923}{1.626} = 567.7 \text{ K}$$

where $T_1 = 650 + 273 = 923$ K.

Hence substituting for unit mass flow rate

$$\dot{W} = 1 \times 1.11(567.7 - 923) + \left(\frac{45^2 - 9^2}{2 \times 10^3}\right)$$

therefore

$$\dot{W} = -394.4 + 0.97 = -393.4 \text{ kW}$$

i.e.   Power output per kilogram per second $= 393.4$ kW

Note that in Example 3.8 the kinetic energy change is small compared with the enthalpy change. This is often the case in problems on flow processes, and the change in kinetic energy can sometimes be taken to be negligible.

For a vapour undergoing a reversible adiabatic flow process the end state is fixed by equating the initial and final entropies (see Ch. 4).

3.5 Irreversible processes

The criteria of reversibility are stated in section 1.4. The equations of sections 3.1, 3.2, and 3.3 can only be used when the process obeys the criteria of reversibility to a close approximation. In processes in which a fluid is enclosed in a cylinder behind a piston, friction effects can be assumed to be negligible. However, in order to satisfy criterion (c) in section 1.4 heat must never be transferred to or from the system through a finite temperature difference. Only in an isothermal process is this conceivable, since in all other processes the temperature of the system is continually changing during the process; in order to satisfy criterion (c) the temperature of the cooling or heating medium external to the system would be required to change correspondingly. Ideally a way of achieving reversibility can be imagined, but in practice it cannot even be approached as an approximation. Nevertheless, if we accept inevitable irreversibilities in the surroundings, we can still have processes which are internally reversible. That is, the system undergoes a process which can be reversed, but the surroundings undergo an irreversible change. Most processes occurring in a cylinder behind a piston can be assumed to be internally reversible to a close approximation, and the equations of sections 3.1, 3.2, and 3.3 can be used where applicable. Certain processes cannot be assumed to be internally reversible, and the important cases will now be considered.

Unresisted, or free, expansion

This process was mentioned in section 1.5 in order to show that in an irreversible process the work done is not given by $-\int p \, dv$. Consider two vessels A and B, interconnected by a short pipe with a valve, and perfectly thermally insulated (see Fig. 3.13). Initially let the vessel A be filled with a fluid at a certain pressure, and let B be completely evacuated. When the valve is opened the fluid in A will expand rapidly to fill both vessels A and B. The pressure finally will be lower than the initial pressure in vessel A. This is known as an unresisted expansion or a free expansion. The process is not reversible, since external work would have to be done to restore the fluid to its initial condition. The non-flow energy equation, (1.4), can be applied between the initial and final states,

i.e. $\qquad Q + W = u_2 - u_1$

Fig. 3.13 Two perfectly insulated interconnected vessels

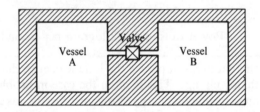

Now in this process no work is done on or by the fluid, since the boundary of the system does not move. No heat flows to or from the fluid since the system is well lagged. The process is therefore adiabatic, but irreversible,

i.e. $\qquad u_2 - u_1 = 0 \quad$ or $\quad u_2 = u_1$

In a free expansion therefore the internal energy initially equals the internal energy finally.

For a perfect gas, we have, from equation (2.14),

$$u = c_v T$$

Therefore for a free expansion of a perfect gas

$$c_v T_1 = c_v T_2$$

i.e. $\qquad T_1 = T_2$

That is, for a perfect gas undergoing a free expansion, the initial temperature is equal to the final temperature.

Example 3.9

Air at 20 bar is initially contained in vessel A of Fig. 3.13, the volume of which can be assumed to be 1 m^3. The valve is opened and the air expands to fill vessels A and B. Assuming that the vessels are of equal volume, calculate the final pressure of the air.

Solution

For a perfect gas for a free expansion, $T_1 = T_2$. Also from equation (2.6), $pV = mRT$, hence $p_1 V_1 = p_2 V_2$.

Now V_2 is the combined volumes of vessels A and B,

i.e. $V_2 = V_A + V_B = 1 + 1 = 2 \text{ m}^3$ and $V_1 = 1 \text{ m}^3$

Therefore we have

$$p_2 = p_1 \times \frac{V_1}{V_2} = 20 \times \tfrac{1}{2} = 10 \text{ bar}$$

i.e. Final pressure = 10 bar

The process is shown on a p–v diagram in Fig. 3.14. State 1 is fixed at 20 bar and 1 m^3 when the mass of gas is known; state 2 is fixed at 10 bar and 2 m^3 for the same mass of gas. The process between these states is irreversible and must be drawn dotted. The points 1 and 2 lie on an isothermal line, but the process between 1 and 2 cannot be called isothermal, since the intermediate temperatures are not the same throughout the process. There is no work done during the process, and the area under the dotted line does not represent work done.

Fig. 3.14 Irreversible process on a p–v diagram for Example 3.9

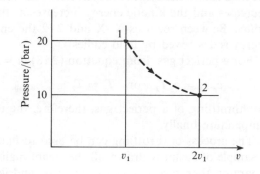

Throttling

A flow of fluid is said to be throttled when there is some restriction to the flow, when the velocities before and after the restriction are either equal or negligbly small, and when there is a negligible heat loss to the surroundings. The restriction to flow can be a partly open valve, an orifice, or any other sudden reduction in the cross-section of the flow.

An example of throttling is shown in Fig. 3.15. The fluid, flowing steadily along a well-lagged pipe, passes through an orifice at section X–X. Since the pipe is well lagged it can be assumed that no heat flows to or from the fluid. The flow equation (1.10) can be applied between any two sections of the flow,

i.e.
$$\dot{m}\left(h_1 + \frac{C_1^2}{2}\right) + \dot{Q} + \dot{W} = \dot{m}\left(h_2 + \frac{C_2^2}{2}\right)$$

Now since $Q = 0$, and $W = 0$, then

$$h_1 + \frac{C_1^2}{2} = h_2 + \frac{C_2^2}{2}$$

Fig. 3.15 Throttling process

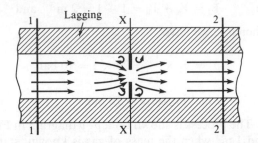

When the velocities C_1 and C_2 are small, or when C_1 is approximately equal to C_2, then the kinetic energy terms may be neglected. (Note that sections 1–1 and 2–2 can be chosen well upstream and well downstream of the disturbance to the flow, so that this latter assumption is justified.) Then $h_1 = h_2$. Therefore for a throttling process, the enthalpy initially is equal to the enthalpy finally.

The process is adiabatic, but is highly irreversible because of the eddying of the fluid round the orifice at X–X. Between sections 1–1 and X–X the enthalpy decreases and the kinetic energy increases as the fluid accelerates through the orifice. Between sections X–X and 2–2 the enthalpy increases as the kinetic energy is destroyed by fluid eddies.

For a perfect gas, from equation (2.18), $h = c_p T$, therefore,

$$c_p T_1 = c_p T_2 \quad \text{or} \quad T_1 = T_2$$

For throttling of a perfect gas, therefore, the temperature initially equals the temperature finally.

The process of throttling can be used to find the dryness fraction of steam. A sample of steam is drawn off the steam main, passed through a mechanical separator, then through a throttle valve, and finally through a condenser; the water separated from the mechanical separator and the water from the condenser are weighed and the dryness fraction calculated as shown in the following example.

Example 3.10   The dryness fraction of wet steam in a main is determined using a separating and throttling calorimeter. The pressure in the main is 5 bar; after throttling, the steam pressure and temperature are 1.013 25 bar and 120 °C; the water collected from the separator is at the rate of 0.5 kg/h, and that from the condenser at the rate of 9 kg/h. Making suitable assumptions, calculate the dryness fraction of the steam in the main.

Solution   The assumptions made are as follows: negligible pressure drop of steam in the separator; no change of kinetic energy across the throttle valve; negligible heat loss in the separator and in the throttling process.

    The processes are shown on a p–v diagram in Fig. 3.16; process 1–2 represents the separating process, process 2–3 the throttling process, and process 3–4 the condensing process. Process 2–3 is shown dotted since the process is irreversible; no work is done during the process and the area under line 2–3 is *not* equal to work done.

Fig. 3.16 Processes on a p–v diagram for Example 3.10

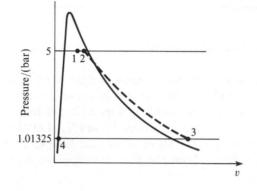

    The enthalpy after throttling is obtained by interpolating from steam tables,

$$h_3 = 2676 + \frac{(120 - 100)}{(150 - 100)} \times (2777 - 2676)$$

$$= 2716.4 \text{ kJ/kg}$$

For an adiabatic throttling process neglecting kinetic energy changes, $h_2 = h_3$, therefore using equation (2.2)

$$h_2 = h_3 = h_{f2} + x_2 h_{fg2}$$

therefore

$$x_2 = \frac{2716.4 - 640}{2109} = 0.985$$

where $h_{f2} = 640 \text{ kJ/kg}$, and $h_{fg2} = 2109 \text{ kJ/kg}$, are read from saturation tables at 5 bar.

    The mass flow rate of water in the steam at state 2 is therefore given by

$$\dot{m}_{w2} = (1 - x_2) \times (\text{mass flow rate of condensate})$$

$$= (1 - 0.985) \times 9 = 0.135 \text{ kg/h}$$

Therefore the total mass flow rate of water in the steam sample from the main is given by the mass flow rate after separation, $\dot{m}_{w2}$ plus the mass of water separated, given as 0.5 kg/h,

i.e. $\dot{m}_{w1} = 0.135 + 0.5 = 0.635 \text{ kg/h}$

The mass flow rate of dry vapour in the sample is therefore the total mass flow rate of $(0.5 + 9)$ kg/h minus $\dot{m}_{w2}$, ($= 0.635$ kg/h)

i.e. Mass flow rate of dry vapour in sample
$$= 0.5 + 9 - 0.635 = 8.865 \text{ kg/h}$$

Then the dryness fraction in the main is the mass flow rate of dry vapour divided by the total mass flow rate,

i.e. $\qquad x_1 = \dfrac{8.865}{(0.5 + 9)} = 0.933$

Adiabatic mixing

The mixing of two streams of fluid is quite common in engineering practice, and can usually be assumed to occur adiabatically. Consider two streams of a fluid mixing as shown in Fig. 3.17. Let the streams have mass flow rates $\dot{m}_1$ and $\dot{m}_2$ and temperatures T_1 and T_2. Let the resulting mixed stream have a temperature T_3. There is no heat flow to or from the fluid, and no work is done, hence from the flow equation, we have, neglecting changes in kinetic energy,

$$\dot{m}_1 h_1 + \dot{m}_2 h_2 = \dot{m}_3 h_3 \tag{3.33}$$

Fig. 3.17 Mixing process

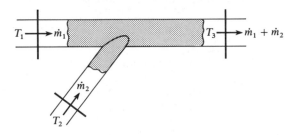

For a perfect gas, from equation (2.18), $h = c_p T$, hence,

$$\dot{m}_1 c_{p_1} T_1 + \dot{m}_2 c_{p_2} T_2 = (\dot{m}_1 c_{p_1} + \dot{m}_2 c_{p_2}) T_3$$

Or, assuming that the two streams 1 and 2 are of the same fluid with the same specific heat capacity,

$$\dot{m}_1 T_1 + \dot{m}_2 T_2 = (\dot{m}_1 + \dot{m}_2) T_3 \tag{3.34}$$

The mixing process is highly irreversible due to the large amount of eddying and churning of the fluid that takes place.

3.6 Nonsteady-flow processes

There are many cases in practice when the rate of mass flow crossing the boundary of a system at inlet is not the same as the rate of mass flow crossing the boundary of the system at outlet. Also, the rate at which work is done on or by the fluid, and the rate at which heat is transferred to or from the system is not necessarily constant with time. In a case of this kind the total energy of the system within the boundary is no longer constant, as it is in a steady-flow process, but varies with time.

Let the total energy of the system within the boundary at any instant be E. During a small time interval let the mass entering the system be δm_1, and let the mass leaving the system be δm_2; let the heat supplied and the work input during the same time be δQ and δW respectively. Consider a similar system to the one shown in Fig. 1.22. Now, as shown in section 1.8 (p. 19), work is done at inlet and outlet in introducing and expelling mass across the system boundaries, i.e. at inlet

$$\text{Energy required} = \delta m_1 p_1 v_1$$

and at outlet

$$\text{Energy required} = \delta m_2 p_2 v_2$$

Also, as before, the energy of unit mass of the flowing fluid is given by $(u_1 + C_1^2/2 + Z_1 g)$ at inlet, and by $(u_2 + C_2^2/2 + Z_2 g)$ at outlet. Hence

Energy entering system

$$= \delta Q + \delta W + \delta m_1(u_1 + C_1^2/2 + Z_1 g) + \delta m_1 p_1$$

and

$$\text{Energy leaving system} = \delta m_2(u_2 + C_2^2/2 + Z_2 g) + \delta m_2 p_2 v_2$$

Then, applying the first law,

Energy entering $-$ energy leaving $=$ increase of energy of the system, δE

therefore

$$\delta Q + \delta W + \delta m_1(u_1 + C_1^2/2 + Z_1 g + p_1 v_1)$$
$$- \delta m_2(u_2 + C_2^2/2 + Z_2 g + p_2 v_2) = \delta E$$

During a finite time the total heat transferred is given by $\sum \delta Q = Q$, and the total work done is given by $\sum \delta W = W$.

Let the initial mass within the system boundaries be m', at a height, Z', and the initial internal energy be u'; let the final mass within the boundaries at the end of the time interval be m'', at a height, Z'', and the final internal energy be u''. Therefore

$$\sum \delta E = m''(u'' + Z'' g) - m'(u' + Z' g)$$

Therefore we have

$$Q + W + \sum [\delta m_1(u_1 + p_1 v_1 + C_1^2/2 + Z_1 g)]$$
$$= \sum [\delta m_2(u_2 + p_2 v_2 + C_2^2/2 + Z_2 g)]$$
$$+ m''(u'' + Z'' g) - m'(u' + Z' g) \tag{3.35}$$

79

or

$$Q + W + \sum \delta m_1 (h_1 + C_1^2/2 + Z_1 g)$$
$$= \sum \delta m_2 (h_2 + C_2^2/2 + Z_2 g) + (m''u'' - m'u')$$
$$+ (m''Z''g - m'Z'g) \tag{3.36}$$

Also, from continuity of mass,

Mass entering − mass leaving
= increase of mass within system boundary

i.e. $\quad \sum \delta m_1 - \sum \delta m_2 = m'' - m' \tag{3.37}$

One of the most commonly occurring problems involving the nonsteady-flow equation is the filling of a bottle or reservoir from a source which is large in comparison with the bottle or reservoir. Figure 3.18 shows a typical example. It is assumed that the condition of the fluid in the pipeline is unchanged during the filling process. In this case there is no work done on the system boundary; also, no mass leaves the system during the process, hence $\delta m_2 = 0$.

Fig. 3.18 Filling a bottle or reservoir from a pipeline

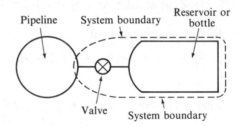

Applying equation (3.36), making the additional assumption that changes in potential energy are zero, and that the kinetic energy, $C_1^2/2$, is small compared with the enthalpy, h_1, we have

$$Q + \sum (\delta m_1 h_1) = m''u'' - m'u'$$

Or, since, h_1 is constant during the process,

$$Q + h_1 \sum (\delta m_1) = m''u'' - m'u'$$

In this case equation (3.37) becomes

$$\sum \delta m_1 = m'' - m'$$

Hence substituting

$$Q + h_1(m'' - m') = m''u'' - m'u' \tag{3.38}$$

It is often possible to assume that the process is adiabatic, and in that case we have

$$h_1(m'' - m') = m''u'' - m'u'$$

Or in words:

Enthalpy of mass which enters the bottle
= increase of internal energy of the system

Example 3.11   A rigid vessel of volume $10 \, \text{m}^3$ containing steam at 2.1 bar and dryness fraction 0.9 is connected to a pipeline, and steam is allowed to flow from the pipeline into the vessel until the pressure and temperature in the vessel are 6 bar and 200 °C respectively. The steam in the pipeline is at 10 bar and 250 °C throughout the process. Calculate the heat transfer to or from the vessel during the process.

Solution   Using the notation previously introduced we have

$$u' = u_f'(1 - 0.9) + (u_g' \times 0.9) = 511 \times 0.1 + 2531 \times 0.9$$

i.e. $u' = 2329 \, \text{kJ/kg}$

Also, $m' = V/v' = 10/0.9v_g = 10/(0.9 \times 0.8461) = 13.13 \, \text{kg}$

The steam is superheated finally at 6 bar and 200 °C, therefore

$$u'' = 2640 \, \text{kJ/kg}$$

and $v'' = 0.3522 \, \text{m}^3/\text{kg}$

i.e. $m'' = V/v'' = 10/0.3522 = 28.4 \, \text{kg}$

The steam in the pipeline is superheated at 10 bar and 250 °C, hence

$$h_1 = 2944 \, \text{kJ/kg}$$

Then using equation (3.38)

$$Q + 2944(28.4 - 13.13) = (28.4 \times 2640) - (13.13 \times 2329)$$

therefore

$$Q = 74\,980 - 30\,590 - 44\,940 = -550 \, \text{kJ}$$

i.e.   Heat rejected from vessel $= 550 \, \text{kJ}$

Another commonly occurring example of the nonsteady-flow process is the case in which a vessel is opened to a large space and fluid is allowed to escape (Fig. 3.19). There is no work done and in this case $\delta m_1 = 0$ since no mass enters the system. Neglecting changes in potential energy and applying equation (3.36):

$$Q = \sum [\delta m_2(h_2 + C_2^2/2)] + (m''u'' - m'u')$$

Fig. 3.19 Fluid escaping from a vessel

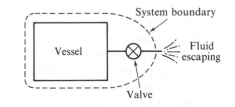

The difficulty arising in this analysis is that the state 2 of the mass leaving the vessel is continually changing, and hence it is impossible to evaluate the term $\sum [\delta m_2(h_2 + C_2^2/2)]$. An approximation can be made in order to find the mass of fluid which leaves the vessel as the pressure drops to a given value. It can be assumed that the fluid remaining in the vessel undergoes a reversible adiabatic expansion. This is a good approximation if the vessel is well lagged, or if the duration of the process is short. Using this assumption the end state of the fluid in the vessel can be found, and hence the mass remaining in the vessel, m'', can be calculated.

Example 3.12   An air receiver of volume 6 m³ contains air at 15 bar and 40.5 °C. A valve is opened and some air is allowed to blow out to atmosphere. The pressure of the air in the receiver drops rapidly to 12 bar when the valve is then closed. Calculate the mass of air which has left the receiver.

Solution   Initially

$$m' = p'V/RT' = \frac{15 \times 10^5 \times 6}{0.287 \times 10^3 \times 313.5} = 100 \text{ kg}$$

Assuming that the mass in the receiver undergoes a reversible adiabatic process, then using equation (3.21)

$$\frac{T'}{T''} = \left(\frac{p'}{p''}\right)^{\gamma-1/\gamma} = \left(\frac{15}{12}\right)^{0.4/1.4} = 1.25^{0.286} = 1.066$$

therefore

$$T'' = 313.5/1.066 = 294.1 \text{ K}$$

Hence $$m'' = p''V/RT'' = \frac{12 \times 10^5 \times 6}{0.287 \times 10^3 \times 294.1} = 85.3 \text{ kg}$$

Therefore

$$\text{Mass of air which left receiver} = 100 - 85.3 = 14.7 \text{ kg}$$

In the case of a vapour undergoing a reversible adiabatic expansion no equation such as (3.21), as used above, holds true. It is necessary to make use of the property entropy, s, which can be shown to remain constant during a reversible adiabatic process, i.e. $s' = s''$. Then using tables the value of v'' can be calculated and hence m'' found (see Problem 4.22).

Example 3.13   At the beginning of the induction stroke of a petrol engine of compression ratio 8/1, the clearance volume is occupied by residual gas at a temperature of 840 °C and pressure 1.034 bar. The volume of mixture induced during the stroke, measured at atmospheric conditions of 1.013 bar and 15 °C, is 0.75 of the cylinder swept volume. The mean pressure and temperature in the induction manifold during induction is 0.965 bar and 27 °C respectively, and the mean pressure in the cylinder during the induction stroke is 0.828 bar. Calculate the temperature and pressure of the mixture at the end of the induction stroke assuming the process to be adiabatic. For the induced mixture and final mixture take $c_v = 0.718$ kJ/kg K and $R = 0.287$ kJ/kg K; for the residual gas take $c_v = 0.840$ kJ/kg K and $R = 0.296$ kJ/kg K.

Solution Let swept volume be V_s and clearance volume be V_c. Then

$$\text{Compression ratio} = \frac{V_s + V_c}{V_c} = 8 \quad \text{(see p. 135)}$$

i.e. $\quad V_s = 7V_c$

Initially the residual gas occupies the volume, $V_c = V_s/7$, therefore

$$m' = \frac{p'V_c}{RT'} = \frac{1.034 \times 10^5 \times V_s}{0.296 \times 1113 \times 7 \times 10^3} = 0.0448V_s \, \text{kg}$$

where $T' = 840 + 273 = 1113$ K.

Also using equation (3.37)

$$m'' - m' = \sum \delta m_1 - \sum \delta m_2$$

and noting that in this example $\sum \delta m_2 = 0$, we have

$$m'' - m' = m_1 = \frac{1.013 \times 10^5 \times 0.75V_s}{0.287 \times 288 \times 10^3} = 0.9192V_s \, \text{kg}$$

therefore

$$m'' = 0.9192V_s + 0.0448V_s = 0.964V_s \, \text{kg}$$

Changes in kinetic and potential energy can be neglected, and the process is adiabatic (i.e. $Q = 0$), hence applying equation (3.36) we have

$$m_1 h_1 + W = m''u'' - m'u'$$

Also, the temperature of the mixture in the induction manifold is constant throughout the stroke, i.e. $h_1 = c_p T_1 = \text{constant}$. Therefore

$$m_1 c_p T_1 + W = m'' c_v T'' - m' c_v T' \tag{1}$$

The work input is given by

$$-W = (\text{mean pressure in cylinder during induction} \times \text{swept volume})$$

$$= 0.828 \times 10^5 \times V_s = 82\,800V_s \, \text{N m} = 82.8V_s \, \text{kJ}$$

therefore

$$W = -82.8V_s \, \text{kJ}$$

For the mixture induced, $c_p = c_v + R = 0.718 + 0.287 = 1.005$ kJ/kg K. Then substituting values in equation [1]

$$(0.9192 \times V_s \times 1.005 \times 300) - 82.8V_s$$

$$= (0.964V_s \times 0.718 \times T'') - (0.0448V_s \times 0.84 \times 1113)$$

therefore

$$T'' = 341.3 \, \text{K} = 68.3\,°\text{C}$$

i.e. $\quad$ Final temperature $= 68.3\,°\text{C}$

Then

$$p'' = \frac{m'' R T''}{(V_s + V_c)} = \frac{0.964 V_s \times 0.287 \times 341.3 \times 10^3}{8 V_s / 7} = 82\,623 \text{ N/m}^2$$

i.e.   Final pressure $= 0.826$ bar

Problems

3.1   1 kg of air enclosed in a rigid container is initially at 4.8 bar and 150 °C. The container is heated until the temperature is 200 °C. Calculate the pressure of the air finally and the heat supplied during the process.

(5.37 bar; 35.9 kJ/kg)

3.2   A rigid vessel of volume 1 m³ contains steam at 20 bar and 400 °C. The vessel is cooled until the steam is just dry saturated. Calculate the mass of steam in the vessel, the final pressure of the steam, and the heat rejected during the process.

(6.62 kg; 13.01 bar; 2355 kJ)

3.3   Oxygen (molar mass 32 kg/kmol) expands reversibly in a cylinder behind a piston at a constant pressure of 3 bar. The volume initially is 0.01 m³ and finally is 0.03 m³; the initial temperature is 17 °C. Calculate the work input and the heat supplied during the expansion. Assume oxygen to be a perfect gas and take $c_p = 0.917$ kJ/kg K.

(-6 kJ; 21.18 kJ)

3.4   Steam at 7 bar, dryness fraction 0.9, expands reversibly at constant pressure until the temperature is 200 °C. Calculate the work input and heat supplied per unit mass of steam during the process.

(-38.2 kJ/kg; 288.7 kJ/kg)

3.5   0.05 m³ of a perfect gas at 6.3 bar undergoes a reversible isothermal process to a pressure of 1.05 bar. Calculate the heat supplied.

(56.4 kJ)

3.6   Dry saturated steam at 7 bar expands reversibly in a cylinder behind a piston until the pressure is 0.1 bar. If heat is supplied continuously during the process in order to keep the temperature constant, calculate the change of internal energy per unit mass of steam.

(37.2 kJ/kg)

3.7   1 kg of air is compressed isothermally and reversibly from 1 bar and 30 °C to 5 bar. Calculate the work input and the heat supplied.

(140 kJ/kg; -140 kJ/kg)

3.8   1 kg of air at 1 bar, 15 °C is compressed reversibly and adiabatically to a pressure of 4 bar. Calculate the final temperature and the work input.

(155 °C; 100.5 kJ/kg)

3.9   Nitrogen (molar mass 28 kg/kmol) expands reversibly in a perfectly thermally insulated cylinder from 3.5 bar, 200 °C to a volume of 0.09 m³. If the initial volume occupied was 0.03 m³, calculate the work input. Assume nitrogen to be a perfect gas and take $c_v = 0.741$ kJ/kg K.

(-9.31 kJ)

3.10 A certain perfect gas is compressed reversibly from 1 bar, $17\,°C$ to a pressure of 5 bar in a perfectly thermally insulated cylinder, the final temperature being $77\,°C$. The work done on the gas during the compression is $45\,kJ/kg$. Calculate γ, c_v, R, and the molar mass of the gas.

$(1.132; 0.75\,kJ/kg\,K; 0.099\,kJ/kg\,K; 84\,kg/kmol)$

3.11 1 kg of air at 1.02 bar, $20\,°C$ is compressed reversibly according to a law $pv^{1.3} = $ constant, to a pressure of 5.5 bar. Calculate the work done on the air and the heat supplied during the compression.

$(133.46\,kJ/kg; -33.3\,kJ/kg)$

3.12 Oxygen (molar mass $32\,kg/kmol$) is compressed reversibly and polytropically in a cylinder from 1.05 bar, $15\,°C$ to 4.2 bar in such a way that one-third of the work input is rejected as heat to the cylinder walls. Calculate the final temperature of the oxygen. Assume oxygen to be a perfect gas and take $c_v = 0.649\,kJ/kg\,K$.

$(113\,°C)$

3.13 A mass of 0.05 kg of carbon dioxide (molar mass $44\,kg/kmol$), occupying a volume of $0.03\,m^3$ at 1.025 bar, is compressed reversibly until the pressure is 6.15 bar. Calculate the final temperature, the work done on the CO_2, and the heat supplied:
  (i) when the process is according to a law $pv^{1.4} = $ constant;
  (ii) when the process is isothermal;
  (iii) when the process takes place in a perfectly thermally insulated cylinder.
Assume carbon dioxide to be a perfect gas, and take $\gamma = 1.3$.

$(270\,°C; 5.135\,kJ; 1.712\,kJ; 52.5\,°C; 5.51\,kJ; -5.51\,kJ; 219\,°C; 5.25\,kJ; 0\,kJ)$

3.14 A refrigerant is compressed reversibly in a cylinder according to a polytropic law from 2.62 bar, dry saturated, to 8.20 bar when the temperature is then $40\,°C$. Using the refrigerant properties given as Table 3.2, calculate:
  (i) the polytropic index;
  (ii) the work input during the compression process;
  (iii) the heat transferred to or from the cylinder walls during the process.

$(1.073; 21.93\,kJ/kg; 6.16\,kJ/kg)$

Table 3.2 Properties of refrigerant for Problem 3.14

Saturation values			Superheat values at 8.2 bar, $40\,°C$	
p_g	v_g	h_g	v	h
(bar)	(m^3/kg)	(kJ/kg)	(m^3/kg)	(kJ/kg)
2.62	0.0757	292.9	0.02615	322.6

3.15 A refrigerant is dry saturated at 2 bar and is compressed reversibly in a cylinder according to a law $pv = $ constant to a pressure of 10 bar. Using the properties of refrigerant given as Table 3.3, calculate:
  (i) the final specific volume and temperature of the refrigerant:
  (ii) the final specific volume and temperature when the working substance is air, compressed between the same pressures and from the same initial temperature.

$(0.024\,m^3/kg, 24\,°C; 0.071\,m^3/kg, -25\,°C)$

Table 3.3 Properties of refrigerant for Problem 3.15

t_g	p_g	v_g
(°C)	(bar)	(m³/kg)
−25	2	0.120
24	10	0.024

3.16 A refrigerant leaves a condenser as a saturated liquid at a temperature of 25 °C, and is throttled to a pressure of 1.83 bar where it enters the evaporator. Using the properties of refrigerant given as Table 3.4, calculate the dryness fraction of the refrigerant vapour entering the evaporator.

(0.236)

Table 3.4 Properties of refrigerant for Problem 3.16

t_g	p_g	h_f	h_g
(°C)	(bar)	(kJ/kg)	
25	6.52	59.7	197.7
−15	1.83	22.3	181.0

3.17 The pressure in a steam main is 12 bar. A sample of steam is drawn off and passed through a throttling calorimeter, the pressure and temperature at exit from the calorimeter being 1 bar and 140 °C respectively. Calculate the dryness fraction of the steam in the main, stating any assumptions made in the throttling process.

(0 .966)

3.18 Air at 6.9 bar, 260 °C is throttled to 5.5 bar before expanding through a nozzle to a pressure of 1.1 bar. Assuming that the air flows reversibly in steady flow through the nozzle, and that no heat is rejected, calculate the velocity of the air at exit from the nozzle when the inlet velocity is 100 m/s.

(636 m/s)

3.19 Air at 40 °C enters a mixing chamber at a rate of 225 kg/s where it mixes with air at 15 °C entering at a rate of 540 kg/s. Calculate the temperature of the air leaving the chamber, assuming steady-flow conditions. Assume that the heat loss is negligible.

(22.4 °C)

3.20 Steam from a superheater at 7 bar, 300 °C is mixed in steady adiabatic flow with wet steam at 7 bar, dryness fraction 0.9. Calculate the mass of wet steam required per kilogram of superheated steam to produce steam at 7 bar, dry saturated.

(1.43 kg)

3.21 A rigid cylinder contains helium (molar mass 4 kg/kmol) at a pressure of 5 bar and a temperature of 15 °C. The cylinder is now connected to a large source of helium at 10 bar and 15 °C, and the valve connecting the cylinder is closed when the cylinder pressure has risen to 8 bar. Calculate the final temperature of the helium in the cylinder assuming that the heat transfer during the process is negligibly small. Take c_v for helium as 3.12 kJ/kg K.

(65.8 °C)

3.22 A well-lagged vessel of volume $1 \, \text{m}^3$, containing 1.25 kg of steam at a pressure of 2.2 bar, is connected via a valve to a large source of steam at 20 bar. The valve is opened and the pressure in the vessel is allowed to rise until the steam in the vessel is just dry saturated at 4 bar and the valve is then closed. Calculate the dryness fraction of the steam supplied.

(0.904)

3.23 An air receiver contains 10 kg of air at 7 bar. A blow-off valve is opened in error and closed again within seconds, but the pressure is observed to drop to 6 bar. Calculate the mass of air which has escaped from the receiver, stating clearly any assumptions made.

Calculate also the pressure of the air in the receiver some time after the valve has been closed such that the air temperature has attained its original value.

(1.04 kg; 6.27 bar)

3.24 A vertical cylinder of cross-sectional area $6450 \, \text{mm}^2$ is open to the atmosphere at one end and connected to a large storage vessel at the other end by means of a pipeline and valve. A frictionless piston, of weight 100 N, is fitted into the cylinder and the initial cylinder volume is negligible. The valve is then opened and air is slowly admitted from the large storage vessel into the cylinder until the piston has moved very slowly a distance of 0.6 m, when the valve is shut. If the temperature of the air in the cylinder is 30 °C at the end of the operation and the temperature of the air in the large storage vessel is constant at 90 °C, calculate:

  (i) the pressure of the air in the cylinder during the process;
 (ii) the work done on the air during the process;
(iii) the work done on the piston;
(iv) the heat supplied to the air in the cylinder during the process.

Take the atmospheric pressure as 1.013 bar.

(1.168 bar; −452 N m; 60 N m; −0.31 kJ)

4

The Second Law

In Chapter 1 it is stated that, according to the First Law of Thermodynamics, when a system undergoes a complete cycle the net heat supplied plus the net work input is zero. This is based on the conservation of energy principle which follows from observation of natural events. The Second Law of Thermodynamics, which is also a natural law, indicates that in any complete cycle the *gross* heat supplied plus the net work input must be greater than zero. Thus for any cycle in which there is a net work output (i.e. W −ve), heat must always be rejected. For any cycle in which heat is supplied at a low temperature and rejected at a higher temperature there must always be a positive work input.

To enable the second law to be discussed more fully the heat engine must be defined and discussed.

4.1 The heat engine

A heat engine is a system which operates continuously and across whose boundaries flow only heat and work.

A diagrammatic representation of a heat engine is given in Fig. 4.1; a forward heat engine is shown in Fig. 4.1(a) and a reversed heat engine in Fig. 4.1(b). (Note: the term 'reservoir' is taken to mean an energy source at a uniform temperature.) In both cases the first law applies as given by equation (1.3), $\sum Q + \sum W = 0$, or referring to Fig. 4.1,

$$Q_1 + Q_2 + W = 0$$

As shown in Fig. 4.1, for a forward heat engine cycle such as that used for power production, Q_1 is positive and W and Q_2 are negative; for a reversed cycle such as that used for a refrigerator or heat pump, Q_1 is negative and W and Q_2 are positive.

An example of a forward heat engine is shown in Fig. 4.2; by the first law the heat supplied of 100 units equates to the work output of 30 units plus the heat rejected of 70 units. It can be demonstrated that it is impossible to construct a forward heat engine in which 100 units of heat are supplied and 100 units of work output are produced; some heat must always be rejected. One statement

Fig. 4.1 Forward and reversed heat engines

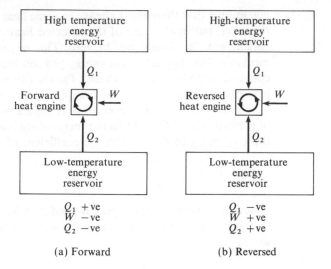

Q_1 +ve
W −ve
Q_2 −ve

Q_1 −ve
W +ve
Q_2 +ve

(a) Forward

(b) Reversed

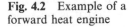

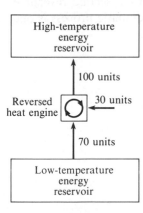

Fig. 4.2 Example of a forward heat engine

Fig. 4.3 Example of a reversed heat engine

of the Second Law of Thermodynamics is therefore as follows:

It is impossible for a heat engine to produce a net work output in a complete cycle if it exchanges heat only with a single energy reservoir.

Using the chosen sign convention for heat and work and referring to Fig. 4.1, we can write this statement of the second law in symbols as follows:

$$Q_1 > -W \tag{4.1}$$

The cycle efficiency of a forward heat engine can then be defined as the ratio of the net work output to the gross heat supplied,

i.e.   Cycle efficiency, $\eta = \dfrac{-W}{Q_1}$ (4.2)

The second law implies that the cycle efficiency must always be less than unity. For the heat engine shown in Fig. 4.2 the cycle efficiency is $\eta = 30/100 = 0.3$ or 30%.

An example of a reversed heat engine is shown in Fig. 4.3; by the first law the 70 units of heat supplied from the low-temperature energy reservoir plus the work input of 30 units is equal to the 100 units of heat rejected to the high-temperature reservoir. It can be demonstrated that it is impossible to transfer 70 units of heat from the low-temperature reservoir to the high-temperature reservoir without a work input. An alternative statement of the second law is therefore as follows:

It is impossible to construct a device that operating in a cycle will produce no effect other than the transfer of heat from a cooler to a hotter body.

Referring to Fig. 4.1(b) this statement of the second law can be expressed as

$$W > 0 \tag{4.3}$$

It is interesting to note that in the case of a forward heat engine the second

89

law implies that there must always be some heat rejected to the low-temperature reservoir, but in the case of the reversed heat engine (Fig. 4.1(b)), there is no reason why Q_2 should not be zero. This means that it is possible to convert mechanical energy into heat energy (e.g. on braking a car the kinetic energy is converted into heat at the wheels). On the other hand it is impossible to convert heat energy continuously and completely into mechanical work.

The effectiveness of a reversed heat engine is defined in terms of a coefficient of performance, COP. When the reversed engine is used mainly as a refrigerator then, referring to Fig. 5.1(b), the coefficient of performance is defined as

$$COP_{ref} = \frac{Q_2}{W} \tag{4.4}$$

When the reversed heat engine is used as a heat pump the heat transferred to the high-temperature reservoir is the useful energy, and referring to Fig. 4.1(b) we have

$$COP_{hp} = \frac{-Q_1}{W} \tag{4.5}$$

(Note: the coefficient of performance of both a refrigerator and a heat pump is always greater than unity; in the example given in Fig. 4.3, $COP_{ref} = 70/30 = 2.333$, and $COP_{hp} = 100/30 = 3.333$.) Refrigeration and heat pumps are considered in more detail in Chapter 14.

4.2 Entropy

In section 1.7, an important property, internal energy, was found to arise as a consequence of the First Law of Thermodynamics. Another important property, *entropy*, follows from the second law.

Consider a reversible adiabatic process for any system on a p–v diagram. This is represented by line AB on Fig. 4.4. Let us suppose that it is possible for the system to undergo a reversible isothermal process at temperature T_1 from B to C and then be restored to its original state by a second reversible adiabatic process from C to A. Now by definition an adiabatic process is one

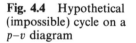

Fig. 4.4 Hypothetical (impossible) cycle on a p–v diagram

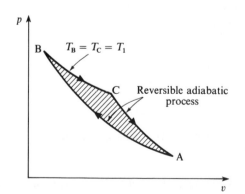

in which no heat flows to or from the system. Hence the only heat transferred is from B to C during the isothermal process. The work done *by* the system is given by the enclosed area (see section 1.6). We therefore have a system undergoing a cycle and developing a net work output while drawing heat from a reservoir at one fixed temperature. This is impossible because it violates the second law, as stated in section 4.1. Therefore the original supposition is wrong, and it is not possible to have two reversible adiabatic processes passing through the same state A.

Now one of the characteristics of a property of a system is that there is one unique line which represents a value of the property on a diagram of properties. (For example, the line BC on Fig. 4.4 represents the isothermal at T_1.) Hence there must be a property represented by a reversible adiabatic process. This property is called entropy, s.

It follows that there is no change of entropy in a reversible adiabatic process. Each reversible adiabatic process represents a unique value of entropy. On a $p-v$ diagram a series of reversible adiabatic processes appear as shown in Fig. 4.5(a), each line representing one value of entropy. This is similar to Fig. 4.5(b) in which a series of isothermals is drawn, each representing one value of temperature.

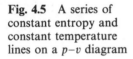

Fig. 4.5 A series of constant entropy and constant temperature lines on a $p-v$ diagram

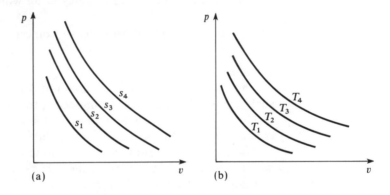

In order to be able to define entropy in terms of the other thermodynamic properties a rigorous approach is necessary; a much simplified approach has been adopted in this book. For a more rigorous approach ref. 4.1 should be consulted.

In section 3.2 a reversible adiabatic process for a perfect gas was shown to follow a law pv^γ = constant. Now the law pv^γ = constant is a unique line on a $p-v$ diagram, so that the proof given in section 3.2 for a perfect gas is a similar proof to that given above (i.e. the proof that a reversible adiabatic process occupies a unique line on a diagram of properties). The proof given above depends on the second law and has been used to introduce entropy as a property. It follows therefore that the proof of pv^γ = constant in section 3.2 must imply the fact that the entropy does not change during a reversible adiabatic process. Referring to the proof in section 3.2, starting with the non-flow energy equation for a reversible process

$$dQ = du + p\, dv$$

and for a perfect gas

$$dQ = c_v \, dT + RT \frac{dv}{v}$$

This equation can be integrated after dividing through by T,

i.e.
$$\frac{dQ}{T} = \frac{c_v \, dT}{T} + \frac{R \, dv}{v}$$

Also for an adiabatic process, $dQ = 0$,

i.e.
$$\frac{dQ}{T} = \frac{c_v \, dT}{T} + \frac{R \, dv}{v} = 0 \tag{4.6}$$

Now apart from mathematical manipulation and the introduction of the relationship between R, c_p, c_v, and γ, there are no other major steps in the proof. This must mean that dividing through by T is the one step which implies the restriction of the second law, and the important fact that the change of entropy is zero. We can say, therefore, $dQ/T = 0$ for a reversible adiabatic process. For any other reversible process $dQ/T \neq 0$.

This result can be shown to apply to all working substances,

i.e.
$$ds = \frac{dQ}{T} \quad \text{for all working substances} \tag{4.7}$$

where s is entropy.

The argument in this section does not constitute a proof of $ds = dQ/T$. For such a proof the reader is recommended to ref. 4.1.

Note that since equation (4.6) is for a reversible process, then dQ in equation (4.7) is the heat added reversibly.

The change of entropy is more important than its absolute value, and the zero of entropy can be chosen quite arbitrarily.

Integrating equation (4.7) gives

$$s_2 - s_1 = \int_1^2 \frac{dQ}{T} \tag{4.8}$$

Considering unit mass of fluid, the units of entropy are given by kilojoules per kilogram divided by K. That is, the units of specific entropy, s, are kJ/kg K. The symbol S will be used for the entropy of mass, m, of a fluid,

i.e.
$$S = ms$$

Rewriting equation (4.7) we have

$$dQ = T \, ds$$

or for any reversible process

$$Q = \int_1^2 T \, ds \tag{4.9}$$

This equation is analogous to equation (1.2),

$$W = - \int_1^2 p \, dv \quad \text{for any reversible process}$$

Thus, as there is a diagram on which areas represent work output in a reversible process, there is also a diagram on which areas represent heat supplied in a reversible process. These diagrams are the p–v and the T–s diagrams respectively, as shown in Figs 4.6(a) and 4.6(b). For a reversible process 1–2 in Fig. 4.6(a), the shaded area $\int_1^2 p \, dv$ represents work output, $-W$; for a reversible process 1–2 in Fig. 4.6(b), the shaded area $\int_1^2 T \, ds$ represents heat supplied, Q. Therefore one great use of the property entropy is that it enables a diagram to be drawn on which areas represent heat flow in a reversible process. In the section 4.3 the T–s diagram will be considered for a vapour and for a perfect gas.

Fig. 4.6 Area under a reversible process on a p–v and on a T–s diagram

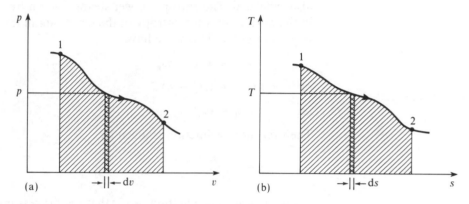

4.3 The *T–s* diagram

For a vapour

The T–s diagram for steam only will be considered here; the diagram for a refrigerant is exactly similar with the important exception of the zeros of entropy and enthalpy which vary according to the source of the tabular information (see Ch. 14). The T–s diagram for steam is shown in Fig. 4.7. Three lines of constant pressure (p_1, p_2, and p_3) are shown (i.e. lines ABCD, EFGH, and JKLM). The pressure lines in the liquid region are practically coincident with the saturated liquid line (i.e. portions AB, EF, and JK), and the difference is usually neglected. The pressure remains constant with temperature when the latent heat is added, hence the pressure lines are horizontal in the wet region (i.e. portions BC, FG, and KL). The pressure lines curve upwards in the superheat region as shown (i.e. portions CD, GH, and LM). Thus the temperature rises as heating continues at constant pressure. One constant volume line (shown as a broken line) is drawn in Fig. 4.7. Lines of constant volume are concave down in the wet region and slope up more steeply than pressure lines in the superheat region.

Fig. 4.7 *T*–s diagram
for a vapour

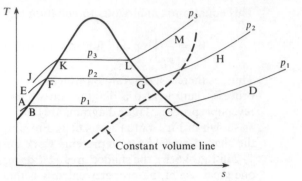

In steam tables the entropy of the saturated liquid and the dry saturated vapour are represented by s_f and s_g respectively. The difference, $s_g - s_f = s_{fg}$, is also tabulated. The entropy of wet steam is given by the entropy of the water in the mixture plus the entropy of the dry steam in the mixture. For wet steam with dryness fraction, x, we have

$$s = (1 - x)s_f + xs_g \qquad (4.10)$$

or $\qquad s = s_f + x(s_g - s_f)$

i.e. $\qquad s = s_f + xs_{fg} \qquad (4.11)$

Then the dryness fraction is given by

$$x = \frac{s - s_f}{s_{fg}} \qquad (4.12)$$

It can be seen from equation (4.12) that the dryness fraction is proportional to the distance of the state point from the liquid line on a T–s diagram. For example, for state 1 on Fig. 4.8 the dryness fraction

$$x_1 = \frac{\text{distance F1}}{\text{distance FG}} = \frac{s_1 - s_{f1}}{s_{fg1}}$$

The area under the line FG on Fig. 4.8 represents the specific enthalpy of vaporization h_{fg}. The area under line F1 is given by $x_1 h_{fg}$.

Fig. 4.8 Dryness
fraction from areas on a
T–s diagram

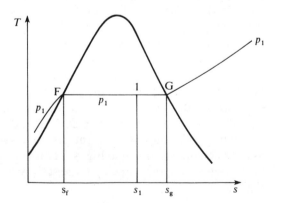

Example 4.1   1 kg of steam at 7 bar, entropy 6.5 kJ/kg K, is heated reversibly at constant pressure until the temperature is 250 °C. Calculate the heat supplied, and show on a *T–s* diagram the area which represents the heat flow.

Solution   At 7 bar, $s_g = 6.709$ kJ/kg K, hence the steam is wet, since the actual entropy, s, is less than s_g.

From equation (4.12)

$$x_1 = \frac{s_1 - s_{f1}}{s_{fg1}} = \frac{6.5 - 1.992}{4.717} = 0.955$$

Then from equation (2.2)

$$h_1 = h_{f1} + x_1 h_{fg1} = 697 + (0.955 \times 2067)$$

i.e. $h_1 = 697 + 1975 = 2672$ kJ/kg

At state 2 the steam is at 250 °C at 7 bar, and is therefore superheated. From superheat tables, $h_2 = 2955$ kJ/kg.

At constant pressure from equation (3.3)

$$Q = h_2 - h_1 = 2955 - 2672 = 283 \text{ kJ/kg}$$

i.e.   Heat supplied = 283 kJ/kg

The *T–s* diagram showing the process is given in Fig. 4.9, the shaded area representing the heat flow.

Fig. 4.9 *T–s* diagram for Example 4.1

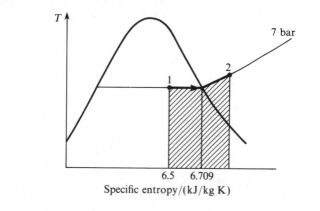

Specific entropy/(kJ/kg K)

Example 4.2   A rigid cylinder of volume 0.025 m³ contains steam at 80 bar and 350 °C. The cylinder is cooled until the pressure is 50 bar. Calculate the state of the steam after cooling and the amount of heat rejected by the steam. Sketch the process on a *T–s* diagram indicating the area which represents the heat flow.

Solution   Steam at 80 bar and 350 °C is superheated, and the specific volume from tables is 0.029 94 m³/kg. Hence the mass of steam in the cylinder is given by

$$m = \frac{0.025}{0.029\,94} = 0.835 \text{ kg}$$

95

For superheated steam above 80 bar the internal energy is found from equation (3.7),

$$u_1 = h_1 - p_1 v_1 = 2990 - \frac{80 \times 10^5 \times 0.02994}{10^3} = 2990 - 239.5$$

i.e. $u_1 = 2750.5 \ \text{kJ/kg}$

At state 2, $p_2 = 50$ bar and $v_2 = 0.029\,94 \ \text{m}^3/\text{kg}$, therefore the steam is wet, and the dryness fraction is given by equation (2.1)

$$x_2 = \frac{v_2}{v_{g_2}} = \frac{0.029\,94}{0.039\,44} = 0.758$$

From equation (2.3)

$$u_2 = (1 - x_2)u_{f_2} + x_2 u_{g_2} = (0.242 \times 1149) + (0.758 \times 2597)$$

i.e. $u_2 = 278 + 1969 = 2247 \ \text{kJ/kg}$

At constant volume from equation (3.2),

$$Q = U_2 - U_1 = m(u_2 - u_1) = 0.835(2247 - 2750.5)$$

i.e. $Q = -0.835 \times 503.5 = -420 \ \text{kJ}$

i.e.   Heat rejected $= 420 \ \text{kJ}$

Figure 4.10 shows the process drawn on a $T\!-\!s$ diagram, the shaded area representing the heat rejected by the steam.

Fig. 4.10 $T\!-\!s$ diagram for Example 4.2

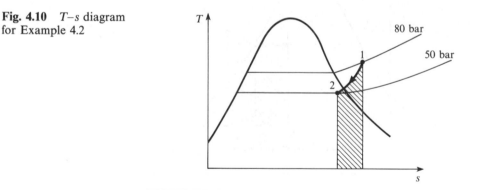

For a perfect gas

It is useful to plot lines of constant pressure and constant volume on a $T\!-\!s$ diagram for a perfect gas. Since changes of entropy are of more direct application than the absolute value, the zero of entropy can be chosen at any arbitrary reference temperature and pressure. In Fig. 4.11 the pressure line p_1 and the volume line v_1 have been drawn passing through the state point 1. Note that a line of constant pressure slopes less steeply than a line of constant volume.

Fig. 4.11 Entropy changes at constant pressure and at constant volume for a perfect gas on a *T–s* diagram

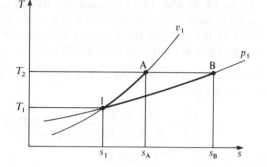

This can be proved easily by reference to Fig. 4.11. Let points A and B be at T_2 and v_1, and T_2 and p_1 respectively as shown. Now between 1 and A from equation (4.8) we have

$$s_A - s_1 = \int_1^A \frac{dQ}{T}$$

Also at a constant volume for 1 kg of gas from equation (2.11), $dQ = c_v \, dT$. Therefore

$$s_A - s_1 = \int_1^A \frac{c_v \, dT}{T} = c_v \ln\left(\frac{T_A}{T_1}\right) = c_v \ln\left(\frac{T_2}{T_1}\right)$$

Similarly, at constant pressure for 1 kg of gas, $dQ = c_p \, dT$. Hence,

$$s_B - s_1 = \int_1^B \frac{c_p \, dT}{T} = c_p \ln\left(\frac{T_B}{T_1}\right) = c_p \ln\left(\frac{T_2}{T_1}\right)$$

Now since c_p is greater than c_v for any perfect gas, then $s_B - s_1$ is greater than $s_A - s_1$. Point A must therefore lie to the left of point B on the diagram, and hence a line of constant pressure slopes less steeply than a line of constant volume. Figure 4.12(a) shows a series of constant pressure lines on a *T–s* diagram, and Fig. 4.12(b) shows a series of constant volume lines on a *T–s* diagram. Note that in Fig. 4.12(a), $p_6 > p_5 > p_4 > p_3$, etc. and in Fig. 4.12(b), $v_1 > v_2 > v_3$, etc. As the pressure rises the temperature rises and the volume decreases; conversely as the pressure and temperature fall the volume increases.

Fig. 4.12 Constant pressure and constant volume lines plotted on a *T–s* diagram for a perfect gas

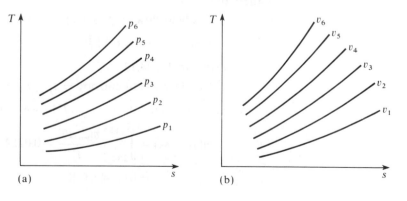

Example 4.3   Air at 15 °C and 1.05 bar occupies 0.02 m³. The air is heated at constant volume until the pressure is 4.2 bar, and then cooled at constant pressure back to the original temperature. Calculate the net heat flow to or from the air and the net entropy change. Sketch the process on a T–s diagram.

Solution   The processes are shown on a T–s diagram in Fig. 4.13. From equation (2.6), for a perfect gas,

$$m = \frac{pV}{RT} = \frac{1.05 \times 10^5 \times 0.02}{0.287 \times 10^3 \times 288} = 0.0254 \text{ kg}$$

where $T_1 = 15 + 273 = 288$ K.

Fig. 4.13 Processes on a T–s diagram for Example 4.3

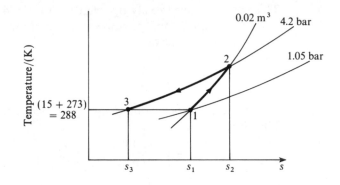

For a perfect gas at constant volume, $p_1/T_1 = p_2/T_2$, hence

$$T_2 = \frac{4.2 \times 288}{1.05} = 1152 \text{ K}$$

From equation (2.13), at constant volume

$$Q = mc_v(T_2 - T_1) = 0.0254 \times 0.718(1152 - 288)$$

i.e. $Q_{1-2} = 15.75$ kJ

From equation (2.12), at constant pressure

$$Q = mc_p(T_3 - T_2) = 0.0254 \times 1.005(288 - 1152)$$

i.e. $Q_{2-3} = -22.05$ kJ

therefore

Net heat flow $= Q_{1-2} + Q_{2-3} = 15.75 - 22.05 = -6.3$ kJ

i.e.   Heat rejected $= 6.3$ kJ

Referring to Fig. 4.13

Net decrease in entropy $= s_1 - s_3 = (s_2 - s_3) - (s_2 - s_1)$

At constant pressure, $dQ = mc_p\,dT$, hence, using equation (4.8),

$$m(s_2 - s_3) = \int_{288}^{1152} \frac{mc_p\,dT}{T} = 0.0254 \times 1.005 \times \ln\left(\frac{1152}{288}\right)$$

$$= 0.0354 \text{ kJ/K}$$

At constant volume, $dQ = mc_v \, dT$, hence, using equation (4.8)

$$m(s_2 - s_1) = \int_{288}^{1152} \frac{mc_v \, dT}{T} = 0.0254 \times 0.718 \times \ln\left(\frac{1152}{288}\right)$$

$$= 0.0253 \text{ kJ/kg}$$

Therefore,

$$m(s_1 - s_3) = 0.0354 - 0.0253 = 0.0101 \text{ kJ/K}$$

i.e.   Decrease in entropy of air $= 0.0101 \text{ kJ/K}$

Note that since entropy is a property, the decrease in entropy in Example 4.3, given by $s_1 - s_3$, is independent of the processes undergone between states 1 and 3. The change $s_1 - s_3$ can also be found by imagining a reversible isothermal process taking place between 1 and 3. The isothermal process on the *T–s* diagram will be considered in section 4.4.

4.4 Reversible processes on the *T–s* diagram

The various reversible processes dealt with in Chapter 3 will now be considered in relation to the *T–s* diagram. The constant volume and constant pressure processes have been represented on the *T–s* diagram in section 4.3, and will therefore not be discussed again in this section.

Reversible isothermal process

A reversible isothermal process will appear as a horizontal line on a *T–s* diagram, and the area under the line must represent the heat flow during the process. For example, Fig. 4.14 shows a reversible isothermal expansion of wet steam into the superheat region. The shaded area represents the heat supplied during the process,

i.e.   Heat supplied $= T(s_2 - s_1)$

Note that the absolute temperature must be used.

Fig. 4.14 Reversible isothermal process for steam on a *T–s* diagram

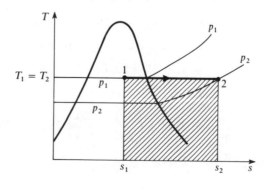

Example 4.4 Dry saturated steam at 100 bar expands isothermally and reversibly to a pressure of 10 bar. Calculate the heat supplied and the work done per kilogram of steam during the process.

Solution The process is shown in Fig. 4.15, the shaded area representing the heat supplied.

Fig. 4.15 Process on a *T–s* diagram for Example 4.4

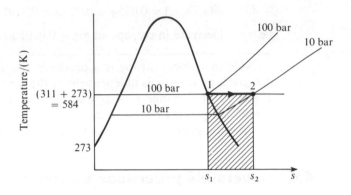

From tables at 100 bar, dry saturated

$$s_1 = s_g = 5.615 \text{ kJ/kg K} \quad \text{and} \quad t_1 = 311\,°\text{C}$$

At 10 bar and 311 °C the steam is superheated, hence interpolating

$$s_2 = 7.124 + \left(\frac{311 - 300}{350 - 300}\right)(7.301 - 7.124)$$

i.e. $s_2 = 7.124 + 0.039 = 7.163 \text{ kJ/kg K}$

Then we have

$$\text{Heat supplied} = \text{shaded area} = T(s_2 - s_1)$$
$$= 584(7.163 - 5.615) = 584 \times 1.548$$

where $T = 311 + 273 = 584$ K.

i.e. Heat supplied $= 904 \text{ kJ/kg}$

To find the work done it is necessary to apply the non-flow energy equation,

i.e. $Q + W = u_2 - u_1 \quad \text{or} \quad W = (u_2 - u_1) - Q$

From tables, at 100 bar, dry saturated,

$$u_1 = u_g = 2545 \text{ kJ/kg}$$

At 10 bar and 311 °C, interpolating,

$$u_2 = 2794 + \left(\frac{311 - 300}{350 - 300}\right)(2875 - 2794) = 2794 + 17.8$$

i.e. $u_2 = 2811.8 \text{ kJ/kg}$

Then

$$W = (u_2 - u_1) - Q$$
$$= (2811.8 - 2545) - 904 = 266.8 - 904 = -637.2 \text{ kJ/kg}$$

i.e.　　Work done *by* the steam = 637.2 kJ/kg

A reversible isothermal process for a perfect gas is shown on a *T–s* diagram in Fig. 4.16. The shaded area represents the heat supplied during the process,

i.e.　　$Q = T(s_2 - s_1)$

Fig. 4.16 Reversible isothermal process for a perfect gas

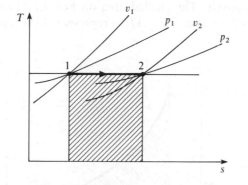

For a perfect gas undergoing an isothermal process it is possible to evaluate $s_2 - s_1$. From the non-flow equation (1.6) we have, for a reversible process,

$$dQ = du + p \, dv$$

Also for a perfect gas from Joule's law $du = c_v \, dT$,

i.e.　　$dQ = c_v \, dT + p \, dv$

For an isothermal process, $dT = 0$, hence

$$dQ = p \, dv$$

Then, since $pv = RT$, we have

$$dQ = RT\frac{dv}{v}$$

Now from equation (4.8)

$$s_2 - s_1 = \int_1^2 \frac{dQ}{T} = \int_{v_1}^{v_2} \frac{RT \, dv}{Tv} = R \int_{v_1}^{v_2} \frac{dv}{v}$$

i.e.　　$s_2 - s_1 = R \ln\left(\frac{v_2}{v_1}\right) = R \ln\left(\frac{p_1}{p_2}\right)$　　　　(4.13)

Therefore the heat supplied is given by

$$Q = T(s_2 - s_1) = RT \ln\left(\frac{v_2}{v_1}\right) = RT \ln\left(\frac{p_1}{p_2}\right)$$

Note that this result is the same as that derived in section 3.1,

i.e. $\quad Q = -W = RT\ln\left(\dfrac{p_1}{p_2}\right) = p_1 v_1 \ln\left(\dfrac{p_1}{p_2}\right)$

Example 4.5 0.03 m^3 of nitrogen (molar mass 28 kg/kmol) contained in a cylinder behind a piston is initially at 1.05 bar and 15 °C. The gas is compressed isothermally and reversibly until the pressure is 4.2 bar. Calculate the change of entropy, the heat flow, and the work done, and sketch the process on a p–v and T–s diagram. Assume nitrogen to act as a perfect gas.

Solution The process is shown on a p–v and a T–s diagram in Figs 4.17(a) and 4.17(b) respectively. The shaded area on Fig. 4.17(a) represents work input, and the shaded area on Fig. 4.17(b) represents heat rejected.

Fig. 4.17 Processes for Example 4.5 on p–v and T–s diagrams

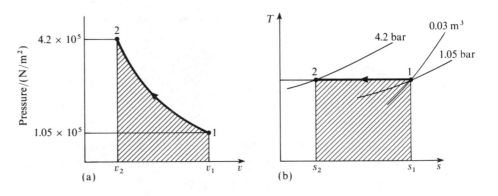

From equation (4.13)

$$s_2 - s_1 = R\ln\left(\frac{p_1}{p_2}\right) = \frac{297}{10^3}\ln\left(\frac{1.05}{4.2}\right)$$

i.e. $\quad s_2 - s_1 = -\dfrac{297}{10^3}\ln\left(\dfrac{4.2}{1.05}\right) = -0.4117 \text{ kJ/kg K}$

From equation (2.9)

$$R = \frac{\tilde{R}}{\tilde{m}} = \frac{8314.5}{28} = 297 \text{ N m/kg K}$$

Then, since $pV = mRT$, we have

$$m = \frac{pV}{RT} = \frac{1.05 \times 10^5 \times 0.03}{297 \times 288} = 0.0368 \text{ kg}$$

where $T = 15 + 273 = 288 \text{ K}$.

Then $\quad S_1 - S_2 = m(s_1 - s_2)$

$$= 0.0368 \times 0.4117 = 0.01516 \text{ kJ/K}$$

Heat rejected $= m$ (shaded area on Fig. 4.17(b)) $= mT(s_1 - s_2)$

$$= 0.01516 \times 288 = 4.37 \text{ kJ}$$

i.e. $Q = -4.37 \text{ kJ}$

Then for an isothermal process for a perfect gas, from equation (3.12),

$$Q + W = 0$$

$$-4.37 + W = 0$$

i.e.   Work input, $W = 4.37 \text{ kJ}$

Reversible adiabatic process (or isentropic process)

For a reversible adiabatic process the entropy remains constant, and hence the process is called an *isentropic process*. Note that for a process to be isentropic it need not be either adiabatic or reversible, but the process will always appear as a vertical line on a *T–s* diagram. Cases in which an isentropic process is not both adiabatic and reversible occur infrequently and will be ignored throughout this book.

An isentropic process for superheated steam expanding into the wet region is shown in Fig. 4.18. When the reversible adiabatic process was considered in section 3.1, it was stated that no simple method was available for fixing the end states. Now, using the fact that the entropy remains constant, the end states can be found easily from tables. This is illustrated in the following example.

Fig. 4.18 Isentropic process for steam on a *T–s* diagram

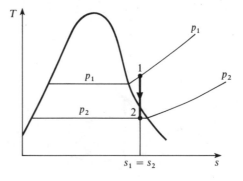

Example 4.6   Steam at 100 bar, 375 °C expands isentropically in a cylinder behind a piston to a pressure of 10 bar. Calculate the work done per kilogram of steam.

Solution   From superheat tables, at 100 bar, 375 °C, we have

$$s_1 = s_2 = 6.091 \text{ kJ/kg K}$$

At 10 bar and $s_2 = 6.091$, the steam is wet, since s_2 is less than s_{g_2}. Then from equation (4.12)

$$x_2 = \frac{s_2 - s_{f_2}}{s_{fg_2}} = \frac{6.091 - 2.138}{4.448} = 0.889$$

103

Then from equation (2.3)

$$u_1 = (1 - x_2)u_{f_2} + x_2 u_{g_2} = (0.111 \times 762) + (0.889 \times 2584)$$

i.e. $u_2 = 84.6 + 2297 = 2381.6 \text{ kJ/kg}$

At 100 bar, 375 °C, we have from tables, $h_1 = 3017 \text{ kJ/kg}$ and $v_1 = 0.024\,53 \text{ m}^3/\text{kg}$. Then using equation (1.9)

$$u_1 = h_1 - p_1 v_1 = 3017 - \frac{100 \times 10^5 \times 0.024\,53}{10^3} = 3017 - 245.3$$

i.e. $u_1 = 2771.7 \text{ kJ/kg}$

For an adiabatic process from equation (3.13),

$$W = u_2 - u_1$$

therefore,

$$W = 2381.6 - 2771.7$$

$$= -390.1 \text{ kJ/kg}$$

i.e. Work done *by* the steam $= 390.1 \text{ kJ/kg}$

For a perfect gas an isentropic process on a T–s diagram is shown in Fig. 4.19. It is shown in section 3.1 that for a reversible adiabatic process for a perfect gas the process follows a law $pv^\gamma = \text{constant}$. Since a reversible adiabatic process occurs at constant entropy and is known as an isentropic process, the index γ is known as the *isentropic index* of the gas.

Fig. 4.19 Isentropic process for a perfect gas on a T–s diagram

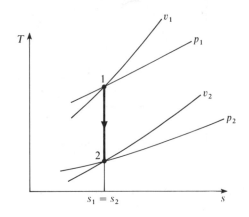

Polytropic process

To find the change of entropy in a polytropic process for a vapour when the end states have been fixed using $p_1 v_1^n = p_2 v_2^n$, the entropy values at the end states can be read straight from tables.

Example 4.7 In a reciprocating compressor of a refrigeration plant the refrigerant is dry saturated at 2.01 bar at the beginning of compression and is compressed reversibly according to a polytropic law $pv^{1.1}$ = constant to a pressure of 10 bar. Calculate the change of specific entropy during the process using the table of properties of refrigerant given in Table 4.1, interpolating where necessary.

Table 4.1 Properties of refrigerant for Example 4.7

	Saturation values			Superheated values at 10 bar	
p_g	v_g	s_g		v	s
(bar)	(m³/kg)	(kJ/kg K)		(m³/kg)	(kJ/kg K)
2.01	0.0978	1.7189		0.0222	1.7564
10	0.0202	1.7033		0.0233	1.7847

Solution   From Table 4.1,

$$s_1 = s_{g_1} = 1.7189 \text{ kJ/kg K}$$

and $$v_1 = v_{g_1} = 0.0978 \text{ m}^3/\text{kg}$$

Then,

$$v_2 = v_1\left(\frac{p_1}{p_2}\right)^{1/1.1} = 0.0978\left(\frac{2.01}{10}\right)^{1/1.1}$$

$$= 0.0228 \text{ m}^3/\text{kg}$$

At 10 bar and a specific volume of 0.0228 m³/kg the steam is superheated. The process is shown in Fig. 4.20.

Fig. 4.20 Isentropic compression process for Example 4.7

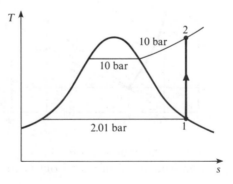

Interpolating from the superheat tables of Table 4.1,

$$s_2 = 1.7564 + \frac{(0.0228 - 0.0222)}{(0.0233 - 0.0222)} \times (1.7847 - 1.7564)$$

$$= 1.7704 \text{ kJ/kg K}$$

Increase of entropy $= 1.7704 - 1.7189 = 0.0515 \text{ kJ/kg K}$

It was shown in section 3.1 that the polytropic process is the general case for a perfect gas. To find the entropy change for a perfect gas in the general case, consider the non-flow energy equation for a reversible process equation (1.6),

$$dQ = du + p\,dv$$

Also for unit mass of a perfect gas from Joule's law $du = c_v\,dT$, and from equation (2.5), $pv = RT$. Therefore

$$dQ = c_v\,RT + \frac{RT\,dv}{v}$$

Then from equation (4.7)

$$ds = \frac{dQ}{T} = \frac{c_v\,dT}{T} + \frac{R\,dv}{v}$$

Hence between any two states 1 and 2

$$s_2 - s_1 = c_v \int_{T_1}^{T_2} \frac{dT}{T} + R \int_{v_1}^{v_2} \frac{dv}{v} = c_v \ln\!\left(\frac{T_2}{T_1}\right) + R \ln\!\left(\frac{v_2}{v_1}\right) \qquad (4.14)$$

This can be illustrated on a $T–s$ diagram as in Fig. 4.21. Since in the process in Fig. 4.21, $T_2 < T_1$, then it is more convenient to write

$$s_2 - s_1 = R \ln\!\left(\frac{v_2}{v_1}\right) - c_v \ln\!\left(\frac{T_1}{T_2}\right) \qquad (4.15)$$

Fig. 4.21 Polytropic process for a perfect gas on a $T–s$ diagram

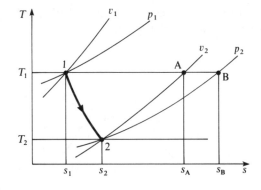

The first part of the expression for $s_2 - s_1$ in equation (4.15) is the change of entropy in an isothermal process from v_1 to v_2, i.e. from equation (4.13)

$$s_A - s_1 = R \ln\!\left(\frac{v_2}{v_1}\right) \quad \text{(see Fig. 4.21)}$$

Also the second part of the expression for $s_2 - s_1$ in equation (4.15) is the change of entropy in a constant volume process from T_1 to T_2, i.e. referring to Fig. 4.21,

$$s_A - s_2 = c_v \ln\!\left(\frac{T_1}{T_2}\right)$$

It can be seen therefore that in calculating the entropy change in a polytropic process from state 1 to state 2 we have in effect replaced the process by two simpler processes: from 1 to A and then from A to 2. It is clear from Fig. 4.21 that

$$s_2 - s_1 = (s_A - s_1) - (s_A - s_2)$$

Any two processes can be chosen to replace a polytropic process in order to find the entropy change. For example, going from 1 to B and then from B to 2 as in Fig. 4.21, we have

$$s_2 - s_1 = (s_B - s_1) - (s_B - s_2)$$

At constant temperature between p_1 and p_2, using equation (4.13),

$$s_B - s_1 = R \ln\left(\frac{p_1}{p_2}\right)$$

and at constant pressure between T_1 and T_2 we have

$$s_B - s_2 = c_p \ln\left(\frac{T_1}{T_2}\right)$$

Hence,

$$s_2 - s_1 = R \ln\left(\frac{p_1}{p_2}\right) - c_p \ln\left(\frac{T_1}{T_2}\right)$$

or $\qquad s_2 - s_1 = c_p \ln\left(\frac{T_2}{T_1}\right) + R \ln\left(\frac{p_1}{p_2}\right)$ $\qquad\qquad$ (4.16)

Equation (4.16) can also be derived easily from equation (4.14).

There are obviously a large number of possible equations for the change of entropy in a polytropic process. Each problem can be dealt with by sketching the *T–s* diagram and replacing the process by two other simpler reversible processes, as in Fig. 4.21.

Example 4.8$\qquad$ Calculate the change of entropy of 1 kg of air expanding polytropically in a cylinder behind a piston from 6.3 bar and 550 °C to 1.05 bar. The index of expansion is 1.3.

Solution$\qquad$ The process is shown on a *T–s* diagram in Fig. 4.22. From equation 3.29,

$$\frac{T_1}{T_2} = \left(\frac{p_1}{p_2}\right)^{(n-1)/n} = \left(\frac{6.3}{1.05}\right)^{(1.3-1)/1.3} = 1.512$$

Fig. 4.22 Process for Example 4.8 on a *T–s* diagram

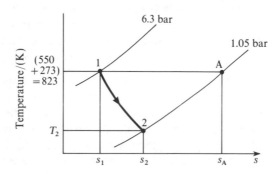

therefore

$$T_2 = \frac{823}{1.512} = 544 \text{ K}$$

where $T_1 = 550 + 273 = 823$ K.

Now replace the process 1 to 2 by two processes, 1 to A and A to 2. Then at constant temperature from 1 to A, from equation (4.13)

$$s_A - s_1 = R \ln\left(\frac{p_1}{p_2}\right) = 0.287 \ln\left(\frac{6.3}{1.05}\right)$$

$$= 0.514 \text{ kJ/kg K}$$

At constant pressure from A to 2

$$s_A - s_2 = c_p \ln\left(\frac{T_1}{T_2}\right) = 1.005 \ln\left(\frac{823}{544}\right)$$

$$= 0.416$$

Then $\quad s_2 - s_1 = 0.514 - 0.416 = 0.098$ kJ/kg K

i.e. $\qquad$ Increase in entropy $= 0.098$ kJ/kg K

Note that if in Example 4.8 $s_A - s_2$ happened to be greater than $s_A - s_1$, this would mean that s_1 was greater than s_2, and the process should appear as in Fig. 4.23.

Fig. 4.23 Alternative *T–s* diagram for Example 4.8

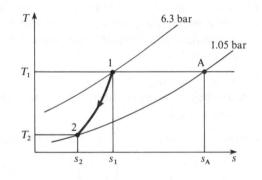

Example 4.9

0.05 kg of carbon dioxide (molar mass 44 kg/kmol) is compressed from 1 bar, 15 °C, until the pressure is 8.3 bar, and the volume is then 0.004 m³. Calculate the change of entropy. Take c_p for carbon dioxide as 0.88 kJ/kg K, and assume carbon dioxide to be a perfect gas.

Solution The two end states are marked on a *T–s* diagram in Fig. 4.24. The process is not specified in the example and no information about it is necessary. States 1 and 2 are fixed and hence $s_2 - s_1$ is fixed. The process between 1 and 2 could be reversible or irreversible; the change of entropy is the same between the end states given. With reference to Fig. 4.24, to find $s_1 - s_2$ we can first find $s_A - s_2$ and then subtract $s_A - s_1$. First of all it is necessary to find R and then T_2.

Fig. 4.24 $T-s$ diagram for Example 4.9

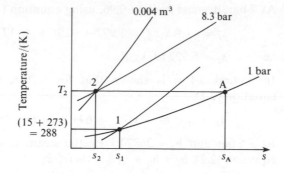

From equation (2.9)

$$R = \frac{\tilde{R}}{\tilde{m}} = \frac{8314.5}{44} = 189 \text{ N m/kg K}$$

From equation (2.6), $pV = mRT$, therefore,

$$T_2 = \frac{p_2 V_2}{mR} = \frac{8.3 \times 10^5 \times 0.004}{0.05 \times 189} = 351 \text{ K}$$

Then from equation (4.13)

$$s_A - s_2 = R \ln\left(\frac{p_2}{p_A}\right) = 0.189 \ln\left(\frac{8.3}{1}\right) = 0.4 \text{ kJ/kg K}$$

Also at constant pressure from 1 to A

$$s_A - s_1 = c_p \ln\left(\frac{T_2}{T_1}\right) = 0.88 \ln\left(\frac{351}{288}\right) = 0.174 \text{ kJ/kg K}$$

where $T_1 = 15 + 273 = 288$ K. Then

$$s_1 - s_2 = 0.4 - 0.174 = 0.226 \text{ kJ/kg K}$$

Hence for 0.05 kg of carbon dioxide

Decrease in entropy $= 0.05 \times 0.226 = 0.0113 \text{ kJ/K}$

4.5 Entropy and irreversibility

In section 4.4 it is pointed out that, since entropy is a property, the change of entropy depends only on the end states and not on the process between the end states. Therefore, provided an irreversible process gives enough information to fix the end states then the change of entropy can be found. This can best be illustrated by some examples.

Example 4.10   Steam at 7 bar, dryness fraction 0.96, is throttled down to 3.5 bar. Calculate the change of entropy per unit mass of steam.

109

Solution   At 7 bar, dryness fraction 0.96, using equation (4.11) we have

$$s_1 = s_{f_1} + x_1 s_{fg_1} = 1.992 + (0.96 \times 4.717)$$

i.e. $s_1 = 6.522 \text{ kJ/kg K}$

In section 3.4 it is shown that for a throttling process, $h_1 = h_2$. From equation (2.2)

$$h_2 = h_1 = h_{f_1} + x_1 h_{fg_1} = 697 + (0.96 \times 2067) = 2682 \text{ kJ/kg}$$

At 3.5 bar and $h_2 = 2682 \text{ kJ/kg}$ the steam is still wet, since $h_{g_2} > h_2$. From equation (2.2), $h_2 = h_{f_2} + x_2 h_{fg_2}$, therefore

$$x_2 = \frac{h_2 - h_{f_2}}{h_{fg_2}} = \frac{2682 - 584}{2148} = 0.977$$

Then from equation (4.11)

$$s_2 = s_{f_2} + x_2 s_{fg_2} = 1.727 + (0.977 \times 5.214) = 6.817 \text{ kJ/kg K}$$

Therefore,

$$\text{Increase of entropy} = 6.817 - 6.522 = 0.295 \text{ kJ/kg K}$$

The process is shown on a T–s diagram in Fig. 4.25. Note that the process is shown dotted, and the area under the line does not represent heat flow; a throttling process assumes no heat flow, but there is a change in entropy because the process is irreversible.

Fig. 4.25 Throttling process for Example 4.10 on a T–s diagram

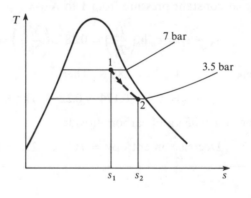

Example 4.11   Two vessels of equal volume are connected by a short length of pipe containing a valve; both vessels are well lagged. One vessel contains air and the other is completely evacuated. Calculate the change of entropy per kg of air in the system when the valve is opened and the air is allowed to fill both vessels.

Solution   Initially the vessel A contains air and the vessel B is completely evacuated, as in Fig. 4.26; finally the air occupies both vessels A and B. In section 3.4 it was shown that in an unresisted expansion for a perfect gas, the initial and final temperatures are equal. In this case the initial volume is V_A and the final volume

Fig. 4.26 Two well-lagged interconnected vessels for Example 4.11

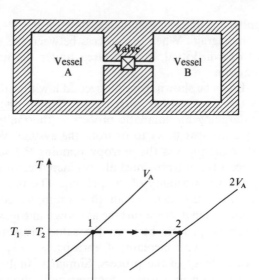

Fig. 4.27 Process on a $T-s$ diagram for Example 4.11

is $V_A + V_B = 2V_A$. The end states can be marked on a $T-s$ diagram as shown in Fig. 4.27. The process 1 to 2 is irreversible and must be drawn dotted. The change of entropy is $s_2 - s_1$, regardless of the path of the process between states 1 and 2. Hence, for the purpose of calculating the change of entropy, imagine the process replaced by a reversible isothermal process between states 1 and 2. Then from equation (4.13)

$$(s_2 - s_1) = R \ln\left(\frac{V_2}{V_1}\right) = 0.287 \ln\left(\frac{2V_A}{V_A}\right)$$

$$= 0.287 \ln 2 = 0.199 \text{ kJ/kg K}$$

i.e.   Increase of entropy = 0.199 kJ/kg K

Note that the process is drawn dotted in Fig. 4.27, and the area under the line has no significance; the process is adiabatic and there is a change in entropy since the process is irreversible.

It is important to remember that the equation (4.7), $ds = dQ/T$, is true only for reversible processes. In the same way the equation $dW = -p\,dv$ is true only for reversible processes. In Example 4.11 the volume of the air increased from V_A to $2V_A$, and yet no work was done by the air during the process,

i.e. $dW = 0$ yet $V_2 - V_1 = 2V_A - V_A = V_A$

Similarly, the entropy in Example 4.11 increased by 0.199 kJ/kg K and yet the heat flow was zero, i.e. $ds \neq dQ/T$. No confusion should be caused if the $T-s$ and/or the $p-v$ diagram is drawn for each problem and the state points marked in their correct positions. Then, when a process between two states is reversible, the lines representing the process can be drawn in as full lines, and the area

111

under the line represents heat flow on the T–s diagram and work done on the p–v diagram. When the process between the states is irreversible the line must be drawn dotted, and the area under the line has no significance on either diagram.

It can be shown from the second law that the entropy of a thermally isolated system must either increase or remain the same. For instance, a system undergoing an adiabatic process is thermally isolated from its surroundings since no heat flows to or from the system. We have seen that in a reversible adiabatic process the entropy remains the same. In an irreversible adiabatic process the entropy must always increase, and the gain of entropy is a measure of the irreversibility of the process. The processes in Examples 4.10 and 4.11 illustrate this fact. As another example, consider an irreversible adiabatic expansion in a steam turbine as shown in Fig. 4.28 as process 1 to 2. A reversible adiabatic process between the same pressures is represented by 1 to 2s in Fig. 4.28. The increase of entropy, $s_2 - s_1 = s_2 - s_{2s}$, is a measure of the irreversibility of the process. Similarly, in Fig. 4.29, an irreversible adiabatic compression in a rotary compressor is shown as process 1 to 2. A reversible adiabatic process between the same pressures is represented by 1 to 2s. As before, the increase of entropy shows the irreversibility of the process.

Fig. 4.28 Irreversible adiabatic process for steam on a T–s diagram

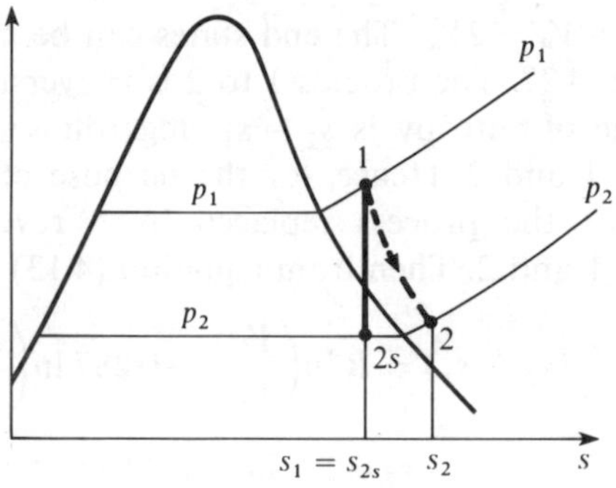

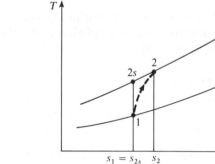

Fig. 4.29 Irreversible adiabatic compression for a perfect gas on a T–s diagram

Example 4.12 In an air turbine the air expands from 6.8 bar and 430 °C to 1.013 bar and 150 °C. The heat loss from the turbine can be assumed to be negligible. Show that the process is irreversible, and calculate the change of entropy per kilogram of air.

Solution   Since the heat loss is negligible the process is adiabatic. For a reversible adiabatic process for a perfect gas, using equation (3.21),

$$\frac{T_1}{T_2} = \left(\frac{p_1}{p_2}\right)^{(\gamma-1)/\gamma}$$

i.e.

$$\frac{703}{T_2} = \left(\frac{6.8}{1.013}\right)^{(1.4-1)/1.4}$$

where $T_1 = 430 + 273 = 703$ K,

$$T_2 = \frac{703}{1.723} = 408 \text{ K} = 408 - 273 = 135\,^\circ\text{C}$$

But the actual temperature is 150 °C at the pressure of 1.013 bar, hence the process is irreversible. The process is shown as 1 to 2 in Fig. 4.30; the ideal isentropic process 1 to 2s is also shown. It is not possible for process 1 to 2 to be reversible, because in that case the area under line 1–2 would represent heat flow and yet the process is adiabatic.

Fig. 4.30 T–s diagram for Example 4.12

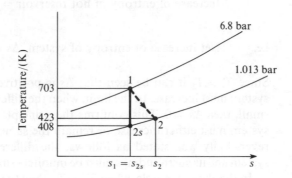

The change of entropy, $s_2 - s_1$, can be found by considering a reversible constant pressure process between 2s and 2. Then from equation (4.7), $ds = dQ/T$, and at constant pressure for 1 kg of a perfect gas we have $dQ = c_p \, dT$ therefore

$$s_2 = s_{2s} = \int_{2s}^{2} \frac{c_p \, dT}{T} = c_p \ln\left(\frac{T_2}{T_{2s}}\right)$$

$$= 1.005 \ln\left(\frac{423}{408}\right) = 0.0363 \text{ kJ/kg K}$$

i.e.   Increase of entropy, $s_2 - s_1 = 0.0363$ kJ/kg K

Consider now the case when a system is not thermally isolated from its surroundings. The entropy of such a system can increase, decrease, or remain the same, depending on the heat crossing the boundary. However, if the boundary is extended to include the source or sink of heat with which the system is in communication, then the entropy of this new system must either increase or remain the same. To illustrate this, consider a hot reservoir at T_1 and a cold

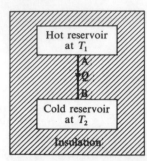

Fig. 4.31 Two thermally-insulated interconnected reservoirs of energy

reservoir at T_2, and assume that the two reservoirs are thermally insulated from the surroundings as in Fig. 4.31. Let the heat flow from the hot to the cold reservoir be Q. There is a continuous temperature gradient from T_1 to T_2 between points A and B, and it can be assumed that heat is transferred reversible from the hot reservoir to point A, and from point B to the cold reservoir. It will be assumed that the reservoirs are such that the temperature of each remains constant. Then we have

$$\text{Heat supplied to cold reservoir} = +Q$$

Hence from equation (4.8)

$$\text{Increase of entropy of cold reservoir} = +\frac{Q}{T_2}$$

Also,

$$\text{Heat supplied to hot reservoir} = -Q$$

therefore

$$\text{Increase of entropy of hot reservoir} = -\frac{Q}{T_1}$$

i.e. $$\text{Net increase of entropy of system, } \Delta s = \left(\frac{Q}{T_2} - \frac{Q}{T_1}\right)$$

Since $T_1 > T_2$ it can be seen that Δs is positive, and hence the entropy of the system must increase. In the limit, when the difference in temperature is infinitely small, then $\Delta s = 0$. This confirms the principle that the entropy of an isolated system must either increase or remain the same. In section 1.4, criterion (c) for reversibility was stated as follows: the difference in temperature between a system and its surroundings must be infinitely small during a reversible process.

In the above example, when $T_1 > T_2$, then the heat flow between the reservoirs is irreversible by the above criterion. Thus the entropy of the system increases when the heat flow process is irreversible, but remains the same when the process is reversible. The increase of entropy is a measure of the irreversibility. The processes occurring in the above example can be drawn on a T–s diagram as in Fig. 4.32. The two processes have been superimposed on the same diagram. Process P–R represents the transfer of Q units of heat from the hot reservoir,

Fig. 4.32 Processes for the hot and cold reservoirs on a T–s diagram

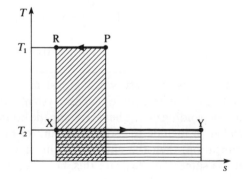

and the area under P–R is equal to Q. Process X–Y represents the transfer of Q units of heat to the cold reservoir, and the area under X–Y is equal to Q. The area under P–R is equal to the area under X–Y, and hence it can be seen from the diagram that the entropy of the cold reservoir must always increase more than the entropy of the hot reservoir decreases. Thus the entropy of the combined system must increase. Note that, since in this example both processes P–R and X–Y are reversible, then the irreversibility occurs between A and B on Fig. 4.31. That is, the irreversibility is caused by the heat transfer process between A and B. Whenever heat is transferred through a finite temperature difference the process is irreversible and there is an increase of entropy of the system and its surroundings.

In certain processes the irreversibility may occur in the surroundings, then the process is internally reversible, and areas on the $p-v$ and $T-s$ diagrams relate closely to the work and heat respectively as before. Internal reversibility was mentioned earlier in section 1.5. In most problems when a process is assumed to be reversible it is internal reversibility which is implied. Conversely, most processes in practice which are said to be irreversible are internally irreversible due to eddying and churning of the working fluid, as in Example 4.13.

Referring to Fig. 4.31, if a heat engine were interposed between the hot and cold reservoirs, some work could be developed. The second law states that heat can never flow unaided from a cold reservoir to a hot reservoir, therefore in order to develop work from the quantity of energy, Q, after it has been transferred to the cold reservoir, it would be necessary to have a third reservoir at lower temperature than the cold reservoir. It is clear that when a quantity of heat is transferred through a finite temperature difference, its usefulness becomes less, and in the limit when the heat has been transferred to the lowest existing temperature reservoir then no more work can be developed. Irreversibility therefore has a degrading effect on the energy available, and entropy can be considered as a measure not only of irreversibility but also of the degradation of energy. Note that, by the principle of conservation of energy, no energy can be destroyed; by the Second Law of Thermodynamics, energy can only become less useful and never more useful. Systems tend naturally to states of lower grade energy; any system moving to a state of higher-grade energy without an external supply of energy would be violating the second law. The second law can be seen to imply a direction or a gradient of usefulness of energy. Work is more useful than heat; the higher the temperature of a reservoir of energy the more useful is the amount of energy available. Applying this latter conclusion to a heat engine it can be deduced that, for a given cold reservoir (e.g. the atmosphere), then the higher the temperature of the hot reservoir, the higher will be the thermal efficiency of the heat engine. This will be discussed more fully in Chapter 5.

4.6 Exergy

The theoretical maximum amount of work that can be obtained from a system at any state p_1 and T_1 when operating with a reservoir at the constant pressure and temperature p_0 and T_0 is called the *exergy* or *availability*.

Non-flow systems

Consider a system consisting of a fluid in a cylinder behind a piston, the fluid expanding reversibly from initial conditions of p_1 and T_1 to final atmospheric conditions of p_0 and T_0. Imagine also that the system works in conjunction with a reversible heat engine which receives heat reversibly from the fluid in the cylinder such that the working substance of the heat engine follows the cycle 01A0 as shown in Figs 4.33(a) and 4.33(b), where $s_1 = s_A$ and $T_0 = T_A$.

Fig. 4.33 Illustration of the exergy of a system

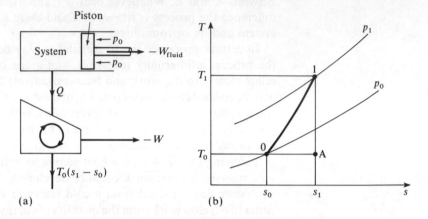

(a)   (b)

Note: The only possible way in which this could occur would be if an infinite number of reversible heat engines were arranged in parallel, each operating on a Carnot cycle (see Ch. 5), each one receiving heat at a different constant temperature and each one rejecting heat at T_0. The work done *by* this engine is given by

$$-W = \sum Q$$

therefore

$$-W = Q - T_0(s_1 - s_0)$$

The heat supplied to the engine is equal to the heat rejected by the fluid in the cylinder. Therefore for the fluid in the cylinder undergoing the process 1 to 0, we have

$$-Q + W_{\text{fluid}} = u_0 - u_1$$

i.e. $W_{\text{fluid}} = (u_0 - u_1) + Q$

Therefore the total work output is given by

$$-W_{\text{fluid}} - W = -(u_0 - u_1) - Q + Q - T_0(s_1 - s_0)$$
$$= (u_1 - u_0) - T_0(s_1 - s_0)$$

The work done by the fluid on the piston is less than the total work done by the fluid, $-W_{\text{fluid}}$, since there is work done on the atmosphere which is at the constant pressure p_0 (see Problem 3.24). That is

Work done by fluid on atmosphere, $-W_{\text{atm}} = p_0(v_0 - v_1)$

Note: when a fluid undergoes a complete cycle then the net work done on or by the atmosphere is zero. Hence

$$\text{Maximum work available} = \text{Work done by fluid on piston}$$
$$+ \text{Work done by engine}$$
$$= -W_{\text{fluid}} + W_{\text{atm}} - W$$
$$= (u_1 - u_0) - T_0(s_1 - s_0) - p_0(v_0 - v_1)$$
$$= (u_1 + p_0 v_1 - T_0 s_1) - (u_0 + p_0 v_0 - T_0 s_0)$$
$$= a_1 - a_0$$

The property $a = u + p_0 v - T_0 s$ is called the *specific non-flow exergy*.

Steady-flow systems

Let fluid flow steadily with a velocity C_1 from a reservoir in which the pressure and temperature remain constant at p_1 and T_1 through an apparatus to atmospheric pressure of p_0. Let the reservoir be at a height Z_1 from the datum, which can be taken at exit from the apparatus, i.e. $Z_0 = 0$. For a maximum work output to be obtained from the apparatus the exit velocity, C_0, must be zero. It can be shown as for non-flow systems above that a reversible heat engine working between the limits would reject $T_0(s_1 - s_0)$ units of heat, where T_0 is the atmospheric temperature.

Therefore we have

$$\text{Specific exergy} = (h_1 + C_1^2/2 + Z_1 g) - h_0 - T_0(s_1 - s_0)$$

In many thermodynamic systems the kinetic and potential energy terms are negligible,

i.e. $$\text{Specific exergy} = (h_1 - T_0 s_1) - (h_0 - T_0 s_0) = b_1 - b_0$$

Effectiveness

Instead of comparing a process to some imaginary ideal process, as is done in the case of isentropic efficiency for instance (see p. 238), it is a better measure of the usefulness of the process to compare the useful output of the process with the loss of exergy of the system. The useful output of a system is given by the increase of exergy of the surroundings,

i.e. Effectiveness, $$\varepsilon = \frac{\text{increase of exergy of surroundings}}{\text{loss of exergy of the system}} \qquad (4.17)$$

For a compression or heating process the effectiveness becomes

$$\varepsilon = \frac{\text{increase of exergy of the system}}{\text{loss of exergy of the surroundings}}$$

Example 4.13   Steam expands adiabatically in a turbine from 20 bar, 400 °C to 4 bar, 250 °C. Calculate:

  (i) the isentropic efficiency of the process;
 (ii) the loss of exergy of the system assuming an atmospheric temperature of 15 °C;
(iii) the effectiveness of the process.

Neglect changes in kinetic and potential energy.

Solution   (i) Initially the steam is superheated at 20 bar and 400 °C, hence from tables,

$$h_1 = 3248 \text{ kJ/kg} \quad \text{and} \quad s_1 = 7.126 \text{ kJ/kg K}$$

Finally the steam is superheated at 4 bar and 250 °C, hence from tables

$$h_2 = 2965 \text{ kJ/kg} \quad \text{and} \quad s_2 = 7.379 \text{ kJ/kg K}$$

The process is shown as 1 to 2 in Fig. 4.34

$$s_1 = s_{2s} = 7.126 \text{ kJ/kg K}$$

Fig. 4.34 *T–s* diagram
for Example 4.13

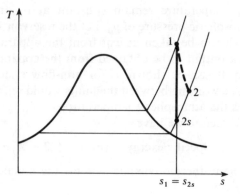

Hence interpolating

$$h_{2s} = 2753 + \left(\frac{7.126 - 6.929}{7.172 - 6.929}\right)(2862 - 2753)$$

$$= 2841.4 \text{ kJ/kg}$$

$$\text{Isentropic efficiency} = \frac{\text{actual work output}}{\text{isentropic work}}$$

$$= \frac{h_1 - h_2}{h_1 - h_{2s}} = \frac{3248 - 2965}{3248 - 2841.4} = \frac{283}{406.6} = 69.6\%$$

(ii)   Loss of exergy $= b_1 - b_2 = h_1 - h_2 + T_0(s_2 - s_1)$

$$= 283 + 288(7.379 - 7.126)$$

$$= 355.9 \text{ kJ/kg}$$

(iii) Effectiveness, $\varepsilon = \dfrac{-W}{b_1 - b_2} = \dfrac{h_1 - h_2}{b_1 - b_2}$

i.e. $\qquad\qquad\qquad \varepsilon = \dfrac{283}{355.9} = 79.6\%$

Note: the effectiveness is greater than the isentropic efficiency; this is because the steam at state 2 has a higher exergy than that at state 2s due to the heating effect of the irreversibilities in the expansion process.

Example 4.14

Air at $15\,°C$ is to be heated to $40\,°C$ by mixing it in steady flow with a quantity of air at $90\,°C$. Assuming that the mixing process is adiabatic and neglecting changes in kinetic and potential energy, calculate the ratio of the mass flow of air initially at $90\,°C$ to that initially at $15\,°C$. Calculate also the effectiveness of the heating process if the atmospheric temperature is $15\,°C$.

Solution Let the ratio of mass flows required be y; let the air at $15\,°C$ be stream 1, the air at $90\,°C$ be stream 2, and the mixed airstream at $40\,°C$ be stream 3.

Then $\qquad c_p T_1 + y c_p T_2 = (1 + y) c_p T_3$

or $\qquad y c_p (T_2 - T_3) = c_p (T_3 - T_1)$

i.e. $\qquad y(90 - 40) = 40 - 15$

therefore

$$y = \frac{25}{50} = 0.5$$

Let the system considered be a stream of air of unit mass, heated from 15 to $40\,°C$.

Increase of exergy of system per unit mass of air initially at $15\,°C$

$$= b_3 - b_1 = (h_3 - h_1) - T_0(s_3 - s_1)$$
$$= 1.005(40 - 15) - 288(s_3 - s_1)$$

$$s_3 - s_1 = c_p \ln\left(\frac{T_3}{T_1}\right) = 1.005 \ln\left(\frac{313}{288}\right) = 0.0831 \text{ kJ/kg K}$$

therefore

Increase of exergy of system per unit mass of air initially at $15\,°C$

$$= (1.005 \times 25) - (288 \times 0.0831)$$
$$= 1.195 \text{ kJ/kg}$$

The system, which is the air being heated, is 'surrounded' by the airstream being cooled. Therefore, the loss of exergy of the surroundings is given by $y(b_2 - b_3)$ per unit mass of air initially at $15\,°C$,
i.e.

Loss of exergy of surroundings per unit mass of air initially at 15°C

$$= 0.5\{(h_2 - h_3) - T_0(s_2 - s_3)\}$$

$$= 0.5\left(1.005(90 - 40) - 288 \times 1.005 \ln \frac{363}{313}\right)$$

$$= 3.65 \text{ kJ/kg}$$

Therefore

$$\text{Effectiveness} = \frac{1.195}{3.65} = 0.327 \text{ or } 32.7\%$$

The low figure for the effectiveness is an indication of the highly irreversible nature of the mixing process.

Example 4.15

A liquid of specific heat 6.3 kJ/kg K is heated at approximately constant pressure from 15 to 70°C by passing it through tubes which are immersed in a furnace. The furnace temperature is constant at 1400°C. Calculate the effectiveness of the heating process when the atmospheric temperature is 10°C.

Solution   Increase of exergy of the liquid is

$$b_2 - b_1 = (h_2 - h_1) - T_0(s_2 - s_1)$$

i.e. $$b_2 - b_1 = 6.3(70 - 15) - 283 \times 6.3 \ln\left(\frac{343}{288}\right)$$

$$= 34.7 \text{ kJ/kg}$$

Now the heat rejected by the furnace is equal to the heat supplied to the liquid $(h_2 - h_1)$. If this quantity of heat were supplied to a heat engine operating on the Carnot cycle its thermal efficiency would be

$$\left(1 - \frac{T_0}{1400 + 273}\right)$$

For the thermal efficiency of a Carnot cycle see p. 126. Therefore work which could be obtained from a heat engine is given by the product of the thermal efficiency and the heat supplied.

i.e.   Possible work of a heat engine $= (h_2 - h_1)\left(1 - \frac{283}{1673}\right)$

The possible work from a heat engine is a measure of the loss of exergy of the furnace. That is

$$\text{Loss of exergy of surroundings} = 6.3(70 - 15)\left(1 - \frac{283}{1673}\right)$$

$$= 288 \text{ kJ/kg}$$

Then   Effectiveness $= \dfrac{34.7}{288} = 0.121$ or 12.1%

The very low value of effectiveness reflects the irreversibility of the transfer of heat through a large temperature difference. If the furnace temperature were much lower the process would be much more effective, although the heat transferred to the liquid would remain the same.

For further reading on this topic see ref. 4.2.

Problems

4.1   1 kg of steam at 20 bar, dryness fraction 0.9, is heated reversibly at constant pressure to a temperature of 300 °C. Calculate the heat supplied, and the change of entropy, and show the process on a $T-s$ diagram, indicating the area which represents the heat flow.

(415 kJ/kg; 0.8173 kJ/kg K)

4.2   Steam at 0.05 bar, 100 °C is to be condensed completely by a reversible constant pressure process. Calculate the heat rejected per kilogram of steam, and the change of specific entropy. Sketch the process on a $T-s$ diagram and shade in the area which represents the heat flow.

(2550 kJ/kg; 8.292 kJ/kg K)

4.3   0.05 kg of steam at 10 bar, dryness fraction 0.84, is heated reversibly in a rigid vessel until the pressure is 20 bar. Calculate the change of entropy and the heat supplied. Show the area which represents the heat supplied on a $T-s$ diagram.

(0.0704 kJ/kg K; 36.85 kJ)

4.4   A rigid cylinder containing 0.006 m³ of nitrogen (molar mass 28 kg/kmol) at 1.04 bar, 15 °C, is heated reversibly until the temperature is 90 °C. Calculate the change of entropy and the heat supplied. Sketch the process on a $T-s$ diagram. Take the isentropic index, γ, for nitrogen as 1.4, and assume that nitrogen is a perfect gas.

(0.001 25 kJ/K; 0.407 kJ)

4.5   1 m³ of air is heated reversibly at constant pressure from 15 to 300 °C, and is then cooled reversibly at constant volume back to the initial temperature. The initial pressure is 1.03 bar. Calculate the net heat flow and the overall change of entropy, and sketch the processes on a $T-s$ diagram.

(101.5 kJ; 0.246 kJ/K)

4.6   1 kg of steam undergoes a reversible isothermal process from 20 bar and 250 °C to a pressure of 30 bar. Calculate the heat flow, stating whether it is supplied or rejected, and sketch the process on a $T-s$ diagram.

(-135 kJ/kg)

4.7   1 kg of air is allowed to expand reversibly in a cylinder behind a piston in such a way that the temperature remains constant at 260 °C while the volume is doubled. The piston is then moved in, and heat is rejected by the air reversibly at constant pressure until the volume is the same as it was initially. Calculate the net heat flow and the overall change of entropy. Sketch the process on a $T-s$ diagram.

(-161.9 kJ/kg; -0.497 kJ/kg K)

4.8 Steam at 5 bar, 25 °C, expands isentropically to a pressure of 0.7 bar. Calculate the final condition of the steam.

(0.967)

4.9 Steam expands reversibly in a cylinder behind a piston from 6 bar dry saturated, to a pressure of 0.65 bar. Assuming that the cylinder is perfectly thermally insulated, calculate the work done during the expansion per kilogram of steam. Sketch the process on a T–s diagram.

(323.8 kJ/kg)

4.10 1 kg of a fluid at 30 bar, 300 °C, expands reversibly and isothermally to a pressure of 0.75 bar. Calculate the heat flow and the work done (i) when the fluid is air, (ii) when the fluid is steam. Sketch each process on a T–s diagram.

(607 kJ/kg; − 607 kJ/kg; 1035 kJ/kg; − 975 kJ/kg)

4.11 1 kg of a fluid at 2.62 bar, −3 °C, is compressed according to a law pv = constant to a pressure of 8.2 bar. Calculate the work input and the heat supplied (i) when the fluid is air, (ii) when the fluid is a refrigerant initially dry saturated with the properties given in Table 4.2. Sketch each process on a T–s diagram.

(88.41 kJ/kg; −88.41 kJ/kg; 22.63 kJ/kg; −6.69 kJ/kg)

Table 4.2 Properties of refrigerant for Problem 4.11

Saturation values				
t_g	p_g	v_g	h_f	h_g
(°C)	(bar)	(m³/kg)	(kJ/kg)	(kJ/kg)
−3.0	2.62	0.0757	96.07	292.94
32.3	8.20	0.0248	144.29	313.05

4.12 1 kg of air at 1.013 bar, 17 °C, is compressed according to a law $pv^{1.3}$ = constant, until the pressure is 5 bar. Calculate the change of entropy and sketch the process on a T–s diagram, indicating the area which represents the heat flow.

(−0.0885 kJ/kg)

4.13 0.06 m³ of ethane (molar mass 30 kg/kmol), at 6.9 bar and 260 °C, is allowed to expand isentropically in a cylinder behind a piston to a pressure of 1.05 bar and a temperature of 107 °C. Calculate γ, R, c_p, c_v, for ethane, and calculate the work done during the expansion. Assume ethane to be a perfect gas.

The same mass of ethane at 1.05 bar, 107 °C, is compressed to 6.9 bar according to a law $pv^{1.4}$ = constant. Calculate the final temperature of the ethane and the heat flow to or from the cylinder walls during the compression. Calculate also the change of entropy during the compression, and sketch both processes on a p–v and a T–s diagram.

(1.219; 0.277 kJ/kg K; 1.542 kJ/kg K; 1.265 kJ/kg K; 54.2 kJ; 377.7 °C; 43.4 kJ; 0.0862 kJ/K)

4.14 At the start of the compression process in the reciprocating compressor of a refrigeration plant the refrigerant is at 1.5 bar, dry saturated. At the end of the compression process, which is according to a reversible polytropic law $pv^{1.2}$ = constant, the pressure is 6.5 bar. Using the properties of refrigerant given as Table 4.3, interpolating where necessary, calculate:

(i) the change of specific entropy during the process;

(ii) the degree of superheat of the refrigerant after compression.

(0.06 kJ/kg K; 35 K)

Table 4.3 Properties of refrigerant for Problem 4.14

Saturation values				Superheated values at 6.5 bar		
t_g	p_g	v_g	s_g	t	v	s
(°C)	(bar)	(m³/kg)	(kJ/kg K)	(°C)	(m³/kg)	(kJ/kg K)
−20	1.5	0.109	1.12	50	0.030	1.15
25	6.5	0.027	1.11	70	0.034	1.21

4.15 A certain perfect gas for which $\gamma = 1.26$ and the molar mass is 26 kg/kmol, expands reversibly from 727 °C, 0.003 m³ to 2 °C, 0.6 m³, according to a linear law on the $T-s$ diagram. Calculate the work done per kilogram of gas and sketch the process on a $T-s$ diagram.

$$(-959.3 \text{ kJ/kg})$$

4.16 1 kg of air at 1.02 bar, 20 °C, undergoes a process in which the pressure is raised to 6.12 bar, and the volume becomes 0.25 m³. Calculate the change of entropy and mark the initial and final states on a $T-s$ diagram.

$$(0.083 \text{ kJ/kg K})$$

4.17 Steam at 15 bar is throttled to 1 bar and a temperature of 150 °C. Calculate the initial dryness fraction and the change of specific entropy. Sketch the process on a $T-s$ diagram and state the assumptions made in the throttling process.

$$(0.992; 1.202 \text{ kJ/kg K})$$

4.18 Two vessels, one exactly twice the volume of the other, are connected by a valve and immersed in a constant temperature bath of water. The smaller vessel contains hydrogen (molar mass 2 kg/kmol), and the other is completely evacuated. Calculate the change of entropy per kilogram of gas when the valve is opened and conditions are allowed to settle. Sketch the process on a $T-s$ diagram. Assume hydrogen to be a perfect gas.

$$(4.567 \text{ kJ/kg K})$$

4.19 A turbine is supplied with steam at 40 bar, 400 °C, which expands through the turbine in steady flow to an exit pressure of 0.2 bar, and a dryness fraction of 0.93. The inlet velocity is negligible, but the steam leaves at high velocity through a duct of 0.14 m² cross-sectional area. If the mass flow is 3 kg/s, and the mechanical efficiency is 90%, calculate the power output of the turbine. Show that the process is irreversible and calculate the change of specific entropy. Heat losses from the turbine are negligible.

$$(2048 \text{ kW}; 0.643 \text{ kJ/kg K})$$

4.20 In a centrifugal compressor the air is compressed through a pressure ratio of 4 to 1, and the temperature of the air increases by a factor of 1.65. Show that the process is irreversible and calculate the change of entropy per kilogram of air. Assume that the process is adiabatic. Sketch the process on a $T-s$ diagram.

$$(0.105 \text{ kJ/kg K})$$

4.21 In a gas turbine unit the gases enter the turbine at 550 °C and 5 bar and leave at 1 bar. The process is approximately adiabatic, but the entropy changes by 0.174 kJ/kg K. Calculate the exit temperature of the gases. Assume the gases to act as a perfect gas, and take $\gamma = 1.333$ and $c_p = 1.11$ kJ/kg K. Sketch the process on a $T-s$ diagram.

$$(370.9 \text{ °C})$$

4.22 A rigid, well-lagged vessel of 0.3 m³ capacity contains 0.7614 kg of steam at 6 bar. A valve is opened and the pressure falls to 1.4 bar before the valve is shut again. Calculate the condition of the steam remaining in the vessel, and the mass of steam which has escaped.

(0.989; 0.516 kg)

4.23 A rigid vessel contains 0.5 kg of a perfect gas of specific heat at constant volume 1.1 kJ/kg K. A stirring paddle is inserted into the vessel and 11 kJ of work are done on the paddle by the stirrer motor. Assuming that the vessel is well lagged and that the gas is initially at the temperature of the surroundings which are at 17 °C, calculate the effectiveness of the process.

(3.3%)

4.24 The identical vessel of Problem 4.23 is heated through the same temperature difference by immersing it in a furnace of constant temperature 100 °C. Calculate the effectiveness of the process.

(14.8%)

4.25 Steam enters a turbine at 70 bar, 500 °C and leaves at 2 bar in a dry saturated state. Calculate the isentropic efficiency and the effectiveness of the process. Neglect changes of kinetic and potential energy and assume that the process is adiabatic. The atmospheric temperature is 17 °C.

(84.4%; 88%)

4.26 In an open-type feed heater (see p. 249), steam enters at 15 bar, 200 °C. The feedwater enters the heater at 130 °C and leaves the heater at the saturation temperature corresponding to the pressure in the heater of 15 bar. Calculate the mass of steam entering per unit mass of feed water entering the heater.

Calculate also the loss of exergy of the steam per unit mass and the effectiveness of the heater. Assume that there is no heat loss from the heater and that the atmospheric temperature is 20 °C. State any other assumptions made.

(0.1536 kg; 734.9 kJ/kg; 88.1%)

References

4.1 ROGERS G F C and MAYHEW Y R 1992 *Engineering Thermodynamics, Work and Heat Transfer*, 4th edn Longman
4.2 HAYWOOD R W 1991 *Analysis of Engineering Cycles* Pergamon

5

The Heat Engine Cycle

In this chapter the heat engine cycle is discussed more fully and gas power cycles are considered. It can be shown that there is an ideal theoretical cycle which is the most efficient conceivable; this cycle is called the Carnot cycle. The highest thermal efficiency possible for a heat engine in practice is only about half that of the ideal theoretical Carnot cycle, between the same temperature limits. This is due to irreversibilities in the actual cycle, and to deviations from the ideal cycle, which are made for various practical reasons. The choice of a power plant in practice is a compromise between thermal efficiency and various factors such as the size of the plant for a given power requirement, mechanical complexity, operating cost, and capital cost.

5.1 The Carnot cycle

It can be shown from the Second Law of Thermodynamics that no heat engine can be more efficient than a reversible heat engine working between the same temperature limits (see ref. 5.1). Carnot showed that the most efficient possible cycle is one in which all the heat supplied is supplied at one fixed temperature, and all the heat rejected is rejected at a lower fixed temperature. The cycle therefore consists of two isothermal processes joined by two adiabatic processes. Since all processes are reversible, then the adiabatic processes in the cycle are also isentropic. The cycle is most conveniently represented on a T–s diagram as shown in Fig. 5.1.

Process 1 to 2 is isentropic expansion from T_1 to T_2.
Process 2 to 3 is isothermal heat rejection.
Process 3 to 4 is isentropic compression from T_2 to T_1.
Process 4 to 1 is isothermal heat supply.

The cycle is completely independent of the working substance used.

The cycle efficiency of a heat engine, defined in section 4.1, is given by the net work output divided by the gross heat supplied

i.e. $$\eta = \frac{-\sum W}{Q_1} = \frac{\sum Q}{Q_1} \qquad (5.1)$$

Fig. 5.1 Carnot cycle on a T–s diagram

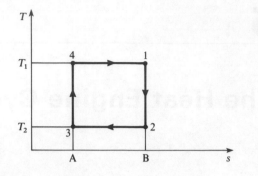

In the Carnot cycle, with reference to Fig. 5.1, it can be seen that the gross heat supplied, Q_1, is given by the area 41BA4,

i.e. $\quad Q_1 = \text{area } 41\text{BA}4 = T_1(s_\text{B} - s_\text{A})$

Similarly the net heat supplied, $\sum Q$, is given by the area 41234,

i.e. $\quad \sum Q = (T_1 - T_2)(s_\text{B} - s_\text{A})$

Hence we have Carnot cycle efficiency

$$\eta_{\text{Carnot}} = \frac{(T_1 - T_2)(s_\text{B} - s_\text{A})}{T_1(s_\text{B} - s_\text{A})}$$

i.e. $\quad \eta_{\text{Carnot}} = 1 - \frac{T_2}{T_1}$ $\hspace{4cm}$ (5.2)

If a sink for heat rejection is available at a fixed temperature T_2 (e.g. a large supply of cooling water), then the ratio T_2/T_1 will decrease as the temperature of the source T_1 is increased. From equation (5.2) it can be seen that as T_2/T_1 decreases, then the thermal efficiency increases. Hence for a fixed lower temperature for heat rejection, the upper temperature at which heat is supplied must be made as high as possible. The maximum possible thermal efficiency between any two temperatures is that of the Carnot cycle.

Example 5.1 What is the highest possible theoretical efficiency of a heat engine operating with a hot reservoir of furnace gases at 2000 °C when the cooling water available is at 10 °C?

Solution From equation (5.2)

$$\eta_{\text{Carnot}} = 1 - \frac{T_2}{T_1} = 1 - \frac{10 + 273}{2000 + 273} = 1 - \frac{283}{2273}$$

i.e. $\quad$ Highest possible efficiency $= 1 - 0.1246$

$$= 0.8754 \quad \text{or} \quad 87.54\%$$

It should be noted that a system in practice operating between similar temperatures (e.g. a steam-generating plant) would have a thermal efficiency of about 30%. The discrepancy is due to losses due to irreversibility in the actual plant, and also because of deviations from the ideal Carnot cycle made for various practical reasons.

It is difficult in practice to devise a system which can receive and reject heat at constant temperature. A wet vapour is the only working substance which can do this conveniently, since for a wet vapour the pressure and temperature remain constant as the specific enthalpy of vaporization is supplied or rejected. A Carnot cycle for a wet vapour is as shown in Fig. 5.2. Although this cycle is the most efficient possible vapour cycle, it is not used in steam plant. The theoretical cycle on which steam cycles are based is known as the Rankine cycle. This will be discussed in detail in Chapter 8, and the reasons for using it in preference to the Carnot cycle will be given.

Fig. 5.2 Carnot cycle for a wet vapour on a T–s diagram

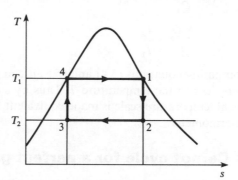

5.2 Absolute temperature scale

In the preceding chapters a temperature scale based on the perfect gas thermometer has been assumed. Using the Second Law of Thermodynamics it is possible to establish a temperature scale which is independent of the working substance.

We have, for any heat engine from equation (5.1),

$$\eta = \frac{\sum Q}{Q_1}$$

Also the efficiency of an engine operating on the Carnot cycle depends only on the temperatures of the hot and cold reservoirs. Denoting temperature on an arbitrary scale by X, we have

$$\eta = \phi(X_1, X_2) \tag{5.3}$$

where ϕ is a function, and X_1 and X_2 are the temperatures of the hot and cold reservoirs.

Combining equations (5.1) and (5.3) we have

$$\frac{\sum Q}{Q_1} = \phi(X_1, X_2)$$

There are a large number of possible temperature scales which are all independent of the working substance. Any working scale can be chosen by suitably selecting the value of the function ϕ. The function can be chosen that

$$1 - \frac{\sum Q}{Q_1} = \frac{X_2}{X_1} \tag{5.4}$$

Also from equation (5.2) we have

$$\eta = 1 - \frac{T_2}{T_1}$$

Hence using equation (5.1)

$$\eta = \frac{\sum Q}{Q_1} = 1 - \frac{T_2}{T_1}$$

or

$$1 - \frac{\sum Q}{Q_1} = \frac{T_2}{T_1} \tag{5.5}$$

Comparing equations (5.4) and (5.5) it can be seen that the temperature X is equivalent to the temperature T. Thus by suitably choosing the function ϕ, the ideal temperature scale is made equivalent to the scale based on the perfect gas thermometer.

5.3 The Carnot cycle for a perfect gas

A Carnot cycle for a perfect gas is shown on a $T–s$ diagram in Fig. 5.3. Note that the pressure of the gas changes continuously from p_4 to p_1 during the isothermal heat supply, and from p_2 to p_3 during the isothermal heat rejection. In practice it is much more convenient to heat a gas at approximately constant pressure or at constant volume, hence it is difficult to attempt to operate an actual heat engine on the Carnot cycle using a gas as working substance. Another important reason for not attempting to use the Carnot cycle in practice is illustrated by drawing the cycle on a $p–v$ diagram, as in Fig. 5.4. The net work output of the cycle is given by the area 12341. This is a small quantity compared with the gross work output of the expansion processes of the cycle, given by area 412BA4. The work of the compression processes (i.e. work done on the gas) is given by the area 234AB2. The ratio of the net work output to the gross work output of the system is called the *work ratio*. The Carnot cycle, despite its high thermal efficiency, has a low work ratio.

Fig. 5.3 Carnot cycle for a perfect gas on a $T–s$ diagram

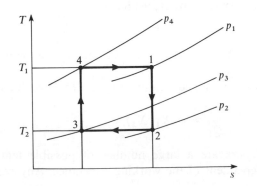

Fig. 5.4 Carnot cycle on a p–v diagram

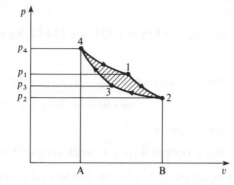

Example 5.2

A hot reservoir at 800 °C and a cold reservoir at 15 °C are available. Calculate the thermal efficiency and the work ratio of a Carnot cycle using air as the working fluid, if the maximum and minimum pressures in the cycle are 210 bar and 1 bar.

Solution

The cycle is shown on a T–s and p–v diagram in Figs 5.5(a) and 5.5(b) respectively.

Using equation (5.2),

$$\eta_{\text{Carnot}} = 1 - \frac{T_2}{T_1} = 1 - \frac{15 + 273}{800 + 273} = 1 - 0.268$$

i.e. $\eta_{\text{Carnot}} = 0.732$ or   73.2%

In order to find the work output and the work ratio it is necessary to find the entropy change $(s_1 - s_4)$.

For an isothermal process from 4 to A, using equation (4.12),

$$s_A - s_4 = R \ln\!\left(\frac{p_4}{p_2}\right) = 0.287 \ln\!\left(\frac{210}{1}\right) = 1.535 \text{ kJ/kg K}$$

At constant pressure from A to 2, we have

Fig. 5.5 Carnot cycle for Example 5.2 on p–v and T–s diagrams

$$s_A - s_2 = c_p \ln\!\left(\frac{T_1}{T_2}\right) = 1.005 \ln\!\left(\frac{1073}{288}\right) = 1.321 \text{ kJ/kg K}$$

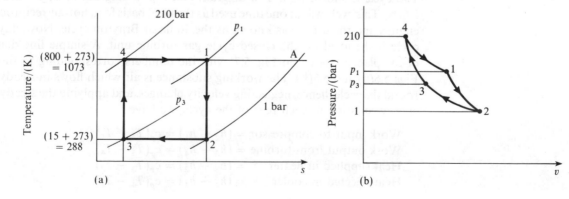

(a)

(b)

129

therefore

$$s_1 - s_4 = 1.535 - 1.321 = 0.214 \text{ kJ/kg K}$$

Then

$$\text{Net work output} = (T_1 - T_2)(s_1 - s_4) = \text{area } 12341$$
$$= (1073 - 288) \times 0.214 = 168 \text{ kJ/kg}$$

Gross work output is

Work output 4 to 1 + work output 1 to 2

From equation (3.12) for an isothermal process, $Q + W = 0$,

i.e. $\quad -W_{4-1} = Q_{4-1} = \text{area under line } 4\text{–}1 \text{ on Fig. } 5.5(a)$

$$= (s_1 - s_4) \times T_1 = 0.214 \times 1073$$
$$= 229.6 \text{ kJ/kg}$$

For an isentropic process from 1 to 2, from equation (3.13), $W = (u_2 - u_1)$, therefore for a perfect gas

$$-W_{2-1} = c_v(T_1 - T_2)$$
$$= 0.718(1073 - 288) = 563.6 \text{ kJ/kg}$$

Therefore

$$\text{Gross work output} = 229.6 + 563.6 = 793.2 \text{ kJ/kg}$$

i.e. $\quad \text{Work ratio} = \dfrac{\text{net work output}}{\text{gross work output}} = \dfrac{168}{793.2} = 0.212$

5.4 The constant pressure cycle

In this cycle the heat supply and heat rejection processes occur reversibly at constant pressure. The expansion and compression processes are isentropic. The cycle is shown on a T–s diagram and a p–v diagram in Figs 5.6(a) and 5.6(b). This cycle was at one time used as the ideal basis for a hot-air reciprocating engine, and the cycle was known as the Joule or Brayton cycle. Nowadays the cycle is the ideal for the closed cycle gas turbine unit. A simple line diagram of the plant is shown in Fig. 5.7, with the numbers corresponding to those of Figs 5.6(a) and 5.6(b). The working substance is air which flows in steady flow round the cycle, hence, neglecting velocity changes, and applying the steady-flow energy equation to each part of the cycle, we have

Work input to compressor $= (h_2 - h_1) = c_p(T_2 - T_1)$
Work output from turbine $= (h_3 - h_4) = c_p(T_3 - T_4)$
Heat supplied in heater $\quad = (h_3 - h_2) = c_p(T_3 - T_2)$
Heat rejected in cooler $\quad = (h_4 - h_1) = c_p(T_4 - T_1)$

Fig. 5.6 Constant
pressure cycle on p–v
and T–s diagrams

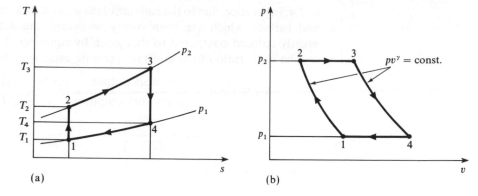

(a) (b)

Fig. 5.7 Closed-cycle
gas turbine unit

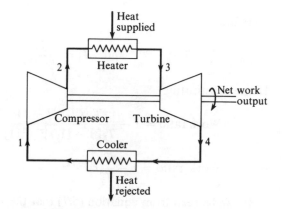

Then from equation (5.1)

$$\eta = \frac{\sum Q}{Q_1} = \frac{c_p(T_3 - T_2) - c_p(T_4 - T_1)}{c_p(T_3 - T_2)} = 1 - \frac{T_4 - T_1}{T_3 - T_2}$$

Now since processes 1 to 2 and 3 to 4 are isentropic between the same pressures p_2 and p_1, we have, using equation (3.21),

$$\frac{T_2}{T_1} = \left(\frac{p_2}{p_1}\right)^{(\gamma-1)/\gamma} = \frac{T_3}{T_4} = r_p^{(\gamma-1)/\gamma}$$

where r_p is the pressure ratio, p_2/p_1.

i.e. $T_3 = T_4 r_p^{(\gamma-1)/\gamma}$ and $T_2 = T_1 r_p^{(\gamma-1)/\gamma}$

$$T_3 - T_2 = r_p^{(\gamma-1)/\gamma}(T_4 - T_1)$$

Hence, substituting in the expression for the efficiency

$$\eta = 1 - \frac{T_4 - T_1}{(T_4 - T_1)r_p^{(\gamma-1)/\gamma}} = 1 - \frac{1}{r_p^{(\gamma-1)/\gamma}} \qquad (5.6)$$

Thus for the constant pressure cycle the cycle efficiency depends only on the pressure ratio. In the ideal case the value of γ for air is constant and equal to 1.4. In practice, due to the eddying of the air as it flows through the compressor and turbine which are both rotary machines, the actual cycle efficiency is greatly reduced compared to that given by equation (5.6).

The work ratio of the constant pressure cycle may be found as follows:

$$\text{Work ratio} = \frac{\text{Net work output}}{\text{Gross work output}} = \frac{c_p(T_3 - T_4) - c_p(T_2 - T_1)}{c_p(T_3 - T_4)}$$

$$= 1 - \frac{T_2 - T_1}{T_3 - T_4}$$

Now, as previously,

$$\frac{T_2}{T_1} = r_p^{(\gamma - 1)/\gamma} = \frac{T_3}{T_4}$$

therefore

$$T_2 = T_1 r_p^{(\gamma - 1)/\gamma} \quad \text{and} \quad T_4 = \frac{T_3}{r_p^{(\gamma - 1)/\gamma}}$$

Hence substituting

$$\text{Work ratio} = 1 - \frac{T_1(r_p^{(\gamma - 1)/\gamma} - 1)}{T_3[1 - (1/r_p^{(\gamma - 1)/\gamma})]}$$

i.e.
$$\text{Work ratio} = 1 - \frac{T_1}{T_3} r_p^{(\gamma - 1)/\gamma} \tag{5.7}$$

It can be seen from equation (5.7) that the work ratio depends not only on the pressure ratio but also on the ratio of the minimum and maximum temperatures. For a given inlet temperature, T_1, the maximum temperature, T_3, must be made as high as possible for a high work ratio.

For an open-cycle gas turbine unit the actual cycle is not such a good approximation to the ideal constant pressure cycle, since fuel is burned with the air, and a fresh charge is continuously induced into the compressor. The ideal cycle provides nevertheless a good basis for comparison, and in many calculations for an ideal open-cycle gas turbine the effects of the mass of fuel and the charge in the working fluid are neglected.

Example 5.3   In a gas turbine unit air is drawn at 1.02 bar and 15 °C, and is compressed to 6.12 bar. Calculate the thermal efficiency and the work ratio of the ideal constant pressure cycle, when the maximum cycle temperature is limited to 800 °C.

Solution   The ideal cycle is shown on a T–s diagram in Fig. 5.8. From equation (5.6)

$$\text{Thermal efficiency, } \eta = 1 - \frac{1}{r_p^{(\gamma - 1)/\gamma}}$$

i.e.
$$\eta = 1 - \left(\frac{1.02}{6.12}\right)^{(1.4 - 1)/1.4} = 1 - 0.599$$

Fig. 5.8 T–s diagram for Example 5.3

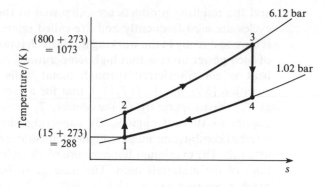

therefore

Thermal efficiency = 0.401 or 40.1%

The net work output of the cycle is given by the work output of the turbine minus the work input in the compressor,

i.e.   Net work output $= c_p(T_3 - T_4) - c_p(T_2 - T_1)$

From equation (3.21)

$$\frac{T_2}{T_1} = \left(\frac{p_2}{p_1}\right)^{(\gamma-1)/\gamma} = \frac{T_3}{T_4} = \left(\frac{6.12}{1.02}\right)^{(1.4-1)/1.4} = 1.669$$

therefore

$$T_2 = 1.669 \times T_1 = 1.669 \times 288 = 480.5 \text{ K}$$

where $T_2 = 15 + 273 = 288$ K and

$$T_4 = \frac{T_3}{1.669} = \frac{1073}{1.669} = 642.9 \text{ K}$$

where $T_3 = 800 + 273 = 1073$ K. Therefore

Net work output $= 1.005(1073 - 642.9) - 1.005(480.5 - 288)$

$= 238.8 \text{ kJ/kg}$

Gross work output = work output of the turbine $= c_p(T_3 - T_4)$

$= 1.005(1073 - 642.9) = 432.3 \text{ kJ/kg}$

Then   Work ratio $= \dfrac{\text{net work output}}{\text{gross work output}} = \dfrac{238.8}{432.3} = 0.553$

5.5 The air standard cycle

Cycles in which the fuel is burned directly in the working fluid are not heat engines in the true meaning of the term since the system is not reduced to its initial state. The working fluid undergoes a chemical change by combustion and the resulting products are exhausted to the atmosphere. In practice such cycles are used frequently and are called internal-combustion cycles. The fuel is burned directly in the working fluid which is normally air. The main advantage of such power units is that high temperatures of the fluid can be attained, since heat is not transferred through metal walls to the fluid. It is seen from equation (5.2), $\eta = 1 - (T_2/T_1)$, that for a given sink for the rejection of heat at T_2, the temperature of the source, T_1, must be as high as possible. This applies to all heat engines. By supplying fuel inside the cylinder as in the internal-combustion engine, higher temperatures for the working fluid can be attained. The maximum temperature of all cycles is limited by the metallurgical limit of the materials used. The fluid in an internal-combustion engine may reach a temperature as high as 3000 K. This is made possible by externally cooling the cylinder by water or air cooling; also, due to the intermittent nature of the cycle, the working fluid reaches its maximum temperature for only an instant during each cycle.

Examples of internal-combustion cycles are the open cycle gas turbine unit, the petrol engine, the diesel engine or oil engine, and the gas engine. The open cycle gas turbine unit, although an internal combustion cycle, is nevertheless in a different category to the other internal-combustion engines; the cycle is a steady-flow cycle in which the working fluid flows from one component to another round the cycle. It will be assumed, therefore, that the gas turbine unit, whether operating on the open or the closed cycle, can be satisfactorily compared with the ideal constant pressure cycle, dealt with in section 5.4. Gas turbine cycles are considered in more detail in Chapter 9.

In the petrol engine a mixture of air and petrol is drawn into the cylinder, compressed by the piston, then ignited by an electric spark. The hot gases expand, pushing the piston back, and are then swept out to exhaust, and the cycle recommences with the induction of a fresh charge of petrol and air. In the diesel or oil engine the oil is sprayed under pressure into the compressed air at the end of the compression stroke, and combustion is spontaneous due to the high temperature of the air after compression. In a gas engine a mixture of gas and air is induced into the cylinder, compressed, and then ignited as in the petrol engine, by an electric spark. Reciprocating internal-combustion engines are considered in more detail in Chapter 13. To give a basis of comparison for the actual internal-combustion engine the air standard cycle is defined. In an *air standard cycle* the working substance is assumed to be air throughout, all processes are assumed to be reversible, and the source of heat supply and the sink for heat rejection are assumed to be external to the air. The cycle can be represented on any diagram of properties, and is usually drawn on the $p–v$ diagram, since this allows a more direct comparison to be made with the actual engine machine cycle. It must be stressed that an air standard cycle on a $p–v$ diagram is a true thermodynamic cycle, whereas a record of pressure variations in an engine cylinder against piston displacement is a machine cycle.

5.6 The Otto cycle

The Otto cycle is the ideal air standard cycle for the petrol engine, the gas engine, and the high-speed oil engine. The cycle is shown on a $p-v$ diagram in Fig. 5.9.

Process 1 to 2 is isentropic compression.
Process 2 to 3 is reversible constant volume heating.
Process 3 to 4 is isentropic expansion.
Process 4 to 1 is reversible constant volume cooling.

Fig. 5.9 Otto cycle on a $p-v$ diagram

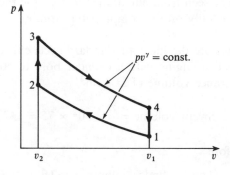

To give a direct comparison with an actual engine the ratio of the specific volumes, v_1/v_2, is taken to be the same as the compression ratio of the actual engine,

i.e.   Compression ratio, $r_v = \dfrac{v_1}{v_2}$

$$= \frac{\text{swept volume} + \text{clearance volume}}{\text{clearance volume}} \tag{5.8}$$

The thermal efficiency of the Otto cycle can be found using equation (5.1),

$$\eta = \frac{\sum Q}{Q_1}$$

The heat supplied, Q_1, at constant volume between T_2 and T_3 is given by equation (2.13) per unit mass of air

$$Q_1 = c_v(T_3 - T_2)$$

Similarly the heat rejected per unit mass at constant volume between T_4 and T_1 is given by equation (2.13), $c_v(T_4 - T_1)$.

The processes 1 to 2 and 3 to 4 are isentropic and therefore there is no heat flow during these processes. Therefore

$$\eta = \frac{\sum Q}{Q_1} = \frac{c_v(T_3 - T_2) - c_v(T_4 - T_1)}{c_v(T_3 - T_2)} = 1 - \left(\frac{T_4 - T_1}{T_3 - T_2}\right)$$

135

Now since processes 1 to 2 and 3 to 4 are isentropic, then using equation (3.20),

$$\frac{T_2}{T_1} = \left(\frac{v_1}{v_2}\right)^{\gamma-1} = \left(\frac{v_4}{v_3}\right)^{\gamma-1} = \frac{T_3}{T_4} = r_v^{\gamma-1}$$

where r_v is the compression ratio from equation (5.8).

Then $\quad T_3 = T_4 r_v^{\gamma-1} \quad$ and $\quad T_2 = T_1 r_v^{\gamma-1}$

Hence substituting

$$\eta = 1 - \frac{T_4 - T_1}{(T_4 - T_1) r_v^{\gamma-1}} = 1 - \frac{1}{r_v^{\gamma-1}} \tag{5.9}$$

It can be seen from equation (5.9) that the thermal efficiency of the Otto cycle depends only on the compression ratio, r_v.

Example 5.4   Calculate the ideal air standard cycle efficiency based on the Otto cycle for a petrol engine with a cylinder bore of 50 mm, a stroke of 75 mm, and a clearance volume of 21.3 cm^3.

Solution   Swept volume $= \dfrac{\pi}{4} \times 50^2 \times 75 = 147\,200 \text{ m}^3 = 147.2 \text{ cm}^3$

Therefore

Total cylinder volume $= 147.2 + 21.3 = 168.5 \text{ cm}^3$

i.e.   Compression ratio, $r_v = \dfrac{168.5}{21.3} = 7.914/1$

Then using equation (5.9)

$$\eta = 1 - \frac{1}{r_v^{\gamma-1}} = 1 - \frac{1}{7.914^{0.4}} = 1 - 0.437 = 0.563 \text{ or } 56.3\%$$

5.7   The diesel cycle

The engines in use today which are called diesel engines are far removed from the original engine invented by Diesel in 1892. Diesel worked on the idea of spontaneous ignition of powdered coal, which was blasted into the cylinder by compressed air. Oil became the accepted fuel used in compression–ignition engines, and the oil was originally blasted into the cylinder in the same way that Diesel had intended to inject the powdered coal. This gave a cycle of operation which has as its ideal counterpart the ideal air standard diesel cycle shown in Fig. 5.10.

As before the compression ratio, r_v, is given by the ratio v_1/v_2.

Process 1 to 2 is isentropic compression.
Process 2 to 3 is reversible constant pressure heating.
Process 3 to 4 is isentropic expansion.
Process 4 to 1 is reversible constant volume cooling.

Fig. 5.10 Diesel cycle on a p–v diagram

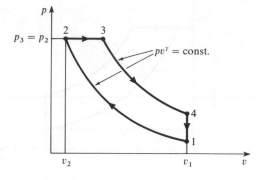

From equation (5.1)

$$\eta = \frac{\sum Q}{Q_1}$$

At constant pressure from equation (2.12) per kg of air

$$Q_1 = c_p(T_3 - T_2)$$

Also at constant volume from equation (2.13), per kilogram of air the heat rejected is $c_v(T_4 - T_1)$.

There is no heat flow in processes 1 to 2 and 3 to 4 since these processes are isentropic. Hence by substituting in the expression for thermal efficiency the following equation may be derived:

$$\eta = 1 - \frac{\beta^\gamma - 1}{(\beta - 1)r_v^{\gamma - 1}\gamma} \tag{5.10}$$

where $\beta = v_3/v_2 = $ cut-off ratio.

Equation (5.10) shows that the thermal efficiency depends not only on the compression ratio but also on the heat supplied between 2 and 3, which fixes the ratio, v_3/v_2. Equation (5.10) is derived by expressing each temperature in terms of T_1 and r_v or β. The derivation is not given here because it is believed that the best method of working out the thermal efficiency is to calculate each temperature individually round the cycle, and then apply equation (5.1), $\eta = \sum Q/Q_1$. This is illustrated in Example 5.5.

Example 5.5

A diesel engine has an inlet temperature and pressure of 15 °C and 1 bar respectively. The compression ratio is 12/1 and the maximum cycle temperature is 1100 °C. Calculate the air standard thermal efficiency based on the diesel cycle.

Solution

Referring to Fig. 5.11, $T_1 = 15 + 273 = 288$ K and $T_3 = 1100 + 273 = 1373$ K. From equation (3.20)

$$\frac{T_2}{T_1} = \left(\frac{v_1}{v_2}\right)^{\gamma - 1} = r_v^{\gamma - 1} = 12^{0.4} = 2.7$$

i.e. $T_2 = 2.7 \times 288 = 778$ K

137

Fig. 5.11 Diesel cycle
on a $p-v$ diagram for
Example 5.5

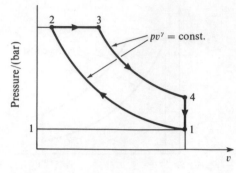

At constant pressure from 2 to 3, since $pv = RT$ for a perfect gas, then

$$\frac{T_3}{T_2} = \frac{v_3}{v_2}$$

i.e. $\dfrac{v_3}{v_2} = \dfrac{1373}{778} = 1.765$

Therefore

$$\frac{v_4}{v_3} = \frac{v_4 v_2}{v_2 v_3} = \frac{v_1 v_2}{v_2 v_3} = 12 \times \frac{1}{1.765} = 6.8$$

Then using equation (3.20)

$$\frac{T_3}{T_4} = \left(\frac{v_4}{v_3}\right)^{\gamma - 1} = 6.8^{0.4} = 2.153$$

i.e. $T_4 = \dfrac{1373}{2.153} = 638 \text{ K}$

Then from equation (2.12), per kilogram of air

$$Q_1 = c_p(T_3 - T_2) = 1.005(1373 - 778) = 598 \text{ kJ/kg}$$

Also, from equation (2.13), per kilogram of air, the heat rejected is

$$c_v(T_4 - T_1) = 0.718(638 - 288) = 251 \text{ kJ/kg}$$

Therefore from equation (5.1)

$$\eta = \frac{\sum Q}{Q_1} = \frac{598 - 251}{598} = 0.58 \text{ or } 58\%$$

5.8 The dual-combustion cycle

Modern oil engines, although still called diesel engines, are more closely derived
from an engine invented by Ackroyd-Stuart in 1888. All oil engines today use
solid injection of the fuel; the fuel is injected by a spring-loaded injector, the

fuel pump being operated by a cam driven from the engine crankshaft (see section 13.10). The ideal cycle used as a basis for comparison is called the dual-combustion cycle or the mixed cycle, and is shown on a p–v diagram in Fig. 5.12.

Process 1 to 2 is isentropic compression.
Process 2 to 3 is reversible constant volume heating.
Process 3 to 4 is reversible constant pressure heating.
Process 4 to 5 is isentropic expansion.
Process 5 to 1 is reversible constant volume cooling.

Fig. 5.12 Dual combustion cycle on a p–v diagram

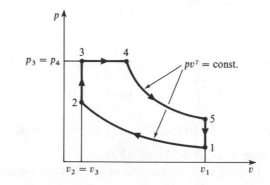

The heat is supplied in two parts, the first part at constant volume and the remainder at constant pressure, hence the name 'dual-combustion'. In order to fix the thermal efficiency completely, three factors are necessary. These are the compression ratio, $r_v = v_1/v_2$, the ratio of pressures, $k = p_3/p_2$, and the ratio of volumes, $\beta = v_4/v_3$.

Then it can be shown that

$$\eta = 1 - \frac{k\beta^\gamma - 1}{[(k-1) + \gamma k(\beta - 1)]r_v^{\gamma-1}} \tag{5.11}$$

Note that when $k = 1$ (i.e. $p_3 = p_2$), then the equation (5.11) reduces to the thermal efficiency of the diesel cycle given by equation (5.10). The efficiency of the dual-combustion cycle depends not only on the compression ratio but also on the relative amounts of heat supplied at constant volume and at constant pressure. Equation (5.11) is much too cumbersome to use, and the best method of calculating thermal efficiency is to evaluate each temperature round the cycle and then use equation (5.1), $\eta = \sum Q/Q_1$. The heat supplied, Q_1, is found by using equations (2.13) and (2.12) for the heat added at constant volume and at constant pressure respectively,

i.e. $\qquad Q_1 = c_v(T_3 - T_2) + c_p(T_4 - T_3)$

The heat rejected is given by $c_v(T_5 - T_1)$.

Example 5.6 An oil engine takes in air at 1.01 bar, 20 °C and the maximum cycle pressure is 69 bar. The compressor ratio is 18/1. Calculate the air standard thermal efficiency based on the dual-combustion cycle. Assume that the heat added at constant volume is equal to the heat added at constant pressure.

Fig. 5.13 Dual combustion cycle for Example 5.6

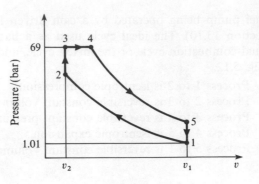

Solution The cycle is shown on a p–v diagram in Fig. 5.13. Using equation (3.20),

$$\frac{T_2}{T_1} = \left(\frac{v_1}{v_2}\right)^{\gamma-1} = 18^{0.4} = 3.18$$

i.e. $\quad T_2 = 3.18 \times T_1 = 3.18 \times 293 = 931 \text{ K}$

where $T_1 = 20 + 273 = 293$ K.

From 2 to 3 the process is at constant volume, hence

$$\frac{p_3}{p_2} = \frac{T_3}{T_2} \quad \text{since} \quad \frac{p_3 v_3}{T_3} = \frac{p_2 v_2}{T_2} \quad \text{and} \quad v_3 = v_2$$

i.e. $\quad T_3 = \dfrac{p_3}{p_2} \times T_2 = \dfrac{69 \times 931}{p_2}$

To find p_2, use equation (3.19),

$$\frac{p_2}{p_1} = \left(\frac{v_1}{v_2}\right)^{\gamma} = 18^{1.4} = 57.2$$

i.e. $\quad p_2 = 57.2 \times 1.01 = 57.8 \text{ bar}$

Then substituting

$$T_3 = \frac{69 \times 931}{57.8} = 1112 \text{ K}$$

Now the heat added at constant volume is equal to the heat added at constant pressure in this example, therefore

$$c_v(T_3 - T_2) = c_p(T_4 - T_3)$$

i.e. $\quad 0.718(1112 - 931) = 1.005(T_4 - 1112)$

therefore

$$T_4 = \frac{0.718 \times 181}{1.005} + 1112$$

i.e. $\quad T_4 = 1241.4 \text{ K}$

To find T_5 it is necessary to know the value of the volume ratio, v_5/v_4. At constant pressure from 3 to 4

$$\frac{v_4}{v_3} = \frac{T_4}{T_3} = \frac{1241.4}{1112} = 1.116$$

Therefore

$$\frac{v_5}{v_4} = \frac{v_1}{v_4} = \frac{v_1 v_3}{v_2 v_4} = 18 \times \frac{1}{1.116} = 16.14$$

Then using equation (3.20)

$$\frac{T_4}{T_5} = \left(\frac{v_5}{v_4}\right)^{\gamma-1} = 16.14^{0.4} = 3.04$$

i.e.
$$T_5 = \frac{1241.4}{3.04} = 408 \text{ K}$$

Now the heat supplied, Q_1, is given by

$$Q_1 = c_v(T_3 - T_2) + c_p(T_4 - T_3) \quad \text{or} \quad Q_1 = 2c_v(T_3 - T_2)$$

since in this example the heat added at constant volume is equal to the heat added at constant pressure. Therefore

$$Q_1 = 2 \times 0.718 \times (1112 - 931) = 260 \text{ kJ/kg}$$

The heat rejected is given by

$$c_v(T_5 - T_1) = 0.718(408 - 293) = 82.6 \text{ kJ/kg}$$

Then from equation (4.3)

$$\eta = \frac{\sum Q}{Q_1} = \frac{260 - 82.6}{260} = 1 - 0.318 = 0.682 \text{ or } 68.2\%$$

It should be mentioned here that the modern high-speed oil engine operates on a cycle for which the Otto cycle is a better basis of comparison. Also, since the Otto cycle calculation for thermal efficiency is much simpler than that of the dual-combustion cycle, then this is another reason for using the Otto cycle as a standard of comparison.

5.9 Mean effective pressure

The term work ratio is defined in section 5.3, and is shown to be a useful criterion for practical power plants. For internal-combustion engines work ratio is not such a useful concept, since the work done on and by the working fluid takes place inside one cylinder. In order to compare reciprocating engines another term is defined called the *mean effective pressure*. The mean effective pressure is defined as the height of a rectangle having the same length and area as the cycle plotted on a $p-v$ diagram. This is illustrated for an Otto cycle in Fig. 5.14. The rectangle ABCDA is the same length as the cycle 12341, and area

Fig. 5.14 Mean effective pressure on a $p-v$ diagram

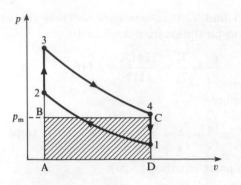

ABCDA is equal to area 12341. Then the mean effective pressure, p_m, is the height AB of the rectangle. The work output per kilogram of air can therefore be written as

$$-W = \text{area ABCDA} = p_m(v_1 - v_2) \tag{5.12}$$

The term $(v_1 - v_2)$ is proportional to the swept volume of the cylinder, hence it can be seen from equation (5.12) that the mean effective pressure gives a measure of the work output per swept volume. It can therefore be used to compare similar engines of different size. The mean effective pressure discussed in this section is for the air standard cycle. It will be shown in Chapter 13 that the indicated mean effective pressure of an actual engine can be measured from an indicator diagram and used to evaluate the indicated work done by the engine.

Example 5.7   Calculate the mean effective pressure for the cycle of Example 5.6.

Solution   In Example 5.6 the heat supplied, Q_1, and the cycle efficiency were found to be 260 kJ/kg and 68.2% respectively. From equation (4.2)

$$\eta = \frac{-W}{Q_1}$$

therefore

$$-W = \eta Q_1 = 0.682 \times 260 = 177 \text{ kJ/kg}$$

Now from the definition of mean effective pressure, and equation (5.12), we have

$$-W = p_m(v_1 - v_2)$$

Using equation (2.5), $pv = RT$ and equation (5.8), $r_v = v_1/v_2 = 18$, then

$$v_1 - v_2 = \left(v_1 - \frac{v_1}{18}\right) = \frac{17}{18}v_1 = \frac{17}{18}\frac{RT_1}{p_1} = \frac{17 \times 287 \times 293}{18 \times 1.01 \times 10^5}$$

i.e. $v_1 - v_2 = 0.786 \text{ m}^3/\text{kg}$

Then substituting,

$$-W = p_m \times 0.786 \quad \text{or} \quad p_m = -W/0.786 \text{ kJ/m}^3$$

i.e.   Mean effective pressure $= \dfrac{177 \times 10^3}{10^5 \times 0.786} = 2.25 \text{ bar}$

5.10 The Stirling and Ericsson cycles

It has been shown that no cycle can have an efficiency greater than that of the Carnot cycle working between given temperature limits T_1 and T_2. Cycles which have an efficiency equal to that of the Carnot cycle have been defined and are known as the Stirling and Ericsson cycles and they are superior to the Carnot cycle in that they have higher work ratios.

The Stirling cycle is shown in the p–v diagram in Fig. 5.15(a) and is represented diagrammatically in Fig. 5.15(b): it must be emphasized that this is not a physical description of a Stirling engine but one which may help to give an understanding of the way the processes which make up the cycle are related.

Fig. 5.15 Stirling engine and the Stirling cycle

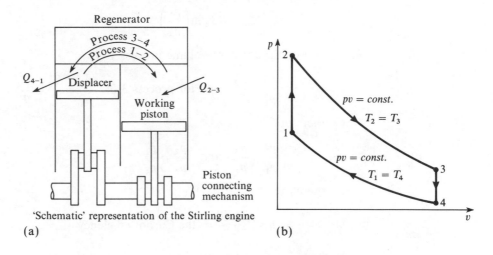

'Schematic' representation of the Stirling engine

(a)

(b)

Heat is supplied to the working fluid from an external source, process 2–3, as the gas expands isothermally ($T_2 = T_3$), and heat is rejected to an external sink, process 4–1, as the gas is compressed isothermally ($T_1 = T_4$). The two isothermals are connected by the reversible constant volume processes 1–2 and 3–4 during which the temperature changes are equal to ($T_2 - T_1$). The heat rejected during process 3–4, $c_v(T_2 - T_1)$, is used to heat the gas during process 1–2, $c_v(T_2 - T_1) = c_v(T_3 - T_4)$ and this is assumed to take place ideally and reversibly in a *regenerator*. The regenerator requires a matrix of material which separates the heating and cooling gases, but allows the temperatures to change progressively by infinitesimal and corresponding amounts during the processes. This regenerative process takes place at constant volume and is internal to the cycle.

The efficiency of the Stirling cycle is obtained by considering the heat transfers between the system and the bodies external to it, i.e. a high-temperature heat supply and a low-temperature sink to which heat is rejected.

Heat supplied from the hot source, using equations (3.11) and (3.12),

$$Q_{2-3} = -W_{2-3} = RT_2 \ln\left(\frac{p_2}{p_3}\right) \text{ per unit mass of gas}$$

143

Similarly

$$\text{Heat } rejected \text{ to the coldsink} = W_{4-1} = RT_1 \ln\left(\frac{p_1}{p_4}\right)$$

therefore

$$\sum Q = RT_2 \ln\left(\frac{p_2}{p_3}\right) - RT_1 \ln\left(\frac{p_1}{p_4}\right)$$

and as the cycle efficiency, $\eta = \sum Q/Q_{2-3}$, therefore

$$\eta = 1 - \frac{RT_1 \ln\left(\dfrac{p_2}{p_3}\right)}{RT_2 \ln\left(\dfrac{p_1}{p_4}\right)}$$

For the constant volume process 1–2,

$$\frac{p_2}{p_1} = \frac{T_2}{T_1} \quad \text{and for process 3–4} \quad \frac{p_3}{p_4} = \frac{T_3}{T_4} = \frac{T_2}{T_1}$$

therefore

$$\frac{p_2}{p_1} = \frac{p_3}{p_4} \quad \text{and} \quad \frac{p_2}{p_3} = \frac{p_1}{p_4}$$

therefore

$$\eta = 1 - \frac{T_1}{T_2} = \text{the Carnot efficiency}$$

This result can be deduced without formal proof as the heat supply and rejection processes take place at constant temperatures.

$$\text{Work ratio} = \frac{\text{net work output}}{\text{gross work output}} = \frac{-W_{2-3} + W_{4-1}}{-W_{2-3}}$$

$$= \frac{\sum Q}{Q_{2-3}} = \text{cycle efficiency, } \eta$$

since $-W_{2-3} = Q_{2-3}$.

The practical interpretation of the ideal cycle will not be described in detail here and the reader is advised to consult the specialist literature for the mechanical arrangements employed and the performance assessments (see refs 5.2 and 5.3). Figure 5.15(b) gives a simplified representation of the engine and shows the necessity for two pistons, a working piston and a displacing piston, which in fact work in different parts of the same cylinder and not as represented. It is necessary to the ideal cycle for the pistons to move discontinuously and this is only approximated to by the mechanisms employed. The result is that the processes of the ideal cycle are not achieved and there is a considerable 'rounding off' of the ideal p–v diagram as the heating and cooling processes merge to depart considerably from the constant volume heating concept.

The attractions of the Stirling engine are that it can utilize any form of heat from conventional or indigenous fuel, solar or nuclear sources, provided the temperature created is high enough. The engines are quiet, with an efficiency equal to or better than the best internal combustion engines and with little vibration due to the nature of the drive needed to give the differential movements between the working and displacing pistons. The possible range of application of the Stirling engine is wide and includes marine use, electricity generation for peak loads and as a stand-by unit, automotive purposes, particularly in comparison with the diesel engine, and for situations when unconventional fuels or heat sources can, or must, be used. The most important applications up to now have been as an air engine and as a refrigerator; with the Stirling cycle reversed, it is capable of reaching the low temperatures of the cryogenic regions.

The Ericsson cycle is similar to the Stirling cycle except that the two isothermals are connected by constant pressure processes, as shown in Fig. 5.16.

Fig. 5.16 Ericsson cycle on a p–v diagram

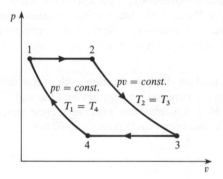

Problems

5.1 What is the highest cycle efficiency possible for a heat engine operating between 800 and 15 °C?

(73.2%)

5.2 Two reversible heat engines operate in series between a source at 527 °C and a sink at 17 °C. If the engines have equal efficiencies and the first rejects 400 kJ to the second, calculate:
  (i) the temperature at which heat is supplied to the second engine;
 (ii) the heat taken from the source;
(iii) the work done by each engine.
Assume that each engine operates on the Carnot cycle.

(208.7 °C; 664.4 kJ; 264.4 kJ; 159.2 kJ)

5.3 In a Carnot cycle operating between 307 and 17 °C the maximum and minimum pressures are 62.4 bar and 1.04 bar. Calculate the cycle efficiency and the work ratio. Assume air to be the working fluid.

(50%; 0.286)

5.4 A closed-cycle gas turbine unit operating with maximum and minimum temperatures of 760 and 20 °C has a pressure ratio of 7/1. Calculate the ideal cycle efficiency and the work ratio.

(42.7%; 0.505)

145

5.5 In an air standard Otto cycle the maximum and minimum temperatures are 1400 and 15 °C. The heat supplied per kilogram of air is 800 kJ. Calculate the compression ratio and the cycle efficiency. Calculate also the ratio of maximum to minimum pressures in the cycle.

(5.27/1; 48.5%; 30.65/1)

5.6 A four-cylinder petrol engine has a swept volume of 2000 cm^3, and the clearance volume in each cylinder is 60 cm^3. Calculate the air standard cycle efficiency. If the introduction conditions are 1 bar and 24 °C, and the maximum cycle temperature is 1400 °C, calculate the mean effective pressure based on the air standard cycle.

(59.1%; 5.28 bar)

5.7 Calculate the cycle efficiency and mean effective pressure of an air standard diesel cycle with a compression ratio of 15/1, and maximum and minimum cycle temperatures of 1650 °C and 15 °C respectively. The maximum cycle pressure is 45 bar.

(59.1%; 8.38 bar)

5.8 In a dual-combustion cycle the maximum temperature is 2000 °C and the maximum pressure is 70 bar. Calculate the cycle efficiency and the mean effective pressure when the pressure and temperature at the start of compression are 1 bar and 17 °C respectively. The compression ratio is 18/1.

(63.6%; 10.46 bar)

5.9 An air standard dual-combustion cycle has a mean effective pressure of 10 bar. The minimum pressure and temperature are 1 bar and 17 °C respectively, and the compression ratio is 16/1. Calculate the maximum cycle temperature when the cycle efficiency is 60%. The maximum cycle pressure is 60 bar.

(1959 °C)

References

5.1 ROGERS G F C and MAYHEW Y R 1992 *Engineering Thermodynamics, Work and Heat Transfer*, 4th edn Longman
5.2 WALKER G 1980 *Stirling Engines* Oxford Univ. Press
5.3 I Mech E Conference 1982 *Stirling Engines – Progress Towards Reality* MEP

6

Mixtures

A pure substance is defined as a substance having a constant and uniform chemical composition, and this definition can be extended to include a homogeneous mixture of gases when there is no chemical reaction taking place. The thermodynamic properties of a mixture of gases can be determined in the same way as for a single gas. The most common example of this is dry air, which is a mixture of oxygen, nitrogen, a small percentage of argon, and traces of other gases. The properties of air have been determined and it is considered as a single substance.

The mixtures to be considered in this chapter are those composed of perfect gases, and perfect gases and vapours. The properties of such mixtures are important in combustion calculations. Air and water vapour mixtures are considered later in the chapter with reference to surface condensers, but for moist atmospheric air there is a special nomenclature and this is considered in Chapter 15 on psychrometry and air-conditioning.

6.1 Dalton's law and the Gibbs–Dalton law

Consider a closed vessel of volume V at temperature T, which contains a mixture of perfect gases at a known pressure. If some of the mixture were removed, then the pressure would be less than the initial value. If the gas removed were the full amount of one of the constituents then the reduction in pressure would be equal to the contribution of that constituent to the initial total pressure. Each constituent contributes to the total pressure by an amount which is known as the *partial pressure* of the constituent. The relationship between the partial pressures of the constituents is expressed by Dalton's law, as follows:

> *The pressure of a mixture of gases is equal to the sum of the partial pressure of the constituents.*
>
> *The partial pressure of each constituent is that pressure which the gas would exert if it occupied alone that volume occupied by the mixture at the same temperature.*

Fig. 6.1 Gas A mixing
with gas B

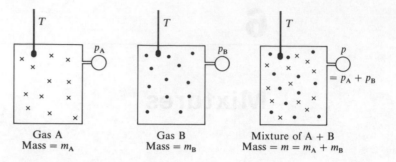

Gas A
Mass = m_A

Gas B
Mass = m_B

Mixture of A + B
Mass = $m = m_A + m_B$

This is expressed diagrammatically in Fig. 6.1. The gases A and B, originally occupying volume V at temperature T, are mixed in the third vessel which is of the same volume and is at the same temperature.

By the conservation of mass

$$m = m_A + m_B \tag{6.1}$$

By Dalton's law

$$p = p_A + p_B \tag{6.2}$$

Dalton's law is based on experiment and is found to be obeyed more accurately by gas mixtures at low pressures. As shown in Fig. 6.1 each constituent occupies the whole vessel. The example given in Fig. 6.1 and the relationships in equations (6.1) and (6.2) refer to a mixture of two gases, but the law can be extended to any number of gases,

i.e. $\quad m = m_A + m_B + m_C + \text{etc.} \quad \text{or} \quad m = \sum m_i \tag{6.3}$

where m_i is the mass of a constituent.

Table 6.1 Analyses of air

| Constituent | Chemical symbol | Analysis | | Molar mass |
		By volume (%)	By mass (%)	(kg/kmol)
Oxygen	O_2	20.95	23.14	31.999
Nitrogen	N_2	78.09	75.53	28.013
Argon	Ar	0.93	1.28	39.948
Carbon dioxide	CO_2	0.03	0.05	44.010

Similarly

$$p = p_A + p_B + p_C + \text{etc.} \quad \text{or} \quad p = \sum p_i \tag{6.4}$$

where p_i is the partial pressure of a constituent.

Air is the most common mixture and since it will be referred to frequently, its composition is as given in Table 6.1. The mean molar mass of air is 28.96 kg/kmol, and the specific gas constant R is 0.2871 kJ/kg K. For approximate calculations the air is said to be composed of oxygen and 'atmospheric nitrogen' (see Table 6.2). Note: volumetric analysis is the analysis by volume; gravimetric analysis is the analysis by mass.

Table 6.2 Approximate
analyses for air

| Constituent | Analysis | | Molar mass |
	By volume (%)	By mass (%)	(kg/kmol)
Oxygen	21.0	23.3	32.0
Nitrogen	79.0	76.7	28.0

Example 6.1 A vessel of volume $0.4\,\text{m}^3$ contains $0.45\,\text{kg}$ of carbon monoxide and $1\,\text{kg}$ of air, at $15\,°\text{C}$. Calculate the partial pressure of each constituent and the total pressure in the vessel. The gravimetric analysis of air is to be taken as 23.3% oxygen and 76.7% nitrogen. Take the molar masses of carbon monoxide, oxygen and nitrogen as 28, 32 and 28 kg/kmol.

Solution
$$\text{Mass of oxygen present} = \frac{23.3}{100} \times 1 = 0.233\,\text{kg}$$

$$\text{Mass of nitrogen present} = \frac{76.7}{100} \times 1 = 0.767\,\text{kg}$$

From equation (2.9)

$$R = \frac{\tilde{R}}{\tilde{m}}$$

and from equation (2.6)

$$pV = mRT$$

Hence $$p = \frac{m\tilde{R}T}{\tilde{m}V}$$

or for a constituent

$$p_i = \frac{m_i \tilde{R} T}{\tilde{m}_i V}$$

The volume V is $0.4\,\text{m}^3$ and the temperature T is $(15 + 273) = 288\,\text{K}$. Therefore we have for O_2

$$p_{O_2} = \frac{0.233 \times 8.3145 \times 288}{32 \times 0.4} = 43.59\,\text{kN/m}^2$$

$$= \frac{43.59 \times 10^3}{10^5} = 0.4359\,\text{bar}$$

for N_2 $$p_{N_2} = \frac{0.767 \times 8.3145 \times 288}{28 \times 0.4} = 163.99\,\text{kN/m}^2$$

$$= \frac{163.99 \times 10^3}{10^5} = 1.6399\,\text{bar}$$

149

for CO $\quad p_{CO} = \dfrac{0.45 \times 8.3145 \times 288}{28 \times 0.4} = 96.21 \text{ kN/m}^2$

$$= \dfrac{96.21 \times 10^3}{10^5} = 0.9621 \text{ bar}$$

The total pressure in the vessel is given by equation (6.4)

$$p = \sum p_i = 0.436 + 1.640 + 0.962 = 3.038 \text{ bar}$$

i.e. Pressure in vessel = 3.038 bar

Dalton's law was reformulated by Gibbs to include a second statement on the properties of mixtures. The combined statement is known as the Gibbs–Dalton law, and is as follows:

The internal energy, enthalpy, and entropy, of a gaseous mixture are respectively equal to the sums of the internal energies, enthalpies, and entropies, of the constituents.

Each constituent has that internal energy, enthalpy, and entropy, which it would have if it occupied alone that volume occupied by the mixture at the temperature of the mixture.

This statement leads to the equations

$$mu = m_A u_A + m_B u_B + \text{etc.} \quad \text{or} \quad mu = \sum m_i u_i \tag{6.5}$$

and $\quad mh = m_A h_A + m_B h_B + \text{etc.} \quad \text{or} \quad mh = \sum m_i h_i \tag{6.6}$

and $\quad ms = m_A s_A + m_B s_B + \text{etc.} \quad \text{or} \quad ms = \sum m_i s_i \tag{6.7}$

6.2 Volumetric analysis of a gas mixture

The analysis of a mixture of gases is often quoted by volume as this is the most convenient for practical determinations.

Consider a volume V of a gaseous mixture at a temperature T, consisting of three constituents A, B, and C as in Fig. 6.2(a). Let each of the constituents be compressed to a pressure p equal to the total pressure of the mixture, and let the temperature remain constant. The partial volumes then occupied by the constituents will be V_A, V_B, and V_C. From equation (2.6) $pV = mRT$, therefore, referring to Fig. 6.2(a)

$$m_A = \dfrac{p_A V}{R_A T}$$

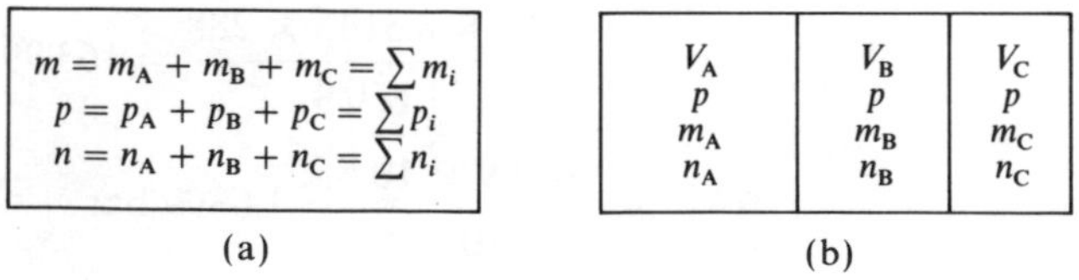

Fig. 6.2 Illustration of partial volume

| $m = m_A + m_B + m_C = \sum m_i$ |
| $p = p_A + p_B + p_C = \sum p_i$ |
| $n = n_A + n_B + n_C = \sum n_i$ |

V_A	V_B	V_C
p	p	p
m_A	m_B	m_C
n_A	n_B	n_C

(a)　　　　　　　　　　　(b)

and referring to Fig. 6.2(b)

$$m_A = \frac{pV_A}{R_A T}$$

Equating the two values for m_A, we have

$$\frac{p_A V}{R_A T} = \frac{pV_A}{R_A T}$$

i.e. $p_A V = pV_A$ or $V_A = \dfrac{p_A}{p} V$

In general therefore,

$$V_i = \frac{p_i}{p} V \tag{6.8}$$

i.e. $\displaystyle\sum V_i = \sum \frac{p_i V}{p} = \frac{V}{p}\sum p_i$

Now from equation (6.4), $p = \sum p_i$, therefore

$$\sum V_i = V \tag{6.9}$$

Therefore the volume of a mixture of gases is equal to the sum of the volumes of the individual constituents when each exists alone at the pressure and temperature of the mixture. This is the statement of another empirical law, the law of partial volumes (sometimes called Amagat's law or Leduc's law).

The amount of substance is defined in section 2.3 and is given by equation (2.7) as $n = m/\tilde{m}$. By Avogadro's law, the amount of substance of any gas is proportional to the volume of the gas at a given pressure and temperature. Referring to Fig. 6.2(a), the volume V contains an amount of substance n of the mixture at p and T. In Fig. 6.2(b), the gas A occupies a volume V_A at p and T, and this volume contains an amount of substance n_A. Similarly there are amounts of substance n_B of gas B in volume V_B, and n_C of gas C in volume V_C. Now from equation (6.9),

$$\sum V_i = V \quad \text{or} \quad V_A + V_B + V_C = V$$

Therefore the total amount of substance in the vessel must equal the sum of the amounts of substance of the individual constituents,

i.e. $n_A + n_B + n_C = n$ or $n = \sum n_i$ \tag{6.10}

6.3 The molar mass and specific gas constant

For any gas in a gas mixture occupying a total volume of V at a temperature T, from equation (2.8) $pV = n\tilde{R}T$, and the definition of partial pressure, we have

$$p_i V = n_i \tilde{R} T \tag{6.11}$$

therefore

$$\sum (p_i V) = \sum (n_i \tilde{R} T)$$

i.e. $V \sum p_i = \tilde{R} T \sum n_i$

From equation (6.4), $p = \sum p_i$, hence,

$$pV = \tilde{R} T \sum n_i$$

Also from equation (6.10), $n = \sum n_i$, therefore

$$pV = n \tilde{R} T$$

The mixture therefore acts as a perfect gas, and this is the characteristic equation for the mixture. A molar mass is defined by the equation, $\tilde{m} = m/n$, where m is the mass of the mixture and n is the amount of substance of the mixture. Similarly, a specific gas constant is defined by the equation $R = \tilde{R}/\tilde{m}$. It can be assumed that a mixture of perfect gases obeys all the perfect gas laws.

To find the specific gas constant for the mixture in terms of the specific gas constants of the constituents, consider equation (2.6) both for the mixture and for a constituent,

i.e. $pV = mRT$ and $p_i V = m_i R_i T$

Then $\sum p_i V = \sum m_i R_i T$

therefore

$$V \sum p_i = T \sum m_i R_i$$

Now from equation (6.4), $p = \sum p_i$, therefore

$$pV = T \sum m_i R_i \quad \text{or} \quad pV = mRT = T \sum m_i R_i$$

i.e. $mR = \sum m_i R_i$ or $R = \sum \dfrac{m_i}{m} R_i$ (6.12)

where m_i/m is the mass fraction of a constituent.

Example 6.2   The gravimetric analysis of air is 23.14% oxygen, 75.53% nitrogen, 1.28% argon, 0.05% carbon dioxide. Calculate the specific gas constant for air and the molar mass. Take the molar masses from Table 6.1 on p. 148.

Solution   From equation (2.9), $R = \tilde{R}/\tilde{m}$, therefore

$$R_{O_2} = \frac{8.3145}{31.999} = 0.2598 \text{ kJ/kg K}$$

$$R_{N_2} = \frac{8.3145}{28.013} = 0.2968 \text{ kJ/kg K}$$

$$R_{A_r} = \frac{8.3145}{39.948} = 0.2081 \text{ kJ/kg K}$$

$$R_{CO_2} = \frac{8.3145}{44.010} = 0.1889 \text{ kJ/kg K}$$

Then using equation (6.12), $R = \sum (m_i/m) R_i$, we have

$$R = (0.2314 \times 0.2598) + (0.7553 \times 0.2968) + (0.0128 \times 0.2081)$$
$$+ (0.0005 \times 0.1889) = 0.2871 \, \text{kJ/kg K}$$

i.e.   Specific gas constant for air = 0.2871 kJ/kg K

From equation (2.9), $\tilde{m} = \tilde{R}/R$, therefore,

$$\tilde{m} = \frac{8.3145}{0.2871} = 28.960 \, \text{kg/kmol}$$

i.e.   Molar mass of air = 28.96 kg/kmol

When the approximate analysis for air is used (i.e. 23.3% O_2 and 76.7% N_2 by mass), it is usual practice to take R as 0.287 kJ/kg K and $\tilde{m}$ as 29 kg/kmol.

From equation (6.11), $p_i V = n_i \tilde{R} T$, and combining this with equation (2.8) applied to the mixture (i.e. $pV = n\tilde{R}T$), we have

$$\frac{p_i V}{pV} = \frac{n_i \tilde{R} T}{n \tilde{R} T}$$

i.e. $$\frac{p_i}{p} = \frac{n_i}{n} \tag{6.13}$$

This can be combined with equation (6.8), to give

$$\frac{p_i}{p} = \frac{n_i}{n} = \frac{V_i}{V} \tag{6.14}$$

This is an important result which means that the molar analysis is identical with the volumetric analysis, and both are equal to the ratio of the partial pressure to the total pressure.

Another method of determining the molar mass is as follows. Applying the characteristic equation, (2.6), to each constituent and to the mixture we have $m_i = p_i V/R_i T$, and $m = pV/RT$.

From equation (6.3), $m = \sum m_i$, therefore

$$\frac{pV}{RT} = \sum \frac{p_i V}{R_i T} \quad \text{or} \quad \frac{p}{R} = \sum \frac{p_i}{R_i}$$

Using equation (2.9), $R = \tilde{R}/\tilde{m}$, and substituting, we have

$$\frac{p\tilde{m}}{\tilde{R}} = \sum \frac{p_i \tilde{m}_i}{\tilde{R}} \quad \text{or} \quad p\tilde{m} = \sum p_i \tilde{m}_i$$

i.e. $$\tilde{m} = \sum \frac{p_i}{p} \tilde{m}_i \tag{6.15}$$

Also using equation (6.14)

$$\tilde{m} = \sum \frac{V_i}{V} \tilde{m}_i \tag{6.16}$$

153

and $\quad \tilde{m} = \sum \dfrac{n_i}{n} \tilde{m}_i$ $\hspace{3cm}$ (6.17)

Example 6.3 $\quad$ The gravimetric analysis of air is 23.14% oxygen, 75.53% nitrogen, 1.28% argon, and 0.05% carbon dioxide. Calculate the analysis by volume and the partial pressure of each constituent when the total pressure is 1 bar.

Solution $\quad$ From equation (6.14) the analysis by volume, V_i/V, is the same as the fraction n_i/n. Also from equation (2.7), $n_i = m_i/\tilde{m}_i$, therefore considering 1 kg of mixture we have the tabular solution shown in Table 6.3.

Table 6.3 Solution for Example 6.3

Constituent	m_i (kg)	$\tilde{m}_i$ (kg/kmol)	$n_i = m_i/\tilde{m}_i$ (kmol)	$n_i/n = V_i/V$ (%)
Oxygen	0.2314	31.999	0.00723	$\dfrac{0.00723}{0.03452} \times 100 = 20.95$
Nitrogen	0.7553	28.013	0.02696	$\dfrac{0.02696}{0.03452} \times 100 = 78.09$
Argon	0.0128	39.948	0.00032	$\dfrac{0.00032}{0.03452} \times 100 = 0.93$
Carbon dioxide	0.0005	44.010	0.00001	$\dfrac{0.00001}{0.03452} \times 100 = 0.03$

$$n = \sum n_i = \underline{\underline{0.03452}}$$

From equation (6.14), $p_i/p = V_i/V = n_i/n$, therefore, $p_i = (n_i/n)p$, hence using the volume fractions from Table 6.3,

for O_2 $\quad p_{O_2} = 0.2095 \times 1 = 0.2095$ bar

for N_2 $\quad p_{N_2} = 0.7809 \times 1 = 0.7809$ bar

for Ar $\quad p_{A_r} = 0.0093 \times 1 = 0.0093$ bar

for CO_2 $\quad p_{CO_2} = 0.0003 \times 1 = 0.0003$ bar

Example 6.4 $\quad$ A mixture of 1 kmol CO_2 and 3.5 kmol of air is contained in a vessel at 1 bar and 15 °C. The volumetric analysis of air can be taken as 21% oxygen and 79% nitrogen. Calculate for the mixture:

(i) the masses of CO_2, O_2, and N_2, and the total mass;
(ii) the percentage carbon content by mass;
(iii) the molar mass and the specific gas constant for the mixture;
(iv) the specific volume of the mixture.

Take the molar masses of carbon, oxygen and nitrogen as 12 kg/kmol, 32 kg/kmol and 28 kg/kmol respectively.

Solution (i) From equation (6.14), $n_i = (V_i/V)n$, we have

$$n_{O_2} = 0.21 \times 3.5 = 0.735 \text{ kmol}$$

and $n_{N_2} = 0.79 \times 3.5 = 2.765 \text{ kmol}$

From equation (2.7), $m_i = n_i \tilde{m}_i$, therefore

$$m_{CO_2} = 1 \times 44 = 44 \text{ kg}$$

$$m_{O_2} = 0.735 \times 32 = 23.55 \text{ kg}$$

and $m_{N_2} = 2.765 \times 28 = 77.5 \text{ kg}$

Total mass, $m = m_{O_2} + m_{N_2} + m_{CO_2}$

$$= 23.55 + 77.5 + 44 = 145.05 \text{ kg}$$

(ii) The molar mass of carbon is 12 kg/kmol, therefore there are 12 kg of carbon present in 1 kmol of carbon dioxide,

i.e. Percentage carbon in mixture $= \dfrac{12 \times 100}{145.05} = 8.27\%$ by mass

(iii) From equation (6.10), $n = \sum n_i$, then

$$n = n_{CO_2} + n_{O_2} + n_{N_2} = 1 + 0.735 + 2.765 = 4.5 \text{ kmol}$$

Then using equation (6.17),

$$\tilde{m} = \sum \left(\frac{n_i}{n} \tilde{m}_i \right)$$

we have

$$\tilde{m} = \left(\frac{1}{4.5} \times 44 \right) + \left(\frac{0.735}{4.5} \times 32 \right) + \left(\frac{2.765}{4.5} \times 28 \right) = 32.2 \text{ kg/kmol}$$

i.e. Molar mass of mixture $= 32.2$ kg/kmol

From equation (2.9), $R = \tilde{R}/\tilde{m}$, we have

$$R = \frac{8.3145}{32.2} = 0.2581 \text{ kJ/kg K}$$

i.e. Specific gas constant for the mixture $= 0.2581$ kJ/kg K

(iv) From equation (2.5), $pv = RT$, therefore

$$v = \frac{RT}{p} = \frac{0.2581 \times 288 \times 10^3}{1 \times 10^5} = 0.7435 \text{ m}^3/\text{kg}$$

where $T = 15 + 273 = 288$ K.
i.e. specific volume of the mixture at 1 bar and 15 °C is 0.7435 m³/kg.

Example 6.5   A mixture of H_2 and O_2 is to be made so that the ratio of H_2 to O_2 is 2 to 1 by volume. Calculate the mass of O_2 required and the volume of the container, per kilogram of H_2, if the pressure and temperature are 1 bar and $15\,^\circ C$ respectively. Take the molar masses of hydrogen and oxygen as 2 kg/kmol and 32 kg/kmol.

Solution   Let the mass of O_2 per kilogram of H_2 be x.
From equation (2.7), $n_i = m_i/\tilde{m}_i$, therefore

$$n_{H_2} = \frac{1}{2} = 0.5\ \text{kmol} \quad \text{and} \quad n_{O_2} = \frac{x}{32}\ \text{kmol}$$

From equation (6.14), $V_i/V = n_i/n$, therefore

$$\frac{V_{H_2}}{V_{O_2}} = \frac{n_{H_2}}{n_{O_2}} \quad \text{and} \quad \frac{V_{H_2}}{V_{O_2}} = 2 \quad \text{(given)}$$

i.e. $\dfrac{0.5}{x/32} = 2$ therefore $x = \dfrac{32 \times 0.5}{2} = 8\ \text{kg}$

i.e.   Mass of oxygen per kilogram of hydrogen $= 8\ \text{kg}$

The total amount of substance in the vessel per kilogram of H_2 is

$$n = n_{H_2} + n_{O_2} = 0.5 + \frac{x}{32} = 0.5 + \frac{8}{32} = 0.5 + 0.25 = 0.75\ \text{kmol}$$

Then from equation (2.8),

$$pV = n\tilde{R}T$$

therefore

$$V = \frac{0.75 \times 8.3145 \times 288 \times 10^3}{1 \times 10^5} = 17.96\ \text{m}^3$$

Example 6.6   A vessel contains a gaseous mixture of composition by volume, 80% H_2 and 20% CO. It is desired that the mixture should be made in the proportion 50% H_2 and 50% CO by removing some of the mixture and adding some CO. Calculate per kilomole of mixture the mass of mixture to be removed, and the mass of CO to be added. The pressure and temperature in the vessel remain constant during the procedure.

Take the molar mass of hydrogen and carbon monoxide as 2 kg/kmol and 28 kg/kmol.

Solution   Since the pressure and temperature remain constant, then the amount of substance in the vessel remains the same throughout. Therefore the amount of substance of mixture removed is equal to the amount of substance of CO added.

Let x kg of mixture be removed and y kg of CO be added.
For the mixture, from equation (6.16)

$$\tilde{m} = \sum \frac{V_i}{V}\tilde{m}_i$$

therefore,

$$\tilde{m} = (0.8 \times 2) + (0.2 \times 28) = 7.2 \text{ kg/kmol}$$

Then using equation (2.7), $n = m/\tilde{m}$, we have

$$\text{amount of substance of mixture removed} = \frac{x}{7.2} \text{ kmol}$$

$$\text{amount of substance of CO added} = \frac{y}{28} \text{ kmol}$$

and $x/7.2 = y/28$

From equation (6.14), $V_i/V = n_i/n$, therefore

$$\text{amount of substance of } H_2 \text{ in the mixture removed} = 0.8 \times \frac{x}{7.2} = \frac{x}{9} \text{ kmol}$$

and   amount of substance of H_2 initially $= 0.8 \times 1 = 0.8 \text{ kmol}$

Hence   amount of substance of H_2 remaining in vessel $= \left(0.8 - \dfrac{x}{9}\right) \text{ kmol}$

But 1 kmol of the new mixture is 50% H_2 and 50% CO, therefore

$$0.8 - \frac{x}{9} = 0.5$$

i.e. $x = (0.8 - 0.5) \times 9 = 2.7 \text{ kg}$

i.e.   Mass of mixture removed $= 2.7 \text{ kg}$

Also since $x/7.2 = y/28$, therefore

$$y = \frac{28}{7.2} \times x = \frac{28 \times 2.7}{7.2} = 10.5 \text{ kg}$$

i.e.   Mass of CO added $= 10.5 \text{ kg}$

6.4 Specific heat capacities of a gas mixture

It was shown in section 6.1 that as a consequence of the Gibbs–Dalton law the internal energy of a mixture of gases is given by equation (6.5), $mu = \sum m_i u_i$. Also for a perfect gas from equation (2.14), $u = c_v T$. Hence substituting we have

$$mc_v T = \sum m_i c_{v_i} T$$

therefore

$$mc_v = \sum m_i c_{v_i}$$

or $$c_v = \sum \frac{m_i}{m} c_{v_i} \tag{6.18}$$

157

Similarly from equation (6.6), $mh = \sum m_i h_i$, and from equation (2.18), $h = c_p T$, therefore

$$mc_p T = \sum m_i c_{p_i} T$$

therefore

$$mc_p = \sum m_i c_{p_i}$$

or
$$c_p = \sum \frac{m_i}{m} c_{p_i} \qquad (6.19)$$

From equations (6.18) and (6.19)

$$c_p - c_v = \sum \frac{m_i}{m} c_{p_i} - \sum \frac{m_i}{m} c_{v_i} = \sum \frac{m_i}{m} (c_{p_i} - c_{v_i})$$

Using equation (2.17), $c_{p_i} - c_{v_i} = R_i$, therefore

$$c_p - c_v = \sum \frac{m_i}{m} R_i$$

Also from equation (6.12), $R = \sum \frac{m_i}{m} R_i$, therefore for the mixture

$$c_p - c_v = R$$

The equations (2.20), (2.21), and (2.22), can be applied to a mixture of gases,

$$\gamma = \frac{c_p}{c_v}; \qquad c_v = \frac{R}{\gamma - 1}; \qquad c_p = \frac{\gamma R}{\gamma - 1}$$

It should be noted that γ must be determined from equation 2.20; there is no weighted mean expression as there is for R, c_v, and c_p.

Example 6.7   The gas in an engine cylinder has a volumetric analysis of 12% CO_2, 11.5% O_2, and 76.5% N_2. The temperature at the beginning of expansion is 1000 °C and the gas mixture expands reversibly through a volume ratio of 7 to 1, according to a law $pv^{1.25}$ = constant. Calculate the work done and the heat flow per unit mass of gas. The values of c_p for the constituents averaged over the temperature are as follows: c_p for CO_2 = 1.271 kJ/kg K; c_p for O_2 = 1.110 kJ/kg K; c_p for N_2 = 1.196 kJ/kg K.

Solution   From equation (2.7) $m_i = n_i \tilde{m}_i$, therefore a conversion from volume fraction to mass fraction is as given in Table 6.4. Then using equation (6.19) and the mass fractions from Table 6.4

$$c_p = \sum \frac{m_i}{m} c_{p_i}$$

therefore

$$c_p = (0.174 \times 1.271) + (0.121 \times 1.110) + (0.705 \times 1.196)$$

$$= 1.199 \text{ kJ/kg K}$$

Table 6.4 Solution for Example 6.7

Constituent	n_i (kmol)	$\tilde{m}_i$ (kg/kmol)	$m_i = n_i \tilde{m}_i$ (kg)	m_i/m
Carbon dioxide	0.120	44	5.28	$5.28/30.36 = 0.174$
Oxygen	0.115	32	3.68	$3.68/30.36 = 0.121$
Nitrogen	0.765	28	21.40	$21.40/30.36 = 0.705$
			$m = \sum m_i = 30.36$	

From equation (6.12), $R = \sum (m_i/m) R_i$, and from equation (2.9), $R_i = \tilde{R}/\tilde{m}_i$, therefore

$$R = \left(0.174 \times \frac{8.3145}{44} \right) + \left(0.121 \times \frac{8.3145}{32} \right) + \left(0.705 \times \frac{8.3145}{28} \right)$$

$$= 0.274 \text{ kJ/kg K}$$

Then from equation (2.17), $c_p - c_v = R$, we have

$$c_v = 1.199 - 0.274 = 0.925 \text{ kJ/kg K}$$

The work done per kg of gas can be obtained from equation (3.20)

$$W = \frac{R(T_2 - T_1)}{n - 1}$$

T_2 can be found using equation (3.28)

$$\frac{T_2}{T_1} = \left(\frac{v_1}{v_2} \right)^{n-1} = \left(\frac{1}{7} \right)^{0.25}$$

$$T_2 = \frac{T_1}{7^{0.25}} = \frac{1273}{1.627} = 782.6 \text{ K}$$

where $T_1 = 1000 + 273 = 1273$ K. Therefore

$$W = \frac{0.274(782.6 - 1273)}{1.25 - 1} = -537.5 \text{ kJ/kg}$$

i.e.   Work done *by* the gas mixture $= +537.5$ kJ/kg

Also from equation (2.16), for unit mass, $u_2 - u_1 = c_v(T_2 - T_1)$, therefore

$$u_2 - u_1 = 0.925(782.6 - 1273) = -453.6 \text{ kJ/kg}$$

Finally, from the non-flow energy equation (1.4), $Q + W = (u_2 - u_1)$,

i.e. $Q - 537.5 = -453.6$ therefore $Q = 83.9$ kJ/kg

i.e.   Heat supplied $= 83.9$ kJ/kg

Example 6.8   Calculate for the data of Example 6.7 the change of entropy per kilogram of mixture.

Fig. 6.3 T–s diagram for Example 6.8

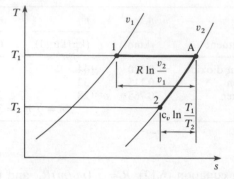

Solution   Referring to Fig. 6.3, the change of entropy between state 1 and state 2 can be found by imagining the process replaced by two other processes, 1 to A and A to 2. This method is described in section 4.4.

For isothermal process 1 to A, from equation (4.12)

$$s_A - s_1 = R \ln\left(\frac{v_2}{v_1}\right) = 0.274 \times \ln 7 = 0.533 \text{ kJ/kg K}$$

For the constant volume process A to 2

$$s_A - s_2 = c_v \int_2^A \frac{dT}{T} = c_v \ln\left(\frac{T_1}{T_2}\right) = 0.925 \times \ln\left(\frac{1273}{782.6}\right)$$

i.e. $s_A - s_2 = 0.450 \text{ kJ/kg K}$

Then by subtraction,

$$s_2 - s_1 = 0.533 - 0.450 = 0.083 \text{ kJ/kg K}$$

It is often convenient to use amount of substance in problems on mixtures, and to define heat capacities expressed in terms of the amount of substance. These are known as *molar heat capacities*, and are denoted by $\tilde{c}_p$ and $\tilde{c}_v$. Molar heat capacities are defined as follows:

$$\tilde{c}_p = \tilde{m}c_p \quad \text{and} \quad \tilde{c}_v = \tilde{m}c_v \tag{6.20}$$

From equation (2.17), $c_p - c_v = R$, therefore

$$\tilde{c}_p - \tilde{c}_v = \tilde{m}c_p - \tilde{m}c_v = \tilde{m}R$$

Also from equation (2.9), $\tilde{m}R = \tilde{R}$, hence

$$\tilde{c}_p - \tilde{c}_v = \tilde{R} \tag{6.21}$$

From equation (2.15)

$$U = mc_v T$$

Also from equation (2.7), $m = n\tilde{m}$, and from equation (6.20), $\tilde{m}c_v = \tilde{c}_v$, therefore,

$$U = n\tilde{c}_v T \tag{6.22}$$

Similarly

$$H = n\tilde{c}_p T \tag{6.23}$$

By the Gibbs–Dalton law,

$$U = \sum U_i \quad \text{and} \quad H = \sum H_i$$

therefore

$$n\tilde{c}_v T = \sum n_i \tilde{c}_{v_i} T \quad \text{and} \quad n\tilde{c}_p T = \sum n_i \tilde{c}_{p_i} T$$

i.e.

$$\tilde{c}_v = \sum \frac{n_i}{n} \tilde{c}_{v_i} \tag{6.24}$$

and

$$\tilde{c}_p = \sum \frac{n_i}{n} \tilde{c}_{p_i} \tag{6.25}$$

Example 6.9   A producer gas has the following volumetric analysis: 29% CO, 12% H_2, 3% CH_4, 4% CO_2, 52% N_2. Calculate the values of $\tilde{c}_p$, $\tilde{c}_v$, c_p, and c_v for the mixture. The values of $\tilde{c}_p$ for the constituents are as follows: for CO, $\tilde{c}_p = 29.27$ kJ/kmol K; for H_2, $\tilde{c}_p = 28.89$ kJ/kmol K; for CH_4, $\tilde{c}_p = 35.80$ kJ/kmol K; for CO_2, $\tilde{c}_p = 37.22$ kJ/kmol K; for N_2, $\tilde{c}_p = 29.14$ kJ/kmol K.

The molar masses may be taken as follows: for H_2, 2 kg/kmol; for CH_4, 16 kg/kmol; for CO_2, 44 kg/kmol; for N_2, 28 kg/kmol.

Solution   From equation (6.25),

$$\tilde{c}_p = \sum \frac{n_i}{n} \tilde{c}_{p_i}$$

Therefore,

$$\tilde{c}_p = (0.29 \times 29.27) + (0.12 \times 23.89) + (0.03 \times 35.80)$$
$$+ (0.04 \times 37.22) + (0.52 \times 29.14)$$

i.e. $\tilde{c}_p = 29.6707$ kJ/kmol K

From equation (6.21),

$$\tilde{c}_p - \tilde{c}_v = \tilde{R}$$

therefore

$$\tilde{c}_v = \tilde{c}_p - \tilde{R} = 29.6707 - 8.3145 = 21.3562 \text{ kJ/kmol K}$$

i.e. $\tilde{c}_v = 21.3562$ kJ/kmol K

The molar mass can be found from equation (6.17), i.e.

$$\tilde{m} = \sum \frac{n_i}{n} \tilde{m}$$

$$= (0.29 \times 28) + (0.12 \times 2) + (0.03 \times 16) + (0.04 \times 44)$$
$$+ (0.52 \times 28)$$

161

Table
6.5 Properties of some common gases at 300K

Gas	c_p (kJ/kg K)	c_v	γ	$\tilde{c}_p$ (kJ/kmol K)	$\tilde{c}_v$	$\tilde{m}$ (kg/kmol)	R (kJ/kg K)
Diatomic							
Carbon monoxide (CO)	1.0410	0.7442	1.399	29.158	20.845	28.010	0.2968
Hydrogen (H_2)	14.3230	10.1987	1.404	28.875	20.561	2.016	4.1243
Nitrogen (N_2)	1.0400	0.7432	1.399	29.134	20.819	28.013	0.2968
Oxygen (O_2)	0.9182	0.6584	1.395	29.382	21.068	31.999	0.2598
Monatomic							
Argon (Ar)	0.5203	0.3122	1.666	20.786	12.472	39.950	0.2081
Helium (He)	5.1930	3.1159	1.666	20.788	12.473	4.003	2.0771
Triatomic							
Carbon dioxide (CO_2)	0.8457	0.6568	1.288	37.219	28.906	44.010	0.1889
Sulphur dioxide (SO_2)	0.6448	0.5150	1.252	41.306	32.991	64.060	0.1298
Hydrocarbons							
Ethane (C_2H_6)	1.7668	1.4903	1.186	53.128	44.813	30.070	0.2765
Methane (CH_4)	2.2316	1.7132	1.303	35.795	27.480	16.040	0.5184
Propane (C_3H_8)	1.6915	1.5029	1.126	74.578	66.263	44.090	0.1886

i.e. $\tilde{m} = 25.16 \text{ kg/kmol}$

Then from equation (6.20)

$$c_p = \frac{\tilde{c}_p}{\tilde{m}} = \frac{29.6707}{25.16} = 1.1793 \text{ kJ/kg K}$$

and

$$c_v = \frac{\tilde{c}_v}{\tilde{m}} = \frac{21.3562}{25.16} = 0.8488 \text{ kJ/kg K}$$

Values of γ, c_p, c_v, $\tilde{c}_p$, $\tilde{c}_v$, $\tilde{m}$, and R at 300 K for some of the more common gases are shown in Table 6.5.

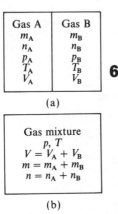

(a)

(b)

Fig. 6.4 Mixing of two gases initially separate

6.5 Adiabatic mixing of perfect gases

Consider two gases A and B separated from each other in a closed vessel by a thin diaphragm, as shown in Fig. 6.4(a). If the diaphragm is punctured or removed, then the gases mix as in Fig. 6.4(b), and each then occupies the total volume, behaving as if the other gas were not present. This process is equivalent to a free expansion of each gas, and is irreversible. The process can be simplified by the assumption that it is adiabatic; this means that the vessel is perfectly thermally insulated and therefore there will be an increase in entropy of the system. In section 4.5 it is shown that there is always an increase in entropy of a thermally isolated system which undergoes an irreversible process.

It is shown in section 3.5 that in a free expansion process the internal energy initially is equal to the internal energy finally. In this case, from equation (6.22)

$$U_1 = n_A \tilde{c}_{v_A} T_A + n_B \tilde{c}_{v_B} T_B$$

and

$$U_2 = (n_A \tilde{c}_{v_A} + n_B \tilde{c}_{v_B}) T$$

Extending this result to any number of gases,

$$U_1 = \sum n_i \tilde{c}_{v_i} T_i \quad \text{and} \quad U_2 = T \sum n_i \tilde{c}_v$$

Then $U_1 = U_2$

i.e. $\sum n_i \tilde{c}_{v_i} T_i = T \sum n_i \tilde{c}_{v_i}$

i.e. $T = \dfrac{\sum n_i \tilde{c}_{v_i} T_i}{\sum n_i \tilde{c}_{v_i}}$ (6.26)

Example 6.10   A vessel of 1.5 m^3 capacity contains oxygen at 7 bar and 40 °C. The vessel is connected to another vessel of 3 m^3 capacity containing carbon monoxide at 1 bar and 15 °C. A connecting valve is opened and the gases mix adiabatically. Calculate:

(i) the final temperature and pressure of the mixture;
(ii) the change in entropy of the system.

For oxygen, $\tilde{c}_v = 21.07$ kJ/kmol K; for carbon monoxide, $\tilde{c}_v = 20.86$ kJ/kmol K.

Solution   (i) From equation (2.8)

$$n = \frac{pV}{\tilde{R}T}$$

Therefore

$$n_{O_2} = \frac{7 \times 10^5 \times 1.5}{8.3145 \times 313 \times 10^3} = 0.4035 \quad \text{where } T_{O_4} = 40 + 273 = 313 \text{ K}$$

and

$$n_{CO} = \frac{1 \times 10^5 \times 3}{8.3145 \times 288 \times 10^3} = 0.1253 \quad \text{where } T_{CO} = 15 + 273 = 288 \text{ K}$$

Before mixing

$$U_1 = \sum (n_i \tilde{c}_{v_i} T_i) = (0.4035 \times 21.07 \times 313) + (0.1253 \times 20.86 \times 288)$$

i.e. $U_1 = 3413.8$ kJ

After mixing

$$U_2 = T \sum (n_i \tilde{c}_{v_i}) = T\{(0.4035 \times 21.07) + (0.1253 \times 20.86)\}$$

i.e. $U_2 = 11.118 \times T$

For adiabatic mixing, $U_1 = U_2$, therefore

$$3413.8 = 11.118 \times T \quad \text{therefore } T = \frac{3413.8}{11.118} = 307 \text{ K}$$

i.e.   Temperature of mixture $= 307 - 273 = 34 \,°\text{C}$

From equation (2.8),

$$p = \frac{n\tilde{R}T}{V}$$

Therefore

$$p = \frac{(0.4035 + 0.1253) \times 8.3145 \times 307 \times 10^3}{(1.5 + 3.0) \times 10^5} = 3 \text{ bar}$$

i.e.   Pressure after mixing $= 3$ bar

(ii) The change of entropy of the system is equal to the change of entropy of the oxygen plus the change of entropy of the carbon monoxide; this follows from the Gibbs–Dalton law.

Referring to Fig. 6.5, the change of entropy of the oxygen can be calculated by replacing the process undergone by the oxygen by the two processes 1 to A and A to 2.

Fig. 6.5 T–s diagram for oxygen for Example 6.10

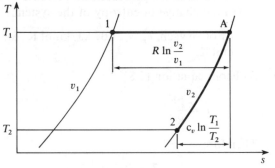

For an isothermal process from 1 to A, from equation (4.13), we have

$$s_A - s_1 = R \ln\left(\frac{V_A}{V_1}\right) \quad \text{or} \quad S_A - S_1 = mR \ln\left(\frac{V_A}{V_1}\right)$$

i.e. $S_A - S_1 = n\tilde{R} \ln\left(\frac{V_A}{V_1}\right) = 0.4035 \times 8.3145 \times \ln\left(\frac{4.5}{1.5}\right) = 3.686 \text{ kJ/K}$

At constant volume from A to 2,

$$s_A - s_2 = c_v \int_2^A \frac{dT}{T} = c_v \ln\left(\frac{T_1}{T_2}\right) \quad \text{or} \quad S_A - S_2 = mc_v \ln\left(\frac{T_1}{T_2}\right)$$

$$S_A - S_2 = n\tilde{c}_v \ln\left(\frac{T_1}{T_2}\right) = 0.4035 \times 21.07 \times \ln\left(\frac{313}{307}\right) = 0.1683 \text{ kJ/K}$$

therefore

$$S_2 - S_1 = 3.686 - 0.168 = 3.518 \text{ kJ/K}$$

Referring to Fig. 6.6, the change of entropy of the carbon monoxide can be found in a similar way to the above,

i.e. $$S_2 - S_1 = (S_B - S_1) + (S_2 - S_B)$$

Fig. 6.6 T–s diagram for carbon monoxide for Example 6.10

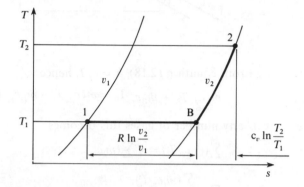

therefore

$$S_2 - S_1 = n\tilde{R} \ln\left(\frac{V_B}{V_1}\right) + n\tilde{c}_v \ln\left(\frac{T_2}{T_1}\right)$$

$$= \left\{0.1253 \times 8.314 \times \ln\left(\frac{4.5}{3}\right)\right\}$$

$$+ \left\{0.1253 \times 20.86 \times \ln\left(\frac{307}{288}\right)\right\}$$

therefore

$$S_2 - S_1 = 0.590 \text{ kJ/K}$$

Hence the change of entropy of the whole system is given by

$$(S_2 - S_1)_{\text{system}} = (S_2 - S_1)_{O_2} + (S_2 - S_1)_{CO}$$

i.e.   Change of entropy of system $= 3.518 + 0.590 = 4.108 \text{ kJ/K}$

Another form of mixing is that which occurs when streams of fluid meet to form a common stream in steady flow. This is shown diagrammatically in Fig. 6.7. The steady-flow energy equation can be applied to the mixing section, and changes in kinetic and potential energy are usually negligible,

i.e. $$\dot{m}_A h_{A_1} + \dot{m}_B h_{B_1} + \dot{Q} + \dot{W} = \dot{m}_A h_{A_2} + \dot{m}_B h_{B_2}$$

For adiabatic flow $Q = 0$, and also $W = 0$ in this case, therefore

$$\dot{m}_A h_{A_1} + \dot{m}_B h_{B_1} = \dot{m}_A h_{A_2} + \dot{m}_B h_{B_2}$$

165

Fig. 6.7 Mixing of two
fluid streams

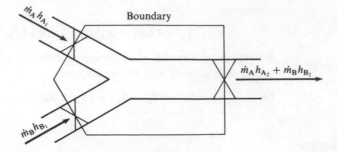

From equation (2.18), $h = c_p T$, hence

$$\dot{m}_A c_{p_A} T_A + \dot{m}_B c_{p_B} T_B = \dot{m}_A c_{p_A} T + \dot{m}_B c_{p_B} T$$

For any number of gases this becomes

$$\sum (\dot{m}_i c_{p_i} T_i) = T \sum (\dot{m}_i c_{p_i})$$

i.e. $$T = \frac{\sum (\dot{m}_i c_{p_i} T_i)}{\sum (\dot{m}_i c_{p_i})} \qquad (6.27)$$

Also, since from equation (6.20), $\tilde{c}_p = \tilde{m} c_p$, and $\tilde{m} = m/n$ from equation (2.7), then

$$n \tilde{c}_p = m c_p$$

Hence, $$T = \frac{\sum (n_i \tilde{c}_{p_i} T_i)}{\sum (n_i \tilde{c}_{p_i})} \qquad (6.28)$$

Equation (6.27) or (6.28) represents one condition which must be satisfied in an adiabatic mixing process of perfect gases in steady flow. In a particular problem some other information must be known (e.g. the final pressure or specific volume) before a complete solution is possible. To find the change of entropy in such a process the procedure is as described above for adiabatic mixing by a free expansion. The entropy change of each gas is found and the results added together.

6.6 Gas and vapour mixtures

Consider a vessel of fixed volume which is maintained at a constant temperature as shown in Fig. 6.8(a). The vessel is evacuated and the absolute pressure is therefore zero. In Fig. 6.8(b) a small quantity of water is introduced into the vessel and it evaporates to occupy the whole volume. For a small quantity of water introduced, the pressure in the vessel will be less than the saturation pressure corresponding to the temperature of the vessel. At this condition of pressure and temperature the vessel will be occupied by superheated vapour.

Fig. 6.8 Liquid introduced into an evacuated vessel

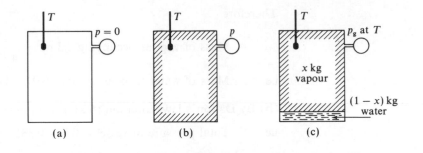

(a) (b) (c)

As more water is introduced the pressure increases and the water continues to evaporate until such a condition is reached that the volume can hold no more vapour. Any additional water introduced into the vessel after this will not evaporate but will exist as water, the condition being as in Fig. 6.8(c), which shows the vapour in contact with its liquid. Per kilogram of water introduced, the vessel can be thought of as containing either $(1 - x)$ kg of water plus x kg of dry saturated vapour, or as containing 1 kg of wet steam of dryness fraction x.

During the entire process of evaporation the temperature remains constant. If the temperature is now raised by the addition of heat, then more vapour will evaporate and the pressure in the vessel will increase. Eventually the vessel will contain a superheated vapour as before, but at a higher pressure and temperature.

The vessel in Fig. 6.8 is considered to be initially evacuated, but the water would evaporate in exactly the same way if the vessel contained a gas or a mixture of gases. As stated in the Gibbs–Dalton law, each constituent behaves as if it occupies the whole vessel at the temperature of the vessel. When a little water is sprayed into a vessel containing a gas mixture, then the vapour formed will exert the saturation pressure corresponding to the temperature of the vessel, and this is the partial pressure of the vapour in the mixture. (It must be remembered that the vapour is only saturated when it is in contact with its liquid.)

When a mixture contains a saturated vapour, then the partial pressure of the vapour can be found from tables at the temperature of the mixture. This assumes that a saturated vapour obeys the Gibbs–Dalton law; this is only a good approximation at low values of the total pressure.

Example 6.11

(a) A vessel of 0.3 m^3 capacity contains air at 0.7 bar and 75 °C. The vessel is maintained at this temperature as water is injected into it. Calculate the mass of water to be injected so that the vessel is just filled with saturated vapour.

(b) If injection now continues until a total mass of 0.7 kg of water is introduced, calculate the new total pressure in the vessel.

(c) The vessel is now heated until all the water in it just evaporates. Calculate:

(i) the total pressure for this condition;
(ii) the heat to be supplied.

Solution

(a) The subscripts s, w, and a will be used for steam, water, and air respectively.

At 75 °C, the saturation pressure $p_g = 0.3855$ bar and $v_g = 4.133$ m^3/kg.

167

Therefore

$$\text{Mass of vapour occupying } 0.3 \text{ m}^3 = \frac{0.3}{4.133} = 0.0726 \text{ kg}$$

i.e.   Mass of water to be injected = 0.0726 kg

(b) By Dalton's law, equation (6.2), $p = p_a + p_s$,

i.e.   Total pressure in vessel $= 0.7 + 0.3855 = 1.0855$ bar

Note that the dry vapour is assumed to act as a perfect gas, hence the vapour and the air are assumed to occupy the same volume while each exerts its partial pressure.

When a total mass of 0.7 kg of water has been injected into the vessel it will exist partly as dry saturated vapour (say m_s kg) and partly as water (say m_w kg, where $m_w = (0.7 - m_s)$) in such proportions that the mixture occupies the total volume of 0.3 m³, therefore

$$(m_s \times 4.133) + (0.7 - m_s) \times 0.001\,026 = 0.3$$

where 0.001 026 m³/kg is the specific volume of water.

i.e. $m_s(4.133 - 0.001\,026) = 0.3 - (0.7 \times 0.001\,026)$

$$4.132 \times m_s = 0.2993 \quad \text{therefore } m_s = \frac{0.2993}{4.132} = 0.0724 \text{ kg}$$

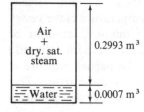

Air
+
dry. sat.
steam

0.2993 m³

Water   0.0007 m³

Fig. 6.9 Conditions in vessel for part (b) of Example 6.11

Note that the volume of water is negligibly small compared to the volume of the air–vapour mixture

$$m_w = 0.7 - 0.0724 = 0.6276 \text{ kg}$$

The volume occupied by the dry vapour $= 0.0724 \times 4.133 = 0.2993$ m³. The vessel may be assumed to contain air, dry saturated steam, and water, as shown in Fig. 6.9.

Since $T_1 = T_2$, we can write

$$p_{a_1} V_{a_1} = p_{a_2} V_{a_2}$$

therefore

$$p_{a_2} = 0.7 \times \frac{0.3}{0.2993} = 0.7017 \text{ bar}$$

i.e.   Total pressure $= p_a + p_s = 0.7017 + 0.3855 = 1.0872$ bar

(c) (i) The water can be completely evaporated by raising the temperature to a value such that the total volume is occupied by saturated steam and air. This condition is reached when the steam has a specific volume v_g, such that, $0.7 \times v_g = 0.3$,

i.e. $$v_g = \frac{0.3}{0.7} = 0.4286 \text{ m}^3/\text{kg}$$

From tables the saturation pressure at $v_g = 0.4286$ m^3/kg is, by interpolating,

$$p = 4 + \left(\frac{0.4623 - 0.4286}{0.4623 - 0.4139}\right) \times (4.5 - 4.0) = 4.35 \text{ bar}$$

The air now occupies the volume of 0.3 m^3 while exerting its partial pressure p_{a_3} at the new temperature. The new temperature is that saturation temperature corresponding to the pressure of 4.35 bar.

From tables by interpolation at 4.35 bar

$$t = 143.6 + \frac{0.35}{0.5} \times 4.3 = 146.6 \,^\circ\text{C} \quad \text{therefore } T = 146.6 + 273 = 419.6 \text{ K}$$

Then for the air

$$\frac{p_{a_3}}{T_{a_3}} = \frac{p_{a_1}}{T_{a_1}} \quad \text{therefore } p_{a_3} = 0.7 \times \frac{419.6}{348} = 0.8439 \text{ bar}$$

where $T_{a_1} = 75 + 273 = 348$ K,

i.e.   Total pressure in vessel = $4.35 + 0.8439 = 5.194$ bar

(ii) From the non-flow energy equation

$$Q + W = U_2 - U_1$$

In this case $W = 0$, therefore $Q = (U_2 - U_1)$. Then

$$U_1 = m_{w_1} u_{w_1} + m_a u_{a_1} + m_{s_1} u_{s_1}$$

and $$U_2 = m_a u_{a_2} + m_{s_2} u_{s_2}$$

For a perfect gas, from equation (2.15), $U = mc_v T$, therefore

$$Q = m_{s_2} u_{s_2} - m_{s_1} u_{s_1} - m_{w_1} u_{w_1} + m_a c_v (T_2 - T_1)$$

Then taking u_s and u_w from tables, and substituting for

$$m_a = \frac{p_a V}{R_a T} = \frac{0.7 \times 10^5 \times 0.3}{0.287 \times 348 \times 10^3} = 0.2102 \text{ kg}$$

we have

$$Q = (0.7 \times 2556.8) - (0.0724 \times 2475.3)$$

$$- (0.6276 \times 313.5) + 0.2102 \times 0.718(419.6 - 348)$$

i.e.   Heat supplied = $1789.8 - 179.2 - 196.8 + 10.8 = 1424.6$ kJ

Example 6.12   The products of combustion of a fuel have a volumetric analysis of CO_2 8%, H_2O 15%, O_2 5.5%, and N_2 71.5%. If the total pressure is 1.4 bar, calculate the temperature to which the gas must be cooled at constant pressure for condensation of the H_2O just to commence.

Solution   From equation (6.14)

$$\text{Partial pressure of } H_2O = \frac{n_i}{n} \times p = 0.15 \times 1.4 = 0.21 \text{ bar}$$

The saturation temperature corresponding to 0.21 bar is 61.15 °C, i.e. the gas must be cooled to 61.15 °C for condensation of the H_2O to commence.

6.7 The steam condenser

The condenser is an essential part of any steam power plant. The temperature at which condensation occurs is in the order of 25 to 40 °C, the corresponding saturation pressures being 0.031 66 and 0.073 75 bar. The shell and tube type condenser is a vessel in which this low pressure is maintained by a pump, and the steam condenses on the outside of tubes through which cold water is flowing. This type is called a surface condenser. There will be some leakage of air into the condenser, both through the glands and from air dissolved in the feedwater which comes out of solution and is carried into the condenser by the steam. This air impairs the condenser performance since it reduces the heat transfer from the steam to the cooling water.

The condenser contains a mixture of steam, air, and water. The air must be pumped out of the condenser continually to maintain the vacuum, and the air which is pumped out carries with it some of the steam. This results in a loss of feedwater to the boiler. This loss has to be made up by the addition of cold water. Another effect of the presence of air is that the condensate is undercooled (i.e. cooled to a temperature below the saturation temperature), which means that more heat has to be supplied to the water in the boiler than if no undercooling had occurred.

The pressure in the condenser is approximately constant throughout and steam and air enter the condenser in fixed proportions when steady conditions prevail. As some of the steam is condensed the partial pressure of the remaining steam decreases, and hence the partial pressure of the air increases to maintain the same total pressure. At reduced partial pressures the steam has a saturation temperature which is below that of the incoming steam. Hence condensation proceeds at progressively lower temperatures.

Some condensers are designed to make up for the deficiencies of the simple type. Two of these are indicated in Figs 6.10(a) and 6.10(b). In Fig. 6.10(a) most of the condensation is carried out on the main bank of tubes and the air is drawn over another, smaller, bank which is shielded from the main bank and is called the air cooler. Here further condensation takes place at a lower temperature with a subsequent saving in feedwater, and a smaller pump is required for the condenser. In Fig. 6.10(b) the air-cooling tubes are in the centre of the condenser and the air is pumped away from this region. The incoming steam passes all round the bank of tubes and some is drawn upwards to the centre. In doing so it meets the undercooled condensate which has been formed and reheats it, hence reducing the amount of undercooling.

Fig. 6.10 Two arrangements for air extraction in a condenser

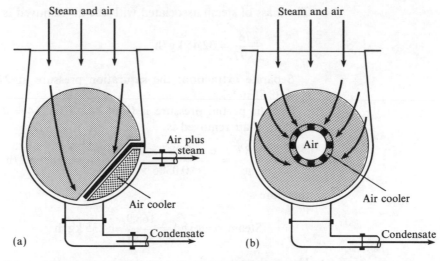

(a)

(b)

Example 6.13

A surface condenser is required to deal with 20 000 kg of steam per hour, and the air leakage is estimated at 0.3 kg per 1000 kg of steam. The steam enters the condenser dry saturated at 38 °C. The condensate is extracted at the lowest point of the condenser at a temperature of 36 °C. The condensate loss is made up with water at 7 °C. It is required to find the saving in condensate and the saving in heat supplied in the boiler, by fitting a separate air extraction pump which draws the air over an air cooler. Assume that the air leaves the cooler at 27 °C. The pressure in the condenser can be assumed to remain constant.

Solution

At entry, mass of air per kilogram of steam = 0.3/1000 kg.

At 38 °C the saturation pressure is 0.066 24 bar and $v_g = 21.63$ m^3/kg.

For 1 kg of steam the volume is 21.63 m^3, and this must be the volume occupied by 0.3/1000 kg of air when exerting its partial pressure,

$$\text{Partial pressure of air} = \frac{m_a R_a T}{V} = \frac{0.3 \times 0.287 \times 311 \times 10^3}{1000 \times 21.63 \times 10^5}$$

$$= 1.2 \times 10^{-5} \text{ bar}$$

This is negligibly small and may be neglected.

Condensate extraction: the saturation pressure at 36 °C is 0.0594 bar, and $v_g = 23.97$ m^3/kg. The total pressure in the condenser is 0.066 24 bar, hence

$$0.066\,24 = 0.0594 + p_a \quad \text{therefore } p_a = 0.006\,84 \text{ bar}$$

The mass of air removed per hour is

$$\frac{20\,000 \times 03}{1000} = 6 \text{ kg/h}$$

Hence the volume of air removed per hour is

$$\frac{mRT}{p} = \frac{6 \times 0.287 \times 309 \times 10^3}{0.006\,84 \times 10^5} = 778 \text{ m}^3/\text{h}$$

171

The mass of steam associated with the air removed is therefore given by

$$\frac{778}{23.97} = 32.45 \text{ kg/h}$$

Separate extraction: the saturation pressure at 27 °C is 0.035 64 bar and $v_g = 38.81 \text{ m}^3/\text{kg}$.

The air partial pressure is $0.066\,24 - 0.035\,64 = 0.0306$ bar. Therefore the volume of air removed is

$$\frac{mRT}{p} = \frac{6 \times 0.287 \times 300 \times 10^3}{0.0306 \times 10^5} = 168.9 \text{ m}^3/\text{h}$$

therefore

$$\text{Steam removed} = \frac{168.9}{38.81} = 4.35 \text{ kg/h}$$

Hence the saving in condensate by using the separate extraction method is given by $32.45 - 4.35 = 28.1 \text{ kg/h}$.

Also, the saving in heat to be supplied in the boiler is approximately $28.1 \times 4.182(36 - 7)/3600 = 0.95 \text{ kW}$, where the mean specific heat of the water is 4.182 kJ/kg K.

Example 6.14

For the data of Example 6.13 calculate the percentage reduction in air pump capacity by using the separate extraction method. If the temperature rise of the cooling water is 5.5 K, calculate the mass flow of cooling water required.

Solution

Air pump capacity without air cooler = 778 m³/h

Air pump capacity with the air cooler = 168.9 m³/h

Therefore

$$\text{Percentage reduction in capacity} = \left(\frac{778 - 168.9}{778}\right) \times 100$$

$$= 78.3\%$$

The system to be analysed is shown in Fig. 6.11. Let suffixes s, a, and c denote steam, air, and condensate respectively. Applying the steady-flow energy equation and neglecting changes in kinetic energy, we have

$$Q + \dot{m}_{s_1} h_{s_1} + \dot{m}_{a_1} h_{a_1} = (\dot{m}_{s_2} h_{s_2} + \dot{m}_{a_2} h_{a_2}) + \dot{m}_c h_c$$

$$\dot{m}_{a_1} = \dot{m}_{a_2} = 6 \text{ kg/h}; \qquad \dot{m}_{s_1} = 20\,000 \text{ kg/h}; \qquad \dot{m}_{s_2} = 4.35 \text{ kg/h}$$

$$\dot{m}_c = 20\,000 - 4.35 = 20\,000 \text{ kg/h approx.}$$

$$h_{a_1} - h_{a_2} = c_p(T_1 - T_2) \quad \text{(from equation (2.18))}$$

i.e. $$Q = (4.35 \times 2550.3) + \{6 \times 1.005(38 - 27)\}$$

$$+ (20\,000 \times 150.7) - (20\,000 \times 2570.1)$$

Fig. 6.11 Condenser system for Example 6.14

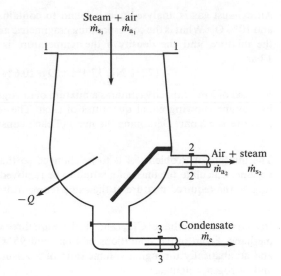

therefore

$$Q = -48.38 \times 10^6 \, \text{kJ/h} = -13\,439 \, \text{kW}$$

where $h_c = h_f$ at $36\,°C = 150.7 \, \text{kJ/kg}$,

i.e. Heat rejected $= +13\,439 \, \text{kW}$

The mass of cooling water required for a 5.5 K rise in temperature is $48.38 \times 10^6/(5.5 \times 4.182) = 2.1 \times 10^6 \, \text{kg/h}$, approximately.

Unless a very large natural supply of cooling water is available for large steam plants, means must be found to cool the cooling water after use. This can be done by passing the cooling water through a cooling tower; cooling towers are considered in Section 15.5.

Problems

(For values of $\tilde{m}$, R, $\tilde{c}_p$, $\tilde{c}_v$, etc. which are necessary in the following problems, refer to Table 6.5 on p. 162; take values of $\tilde{m}$ to the nearest whole number.)

6.1 A mixture of carbon monoxide and oxygen is to be prepared in the proportion of 7 kg to 4 kg in a vessel of 0.3 m³ capacity. If the temperature of the mixture is $15\,°C$, determine the pressure to which the vessel is subject. If the temperature is raised to $40\,°C$, what will then be the pressure in the vessel?

(29.94 bar; 32.54 bar)

6.2 For the mixture of Problem 6.1 calculate the volumetric analysis, the molar mass and the characteristic gas constant. Calculate also the total amount of substance in the mixture.

(33.3% O_2; 66.7% CO; 29.3 kg/kmol; 0.283 kJ/kg K; 0.375 kmol)

6.3 An exhaust gas is analysed and is found to contain, by volume, 78% N_2, 12% CO_2, and 10% O_2. What is the corresponding gravimetric analysis? Calculate the molar mass of the mixture, and the density if the temperature is 550 °C and the total pressure is 1 bar.

(72% N_2, 17.4% CO_2, 10.6% O_2; 30.33 kg/kmol; 0.443 kg/m^3)

6.4 A vessel of 3 m^3 capacity contains a mixture of nitrogen and carbon dioxide, the analysis by volume showing equal quantities of each. The temperature is 15 °C and the total pressure is 3.5 bar. Determine the mass of each constituent.

(6.14 kg N_2; 9.65 kg CO_2)

6.5 The mixture of Problem 6.4 is to be changed so that it is 70% CO_2 and 30% N_2 by volume. Calculate the mass of mixture to be removed and the mass of CO_2 to be added to give the required mixture at the same temperature and pressure as before.

(6.31 kg; 7.72 kg CO_2)

6.6 In a mixture of methane (CH_4) and air there are three volumes of oxygen to one volume of methane. From initial conditions of 1 bar and 95 °C the gas is compressed reversibly and adiabatically through a volume ratio of 5. Assuming that air contains only oxygen and nitrogen, calculate:
(i) the values of c_p, c_v, $\tilde{c}_p$, $\tilde{c}_v$, R and γ for the mixture;
(ii) the final pressure and temperature of the mixture;
(iii) the work input per unit mass of mixture.

(1.057 kJ/kg K, 0.761 kJ/kg K, 29.60 kJ/kmol K, 21.31 kJ/kmol K, 0.297 kJ/kg K, 1.389; 9.35 bar, 415.3 °C; 243.8 kJ/kg)

6.7 A mixture is made up of 25% N_2, 35% O_2, 20% CO_2, and 20% CO by volume. Calculate:
(i) the molar mass of the mixture;
(ii) $\tilde{c}_p$ and $\tilde{c}_v$ for the mixture;
(iii) γ for the mixture;
(iv) the partial pressure of each constituent when the total pressure is 1.5 bar;
(v) the density of the mixture at 1.5 bar and 15 °C.

(32.6 kg/kmol; 30.84, 22.53 kJ/kmol K; 1.37; 0.375, 0.525, 0.3, 0.3 bar; 2.04 kg/m^3)

6.8 Two vessels are connected by a pipe in which there is a valve. One vessel of 0.3 m^3 contains air at 7 bar and 32 °C, and the other of 0.03 m^3 contains oxygen at 21 bar and 15 °C. The valve is opened and the two gases are allowed to mix. Assuming that the system is well lagged, calculate:
(i) the final temperature of the mixture;
(ii) the final pressure of the mixture;
(iii) the partial pressure of each constituent;
(iv) the volumetric analysis of the mixture;
(v) the values of c_p, c_v, R, $\tilde{m}$, and γ for the mixture;
(vi) the increase of entropy of the system per kilogram of mixture;
(vii) the change in internal energy and enthalpy of the mixture per kilogram if the vessel is cooled to 10 °C.
Assume that air consists only of oxygen and nitrogen.

(27.9 °C; 8.27 bar; 3.31, 4.96 bar; 60% N_2, 40% O_2; 0.987, 0.709 kJ/kg K; 0.278 kJ/kg K; 29.91 kg/mol; 1.392; 0.183 kJ/kg K; 12.69, 17.67 kJ/kg)

6.9 Air and carbon monoxide are mixed in the proportion 3 to 1 by mass. The CO is supplied at 4 bar and 15 °C, and the air is supplied at 7 bar and 32 °C. The two constituents are passed in steady flow through non-return valves to mix adiabatically at a pressure of 1 bar. Calculate:
(i) the final temperature of the mixture;

  (ii) the partial pressure of each constituent of the mixture;
 (iii) the increase of entropy per kilogram of mixture;
 (iv) the volume flow of mixture for a flow of 1 kg/min of CO;
  (v) the velocity of the mixture if the area of the pipe downstream of the mixing section
      is 0.1 m²

$$(27.7\,°C;\ 0.256,\ 0.156,\ 0.588\ bar;\ 0.689\ kJ/kg\ K;\ 3.49\ m^3/min;\ 0.582\ m/s)$$

6.10 Ammonia in air is a toxic mixture when the ammonia is 0.55% by volume. Calculate how much leakage from an ammonia compressor can be tolerated per 1000 m³ of space. The pressure is 1 bar and the temperature is 15 °C. The molar mass of ammonia (NH_3) is 17 kg/kmol, and it may be assumed to act as a perfect gas in this case.

$$(3.91\ kg)$$

6.11 A vessel of 0.3 m³ capacity contains a mixture of air and steam which is 0.75 dry. If the pressure is 7 bar and the temperature is 116.9 °C, calculate the mass of water present, the mass of dry saturated vapour, and the mass of air.

$$(0.102\ kg;\ 0.307\ kg;\ 1.394\ kg)$$

6.12 If the vessel of problem 6.11 is cooled to 100 °C calculate:
  (i) the mass of vapour condensed;
 (ii) the final pressure in the vessel;
 (iii) the heat rejected.

$$(0.128\ kg;\ 5.99\ bar;\ 297\ kJ)$$

6.13 A closed vessel of volume 3 m³ contains air saturated with water vapour at 38 °C and a vacuum pressure of 660 mm of mercury. The vacuum falls to 560 mm of mercury and the temperature falls to 26.7 °C. Calculate the mass of air that has leaked in and the quantity of vapour that has condensed. Take the barometric pressure as 760 mm Hg.

$$(0.583\ kg;\ 0.0627\ kg)$$

6.14 The air in a cylinder fitted with a piston is saturated with water vapour. The volume is 0.3 m³, the pressure is 3.5 bar and the temperature is 60.1 °C. The mixture is compressed to 5.5 bar, the temperature remaining constant. Calculate:
  (i) the masses of air and vapour present initially;
 (ii) the mass of vapour condensed on compression.

$$(1.036\ kg;\ 0.0392\ kg;\ 0.0148\ kg)$$

6.15 The temperature in a vessel is 36 °C and the proportion by mass of air to dry saturated steam is 0.1. What is the pressure in the vessel in bar and in mm of mercury vacuum? The barometric pressure is 760 mm Hg.

$$(0.0631\ bar;\ 712.7\ mm\ Hg)$$

6.16 A surface condenser is fitted with separate air and condensate outlets. A portion of the cooling surface is screened from the incoming steam and the air passes over these screened tubes to the air extraction and becomes cooled below the condensate temperature. The condenser receives 20 000 kg/h of steam dry saturated at 36.2 °C. At the condensate outlet the temperature is 34.6 °C, and at the air extraction the temperature is 29 °C. The volume of air plus vapour leaving the condenser is 3.8 m³/min. Assuming constant pressure throughout the condenser calculate:
  (i) the mass of air removed per 10 000 kg of steam;
 (ii) the mass of steam condensed in the air cooler per minute;
 (iii) the heat rejected to the cooling water.
Neglect the partial pressure of the air at inlet to the condenser.

$$(2.63\ kg;\ 0.5\ kg/min;\ 13\,451\ kW)$$

7

Combustion

The ideal cycles previously considered use fluids which remain unchanged chemically as they pass through the various processes of the cycle. In practical engines and power plants the source of heat is the chemical energy of substances called fuels. This energy is released during the chemical reaction of the fuel with oxygen. The fuel elements combine with oxygen in an oxidation process which is rapid and is accompanied by the evolution of heat.

The combustion process takes place in a controlled manner in some form of combustion chamber after initiation of combustion by some means (e.g. in a petrol engine the combustion is started by an electric spark). The most convenient source of oxygen supply is that of the atmosphere which contains oxygen and nitrogen and traces of other gases. Normally no attempt is made to separate out the oxygen from the atmosphere, and the nitrogen, etc. accompanies the oxygen into the combustion chamber.

Nitrogen does not oxidize easily and is inert as far as the combustion process is concerned, but it acts as a moderator in that it absorbs some of the heat of combustion and so limits the maximum temperature reached. As combustion proceeds the oxygen is progressively used up and the proportion of nitrogen plus products of combustion to the available oxygen increases. For a given amount of fuel there is a definite amount of oxygen, and therefore air, which is required for the complete combustion of a given fuel. To ensure complete combustion it is usual to supply air in excess of the amount required for chemically correct combustion. The oxygen not consumed in the reaction passes into the exhaust with the products of combustion.

Internal-combustion engines are run on liquid fuels which are grouped as 'petrols' (known as gasoline in the USA), and diesel oils, or gaseous fuels, commonly used in combined heat and power plant; gas turbines are run mainly on kerosene although natural gas is now commonly used. Engines burning solid fuels have been built but are mainly experimental. In the many and diverse applications in industry, solid, liquid, and gaseous fuels are used. Generalization is not possible on the selection of fuels, since the fuel used and its necessary firing equipment depend on the particular application, the practical circumstances, and economic considerations.

7.1 Basic chemistry

It is necessary to understand the construction and use of chemical formulae, before combustion problems can be considered. This involves elementary concepts which have been met before by most students, but a brief explanation will be given here.

Atoms. Chemical elements cannot be divided indefinitely and the smallest particle which can take part in a chemical change is called an atom. If an atom is split as in a nuclear reaction, the divided atom does not retain the original chemical properties.

Molecules. Elements are seldom found to exist naturally as single atoms. Some elements have atoms which exist in pairs, each pair forming a molecule (e.g. oxygen), and the atoms of each molecule are held together by strong inter-atomic forces. The isolation of a molecule of oxygen would be tedious, but possible; the isolation of an atom of oxygen would be a different prospect.

The molecules of some substances are formed by the mating up of atoms of different elements. For example, water (which is chemically the same as ice and steam) has a molecule which consists of two atoms of hydrogen and one atom of oxygen.

The atoms of different elements have different masses and these values are important when a quantitative analysis is required. The actual masses are infinitesimally small, and the ratios of the masses of atoms are used. These ratios are given by the relative atomic masses quoted on a scale which defines the atomic mass of isotope 12 of carbon as 12 (see Ch. 2, p. 40). The *relative atomic mass* of a substance is the mass of a single entity of the substance relative to a single entity of carbon-12. Table 7.1 gives the relative atomic masses of some common elements rounded off to give values accurate enough for most purposes.

Table 7.1 Relative atomic and molecular masses of some common substances

Element	Oxygen	Hydrogen	Carbon	Sulphur	Nitrogen
Atomic symbol	O	H	C	S	N
Relative atomic mass	16	1	12	32	14
Molecular grouping	O_2	H_2	C	S	N_2
Relative molecular mass (rounded)	32	2	12	32	28
Accurate values	31.999	2.016	12	32.030	28.013

Relative molecular masses are based on the relative masses of the atoms which constitute the molecule. In chemical formulae one atom of an element is represented by the symbol for the element, i.e. an atom of hydrogen is written as H, and other examples are given in Table 7.1. If a substance exists as a molecule containing, say, two atoms, as for hydrogen, it is written as H_2. Two molecules of hydrogen is written as $2H_2$, etc. Table 7.1 includes relative molecular masses, rounded off, and, for comparison, the accurate values.

Table 7.2 Compounds and their relative molecular masses

Compound	Formula	Relative molecular mass	
Water, steam	H_2O	$(2 \times 1) + (1 \times 16)$ =	18
Carbon monoxide	CO	$(1 \times 12) + (1 \times 16)$ =	28
Carbon dioxide	CO_2	$(1 \times 12) + (2 \times 16)$ =	44
Sulphur dioxide	SO_2	$(1 \times 32) + (2 \times 16)$ =	64
Methane	CH_4	$(1 \times 12) + (4 \times 1)$ =	16
Ethane	C_2H_6	$(2 \times 12) + (6 \times 1)$ =	30
Propane	C_3H_8	$(3 \times 12) + (8 \times 1)$ =	44
n-Butane	C_4H_{10}	$(4 \times 12) + (10 \times 1)$ =	58
Ethylene	C_2H_4	$(2 \times 12) + (4 \times 1)$ =	28
Propylene	C_3H_6	$(3 \times 12) + (6 \times 1)$ =	42
n-Pentane	C_5H_{12}	$(5 \times 12) + (12 \times 1)$ =	72
Benzene	C_6H_6	$(6 \times 12) + (6 \times 1)$ =	78
Toluene	C_7H_8	$(7 \times 12) + (8 \times 1)$ =	92
n-Octane	C_8H_{18}	$(8 \times 12) + (18 \times 1)$ =	114

Some of the other substances met in combustion work are given in Table 7.2 to illustrate the calculations of the relative molecular mass from the relative atomic masses of the elements.

7.2 Fuels

The most important fuel elements are carbon and hydrogen, and most fuels consist of these and sometimes a small amount of sulphur. The fuel may contain some oxygen and a small quantity of incombustibles (e.g. water vapour, nitrogen, or ash).

Coal is the most important solid fuel and the various types are divided into groups according to their chemical and physical properties. An accurate chemical analysis by mass of the important elements in the fuel is called the *ultimate analysis*, the elements usually included being carbon, hydrogen, nitrogen, and sulphur. The main groups are shown in Table 7.3, and their ultimate analyses are given. The analyses are typical but may vary from one sample to another within the group, and hence can be taken only as a guide. Another analysis of coal, also shown in Table 7.3 called the *proximate analysis*, gives the percentages of inherent moisture, volatile matter, and combustible solid (called fixed carbon). The fixed carbon is found as a remainder by deducting the percentages of the other quantities. The volatile matter includes the water derived from the chemical decomposition of the coal (not to be confused with free, or inherent moisture), the combustible gases (e.g. hydrogen, methane, ethane, etc.), and tar (i.e. a complex mixture of hydrocarbons and other organic compounds). The procedures for both analyses are given in ref. 7.1; see also ref. 7.2 and a concise treatment in ref. 7.3.

Most liquid fuels are hydrocarbons which exist in the liquid phase at atmospheric conditions. Petroleum oils are complex mixtures of sometimes hundreds of different fuels, but the necessary information to the engineer is the relative proportions of carbon, hydrogen, etc. as given by the ultimate analysis. Table 7.4 gives the ultimate analyses of some liquid fuels.

Table 7.3 Analysis of solid fuels

Ultimate analysis

		Percentage by mass of dry fuel					
Fuel	Rank	C	H	O	N	S	Mineral matter
Anthracite	101	88.2	2.7	1.7	1.0	1.2	5.2
Medium-rank coal	401	81.8	4.9	4.4	1.8	1.9	5.2
Low-rank coal	902	75.0	4.6	10.7	1.6	2.1	6.0
Coke	—	90.0	0.4	1.9	—	—	7.7

Proximate analysis on a mineral matter-free basis

	Percentage by mass of fuel		
Fuel	Inherent moisture	Volatile matter	Fixed carbon
Anrhtacite	2	6	92
Medium-rank coal	3	39	58
Low-rank coal	10	42	48

Table 7.4 Analyses of liquid fuels

Fuel	Carbon	Hydrogen	Sulphur	Ash, etc.
100-octane petrol	85.1	14.9	0.01	—
Motor petrol	85.5	14.4	0.1	—
Benzole	91.7	8.0	0.3	—
Kerosene (paraffin)	86.3	13.6	0.1	—
Diesel oil	86.3	12.8	0.9	—
Light fuel oil	86.2	12.4	1.4	—
Heavy fuel oil	86.1	11.8	2.1	—
Residual fuel oil	88.3	9.5	1.2	1.0

Table 7.5 Analysis by volume of a typical natural gas

Methane CH_4	Ethane C_2H_6	Propane C_3H_8	Butane C_4H_{10}	Nitrogen N_2	Carbon dioxide CO_2
92.6%	3.6%	0.8%	0.3%	2.6%	0.1%

Gaseous fuels are chemically the simplest of the three groups. The main gaseous fuel in use occurs naturally but other gaseous fuels may be manufactured by the various treatments of coal. Carbon monoxide is an important gaseous fuel which is a constituent of other gas mixtures, and is also a product of the incomplete combustion of carbon. A typical analysis of a natural gas is given in Table 7.5. The table gives the analyses by volume, each constituent having been measured by volume at atmospheric pressure and temperature. The volumetric analysis is the same as the molar analysis (see equation (6.14)).

Fuels are tested according to standardized procedures and for further information ref. 7.2 should be consulted.

7.3 Combustion equations

Proportionate masses of air and fuel enter the combustion chamber where the chemical reaction takes place, and from which the products of combustion pass to the exhaust. By the conservation of mass the mass flow remains constant (i.e. total mass of products equals total mass of reactants), but the reactants are chemically different from the products, and the products leave at a higher temperature. The total number of atoms of each element concerned in the combustion remains constant, but the atoms are rearranged into groups having different chemical properties. This information is expressed in the chemical equation which shows:

(i) the reactants and the products of combustion;
(ii) the relative quantities of the reactants and products.

The two sides of the equation must be consistent, each having the same number of atoms of each element involved. It should not be assumed that if an equation can be written, that the reaction it represents is inevitable or even possible. For possibility and direction the reaction has to be considered with reference to the Second Law of Thermodynamics. For the present the only concern is known combustion equations.

The equation shows the number of molecules of each reactant and product. The amount of substance, introduced in section 2.3, is proportional to the number of molecules, hence the relative numbers of molecules of the reactants and the products give the molar, and therefore the volumetric, analysis of the gaseous constituents.

As stated earlier the oxygen supplied for combustion is usually provided by atmospheric air, and it is necessary to use accurate and consistent analyses of air by mass and by volume. It is usual in combustion calculations to take air as 23.3% O_2, 76.7% N_2 by mass, and 21% O_2, 79% N_2 by volume. The small traces of other gases in dry air are included in the nitrogen, which is sometimes called 'atmospheric nitrogen'.

Consider the combustion equation for hydrogen:

$$2H_2 + O_2 \rightarrow 2H_2O \tag{7.1}$$

This tells us that

(i) hydrogen reacts with oxygen to form steam or water;
(ii) two molecules of hydrogen react with one molecule of oxygen to give two molecules of steam or water,

i.e.   2 volumes H_2 + 1 volume $O_2 \rightarrow$ 2 volumes H_2O

The H_2O may be a liquid or a vapour depending on whether the product has been cooled sufficiently to cause condensation. The proportions by mass are obtained by using relative atomic masses,

i.e. $2H_2 + O_2 \rightarrow 2H_2O$

therefore

$$2 \times (2 \times 1) + (2 \times 16) \rightarrow 2 \times \{(2 \times 1) + 16\}$$

i.e. $4 \text{ kg H}_2 + 32 \text{ kg O}_2 \rightarrow 36 \text{ kg H}_2\text{O}$

or $1 \text{ kg H}_2 + 8 \text{ kg O}_2 \rightarrow 9 \text{ kg H}_2\text{O}$

The same proportions are obtained by writing equation (7.1) as $H_2 + \frac{1}{2}O_2 \rightarrow H_2O$, and this is sometimes done.

It will be noted from equation (7.1) that the total volume of the reactants is 2 volumes H_2 + 1 volume O_2 = 3 volumes. The total volume of the product is only 2 volumes. There is therefore a volumetric contraction on combustion.

Since oxygen is accompanied by nitrogen if air is supplied for the combustion, then this nitrogen should be included in the equation. As nitrogen is inert as far as the chemical reaction is concerned, it will appear on both sides of the equation.

With 1 kmol of oxygen there are 79/21 kmol of nitrogen, hence equation (7.1) becomes

$$2H_2 + O_2 + \frac{79}{21}N_2 \rightarrow 2H_2O + \frac{79}{21}N_2 \tag{7.2}$$

Similar equations can be found for the combustion of carbon. There are two possibilities to consider:

(i) The complete combustion of carbon to carbon dioxide

$$C + O_2 \rightarrow CO_2 \tag{7.3}$$

and including the nitrogen

$$C + O_2 + \frac{79}{21}N_2 \rightarrow CO_2 + \frac{79}{21}N_2 \tag{7.4}$$

Considering the volumes of reactants and products

0 volume C + 1 volume $O_2 + \dfrac{79}{21}$ volumes N_2

$\rightarrow 1$ volume $CO_2 + \dfrac{79}{21}$ volumes N_2

The volume of carbon is written as zero since the volume of a solid is negligible in comparison with that of a gas.

By mass

$$12 \text{ kg C} + (2 \times 16) \text{ kg O}_2 + \frac{79}{21}(2 \times 14) \text{ kg N}_2$$

$$\rightarrow \{12 + (2 \times 16)\} \text{ kg CO}_2 + \frac{79}{21}(2 \times 14) \text{ kg N}_2$$

i.e. $12 \text{ kg C} + 32 \text{ kg O}_2 + 105.3 \text{ kg N}_2 \rightarrow 44 \text{ kg CO}_2 + 105.3 \text{ kg N}_2$

or $1 \text{ kg C} + \dfrac{32}{12} \text{ kg O}_2 + \dfrac{105.3}{12} \text{ kg N}_2 \rightarrow \dfrac{44}{12} \text{ kg CO}_2 + \dfrac{105.3}{12} \text{ kg N}_2$

(ii) The incomplete combustion of carbon. This occurs when there is an insufficient supply of oxygen to burn the carbon completely to carbon dioxide,

i.e.
$$2C + O_2 \rightarrow 2CO \tag{7.5}$$

$$2C + O_2 + \frac{79}{21}N_2 \rightarrow 2CO + \frac{79}{21}N_2 \tag{7.6}$$

By mass

$$(2 \times 12)\,kg\,C + (2 \times 16)\,kg\,O_2 + \frac{79}{21}(2 \times 14)\,kg\,N_2$$

$$\rightarrow 2(12 + 16)\,kg\,CO + \frac{79}{21}(2 \times 14)\,kg\,N_2$$

i.e. $24\,kg\,C + 32\,kg\,O_2 + 105.3\,kg\,N_2 \rightarrow 56\,kg\,CO + 105.3\,kg\,N_2$

or $1\,kg\,C + \dfrac{32}{24}\,kg\,O_2 + \dfrac{105.3}{24}\,kg\,N_2 \rightarrow \dfrac{56}{24}\,kg\,CO + \dfrac{105.3}{24}\,kg\,N_2$

If a further supply of oxygen is available then the combustion can continue to completion

$$2CO + O_2 + \frac{79}{21}N_2 \rightarrow 2CO_2 + \frac{79}{21}N_2 \tag{7.7}$$

By mass,

$$56\,kg\,CO + 32\,kg\,O_2 + 105.3\,kg\,N_2 \rightarrow 88\,kg\,CO_2 + 105.3\,kg\,N_2$$

or $1\,kg\,CO + \dfrac{32}{56}\,kg\,O_2 + \dfrac{105.3}{56}\,kg\,N_2 \rightarrow \dfrac{88}{56}\,kg\,CO_2 + \dfrac{105.3}{56}\,kg\,N_2$

7.4 Stoichiometric air–fuel ratio

A stoichiometric mixture of air and fuel is one that contains just sufficient oxygen for the complete combustion of the fuel. A mixture which has an excess of air is termed a *weak mixture*, and one which has a deficiency of air is termed a *rich mixture*. The percentage of excess air is given by the following:

Percentage excess air
$$= \frac{\text{actual A/F ratio} - \text{stoichiometric A/F ratio}}{\text{stoichiometric A/F ratio}} \tag{7.8}$$

where A denotes air and F denotes fuel.

For gaseous fuels the ratios are expressed by volume and for solid and liquid fuels the ratios are expressed by mass. Equation (7.8) gives a positive result when the mixture is weak, and a negative result when the mixture is rich. For boiler plant the mixture is usually greater than 20% weak; for gas turbines it can be as much as 300% weak. Petrol engines have to meet various conditions of load and speed, and operate over a wide range of mixture strengths. The following definition is used:

$$\text{Mixture strength} = \frac{\text{stoichiometric A/F ratio}}{\text{actual A/F ratio}} \qquad (7.9)$$

The working values range between 80% (weak) and 120% (rich) (see section 13.6).

Where fuels contain some oxygen (e.g. ethyl alcohol C_2H_6O) this oxygen is available for the combustion process, and so the fuel requires a smaller supply of air.

7.5 Exhaust and flue gas analysis

The products of combustion are mainly gaseous. When a sample is taken for analysis it is usually cooled down to a temperature which is below the saturation temperature of the steam present. The steam content is therefore not included in the analysis, which is then quoted as the analysis of the dry products. Since the products are gaseous, it is usual to quote the analysis by volume. An analysis which includes the steam in the exhaust is called a wet analysis. The following examples illustrate the principles covered in this chapter up to this point.

Example 7.1   A sample of dry anthracite has the following composition by mass.

C 90%; H 3%; O 2.5%; N 1%; S 0.5%; ash 3%

Calculate:
 (i) the stoichiometric A/F ratio;
 (ii) the A/F ratio and the dry and wet analysis of the products of combustion by mass and by volume, when 20% excess air is supplied.

Solution   (i) Each constituent is taken separately and the amount of oxygen required for complete combustion is found from the relevant chemical equation. Carbon:

$$C + O_2 \rightarrow CO_2 \qquad 12\,\text{kg C} + 32\,\text{kg O}_2 \rightarrow 44\,\text{kg CO}_2$$

i.e.   Oxygen required $= 0.9 \times \dfrac{32}{12} = 2.4\,\text{kg/kg coal}$

where the carbon content is 0.9 kg per kilogram of coal

$$\text{Carbon dioxide produced} = 0.9 \times \frac{44}{12} = 3.3\,\text{kg CO}_2$$

Hydrogen (H):

$$H_2 + \tfrac{1}{2}O_2 \rightarrow H_2O \qquad 2\,\text{kg H}_2 + 16\,\text{kg O}_2 \rightarrow 18\,\text{kg H}_2O$$

or $1\,\text{kg H}_2 + 8\,\text{kg O}_2 \rightarrow 9\,\text{kg H}_2O$

i.e.   Oxygen required $= 0.03 \times 8 = 0.24\,\text{kg/kg coal}$

and   Steam produced $= 0.03 \times 9 = 0.27\,\text{kg/kg coal}$

Sulphur:

$$S + O_2 \rightarrow SO_2 \qquad 32\,kg\,S + 32\,kg\,O_2 \rightarrow 64\,kg\,SO_2$$

or $$1\,kg\,S + 1\,kg\,O_2 \rightarrow 2\,kg\,SO_2$$

i.e. Oxygen required = 0.005 kg/kg coal

Sulphur dioxide produced = 2 × 0.005 = 0.01 kg/kg coal

These results are tabulated in Table 7.6; the oxygen in the fuel is shown as a negative quantity in the column 'oxygen required'.

Table 7.6 Solution for Example 7.1(i)

Constituent	Mass fraction	Oxygen required (kg/kg coal)	Product mass (kg/kg coal)
Carbon (C)	0.900	2.400	3.30 (CO_2)
Hydrogen (H)	0.030	0.240	0.27 (H_2O)
Sulphur (S)	0.005	0.005	0.01 (SO_2)
Oxygen (O)	0.025	−0.025	—
Nitrogen (N)	0.010	—	0.01 (N_2)
Ash	0.030	—	—
		2.620	

From Table 7.6:

$$O_2 \text{ required per kilogram of coal} = 2.62\,kg$$

therefore

$$\text{Air required per kilogram of coal} = \frac{2.62}{0.233} = 11.245\,kg$$

where air is assumed to contain 23.3% O_2 by mass.

i.e. stoichiometric air–fuel ratio = 11.245.

(ii) For an air supply which is 20% in excess, using equation (7.8),

$$\text{Actual A/F ratio} = 11.245 + \left(\frac{20}{100} \times 11.245 \right) = 1.2 \times 11.245$$

$$= 13.494/1$$

Therefore

$$N_2 \text{ supplied} = 0.767 \times 13.494 = 10.350\,kg$$

Also O_2 supplied = 0.233 × 13.494 = 3.144 kg

In the products, then, we have

$$N_2 = 10.350 + 0.01 = 10.360\,kg$$

and excess O_2 = 3.144 − 2.620 = 0.524 kg

The products are entered in Table 7.7 and the analysis by volume is obtained. In column 3 the percentage by mass is given by 100 times the mass of each product divided by the total mass of 14.464 kg. In column 5 the amount of substance per kilogram of coal is given by equation (2.7), $n_i = m_i / \tilde{m}_i$. The total of 0.4764 in column 5 gives the total amount of substance of wet products per kilogram of coal, and by subtracting the amount of substance of H_2O from this total, the total amount of substance of the dry products is obtained as 0.4616. Column 6 gives the proportion of each constituent of column 5 expressed as a percentage of the total amount of substance of the wet products. Similarly column 7 gives the percentage by volume of the dry products.

Table 7.7 Solution for Example 7.1(ii)

Product	m_i (kg/kg coal)	m_i/m (%)	$\tilde{m}_i$ (kg/kmol)	$n_i = m_i / \tilde{m}_i$	Wet n_i/n (%)	Dry n_i/n (%)
1	2	3	4	5	6	7
CO_2	3.300	22.82	44	0.0750	15.74	16.25
H_2O	0.270	1.87	18	0.0150	3.15	—
SO_2	0.010	0.07	64	0.0002	0.04	0.04
O_2	0.524	3.62	32	0.0164	3.44	3.55
N_2	10.360	71.63	28	0.3700	77.63	80.16
	14.464	100.01		0.4766 (wet) (0.4616) (dry)	100.00	100.00

Example 7.2

The analysis of a supply of gas is as follows: H_2 49.4%; CO 18%; CH_4 20%; C_4H_8 2%; O_2 0.4% N_2 6.2%; CO_2 4%. Calculate:
(i) the stoichiometric A/F ratio;
(ii) the wet and dry analysis of the products of combustion if the actual mixture is 20% weak.

Solution

(i) The example is solved by a similar tabular method to Example 7.1; a specimen calculation is shown more fully as follows.

$$CH_4 + 2O_2 \rightarrow CO_2 + 2H_2O$$

i.e. $1 \text{ kmol } CH_4 + 2 \text{ kmol } O_2 \rightarrow 1 \text{ kmol } CO_2 + 2 \text{ kmol } H_2O$

There are 0.2 kmol of CH_4 for 1 kmol gas, hence

$$0.2 \text{ kmol } CH_4 + (0.2 \times 2) \text{ kmol } O_2 \rightarrow 0.2 \text{ kmol } CO_2 + (0.2 \times 2) \text{ kmol } H_2O$$

Therefore the oxygen required for the CH_4 in the gas is 0.4 kmol/kmol gas. The results are summarized in Table 7.8; the oxygen in the gas, (0.004 kmol/kmol of gas) is included in column 4 as a negative quantity. From Table 7.8,

$$\text{Air required} = \frac{0.853}{0.21} = 4.062 \text{ kmol/kmol gas}$$

185

Table 7.8 Results for Example 7.2

1	2	3	4	5	6
	kmol/ kmol fuel	Combustion equation	O_2 kmol/ kmol fuel	Products kmol/kmol fuel	
				CO_2	H_2O
H_2	0.494	$2H_2 + O_2 \rightarrow 2H_2O$	0.247	—	0.494
CO	0.18	$2CO + O_2 \rightarrow 2CO_2$	0.09	0.18	—
CH_4	0.2	$CH_4 + 2O_2 \rightarrow CO_2 + 2H_2O$	0.4	0.2	0.4
C_4H_8	0.02	$C_4H_8 + 6O_2 \rightarrow 4CO_2 + 4H_2O$	0.12	0.08	0.08
O_2	0.004	—	−0.004	—	—
N_2	0.062	—			
CO_2	0.04	—		0.04	
		Total	0.853	0.50	0.974

where air is assumed to contain 21% oxygen by volume,

i.e. Stoichiometric A/F ratio = 4.062 by volume

(ii) For a mixture which is 20% weak, using equation (7.8),

Actual A/F ratio = 4.062 + (0.2 × 4.062)

= 1.2 × 4.062 = 4.874 by volume

Associated nitrogen = 0.79 × 4.874 = 3.851 kmol/kmol gas

Excess oxygen = (0.21 × 4.874) − 0.853

= 0.1706 kmol/kmol gas

Total amount of nitrogen in products
= 3.851 + 0.062 = 3.913 kmol/kmol gas

The analysis by volume of the wet and dry products is then as shown in Table 7.9

Table 7.9 Solution for Example 7.2

Product	kmol/kmol fuel	% by vol. (dry)	% by vol. (wet)
CO_2	0.50	10.90	9.0
H_2O	0.974	—	17.5
O_2	0.171	3.72	3.08
N_2	3.912	85.4	70.4
Total wet	5.557	100.02	99.98
−H_2O	0.974		
Total dry	4.583		

Example 7.3 Ethyl alcohol is burned in a petrol engine. Calculate:

(i) the stoichiometric A/F ratio;

(ii) the A/F ratio and the wet and dry analyses by volume of the exhaust gas for a mixture strength of 90%;

(iii) the A/F ratio and the wet and dry analyses by volume of the exhaust gas for a mixture strength of 120%.

Solution   (i) The equation for the combustion of ethyl alcohol is as follows:

$$C_2H_6O + 3O_2 \rightarrow 2CO_2 + 3H_2O$$

Since there are two atoms of carbon in each mole of C_2H_6O then there must be 2 mol of CO_2 in the products, giving two atoms of carbon on each side of the equation. Similarly, since there are six atoms of hydrogen in each mole of ethyl alcohol then there must be 3 mol of H_2O in the products, giving six atoms of hydrogen on each side of the equation. Then balancing the atoms of oxygen, it is seen that there are $\{(2 \times 2) + 3\} = 7$ atoms on the right-hand side of the equation, hence seven atoms must appear on the left-hand side of the equation. There is one atom of oxygen in the ethyl alcohol, therefore a further six atoms of oxygen must be supplied, and hence 3 mol of oxygen are required as shown. Since the O_2 is supplied as air, the associated N_2 must appear in the equation,

i.e. $$C_2H_6O + 3O_2 + \left(3 \times \frac{79}{21}\right)N_2 \rightarrow 2CO_2 + 3H_2O + \left(3 \times \frac{79}{21}\right)N_2$$

1 kmol of fuel has a mass of $(2 \times 12) + (6 + 16) = 46$ kg; 3 kmol of oxygen have a mass of $(3 \times 32) = 96$ kg.
   Therefore

$$O_2 \text{ required per kg of fuel} = \frac{96}{46} = 2.087 \text{ kg}$$

Therefore

$$\text{Stoichiometric A/F ratio} = \frac{2.087}{0.233} = 8.957/1$$

(ii) Considering a mixture strength of 90% then, from equation (7.9),

$$0.9 = \frac{\text{stoichiometric A/F ratio}}{\text{actual A/F ratio}}$$

Therefore

$$\text{Actual A/F ratio} = \frac{8.957}{0.9} = 9.952/1$$

This means that $1/0.9$ times as much air is supplied as is necessary for complete combustion. The exhaust will therefore contain $(1/0.9) - 1 = 0.1/0.9$ of the stoichiometric oxygen,

$$C_2H_6O + \frac{1}{0.9}\left\{3O_2 + \left(3 \times \frac{79}{21}\right)N_2\right\}$$

$$\rightarrow 2CO_2 + 3H_2O + \left(\frac{0.1}{0.9} \times 3\right)O_2 + \left(\frac{1}{0.9} \times 3 \times \frac{79}{21}\right)N_2$$

i.e. the products are

$$2 \text{ kmol } CO_2 + 3 \text{ kmol } H_2O + 0.333 \text{ kmol } O_2 + 12.540 \text{ kmol } N_2$$

The total amount of substance $= 2 + 3 + 0.333 + 12.540$

$$= 17.873 \text{ kmol}$$

Hence wet analysis is

$$\frac{2}{17.873} \times 100 = 11.19\% \ CO_2; \qquad \frac{3}{17.873} \times 100 = 16.79\% \ H_2O$$

$$\frac{0.333}{17.873} \times 100 = 1.86\% \ O_2; \qquad \frac{12.540}{17.873} \times 100 = 70.16\% \ N_2$$

The total dry amount of substance $= 2 + 0.333 + 12.540$

$$= 14.873 \text{ kmol}$$

Hence the dry analysis is

$$\frac{2}{14.873} \times 100 = 13.45\% \ CO_2; \qquad \frac{0.333}{14.873} \times 100 = 2.24\% \ O_2;$$

$$\frac{12.540}{14.873} \times 100 = 84.31\% \ N_2$$

(iii) Considering a mixture strength of 120%, then from equation (7.9),

$$1.2 = \frac{\text{stoichiometric A/F ratio}}{\text{actual A/F ratio}}$$

Therefore

$$\text{Actual A/F ratio} = \frac{8.957}{1.2} = 7.47/1$$

This means that $1/1.2$ of the stoichiometric air is supplied. The combustion cannot be complete, as the necessary oxygen is not available. It is usual to assume that all the hydrogen is burned to H_2O, since hydrogen atoms have a greater affinity for oxygen than carbon atoms. The carbon in the fuel will burn to CO and CO_2, but the relative proportions have to be determined. Let there be a kmol CO_2 and b kmol CO in the products. Then the combustion equation is as follows:

$$C_2H_6O + \frac{1}{1.2}\left\{ 3O_2 + \left(3 \times \frac{79}{21} \right)N_2 \right\}$$

$$\rightarrow aCO_2 + bCO + 3H_2O + \left(\frac{1}{1.2} \times 3 \times \frac{79}{21} \right)N_2$$

To find a and b a balance of the carbon and the oxygen atoms can be made, i.e.   Carbon balance: $2 = a + b$

Oxygen balance: $1 + \left(2 \times \dfrac{1}{1.2} \times 3\right) = 2a + b + 3$

Subtracting the equations gives

$a = 1$ and then $b = 2 - 1 = 1$

i.e. the products are

$$(1 \text{ kmol CO}_2) + (1 \text{ kmol CO}) + (3 \text{ kmol H}_2) + (9.405 \text{ kmol N}_2)$$

$$= 1 + 1 + 3 + 9.405 = 14.405 \text{ kmol}$$

Hence wet analysis is

$$\frac{1}{14.405} \times 100 = 6.94\% \text{ CO}_2; \qquad \frac{1}{14.405} \times 100 = 6.94\% \text{ CO}$$

$$\frac{3}{14.405} \times 100 = 20.83\% \text{ H}_2; \qquad \frac{9.405}{14.405} \times 100 = 65.29\% \text{ N}_2$$

The total dry amount of substance $= 1 + 1 + 9.405$

$$= 11.405 \text{ kmol}$$

Hence dry analysis is

$$\frac{1}{11.405} \times 100 = 8.77\% \text{ CO}_2; \qquad \frac{1}{11.405} \times 100 = 8.77\% \text{ CO}$$

$$\frac{9.405}{11.405} \times 100 = 82.46\% \text{ N}_2$$

Example 7.4   For the stoichiometric mixture of Example 7.3, calculate:

(i) the volume of the mixture per kilogram of fuel at a temperature of 65 °C and a pressure of 1.013 bar;

(ii) the volume of the products of combustion per kilogram of fuel after cooling to a temperature of 120 °C at a pressure of 1 bar.

Solution   (i) As before

$$C_2H_6O + 3O_2 + \left(3 \times \frac{79}{21}\right)N_2 \rightarrow 2CO_2 + 3H_2O + \left(3 \times \frac{79}{21}\right)N_2$$

Therefore

$$\text{Total amount of substance reactants} = 1 + 3 + \left(3 \times \frac{79}{21}\right)$$

$$= 15.286 \text{ kmol}$$

189

From equation (2.8), $p V = n \tilde{R} T$, therefore

$$V = \frac{15.286 \times 10^3 \times 8.3145 \times 338}{10^5 \times 1.013} = 424.06 \text{ m}^3/\text{kmol of fuel}$$

where $T = 65 + 273 = 338$ K.

In 1 kmol of fuel there are $(2 \times 12) + (6 + 16) = 46$ kg, therefore

$$\text{Volume of reactants per kilogram of fuel} = \frac{424.06}{46} = 9.219 \text{ m}^3$$

(ii) When the products are cooled to $120\,°C$ the H_2O exists as steam, since the temperature is well above the saturation temperature corresponding to the partial pressure of the H_2O. (This must be so since the saturation temperature corresponding to the total pressure is $99.6\,°C$, and saturation temperature decreases with pressure.) The total amount of substance of the products is

$$2 + 3 + \left(3 \times \frac{79}{21} \right) = 16.286 \text{ kmol}$$

From equation (2.8), $p V = n \tilde{R} T$, therefore

$$V = \frac{16.286 \times 10^3 \times 8.3145 \times 393}{10^5 \times 1} = 532.15 \text{ m}^3/\text{kmol of fuel}$$

where $T = 120 + 273 = 393$ K.

i.e.　　Volume of products per kg of fuel $= \dfrac{532.15}{46} = 11.57 \text{ m}^3$

Example 7.5　　If the products in Example 7.4 are cooled to $15\,°C$ at constant pressure, calculate the amount of water which will condense per kilogram of fuel.

Solution　At $15\,°C$, since some condensation has taken place, the steam remaining is dry saturated, being in contact with its liquid. The saturation pressure at $15\,°C$ is $0.017\,04$ bar, and this is the partial pressure of the dry saturated steam.

Then using equation (6.14)

$$\frac{V_i}{V} = \frac{n_i}{n} = \frac{p_i}{p}$$

For the steam

$$\frac{n_s}{n} = \frac{0.017\,04}{1} = 0.017\,04$$

From Example 7.4 the total amount of substance of dry products is $(16.286 - 3) = 13.286$ kmol, therefore

$$\frac{n_s}{n_s + 13.286} = 0.017\,04 \quad \text{therefore } n_s = \left(\frac{0.017\,04 \times 13.286}{1 - 0.017\,04} \right) = 0.2303$$

i.e. amount of substance of dry saturated steam remaining at $15\,°C$ is 0.2303 kmol, therefore amount of substance of water condensed is $(3 - 0.2303) = 2.77$ kmol. Also 1 kmol of H_2O contains $(2 + 16) = 18$ kg, therefore mass of water condensed is (2.77×18) kg/kmol fuel

i.e.   Mass of water condensed per kilogram of fuel $= \dfrac{2.77 \times 18}{46} = 1.084$ kg

Any problem in combustion can be solved using the amount of substance. The following examples illustrate the method for a solid and for a gaseous fuel; these examples should be compared with the method used in Examples 7.1 and 7.2.

Example 7.6   The gravimetric analysis of a sample of coal is given as 80% C, 12% H, and 8% ash. Calculate the stoichiometric A/F ratio and the analysis of the products by volume.

Solution   1 kg of coal contains 0.8 kg C and 0.12 kg H

1 kg of coal contains $\dfrac{0.8}{12}$ kmol C and 0.12 kmol H

Let the oxygen required for complete combustion be x kmol, the nitrogen supplied with the oxygen is then $x \times 79/21 = 3.76x$ kmol.

For 1 kg of coal the combustion equation is therefore as follows:

$$\frac{0.8}{12}C + 0.12H + xO_2 + 3.76xN_2 \rightarrow aCO_2 + bH_2O + 3.76xN_2$$

Then   Carbon balance: $\dfrac{0.8}{12} = a$ or $a = 0.067$ kmol

Hydrogen balance: $0.12 = 2b$ or $b = 0.060$ kmol

Oxygen balance: $2x = 2a + b$

i.e. $x = a + b/2 = 0.067 + 0.030 = 0.097$ kmol

The mass of 1 kmol of oxygen is 32 kg, therefore the mass of O_2 supplied per kilogram of coal is 32×0.097 kg,

Stoichiometric A/F ratio $= \dfrac{32 \times 0.097}{0.233} = 13.3/1$

Total amount of substance of products $= a + b + 3.76x$

$$= 0.067 + 0.06 + (3.76 \times 0.097)$$

$$= 0.492 \text{ kmol}$$

Hence wet analysis is

$\dfrac{0.067}{0.492} \times 100 = 13.6\%$ CO_2; $\dfrac{0.06}{0.492} \times 100 = 12.2\%$ H_2

$\dfrac{0.365}{0.492} \times 100 = 74.2\%$ N_2

Example 7.7

A gas engine is supplied with natural gas of the following composition: CH_4 93%; C_2H_6 3%; N_2 3%; CO 1%. If the A/F ratio is 30 by volume, calculate the analysis of the dry products of combustion. It can be assumed that the stoichiometric A/F ratio is less than 30.

Solution

Since we are told that the actual air–fuel ratio is greater than the stoichiometric it follows that excess air has been supplied. The products will therefore consist of CO_2, H_2O, O_2, and N_2. The combustion equation can be written as follows:

$$0.93CH_4 + 0.03C_2H_6 + 0.01CO + 0.03N_2$$
$$+ (0.21 \times 30)O_2 + (0.79 \times 30)N_2$$
$$\rightarrow aCO_2 + bH_2O + cO_2 + dN_2$$

Then   Carbon balance: $0.93 + (2 \times 0.03) + 0.01 = a$ or $a = 1$

Hydrogen balance: $(4 \times 0.93) + (6 \times 0.03) = 2b$ or $b = 1.95$

Oxygen balance: $0.01 + (0.21 \times 30) = 2a + b + 2c$

Therefore

$$c = \{6.31 - 2 - (2 \times 1.95)\}/2 = 0.205$$

Nitrogen balance: $0.03 + (0.79 \times 30) = d$ or $d = 23.73$

i.e.   Total amount of substance of dry products $= 1 + 0.205 + 23.73$
$$= 24.935 \text{ kmol}$$

Then analysis by volume is

$$\frac{1}{24.935} \times 100 = 4.01 \text{ CO}_2; \qquad \frac{0.205}{24.935} \times 100 = 0.82\% \text{ O}_2$$

$$\frac{23.73}{24.935} \times 100 = 95.17\% \text{ N}_2$$

7.6 Practical analysis of combustion products

The experimental investigation of a combustion process requires the analysis of the products of combustion. The most basic method is to take a sample of the gas and analyse it chemically as in an Orsat apparatus for example (see ref. 7.2). Modern instrumentation now provides a quicker, more accurate, and continuous means of analysis. Some of the methods used are summarized below, but manufacturers' catalogues should be consulted for details of actual instruments since improvements are continually being made to the measuring systems.

Non-dispersive infra-red (NDIR)

In this method the concentration of the constituent gas is measured by recording its 'optical' absorption in the infra-red spectrum. The mixture is continuously examined by being drawn through a tube, the inside of which is subject to radiation from an infra-red source, each gas absorbing on a particular waveband of the radiation. Full descriptions of infra-red analysis and other methods are given in ref. 7.4.

One particular instrument is shown diagrammatically in Fig. 7.1. The gas being analysed is passed through tube A and dry air is passed through tube B. The chambers C and D contain pure samples of the constituents to be detected. Radiation from nichrome elements is passed through tubes A and B and thence to chambers C and D where it is absorbed. The subsequent heating of the gases in C and D causes an increase in pressure in the two chambers; C and D are separated by a thin metal diaphragm which, together with an insulated, perforated plate, forms a capacitor.

Fig. 7.1 Non-dispersive infra-red gas analyser

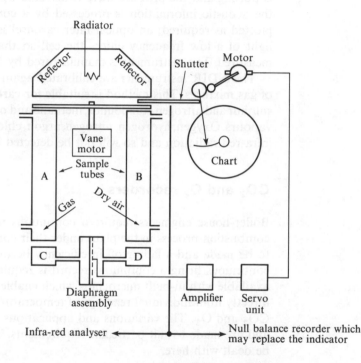

If the constituent being sought is not present in tube A then equal amounts of radiation will be absorbed in C and D, which will be heated equally and there will be no pressure difference across the diaphragm. If the constituent is present in A, it will absorb some of the radiation admitted, and the rest will pass on to chamber C. The radiation absorbed in C will then be less than in D so that a greater pressure will be reached in D than in C, and the diaphragm will be displaced. This displacement causes a change in capacitance of the capacitor and a current is produced which is amplified to give a reading on a microammeter. The microammeter scale is calibrated to give the corresponding concentration of the constituent in the gas being analysed.

To avoid zero errors the radiation is cut off from both tubes simultaneously and allowed to fall on them simultaneously, by means of a vane which rotates at a low frequency. The pressure changes are then related to the temperature changes produced by the differential absorption in C and D.

To provide a continuously recording instrument as shown in the right-hand part of Fig. 7.1, a 'null' balance recorder is employed. The principle is that the pressure difference created should be nullified by cutting off from the vessel B a sufficient amount of radiation to balance that absorbed in A. This is done by means of a shutter driven by a servo-motor which receives a signal from the detector unit. A recording pen is linked to the balancing shutter mechanism and records on a circular chart.

Another NDIR instrument uses photoacoustic detection: a sample is drawn into a closed cell where infra-red light of the correct frequency is absorbed by the gas to be measured; the pressure increases and decreases because the light is pulsing and the pressure wave is detected by a microphone in the cell wall; the acoustic information is processed by a computer and can be printed or plotted as required; an optical filter carousel is then turned so that infra-red light of a low frequency enters the cell so that the next constituent can be measured. The instrument is manufactured by Brüel and Kjaer of Denmark.

The NDIR instruments are calibrated against accurately prepared samples of gas mixtures. This method is suitable for carbon monoxide, carbon dioxide, sulphur and nitrogen compounds, methane and other hydrocarbons, and organic vapours. Oxygen, hydrogen, nitrogen, argon, chlorine, and helium, do not absorb infra-red radiation and so will not be detected by this type of instrument.

CO_2 and O_2 recorders

Boiler-house engineers require a continuous indication of the quality of the combustion process in the plant under their control. This enables comparisons to be made and a falling-off in efficiency becomes immediately apparent. For continuous firing a continuous record is required, and digital instruments are available with in-built microchips which enable boiler efficiency to be obtained directly from individual readings of temperature, and percentage by volume of CO_2 and O_2. The variations and applications of these instruments are many, and are made to suit particular requirements. The general principles only will be dealt with here.

CO_2 measurement by thermal conductivity variations

The CO_2 content of a flue gas is an important criterion of efficient and economic combustion, and is important in observing the regulations governing smoke emission.

When a heated wire is placed in a gaseous atmosphere it loses heat by radiation, convection and conduction. If the losses by radiation and convection are kept constant, the total heat loss is dependent on the heat loss by conduction, which varies with the constituents of the gas since each has a different and characteristic thermal conductivity. If a constant heat input is supplied to the wire there is an equilibrium temperature for each mixture, and if the CO_2 content alone is varied, then its concentration will be indicated by a measurement of the temperature of the wire. In an actual instrument the heat loss from the wire is mainly by conduction, the other means (e.g. convection, radiation, end cooling, diffusion) account for about 1% each of the total loss. The convection loss is reduced by mounting the wire vertically. The instrument is calibrated against mixtures of known composition.

The sample of gas is passed over an electrically heated platinum wire in a cell which forms one arm of a Wheatstone bridge. In another arm of the bridge is a similar cell containing air. A difference in CO_2 content between the two cells causes a difference in temperature between the two wires and hence a difference in resistance. The out-of-balance potential of the bridge is measured by a recording potentiometer, calibrated to give the CO_2 content directly.

Oxygen measurement by magnetic means

Gases may be classified in two groups:

(i) diamagnetic gases which seek the weakest part of a magnetic field;
(ii) paramagnetic gases which seek the strongest part of a magnetic field.

Most gases are diamagnetic, but oxygen is paramagnetic, and this property of oxygen can be utilized in measuring the oxygen content of gas mixtures. Referring to Fig. 7.2 the gas sample is introduced into the analysis cell and passes through the annulus as shown. The horizontal cross-tube carries two identical platinum windings, coils 1 and 2, which are connected in adjacent arms of a Wheatstone bridge, and are heated by the applied voltage. The winding at A is traversed by a magnetic field of high intensity from a large permanent magnet. When the gas sample passes the end of the tube the oxygen is drawn into the cross-tube. It is heated and its paramagnetic property decreases due

Fig. 7.2 Paramagnetic analyser

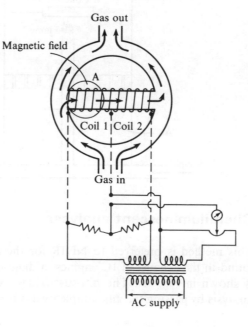

to the increase in temperature as it passes through the tube. The induction of fresh cool oxygen continues, hence a continuous flow is established. The result is that the two windings are cooled by different amounts, the resistance changes and the bridge goes out of balance. The resulting electromotive force (emf) is measured by a potentiometer, and since this is proportional to the oxygen content, the reading gives the oxygen content of the mixture.

Zirconia cell

An absolute method of measurement of oxygen may be obtained using a zirconia cell; the output of the cell is directly related to the oxygen concentration of the gas sample (see ref. 7.5).

Flame ionization detector (FID)

This is a method used for detecting the hydrocarbon (HC) content of the exhaust from internal combustion (IC) engines; Fig. 7.3 shows a diagram of a typical system. The sample to be analysed is mixed with a special burner fuel which may be hydrogen, hydrogen plus helium, or hydrogen plus nitrogen. A polarizing voltage exists between the burner and the collector which causes a migration of ions in the flame and so a current to flow in the collector circuit. The current is proportional to the rate of ion formation which depends on the HC concentration in the gases and is detected by a suitable electrometer and displayed as an analogue output. The FID gives a rapid, accurate, and continuous reading of total HC concentration for levels as low as one part in 10^9.

Fig. 7.3 Flame ionization detector

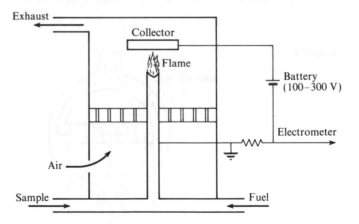

Chemiluminescent analyser

This method is preferred to NDIR for the analysis of the oxides of nitrogen found in the exhaust of IC engines; a diagram of a chemiluminescent analyser is shown in Fig. 7.4. The nitrous oxides, NO_x, are converted to NO before analysis by passing the gas sample over a heated catalyst such as stainless steel,

Fig. 7.4
Chemiluminescent
analyser

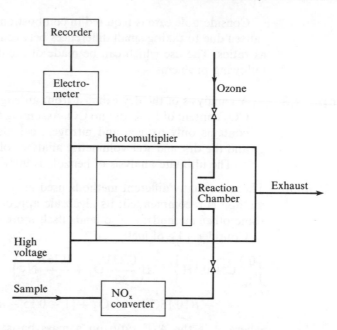

graphite, or molybdenum at about 600 °C. The sample is now mixed with ozone where the reaction gives NO_2 and O_2; some of the NO_2 contains an excess of energy within its atoms and this NO_2 then changes to the ground state with the emission of light (chemiluminescence) which is amplified then measured by a photomultiplier tube and the signal displayed as an analogue reading of the NO_x content of the original sample.

Results of the analysis

An analysis will show whether or not the combustion is complete. For instance the presence of CO will indicate that the combustion is not complete, and if an oxygen reading is obtained this will mean that excess air has been supplied. Both CO and O_2 may appear in the analysis as a result of incomplete combustion and dissociation (see section 7.7). Quantitatively the dry product analysis can be used to calculate the A/F ratio. This method of obtaining the A/F ratio is not so reliable as direct measurement of the air consumption and fuel consumption of the engine. More caution is required when analysing the products of combustion of a solid fuel since some of the products do not appear in the flue gases (e.g. ash and unburnt carbon). The residual solid must be analysed as well in order to determine the carbon content, if any. With an engine using petrol or diesel fuel the exhaust may include unburnt particles of carbon and this quantity will not appear in the analysis. The exhaust from internal combustion engines may contain also some CH_4 and H_2 due to incomplete combustion. The CH_4 content is approximately 0.22% of the dry products and the H_2 content is of the order of half the CO content.

Considerable care is required in combustion calculations as inaccuracies are caused due to taking small differences between quantities and then using these as ratios. The use which can be made of the dry analysis is illustrated by the following problems.

Example 7.8

An analysis of the dry exhaust from an engine running on benzole shows a CO_2 content of 15%, but no CO. Assuming that the remainder of the exhaust contains only oxygen and nitrogen, calculate the A/F ratio of the engine and the dry and wet volumetric analyses of the products of combustion.

The ultimate analysis of benzole is 90% C and 10% H.

Solution

There are many different methods used in combustion analysis. The following method is recommended; its algebraic approach makes it less confusing than some other methods; it also lends itself more easily to a computer solution.

Consider 1 kg of fuel:

$$\left\{ \frac{0.9}{12}C + 0.1H \right\} + A\left(\frac{0.233}{32}O_2 + \frac{0.767}{28}N_2 \right)$$

$$\rightarrow B\{0.15CO_2 + aO_2 + (1 - 0.15 - a)N_2\} + bH_2O$$

where A is the A/F ratio on a mass basis; B the amount of substance of dry exhaust gas per kilogram of fuel, kmol/kg; a the fraction by volume of oxygen in the dry exhaust gas; b the amount of substance of steam per kilogram of fuel, kmol/kg. A brief explanation of the above equation is given below.

Left-hand side of the equation. There are $(0.9/12)$ kmol C and 0.1 kmol H in 1 kg of fuel. A kg of air per kg of fuel contains $0.233A$ kg of oxygen and hence contains $(0.233/32)$ kmol of oxygen; similarly there are $(0.767/28)$ kmol of nitrogen in A kg of air for every kilogram of fuel.

Right-hand side of the equation. There are 0.15 kmol CO_2 for 1 kmol of dry exhaust gas (given in the example), hence there are $0.15B$ kmol CO_2 per kilogram of fuel since there are B kmol dry exhaust gas per kilogram of fuel.

Similarly, there are Ba kmol oxygen in the products per kilogram of fuel; the nitrogen in the dry exhaust gas is obtained by difference, i.e. $(1 - 0.15 - a)$.

There are four unknowns in the above equation and four equations may be obtained from the mass balances on carbon, oxygen, nitrogen and hydrogen.

Carbon balance:	$0.9/12 = 0.15B$ therefore $B = 0.5$	
Hydrogen balance:	$0.1 = 2b$ therefore $b = 0.05$	
Oxygen balance:	$0.233A/32 = 0.15B + Ba + b/2$	
	i.e. $a = 0.01456A - 0.2$	[1]
Nitrogen balance:	$0.767A/28 = B(0.85 - a)$	
	i.e. $a = 0.85 - 0.05479A$	[2]

Solving equations [1] and [2] for A we have

$$0.01456A - 0.2 = 0.85 - 0.05479A$$

therefore

$$A = 15.14$$

i.e.   A/F ratio = 15.14

The value of a can then be found from either equation [1] or equation [2] as 0.02. Hence the complete dry exhaust gas volumetric analysis is given by 15% CO_2, 20% O_2, 65% N_2. The wet volumetric analysis can be found by totalling the amount of substance of wet products and taking the appropriate ratios,

i.e.   Total amount of substance of wet products

$$= (0.5 \times 0.15)CO_2 + (0.5 \times 0.2)O_2 + (0.5 \times 0.65)N_2 + 0.05 \, H_2$$

$$= 0.55 \, \text{kmol/kg fuel}$$

Then the wet volumetric analysis is as follows:

$$\frac{0.5 \times 0.15}{0.55} \times 100 = 13.64\% \, CO_2, \qquad \frac{0.5 \times 0.2}{0.55} \times 100 = 18.18\% \, O_2,$$

$$\frac{0.5 \times 0.65}{0.55} \times 100 = 59.09\% \, N_2, \qquad \frac{0.05}{0.55} \times 100 = 9.09\% \, H_2$$

Example 7.9   The analysis of the dry exhaust gas from an internal combustion engine gave 12% CO_2, 2% CO, 4% CH_4, 1% H_2, 4.5% O_2, 76.5% N_2. Calculate the proportions by mass of carbon to hydrogen in the fuel, assuming it to be a pure hydrocarbon, and the A/F ratio used.

Solution   Let the unknown mass fraction of carbon in the fuel be x kg per kilogram of fuel. Then we can write the combustion equation as follows:

$$\frac{x}{12}C + (1 - x)H + A\left(\frac{0.233}{32}O_2 + \frac{0.767}{28}N_2\right)$$

$$\rightarrow B\{0.12CO_2 + 0.02CO + 0.04CH_4$$

$$+ 0.01H_2 + 0.045O_2 + 0.765N_2\} + aH_2O$$

As in Example 7.8 there are four unknowns and four possible equations.

Carbon balance:	$x/12 = B(0.12 + 0.02 + 0.04)$	
	$x = 2.16B$	[1]
Hydrogen balance:	$(1 - x) = (4 \times 0.04)B + (2 \times 0.01)B + 2a$	
	$a = 0.5 - 0.5x - 0.09B$	[2]
Oxygen balance:	$\dfrac{0.233A}{32} = B\left(0.12 + \dfrac{0.02}{2} + 0.045\right) + \dfrac{a}{2}$	
	$a = 0.01456A - 0.35B$	[3]
Nitrogen balance:	$0.767A/28 = 0.765B$	
	$A = 27.927B$	[4]

Equating [2] and [3] we have

$$0.5 - 0.5x - 0.09B = 0.01456A - 0.35B$$
$$A = 34.341 - 34.341x + 17.857B$$

Then using equation [4]

$$34.341 - 34.341x + 17.857B = 27.927B$$
$$B = 3.41 - 3.41x$$

Substituting in [1]

$$x = 7.366 - 7.366x$$

i.e. $x = 0.8805$

The composition of the fuel is therefore 88.05% C, 11.95% H.

Also, $B = 3.41 \times 0.1195 = 0.4075$

Then in equation [4],

$$A = 27.927 \times 0.4075 = 11.38$$

i.e.   A/F ratio $= 11.38$

7.7 Dissociation

It is found that during adiabatic combustion the maximum temperature reached is lower than that expected on the basis of elementary calculation. One important reason for this is that the *exothermic* combustion process can be reversed to some extent if the temperature is high enough. The reversed process is an *endothermic* one, i.e. energy is absorbed. In a real process the reaction proceeds in both directions simultaneously and chemical equilibrium is reached when the rate of break-up of product molecules is equal to their rate of formation. This is represented for the combustion of carbon monoxide and hydrogen respectively by the equations

$$2CO + O_2 \rightleftharpoons 2CO_2 \quad \text{and} \quad 2H_2 + O_2 \rightleftharpoons 2H_2O$$

Both of these reactions can take place simultaneously in the same combustion process. The proportions of the constituents adjust themselves to satisfy the equilibrium conditions and their actual values depend on the particular pressure and temperature. The presence of CO and H_2 means that there is further energy to be released on their reaction with O_2 so the maximum temperature reached cannot be as high as that expected on the basis of complete combustion. As the combustion proceeds and the temperature level falls, due to expansion and/or subsequent heat loss, the amount of dissociation decreases (it is significant at temperatures >1500 K) and combustion proceeds to completion. However, since the energy is released at lower temperatures and, in a positive expansion cylinder, at a lower effective compression ratio the efficiency of the process is reduced.

The condition of equilibrium during a reversible combustion process can be studied by means of a conceptual device known as the 'Van't Hoff equilibrium box' as shown in Fig. 7.5. Consider the general reversible combustion process

$$a \text{ kmol } A + b \text{ kmol } B \rightleftharpoons c \text{ kmol } C + d \text{ kmol } D$$

Fig. 7.5 Van't Hoff equilibrium box

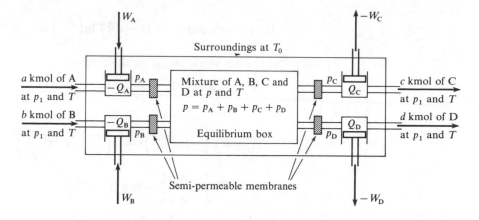

which occurs at a fixed temperature T and a pressure, p, in the equilibrium box. The reactants A and B are each initially at p_1 and T and the products C and D are each finally at p_1 and T. As the process is reversible some energy transfer will take place in the form of work and this is allowed for by the inclusion of isothermal compressors and expanders. The equilibrium box contains a mixture of gases A, B, C and D at total pressure p and temperature T and to allow reversible mass flow of the constituents the pressure of each constituent at entry to the box must be equal to its partial pressure in the box. The pressure adjustments are made by the isothermal expanders and compressors and each constituent enters or leaves through a *semi-permeable membrane*. Some substances permit one gas to pass through but prevent other gases, e.g. a glowing aluminium sheet allows hydrogen to pass through but not other gases. It is assumed here that such substances are available for gases A, B, C and D.

The process may proceed equally well in either direction but it is illustrated here as going from left to right in the equation and in Fig. 7.5. With a reversal of the process the heat and work transfers would be reversed in direction.

The work input during an isothermal expansion by a perfect gas between states 1 and 2 is given by

$$W = mRT \ln\left(\frac{p_2}{p_1}\right) = n\tilde{R}T \ln\left(\frac{p_2}{p_1}\right)$$

and this can be applied to each of the compressors and expanders in the system of Fig. 7.5

$$W_A = \text{work input on A} = a\tilde{R}T \ln\left(\frac{p_A}{p_1}\right) = \tilde{R}T \ln\left(\frac{p_A}{p_1}\right)^a$$

$$W_B = \text{work input on B} = b\tilde{R}T \ln\left(\frac{p_B}{p_1}\right) = \tilde{R}T \ln\left(\frac{p_B}{p_1}\right)^b$$

$$W_C = \text{work input on C} = c\tilde{R}T \ln\left(\frac{p_1}{p_C}\right) = -\tilde{R}T \ln\left(\frac{p_C}{p_1}\right)^c$$

$$W_D = \text{work input on D} = d\tilde{R}T\ln\left(\frac{p_1}{p_D}\right) = -\tilde{R}T\ln\left(\frac{p_D}{p_1}\right)^d$$

Therefore the net work *output* of the system

$$-W = -W_A - W_B - W_C - W_D$$

i.e.

$$-W = \tilde{R}T\left\{-\ln\left(\frac{p_A}{p_1}\right)^a - \ln\left(\frac{p_B}{p_1}\right)^b + \ln\left(\frac{p_C}{p_1}\right)^c + \ln\left(\frac{p_D}{p_1}\right)^d\right\}$$

$$= \tilde{R}T\left\{\ln\frac{p_C^c p_D^d}{p_A^a p_B^b} + \ln p_1^{a+b-c-d}\right\}$$

Suppose that in a second similar system in the same surroundings the pressure in the equilibrium box is p' then it will have a net work output, $-W'$, given by

$$-W' = \tilde{R}T\left\{\ln\frac{(p'_C)^c (p'_D)^d}{(p'_A)^a (p'_B)^b} + \ln p_1^{a+b-c-d}\right\}$$

where $p' = p'_A + p'_B + p'_C + p'_D$

It is proposed that $-W \neq -W'$ and this statement is to be investigated. Suppose $-W > -W'$, then the second system can be reversed, as shown in Fig. 7.6, and a single system formed by using the work output from the first system, $-W$, to provide the work input for the second system reversed.

Fig. 7.6 Hypothetical (impossible) combination of two systems

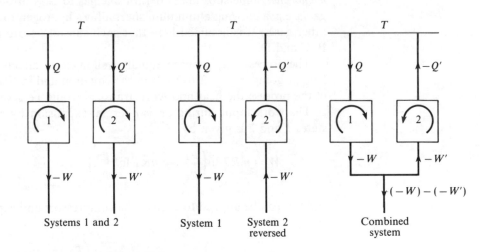

The result is a single system giving a net output of work $(-W) - (-W')$ while exchanging heat with a single source at temperature T. This is a contradiction of the Second Law of Thermodynamics, thus the proposition that $W \neq W'$ is not true so that $W = W'$, therefore

$$\frac{p_C^c p_D^d}{p_A^a p_B^b} = \frac{(p'_C)^c (p'_D)^d}{(p'_A)^a (p'_B)^b} = K$$

where K is the *thermal equilibrium* or *dissociation* constant and has been shown to be independent of the pressure in the equilibrium box. A standard thermodynamic equilibrium constant, $K^\ominus$, can be defined in dimensionless form by referring each partial pressure to a pressure of 1 bar,

i.e.
$$K^\ominus = \ln\left(\frac{p_C}{p^\ominus}\right)^c + \ln\left(\frac{p_D}{p^\ominus}\right)^d - \ln\left(\frac{p_A}{p^\ominus}\right)^a - \ln\left(\frac{p_B}{p^\ominus}\right)^b$$

or in general

$$\ln K^\ominus = \sum_i \ln(p_i/p^\ominus)^{\gamma_i} \tag{7.10}$$

where γ_i is the stoichiometric coefficient, taken as positive for the products and negative for the reactants. The thermodynamic equilibrium constant $K^\ominus$ is a function of temperature and values of $\ln K^\ominus$ are tabulated against temperature for each reaction equation, see ref. 7.6. As the partial pressures of the constituents are proportional to the molar proportions then $K^\ominus$ is an indication of the ratio of products to reactants and so is a measure of the amount of dissociation. If $K^\ominus$ is large then the proportion of product is high and the amount of dissociation is small.

The above general expression for $K^\ominus$, equation (7.10), will now be applied to particular important reactions. For example the combustion of carbon monoxide to carbon dioxide can be written as

$$CO + \tfrac{1}{2}O_2 \rightleftharpoons CO_2$$

with molar proportions for CO, O_2 and CO_2 of 1, 0.5, and 1. Thus K can be written as

$$\ln K^\ominus = \ln\left\{\frac{p_{CO_2}(p^\ominus)^{1/2}}{p_{CO}(p_{O_2})^{1/2}}\right\} \tag{7.11}$$

For the combustion of hydrogen the equation is

$$H_2 + \tfrac{1}{2}O_2 \rightleftharpoons H_2O$$

with molar proportions 1, 0.5, and 1. The equilibrium constant then becomes

$$K^\ominus = \frac{p_{H_2O}(p^\ominus)^{1/2}}{p_{H_2}(p_{O_2})^{1/2}} \tag{7.12}$$

In the combustion of hydrocarbon fuels both of the above reactions may occur simultaneously and another equilibrium constant can be defined by dividing

$$\frac{p_{H_2O}(p^\ominus)^{1/2}}{p_{H_2}(p_{O_2})^{1/2}} \quad \text{by} \quad \frac{p_{CO_2}(p^\ominus)^{1/2}}{p_{CO}(p_{O_2})^{1/2}}$$

giving
$$K^\ominus = \frac{p_{H_2O}\,p_{CO}}{p_{H_2}\,p_{CO_2}} \tag{7.13}$$

which is also tabulated, see ref. 7.6, and can be used to form another equation in the analysis.

It is readily seen that dissociation as described introduces an added complexity into the analysis of the combustion process but the complication does not end there. The conditions of equilibrium must be satisfied at any particular temperature and also the energy balance for the process must be satisfied. The temperature reached will depend on the amount of fuel burned, the proportions of the constituents and their thermodynamic properties, all of which are also dependent on pressure or temperature. Thus several conditions have to be satisfied before any particular state in the process can be determined and it is necessary to establish a sufficient number of equations to complete the analysis. This is discussed more fully in section 7.8 and for the immediate purpose it will be assumed that the energy requirements have been met in the process and only the dissociation effects are required.

The analysis of combustion may be extended to include the formation of nitric oxide, $\frac{1}{2}N_2 + \frac{1}{2}O_2 \rightleftharpoons NO$, which occurs at high temperature and the dissociation of H_2O vapour into hydrogen and hydroxyl, $\frac{1}{2}H_2 + OH \rightleftharpoons H_2O$, as well as into its constituent gases. There may also be the dissociation of molecules of oxygen, hydrogen and nitrogen into atoms. These aspects of combustion are detailed and involve small proportions of the charge. It has been stated previously that recombination occurs as the temperature falls so that combustion is completed with a loss in efficiency of the cycle such that less work output is obtained than expected. The completion of combustion requires the absence of low-temperature quenching conditions, which may arrest the process, and a sufficiency of time for the reaction to be completed.

The importance of the analysis of combustion increases with advancing engine technology as typified in the petrol engine. This engine includes the extremes of combustion conditions starting from a complex chemical fuel and a mixture of air, water vapour and residual exhaust gas. The pressure and temperature levels passed through are large and the duration of a cycle is only a fraction of a second. The cylinders are water-cooled usually and sudden exhausting of gas is provided for. Under these conditions the exhaust gas can have a complex analysis and some of the constituents are responsible for the polluting of the atmosphere with undesirable and irritating results.

Example 7.10 A combustible mixture of carbon monoxide and air which is 10% rich is compressed to a pressure of 8.28 bar and a temperature of 282 °C. The mixture is ignited and combustion occurs adiabatically at constant volume. When the maximum temperature is attained analysis shows 0.228 kmol of CO present for 1 kmol of CO supplied. Show that the maximum temperature reached is 2695 °C.

Solution For stoichiometric conditions

$$CO + \tfrac{1}{2}O_2 + (3.76 \times \tfrac{1}{2}N_2) \rightarrow CO_2 + (3.76 \times \tfrac{1}{2}N_2)$$

$$CO + 0.5O_2 + 1.88N_2 \rightarrow CO_2 + 1.88N_2$$

$$\text{Actual A/F ratio} = \text{stoichiometric A/F ratio} \times \frac{100}{110}$$

$$= \text{stoichiometric A/F ratio}/1.1$$

Therefore the actual reactants are

$$CO + (0.5O_2 + 1.88N_2)/1.1$$

With dissociation there will be some break up of CO_2 giving CO and O_2 in the products such that

$$CO + (0.5O_2 + 1.88N_2)/1.1 = aCO_2 + bCO + cO_2 + (1.88/1.1)N_2$$

The question states that $b = 0.228$, therefore

$$CO + 0.455O_2 + 1.709N_2 \rightarrow aCO_2 + 0.228CO + cO_2 + 1.709N_2$$

Carbon balance: $1 = a + 0.228$ therefore $a = 0.772$
Oxygen balance: $1 + (2 \times 0.455) = 2a + 0.228 + 2c$ therefore $c = 0.069$
For the reaction $CO + \frac{1}{2}O_2 \rightleftharpoons CO_2$

$$K^\ominus = \frac{p_{CO_2}(p^\ominus)^{1/2}}{p_{CO}(p_{O_2})^{1/2}}$$

and $\quad p_{CO_2} = \dfrac{a}{n_2} p_2, \qquad p_{CO} = \dfrac{b}{n_2} p_2, \qquad p_{O_2} = \dfrac{c}{n_2} p_2$

therefore

$$K^\ominus = \frac{a}{b} \sqrt{\left\{ \frac{n_2 p^\ominus}{c p_2} \right\}} \qquad\qquad [1]$$

where $p_2 =$ the total pressure at the required temperature and $n_2 =$ total amount of substance of products

$$n_2 = a + b + c + 1.709 = 0.772 + 0.228 + 0.069 + 1.709$$

$$= 2.778 \text{ kmol}$$

At ignition

$$p_1 = 8.28 \text{ bar} \quad \text{and} \quad T_2 = 273 + 282 = 555 \text{ K}$$

$$p_1 V = n_1 \tilde{R} T_1 \quad \text{and} \quad p_2 V = n_2 \tilde{R} T_2 \quad \text{and} \quad V = \text{constant}$$

therefore

$$p_2 = p_1 \frac{n_2}{n_1} \frac{T_2}{T_1}$$

where $n_1 =$ amount of substance of reactants $= 1 + 0.455 + 1.709 = 3.164$. Then, assuming that $T_2 = 273 + 2695 = 2968$ K, we have

$$p_2 = 8.28 \times \frac{2.778}{3.164} \times \frac{2968}{555} = 38.88 \text{ bar}$$

Substituting in equation [1],

$$K^\ominus = \frac{0.772}{0.228} \sqrt{\left\{ \frac{2.778 \times 1}{0.069 \times 38.88} \right\}} = 3.446$$

therefore

$$\ln K^{\ominus} = 1.237$$

From tables, ref. 7.6, it is seen by interpolation that $\ln K^{\ominus} = 1.235$ for this reaction at 2968 K showing the assumed value to be true. The corresponding pressure was calculated at 38.88 bar.

Example 7.11   A mixture of heptane (C_7H_{16}) and air which is 10% rich is initially at a pressure of 1 bar and temperature 100 °C, and is compressed through a volumetric ratio of 6 to 1. It is ignited and adiabatic combustion proceeds at constant volume. The maximum temperature reached is 2627 °C and at this temperature the equilibrium constants are

$$\frac{p_{H_2O}p_{CO}}{p_{CO_2}p_{H_2}} = 7.01 \quad \text{and} \quad \frac{p_{CO_2}(p^{\ominus})^{1/2}}{p_{CO}(p_{O_2})^{1/2}} = 4.49$$

If the constituents of the gas are CO_2, CO, H_2O, H_2, O_2 and N_2 show that approximately 30% of the carbon has burned incompletely.

Solution   The processes are shown in Fig. 7.7 and for a 10% rich mixture

$$C_7H_{16} + \frac{10}{11}\{11O_2 + (11 \times 3.76)N_2\}$$
$$\rightarrow aCO_2 + bCO + cH_2O + dH_2 + eO_2 + (10 \times 3.76)N_2$$

$$C_7H_{16} + 10O_2 + 37.6N_2$$
$$\rightarrow aCO_2 + bCO + cH_2O + dH_2 + eO_2 + 37.6N_2$$

Fig. 7.7
Pressure–volume
diagram for
Example 7.11

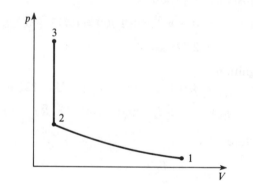

Then,

Carbon balance:	$a + b = 7$	[1]
Hydrogen balance:	$c + d = 8$	[2]
Oxygen balance:	$a + \dfrac{b}{2} + \dfrac{c}{2} + e = 10$	[3]

Amount of substance initially, $n_1 = 1 + 10 + 37.6 = 48.6$; amount of substance finally, $n_3 = a + b + c + d + e + 37.6$. Also,

$$p_1 V_1 = n_1 \tilde{R} T_1 \quad \text{and} \quad p_3 V_3 = n_3 \tilde{R} T_3$$

and combining these gives

$$\frac{p_3}{n_3} = \frac{p_1 \, V_1 \, T_3}{n_1 \, V_3 \, T_1} = \frac{1}{48.6} \times \frac{6}{1} \times \frac{2900}{373} = 0.958 \text{ bar/kmol}$$

where $T_1 = 273 + 100 = 373$ K, and $T_2 = 273 + 2627 = 2900$ K.
The partial pressures are given by equation 6.14 as

$$p_{CO_2} = \frac{a}{n_3} p_3 \qquad p_{CO} = \frac{b}{n_3} p_3$$

$$p_{H_2O} = \frac{c}{n_3} p_3 \qquad p_{H_2} = \frac{d}{n_3} p_3 \qquad p_{O_2} = \frac{e}{n_3} p_3$$

The example gives

$$\frac{p_{CO} \, p_{H_2O}}{p_{CO_2} \, p_{H_2}} = 7.01 \qquad \frac{p_{CO_2}(p^{\ominus})^{1/2}}{p_{CO}(p_{O_2})^{1/2}} = 4.49$$

therefore

$$\frac{bc}{ad} = 7.01 \qquad\qquad\qquad\qquad\qquad\qquad [4]$$

and $\qquad \dfrac{a}{b\,e^{1/2}} \left(\dfrac{n_3}{p_3}\right)^{1/2} = 4.49 \qquad\qquad\qquad\qquad [5]$

The proportion of carbon burned incompletely is given as 0.3, so that 0.3 of the initial 7 atoms of carbon burn to give b CO.

i.e. $\qquad b/7 = 0.3 \quad$ therefore $b = 2.1$

Then from equation [1], $a = 7 - 2.1 = 4.9$. Substituting in [4] gives

$$\frac{c}{d} = 7.01 \times \frac{a}{b} = 7.01 \times \frac{4.9}{2.1} = 16.36$$

From equation [2]

$$c + d = 8 \quad \text{therefore } 16.36d + d = 8 \qquad d = \frac{8}{17.36} = 0.461$$

and $\qquad c = 16.36 \times 0.461 = 7.539$

From equation [3]

$$e = 10 - a - \frac{b}{2} - \frac{c}{2} = 10 - 4.9 - \frac{2.1}{2} - \frac{7.539}{2} = 0.280$$

substituting in equation [5] gives

$$\frac{a}{b\,e^{1/2}}\left(\frac{n_3}{p_3}\right)^{1/2} = \frac{4.9}{2.1}\left(\frac{1}{0.28 \times 0.958}\right)^{1/2} = 4.51$$

which gives sufficient agreement to the 4.49 quoted showing that approximately 30% of carbon was burned to CO.

7.8 Internal energy and enthalpy of reaction

Previous consideration of the combustion process has not included the energy released during the process and final temperatures attained. It is evident, however, that such a process must obey the First Law of Thermodynamics. Applications of this law to other processes have been for pure substances, or those that can be considered to be so, with the stipulation that their thermodynamic state is defined by two independent properties. In the type of process now considered there is the potential chemical energy of the fuel to be included which is released during the change from reactants to products.

It is an experimental fact that the energy released on the complete combustion of unit mass of a fuel depends on the temperature at which the process is carried out. Thus such quantities quoted are related to temperature. It will be shown that if the energy release is known for a fuel at one temperature it can be calculated at other temperatures.

The combustion process is defined as taking place from reactants at a state identified by the reference temperature T_0 and another property, either pressure or volume, to products at the same state. If the process is carried out at constant volume then the non-flow energy equation, $Q + W = (U_2 - U_1)$, can be applied to give

$$Q = U_{P_0} - U_{R_0} \quad \text{or} \quad -Q = U_{R_0} - U_{P_0} \tag{7.14}$$

where $W = 0$ for constant volume combustion, $U_1 = U_{R_0}$ the internal energy of the reactants which is a mixture of fuel and air at T_0, and $U_2 = U_{P_0}$ the internal energy of the products of combustion at T_0.

The change in internal energy does not depend on the path between the two states but only on the initial and final values and is given by the quantity, $-Q$, the heat transferred to the surroundings during the process. This is illustrated in Fig. 7.8 and also the property diagram of Fig. 7.9.

Fig. 7.8 Combustion at reference conditions

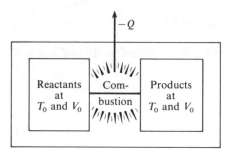

The heat supplied, Q, is called the *internal energy* of *reaction at* T_0 and is denoted by ΔU_0,

i.e. $\qquad Q = \Delta U_0 = U_{P_0} - U_{R_0}$ (7.15)

The molar internal energy of reaction at a standard pressure of 1 bar is defined as $\Delta \tilde{u}^\ominus = \Delta U_0/n$, where n is the amount of substance of the fuel.

As the internal energy of the reactants includes the potential chemical energy,

Fig. 7.9 $U-T$ diagram for combustion at a reference temperature

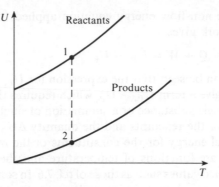

and since heat is transferred *from* the system, it is evident that as defined ΔU_0 is a negative quantity.

For real constant volume combustion processes the initial and final temperatures will be different from the reference temperature T_0. For analytical purposes the change in internal energy between reactants at state 1 to products at state 2 can be considered in three stages:

(a) the change for the reactants from state 1 to the reference temperature T_0;
(b) the constant volume combustion process from reactants to products at T_0;
(c) the change for the products from T_0 to state 2.

The complete process can be conceived as taking place in a piston-cylinder device as indicated in Fig. 7.10.

Fig. 7.10 Combustion between states 1 and 2

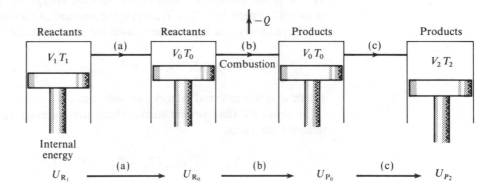

Thus the change in internal energy between states 1 and 2, $(U_2 - U_1)$, can be written more explicitly as, $(U_{P_2} - U_{R_1})$, to show the chemical change involved and this can be further expanded for analytical purposes:

$$U_{P_2} - U_{R_1} = (U_{P_2} - U_{P_0}) + (U_{P_0} - U_{R_0}) + (U_{R_0} - U_{R_1})$$

therefore

$$U_{P_2} - U_{R_1} = \underset{\substack{\text{Products} \\ \text{(c)}}}{(U_{P_2} - U_{P_0})} + \underset{\text{(b)}}{\Delta U_0} + \underset{\substack{\text{Reactants} \\ \text{(a)}}}{(U_{R_0} - U_{R_1})} \qquad (7.16)$$

The non-flow energy equation applied to a process involving combustion and work gives

$$Q + W = U_{P_2} - U_{R_1}$$

It can be seen that the expression for $U_{P_2} - U_{R_1}$ has been conveniently split up to give a term $U_{P_2} - U_{P_0}$ which requires the product mixture to be regarded as a single substance or a summation of single substance constituents, a similar term for the reactants and the quantity ΔU_0 previously defined. The values of internal energy for the constituents of the mixtures remain to be determined. These are functions of temperature and the most accurate method is to use tabulated values such as those of ref. 7.6. In some cases it is a good approximation to calculate changes in internal energy assuming the gaseous constituents to be perfect using an average value of c_v for the temperature range involved. If the temperature range is T_1 to T_2 then the value of c_v at $T = (T_1 + T_2)/2$ can be used, this assumes a linear change in c_v with temperature, but if the temperature range is large the result may not be accurate enough and tabulated values of the properties are required. The tables of ref. 7.6 give the values of $\tilde{u}$ and $\tilde{h}$, with $\tilde{h} = 0$ at the normal reference temperature of $25\,°C = 298.15$ K; $\tilde{u}$ and $\tilde{h}$ are virtually independent of pressure.

The changes in internal energy for the reactants ($U_{R_0} - U_{R_1}$) and for the products ($U_{P_2} - U_{P_0}$) can be calculated from the following expressions:

$$U_{R_0} - U_{R_1} = \sum_{R} n_i(\tilde{u}_{i_0} - \tilde{u}_{i_1}) \tag{7.17}$$

where Σ_R denotes the summation for all the constituents of the reactants denoted by i, $\tilde{u}_i$ is the tabulated value of the internal energy for the constituent at the required temperature T_0 or T_1, and n_i the amount of substance of the constituent.

Alternatively if a mass base is used for the tabulated values or calculation

$$U_{R_0} - U_{R_1} = \sum_{R} m_i(u_{i_0} - u_{i_1}) \tag{7.18}$$

where u_i is the internal energy per unit mass.

In terms of the specific heats which are average values for the required temperature range

$$U_{R_0} - U_{R_1} = \sum_{R} m_i c_{v_i}(T_0 - T_1) = (T_0 - T_1) \sum_{R} m_i c_{v_i} \tag{7.19}$$

and similar expressions for the products are

$$U_{P_2} - U_{P_0} = \sum_{P} n_i(\tilde{u}_{i_2} - \tilde{u}_{i_0}) \tag{7.20}$$

$$U_{P_2} - U_{P_0} = \sum_{P} m_i(u_{i_2} - u_{i_0}) \tag{7.21}$$

$$U_{P_2} - U_{P_0} = \sum_{P} m_i c_{v_i}(T_2 - T_0) = (T_2 - T_0) \sum_{P} m_i c_{v_i} \tag{7.22}$$

Note that $n_i \tilde{c}_{v_i} = m_i c_{v_i}$.

The process has been analysed on the basis of a non-flow process which involves combustion at constant volume. A similar analysis can be made for a steady-flow or constant pressure combustion process in which the changes in enthalpy are important

$$H_{P_2} - H_{R_1} = (H_{P_2} - H_{P_0}) + \Delta H_0 + (H_{R_0} - H_{R_1}) \qquad (7.23)$$
$$\text{Products} \qquad\qquad\qquad \text{Reactants}$$

where $\Delta H_0 = $ *enthalpy of reaction* at T_0 and

$$\Delta H_0 = H_{P_0} - H_{R_0} \text{ and is always negative} \qquad (7.24)$$

The molar enthalpy of reaction is defined as $\Delta \tilde{h}^{\ominus} = \Delta H_0 / n$ at 1 bar where n is the amount of substance of the fuel.

The expressions for the change in enthalpy of reactants and products are

$$H_{R_0} - H_{R_1} = \sum_R n_i (\tilde{h}_{i_0} - \tilde{h}_{i_1}) \qquad (7.25)$$

$$H_{R_0} - H_{R_1} = \sum_R m_i (h_{i_0} - h_{i_1}) \qquad (7.26)$$

and if mean specific heats are used

$$H_{R_0} - H_{R_1} = \sum_R m_i c_{p_i} (T_0 - T_1) = (T_0 - T_1) \sum_R m_i c_{p_i} \qquad (7.27)$$

$$H_{P_2} - H_{P_0} = \sum_P n_i (\tilde{h}_{i_2} - \tilde{h}_{i_0}) \qquad (7.28)$$

$$H_{P_2} - H_{P_0} = \sum_P m_i (h_{i_2} - h_{i_0}) \qquad (7.29)$$

and if mean specific heats are used

$$H_{P_2} - H_{P_0} = \sum_P m_i c_{p_i} (T_2 - T_0) = (T_2 - T_0) \sum_P m_i c_{p_i} \qquad (7.30)$$

Note that $n_i \tilde{c}_{p_i} = m_i c_{p_i}$.

Values of molar enthalpy, $\Delta \tilde{h}^{\ominus}$, referred to a pressure of 1 bar are given in ref. 7.6 for some common reactions.
For example,

Reaction	$\Delta \tilde{h}^{\ominus}$ at 25 °C (298.15 K) (kJ/kmol)
$C(sol) + O_2 \rightarrow CO_2$	$-393\,520$
$CO + \frac{1}{2}O_2 \rightarrow CO_2$	$-282\,990$
$C_6H_6(vap) + 7\frac{1}{2}O_2 \rightarrow 6CO_2 + 3H_2O(liq)$	$-3\,301\,397$

It will be noted that the state of the fuel is given if it is solid (sol), liquid (liq) or vapour (vap) if this is required. If H_2O is a product of the combustion then it is necessary to know the state, liquid or vapour, at the end of the process by which $\Delta \tilde{h}^{\ominus}$ was determined. If $\Delta \tilde{h}_0$ is known for a particular fuel with the H_2O formed in the liquid state the value of $\Delta \tilde{h}_0$ with the H_2O in the vapour state can be calculated.

Example 7.12

For benzene vapour (C_6H_6) at 25 °C $\Delta\tilde{h}_0$ is $-3\,301\,397$ kJ/kmol with the H_2O in the liquid phase. Calculate $\Delta\tilde{h}_0$ for the H_2O in the vapour phase.

Solution

If the H_2O remains as a vapour the heat transferred to the surroundings will be less than that when the vapour condenses, by the amount due to the change in enthalpy of the vapour during condensation at the reference temperature.

$$\Delta\tilde{h}_0(\text{vap}) = \Delta\tilde{h}_0(\text{liq}) + m_s h_{fg_0}$$

where m_s is the mass of H_2O formed for 1 kmol of fuel; h_{fg_0} is the change in enthalpy of steam between saturated liquid and saturated vapour at the reference temperature T_0 and is 2441.8 kJ/kg at 25 °C.

For the reaction

$$C_6H_6 + 7\tfrac{1}{2}O_2 \rightarrow 6CO_2 + 3H_2O$$

Therefore 3 kmol of H_2O are formed on combustion of 1 kmol of C_6H_6; 3 kmol of $H_2O = 3 \times 18 = 54$ kg H_2O.

$$\Delta\tilde{h}_0(\text{vap}) = -3\,301\,397 + (54 \times 2441.8)$$

$$= -3\,169\,540 \text{ kJ/kmol}$$

In the equations for the change in internal energy and enthalpy for reactants and products, and for ΔU_0 and ΔH_0, if a change in state takes place (e.g. from liquid fuel to vapour), a term describing this process must be included.

Nothing has been said of the air–fuel ratio for combustion during the determination of ΔU_0 or ΔH_0. Consideration will show that this does not matter provided there is sufficient air to ensure complete combustion. Excess oxygen, like the nitrogen present, starts and finishes at the reference temperature T_0 and so suffers no change in internal energy or enthalpy, thus not affecting ΔU_0 or ΔH_0.

From the definition of the enthalpy of a perfect gas

$$H = U + pV = U + n\tilde{R}T$$

So if we are concerned only with gaseous mixtures in the reaction then for products and reactants

$$H_{P_0} = U_{P_0} + n_P \tilde{R}_0 T_0 \quad \text{and} \quad H_{R_0} = U_{R_0} + n_R \tilde{R}_0 T_0$$

where n_P and n_R are the amounts of substance of products and reactants respectively, and the temperature is the reference temperature T_0.

Then, using equations (7.15) and (7.24), we have

$$\Delta H_0 = \Delta U_0 + (n_P - n_R)\tilde{R}T_0 \tag{7.31}$$

If there is no change in the amount of substance during the reaction, or if the reference temperature is absolute zero, then ΔH_0 and ΔU_0 will be equal.

Example 7.13

Calculate the specific internal energy of reaction for the combustion of benzene (C_6H_6) vapour at 25 °C given that $\Delta\tilde{h}_0 = -3\,169\,540$ kJ/kmol and the H_2O is in the vapour phase.

Solution The combustion equation is

$$C_6H_6 + 7\tfrac{1}{2}O_2 \rightarrow 6CO_2 + 3H_2O(vap)$$

$$n_R = 1 + 7.5 = 8.5, \qquad n_P = 6 + 3 = 9$$

From equation (7.31)

$$\Delta U_0 = \Delta H_0 - (n_P - n_R)\tilde{R}T_0$$

$$= -(3\,169\,540 \times 1) - \left\{(9 - 8.5) \times 8.3145 \times 298\right\}$$

where $T_0 = 273 + 25 = 298$ K

i.e. $\Delta U_0 = -3\,169\,540 - 1239 = -3\,170\,779$ kJ

(note that ΔU_0 is negligibly different from ΔH_0)

$$1\ \text{kmol of}\ C_6H_6 = (6 \times 12) + (6 \times 1) = 78\ \text{kg}$$

therefore

$$\Delta u_0 = -\frac{3\,170\,779}{78} = -40\,651\ \text{kJ/kg}$$

Change in reference temperature

It has already been mentioned that the internal energy and enthalpy of reaction depend on the temperature at which the reaction occurs. This is due to the change in enthalpy and internal energy of the reactants and products with temperature.

It can be seen from the property diagram of Fig. 7.11 that the enthalpy of reaction at temperature T, ΔH_T can be obtained from ΔH_0 at T_0 by the relationship

$$-\Delta H_T = -\Delta H_0 + (H_{R_T} - H_{R_0}) - (H_{P_T} - H_{P_0}) \qquad (7.32)$$

where $H_{R_T} - H_{R_0}$ = increase in enthalpy of the reactants from T_0 to T

and $H_{P_T} - H_{P_0}$ = increase in enthalpy of the products from T_0 to T

Fig. 7.11 $H-T$ diagram for combustion

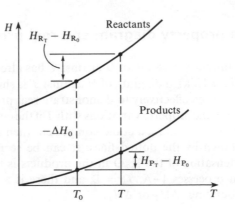

213

Example 7.14 For carbon monoxide at 60 °C, $\Delta\tilde{h}$ is given as $-282\,990$ kJ/kmol. Calculate $\Delta\tilde{h}$ at 2800 K given that the molar enthalpies of the gases concerned are as given in Table 7.10.

Table 7.10 Molar enthalpies of gases for Example 7.14

Gas	Molar enthalpy/(kJ/kmol)	
	At 60 °C	At 2800 K
Carbon monoxide	1018	86 115
Oxygen	1036	90 144
Carbon dioxide	1368	140 440

Solution The reaction equation is

$$CO + \tfrac{1}{2}O_2 \rightarrow CO_2$$

Also, taking values from Table 7.10, and using equations (7.25) and (7.28)

$$H_{R_0} = (1 \times 1018) + (0.5 \times 1036) = 1536 \text{ kJ}$$

$$H_{R_T} = (1 \times 86\,115) + (0.5 \times 90\,144) = 131\,187 \text{ kJ}$$

$$H_{P_0} = 1 \times 1368 = 1368 \text{ kJ}$$

$$H_{P_T} = 1 \times 140\,440 = 140\,440 \text{ kJ}$$

Then using equation (7.32)

$$-\Delta H_T = -\Delta H_0 + (H_{R_T} - H_{R_0}) - (H_{P_T} - H_{P_0})$$

$$= (1 \times 282\,990) + (131\,187 - 1536) - (140\,440 - 1368)$$

$$= 273\,569 \text{ kJ}$$

i.e. $\Delta\tilde{h} = -273\,569$ kJ/kmol CO

The property diagram and real processes

The property diagram of U against T has already been referred to and is shown in Fig. 7.12(a); a diagram of H against T is shown in Fig. 7.12(b). The diagrams can be used effectively to demonstrate real processes of interest. The solid lines indicate the property variations with T if the constituents are gaseous. If reactants or products contain a liquid component then the property lines will be modified as shown by the dotted lines. It can be seen by inspection that the effect of condensation of the H_2O in the products is to increase ΔU_0 and ΔH_0.

In processes 1–A, $T_A = T_0$ and there is a maximum energy transfer to the surroundings ΔU_T or ΔH_T.

Fig. 7.12 $U-T$ and $H-T$ diagrams for combustion

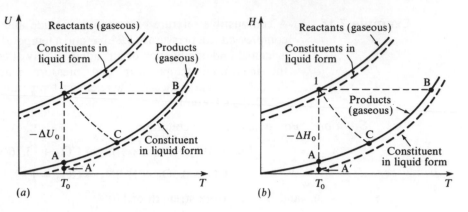

(a)

(b)

In processes 1–B the internal energy, or enthalpy, initially and finally is the same so that the increase in temperature is a maximum and the combustion process is adiabatic.

In 1–C the processes are general with heat transfer and possibly work transfer.

Process 1–A′ in Fig. 7.12(a) corresponds to the constant volume bomb calorimeter test and in Fig. 7.12(b) 1–A′ corresponds to the steady-flow combustion Boys' calorimeter test. Both of these tests are discussed later in section 7.12.

In a general non-flow or steady-flow process the initial state (1) and final state (2) will be different and neither will be at the reference temperature T_0. The quantities of interest are $U_2 - U_1$, and $H_2 - H_1$ for the respective processes. It can be seen readily from Fig. 7.13 that

$$U_2 - U_1 = -(U_1 - U_2) = -\{-\Delta U_0 - (U_{R_0} - U_{R_1}) - (U_{P_2} - U_{P_0})\}$$

$$U_2 - U_1 = U_{P_2} - U_{R_1} = (U_{P_2} - U_{P_0}) + \Delta U_0 + (U_{R_0} - U_{R_1})$$

Fig. 7.13
Energy-absolute temperature diagram between stages 1 and 2

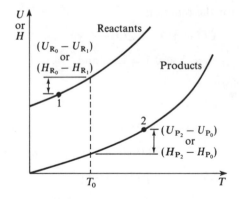

This is equation (7.16) repeated. A corresponding expression is obtained for $H_2 - H_1$.

The principles dealt with in this section will now be applied to typical problems.

215

Example 7.15   A combustible mixture of carbon monoxide and air which is 10% rich is compressed to a pressure of 8.28 bar and a temperature of 282 °C. The mixture is ignited and combustion occurs adiabatically at constant volume. Calculate the maximum temperature and pressure reached assuming that no dissociation takes place. At the reference temperature of 25 °C, $\Delta \tilde{h}_0^\ominus$ for CO is $-282\,990$ kJ/kmol.

Solution   For stoichiometric conditions

$$CO + \tfrac{1}{2}O_2 + (3.76 \times \tfrac{1}{2})N_2 \rightarrow CO_2 + (3.76 \times \tfrac{1}{2})N_2$$

$$CO + 0.5O_2 + 1.88N_2 \rightarrow CO_2 + 1.88N_2$$

and for a mixture strength of 110%

$$\text{Actual A/F ratio} = \text{stoichiometric A/F ratio} \times \frac{100}{110}$$

$$= \text{stoichiometric A/F ratio}/1.1$$

Therefore, for the actual conditions,

$$CO + (0.5O_2 + 1.88N_2)/1.1 \rightarrow aCO_2 + bCO + 1.71N_2$$

Then,

Carbon balance: $1 = a + b$ [1]
Oxygen balance: $1 + (0.5 \times 2)/1.1 = 2a + b$
          i.e. $1.909 = 2a + b$ [2]

From equation [1] and [2]

$$a = 0.909 \quad \text{and} \quad b = 0.091$$

therefore

$$CO + 0.455O_2 + 1.709N_2 \rightarrow 0.909CO_2 + 0.091CO + 1.709N_2$$

Also for the reaction

$$CO + \tfrac{1}{2}O_2 \rightarrow CO_2$$

$$n_R = 1 + 0.5 = 1.5 \quad \text{and} \quad n_P = 1$$

therefore

$$\Delta U_0 = \Delta H_0 - (n_P - n_R)\tilde{R}T_0$$

$$= (-282\,990 \times 1) - \{(1 - 1.5) \times 8.3145 \times 298\}$$

$$= -282\,990 + 1239 = -281\,751 \text{ kJ}$$

i.e. $\Delta \tilde{u}_0^\ominus = -281\,751$ kJ/kmol

The non-flow process is defined by

$$Q + W = (U_2 - U_1)$$

Also $Q = 0$; $W = 0$ at constant volume; $U_1 = U_R$, $U_2 = U_{P_2}$, therefore

$$(U_{P_2} - U_{R_1}) = 0$$

$$(U_{P_2} - U_{P_0}) + (U_{P_0} - U_{R_0}) + (U_{R_0} - U_{R_1}) = 0$$

$$(U_{P_2} - U_{P_0}) + \Delta U_0 + (U_{R_0} - U_{R_1}) = 0 \qquad [3]$$

The values of molar internal energies for the various gases can be read from the tables of ref. 7.6. Then

At 298.15 K $\quad U_{R_0} = -2479 - (0.455 \times 2479) - (1.709 \times 2479)$

$$= -7844 \text{ kJ}$$

At 555 K $\qquad U_{R_1} = 2984.6 + (0.455 \times 3233.1) + (1.709 \times 2944.2)$

$$= 9487 \text{ kJ}$$

At 298.15 K $\quad U_{P_0} = -(0.909 \times 2479) - (0.091 \times 2479) - (1.709 \times 2479)$

$$= -6716 \text{ kJ}$$

Substituting in [3],

$$(U_{P_2} + 6716) - (0.909 \times 281\,751) + (-7844 - 9487) = 0$$

Note: there are 0.091 kmol CO in the products therefore only $(1 - 0.091) = 0.909$ kmol of CO releases energy of reaction,

i.e. $\qquad U_{P_2} = 266\,727 \text{ kJ}$

To find the temperature of products a trial-and-error method is now necessary.

At T_2 $\quad U_{P_2} = \{0.909 \times (\tilde{u}_2 \text{ for } CO_2)\} + \{0.091 \times (\tilde{u}_2 \text{ for } CO)\}$

$$+ \{1.709 \times (\tilde{u}_2 \text{ for } N_2)\}$$

Try $T_2 = 3200 \text{ K}$

$$U_{P_2} = (0.909 \times 138\,720) + (0.091 \times 74\,391) + (1.709 \times 73\,555)$$

$$= 258\,572 \text{ kJ}$$

This compares with the figure of 266 727 kJ calculated above, hence the actual temperature of the products is slightly greater than 3200 K; a second guess from tables gives a value of 3280 K.

The temperature calculated in Example 7.15 is that which would be obtained by adiabatic combustion of the mixture. This temperature would not be attained in practice due to the effect of dissociation; this was discussed in section 7.7 and the inclusion of this effect is illustrated in Example 7.16.

Example 7.16 Calculate the final temperature for the mixture defined in Example 7.15 including the effect of dissociation.

Solution The proportions of the constituents depend on the temperature so the combustion equation is written as

$$CO + 0.455O_2 + 1.709N_2 \rightarrow aCO_2 + bCO + cO_2 + 1.709N_2$$

and b and c can be expressed in terms of a by an atomic balance.

217

Carbon balance: $\quad 1 = a + b \quad$ therefore $b = 1 - a$

Oxygen balance: $\quad 1 + (0.455 \times 2) = 2a + b + 2c$

$$1.91 = 2a + b + 2c$$

i.e. $\qquad 2c = 1.91 - 2a - (1 - a)$

therefore

$$c = 0.455 - 0.5a$$

and

$$CO + 0.455O_2 + 1.709N_2$$
$$\rightarrow aCO_2 + (1 - a)CO + (0.455 - 0.5a)O_2 + 1.709N_2$$

The energy equation can be written as previously

$$(U_{P_2} - U_{P_0}) + \Delta U_0 + (U_{R_0} - U_{R_1}) = 0$$

Using the values calculated in Example 7.15,

$$U_{P_0} = -(a \times 2479) - \{(1 - a) \times 2479\}$$
$$- \{(0.455 - 0.5a) \times 2479\} - [1.709 \times 2479]$$
$$= -(3.164 - 0.5a) \times 2479 \text{ kJ}$$

$$\Delta U_0 = -a \times 281\,751 \text{ kJ}$$

$$U_{R_0} = -7844 \text{ kJ} \qquad U_{R_1} = 9487 \text{ kJ}$$

Substituting,

$$U_{P_2} + \{(3.164 - 0.5a) \times 2479\} - 281\,751a - 17\,331 = 0$$

therefore

$$U_{P_2} = 9487 + 282\,991a \qquad\qquad [1]$$

A second equation is derived using the equilibrium constant for the reaction

$$K^{\ominus} = \frac{p_{CO_2}}{p_{CO}}\left\{\frac{p}{p_{O_2}}\right\}_{1/2}^{1/2} = \frac{a}{(1 - a)}\left\{\frac{n_2}{(0.455 - 0.5a)p_2}\right\}^{1/2}$$

where p_2 is the pressure of the mixture after combustion and n_2 the number of kmol of the products. Now

$$p_1 V = n_1 R T_1 \quad \text{and} \quad p_2 V = n_2 R T_2$$

$$\frac{n_2}{p_2} = \frac{T_1 n_1}{T_2 p_1} = \frac{555}{T_2} \times \frac{(1 + 0.455 + 1.709)}{8.28} = \frac{212.08}{T_2} \text{ kmol/bar}$$

Substituting in the expression for K gives

$$K^{\ominus} = \frac{a}{(1 - a)}\left\{\frac{212.08}{(0.455 - 0.5a)T_2}\right\}^{1/2} \qquad\qquad [2]$$

When the value of T_2 is known then U_{P_2} can be found by reading off $\tilde{u}$ for each of the products and multiplying by the amount of substance of each; equating this to the value of U_{P_2} in [1] allows a value of a to be calculated.

Similarly, when T_2 is known a value of $\ln(K^\ominus)$ can be read from tables, and from [2] a value of a is then found.

Therefore, one convenient trial-and-error method is to choose a value of T_2, calculate a value of a from [1], and then find a value of $K^\ominus$ from [2]. Compare this value of $K^\ominus$ with that found from tables at T_2. When the two values of $K^\ominus$ are the same then the correct value of T_2 has been found.

For example, at $T_2 = 3000$ K, from tables:

$$U_{P_2} = 127\,920a + 68\,598(1 - a)$$
$$+ 73\,155(0.455 - 0.5a) + (67\,795 \times 1.709)$$
$$= 217\,745 + 22\,744.5a$$

i.e. in [1],

$$217\,745 + 22\,744.5a = 9487 + 282\,991a$$

therefore

$$a = 0.8$$

Substituting for a and T_2 in [2]

$$K^\ominus = \frac{0.8}{0.2}\left\{\frac{212.08}{(0.455 - 0.5a) \times 3000}\right\}^{1/2}$$
$$= 4.54$$

From tables at $T_2 = 3000$ K,

$$\ln(K^\ominus) = 1.110 \quad \text{therefore } K^\ominus = 3.03 \text{ (too low)}$$

By such a method the value of T_2 is found to be 2949 K to the nearest degree. This value should be compared with the value of 3280 K from Example 7.15 when dissociation was ignored.

7.9 Enthalpy of formation, $\Delta \tilde{h}_f$

The combustion reaction is a particular kind of chemical reaction in which products are formed from reactants with the release or absorption of energy as heat is transferred to or from the surroundings. As some substances, for instance hydrocarbon fuels, may be many in number and complex in structure the enthalpy of reaction may be calculated on the basis of known values of the molar enthalpy of formation, $\Delta \tilde{h}_f^\ominus$ of the constituents of the reactants and products at the reference temperature T_0.

The molar enthalpy of formation $\Delta \tilde{h}_f$ is the increase in enthalpy when a compound is formed from its constituent elements in their natural form and in a standard state. Something needs to be said about the standard form. The normal forms of oxygen (O_2) and hydrogen (H_2) are gaseous, so $\Delta \tilde{h}_f^\ominus$ for these can be put equal to zero. The normal form of carbon (C) is graphite, so $\Delta \tilde{h}_f^\ominus$ for solid carbon is put to zero. Carbon in another form, e.g. diamond or gas, is not 'normal' and $\Delta \tilde{h}_f^{\ominus}$ is quoted. The standard state is 25 °C, and 1 bar pressure, but it must be borne in mind that not all substances can exist in the natural form, e.g. H_2O cannot be a vapour at 1 bar and 25 °C.

For calculation purposes, for a particular reaction

$$\Delta \tilde{h}^{\ominus} = \sum_{P} n_i \Delta \tilde{h}_{f_i}^{\ominus} - \sum_{R} n_i \Delta \tilde{h}_{f_i}^{\ominus} \qquad (7.33)$$

Typical values of $\Delta \tilde{h}_f^{\ominus}$ are quoted for different substances in Table 7.11.

Table 7.11 Typical heats of formation $\Delta \tilde{h}_f^{\ominus}$ of various species at 25 °C (298 K) and 1 bar

Substance	Formula	State	$\Delta \tilde{h}_f^{\ominus}$ (kJ/kmol)
	O	Gas	249 170
Oxygen	O_2	Gas	0
Water	H_2O	Liquid	−285 820
Water	H_2O	Vapour	−241 830
Carbon	C	Gas	714 990
Carbon	C	Diamond	1 900
Carbon	C	Graphite	0
Carbon monoxide	CO	Gas	−110 530
Carbon dioxide	CO_2	Gas	−393 520
Methane	CH_4	Gas	−74 870
Methyl alcohol	CH_3OH	Vapour	−240 532
Ethyl alcohol	C_2H_5OH	Vapour	−281 102
Ethane	C_2H_6	Gas	−83 870
Ethene	C_2H_4	Gas	52 470
Propane	C_3H_8	Gas	−102 900
Butane	C_4H_{10}	Gas	−125 000
Octane	C_8H_{18}	Liquid	−247 600

Example 7.17   Calculate the molar enthalpy of reaction at 25 °C of ethyl alcohol, C_2H_5OH, using the data of Table 7.11.

Solution   Using equation (7.33)

$$\Delta \tilde{h}_f^{\ominus} = \sum_{P} n_i \Delta \tilde{h}_{f_i}^{\ominus} - \sum_{R} n_i \Delta \tilde{h}_{f_i}^{\ominus}$$

and the combustion equation

$$C_2H_5OH + 3O_2 \rightarrow 2CO_2 + 3H_2O$$

$$\sum_{R} n_i \Delta \tilde{h}_{f_i}^{\ominus} = 1 \times (-281\,102) + 3 \times 0 = -281\,102$$

$$\sum_{P} n_i \Delta \tilde{h}_{f_i}^{\ominus} = 2 \times (-393\,520) + 3 \times (-241\,830) = -1\,512\,530$$

therefore

$$\Delta \tilde{h}^{\ominus} = -1\,512\,530 - (-281\,102)$$

$$= -1\,231\,428 \text{ kJ/kmol}$$

7.10   Calorific value of fuels

The quantities $\Delta h^{\ominus}$ and $\Delta u^{\ominus}$ are approximated to in fuel specifications by quantities called *calorific values* which are obtained by the combustion of the fuels in suitable apparatus. This may be of the constant volume type (e.g. bomb calorimeter) or constant pressure, steady-flow type (e.g. Boys' calorimeter). Both of these are described in section 7.1. Definitions of calorific value are as follows:

(1) The energy transferred as heat to the surroundings (cooling water) per unit quantity of fuel when burned at constant volume with the H_2O product of combustion in the liquid phase is called the gross (or higher) calorific value (GCV) at constant volume $Q_{gr,v}$. This approximates to $-\Delta u^{\ominus}$ at a reference temperature of 25 °C with the H_2O in the liquid phase.

If the H_2O products are in the vapour phase the energy released per unit quantity is called the net (or lower) calorific value (NCV) at constant volume, $Q_{net,v}$. This approximates to $-\Delta u^{\ominus}$ at 25 °C with the H_2O in the vapour phase.

(2) The energy transferred as heat to the surroundings (cooling water) per unit quantity of fuel when burned at constant pressure with the H_2O products of combustion in the liquid phase is called the gross (or higher) calorific value (GCV) at constant pressure, $Q_{gr,p}$. This approximates to $-\Delta h^{\ominus}$ at a reference temperature of 25 °C with the H_2O in the liquid phase.

If the H_2O products are in the vapour phase the energy released is called the net (or lower) calorific value (NCV) at constant pressure, $Q_{net,p}$. This approximates to $-\Delta h^{\ominus}$ at 25 °C with the H_2O in the vapour phase.

Contrary to the definition of $\Delta u^{\ominus}$ and $\Delta h^{\ominus}$ it is usual to quote calorific values as positive quantities. If $Q_{gr,p}$ and the fuel composition is known, the other quantities can be calculated. The above quantities are related as follows. For the constant volume process

$$Q_{gr,v} = Q_{net,v} + m_c u_{fg} \qquad (7.34)$$

For the constant pressure process

$$Q_{gr,p} = Q_{net,p} + m_c h_{fg} \qquad (7.35)$$

where m_c is the mass of condensate per unit quantity of fuel.

$$u_{fg} = u_{fg} \text{ at 25 °C for } H_2O = 2304.4 \text{ kJ/kg}$$

$$h_{fg} = h_{fg} \text{ at 25 °C for } H_2O = 2441.8 \text{ kJ/kg}$$

The calorific values differ from $\Delta h^{\ominus}$ and $\Delta u^{\ominus}$ due to the departure of experimental conditions from ideal with regard to temperatures of products and reactants and also heat transfer conditions. This topic is discussed further in section 7.12.

7.11   Power plant thermal efficiency

The purpose of any power plant is the power output which should be obtained as economically as possible consistent with capital cost and running conditions.

It is necessary to assess the overall performance of a plant for comparison purposes and an important criterion is the *overall thermal efficiency* η_0. This is defined as

$$\eta_0 = \frac{\text{work output}}{\text{fuel energy supplied}}$$

It is necessary to decide on the denominator for this definition. It is desirable, if we consider a steady-flow process, to relate the plant conditions to those for the steady-flow calorimeter in which the products are cooled to atmospheric temperature giving $Q_{gr,p}$. However, it is not possible, or desirable, to cool the products of combustion in a real plant to atmospheric temperature so there is a substantial energy loss to atmosphere. Complete cooling of the products would require large heat transfer surfaces which are expensive and the condensate produced would form corrosive acids. As achieving these conditions is not even attempted it seems that the use of $Q_{gr,p}$ is unsatisfactory and $Q_{net,p}$ is more appropriate and this is often preferred. It does not matter for comparison purposes except that both values are in use making the definition of η_0 somewhat arbitrary as

$$\eta_0 = \frac{-W}{Q_{gr,p}} \quad \text{or} \quad \frac{-W}{Q_{net,p}} \tag{7.36}$$

It is necessary to state with the definition whether $Q_{gr,p}$ or $Q_{net,p}$ is being used.

In a plant it may be required to assess the boiler or steam generator performance only, in which case the *boiler efficiency*, given similarly in equation (8.16), is defined as

$$\text{Boiler efficiency} = \frac{\text{heat transferred to working fluid}}{\text{fuel energy supplied}}$$

therefore

$$\eta_B = \frac{\text{heat transferred to working fluid}}{Q_{gr,p} \text{ or } Q_{net,p}} \tag{7.37}$$

It is again necessary to state whether $Q_{fr,p}$ or $Q_{net,p}$ is being used when boiler efficiency is quoted.

Example 7.18　　A medium-size steam boiler required to supply a generator of output 25 000 kW has a performance specification as follows: steam output 31.6 kg/s; steam pressure 60 bar; steam temperature 500 °C; feed water temperature 100 °C; fuel, natural gas (96.5% CH_4, 0.5% C_2H_6, remainder incombustible); gross calorific value 38 700 kJ/m³ at 1.013 bar and 15 °C; fuel consumption 2.85 m³/s.

Calculate the boiler efficiency and the overall thermal efficiency based on the net calorific value of the fuel.

Solution　　1 kmol of CH_4 burns to give 2 kmol H_2O, therefore 0.965 kmol of CH_4 burns to give $2 \times 0.965 \times 18 = 34.74$ kg H_2O.

1 kmol of C_2H_6 burns to give 3 kmol H_2O, therefore 0.005 kmol of C_2H_6 burns to give $3 \times 0.005 \times 18 = 0.27$ kg H_2O. Therefore 1 kmol of gas produces $34.74 + 0.27 = 35.01$ kg H_2O, and 1 kmol of gas occupies a volume of $(\tilde{R}T/p) = (8314.5 \times 288/1.013 \times 10^5) = 23.64$ m^3 at 1.013 bar and 15 °C, therefore

$$\text{Steam formed per m}^3 \text{ of gas} = \frac{35.01}{23.64} = 1.481 \text{ kg}$$

Using equation (7.35)

$$Q_{gr,p} = Q_{net,p} + m_c h_{fg}$$

therefore

$$Q_{net,p} = 38\,700 - (1.481 \times 2441.8) = 35\,084 \text{ kJ/m}^3$$

$$\text{Heat to working fluid} = h_{\text{supply steam}} - h_{\text{feedwater}}$$

$$= 3421 - 419.1 = 3001.9 \text{ kJ/kg}$$

Using equation (7.37), therefore

$$\eta_B = \frac{3001.9 \times 31.6}{2.85 \times 35\,084} = 0.95(95\%)$$

and using equation (7.36)

$$\eta_0 = \frac{\text{work output}}{Q_{net,p}} = \frac{25\,000}{2.85 \times 35\,084} = 0.25(25\%)$$

7.12 Practical determination of calorific values

The methods for determining calorific value depend on the type of fuel; solid and liquid fuels are usually tested in a bomb calorimeter, and gaseous fuels in a continuous-flow apparatus such as Boys' calorimeter. The apparatus in every case is required to meet with a standard specification and the procedure to be adopted is also laid down.

Solid and liquid fuel

In the bomb calorimeter combustion occurs at constant volume and is a non-flow process; in Boys' calorimeter the gas is burnt continuously under steady-flow conditions.

The bomb is a small stainless steel vessel in which a small mass of the fuel is held in a crucible (see Fig. 7.14). If the fuel is solid, it is usually crushed, passed through a sieve, and then pressed into the form of a pellet in a special press. The size of pellet is estimated from the expected heat release, and is such that the temperature rise to be measured does not exceed 2–3 K. The pellet is ignited

Fig. 7.14 Bomb calorimeter

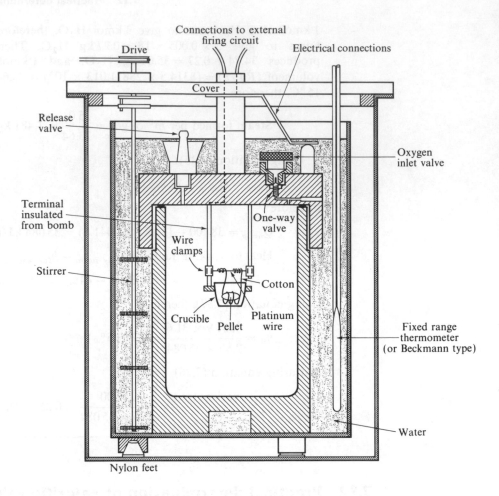

by fusing a piece of platinum or nichrome wire which is in contact with it. The wire forms part of an electric circuit which can be completed by a firing button, which is situated in a position remote from the bomb. With a special form of press the pellet can be formed with the fuse wire passing through it. This facilitates the firing, particularly with some of the more difficult fuels. The crucible carrying the pellet is located in the bomb, a small quantity of distilled water is put into the bomb to absorb the vapours formed by combustion and to ensure that the water vapour produced is condensed, and the top of the bomb is screwed down. Oxygen is then admitted slowly until the pressure is above 23 atm. The bomb is located in the calorimeter and a measured quantity of water is poured into the calorimeter. The calorimeter is closed, the external connections to the circuit are made, and an accurate thermometer of the fixed range or Beckman type is immersed to the proper depth in the water. The water is stirred in a regular manner by a motor-driven stirrer, and temperature observations are taken every minute. At the end of the fifth minute the charge is fired and temperature readings are taken every 10 s during this period. When the temperature readings begin to fall the frequency of readings can be reduced to every minute.

The measured temperature rise is corrected for various losses. The cooling loss is the largest, but corrections are also necessary for the heat released by the combustion of the wire itself, and for the formation of acids on combustion. The cooling correction can be determined graphically or by use of a formula developed by Regnault and Pfaundler. The allowance for the combustion of the wire is determined from its weight and known calorific value. The allowance for acids present is determined by a chemical titration. For most purposes only the correction for cooling need be applied.

If a liquid fuel is being tested, it is contained in a gelatine capsule and the firing may be assisted by including in the crucible a little paraffin of known calorific value.

The water equivalent of the calorimeter is determined by burning a fuel of known calorific value (e.g. benzoic acid) in the bomb. The calculation for the test is then as follows:

Mass of fuel × calorific value

= (mass of water + water equivalent of bomb)

× corrected temperature rise × specific heat capacity of water

From this equation the calorific value of the fuel tested can be found.

Example 7.19

Table 7.12 gives the results of a bomb calorimeter test on a sample of coal. The mass of coal burned was 0.825 g and the total water equivalent of the apparatus was determined as 2500 g. Calculate the calorific value of the coal in kilojoules per kilogram. The temperature rise is to be corrected according to the formula by Regnault and Pfaundler, but no correction need be made for the acids formed.

Table 7.12 Results for a bomb calorimeter test

Pre-firing period time	Temp.	Heating time	Temp.	Cooling period time	Temp.
(min.)	(°C)	(min.)	(°C)	(min.)	(°C)
0	25.730	t_1 6	27.340	t_n 10	27.880
1	25.732	t_2 7	27.880	11	27.878
2	25.734	t_3 8	27.883	12	27.876
3	25.736	t_4 9	27.885	13	27.874
4	25.738			14	27.872
t_0 5	25.740			15	27.870

Solution   The Regnault and Pfaundler cooling correction is as follows:

$$\text{Correction} = nv + \left(\frac{v_1 - v}{t_1 - t} \right) \left\{ \sum_1^{(n-1)} (t) + \tfrac{1}{2}(t_0 + t_n) - nt \right\}$$

where n is the number of minutes between the time of firing and the first reading after the temperature begins to fall from the maximum, v the rate of fall of temperature per minute during the pre-firing period, v_1 the rate of fall of

225

temperature per minute after the maximum temperature, t and t_1 the average temperatures during the pre-firing and final periods respectively, $\Sigma_1^{(n-1)}(t)$ the sum of the readings during the period between firing and the start of cooling, and $\frac{1}{2}(t_0 + t_n)$ the mean of the temperature at the moment of firing and the first temperature after the rate of change of temperature becomes constant. The pre-firing and final periods are of the same duration.

In this example

$$n = 10 - 5 = 5 \text{ min}$$

$$v = -\left(\frac{25.740 - 25.730}{5}\right) = -0.002 \text{ K/min}$$

the negative sign indicates that the temperature was rising in the pre-firing period.

$$v_1 = \frac{27.880 - 27.870}{5} = 0.002 \text{ K/min}$$

$$t = 25.735\,°C \quad \text{and} \quad t_1 = 27.875\,°C$$

$$\sum_1^{(n-1)} (t) = 110.988\,°C \quad \text{and}$$

$$\tfrac{1}{2}(t_0 + t_n) = \frac{25.740 + 27.880}{2} = 26.81\,°C$$

Substituting the values in the equation gives

$$\text{Correction} = -5 \times 0.002 + \left(\frac{0.002 + 0.002}{27.875 - 25.735}\right)$$

$$\times (110.988 + 26.81 - 5 \times 25.735)$$

i.e. $\text{Correction} = 0.00705 \text{ K}$

$\text{Uncorrected temperature rise} = -t_n - t_0$

$$= 27.880 - 25.740 = 2.14 \text{ K}$$

therefore

$\text{Correct temperature rise} = 2.14 + 0.007\,05 = 2.147 \text{ K}$

Then   Heat released from $0.825 \text{ g coal} = 2.147 \times 2500 \times 4.187 \times 10^{-3}$

$$= 22.5 \text{ kJ}$$

i.e. $\text{Calorific value} = \dfrac{22.5}{0.825 \times 10^{-3}} = 27\,250 \text{ kJ/kg}$

i.e. $\text{Calorific value of fuel} = 27\,250 \text{ kJ/kg}$

Gaseous fuels

For a gaseous fuel a continuous supply of the gas is metered and passed at constant pressure into the calorimeter, where it is burned in an ample supply of air (see Fig. 7.15). The products of combustion are cooled as nearly as possible to the initial temperature of the reactants by a continuously circulating supply of cooling water. The gas pressure and temperature are measured and the amount of gas burned is referred to 1.013 bar and 15 °C. The temperature rise of the circulating water is measured, and the condensate from the products of combustion is collected. A test is carried out over a fixed time period. The water flow rate is measured and the condensate is weighed. Then we have

(Volume of fuel at 1.013 bar and 15 °C) × (calorific value)

= (mass of water circulated) × specific heat capacity

× (temperature rise of water)

Fig. 7.15 Boys' gas calorimeter

A – Cooling coil
B – Burners

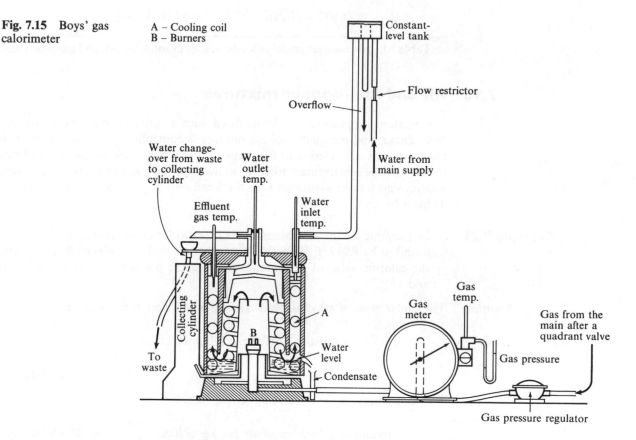

The calorific value of the fuel is obtained in megajoules per cubic metre of gas. The correct procedure for this determination is given in ref. 7.7.

In both the bomb calorimeter test and Boys' calorimeter test the steam formed on combustion is condensed, and so the heat released by the steam on condensing is transferred to the cooling water.

Example 7.20   In a bomb calorimeter test on petrol the GCV was determined and found to be 46 900 kJ/kg. If the fuel contains 14.4% H by mass, calculate the NCV.

Solution   From equation (7.1),

$$2H_2 + O_2 \rightarrow 2H_2O$$

therefore

$$4 \text{ kg H} + 32 \text{ kg O}_2 \rightarrow 36 \text{ kg H}_2O$$

or 1 kg H gives 9 kg H_2O, therefore

$$0.144 \text{ kg H gives } 0.144 \times 9 = 1.296 \text{ kg H}_2O$$

Then using equation (7.34)

$$Q_{net,v} = Q_{gr,v} - m_c u_{fg}$$

therefore

$$Q_{net,v} = 46\,900 - (1.296 \times 2304.4) = 43\,910 \text{ kJ/kg}$$

Table 7.13 gives some typical calorific values of solid, liquid, and gaseous fuels.

7.13 Air and fuel–vapour mixtures

The mixture supplied to an engine fitted with a carburettor is one of air and fuel vapour, and the quality of the mixture is controlled by the carburettor. If the mixture is saturated with fuel vapour then the relative proportions of fuel to air can be determined from a knowledge of the temperature–pressure relationship for the saturated fuel. Such values are given for ethyl alcohol in Table 7.14.

Example 7.21   In Example 7.3 the stoichiometric A/F ratio for ethyl alcohol, C_2H_6O, was found to be 8.957/1. If the NCV of ethyl alcohol is 27 800 kJ/kg, calculate the calorific value of the combustion mixture per cubic metre at 1.013 bar and 15 °C.

Solution   The molar mass of ethyl alcohol is 46 kg/kmol, therefore we have

$$\text{amount of substance of fuel per kilogram of fuel} = \frac{1}{46}$$

$$= 0.021\,74 \text{ kmol/kg}$$

The molar mass of air is 28.96 kg/kmol, therefore

$$\text{amount of substance of air per kg of fuel} = \frac{8.957}{28.96} = 0.3093 \text{ kmol/kg}$$

i.e.

$$\text{Total amount of substance of mixture} = 0.021\,74 + 0.3093$$

$$= 0.3310 \text{ kmol/kg}$$

Table 7.13 Typical calorific values of some fuels

Fuel	Calorific value at 15 °C (kJ/kg)	
	Gross	Net
Solid		
Anthracite	34 600	33 900
Bituminous coal	33 500	32 450
Coke	30 750	30 500
Lignite	21 650	20 400
Peat	15 900	14 500
Liquid		
100-octane petrol	47 300	44 000
Motor petrol	46 800	43 700
Benzole	42 000	40 200
Kerosene	46 250	43 250
Diesel oil	46 000	43 250
Light fuel oil	44 800	42 100
Heavy fuel oil	44 000	41 300
Residual fuel oil	42 100	40 000

Gas	Calorific value at 15 °C and 1 bar (MJ/m³)	
	Gross	Net
Butane	122.00	113.00
Propane	96.00	86.00
Natural gas	38.20	35.20
Coal gas	20.00	17.85
Hydrogen	11.85	10.00
Producer gas	6.04	6.00
Blast-furnace gas	3.41	3.37

Table 7.14 Approximate saturation temperatures and pressures for ethyl alcohol (C_2H_6O)

Temp./(°C)	0	10	20	30	40	50	60
Pressure/(bar)	0.0162	0.0314	0.0584	0.1049	0.1800	0.2960	0.4690

From equation (2.8)

$$pV = n\tilde{R}T$$

therefore

$$V = \frac{0.331 \times 8314.5 \times 288}{1.013 \times 10^5} = 7.824 \text{ m}^3 \text{ per kg of fuel}$$

Now   NCV of fuel $= Q_{net,v} = 27.8$ MJ/kg

therefore

$$\text{NCV of mixture} = \frac{27.8}{7.824} = 3.55 \text{ MJ/m}^3 \text{ of mixture}$$

Example 7.22   For a stoichiometric mixture of ethyl alcohol and air calculate the temperature above which there will be no liquid fuel in the mixture. The pressure of the mixture is 1.013 bar.

Solution   Using the results of Example 7.21, we have

amount of substance of ethyl alcohol $= 0.021\ 74$ kmol/kg

and   Total amount of substance $= 0.331$ kmol

Then using equation (6.14), $p_i = (n_i/n)p$, we have

$$\text{Partial pressure of ethyl alcohol vapour} = \frac{0.021\ 74}{0.331} \times 1.013$$

$$= 0.0665 \text{ bar}$$

From Table 7.14, the saturation temperature corresponding to this pressure lies between 20 and 30 °C. Therefore interpolating

$$t = 20 + \left(\frac{0.0665 - 0.0584}{0.1049 - 0.0584} \right) \times (30 - 20) = 21.74 \,°\text{C}$$

Hence the minimum temperature of the mixture is 21.74 °C for complete evaporation of the liquid fuel.

Problems

7.1   A sample of bituminous coal gave the following ultimate analysis by mass: C 81.9%; H 4.9%; O 6%; N 2.3%; ash 4.9%. Calculate:
  (i) the stoichiometric A/F ratio;
  (ii) the analysis by volume of the wet and dry products of combustion when the air supplied is 25% in excess of that required for complete combustion.
    (10.8/1; CO_2 14.14%; H_2O 5.07%; O_2 4.08%; N_2 76.71%; CO_2 14.89%;
    O_2 4.30%; N_2 80.81%)

7.2   An analysis of natural gas gave the following values: CH_4 93%; C_2H_6 4%; N_2 3%. Calculate the stoichiometric A/F ratio and the analysis of the wet and dry products of combustion when the A/F ratio is 12/1.
    (9.524/1; CO_2 7.76%; H_2O 15.21%; O_2 3.99%; N_2 73.04%; CO_2 9.15%;
    O_2 4.71%; N_2 86.14%)

7.3   Calculate the stoichiometric A/F ratio for benzene (C_6H_6), and the wet and dry analysis of the combustion products.
    (13.2/1; CO_2 16.13%; H_2O 8.06%; N_2 75.81%; CO_2 17.54%; N_2 82.46%)

7.4 In the actual combustion of benzene in an engine the A/F ratio was 12/1. Calculate the analysis of the wet products of combustion.

$$(CO_2\ 13.38\%;\ CO\ 3.94\%;\ H_2O\ 8.66\%;\ N_2\ 74.03\%)$$

7.5 The ultimate analysis of a sample of petrol was 85.5% C and 14.5% H. Calculate:
  (i) the stoichiometric A/F ratio;
  (ii) the A/F ratio when the mixture strength is 90%;
  (iii) the A/F ratio when the mixture strength is 120%;
  (iv) the analyses of the dry products for (ii) and (iii);
  (v) the volume flow rate of the products through the engine exhaust per unit rate of fuel consumption for (iii) when the pressure is 1.013 bar and the temperature is 110 °C.

$$(14.76/1;\ 16.4/1;\ 12.3/1;\ CO_2\ 13.38\%;\ O_2\ 2.24\%;\ N_2\ 84.38\%;\ CO_2\ 8.67\%;$$
$$CO\ 8.79\%;\ N_2\ 82.54\%;\ 15.11\ m^3/s\ per\ kg/s)$$

7.6 The ultimate analysis of a sample of petrol was C 85.5% and H 14.5%. The analysis of the dry products gave 14% CO_2 and some O_2. Calculate the A/F ratio supplied to the engine, and the mixture strength.

$$(15.72/1;\ 94\%)$$

7.7 In an engine test the dry product analysis was CO_2 15.5%; O_2 2.3% and the remainder N_2. Assuming that the fuel burned was a pure hydrocarbon, calculate the ratio of carbon to hydrogen in the fuel, the A/F ratio used, and the mixture strength.

$$(11.5;\ 14.84/1;\ 89.5\%)$$

7.8 The ultimate analysis of a sample of petrol was 85% C and 15% H. The analysis of the dry products showed 13.5% CO_2, some CO and the remainder N_2. Calculate:
  (i) the actual A/F ratio;
  (ii) the mixture strength;
  (iii) the mass of H_2O vapour carried by the exhaust gas per kilogram of total exhaust gas;
  (iv) the temperature to which the gas must be cooled before condensation of the H_2O vapour begins, if the pressure in the exhaust pipe is 1.013 bar.

$$(14.31/1;\ 104\%;\ 0.088\ kg/kg;\ 52.7\ °C)$$

7.9 A quantity of coal used in a boiler had the following analysis: 82% C; 5% H; 6% O; 2% N; 5% ash. The dry flue gas analysis showed 14% CO_2 and some oxygen. Calculate:
  (i) the oxygen content of the dry flue gas;
  (ii) the A/F ratio and the excess air supplied.

$$(5.52\%;\ 14.29/1;\ 31.8\%)$$

7.10 For the mixture of Problem 7.4 calculate the calorific value per cubic metre of mixture at 1.013 bar and 38 °C. The calorific value of benzene is 40 700 kJ/kg.

$$(3.73\ MJ/m^3)$$

7.11 The lower explosive limit of ethyl alcohol in air is 3.56% by volume at a pressure of 1013 mbar. If the pressure in a room is 1013 mbar calculate the lowest temperature at which the explosive mixture would be formed. What quantity of ethyl alcohol in litres would be needed in a room of volume 115 m^3 to produce this mixture. The specific gravity of liquid ethyl alcohol is 0.794. Use the data of Table 7.14 for this problem.

$$(11.73\ °C;\ 10.15/1)$$

7.12 The products of combustion of a hydrocarbon fuel, of carbon to hydrogen ratio 0.85 : 0.15, are found to be CO_2 8%, CO 1%, O_2 8.5%. Calculate the A/F ratio for the process by two methods and hence check the consistency of the data.

$$(23.70,\ 23.53)$$

7.13 A stoichiometric mixture of CO and air was burned adiabatically at constant volume and at the peak pressure of 7.2 bar the temperature was $2469\,°C$; analysis showed the volume of CO present in the products to be 0.192 of the volume of CO supplied. Show by calculating the equilibrium constant for $CO + \frac{1}{2}O_2 \rightleftharpoons CO_2$ that the data are consistent. The value of $\ln K^{\ominus}$ for this reaction at 2600 K is 2.800 and at 2800 K it is 1.893.

7.14 The products of a fuel when measured at a high temperature gave the following analysis: $9.27\%\ CO_2$; $4.00\%\ CO$; $14.20\%\ H_2O$; $0.90\%\ H_2$; $71.03\%\ N_2$. Using appropriate values of equilibrium constants from tables such as those of ref. 7.3, estimate the temperature and pressure of the products.

(2800 K; 20.31 bar)

7.15 A stoichiometric mixture of benzene (C_6H_6) and air is induced into an engine of volumetric compression ratio 5 to 1. The pressure and temperature at the beginning of compression are 1 bar and $100\,°C$. The estimated maximum temperature reached, allowing for dissociation, after adiabatic combustion at constant volume is $2727\,°C$ and at this temperature the standard equilibrium constants are

$$\frac{p_{CO_2}(p^{\ominus})^{1/2}}{p_{CO}(p_{O_2})^{1/2}} = 3.034 \qquad \frac{p_{H_2O}p_{CO}}{p_{H_2}p_{CO_2}} = 7.214$$

Show that about 74.4% of the carbon in the fuel is burned to CO_2 and calculate the maximum pressure reached.

(41.7 bar)

7.16 The enthalpy of combustion of propane gas, C_3H_8, at $25\,°C$ with the H_2O in the products in the liquid phase is $-50\,360\ kJ/kg$. Calculate the enthalpy of combustion with the H_2O in the vapour phase per unit mass of fuel and per unit amount of substance of fuel.

($-46\,364\ kJ/kg$; $-2\,040\,030\ kJ/kmol$)

7.17 Calculate, for propane liquid, C_3H_8, at $25\,°C$ the enthalpy of combustion with the H_2O in the products in the vapour phase. Use the data of Problem 7.16 and take h_{fg} at $25\,°C$ for propane as $372\ kJ/kg$.

($-45\,992\ kJ/kg$)

7.18 Calculate the internal energy of combustion per unit mass of propane vapour, C_3H_8, at $25\,°C$ with the H_2O in the vapour phase from the corresponding value of $\Delta h^{\ominus} = -46\,364\ kJ/kg$.

($-46\,420\ kJ/kg$)

7.19 Calculate the internal energy of combustion per unit mass for gaseous propane, C_3H_8, at $25\,°C$ with the H_2O of combustion in the liquid phase from the corresponding value of $\Delta h^{\ominus} = -50\,360\ kJ/kg$.

($-50\,191\ kJ/kg$)

7.20 Calculate the internal energy of combustion per unit mass for liquid propane, C_3H_8, at $25\,°C$ with the H_2O of combustion in the vapour phase from the corresponding value of $\Delta h^{\ominus} = -45\,992\ kJ/kg$.

($-46\,105\ kJ/kg$)

7.21 $\Delta \tilde{h}^{\ominus}$ for hydrogen at $60\,°C$ is given as $-242\,400\ kJ/kmol$. Calculate $\Delta \tilde{h}^{\ominus}$ at $1950\,°C$ given that the molar enthalpies of the gases concerned are as in Table 7.15.

($-252\,087\ kJ/kmol$)

7.22 A stoichiometric mixture of hydrogen and air at $25\,°C$ is ignited and combustion takes place adiabatically at a constant pressure of 1 bar. $\Delta \tilde{h}^{\ominus}$ for hydrogen at $25\,°C$ with the

Table 7.15 Molar enthalpies of gases for problem 7.21

Gas	Molar enthalpy/(kJ/kmol)	
	60 °C	1950 °C
H_2	9 492	69 250
O_2	9 697	76 500
H_2O	11 147	94 620

H_2O in the liquid phase is $-286\,000$ kJ/kmol. Calculate, neglecting changes in kinetic energy, and using the tables of ref. 7.6:

 (i) the temperature reached if the process is assumed to be adiabatic and dissociation is neglected;

(ii) the temperature reached after adiabatic combustion if the constituents are H_2, O_2, H_2O, and N_2.

At 25 °C h_{fg} for H_2O is 2441.8 kJ/kg

(2254 °C; 2200 °C)

7.23 Octane vapour, C_8H_{18}, is to be burned in air in a steady-flow process. Both the fuel and air are supplied at 25 °C and the product temperature is to be 760 °C. Dissociation is negligible and the heat loss is to be taken as 10% of the increase in enthalpy of the products above the reference temperature.

$\Delta \tilde{h}^{\ominus}$ for octane vapour is $-5\,510\,294$ kJ/kmol with the products in the liquid phase. Calculate the A/F ratio by mass required; h_{fg} at 25 °C for H_2O is 2441.8 kJ/kg. Use the molar enthalpies in ref. 7.6.

(49.2)

7.24 Calculate the molar enthalpy of reaction of methane CH_4, at 25 °C and 1 bar pressure. Use Table 7.11 (p. 220) and assume the H_2O of combustion to be in the liquid phase.

($-890\,290$ kJ/kmol)

7.25 Calculate the molar enthalpy of combustion at 25 °C and 1 bar pressure of a gas of volumetric analysis 12% H_2, 29% CO, 2.6% CH_4, 0.4% C_2H_4, 4% CO_2 and 52% N_2 for the H_2O in the vapour phase and in the liquid phase.

($-137\,240$ kJ/kmol; $-145\,158$ kJ/kmol)

References

7.1 BS 1016: 1989 *Methods for Analysis and Testing of Coal and Coke*

7.2 HARKER J H and BACKHURST J R 1981 *Fuel and Energy* Academic Press

7.3 EASTOP T D and CROFT D R 1990 *Energy Efficiency* Longman

7.4 SKOOG D A and LEARY J L 1992 *Principles of Instrumental Analysis* 4th edn Saunders College Publishing

7.5 WILLARD H H, MERRITT (jr) L L, DEAN J A and SETTLE (jr) F A 1988 *Instrumental Methods of Analysis* 7th edn Wadsworth

7.6 ROGERS G F C and MAYHEW Y R 1987 *Thermodynamic Properties of Fluids and Other Data* 4th edn Blackwell

7.7 BS 3804: Part 1 1964 *Methods for the Determination of the Calorific Value of Fuel Gases*

8

Steam Cycles

The heat engine cycle is discussed in Chapter 5, and it is shown that the most efficient cycle is the Carnot cycle for given temperatures of source and sink. This applies to both gases and vapours, and the cycle for a wet vapour is shown in Fig. 8.1. A brief summary of the essential features is as follows:

4 to 1: heat is supplied at constant temperature and pressure.

1 to 2: the vapour expands isentropically from the high pressure and temperature to the low pressure. In doing so it does work on the surroundings, which is the purpose of the cycle.

2 to 3: the vapour, which is wet at 2, has to be cooled to state point 3 such that isentropic compression from 3 will return the vapour to its original state at 4. From 4 the cycle is repeated.

The cycle described shows the different types of processes involved in the complete cycle and the changes in the thermodynamic properties of the vapour as it passes through the cycle. The four processes are physically very different from each other and thus they each require particular equipment. The heat supply, 4–1, can be made in a boiler. The work output, 1–2, can be obtained by expanding the vapour through a turbine. The vapour is condensed, 2–3, in a condenser, and to raise the pressure of the wet vapour, 3–4, requires a pump or compressor.

Thus the components of the plant are defined, but before these are discussed

Fig. 8.1 Carnot cycle for a wet vapour on a T–s diagram

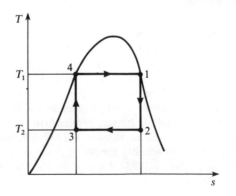

further, the deficiencies of the Carnot cycle as the ideal cycle for a vapour must be considered.

8.1 The Rankine cycle

It is stated in section 5.3 that, although the Carnot cycle is the most efficient cycle, its work ratio is low. Further, there are practical difficulties in following it. Consider the Carnot cycle for steam as shown in Fig. 8.1: at state 3 the steam is wet at T_2 but it is difficult to stop condensation at the point 3 and then compress it just to state 4. It is more convenient to allow the condensation process to proceed to completion, as in Fig. 8.2. The working fluid is water at the new state point 3 in Fig. 8.2, and this can be conveniently pumped to boiler pressure as shown at state point 4. The pump has much smaller dimensions than it would have if it had to pump a wet vapour, the compression process is carried out more efficiently, and the equipment required is simpler and less expensive. One of the features of the Carnot cycle has thus been departed from by the modification to the condensation process. At state 4 the water is not at the saturation temperature corresponding to the boiler pressure. Thus heat must be supplied to change the state from water at 4 to saturated water at 5; this is a constant pressure process, but is not at constant temperature. Hence the efficiency of this modified cycle is not as high as that of the Carnot cycle. This ideal cycle, which is more suitable as a criterion for actual steam cycles than the Carnot cycle, is called the *Rankine cycle*.

Fig. 8.2 Rankine cycle using wet steam on a T–s diagram

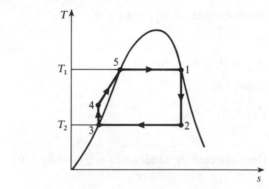

The plant required for the Rankine cycle is shown in Fig. 8.3, and the numbers refer to the state points of Fig. 8.2. The steam at inlet to the turbine may be wet, dry saturated, or superheated, but only the dry saturated condition is shown in Fig. 8.2. The steam flows round the cycle and each process may be analysed using the steady-flow energy equation; changes in kinetic energy and potential energy may be neglected, then for unit mass flow rate

$$Q + W = \mathrm{d}h$$

Each process in the cycle can be considered in turn as follows. Boiler:

$$Q_{451} + W = h_1 - h_4$$

Fig. 8.3 Basic steam plant

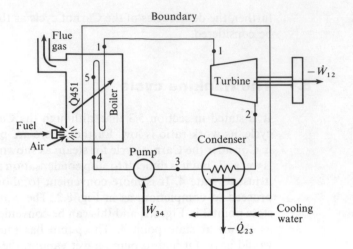

Therefore, since $W = 0$,

$$Q_{451} = h_1 - h_4 \tag{8.1}$$

Turbine: the expansion is adiabatic (i.e. $Q = 0$), and isentropic (i.e. $s_1 = s_2$), and h_2 can be calculated using this latter fact. Then

$$Q_{12} + W_{12} = h_2 - h_1$$

therefore

$$W_{12} = h_2 - h_1$$

or   Work output, $-W_{12} = h_1 - h_2$ $\tag{8.2}$

Condenser:

$$Q_{23} + W = h_3 - h_2$$

Therefore, since $W = 0$

$$Q_{23} = h_3 - h_2$$

therefore

Heat rejected in condenser, $-Q_{23} = h_2 - h_3$ $\tag{8.3}$

Pump:

$$Q_{34} + W_{34} = h_4 - h_3$$

The compression is isentropic (i.e. $s_3 = s_4$), and adiabatic (i.e. $Q = 0$). Therefore

$$W_{34} = (h_4 - h_3)$$

i.e.   Work input to pump, $W_{34} = h_4 - h_3$ $\tag{8.4}$

This is the feed-pump term, and as it is a small quantity in comparison with the turbine work output, $-W_{12}$, it is usually neglected, especially when boiler pressures are low.

Net work input for the cycle $\sum W = W_{12} + W_{34}$

i.e. $\qquad \sum W = (h_2 - h_1) + (h_4 - h_3)$

or $\qquad$ Net work output, $-\sum W = (h_1 - h_2) - (h_4 - h_3)$ $\qquad$ (8.5)

Or, if the feed-pump work is neglected,

$$\text{Net work output,} \quad -\sum W = h_1 - h_2 \qquad (8.6)$$

The heat supplied in the boiler, $Q_{451} = h_1 - h_4$. Then we have

$$\text{Rankine efficiency,} \quad \eta_R = \frac{\text{net work output}}{\text{heat supplied in the boiler}} \qquad (8.7)$$

i.e. $\qquad \eta_R = \dfrac{(h_1 - h_2) - (h_4 - h_3)}{h_1 - h_4}$

or $\qquad \eta_R = \dfrac{(h_1 - h_2) - (h_4 - h_3)}{(h_1 - h_3) - (h_4 - h_3)}$ $\qquad$ (8.8)

If the feed-pump term, $h_4 - h_3$, is neglected equation (8.8) becomes

$$\eta_R = \frac{h_1 - h_2}{h_1 - h_3} \qquad (8.9)$$

When the feed-pump term is to be included it is necessary to evaluate the quantity, W_{34}.

From equation (8.4)

$$\text{Pump work} = W_{34} = h_4 - h_3$$

It can be shown that for a liquid, which is assumed to be incompressible (i.e. $v = $ constant), the increase in enthalpy for isentropic compression is given by

$$(h_4 - h_3) = v(p_4 - p_3)$$

The proof is as follows. For a reversible adiabatic process, from equation (3.15),

$$dQ = dh - v\,dp = 0$$

therefore

$$dh = v\,dp$$

i.e. $\qquad \displaystyle\int_3^4 dh = \int_3^4 v\,dp$

For a liquid, since v is approximately constant, we have

$$h_4 - h_3 = v \int_3^4 dp = v(p_4 - p_3)$$

i.e. $\qquad h_4 - h_3 = v(p_4 - p_3)$

therefore

$$\text{Pump work input} = h_4 - h_3 = v(p_4 - p_3) \qquad (8.10)$$

where v can be taken from tables for water at the pressure p_3.

237

The *efficiency ratio* of a cycle is the ratio of the actual efficiency to the ideal efficiency. In vapour cycles the efficiency ratio compares the actual cycle efficiency to the Rankine cycle efficiency,

i.e. $$\text{Efficiency ratio} = \frac{\text{cycle efficiency}}{\text{Rankine efficiency}} \qquad (8.11)$$

The actual expansion process is irreversible, as shown by line 1–2 in Fig. 8.4. Similarly the actual compression of the water is irreversible, as indicated by line 3–4. The *isentropic efficiency* of a process is defined by

$$\text{Isentropic efficiency} = \frac{\text{actual work}}{\text{isentropic work}} \quad \text{for an expansion process}$$

Fig. 8.4 Rankine cycle showing real processes on a *T*–*s* diagram

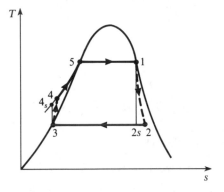

and

$$\text{Isentropic efficiency} = \frac{\text{isentropic work}}{\text{actual work}} \quad \text{for a compression process}$$

Hence

$$\text{Turbine isentropic efficiency} = \frac{-W_{12}}{-W_{12s}} = \frac{h_1 - h_2}{h_1 - h_{2s}} \qquad (8.12)$$

It has been stated that the efficiency of the Carnot cycle is the maximum possible, but that the cycle has a low work ratio. Both efficiency and work ratio are criteria of performance. By the definition of work ratio in section 5.3,

$$\text{Work ratio} = \frac{\text{net work output}}{\text{gross work output}} \qquad (8.13)$$

Another criterion of performance in steam plant is the *specific steam consumption* (ssc). It relates the power output to the steam flow necessary to produce it. The steam flow indicates the size of plant and its component parts, and the ssc is a means whereby the relative sizes of different plants can be compared.

The ssc is the steam flow required to develop unit power output. The power output is $-\dot{m}\sum W$, therefore

$$\text{ssc} = \frac{\dot{m}}{-\dot{m}\sum W} = \frac{1}{-\sum W} \qquad (8.14)$$

Neglecting the feed pump work we have

$$-\sum W = h_1 - h_2$$

therefore

$$\text{ssc} = \frac{1}{(h_1 - h_2)}$$

Note that when h_1 and h_2 are expressed in kilojoules per kilogram then the units of ssc are kg/kJ or kg/kW h.

Example 8.1   A steam power plant operates between a boiler pressure of 42 bar and a condenser pressure of 0.035 bar. Calculate for these limits the cycle efficiency, the work ratio, and the specific steam consumption:

(i) for a Carnot cycle using wet steam;
(ii) for a Rankine cycle with dry saturated steam at entry to the turbine;
(iii) for the Rankine cycle of (ii), when the expansion process has an isentropic efficiency of 80%.

Solution   (i) A Carnot cycle is shown in Fig. 8.5.

Fig. 8.5 Carnot cycle for Example 8.1(a)

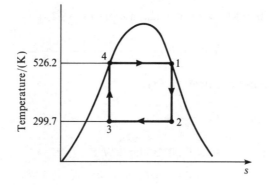

T_1 = saturation temperature at 42 bar

  $= 253.2 + 273 = 526.2$ K

T_2 = saturation temperature at 0.035 bar

  $= 26.7 + 273 = 299.7$ K

Then from equation (5.1)

$$\eta_{\text{Carnot}} = \frac{T_1 - T_2}{T_1} = \frac{526.2 - 299.7}{526.2} = 0.432 \quad \text{or} \quad 43.2\%$$

239

Also   Heat supplied $= h_1 - h_4 = h_{fg}$ at 42 bar $= 1698$ kJ/kg

Then $\eta_{\text{Carnot}} = \dfrac{\text{Net work output}, -\sum W}{\text{Gross heat supplied}} = 0.432$

Therefore $-\sum W = 0.432 \times 1698$,

i.e.   Net work output, $-\sum W = 734$ kJ/kg

To find the gross work of the expansion process it is necessary to calculate h_2, using the fact that $s_1 = s_2$.

From tables

$$h_1 = 2800 \text{ kJ/kg} \quad \text{and} \quad s_1 = s_2 = 6.049 \text{ kJ/kg K}$$

Using equation (4.10)

$$s_2 = 6.049 = s_{f_2} + x_2 s_{fg_2} = 0.391 + x_2 8.13$$

therefore

$$x_2 = \frac{6.049 - 0.391}{8.13} = 0.696$$

Then using equation (2.2)

$$h_2 = h_{f_2} + x_2 h_{fg_2} = 112 + (0.696 \times 2438) = 1808 \text{ kJ/kg}$$

Hence, from equation (8.2)

$$-W_{12} = h_1 - h_2 = 2800 - 1808 = 992 \text{ kJ/kg}$$

Therefore we have, using equation (8.13),

$$\text{Work ratio} = \frac{\text{net work output}}{\text{gross work output}} = \frac{734}{992} = 0.739$$

Using equation (8.14)

$$\text{ssc} = \frac{1}{734}$$

i.e. $\text{ssc} = 0.001\,36$ kg/kW s

      $= 4.91$ kg/kW h

(ii) The Rankine cycle is shown in Fig. 8.6.

As in part (i)

$$h_1 = 2800 \text{ kJ/kg} \quad \text{and} \quad h_2 = 1808 \text{ kJ/kg}$$

Also, $h_3 = h_f$ at 0.035 bar $= 112$ kJ/kg

Using equation (8.10), with $v = v_f$ at 0.035 bar

$$\text{Pump work} = v_f(p_4 - p_3) = 0.001 \times (42 - 0.035) \times \frac{10^5}{10^3}$$

$$= 4.2 \text{ kJ/kg}$$

Fig. 8.6 $T–s$ diagram for Example 8.1(b)

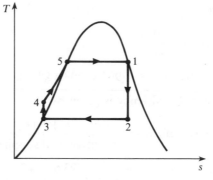

Using equation (8.2)

$$-W_{12} = h_1 - h_2 = 2800 - 1808 = 992 \text{ kJ/kg}$$

Then using equation (8.8)

$$\eta_R = \frac{(h_1 - h_2) - (h_4 - h_3)}{(h_1 - h_3) - (h_4 - h_3)} = \frac{992 - 4.2}{(2800 - 112) - 4.2} = 0.368$$

i.e. $\eta_R = 36.8\%$

Using equation (8.13)

$$\text{Work ratio} = \frac{\text{net work output}}{\text{gross work output}} = \frac{992 - 4.2}{992} = 0.996$$

Using equation (8.14)

$$\text{ssc} = \frac{1}{-\sum W}$$

i.e. $\text{ssc} = \dfrac{1}{(992 - 4.2)} = 0.001\,01 \text{ kg/kW s} = 3.64 \text{ kg/kW h}$

(iii) The cycle with an irreversible expansion process is shown in Fig. 8.7. Using equation (8.12)

$$\text{Isentropic efficiency} = \frac{h_1 - h_2}{h_1 - h_{2s}} = \frac{-W_{12}}{-W_{12s}}$$

Fig. 8.7 $T–s$ diagram for Example 8.1(c)

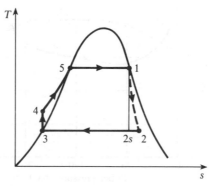

therefore

$$0.8 = \frac{-W_{12}}{992}$$

i.e. $-W_{12} = 0.8 \times 992 = 793.6 \text{ kJ/kg}$

Then the cycle efficiency is given by

$$\text{Cycle efficiency} = \frac{(h_1 - h_2) - (h_4 - h_3)}{\text{gross heat supplied}}$$

$$= \frac{793.6 - 4.2}{(2800 - 112) - 4.2} = 0.294$$

i.e.   Cycle efficiency = 29.4%

$$\text{Work ratio} = \frac{-W_{12} - \text{pump work}}{-W_{12}} = \frac{793.6 - 4.2}{793.6} = 0.995$$

Also

$$\text{ssc} = \frac{1}{793.6 - 4.2} = 0.001\,267 \text{ kg/kW s} = 4.56 \text{ kg/kW h}$$

The feed-pump term has been included in the above calculations, but an inspection of the comparative values shows that it could have been neglected without having a noticeable effect on the results.

It is instructive to carry out these calculations for different boiler pressures and to represent the results graphically against boiler pressure, as in Fig. 8.8. As the boiler pressure increases the specific enthalpy of vaporization decreases, thus less heat is transferred at the maximum cycle temperature. Although the efficiency increases with boiler pressure over the first part of the range, due to the maximum cycle temperature being raised, it is affected by the lowering of the mean temperature at which heat is transferred. Therefore the graph for this efficiency rises, reaches a maximum, and then falls.

Fig. 8.8 Steam cycle efficiency and specific steam consumption against boiler pressure

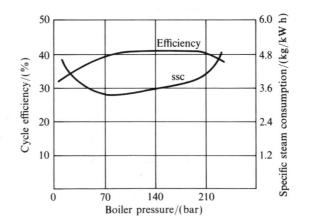

8.2 Rankine cycle with superheat

The average temperature at which heat is supplied in the boiler can be increased by superheating the steam. Usually the dry saturated steam from the boiler drum is passed through a second bank of smaller bore tubes within the boiler. This bank is situated such that it is heated by the hot gases from the furnace until the steam reaches the required temperature.

The Rankine cycle with superheat is shown in Fig. 8.9(a) and 8.9(b). Figure 8.9(a) includes a *steam receiver* which can receive steam from other boilers. In modern plant a receiver is used with one boiler and is placed between the boiler and the turbine. Since the quantity of feedwater varies with the different demands on the boiler, it is necessary to provide a storage of condensate between the condensate and boiler feed pumps. This storage may be either a *surge tank* or *hot well*. A hot well is shown dotted in Fig. 8.9(a).

Fig. 8.9 Steam plant with a superheater (a) and the cycle on a *T–s* diagram (b)

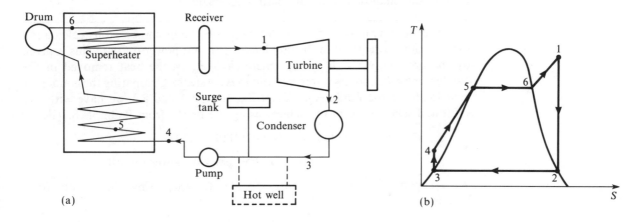

(a)   (b)

Example 8.2   Compare the Rankine cycle performance of Example 8.1 with that obtained when the steam is superheated to 500 °C. Neglect the feed-pump work.

Solution   From tables, by interpolation, at 42 bar:

$$h_1 = 3442.6 \text{ kJ/kg} \quad \text{and} \quad s_1 = s_2 = 7.066 \text{ kJ/kg K}$$

Using equation (4.10)

$$s_2 = s_{f_2} + x_2 s_{fg_2} \quad \text{therefore} \quad 0.391 + x_2 8.13 = 7.066$$

i.e. $x_2 = 0.821$

Using equation (2.2)

$$h_2 = h_{f_2} + x_2 h_{fg_2} = 112 + (0.821 \times 2438) = 2113 \text{ kJ/kg}$$

From tables:

$$h_3 = 112 \text{ kJ/kg}$$

Then, using equation (8.2)

$$-W_{12} = h_1 - h_2 = 3442.6 - 2113 = 1329.6 \text{ kJ/kg}$$

243

Neglecting the feed-pump term, we have

Heat supplied $= h_1 - h_3 = 3442.6 - 112 = 3330.6 \text{ kJ/kg}$

Using equation (8.9)

$$\text{Cycle efficiency} = \frac{h_1 - h_2}{h_1 - h_3} = \frac{1329.6}{3330.6} = 0.399 \text{ or } 39.9\%$$

Also, using equation (8.14)

$$\text{ssc} = \frac{1}{h_1 - h_2} = \frac{1}{1329.6} = 0.000\,752 \text{ kg/kW s} = 2.71 \text{ kg/kW h}$$

The cycle efficiency has increased due to superheating and the improvement in specific steam consumption is even more marked. This indicates that for a given power output the plant using superheated steam will be of smaller proportions than that using dry saturated steam.

The condenser heat loads for different plants can be compared by calculating the rate of heat removal in the condenser, per unit power output. This is given by the product, $\text{ssc} \times (h_2 - h_3)$, where $(h_2 - h_3)$ is the heat removed in the condenser by the cooling water, per unit mass of steam. Comparing the condenser heat loads for the Rankine cycles of Examples 8.1 and 8.2, we have with dry saturated steam at entry to turbine, using the results from Example 8.1(ii):

Condenser heat load $= 0.001\,01(1808 - 112)$

$= 1.713 \text{ kW per kW power output}$

With superheated steam at entry to the turbine, using the results from Example 8.2:

Condenser heat load $= 0.000\,752(2113 - 112)$

$= 1.505 \text{ kW per kW power output}$

For given boiler and condenser pressures, as the superheat temperature increases, the Rankine cycle efficiency increases, and the specific steam consumption decreases, as shown in Fig. 8.10.

Fig. 8.10 Steam cycle efficiency and specific steam consumption against steam temperature at turbine entry

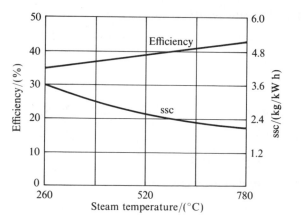

There is also a practical advantage in using superheated steam. For the data of the Rankine cycle of Examples 8.1 and 8.2, the steam leaves the turbine with dryness fractions of 0.696 and 0.821 respectively. The presence of water during the expansion is undesirable, since the droplets are denser than the remainder of the working fluid and therefore have different flow characteristics. The result is the physical erosion of the turbine blades, and a reduction in isentropic efficiency.

The modern tendency is to use higher boiler pressures, and a comparison of cycles on the T–s diagram shows that for a given steam temperature at turbine inlet, the higher pressure plant will have the wetter steam at turbine exhaust (see Fig. 8.11 in which $p_1 > p_2$). It is usual to design for a dryness fraction of not less than 0.9 at the turbine exhaust.

Fig. 8.11 T–s diagram showing the effect of a higher boiler pressure on the steam condition in the turbine

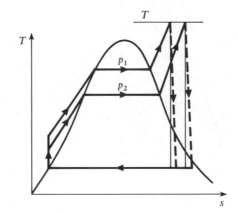

8.3 The enthalpy–entropy chart

In this chapter, and in later ones, we are concerned with changes in enthalpy. It is convenient to have a chart on which enthalpy is plotted against entropy. The h–s chart recommended is that of ref. 8.1, which covers a pressure range of 0.01–1000 bar, and temperatures up to 800 °C. Lines of constant dryness fraction are drawn in the wet region to values less than 0.5, and lines of constant temperature are drawn in the superheat region. In general h–s charts do not show values of specific volume, nor do they show the enthalpies of saturated water at pressures which are of the order of those experienced in steam condensers. Hence the chart is useful only for the enthalpy change in the expansion process of the steam cycle; the methods used in Examples 8.1 and 8.2 are recommended for problems on the Rankine cycle.

A sketch for the h–s chart is shown in Fig. 8.12. Lines of constant pressure are indicated by p_1, p_2, etc.; lines of constant temperature by T_1, T_2, etc. Any two independent properties which appear on the chart are sufficient to define the state (e.g. p_1 and x_1 define state A, and h_A can be read off the vertical axis). In the superheat region, pressure and temperature can define the state (e.g. p_3 and T_4 define the state B, and h_B can be read off). A line of constant entropy between two state points B and C defines the properties at all points during an isentropic process between the two states.

245

Fig. 8.12 Sketch of an enthalpy–entropy chart for steam

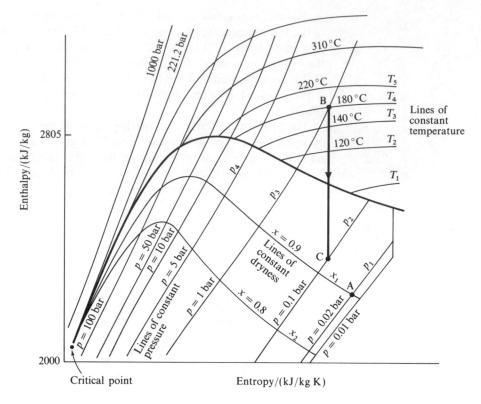

8.4 The reheat cycle

It is desirable to increase the average temperature at which heat is supplied to the steam, and also to keep the steam as dry as possible in the lower pressure stages of the turbine. The wetness at exhaust should be no greater than 10%. The considerations of section 8.2 show that high boiler pressures are required for high efficiency, but that expansion in one stage can result in exhaust steam which is wet. This is a condition which is improved by superheating the steam. The exhaust steam condition can be improved most effectively by reheating the steam, the expansions being carried out in two stages. Referring to Fig. 8.13, 1–2 represents isentropic expansion in the high-pressure turbine, and 6–7 represents isentropic expansion in the low-pressure turbine. The steam is reheated at constant pressure in process 2–6. The reheat can be carried out by returning the steam to the boiler, and passing it through a special bank of tubes, the reheat bank of tubes being situated in the proximity of the superheat tubes. Alternatively, the reheat may take place in a separate reheater situated near the turbine; this arrangement reduces the amount of pipe work required. The use of reheat cycles has encouraged the development of higher pressure, forced circulation boilers, since the specific steam consumption is improved, and the dryness fraction of the exhaust steam is increased.

The analysis is as follows:

$$\text{Heat supplied} = Q_{451} + Q_{26}$$

Fig. 8.13 $T-s$ diagram showing a reheat steam cycle

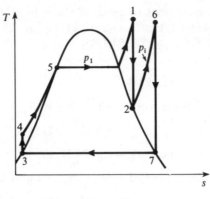

Neglecting the feed-pump work

$$Q_{451} = h_1 - h_3$$

Also, for the reheat process

$$Q_{26} = h_6 - h_2$$

Work output $= -W_{12} - W_{67}$

and $\quad -W_{12} = h_1 - h_2 \quad$ and $\quad -W_{67} = h_6 - h_7$

$$\text{Cycle efficiency} = \frac{-W_{12} - W_{67}}{Q_{451} + Q_{26}}$$

$$= \frac{(h_1 - h_2) + (h_6 - h_7)}{(h_1 - h_3) + (h_6 - h_2)}$$

Example 8.3 Calculate the new cycle efficiency and specific steam consumption if reheat is included in the plant of Example 8.2. The steam conditions at inlet to the turbine are 42 bar and 500 °C, and the condenser pressure is 0.035 bar as before. Assume that the steam is just dry saturated on leaving the first turbine, and is reheated to its initial temperature. Neglect the feed-pump term.

Solution The cycle is shown on a $T-s$ diagram in Fig. 8.14.

Fig. 8.14 $T-s$ diagram for Example 8.3

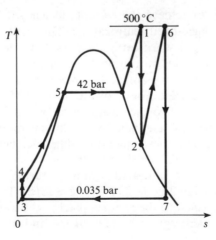

It is convenient to read off the values of enthalpy from the h–s chart, i.e. $h_1 = 3442.6 \text{ kJ/kg}$; $h_2 = 2713 \text{ kJ/kg}$ (at 2.3 bar); $h_6 = 3487 \text{ kJ/kg}$ (at 2.3 bar and 500 °C); $h_7 = 2535 \text{ kJ/kg}$.

From tables

$$h_3 = 112 \text{ kJ/kg}$$

Then   Turbine work $= (h_1 - h_2) + (h_6 - h_7)$

$$= (3443 - 2713) + (3487 - 2535)$$

i.e.   Turbine work $= 1682 \text{ kJ/kg}$

Heat supplied $= (h_1 - h_3) + (h_6 - h_2)$

$$= (3443 - 112) + (3487 - 2713)$$

i.e.   Heat supplied $= 4105 \text{ kJ/kg}$

therefore

$$\text{Cycle efficiency} = \frac{1682}{4105} = 0.41 \text{ or } 41\%$$

Also $\text{ssc} = \dfrac{1}{1682} = 0.000\,595 \text{ kg/kJ}$

i.e. $\text{ssc} = 2.14 \text{ kg/kW h}$

Comparing these answers with the results of Example 8.2 it can be seen that the specific steam consumption has been improved considerably by reheating (i.e. reduced from 2.71 kg/kW h to 2.14 kg/kW h). The efficiency is greater (i.e. increased from 39.9 to 41%). The cycle efficiency will be increased by reheating only if the mean temperature of the heat supply is increased; this will not be the case if the reheat pressure is too low.

8.5   The regenerative cycle

In order to achieve the Carnot efficiency it is necessary to supply and reject heat at single fixed temperatures. One method of doing this, and at the same time having a work ratio comparable to the Rankine cycle, is by raising the feedwater to the saturation temperature corresponding to the boiler pressure before it enters the boiler. This method is not a practical proposition, but is of academic interest. The feedwater is passed from the pump through the turbine in counter-flow to the steam, as shown in Fig. 8.15(a). The feedwater enters the turbine at t_3 and is heated to the steam temperature at inlet to the turbine. If at all points the temperature difference between the steam and the water is negligibly small, then the heat transfer takes place in an ideal reversible manner. Assuming dry saturated steam at turbine inlet, the expansion process is represented by line 1–2–2' in Fig. 8.15(b). The heat rejected by the steam, area 12561, is equal to the heat supplied to the water, area 34783. The heat supplied in the boiler is given by area 41674, and the heat rejected in the condenser is

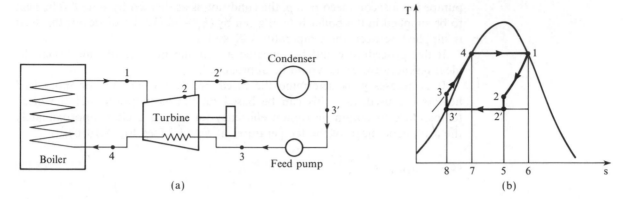

Fig. 8.15 Steam plant operating (a) on a regenerative cycle and (b) the cycle on a $T-s$ diagram

given by area 3'2'583'. This *regenerative* cycle has an efficiency equal to the Carnot cycle, since the heat supplied and rejected externally is done at constant temperature.

This cycle is clearly not a practical proposition, and in addition it can be seen that the turbine operates with wet steam which is to be avoided if possible. However, the Rankine efficiency can be improved upon in practice by bleeding off some of the steam at an intermediate pressure during the expansion, and mixing this steam with feedwater which has been pumped to the same pressure. The mixing process is carried out in a *feed heater*, and the arrangement is represented in Figs 8.16(a) and (b). Only one feed heater is shown but several could be used.

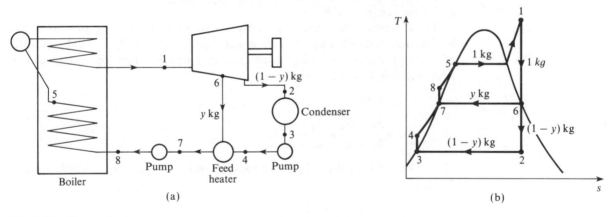

Fig. 8.16 Steam plant with (a) one open feed heater and (b) the cycle on a $T-s$ diagram

The steam expands from condition 1 through the turbine. At the pressure corresponding to point 6, a quantity of steam, say y kg per kilogram of steam supplied from the boiler, is bled off for feed heating purposes. The rest of the steam, $(1 - y)$ kg, completes the expansion and is exhausted at state 2. This amount of steam is then condensed and pumped to the same pressure as the bleed steam. The bleed steam and the feedwater are mixed in the feed heater, and the quantity of bleed steam, y kg, is such that, after mixing and being

249

pumped in a second feed pump, the condition is as defined by state 8. The heat to be supplied in the boiler is then given by $(h_1 - h_8)$ kJ/kg of steam; this heat is supplied between the temperatures T_8 and T_1.

If this procedure could be repeated an infinite number of times, then the ideal regenerative cycle would be approached.

It is necessary to determine the bleed pressure when one or more feed heaters are used, and this can be based on the assumption that the bleed temperature to obtain maximum efficiency for such a cycle is approximately the arithmetic mean of the temperatures at 5 and 2 (see Fig. 8.16(b)),

i.e. $\qquad t_{\text{bleed}} = \dfrac{t_5 + t_2}{2}$ $\qquad\qquad\qquad\qquad\qquad\qquad$ (8.15)

Example 8.4   If the Rankine cycle of Example 8.1 is modified to include one feed heater, calculate the cycle efficiency and the specific steam consumption.

Solution   The steam enters the turbine at 42 bar, dry saturated, and the condenser pressure is 0.035 bar.

At 42 bar, $t_1 = 253.2\,°C$; and at 0.035 bar, $t_2 = 26.7\,°C$.
Therefore by equation (8.15)

$$t_6 = \frac{253.2 + 26.7}{2} = 140\,°C$$

Selecting the nearest saturation pressure from the tables gives the bleed pressure as 3.5 bar (i.e. $t_6 = 138.9\,°C$).

To determine the fraction y, consider the adiabatic mixing process at the feed heater, in which y kg of steam of enthalpy h_6, mix with $(1 - y)$ kg of water of enthalpy h_3, to give 1 kg of water of enthalpy h_7. The feed pump can be neglected (i.e. $h_4 = h_3$). Therefore

$$yh_6 + (1 - y)h_4 = h_7$$

i.e. $\qquad y = \dfrac{h_7 - h_4}{h_6 - h_4} = \dfrac{h_7 - h_3}{h_6 - h_3}$

Now, $h_7 = 584$ kJ/kg; $h_3 = 112$ kJ/kg; and $s_1 = s_6 = s_2 = 6.049$ kJ/kg K. Therefore

$$x_6 = \frac{6.049 - 1.727}{5.214} = 0.829$$

and $\qquad x_2 = \dfrac{6.049 - 0.391}{8.130} = 0.696$

Hence

$$h_6 = h_{f_6} + x_6 h_{fg_6} = 584 + (0.829 \times 2148) = 2364 \text{ kJ/kg}$$

and $\qquad h_2 = h_{f_2} + x_2 h_{fg_2} = 112 + (0.696 \times 2438) = 1808 \text{ kJ/kg}$

therefore

$$y = \frac{584 - 112}{2364 - 112} = 0.21 \text{ kg}$$

Neglecting the second feed-pump term (i.e. $h_7 = h_8$), we have

$$\text{Heat supplied in boiler} = (h_1 - h_7) = 2800 - 584$$
$$= 2216 \text{ kJ/kg}$$

Total work output, $-W = -W_{16} - W_{62} = (h_1 - h_6) + (1 - y)(h_6 - h_2)$

i.e. Work output $= (2800 - 2364) + (1 - 0.21)(2364 - 1808)$
$$= 876 \text{ kJ per kilogram of steam delivered by the boiler}$$

Therefore

$$\text{Cycle efficiency} = \frac{-W}{Q} = \frac{876}{2216} = 0.396 \text{ or } 39.6\%$$

i.e. $\text{ssc} = \dfrac{1}{876} = 0.001\,142 \text{ kg/kJ} = 4.11 \text{ kg/kW h}$

Comparing these results with those of Example 8.1, it can be seen that the addition of one feed heater has increased the cycle efficiency from 36.8 to 39.6%, but the specific steam consumption has increased from 3.64 kg/kW h to 4.11 kg/kW h. The cycle efficiency continues to be increased with the addition of further heaters, but the capital expenditure is also increased considerably. Because of the number of feed pumps required, the heating of the feed water by mixing is dispensed with and *closed heaters* are used. The method is indicated in Fig. 8.17 for two feed heaters, but the number used could be as high as eight. Referring to Fig. 8.17, the feedwater is passed at boiler pressure through the feed heaters 2 and 1 in series. An amount of bleed steam, y_1, is passed to feed heater 1, and the feedwater receives heat from it by the transfer of heat through the separating tubes. The condensed steam is then throttled to the next feed heater which is also supplied with a second quantity of bleed steam, y_2, and a lower temperature heating of the feedwater is carried out. After passing through the various feed heaters the condensed steam is then fed to the condenser. The temperature differences between successive heaters

Fig. 8.17 Steam plant with two closed feed heaters

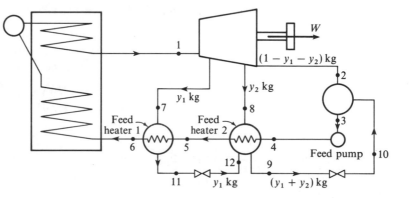

are constant, and in the ideal case the heating process at each is considered to be complete (i.e. the feedwater leaves the feed heater at the temperature of the bleed steam supplied to it).

Example 8.5   In a regenerative steam cycle employing two closed-feed heaters the steam is supplied to the turbine at 40 bar and 500 °C and is exhausted to the condenser at 0.035 bar. The intermediate bleed pressures are obtained such that the saturation temperature intervals are approximately equal, giving pressures of 10 and 1.1 bar.

Calculate the amount of steam bled at each stage, the work output of the plant per kilogram of boiler steam and the cycle efficiency of the plant. Assume ideal processes where required.

Solution   Referring to Fig. 8.17 and the T–s diagram of Fig. 8.18, from tables:

$$h_1 = 3445.8 \text{ kJ/kg} \quad \text{and} \quad s_1 = 7.089 \text{ kJ/kg K} = s_2$$

Fig. 8.18 *T–s* diagram
for Example 8.5

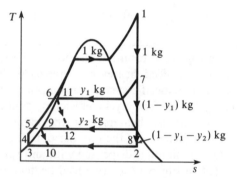

At state 2,

$$0.391 + (x_2 \times 8.13) = 7.089$$

therefore

$$x_2 = \frac{6.698}{8.13} = 0.824$$

i.e. $h_2 = 112 + (0.824 \times 2348) = 2117 \text{ kJ/kg}$

Also $h_3 = h_f$ at 0.035 bar $= 112 \text{ kJ/kg}$

For the first stage of expansion, 1–7, $s_7 = s_1 = 7.089 \text{ kJ/kg K}$, and from tables at 10 bar $s_g < 7.089$, hence the steam is superheated at state 7. By interpolation between 250 and 300 °C at 10 bar we have

$$h_7 = 2944 + \left(\frac{7.089 - 6.926}{7.124 - 6.926}\right)(3052 - 2944) = 2944 + \frac{0.163}{0.198} \times 108$$

i.e. $h_7 = 3032.9 \text{ kJ/kg}$

For the throttling process, 11–12, we have

$$h_6 = h_{11} = h_{12} = 763 \text{ kJ/kg}$$

For the second stage of expansion, 7–8, $s_7 = s_8 = s_1 = 7.089$ kJ/kg K, and from tables at 1.1 bar $s_g > 7.089$ kJ/kg K, hence the steam is wet at state 8. Therefore,

$$1.333 + (x_8 \times 5.994) = 7.089$$

therefore

$$x_8 = 0.961$$

i.e. $\quad h_8 = 429 + (0.961 \times 2251) = 2591$ kJ/kg

For the throttling process, 9–10:

$$h_5 = h_9 = h_{10} = 429 \text{ kJ/kg}$$

Applying an energy balance to the first feed heater, remembering that there is no work or heat transfer:

$$y_1 h_7 + h_5 = y_1 h_{11} + h_6$$

i.e. $\quad y_1 = \dfrac{h_6 - h_5}{h_7 - h_{11}} = \dfrac{763 - 429}{3032.9 - 763} = 0.147$

Similarly for the second heater, taking $h_4 = h_3$:

$$y_2 h_8 + y_1 h_{12} + h_4 = h_5 + (y_1 + y_2)h_9$$

i.e. $\quad y_2(h_8 - h_9) + y_1 h_{12} + h_4 = h_5 + y_1 h_9$

$$y_2(2591 - 429) + (0.147 \times 763) + 112 = 429 + (0.147 \times 429)$$

therefore

$$y_2 = \frac{267.8}{2162} = 0.124$$

The heat supplied to the boiler, Q_1, per kilogram of boiler steam is given by

$$Q_1 = h_1 - h_6 = 3445 - 763 = 2682 \text{ kJ/kg}$$

The work output, neglecting pump work, is given by

$$\begin{aligned}
-W &= (h_1 - h_7) + (1 - y_1)(h_7 - h_8) + (1 - y_1 - y_2)(h_8 - h_2) \\
&= (3445 - 3032.9) + (1 - 0.147)(3032.9 - 2591) \\
&\quad + (1 - 0.147 - 0.124)(2591 - 2117) \\
&= 412.1 + 376.9 + 345.5 = 1134.5 \text{ kJ/kg}
\end{aligned}$$

Then $\quad$ Cycle efficiency $= \dfrac{-W}{Q_1} = \dfrac{1134.5}{2682} = 0.423$ or 42.3%

8.6 Further considerations of plant efficiency

Up to now the considerations of efficiency have been based on the heat which is actually supplied to the steam, and not the heat which has been produced by the combustion of fuel in the boiler. The heat is transferred to the steam from gases which are at a higher temperature than the steam, and the exhaust gases pass to the atmosphere at a high temperature.

To utilize some of the energy in the flue gas an *economizer* can be fitted. This consists of a coil situated in the flue gas stream. The cold feedwater enters at the top of the coil, and as it descends it is heated, and continues to meet higher temperature gas. For the Carnot, the ideal regenerative, and complete feed heating cycles, no use can be made of an economizer since the feedwater enters the boiler at the saturation temperature corresponding to the boiler pressure.

To cool the flue gas even further and improve the plant efficiency, the air which is required for the combustion of the fuel can be pre-heated. For a given temperature of combustion gases, the higher the initial temperature of the air then the less will be the energy input required, and hence less fuel will be used.

Plants which have both economizer and pre-heater coils in the boiler usually require a forced draught for the flue gas, and the power input to the fan, which is a comparatively small quantity, must be taken into account in the energy balance for the plant. Figure 8.19 represents diagrammatically a plant with economizer, pre-heater, and a re-heater.

The boiler efficiency is the heat supplied to the steam in the boiler expressed as a percentage of the chemical energy of the fuel which is available on combustion,

i.e.   Boiler efficiency $= \dfrac{h_1 - (\text{enthalpy of the feedwater})}{m_f \times (\text{GCV or NCV})}$ (8.16)

where h_1 is the enthalpy of the steam entering the turbine and m_f the mass of fuel burned per kilogram of steam delivered from the boiler.

The GCV and NCV are the higher and lower calorific values of the fuel, and the determination of these quantities was considered in Chapter 7.

Fig. 8.19 Steam plant with economizer and air pre-heater

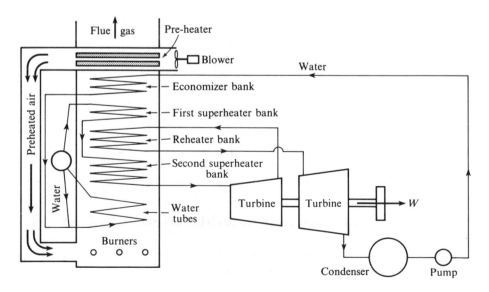

The size of the boiler, or its *capacity*, is quoted as the rate in kilogram per hour at which the steam is generated. A comparison is sometimes made by an *equivalent evaporation*, which is defined as the quantity of steam produced per unit quantity of fuel burned when the evaporation process takes place from and at 100 °C.

8.7 Steam for heating and process use

When steam is required for heating or for a process it may be raised in a boiler and used directly, or passed to a calorifier to heat water which is then circulated. In factory complexes where power and process steam are both required it is usually more efficient to use a plant combining both requirements. In reaching the compromise between power and process demands two main possibilities are available as discussed below.

Back pressure turbine

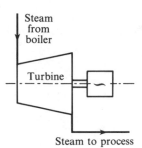

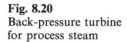

The turbine works with an exhaust pressure which is appropriate to the process steam requirements; the steam leaving the turbine is not condensed but is passed to the process. A typical example is shown in Fig. 8.20.

One of the disadvantages of the back pressure turbine is that if the demand for process steam falls off but the power requirement is unchanged then some steam must be blown off to waste at the process steam pressure. If the power requirement increases with the process demand unchanged then excess power requirements can be bought from the grid. If the power requirements falls off with the process demand unchanged the best solution is to arrange to sell excess power to the grid.

A back pressure turbine used for process steam is suitable for high values of the ratio of process energy to power, say 10 or above.

Fig. 8.20
Back-pressure turbine for process steam

Pass-out turbine

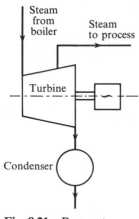

Steam is bled from the turbine at some point or points between inlet and exhaust and is passed to process work. A typical example is shown in Fig. 8.21.

In this system the boiler supply conditions and the condenser pressure can be fixed and the process steam load varied by varying the mass flow rate of process steam bled off the turbine. If the rate of process steam flow is much less than the design value then the excess power generated by the second stage expansion can be solid to the grid; alternatively process steam can be blown off or the boiler operated at part load, but both of these alternatives are wasteful of energy. If the power requirement falls off with the process demand unchanged, the best solution is to sell the excess power to the grid; a more wasteful solution is to blow off excess steam at the bleed point to reduce the power from the second stage of expansion. If the rate of process steam demand increases above the design value the best solution is to use a stand-by boiler to raise the excess

Fig. 8.21 Pass-out turbine for process steam

steam direct. The pass-out turbine system for satisfying process and power requirements is most suitable for low process energy to power ratios, say in the range from 4 to 10.

When the process steam energy to power ratio falls below about 4 it becomes more efficient to generate power and steam for process use separately. Heat–power ratios using various prime movers are discussed in more detail in ref. 7.2.

Example 8.6   A pass-out two-stage turbine receives steam at 50 bar and 350 °C. At 1.5 bar the high-pressure stage exhausts and 12 000 kg of steam per hour are taken at this stage for process purposes. The remainder is reheated at 1.5 bar to 250 °C and then expanded through the low-pressure turbine to a condenser pressure of 0.05 bar. The power output from the turbine unit is to be 3750 kW. The relevant values should be taken from an h–s chart. Take the isentropic efficiency of the high-pressure stage as 0.84, and that of the low-pressure stage as 0.81. Calculate the boiler capacity.

Solution   The processes are shown on an h–s chart in Fig. 8.22. High-pressure stage:

$$\text{Actual work output} = \eta_{\text{isentropic}} \times (h_1 - h_{2s})$$

i.e. $$(h_1 - h_2) = 0.84 \times (3070 - 2397) = 565.3 \text{ kJ/kg}$$

Fig. 8.22 Processes on the h–s chart for Example 8.6

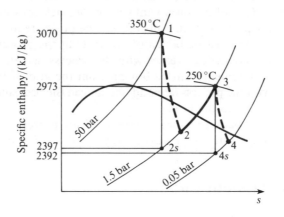

Low-pressure stage:

$$(h_3 - h_4) = \eta_{\text{isentropic}} \times (h_3 - h_{4s})$$

$$= 0.81 \times (2973 - 2392) = 470.6 \text{ kJ/kg}$$

$$\text{Process steam flow} = \frac{12\,000}{3600} = 3.33 \text{ kg/s}$$

Steam flow through the boiler $= \dot{m}$ kg/s

Steam flow through low-pressure stage $= (\dot{m} - 3.33)$ kg/s

Turbine power output $= 3750$ kW

therefore

$$\dot{m}(h_1 - h_2) + (\dot{m} - 3.33)(h_3 - h_4) = 3750$$

i.e. $(\dot{m} \times 565.3) + (\dot{m} - 3.33) \times 470.6 = 3750$

therefore

$$\dot{m} = 5.14\ \text{kg/s}$$

i.e.   Boiler capacity $= 18\,500$ kg of steam per hour

Problems

8.1   (a) Steam is supplied, dry saturated at 40 bar to a turbine and the condenser pressure is 0.035 bar. If the plant operates on the Rankine cycle, calculate, per kilogram of steam:
   (i) the work output neglecting the feed-pump work;
   (ii) the work required for the feed pump;
   (iii) the heat transferred to the condenser cooling water, and the amount of cooling water required through the condenser if the temperature rise of the water is assumed to be 5.5 K;
   (iv) the heat supplied;
   (v) the Rankine efficiency;
   (vi) the specific steam consumption.
   (b) For the same steam conditions calculate the efficiency and the specific steam consumption for a Carnot cycle operating with wet steam.

(986 kJ; 4 kJ; 1703 kJ; 74 kg; 2685 kJ; 36.6%; 3.67 kg/kW h; 42.7%; 4.92 kg/kW h)

8.2   Repeat Problem 8.1(a) for a steam supply condition of 40 bar and 350 °C and the same condenser pressure of 0.035 bar.

(1125 kJ; 4 kJ; 1857 kJ; 80.7 kg; 2978 kJ; 37.6%; 3.21 kg/kW h)

8.3   Steam is supplied to a two-stage turbine at 40 bar and 350 °C. It expands in the first turbine until it is just dry saturated, then it is re-heated to 350 °C and expanded through the second-stage turbine. The condenser pressure is 0.035 bar. Calculate the work output and the heat supplied per kilogram of steam for the plant, assuming ideal processes and neglecting the feed-pump term. Calculate also the specific steam consumption and the cycle efficiency.

(1290 kJ; 3362 kJ; 2.79 kg/kW h; 38.4%)

8.4   If the expansion processes in the turbines of Problem 8.3 have isentropic efficiencies of 84% and 78% respectively, in the first and second stages, calculate the work output and the heat supplied per kilogram of steam, the cycle efficiency, and the specific steam consumption.

   Compare the efficiencies and specific steam consumptions obtained from Problems 8.1, 8.2, 8.3, and 8.4. Compare also the wetness of the steam leaving the turbines in each case.

(1028 kJ; 3311 kJ; 31.1%; 3.5 kg/kW h)
(Dryness fractions at condenser in each case: 0.699, 0.762, 0.85, and 0.94.)

8.5   A generating station is to give a power output of 200 MW. The superheat outlet pressure of the boiler is to be 170 bar and the temperature 600 °C. After expansion through the first-stage turbine to a pressure of 40 bar, 15% of the steam is extracted for feed heating. The remainder is reheated at 600 °C and is then expanded through the second turbine stage to a condenser pressure of 0.035 bar. For preliminary calculations it is assumed that the actual cycle will have an efficiency ratio of 70% and that the generator mechanical and electrical efficiency is 95%. Calculate the maximum continuous rating of the boiler in kilograms per hour.

(632 000 kg/h)

8.6 A steam turbine is to operate on a simple regenerative cycle. Steam is supplied dry saturated at 40 bar, and is exhausted to a condenser at 0.07 bar. The condensate is pumped to a pressure of 3.5 bar at which it is mixed with bleed steam from the turbine at 3.5 bar. The resulting water which is at saturation temperature is then pumped to the boiler. For the ideal cycle calculate, neglecting feed-pump work,
  (i) the amount of bleed steam required per kilogram of supply steam;
 (ii) the cycle efficiency of the plant;
(iii) the specific steam consumption.

(0.1906; 37%; 4.39 kg/kW h)

8.7 Steam is supplied to a two-stage turbine at 40 bar and 500 °C. In the first stage the steam expands isentropically to 3.0 bar at which pressure 2500 kg/h of steam is extracted for process work. The remainder is reheated to 500 °C and then expanded isentropically to 0.06 bar. The by-product power from the plant is required to be 6000 kW. Calculate the amount of steam required from the boiler, and the heat supplied. Neglect feed-pump terms, and assume that the process condensate returns at the saturation temperature to mix adiabatically with the condensate from the condenser.

(14 950 kg/h; 15 880 kW)

8.8 For the plant of Problem 8.7 it is required to improve the efficiency by employing regenerative feed heating by taking off the necessary bleed steam at the same point as the process steam. The process steam is not returned to the boiler, but make-up water at 15 °C is supplied. The bleed steam is mixed with the condensate and make-up water at 3.0 bar such that the resultant water is at the saturation temperature corresponding to 3.0 bar. Calculate:
  (i) the steam supply necessary to meet the same power and process requirements;
 (ii) the amount of bleed steam;
(iii) the heat supplied in kW.
Neglect feed-pump terms.

(16 480 kg/h; 2660 kg/h; 15 460 kW)

8.9 In a regenerative steam cycle employing three closed feed heaters the steam is supplied to the turbine at 42 bar and 500 °C and is exhausted to the condenser at 0.035 bar. The bleed steam for feed heating is taken at pressures of 15, 4, and 0.5 bar. Assuming ideal processes and neglecting pump work, calculate:
  (i) the fraction of the boiler steam bled at each stage;
 (ii) the power output of the plant per unit mass flow rate of boiler steam;
(iii) the cycle efficiency.

(0.0952, 0.0969, 0.0902; 1133.6 kW per kg/s; 43.6%)

8.10 A boiler plant, see Fig. 8.19 (p. 254), incorporates an economizer and an air pre-heater, and generates steam at 40 bar and 300 °C with fuel of calorific value 33 000 kJ/kg burned at a rate of 500 kg/h. The temperature of the feedwater is raised from 40 to 125 °C in the economizer, and the flue gases are cooled at the same time from 395 to 225 °C. The flue gases then enter the air pre-heater in which the temperature of the combustion air is raised by 75 K. A forced-draught fan delivers the air to the pre-heater at a pressure of 1.02 bar and a temperature of 16 °C with a pressure rise across the fan of 180 mm of water. The power input to the fan is 5 kW and it has a mechanical efficiency of 78%. Neglecting heat losses, and taking c_p as 1.01 kJ/kg K for the flue gases, calculate:

 (i) the mass flow rate of air;
 (ii) the temperature of the flue gases leaving the plant;
(iii) the mass flow rate of steam;
(iv) the efficiency of the boiler.

The power required to drive the fan is given by

$$\dot{W} = \frac{h\rho_{\mathrm{w}}g\dot{V}}{\eta_{\mathrm{M}}}$$

where h is the pressure rise across the fan expressed as a head of water, ρ_{w} the density of water, g the acceleration due to gravity, $\dot{V}$ the volume flow rate of air, and η_{M} is the mechanical efficiency of the fan.

(2.72 kg/s; 154 °C; 1.37 kg/s; 83.6%)

References

8.1 HICKSON D C and TAYLOR F R 1980 *Enthalpy–Entropy Diagram for Steam* Basil Blackwell
8.2 EASTOP T D and WATSON W E 1992 *Mechanical Services for Buildings* Longman

9

Gas Turbine Cycles

The simple constant pressure cycle and the open- and closed-cycle gas turbine units have been considered briefly in Chapter 5. In this chapter the various parts of the cycle will be considered in more detail and the practical limitations and modifications to the ideal cycle will be discussed.

The main use for the gas turbine at the present day is in the aircraft field, although gas turbine units for electric power generation are being used increasingly, usually using natural gas as fuel. Gas turbines are used in marine propulsion, but the oil engine and steam turbine are more frequently used, particularly for larger ships. The gas turbine is also used in conjunction with the oil engine, and as part of total energy schemes in combination with steam plant; this is discussed more fully in Chapter 17.

The inefficiencies in the compression and expansion processes become greater for smaller stand-alone gas turbine units and a heat exchanger is frequently used in order to improve the cycle efficiency. A compact effective heat exchanger is necessary before the small gas turbine can compete for economy with the small oil engine or petrol engine.

The use of constant pressure combustion with a rotary compressor driven by a rotary turbine, mounted on a common shaft, gives a combination which is ideal for conditions of steady mass flow over a wide operating range.

9.1 The practical gas turbine cycle

The most basic gas turbine unit is one operating on the open cycle in which a rotary compressor and a turbine are mounted on a common shaft, as shown diagrammatically in Fig. 9.1. Air is drawn into the compressor, C, and after compression passes to a combustion chamber, CC. Energy is supplied in the combustion chamber by spraying fuel into the airstream, and the resulting hot gases expand through the turbine, T, to the atmosphere. In order to achieve net work output from the unit, the turbine must develop more gross work output than is required to drive the compressor and to overcome mechanical losses in the drive.

The compressor used is either a centrifugal or an axial flow compressor and

Fig. 9.1 Open-cycle gas turbine unit

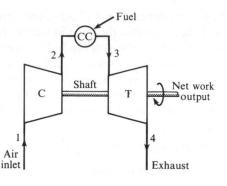

the compression process is therefore irreversible but approximately adiabatic. Similarly the expansion process in the turbine is irreversible but adiabatic. Due to these irreversibilities, more work is required in the compression processes for a given pressure ratio, and less work is developed in the expansion process. It is possible that the compressor and turbine may be so inefficient that the unit is not self-sustaining, and in fact it was the difficulties in improving the compressor and turbine design to cut down irreversibilities that retarded the development of the gas turbine unit.

As stated in section 5.5 the open-cycle gas turbine cannot be compared directly with the ideal constant pressure cycle. The actual cycle involves a chemical reaction in the combustion chamber which results in high-temperature products which are chemically different from the reactants (see section 7.8). During combustion there is no energy exchange with the surroundings, the effect being a gradual decrease in chemical energy with a corresponding increase in enthalpy of the working fluid. The combustion reaction will not be considered in detail here, and a simplification will be made by assuming that the chemical energy released on a combustion is equivalent to a transfer of heat at constant pressure to a working fluid of constant mean specific heat. This simplified approach allows the actual process to be compared with the ideal and to be represented on a T–s diagram.

Neglecting the pressure loss in the combustion chamber the cycle may be drawn on a T–s diagram as shown in Fig. 9.2. Line 1–2 represents irreversible adiabatic compression; line 2–3 represents constant pressure heat supply in the

Fig. 9.2 Gas turbine cycle on a T–s diagram

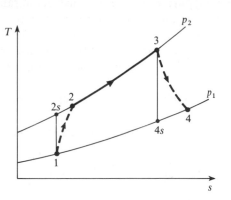

combustion chamber; line 3–4 represents irreversible adiabatic expansion. The process 1–2s represents the ideal isentropic process between the same pressures p_1 and p_2. Similarly the process 3–4s represents the ideal isentropic expansion process between the pressures p_2 and p_1. For the moment it will be assumed that the change in kinetic energy between the various points in the cycle is negligibly small compared with the enthalpy changes. Then applying the flow equation to each part of the cycle, we have the following for unit mass. For the compressor:

$$\text{Work input} = c_p(T_2 - T_1)$$

For the combustion chamber:

$$\text{Heat supplied} = c_p(T_3 - T_2)$$

For the turbine:

$$\text{Work output} = c_p(T_3 - T_4)$$

Then   Net work output $= c_p(T_3 - T_4) - c_p(T_2 - T_1)$

and   Thermal efficiency $= \dfrac{\text{net work output}}{\text{heat supplied}}$

$$= \frac{c_p(T_3 - T_4) - c_p(T_2 - T_1)}{c_p(T_3 - T_2)}$$

The value of the specific heat capacity of a real gas varies with temperature; also, in the open cycle, the specific heat capacity of the gases in the combustion chamber and in the turbine is different from that in the compressor because fuel has been added and a chemical change has taken place. Curves showing the variation of c_p with temperature and air–fuel ratio can be used, and a suitable mean value of c_p and hence γ can be found. It is usual in the gas turbine practice to assume fixed mean values of c_p and γ for the expansion process, and fixed mean values of c_p and γ for the compression process. For the combustion process, curves as shown in Fig. 9.18 (p.282) are used; for simple calculations a mean value of c_p can be assumed. In an open-cycle gas turbine unit the mass flow of gases in the turbine is greater than that in the compressor due to the mass of fuel burned, but it is possible to neglect the mass of fuel, since the air–fuel ratios used are large. Also, in many cases, air is bled from the compressor for cooling purposes, or in the case of aircraft at high altitude, bleed air is used for de-icing and cabin air-conditioning. This amount of air bleed is approximately the same as the mass of fuel injected.

The isentropic efficiency of the compressor is defined as the ratio of the work input required in isentropic compression between p_1 and p_2 to the actual work required.

Neglecting changes in kinetic energy, we have

$$\text{Compressor isentropic efficiency}, \eta_C = \frac{c_p(T_{2s} - T_1)}{c_p(T_2 - T_1)}$$

$$= \frac{T_{2s} - T_1}{T_2 - T_1} \tag{9.1}$$

Similarly the isentropic efficiency of the turbine is defined as the ratio of the actual work output to the isentropic work output between the same pressures.
Neglecting kinetic energy changes

$$\text{Turbine isentropic efficiency}, \eta_{\text{T}} = \frac{c_p(T_3 - T_4)}{c_p(T_3 - T_{4s})}$$

$$= \frac{T_3 - T_4}{T_3 - T_{4s}} \tag{9.2}$$

Example 9.1 A gas turbine unit has a pressure ratio of 10/1 and a maximum cycle temperature of 700 °C. The isentropic efficiencies of the compressor and turbine are 0.82 and 0.85 respectively. Calculate the power output of an electric generator geared to the turbine when the air enters the compressor at 15 °C at the rate of 15 kg/s. Take $c_p = 1.005$ kJ/kg K and $\gamma = 1.4$ for the compression process, and take $c_p = 1.11$ kJ/kg K and $\gamma = 1.333$ for the expansion process.

Solution A line diagram of the unit is shown in Fig. 9.3(a), and the cycle is shown on a T–s diagram in Fig. 9.3(b). In order to evaluate the net work output it is necessary to calculate the temperatures T_2 and T_4. To calculate T_2 we must first calculate T_{2s} and then use the isentropic efficiency.

Fig. 9.3 Gas turbine unit (a) and T–s diagram (b) for Example 9.1

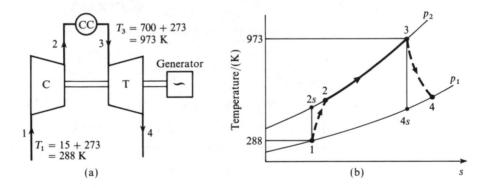

From equation (3.21) for an isentropic process

$$\frac{T_{2s}}{T_1} = \left(\frac{p_2}{p_1}\right)^{(\gamma-1)/\gamma}$$

therefore

$$T_{2s} = 288 \times (10)^{0.4/1.4} = 288 \times 1.931 = 556 \text{ K}$$

Then using equation (9.1)

$$\eta_{\text{C}} = \frac{T_{2s} - T_1}{T_2 - T_1} = \frac{556 - 288}{T_2 - 288} = 0.82$$

i.e. $$(T_2 - 288) = \frac{268}{0.82} = 326.8 \text{ K}$$

263

therefore

$$T_2 = 288 + 326.8 = 614.8 \text{ K}$$

Similarly for the turbine

$$\frac{T_3}{T_{4s}} = \left(\frac{p_2}{p_1}\right)^{(\gamma-1)/\gamma}$$

therefore

$$T_{4s} = \frac{973}{(10)^{0.333/1.333}} = \frac{973}{1.778} = 547.4 \text{ K}$$

Then from equation (9.2)

$$\eta_T = \frac{T_3 - T_4}{T_3 - T_{4s}} = \frac{973 - T_4}{973 - 547.4} = 0.85$$

i.e. $(973 - T_4) = 425.6 \times 0.85 = 361.8 \text{ K}$

therefore

$$T_4 = 973 - 361.8 = 611.2 \text{ K}$$

Hence   Compressor work input $= c_p(T_2 - T_1) = 1.005 \times 326.8$

$$= 328.4 \text{ kJ/kg}$$

Turbine work output $= c_p(T_3 - T_4) = 1.11 \times 361.8$

$$= 401.6 \text{ kJ/kg}$$

therefore

Net work output $= (401.6 - 328.4) = 73.2 \text{ kJ/kg}$

i.e.   Power output $= 73.2 \times 15 = 1098 \text{ kW}$

Example 9.2   Calculate the cycle efficiency and the work ratio of the plant in Example 9.1, assuming that c_p for the combustion process is 1.11 kJ/kg K

Solution   Heat supplied $= c_p(T_3 - T_2)$

$$= 1.11(973 - 614.8) = 1.11 \times 358.2 \text{ kJ/kg}$$

i.e.   Heat supplied $= 397.6 \text{ kJ/kg}$

Therefore

$$\text{Cycle efficiency} = \frac{\text{net work output}}{\text{heat supplied}} = \frac{73.2}{397.6}$$

i.e.   Cycle efficiency $= 0.184$ or 18.4%

From the definition of work ratio given in section 5.3, we have

$$\text{Work ratio} = \frac{\text{net work output}}{\text{gross work output}} = \frac{73.2}{401.6} = 0.182$$

Use of a power turbine

In Examples 9.1 and 9.2 the turbine is arranged to drive the compressor and to develop net work. It is sometimes more convenient to have two separate turbines, one of which drives the compressor while the other provides the power output. The first, or high-pressure (HP) turbine, is then known as the compressor turbine, and the second, or low-pressure (LP) turbine, is called the power turbine. The arrangement is shown in Fig. 9.4(a). Assuming that each turbine has its own isentropic efficiency, the cycle is as shown on a T–s diagram in Fig. 9.4(b). The numbers on Fig. 9.4(b) correspond to those of Fig. 9.4(a). Neglecting kinetic energy changes, we have

$$\text{work from HP turbine} = \text{work input to compressor}$$

i.e. $\qquad c_{p_g}(T_3 - T_4) = c_{p_a}(T_2 - T_1)$

Fig. 9.4 Gas turbine unit with separate power turbine (a) and the cycle on a T–s diagram (b)

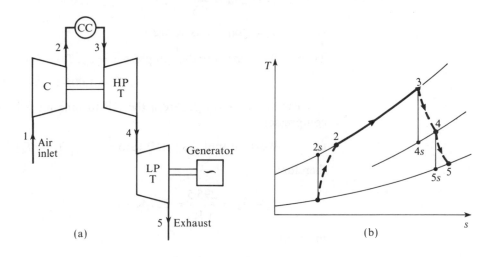

where c_{p_g} and c_{p_a} are the specific heat capacities at constant pressure of the gases in the turbine and the air in the compressor respectively. The net work output is then given by the LP turbine,

i.e. $\qquad \text{Net work output} = c_{p_g}(T_4 - T_5)$

Example 9.3 A gas turbine unit takes in air at $17\,°C$ and 1.01 bar and the pressure ratio is $8/1$. The compressor is driven by the HP turbine and the LP turbine drives a separate power shaft. The isentropic efficiencies of the compressor, and the HP and LP turbines are 0.8, 0.85, and 0.83 respectively. Calculate the pressure and temperature of the gases entering the power turbine, the net power developed by the unit per kg/s mass flow rate, the work ratio and the cycle efficiency of the unit. The maximum cycle temperature is 650 °C. For the compression process take $c_p = 1.005$ kJ/kg K and $\gamma = 1.4$; for the combustion process, and for the expansion process take $c_p = 1.15$ kJ/kg K and $\gamma = 1.333$. Neglect the mass of fuel.

Solution The unit is as shown in Figs 9.4(a) and 9.4(b).

From equation (3.21), for an isentropic process,

$$\frac{T_{2s}}{T_1} = \left(\frac{p_2}{p_1}\right)^{(\gamma-1)/\gamma}$$

i.e. $T_{2s} = 290 \times 8^{0.4/1.4} = 290 \times 1.811 = 525$ K

Then, using equation (9.1)

$$\eta_C = \frac{T_{2s} - T_1}{T_2 - T_1} = \frac{525 - 290}{T_2 - 290} = 0.8$$

therefore

$$T_2 - 290 = \frac{235}{0.8}$$

i.e. $T_2 = 290 + 294 = 584$ K

Then Work input to the compressor $= c_{p_a}(T_2 - T_1)$

$$= 1.005 \times 294 = 295.5 \text{ kJ/kg}$$

Now the work output from the HP turbine must be sufficient to drive the compressor,

i.e. Work output from HP turbine $= c_{p_g}(T_3 - T_4) = 295.5$ kJ/kg

therefore

$$T_3 - T_4 = \frac{295.5}{1.15} = 257 \text{ K}$$

therefore

$$T_4 = T_3 - 257 = 923 - 257 = 666 \text{ K}$$

Then, using equation (9.2),

$$\eta_T \text{ for HP turbine} = \frac{T_3 - T_4}{T_3 - T_{4s}} = \frac{923 - 666}{923 - T_{4s}} = 0.85$$

i.e. $923 - T_{4s} = \dfrac{257}{0.85} = 302.5$ K

therefore

$$T_{4s} = 923 - 302.5 = 620.5 \text{ K}$$

Then from equation (3.21) for an isentropic process,

$$\frac{T_3}{T_{4s}} = \left(\frac{p_3}{p_4}\right)^{(\gamma-1)/\gamma}$$

or $\qquad \dfrac{p_3}{p_4} = \left(\dfrac{T_3}{T_{4s}}\right)^{\gamma/(\gamma-1)} = \left(\dfrac{923}{620.5}\right)^{1.333/0.333} = 4.9$

i.e. $\qquad p_4 = \dfrac{p_3}{4.9} = \dfrac{8 \times 1.01}{4.9} = 1.65 \text{ bar}$

Hence the pressure and temperature at entry to the LP turbine are 1.65 bar and 393 °C, where $t_4 = 666 - 273 = 393$ °C.

To find the power output it is now necessary to evaluate T_5. The pressure ratio, p_4/p_5, is given by $(p_4/p_3) \times (p_3/p_5)$,

i.e. $\qquad \dfrac{p_4}{p_5} = \dfrac{p_4}{p_3} \times \dfrac{p_2}{p_1}$ (since $p_2 = p_3$ and $p_5 = p_1$)

therefore

$$\frac{p_4}{p_5} = \frac{8}{4.9} = 1.63$$

Then $\qquad \dfrac{T_4}{T_{5s}} = \left(\dfrac{p_4}{p_5}\right)^{(\gamma-1)/\gamma} = 1.63^{0.333/1.333} = 1.131$

therefore

$$T_{5s} = \frac{666}{1.131} = 588 \text{ K}$$

Then, using equation (9.2)

$$\eta_T \text{ for the LP turbine} = \frac{T_4 - T_5}{T_4 - T_{5s}}$$

i.e. $\qquad T_4 - T_5 = 0.83(666 - 588) = 0.83 \times 78 = 64.8 \text{ K}$

Then $\qquad$ Work output from LP turbine $= c_{p_g}(T_4 - T_5)$

$$= 1.15 \times 64.8 = 74.5 \text{ kJ/kg}$$

i.e. $\qquad$ Net power output $= 74.5 \times 1 = 74.5 \text{ kW}$

$$\text{Work ratio} = \frac{\text{net work output}}{\text{gross work output}} = \frac{74.5}{74.5 + 295.5} = \frac{74.5}{370} = 0.201$$

Heat supplied $= c_{p_g}(T_3 - T_2) = 1.15(923 - 584)$

i.e. $\qquad$ Heat supplied $= 1.15 \times 339 = 390 \text{ kJ/kg}$

Then $\qquad$ Cycle efficiency $= \dfrac{\text{net work output}}{\text{heat supplied}} = \dfrac{74.5}{390}$

$$= 0.191 \text{ or } 19.1\%$$

Gas Turbine Cycles

Aircraft engines

In a *jet engine* the propulsion nozzle takes the place of the LP stage turbine, as shown diagrammatically in Fig. 9.5(a). The cycle is shown on a *T–s* diagram in Fig. 9.5(b), and it can be seen to be identical with Fig. 9.4(b). The aircraft is powered by the reactive thrust of the jet of gases leaving the nozzle, and this high-velocity jet is obtained at the expense of the enthalpy drop from 4 to 5. The turbine develops just enough work to drive the compressor and overcome mechanical losses.

Fig. 9.5 Simple jet engine (a) with the cycle on a *T–s* diagram (b)

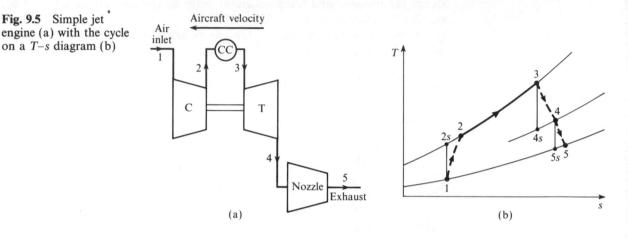

In a *turbo-prop* engine the turbine drives the compressor and also the airscrew, or propeller, as shown in Figs 9.6(a) and 9.6(b). The net work output available to drive the propeller is given by

$$\text{Net work output} = c_{p_g}(T_3 - T_4) - c_{p_a}(T_2 - T_1)$$

(neglecting mechanical losses).

Fig. 9.6 Turboprop engine (a) with the cycle on a *T–s* diagram (b)

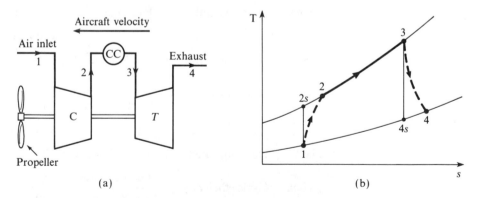

In practice there is also a small jet thrust developed in a turbo-prop aircraft. Jet engines and turbo-prop engines are considered again in Section 10.9.

Parallel flow units

In some industrial and marine gas turbine units, the air flow is split into two streams after the compression process is completed. Some air is then passed to a combustion chamber which supplies hot gases to the turbine driving the compressor, while the rest of the air is passed to a second combustion chamber and from thence to the power turbine. The system is shown diagrammatically in Fig. 9.7, and is called a *parallel flow* unit. In this system each turbine expands the gases received by it through the full pressure ratio. The advantage of this system is that the net power output can be varied using the second combustion chamber, and the power turbine operates independently of the compressor turbine.

Fig. 9.7 Parallel-flow gas turbine unit

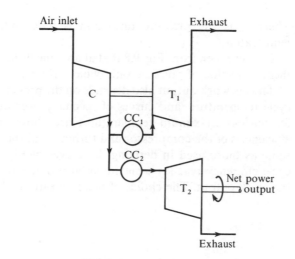

9.2 Modifications to the basic cycle

It can be seen from Examples 9.1, 9.2, and 9.3 that the work ratio and the cycle efficiency of the basic gas turbine cycle are low. These can be improved by increasing the isentropic efficiencies of the compressor and turbine, and this is a matter of blade design and manufacture.

In a practical cycle with irreversibilities in the compression and expansion processes the cycle efficiency depends on the maximum cycle temperatures as well as on the pressure ratio. For fixed values of the isentropic efficiencies of the compressor and turbine, the cycle efficiency can be plotted against pressure ratio for various values of maximum temperature. This is illustrated in Fig. 9.8, for a cycle in which the compressor isentropic efficiency is 0.89, the turbine isentropic efficiency is 0.92, and the air inlet temperature is 20 °C. The ideal air standard cycle thermal efficiency is shown chain-dotted. In section 5.4 it is shown that the ideal constant pressure cycle efficiency is given by

$$\eta = 1 - \left(\frac{1}{r_p}\right)^{(\gamma - 1)/\gamma}$$

269

Fig. 9.8 Gas turbine cycle efficiency against pressure ratio for different maximum cycle temperatures

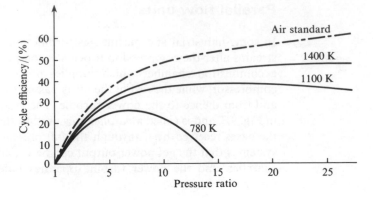

where r_p is the pressure ratio and is independent of the maximum cycle temperature.

It can be seen from Fig. 9.8 that at any one fixed maximum cycle temperature there is a value of pressure ratio which will give maximum cycle efficiency.

The net work output also depends on the pressure ratio and on the maximum cycle temperature, and curves of specific power output against pressure ratio for various maximum temperatures are shown in Fig. 9.9. The isentropic efficiencies of the compressor and turbine, and the air inlet temperature are the same as those used in deriving the curves of Fig. 9.8. It can be seen that the cycle efficiency reaches a maximum at a different value of pressure ratio than the work output. The choice of pressure ratio is therefore a compromise.

Fig. 9.9 Specific power against pressure ratio for different maximum cycle temperatures

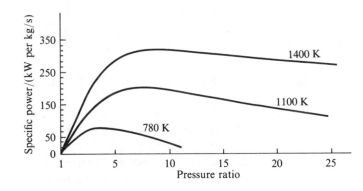

The maximum cycle temperature is limited by metallurgical considerations. The blades of the turbine are under great mechanical stress and the temperature of the blade material must be kept to a safe working value. The temperature of the gases entering the turbine can be raised, provided a means of blade cooling is available. Various methods of blade cooling have been investigated and a discussion of these will be found in ref. 9.1. In aircraft practice where the life expectancy of the engine is shorter, the maximum temperatures used are usually higher than those used in industrial and marine gas turbine units; more expensive alloys and blade cooling allow maximum temperatures of above 1600 K.

It is important to have as high a work ratio as possible, and methods of increasing the work ratio, such as intercooling between compressor stages, and reheating between turbine stages, will be considered in this section. Intercooling and reheating, while increasing the work ratio, can cause a decrease in the cycle efficiency, but when they are used in conjunction with a heat exchanger then intercooling and reheating increase both the work ratio and the cycle efficiency.

Intercooling

When the compression is performed in two stages with an intercooler between the stages, then the work input for a given pressure ratio and mass flow is reduced. Consider a system as shown in Fig. 9.10(a); the T–s diagram for the unit is shown in Fig. 9.10(b). The actual cycle processes are 1–2 in the LP compressor, 2–3 in the intercooler, 3–4 in the HP compressor, 4–5 in the combustion chamber, and 5–6 in the turbine. The ideal cycle for this arrangement is 1–2s–3–4s–5–6s; the compression process without intercooling is shown as 1–A in the actual case, and 1–As in the ideal isentropic case.

Fig. 9.10 Gas turbine unit with intercooling (a) and the cycle on the T–s diagram (b)

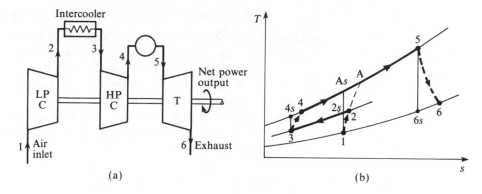

(a) (b)

The work input with intercooling is given by

$$\text{Work input (with intercooling)} = c_p(T_2 - T_1) + c_p(T_4 - T_3) \qquad (9.3)$$

The work input with no intercooling is given by

$$\text{Work input (no intercooling)} = c_p(T_A - T_1)$$
$$= c_p(T_2 - T_1) + c_p(T_A - T_2)$$

Comparing this equation with equation (9.3), it can be seen that the work input with intercooling is less than the work input with no intercooling, when $c_p(T_4 - T_3)$ is less than $c_p(T_A - T_2)$. This is so if it is assumed that the isentropic efficiencies of the two compressors, operating separately, are each equal to the isentropic efficiency of the single compressor which would be required if no intercooling were used. Then $(T_4 - T_3) < (T_A - T_2)$ since the pressure lines diverge from left to right on the T–s diagram.

It can be shown that the best interstage pressure is the one which gives equal pressure ratios in each stage of compression; referring to Fig. 9.10(b) this means that $p_2/p_1 = p_4/p_3$. The work input required is a minimum when the pressure ratio in each stage is the same, and when the temperature of the air is cooled in the intercooler, back to the value at inlet to the unit (i.e. referring to Fig. 9.10(b), $T_3 = T_1$).

Now

$$\text{Work ratio} = \frac{\text{net work output}}{\text{gross work output}}$$

$$= \frac{\text{work of expansion} - \text{work of compression}}{\text{work of expansion}}$$

It follows, therefore, that when the compressor work input is reduced then the work ratio is increased. However, referring to Fig. 9.10(b), the heat supplied in the combustion chamber when intercooling is used in the cycle is given by

$$\text{Heat supplied (with intercooling)} = c_p(T_5 - T_4)$$

whereas the heat supplied when intercooling is not used, with the same maximum cycle temperature T_5, is given by

$$\text{Heat supplied (no intercooling)} = c_p(T_5 - T_A)$$

Hence the heat supplied when intercooling is used is greater than with no intercooling. Although the net work output is increased by intercooling it is found in general that the increase in the heat to be supplied causes the cycle efficiency to decrease. It will be shown later that this disadvantage is offset when a heat exchanger is also used.

When intercooling is used a supply of cooling water must be readily available. The additional bulk of the unit may offset the advantage to be gained by increasing the work ratio.

Reheat

As stated earlier, the expansion process is very frequently performed in two separate turbine stages, the HP turbine driving the compressor and the LP turbine providing the useful power output. The work output of the LP turbine can be increased by raising the temperature at inlet to this stage. This can be done by placing a second combustion chamber between the two turbine stages in order to heat the gases leaving the HP turbine. The system is shown diagrammatically in Fig. 9.11(a), and the cycle is represented on a T–s diagram in Fig. 9.11(b). The line 4–A represents the expansion in the LP turbine if reheating is not used.

As before, the work output of the HP turbine must be exactly equal to the work input required for the compressor (neglecting mechanical losses),

i.e. $\quad c_{p_a}(T_2 - T_1) = c_{p_g}(T_3 - T_4)$

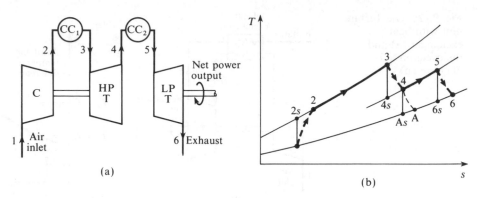

Fig. 9.11 Gas turbine unit with reheating (a) and the cycle on a T–s diagram (b)

The net work output, which is the work output of the LP turbine, is given by

$$\text{Net work output} = c_{p_g}(T_5 - T_6)$$

If reheating is not used, then the work of the LP turbine is given by

$$\text{Net work output (no reheat)} = c_{p_g}(T_4 - T_A)$$

Since pressure lines diverge to the right on the T–s diagram, it can be seen that the temperature difference $(T_5 - T_6)$ is always greater than $(T_4 - T_A)$, so that reheating increases the net work output. Also

$$\text{Work ratio} = \frac{\text{work of expansion} - \text{work of compression}}{\text{work of expansion}}$$

i.e.

$$\text{Work ratio} = 1 - \frac{\text{work of compression}}{\text{work of expansion}}$$

Therefore, when the work of expansion is increased and the work of compression is unchanged, then the work ratio is increased.

Although the net work is increased by reheating, the heat to be supplied is also increased, and the net effect can be to reduce the thermal efficiency,

i.e.

$$\text{Heat supplied} = c_{p_g}(T_3 - T_2) + c_{p_g}(T_5 - T_4)$$

However, the exhaust temperature of the gases leaving the LP turbine is much higher when reheating is used (i.e. T_6 as compared with T_A), and a heat exchanger can be used to enable some of the energy of the exhaust gases to be used.

Heat exchanger

The exhaust gases leaving the turbine at the end of expansion are still at a high temperature, and therefore a high enthalpy (e.g. in Example 9.3, $t_5 = 328.2\,°\text{C}$). If these gases are allowed to pass into the atmosphere, then this represents a loss of available energy. Some of this energy can be recovered by passing the gases from the turbine through a heat exchanger, where the heat transferred from the gases is used to heat the air leaving the compressor. The simple unit

273

Fig. 9.12 Gas turbine unit with heat exchanger (a) and the cycle on a T–s diagram (b)

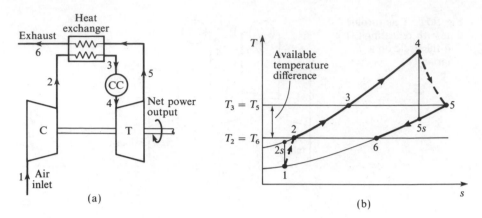

(a)

(b)

with a heat exchanger added is shown diagrammatically in Fig. 9.12(a), and the cycle is represented on a T–s diagram in Fig. 9.12(b). In the ideal heat exchanger the air would be heated from T_2 to $T_3 = T_5$ and the gases would be cooled from T_5 to $T_6 = T_2$. This ideal case is shown in Fig. 9.12(b). In practice this is impossible, since a finite temperature difference is required at all points in the heat exchanger in order to overcome the resistance to the heat transfer. Referring to Fig. 9.13, the required temperature difference between the gases and the air entering the heat exchanger is $(T_6 - T_2)$, and the required temperature difference between the gases and the air leaving the heat exchanger is $(T_5 - T_3)$.

Fig. 9.13 T–s diagram for a gas turbine unit with a heat exchanger showing temperature differences for heat transfer

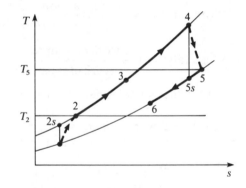

If no heat is lost from the heat exchanger to the atmosphere, then the heat given up by the gases must be exactly equal to the heat taken up by the air,

i.e. $\quad \dot{m}_a c_{p_a}(T_3 - T_2) = \dot{m}_g c_{p_g}(T_5 - T_6)$ (9.4)

The assumption that no heat is lost from the heat exchanger is sufficiently accurate in most practical cases. Equation (9.4) is therefore true whatever the temperatures T_3 and T_6 may be.

A heat exchanger *effectiveness* is defined to allow for the temperature difference necessary for the transfer of heat,

i.e. $\quad \text{Effectiveness} = \dfrac{\text{heat received by the air}}{\substack{\text{maximum possible heat which could be transferred} \\ \text{from the gases in the heat exchanger}}}$

therefore

$$\text{Effectiveness} = \frac{\dot{m}_a c_{p_a}(T_3 - T_2)}{\dot{m}_g c_{p_g}(T_5 - T_2)} \tag{9.5}$$

A more convenient way of assessing the performance of the heat exchanger is to use a *thermal ratio*, defined as

$$\text{Thermal ratio} = \frac{\text{temperature rise of the air}}{\text{maximum temperature difference available}}$$

i.e. $$\text{Thermal ratio} = \frac{T_3 - T_2}{T_5 - T_2} \tag{9.6}$$

Comparing equations (9.5) and (9.6) it can be seen that the thermal ratio is equal to the effectiveness when the product, $\dot{m}_a c_{p_a}$, is equal to the product, $\dot{m}_g c_{p_g}$.

When a heat exchanger is used then the heat to be supplied in the combustion chamber is reduced, assuming that the maximum cycle temperature is unchanged. The net work output is unchanged and hence the cycle efficiency is increased.

Referring to Fig. 9.13

$$\text{Heat supplied by the fuel (without heat exchanger)} = c_{p_g}(T_4 - T_2)$$

$$\text{Heat supplied by the fuel (with heat exchanger)} = c_{p_g}(T_4 - T_3)$$

A heat exchanger can be used only if there is a sufficiently large temperature difference between the gases leaving the turbine and the air leaving the compressor. For example, in the cycle shown in Fig. 9.14 a heat exchanger could not possibly be used because the temperature of the exhaust gases, T_4, is lower than the temperature of the air leaving the compressor, T_2. In practice, although the gas temperature may be higher than the temperature of the air leaving the compressor, the difference in temperature may not be sufficiently large to warrant the additional capital cost and subsequent maintenance required for a heat exchanger. Also, when the temperature difference is small in a heat exchanger, then the surface areas for the heat transfer must be made large in order to achieve a reasonably high value of the thermal ratio. For small gas turbine units (e.g. for pumping sets or for motor cars) a compact heat exchanger must be designed before such units can hope to become competitive for economy

Fig. 9.14 Example of a cycle where a heat exchanger is not feasible

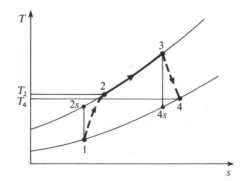

with conventional internal combustion engines of equivalent power. In large gas turbine units for marine propulsion or industrial power, a heat exchanger may be used, although the trend now is towards combined cycles using the turbine exhaust to generate steam or heat water (see Chapter 17).

Example 9.4

A 5000 kW gas turbine generating set operates with two compressor stages with intercooling between stages; the overall pressure ratio is 9/1. A HP turbine is used to drive the compressors, and a LP turbine drives the generator. The temperature of the gases at entry to the HP turbine is 650 °C and the gases are reheated to 650 °C after expansion in the first turbine. The exhaust gases leaving the LP turbine are passed through a heat exchanger to heat the air leaving the HP stage compressor. The compressors have equal pressure ratios and intercooling is complete between stages. The air inlet temperature to the unit is 15 °C. The isentropic efficiency of each compressor stage is 0.8 and the isentropic efficiency of each turbine stage is 0.85; the heat exchanger thermal ratio is 0.75. A mechanical efficiency of 98% can be assumed for both the power shaft and the compressor turbine shaft. Neglecting all pressure losses and changes in kinetic energy, calculate:

  (i) the cycle efficiency;
 (ii) the work ratio;
(iii) the mass flow rate.

For air take $c_p = 1.005 \text{ kJ/kg K}$ and $\gamma = 1.4$, and for the gases in the combustion chamber and in the turbines and heat exchanger take $c_p = 1.15 \text{ kJ/kg K}$ and $\gamma = 1.333$. Neglect the mass of fuel.

Solution

(i) The plant is shown diagrammatically in Fig. 9.15(a), and the cycle is represented on a T–s diagram in Fig. 9.15(b).

Since the pressure ratio and the isentropic efficiency of each compressor is the same, then the work input required for each compressor is the same since both compressors have the same air inlet temperature, i.e. $T_1 = T_3$ and $T_2 = T_4$.

Fig. 9.15 Gas turbine plant (a) and T–s diagram (b) for Example 9.4

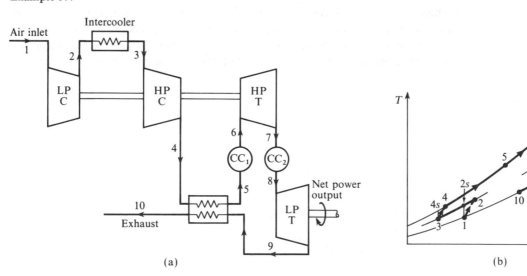

(a)

(b)

From equation (3.21)

$$\frac{T_{2s}}{T_1} = \left(\frac{p_2}{p_1}\right)^{(\gamma-1)/\gamma} \quad \text{and} \quad \frac{p_2}{p_1} = \sqrt{9} = 3$$

therefore

$$T_{2s} = 288 \times 3^{0.4/1.4} = 394 \text{ K}$$

Then from equation (9.1),

$$\eta_C, \text{ LP compressor} = \frac{T_{2s} - T_1}{T_2 - T_1} = 0.8$$

therefore

$$T_2 - T_1 = \frac{394 - 288}{0.8} = \frac{106}{0.8} = 132.5 \text{ K}$$

i.e. $\quad T_2 = 288 + 132.5 = 420.5 \text{ K}$

Also $\quad$ Work input per compressor stage $= c_{p_a}(T_2 - T_1)$

$$= 1.005 \times 132.5 = 133.1 \text{ kJ/kg}$$

The HP turbine is required to drive both compressors and to overcome mechanical friction,

i.e. $\quad$ Work output of HP turbine $= \dfrac{2 \times 133.1}{0.98} = 272 \text{ kJ/kg}$

therefore

$$c_{p_g}(T_6 - T_7) = 272$$

i.e. $\quad 1.15(923 - T_7) = 272$

therefore

$$923 - T_7 = \frac{272}{1.15} = 236.5 \text{ K}$$

i.e. $\quad T_7 = 923 - 236.5 = 686.5 \text{ K}$

From equation (9.2)

$$\eta_T, \text{ HP turbine} = \frac{T_6 - T_7}{T_6 - T_{7s}} = 0.85$$

therefore

$$T_6 - T_{7s} = \frac{236.5}{0.85} = 278 \text{ K}$$

i.e. $\quad T_{7s} = 923 - 278 = 645 \text{ K}$

Then using equation (3.21)

$$\frac{p_6}{p_7} = \left(\frac{T_6}{T_{7s}}\right)^{\gamma/(\gamma-1)} = \left(\frac{923}{645}\right)^{1.333/0.333} = 4.19$$

Then $\dfrac{p_8}{p_9} = \dfrac{9}{4.19} = 2.147$

Using equation (3.21)

$$\frac{T_8}{T_{9s}} = \left(\frac{p_8}{p_9}\right)^{(\gamma-1)/\gamma} = 2.147^{0.333/1.333} = 1.211$$

therefore

$$T_{9s} = \frac{923}{1.211} = 762.6 \text{ K}$$

Then using equation (9.2)

$$\eta_T, \text{ LP turbine} = \frac{T_8 - T_9}{T_8 - T_{9s}} = 0.85$$

therefore

$$T_8 - T_9 = 0.85 \times (923 - 762.6) = 136.3 \text{ K}$$

i.e. $T_9 = 923 - 136.3 = 786.7 \text{ K}$

Therefore

$$\text{Net work output} = c_{p_g}(T_8 - T_9) \times 0.98$$
$$= 1.15 \times 136.3 \times 0.98 = 153.7 \text{ kJ/kg}$$

From equation (9.6)

$$\text{Thermal ratio of heat exchanger} = \frac{T_5 - T_4}{T_9 - T_4} = 0.75$$

i.e.

$$T_5 - 420.5 = 0.75(786.7 - 420.5) = 274.7 \text{ K}$$

therefore

$$T_5 = 420.5 + 274.7 = 695.2 \text{ K}$$

Now Heat supplied $= c_{p_g}(T_6 - T_5) + c_{p_g}(T_8 - T_7)$
$$= 1.15\{(923 - 695.2) + (923 - 686.5)\} = 534 \text{ kJ/kg}$$

Then, from equation (5.2)

$$\text{Cycle efficiency} = \frac{-\dot{W}}{\dot{Q}} = \frac{153.7}{534} = 0.288 \text{ or } 28.8\%$$

(ii) Gross work output
    = work output of HP turbine + work output of LP turbine

i.e.   Gross work output $= 272 + \dfrac{153.7}{0.98} = 429\,\text{kJ/kg}$

Therefore

$$\text{Work ratio} = \frac{\text{net work output}}{\text{gross work output}} = \frac{153.7}{429} = 0.358$$

(iii) The electrical output is 5000 kW. Let the mass flow rate be $\dot{m}$ kg/s, then

$$5000 = \dot{m} \times 153.7$$

$$\dot{m} = \frac{5000}{153.7} = 32.6\,\text{kg/s}$$

i.e.   Rate of flow of air $= 32.6\,\text{kg/s}$

Effect of pressure loss

In Example 9.4 all pressure losses were neglected. In an actual gas turbine unit there are pressure losses due to friction and turbulence in the intercooler, in the air side of the heat exchanger, in both combustion chambers, and in the gas side of the heat exchanger, and in the exhaust duct. The high heat transfer rate in a combustion chamber leading to an appreciable velocity increase in a duct of approximately constant cross-sectional area causes a further pressure loss in addition to that due to friction and turbulence.

Example 9.5   For the gas turbine generating set of Example 9.4 recalculate the cycle efficiency and work ratio, taking the following pressure losses into account, but assuming all other assumptions still apply: air side of heat exchanger, 0.3 bar; gas side of heat exchanger and exhaust duct, 0.05 bar; intercooler, 0.15 bar; each combustion chamber, 0.2 bar.

Take an ambient pressure of 1.01 bar, a pressure ratio for each compressor of 3 : 1 as previously calculated, and find a new overall pressure ratio for the compression. All other data are unchanged.

Solution   Referring to the T–s diagram shown in Fig. 9.16, as before:

$$T_{2s} = 288(3)^{0.286} = 394\,\text{K}$$

Fig. 9.16 *T–s* diagram showing pressure losses for Example 9.5

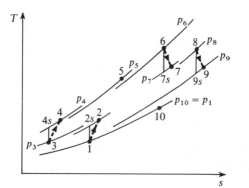

and using the isentropic efficiency
$$T_2 = 420.5 \text{ K} = T_4$$

Also as before

Work output per compressor stage $= 133.1 \text{ kJ/kg}$

The pressure at inlet to the HP compressor, p_3, is now given by $(3 \times 1.01) - 0.15 = 2.88$ bar, and at outlet from the HP compressor, p_4, is $3 \times 2.88 = 8.64$ bar. The new overall pressure ratio is therefore $8.64/1.01 = 8.555$, compared with 9 previously. The pressure at entry to the HP turbine, p_6, is now $8.64 - 0.3 - 0.2 = 8.14$ bar.

The work output of the HP turbine is given as before by $(2 \times 133.1/0.98) = 272 \text{ kJ/kg}$, and hence the temperatures T_7 and T_{7s} are the same as before and hence the ratio $p_6/p_7 = 4.19$ is also the same as before,

i.e. $\quad p_7 = p_6/4.19 = 8.14/4.19 = 1.943 \text{ bar}$

Therefore
$$p_8 = (p_7 - 0.2) = 1.743 \text{ bar}$$

Now $p_{10} = p_1 = 1.01$ bar and therefore
$$p_9 = 1.01 + 0.05 = 1.06 \text{ bar}$$
$$p_8/p_9 = 1.743/1.06 = 1.644$$

Then $\quad T_{9s} = T_8/(1.644)^{0.333/1.333} = 923/1.132 = 815.4 \text{ K}$

and $\quad T_9 = 923 - (923 - 815.4) \times 0.85 = 831.5 \text{ K}$

Therefore
$$\text{Net work output} = c_{p_g}(T_8 - T_9) = 1.15(923 - 831.5)$$
$$= 105.2 \text{ kJ/kg}$$

Then using equation (9.6) for the thermal ratio of the heat exchanger as before, we have
$$T_5 = 420.5 + 0.75(831.5 - 420.5)$$
$$= 728.8 \text{ K}$$

Then $\quad$ Heat supplied $= c_{p_g}(T_6 - T_5) + c_{p_g}(T_8 - T_7)$
$$= 1.15(923 - 728.8) + 1.15(923 - 686.5) = 495.3 \text{ kJ/kg}$$

Hence $\quad$ Cycle efficiency $= 105.2/495.3 = 21.2\%$

This compares with the previous value of 28.8% when pressure losses are neglected.

The gross work of the plant is $(105.2/0.98) + 277 = 384.3 \text{ kJ/kg}$. Therefore,

Work ratio $= 105.2/384.3 = 0.274$

This compares with the previous value of 0.358 when pressure losses were neglected.

9.3 Combustion

In the closed-cycle gas turbine unit heat is transferred to the air in a heat exchanger, but in the open-cycle unit the fuel must be sprayed into the air continuously, and combustion is a continuous process unlike the cyclic combustion of the IC engine.

There are two main combustion systems for open cycles: one in which the air leaving the compressor is split into several streams and each stream is supplied to a separate cylindrical 'can'-type combustion chamber, and the other in which the air flows from the compressor through an annular combustion chamber. The annular type would appear to be more suitable for a unit using an axial flow compressor, but it is difficult to obtain good fuel–air distribution and research and development work on this type is harder than with the simpler can type. The annular type can be modified by having a series of interconnected cans placed in a ring; this is known as the cannular type. In aircraft practice at present the majority of engines use either the cannular or the can type of combustion chamber.

In industrial plants where space is not important the combustion may be arranged to take place in one or two large cylindrical combustion chambers with ducting to convey the hot gases to the turbine; this system gives better control over the combustion process.

In all types of combustion chamber, combustion is initiated by electrical ignition, and once the fuel starts burning, a flame is stabilized in the chamber. In the can type it is usual to have interconnecting pipes between cans, to stabilize the pressure and to allow combustion to be initiated by a spark in one chamber on starting up. A typical can-type chamber is shown diagrammatically in Fig. 9.17. Some of the air from the compressor is introduced directly to the fuel burner; this is called primary air, and represents about 25% of the total airflow. The remaining air enters the annulus round the flame tube, thus cooling the upper portion of the flame tube, and then enters the combustion zone through dilution holes as shown in Fig. 9.17. The primary air forms a comparatively rich mixture and the temperature is high in this zone. The air entering the dilution holes completes the combustion and helps to stabilize the flame in the high-temperature region of the chamber. In some combustion chambers the fuel is injected upstream into the airflow, and a sheet metal cone and perforated baffle plate ensure the necessary mixing of the fuel and air.

Fig. 9.17 Gas turbine can-type combustion chamber

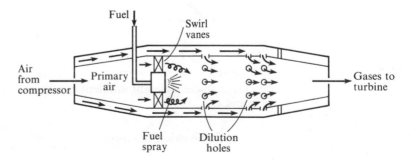

The air–fuel ratio overall is of the order of 60/1 to 120/1, and the air velocity at entry to the combustion chamber is usually not more than 75 m/s. There is a rich and a weak limit for flame stability, and the limit is usually taken at flame blow-out. Instability of the flame results in rough running with consequent effect on the life of the combustion chamber.

It should be noted that because of the high air–fuel ratios used, the gases entering the HP turbine contain a high percentage of oxygen, and therefore if reheating is performed between turbine stages, the additional fuel can be burned satisfactorily in the exhaust gas from the HP turbine.

A combustion efficiency may be defined as follows:

Combustion efficiency

$$= \frac{\text{theoretical fuel–air ratio for actual temperature rise}}{\text{actual fuel–air ratio for actual temperature rise}} \quad (9.7)$$

The theoretical temperature rise is a function of the calorific value of the fuel used, the fuel–air ratio, and the initial temperature of the air. The theoretical temperature rise for any one fuel of known calorific value can be plotted against the fuel–air ratio for various values of air inlet temperature to the chamber, and curves of the form shown in Fig. 9.18 obtained. The combustion efficiency can be evaluated by testing the chamber; sections are traversed to obtain true mean readings of the inlet and outlet temperatures, and the fuel and air mass flow rates are also measured. The fuel used in aircraft gas turbine practice is a light petroleum distillate known as kerosene with a gross calorific value of about 46 400 kJ/kg; for turbines used in power production or as part of a combined heat and power unit the fuel used can also be natural gas; for some process plants a gas turbine unit is used for power production using waste gases as fuel. In cases where kerosene or gas is to be burned a dual-fuel burner is used.

Fig. 9.18 Theoretical temperature rise against fuel–air ratio

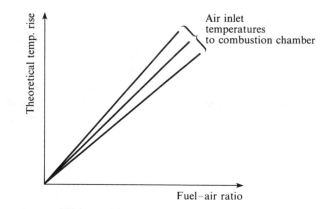

In order to give a comparison of combustion chambers of different size operating under different ambient conditions, a *combustion intensity* is defined as follows:

$$\text{Combustion intensity} = \frac{\text{heat release rate}}{(\text{volume of chamber} \times \text{inlet pressure})} \quad (9.8)$$

The lower the combustion intensity the better the design. In aircraft practice a figure of about 2 kW/m^3 atm would be normal, whereas in the larger industrial plant a figure of about 0.2 kW/m^3 atm is usually achievable.

The pressure loss in the combustion chamber is mainly due to friction and turbulence. There is also a small drop in pressure due to non-adiabatic flow in a duct of approximately constant cross-sectional area. The loss due to friction can be found experimentally by blowing air through the combustion chamber without initiating combustion and measuring the change in pressure. This friction loss in pressure is therefore called the *cold loss*. The loss due to the heating process alone is called the *fundamental loss*. For a more extensive treatment of the combustion process ref. 9.1 should be consulted.

9.4 Additional factors

In considering aircraft propulsion it is necessary first to study the theory of flow in nozzles, and to introduce the concept of total head, or stagnation, pressures and temperatures. Gas turbine cycles for aircraft propulsion are therefore considered again in section 10.9. Another useful concept is small stage, or polytropic, efficiency; this is considered, and cycles analysed using polytropic efficiency in section 11.8 after blading design and the concept of a stage have been introduced.

Problems

(For all problems c_p and γ may be taken as 1.005 kJ/kg K and 1.4 for air, and as 1.15 kJ/kg K and 1.333 for combustion and expansion processes.)

9.1 A gas turbine has an overall pressure ratio of 5 and a maximum cycle temperature of 550 °C. The turbine drives the compressor and an electric generator, the mechanical efficiency of the drive being 97%. The ambient temperature is 20 °C and air enters the compressor at a rate of 15 kg/s; the isentropic efficiencies of the compressor and turbine are 80 and 83%. Neglecting changes in kinetic energy, the mass flow rate of fuel, and all pressure losses, calculate:
  (i) the power output;
  (ii) the cycle efficiency;
  (iii) the work ratio.

(660.3 kW; 12.1%; 0.169)

9.2 In a marine gas turbine unit a HP stage turbine drives the compressor, and a LP stage turbine drives the propeller through suitable gearing. The overall pressure ratio is 4/1, the mass flow rate is 60 kg/s, the maximum temperature is 650 °C, and the air intake conditions are 1.01 bar and 25 °C. The isentropic efficiencies of the compressor, HP turbine, and LP turbine, are 0.8, 0.83, and 0.85 respectively, and the mechanical efficiency of both shafts is 98%. Neglecting kinetic energy changes, and the pressure loss in combustion, calculate:
  (i) the pressure between turbine stages;
  (ii) the cycle efficiency;
  (iii) the shaft power.

(1.57 bar; 14.9%; 4560 kW)

9.3 For the unit of Problem 9.2, calculate the cycle efficiency obtainable when a heat exchanger is fitted. Assume a thermal ratio of 0.75.

(23.4%)

9.4 In a gas turbine generating set two stages of compression are used with an intercooler between stages. The HP turbine drives the HP compressor, and the LP turbine drives the LP compressor and the generator. The exhaust from the LP turbine passes through a heat exchanger which transfers heat to the air leaving the HP compressor. There is a reheat combustion chamber between turbine stages which raises the gas temperature to 600 °C, which is also the gas temperature at entry to the HP turbine. The overall pressure ratio is 10/1, each compressor having the same pressure ratio, and the air temperature at entry to the unit is 20 °C. The heat exchanger thermal ratio may be taken as 0.7, and intercooling is complete between compressor stages. Assume isentropic efficiencies of 0.8 for both compressor stages, and 0.85 for both turbine stages, and that 2% of the work of each turbine is used in overcoming friction. Neglecting all losses in pressure, and assuming that velocity changes are negligibly small, calculate:
 (i) the power output in kilowatts for a mass flow of 115 kg/s;
 (ii) the overall cycle efficiency of the plant.

(14 460 kW; 25.7%)

9.5 A motor car gas turbine unit has two centrifugal compressors in series giving an overall pressure ratio of 6/1. The air leaving the HP compressor passes through a heat exchanger before entering the combustion chamber. The expansion is in two turbine stages, the first stage driving the compressors and the second stage driving the car through gearing. The gases leaving the LP turbine pass through the heat exchanger before exhausting to atmosphere. The HP turbine inlet temperature is 800 °C and the air inlet temperature to the unit is 15 °C. The isentropic efficiency of the compression is 0.8, and that of each turbine is 0.85; the mechanical efficiency of each shaft is 98%. The heat exchanger thermal ration may be assumed to be 0.65. Neglecting pressure losses and changes in kinetic energy, calculate:
 (i) the overall cycle efficiency;
 (ii) the power developed when the air mass flow is 0.7 kg/s;
 (iii) the specific fuel consumption when the calorific value of the fuel used is 42 600 kJ/kg, and the combustion efficiency is 97%.

(29.4%; 94.7 kW; 0.302 kg/kW h)

9.6 In a gas turbine generating station the overall compression ratio is 12/1, performed in three stages with pressure ratios of 2.5/1, 2.4/1, and 2/1 respectively. The air inlet temperature to the plant is 25 °C and intercooling between stages reduces the temperature to 40 °C. The HP turbine drives the HP and intermediate-pressure compressor stages; the LP turbine drives the LP compressor and the generator. The gases leaving the LP turbine are passed through a heat exchanger which heats the air leaving the HP compressor. The temperature at inlet to the HP turbine is 650 °C, and reheating between turbine stages raises the temperature to 650 °C. The gases leave the heat exchanger at a temperature of 200 °C. The isentropic efficiency of each compressor stage is 0.83, and the isentropic efficiencies of the HP and LP turbines are 0.85 and 0.88 respectively. Take the mechanical efficiency of each shaft as 98%. The air mass flow is 140 kg/s. Neglecting pressure losses and changes in kinetic energy, and taking the specific heat of water as 4.19 kJ/kg K, calculate:
 (i) the power output in kilowatts;
 (ii) the cycle efficiency;
 (iii) the flow of cooling water required for the intercoolers when the rise in water temperature must not exceed 30 K;
 (iv) the heat exchanger thermal ratio.

(25 540 kW; 33.4%; 224 kg/s; 0.82)

9.7 In a gas turbine plant air enters a compressor at atmospheric conditions of 15 °C, 1.0133 bar and is compressed through a pressure ratio of 10. The air leaving the compressor passes through a heat exchanger before entering the combustion chamber. The hot gases leave the combustion chamber at 800 °C and expand through an HP turbine which drives the compressor. On leaving the HP turbine the gases pass through a reheat combustion chamber which raises the temperature of the gases to 800 °C before they expand through the power turbine, and thence to the heat exchanger where they flow in counter-flow to the air leaving the compressor. Using the data below, neglecting the mass flow rate of fuel and changes of velocity throughout, calculate:

  (i) the airflow rate required for a net power output of 10 MW;

  (ii) the work ratio of the cycle;

  (iii) the temperature of the air entering the first combustion chamber;

  (iv) the overall cycle efficiency.

Data Isentropic efficiency of compressor, 80%; isentropic efficiencies of HP and power turbine, 87 and 85%; mechanical efficiency of HP turbine-compressor drive, 92%; mechanical efficiency of power turbine drive, 94%; thermal ratio of heat exchanger, 0.75; pressure drop on air side of heat exchanger, 0.125 bar; pressure drop in first combustion chamber, 0.100 bar; pressure drop in reheat combustion chamber, 0.080 bar; pressure drop on gas side of heat exchanger, 0.100 bar.

(91.0 kg/s; 0.25; 611 °C; 18.9%)

9.8 An open-cycle gas turbine plant is used to generate power in an oil refinery. The gas turbine unit drives a generator which supplies electric motors of 2400 kW; the overall mechanical and electrical efficiency is 92%. Some of the exhaust gas from the turbine at 530 °C is supplied to a furnace in the refinery at a rate of 2 kg/s; the remainder of the exhaust gas is passed in counter-flow through a heat exchanger where it heats the air leaving the compressor, and then passes to exhaust at 400 °C. The compressor has a pressure ratio of 8 and the air at entry is at 1.013 bar and 20 °C. The pressure loss in the air side of the heat exchanger is 0.16 bar, the pressure loss in the combustion chamber is 0.12 bar, and the pressure loss in the gas side of the heat exchanger is 0.05 bar. The isentropic efficiencies of the compressor and turbine are 0.85 and 0.92 respectively. Neglecting heat losses in the heat exchanger, and the mass flow rate of fuel, calculate:

  (i) the mass flow rate of air entering the compressor;

  (ii) the temperature of the air entering the combustion chamber;

  (iii) the overall cycle efficiency.

(10.82 kg/s; 421.0 °C; 34.2%)

9.9 A closed-cycle gas turbine plant using helium as the working fluid is proposed for an experimental nuclear reactor. The helium is compressed in two stages with an intercooler between stages. Before passing through a heater where it is heated externally by the reactor coolant, the helium is pre-heated in a heat exchanger where it is in counter-flow with the helium leaving the turbine. The helium leaving the turbine is cooled in the heat exchanger before passing through a cooler where it is cooled by cooling water to the required inlet temperature to the compressor, and the cycle is complete. Using the data below, calculate the overall cycle efficiency.

Data Pressure and temperature at entry to the first compressor, 18 bar and 30 °C; pressure ratio for each compressor, 2; temperature of helium leaving the intercooler, 30 °C; temperature of helium, entering the turbine, 800 °C; isentropic efficiency of each compressor, 0.83; isentropic efficiency of the turbine, 0.86; effectiveness of the heat exchanger, 0.80; pressure loss as a percentage of the inlet pressure to each component: intercooler and external cooler, 1%; each side of heat exchanger, 2%; external heater, 3%. Take γ for helium as 1.666.

(32.6%)

References

9.1 COHEN H, ROGERS G F C, and SARAVANAMUTTOO H I H 1987 *Gas Turbine Theory* 3rd edn Longman

10

Nozzles and Jet Propulsion

A nozzle is a duct of smoothly varying cross-sectional area in which a steadily flowing fluid can be made to accelerate by a pressure drop along the duct. There are many applications in practice which require a high-velocity stream of fluid, and the nozzle is the best means of obtaining this. For example, nozzles are used in steam and gas turbines, in jet engines, in rocket motors, in flow measurement, and in many other applications. When a fluid is decelerated in a duct, causing a rise in pressure along the stream, then the duct is called a *diffuser*; two applications in practice in which a diffuser is used are the centrifugal compressor and the ramjet.

The analysis presented in this chapter will be restricted to *one-dimensional flow*. In one-dimensional flow it is assumed that the fluid velocity, and the fluid properties, change only in the direction of the flow. This means that the fluid velocity is assumed to remain constant at a mean value across the cross-section of the duct. The effects of friction will not be analysed fundamentally, suitable efficiencies or coefficients being adopted to allow for the departure from the ideal frictionless case. The analysis of fluid flow involving friction has become of increasing importance due to the development of the turbojet, the ramjet, and the rocket, and the introduction of high-speed flight. For a fundamental approach to the topic a study of fluid dynamics is required, and the reader is recommended to books on *gas dynamics*, such as ref. 10.1.

10.1 Nozzle shape

Consider a stream of fluid at pressure p_1, enthalpy h_1, and with a low velocity C_1. It is required to find the shape of duct which will cause the fluid to accelerate to a high velocity as the pressure falls along the duct. It can be assumed that the heat loss from the duct is negligibly small (i.e. adiabatic flow, $Q = 0$), and it is clear that no work is done on or by the fluid (i.e. $W = 0$). Applying the steady-flow energy equation, (1.10), between section 1 and any other section X–X where the pressure is p, the enthalpy is h, and the velocity is C, we have

$$h_1 + \frac{C_1^2}{2} = h + \frac{C^2}{2}$$

i.e. $\qquad C^2 = 2(h_1 - h) + C_1^2$

or $\qquad C = \sqrt{\{2(h_1 - h) + C_1^2\}}$ $\qquad\qquad$ (10.1)

If the area at the section X–X is A, and the specific volume is v, then, using equation (1.11)

$$\text{Mass flow, } \dot{m} = \frac{CA}{v}$$

or $\qquad$ Area per unit mass flow, $\dfrac{A}{\dot{m}} = \dfrac{v}{C}$ $\qquad\qquad$ (10.2)

Then substituting for the velocity C, from equation (10.1)

$$\text{Area per unit mass flow} = \frac{v}{\sqrt{\{2(h_1 - h) + C_1^2\}}} \qquad (10.3)$$

It can be seen from equation (10.3) that in order to find the way in which the area of the duct varies it is necessary to be able to evaluate the specific volume, v, and the enthalpy, h, at any section X–X. In order to do this, some information about the process undergone between section 1 and section X–X must be known. For the ideal frictionless case, since the flow is adiabatic and reversible, the process undergone is an isentropic process, and hence

$$s_1 = (\text{entropy at any section X–X}) = s, \text{ say}$$

Now using equation (10.2) and the fact that $s_1 = s$, it is possible to plot the variation of the cross-sectional area of the duct against the pressure along the duct. For a vapour this can be done using tables; for a perfect gas the procedure is simpler, since we have $pv^\gamma = \text{constant}$, for an isentropic process. In either case, choosing fixed inlet conditions, then the variation in the area, A, the specific volume, v, and the velocity, C, can be plotted against the pressure along the duct. Typical curves are shown in Fig. 10.1. It can be seen that the area decreases initially, reaches a minimum, and then increases again. This can also be seen from equation (10.2),

i.e. $\qquad$ Area per unit mass flow $= \dfrac{v}{C}$

Fig. 10.1
Cross-sectional
area, velocity, and
specific volume
variations with pressure
through a nozzle

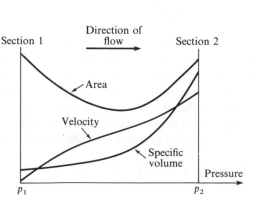

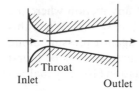

Fig. 10.2 Cross-section through a convergent–divergent nozzle

When v increases less rapidly than C, then the area decreases; when v increases more rapidly than C, then the area increases.

A nozzle, the area of which varies as in Fig. 10.1, is called a *convergent–divergent* nozzle. A cross-section of a typical convergent–divergent nozzle is shown in Fig. 10.2. The section of minimum area is called the *throat* of the nozzle. It will be shown later in section 10.2 that the velocity at the throat of a nozzle operating at its designed pressure ratio is the velocity of sound at the throat conditions. The flow up to the throat is *subsonic*; the flow after the throat is *supersonic*. It should be noted that a sonic or a supersonic flow requires a diverging duct to accelerate it.

The specific volume of a liquid is constant over a wide pressure range, and therefore nozzles for liquids are always convergent, even at very high exit velocities (e.g. a fire-hose uses a convergent nozzle).

10.2 Critical pressure ratio

It has been stated in section 10.1 that the velocity at the throat of a correctly designed nozzle is the velocity of sound. In the same way, for a nozzle that is convergent only, then the fluid will attain sonic velocity at exit if the pressure drop across the nozzle is large enough. The ratio of the pressure at the section where sonic velocity is attained to the inlet pressure of a nozzle is called the *critical pressure ratio*. Cases in which nozzles operate off the design conditions of pressure will be considered in section 10.4; in what follows it will be assumed that the nozzle always operates with its designed pressure ratio.

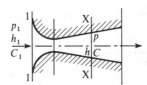

Fig. 10.3 Convergent–divergent nozzle

Consider a convergent–divergent nozzle as shown in Fig. 10.3 and let the inlet conditions be pressure p_1, enthalpy h_1, and velocity C_1. Let the conditions at any other section X–X be pressure p, enthalpy h, and velocity C.

In most practical applications the velocity at the inlet to a nozzle is negligibly small in comparison with the exit velocity. It can be seen from equation (10.2), $A/\dot{m} = v/C$, that a negligibly small velocity implies a very large area, and most nozzles are in fact shaped at inlet in such a way that the nozzle converges rapidly over the first fraction of its length; this is illustrated in the diagram of a nozzle inlet shown in Fig. 10.4.

Now from equation (10.1) we have

$$C = \sqrt{\{2(h_1 - h) + C_1^2\}}$$

and neglecting C_1 this gives

$$C = \sqrt{\{2(h_1 - h)\}} \tag{10.4}$$

Since enthalpy is usually expressed in kilojoules per kilogram, then an additional constant of 10^3 will appear within the root sign if C is to be expressed in metres per second.

Then, substituting from equation (10.4) in equation (10.2), we have

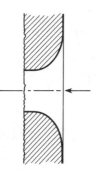

Fig. 10.4 Inlet section of a nozzle

$$\text{Area per unit mass flow,} \quad \frac{A}{\dot{m}} = \frac{v}{C} = \frac{v}{\sqrt{\{2(h_1 - h)\}}} \tag{10.5}$$

As stated in section 10.1 the area can be evaluated at any section where the pressure is p, by assuming that the process is isentropic (i.e. $s_1 = s$). When this is done for a series of pressures, the area can be plotted against pressure along the duct, or against pressure ratio, and the critical pressure can thus be found graphically. For a perfect gas it is possible to simplify equation (10.5) by making use of the perfect gas laws.

From equation (2.18), $h = c_p T$ for a perfect gas, therefore

$$\text{Area per unit mass flow rate} = \frac{v}{\sqrt{\{2c_p(T_1 - T)\}}}$$

$$= \frac{v}{\sqrt{\left\{2c_p T_1\left(1 - \frac{T}{T_1}\right)\right\}}}$$

From equation (2.5), $v = RT/p$, therefore

$$\text{Area per unit mass flow rate} = \frac{RT/p}{\sqrt{\left\{2c_p T_1\left(1 - \frac{T}{T_1}\right)\right\}}}$$

Let the pressure ratio, p/p_1, be x. Then using equation (3.21), for an isentropic process for a perfect gas

$$\frac{T}{T_1} = \left(\frac{p}{p_1}\right)^{(\gamma-1)/\gamma} = x^{(\gamma-1)/\gamma}$$

Substituting for $p = xp_1$, for $T = T_1 x^{(\gamma-1)/\gamma}$, and for $T/T_1 = x^{(\gamma-1)/\gamma}$, we have

$$\text{Area per unit mass flow rate} = \frac{RT_1 x^{(\gamma-1)/\gamma}}{p_1 x\sqrt{\{2c_p T_1(1 - x^{(\gamma-1)/\gamma})\}}}$$

For fixed inlet conditions (i.e. p_1 and T_1 fixed), we have

$$\text{Area per unit mass flow rate} = \text{constant} \times \frac{x^{(\gamma-1)/\gamma}}{x\sqrt{(1 - x^{(\gamma-1)/\gamma})}}$$

$$= \text{constant} \times \frac{1}{x^{1/\gamma}\sqrt{(1 - x^{(\gamma-1)/\gamma})}}$$

$$= \frac{\text{constant}}{\sqrt{(x^{2/\gamma} - x^{2/\gamma}x^{(\gamma-1)/\gamma})}}$$

therefore

$$\text{Area per unit mass flow rate} = \frac{\text{constant}}{\sqrt{(x^{2/\gamma} - x^{(\gamma+1)/\gamma})}} \tag{10.6}$$

To find the value of the pressure ratio, x, at which the area is a minimum it is necessary to differentiate equation (10.6) with respect to x and equate the result to zero, i.e. for minimum area

$$\frac{d}{dx}\left\{\frac{1}{(x^{2/\gamma} - x^{(\gamma+1)/\gamma})^{1/2}}\right\} = 0$$

i.e.
$$-\frac{\left\{\frac{2}{\gamma}x^{(2/\gamma)-1} - \left(\frac{\gamma+1}{\gamma}\right)x^{\{(\gamma+1)/\gamma\}-1}\right\}}{2(x^{2/\gamma} - x^{(\gamma+1)/\gamma})^{3/2}} = 0$$

Hence the area is a minimum when

$$\frac{2}{\gamma}x^{(2/\gamma)-1} = \left(\frac{\gamma+1}{\gamma}\right)x^{\{(\gamma+1)/\gamma\}-1}$$

$$x^{\{(\gamma+1)/\gamma\}-1-(2/\gamma)+1} = \frac{2}{\gamma+1}$$

therefore

$$x = \left(\frac{2}{\gamma+1}\right)^{\gamma/(\gamma-1)}$$

i.e. Critical pressure ratio, $\dfrac{p_c}{p_1} = \left(\dfrac{2}{\gamma+1}\right)^{\gamma/(\gamma-1)}$ (10.7)

It can be seen from equation (10.7) that for a perfect gas the pressure ratio required to attain sonic velocity in a nozzle depends only on the value of γ for the gas. For example, for air $\gamma = 1.4$, therefore

$$\frac{p_c}{p_1} = \left(\frac{2}{1.4+1}\right)^{1.4/0.4} = 0.5283$$

Hence for air at 10 bar, say, a convergent nozzle requires a back pressure of 5.283 bar, in order that the flow should be sonic at exit and for a correctly designed convergent–divergent nozzle with inlet pressure 10 bar, the pressure at the throat is 5.283 bar. For carbon dioxide, $\gamma = 1.3$, therefore

$$\frac{p_c}{p_1} = \left(\frac{2}{1.3+1}\right)^{1.3/0.3} = 0.5457$$

Hence for carbon dioxide at 10 bar, a convergent nozzle requires a back pressure of 5.457 bar for sonic flow at exit, and the pressure at the throat of a convergent–divergent nozzle with inlet pressure 10 bar is 5.457 bar.

The ratio of the temperature at the section of the nozzle where the velocity is sonic to the inlet temperature is called the *critical temperature ratio*,

$$\text{Critical temperature ratio, } \frac{T_c}{T_1} = \left(\frac{p_c}{p_1}\right)^{(\gamma-1)/\gamma} = \frac{2}{\gamma+1}$$

i.e. $\dfrac{T_c}{T_1} = \dfrac{2}{\gamma+1}$ (10.8)

Equations (10.7) and (10.8) apply to perfect gases only, and not to vapours.

However, it is found that a sufficiently close approximation is obtained for a steam nozzle if it is assumed that the expansion follows a law pv^k = constant. The process is assumed to be isentropic, and therefore the index k is an approximate isentropic index for steam. When the steam is initially dry saturated then $k = 1.135$; when the steam is initially superheated then $k = 1.3$. Note that equation (10.8) cannot be used for a wet vapour, since no simple relationship between p and T is known for a wet vapour undergoing an isentropic process. More will be said of this in section 10.6.

The critical velocity at the throat of a nozzle can be found for a perfect gas by substituting in equation (10.1),

i.e. $\qquad C_c = \sqrt{\{2(h_1 - h) + C_1^2\}}$

Putting $C_1 = 0$, as before, and using equation (2.18) for a perfect gas, $h = c_p T$, we have

$$C_c = \sqrt{\{2c_p(T_1 - T_c)\}} = \sqrt{\left\{2c_p T_c\left(\frac{T_1}{T_c} - 1\right)\right\}}$$

From equation (10.8), $T_c/T_1 = 2/(\gamma + 1)$, hence

$$C_c = \sqrt{\left[2c_p T_c\left\{\left(\frac{\gamma + 1}{2}\right) - 1\right\}\right]} = \sqrt{\{c_p T_c(\gamma - 1)\}}$$

Also, from equation (2.22),

$$c_p = \frac{\gamma R}{(\gamma - 1)} \quad \text{or} \quad c_p(\gamma - 1) = \gamma R$$

Hence substituting,

$$C_c = \sqrt{(\gamma R T_c)}$$

i.e. $\qquad$ Critical velocity, $C_c = \sqrt{(\gamma R T_c)}$ $\qquad\qquad$ (10.9)

The critical velocity given by equation (10.9) is the velocity at the throat of a correctly designed convergent–divergent nozzle, or the velocity at the exit of a convergent nozzle when the pressure ratio across the nozzle is the critical pressure ratio.

It can be shown that the critical velocity is the velocity of sound at the critical conditions.

The velocity of sound, a, is defined by the equation

$$a^2 = \frac{dp}{d\rho} \text{ at constant entropy}$$

where p is pressure and ρ is the density. A proof of this expression can be found in ref. 10.2.

Now $\rho = 1/v$, where v is the specific volume.

$$d\rho = d(1/v) = -\frac{1}{v^2} dv$$

Hence $a^2 = -\dfrac{dp}{dv} v^2$

For a perfect gas undergoing an isentropic process, $pv^\gamma = \text{constant}$,

i.e. $p = \dfrac{K}{v^\gamma}$

where K is constant, therefore

$$\frac{dp}{dv} = -\frac{\gamma K}{v^{\gamma+1}}$$

Therefore substituting

$$a^2 = \frac{v^2 \gamma K}{v^{\gamma+1}}$$

Also, $K = pv^\gamma$, hence

$$a^2 = \frac{\gamma p v^\gamma v^2}{v^{\gamma+1}} = \gamma p v$$

i.e. Velocity of sound, $a = \sqrt{(\gamma p v)} = \sqrt{(\gamma R T)}$ (10.10)

It can be seen that the critical velocity of a perfect gas in a nozzle, as given by equation (10.9), is the velocity of sound in the gas at the critical temperature.

Equations (10.9) and (10.10) cannot be applied to a vapour; however, if an approximate isentropic law, $pv^k = \text{constant}$, is assumed for a vapour, then the critical velocity can be taken as $C_c = \sqrt{(kpv)}$. (Note that for a vapour the critical velocity cannot be expressed in terms of the temperature.) It is usually more convenient to evaluate the critical velocity of a vapour using equation (10.4), $C_c = \sqrt{\{2(h_1 - h_c)\}}$, where $(h_1 - h_c)$ is the enthalpy drop from the inlet to the throat, which can be evaluated from tables or by using an h–s chart.

Example 10.1 Air at 8.6 bar and 190 °C expands at the rate of 4.5 kg/s through a convergent–divergent nozzle into a space at 1.03 bar. Assuming that the inlet velocity is negligible, calculate the throat and the exit cross-sectional areas of the nozzle.

Solution The nozzle is shown diagrammatically in Fig. 10.5. From equation (10.7) the critical pressure ratio is given by

$$\frac{p_c}{p_1} = \left(\frac{2}{\gamma+1}\right)^{\gamma/(\gamma-1)} = \left(\frac{2}{2.4}\right)^{1.4/0.4} = 0.5283$$

i.e. $p_c = 0.5283 \times 8.6 = 4.543 \text{ bar}$

Fig. 10.5
Convergent–divergent
nozzle for Example 10.1

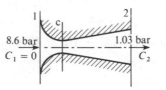

8.6 bar
$C_1 = 0$
1.03 bar
C_2

Also, from equation (10.8)

$$\frac{T_c}{T_1} = \frac{2}{\gamma + 1} = \frac{1}{1.2}$$

$$T_c = \frac{190 + 273}{1.2} = 385.8 \text{ K}$$

From equation (2.5)

$$v_c = \frac{RT_c}{p_c} = \frac{287 \times 385.8}{10^5 \times 4.543} = 0.244 \text{ m}^3/\text{kg}$$

Also, from equation (10.9)

$$C_c = \sqrt{(\gamma RT_c)} = \sqrt{(1.4 \times 287 \times 385.8)} = 393.7 \text{ m/s}$$

[or from equation (10.4)

$$C_c = \sqrt{\{2(h_1 - h_c)\}} = \sqrt{\{2c_p(T_1 - T_c)\}}$$

i.e. $\quad C_c = \sqrt{\{2 \times 1.005 \times 10^3(463 - 385.8)\}} = 393.8 \text{ m/s}]$

To find the area of the throat, using equation (1.11), we have

$$A_c = \frac{\dot{m}v_c}{C_c} = \frac{4.5 \times 0.244}{393.7} = 0.002\,79 \text{ m}^2$$

i.e. $\qquad$ Area of throat $= 0.002\,79 \times 10^6 = 2790 \text{ mm}^2$

Using equation (3.21) for a perfect gas

$$\frac{T_1}{T_2} = \left(\frac{p_1}{p_2}\right)^{(\gamma - 1)/\gamma} = \left(\frac{8.6}{1.03}\right)^{0.4/1.4} = 1.834$$

i.e. $\qquad T_2 = \frac{463}{1.834} = 252.5 \text{ K}$

Then from equation (2.5)

$$v_2 = \frac{RT_2}{p_2} = \frac{287 \times 252.5}{10^5 \times 1.03} = 0.7036 \text{ m}^3/\text{kg}$$

Also, from equation (10.4)

$$C_2 = \sqrt{\{2(h_1 - h_2)\}} = \sqrt{\{2c_p(T_1 - T_2)\}}$$

i.e. $\qquad C_2 = \sqrt{\{2 \times 1.005 \times 10^3(463 - 252.5)\}} = 650.5 \text{ m/s}$

Then to find the exit area, using equation (1.11)

$$A_2 = \frac{\dot{m}v_2}{C_2} = \frac{4.5 \times 0.7036}{650.5} = 0.004\,87 \text{ m}^2$$

i.e. $\qquad$ Exit area $= 0.004\,87 \times 10^6 = 4870 \text{ mm}^2$

10.3 Maximum mass flow

Consider a convergent nozzle expanding into a space, the pressure of which can be varied, while the inlet pressure remains fixed. The nozzle is shown diagrammatically in Fig. 10.6. When the back pressure, p_b, is equal to p_1, then no fluid can flow through the nozzle. As p_b is reduced the mass flow through the nozzle increases, since the enthalpy drops, and hence the velocity, increases. However, when the back pressure reaches the critical value, it is found that no further reduction in back pressure can affect the mass flow. When the back pressure is exactly equal to the critical pressure, p_c, then the velocity at exit is sonic and the mass flow through the nozzle is at a maximum. If the back pressure is reduced below the critical value then the mass flow remains at the maximum value, the exit pressure remains at p_c, and the fluid expands violently outside the nozzle down to the back pressure. It can be seen that the maximum mass flow through a convergent nozzle is obtained when the pressure ratio across the nozzle is the critical pressure ratio. Also, for a convergent–divergent nozzle, with sonic velocity at the throat, the cross-sectional area of the throat fixes the mass flow through the nozzle for fixed inlet conditions.

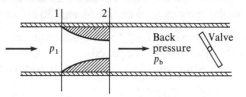

Fig. 10.6 Convergent nozzle with back-pressure variation

When a nozzle operates with the maximum mass flow it is said to be *choked*. A correctly designed convergent–divergent nozzle is always choked.

An attempt can be made to explain the phenomenon of choking, by considering the velocity of any small disturbance in the stream. Any small disturbance in the flow is propagated as small pressure waves travelling at the velocity of sound in the fluid in all directions from the centre of the disturbance. This is illustrated in Fig. 10.7; the pressure waves emanate from point Q at the velocity of sound relative to the fluid, a, while the fluid moves with a velocity, C. The absolute velocity of the pressure waves travelling back upstream is therefore given by $(a - C)$. Now when the fluid velocity is subsonic, then $C < a$,

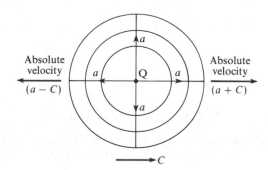

Fig. 10.7 Propagation of a small disturbance in a flowing fluid

and the pressure waves can move back upstream; however, when the flow is sonic, or supersonic (i.e. $C = a$ or $C > a$), then the pressure waves cannot be transmitted back upstream. It follows from this reasoning that in a nozzle in which sonic velocity has been attained no alteration in the back pressure can be transmitted back upstream. For example, when air at 10 bar expands in a nozzle, the critical pressure can be shown to be 5.283 bar. When the back pressure of the nozzle is 4 bar, say, then the nozzle is choked and is passing the maximum mass flow. If the back pressure is reduced to 1 bar, say, the mass flow through the nozzle remains unchanged. Even if the air were allowed to expand into an evacuated space, the mass flow would be no greater than that through the nozzle when the back pressure is 5.283 bar.

Example 10.2   A fluid at 6.9 bar and 93 °C enters a convergent nozzle with negligible velocity, and expands isentropically into a space at 3.6 bar. Calculate the mass flow per square metre of exit area:

(i) when the fluid is helium ($c_p = 5.19$ kJ/kg K);
(ii) when the fluid is ethane ($c_p = 1.88$ kJ/kg K).

Assume that both helium and ethane are perfect gases, and take the respective molar masses as 4 kg/kmol and 30 kg/kmol.

Solution   (i) It is necessary first to calculate the critical pressure in order to discover whether the nozzle is choked.

From equation (2.9), $R = \tilde{R}/\tilde{m}$, therefore for helium,

$$R = \frac{8314.5}{4} = 2079 \text{ N m/kg K}$$

Then from equation (2.22)

$$c_p = \frac{\gamma R}{(\gamma - 1)}$$

i.e. $$\frac{\gamma - 1}{\gamma} = \frac{R}{c_p} = \frac{2079}{10^3 \times 5.19} = 0.4$$

therefore

$$\gamma = \frac{1}{1 - 0.4} = 1.667$$

Then using equation (10.7)

$$\frac{p_c}{p_1} = \left(\frac{2}{\gamma + 1}\right)^{\gamma/(\gamma - 1)} = \left(\frac{2}{2.667}\right)^{1.667/0.667} = 0.487$$

i.e. $p_c = 0.487 \times 6.9$ bar

i.e.   Critical pressure $p_c = 3.36$ bar

The actual back pressure is 3.6 bar, hence in this case the fluid does not reach the critical conditions and the nozzle is not choked.

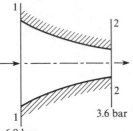

Fig. 10.8 Convergent nozzle with helium for Example 10.2

6.9 bar

3.6 bar

The nozzle is shown diagrammatically in Fig. 10.8.
Using equation (3.21)

$$\frac{T_1}{T_2} = \left(\frac{p_1}{p_2}\right)^{(\gamma-1)/\gamma} = \left(\frac{6.9}{3.6}\right)^{0.4} = 1.297$$

i.e.
$$T_2 = \frac{93 + 273}{1.297} = 282.2 \text{ K}$$

Then from equation (10.4)

$$C_2 = \sqrt{\{2(h_1 - h_2)\}} = \sqrt{2c_p(T_1 - T_2)\}}$$

i.e.
$$C_2 = \sqrt{\{2 \times 5.19 \times 10^3(366 - 282.2)\}} = 932.7 \text{ m/s}$$

Also, from equation (2.5)

$$v_2 = \frac{RT_2}{p_2} = \frac{2079 \times 282.2}{10^5 \times 3.6} = 1.63 \text{ m}^3/\text{kg}$$

Hence from equation (1.11)

$$\dot{m} = \frac{A_2 C_2}{v_2} = \frac{1 \times 932.7}{1.63} = 572.3 \text{ kg/s}$$

i.e.   Mass flow per square metre of exit area = 572.3 kg/s

(ii) Using the same procedure for ethane, we have

$$R = \frac{\tilde{R}}{\tilde{m}} = \frac{8314.5}{30} = 277.1 \text{ N m/kg K}$$

and
$$\frac{\gamma - 1}{\gamma} = \frac{R}{c_p} = \frac{277.1}{10^3 \times 1.88} = 0.147$$

i.e.
$$\gamma = \frac{1}{1 - 0.147} = 1.172$$

Then
$$\frac{p_c}{p_1} = \left(\frac{2}{\gamma + 1}\right)^{\gamma/(\gamma-1)} = \left(\frac{2}{2.172}\right)^{1.172/0.172} = 0.57$$

$$p_c = 0.57 \times 6.9 \text{ bar}$$

i.e.   Critical pressure, $p_c = 3.93$ bar

The actual back pressure is 3.6 bar, hence in this case the fluid reaches critical conditions at exit and the nozzle is choked. The expansion from the exit pressure of 3.93 bar down to the back pressure of 3.6 bar must take place outside the nozzle.

The nozzle is shown diagrammatically in Fig. 10.9.
Since the nozzle is choked, from equation (10.8) we have

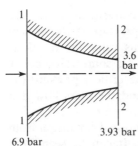

Fig. 10.9 Convergent nozzle with ethane for Example 10.2

6.9 bar

3.6 bar

3.93 bar

$$\frac{T_c}{T_1} = \frac{2}{\gamma + 1} = \frac{2}{2.172}$$

i.e. $\qquad T_2 = T_c = \dfrac{2 \times 366}{2.172} = 337 \text{ K}$

Also, from equation (10.9)

$$C_2 = C_c = \sqrt{(\gamma R T_c)} = \sqrt{(1.172 \times 277.1 \times 337)} = 331 \text{ m/s}$$

From equation (2.5),

$$v_2 = \frac{R T_2}{p_2} = \frac{277.1 \times 337}{10^5 \times 3.93} = 0.238 \text{ m}^3/\text{kg}$$

Then using equation (1.11)

$$\dot{m} = \frac{A_2 C_2}{v_2} = \frac{1 \times 331}{0.238} = 1391 \text{ kg/s}$$

i.e. $\qquad$ Mass flow per square metre of exit area = 1391 kg/s

10.4 Nozzles off the design pressure ratio

When the back pressure of a nozzle is below the design value the nozzle is said to *underexpand*. In underexpansion the fluid expands to the design pressure in the nozzle and then expands violently and irreversibly down to the back pressure on leaving the nozzle (e.g. the nozzle in Example 10.2(ii) shown in Fig. 10.9 is underexpanding).

When the back pressure of a nozzle is above the design value the nozzle is said to *overexpand*. In overexpansion in a convergent nozzle the exit pressure is greater than the critical pressure and the effect is to reduce the mass flow through the nozzle. In overexpansion in a convergent–divergent nozzle there is always an expansion followed by a recompression. The two types of nozzle can be considered separately.

Convergent nozzle

The pressure variations of a fluid flowing through a convergent nozzle are shown in Fig. 10.10. Assuming that the design back pressure is the critical pressure, p_c, then when the back pressure is above this value the nozzle is overexpanding as shown by line (a), and the mass flow is some value below the maximum. When the back pressure is equal to the critical pressure the expansion follows the line (b), the nozzle is choked, and the mass flow is a maximum. When the back pressure is below the critical pressure the expansion in the nozzle still follows the line (b), but there is an additional expansion from p_c down to the back pressure, p_b, outside the nozzle. It can be seen from Fig. 10.10 that in the expansion outside the nozzle the pressure oscillates violently, and in fact a shock wave is formed. In this latter case the nozzle is underexpanding.

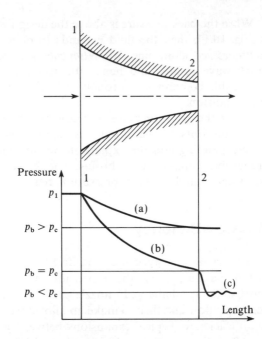

Fig. 10.10 Pressure variations for flow through a convergent nozzle

Convergent–divergent nozzle

The pressure variations of a fluid flowing through a convergent–divergent nozzle are shown in Fig. 10.11. When the mass flow through the nozzle is very low, the pressure at the throat of the nozzle is well above the critical pressure and therefore the divergent portion acts as a diffuser, as shown by line (a) in Fig. 10.11. The nozzle is then acting as a venturimeter.

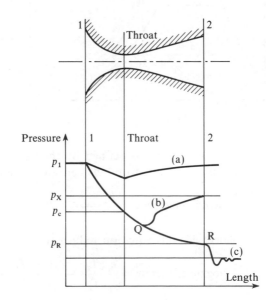

Fig. 10.11 Pressure variations for flow through a convergent–divergent nozzle

When the back pressure is above the design value at some value, p_x, as shown in Fig. 10.11, then the fluid expands from p_1 down to point Q and is then recompressed along line (b). Whenever a supersonic stream is decelerated a shock wave results, and hence the recompression process of line (b) is an irreversible compression through a shock wave. Both (a) and (b) are cases of overexpansion.

When the back pressure is below the design value, p_R, then there is an expansion outside the nozzle as shown by line (c). The nozzle is then underexpanding and the expansion outside the nozzle consists of a series of irreversible compressions through shock waves, alternated with irreversible expansions, until the back pressure is reached.

10.5 Nozzle efficiency

Due to friction between the fluid and the walls of the nozzle, and to friction within the fluid itself, the expansion process is irreversible, although still approximately adiabatic. In nozzle design it is usual to base all calculations on isentropic flow and then to make an allowance for friction by using a coefficient or an efficiency. Typical expansions between p_1 and p_2 in a nozzle are shown on a T–s diagram in Fig. 10.12(a) and (b) for a vapour and for a perfect gas respectively. The line 1–2s on each diagram represents the ideal isentropic expansion, and the line 1–2 represents the actual irreversible adiabatic expansion.

Fig. 10.12 Nozzle expansion processes for a vapour (a) and a perfect gas (b) on a T–s diagram

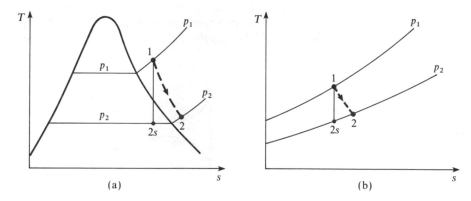

(a) (b)

The *nozzle efficiency* is defined by the ratio of the actual enthalpy drop to the isentropic enthalpy drop between the same pressures,

i.e. $$\text{Nozzle efficiency} = \frac{h_1 - h_2}{h_1 - h_{2s}} \tag{10.11}$$

For a perfect gas this equation reduces to

$$\text{Nozzle efficiency} = \frac{c_p(T_1 - T_2)}{c_p(T_1 - T_{2s})} = \frac{T_1 - T_2}{T_1 - T_{2s}} \tag{10.12}$$

If the actual velocity at exit from the nozzle is C_2, and the velocity at exit when the flow is isentropic is C_{2s}, then using the steady-flow energy equation in each case we have

$$h_1 + \frac{C_1^2}{2} = h_{2s} + \frac{C_{2s}^2}{2} \quad \text{or} \quad h_1 - h_{2s} = \frac{C_{2s}^2 - C_1^2}{2}$$

$$h_1 + \frac{C_1^2}{2} = h_2 + \frac{C_2^2}{2} \quad \text{or} \quad h_1 - h_2 = \frac{C_2^2 - C_1^2}{2}$$

Therefore substituting in equation (10.12)

$$\text{Nozzle efficiency} = \frac{C_2^2 - C_1^2}{C_{2s}^2 - C_1^2} \tag{10.13}$$

When the inlet velocity, C_1, is negligibly small then

$$\text{Nozzle efficiency} = \frac{C_2^2}{C_{2s}^2} \tag{10.14}$$

Sometimes a velocity coefficient is defined as the ratio of the actual exit velocity to the exit velocity when the flow is isentropic between the same pressures,

i.e. $\qquad$ Velocity coefficient $= \dfrac{C_2}{C_{2s}} \tag{10.15}$

It can be seen from equations (10.14) and (10.15) that the velocity coefficient is the square root of the nozzle efficiency, when the inlet velocity is assumed to be negligible.

Another coefficient which is frequently used is the ratio of the actual mass flow through the nozzle, $\dot{m}$, to the mass flow which would be passed if the flow were isentropic, $\dot{m}_s$; this is called the *coefficient of discharge*,

i.e. $\qquad$ Coefficient of discharge $= \dfrac{\dot{m}}{\dot{m}_s} \tag{10.16}$

If the angle of divergence of a convergent–divergent nozzle is made too large, then breakaway of the fluid from the duct walls is liable to occur, with consequent increased friction losses. The included angle of a divergent duct is usually kept below about 20°. It follows that for a given pressure ratio across a convergent–divergent nozzle the divergent portion must be long compared to the convergent portion. Now because the divergent portion of the nozzle is comparatively long, and since a diverging flow is more susceptible to losses, and the velocities in this portion are higher, it follows that the bulk of the friction losses occur in the divergent portion. In fact it is sometimes assumed that all the friction losses occur after the throat of the nozzle. This latter assumption implies that the coefficient of discharge is unity, since any friction after the throat cannot affect the mass flow through a nozzle which is choked.

Nozzles in practice are used with a variety of shapes and cross-sections. The cross-section can be either circular or rectangular, and the axis of the nozzle can be straight or curved. A typical circular section, straight axis nozzle is

Fig. 10.13 Typical
circular section
nozzle (a) and curved
axis steam nozzles (b)

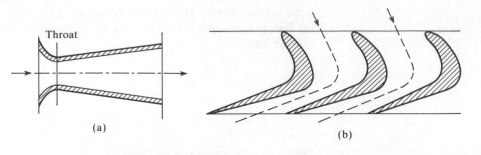

(a)

(b)

shown in Fig. 10.13(a), and a series of typical plate-type, curved-axis steam nozzles is shown in Fig. 10.13(b).

Example 10.3

Gases expand in a propulsion nozzle from 3.5 bar and 425 °C down to a back pressure of 0.97 bar, at the rate of 18 kg/s. Taking a coefficient of discharge of 0.99 and a nozzle efficiency of 0.94, calculate the required throat and exit areas of the nozzle. For the gases take $\gamma = 1.333$ and $c_p = 1.11$ kJ/kg K. Assume that the inlet velocity is negligible.

Solution

The critical pressure is given by equation (10.7)

$$\frac{p_c}{p_1} = \left(\frac{2}{\gamma + 1}\right)^{\gamma/(\gamma - 1)} = \left(\frac{2}{2.333}\right)^{1.333/0.333} = 0.54$$

i.e. $p_c = 0.54 \times 3.5$ bar

i.e.   Critical pressure = 1.89 bar

The nozzle is therefore choking and a convergent–divergent nozzle is required. The mass flow is determined by the throat of the nozzle. Using equation (10.8)

$$\frac{T_c}{T_1} = \frac{2}{\gamma + 1} = \frac{1}{1.1665}$$

i.e. $T_c = \dfrac{425 + 273}{1.1665} = 598.4$ K

Then from equation (10.4)

$$C_c = \sqrt{\{2(h_1 - h_c)\}} = \sqrt{2\{c_p(T_1 - T_c)\}}$$

i.e. $C_c = \sqrt{\{2 \times 1.11 \times 10^3(698 - 598.4)\}} = 470.3$ m/s

(Note that C_c can also be found using equation (10.9) $C_c = \sqrt{(\gamma R T_c)}$.)
    The specific gas constant, R, for the gases can be found from equation (2.22)

$$c_p = \frac{\gamma R}{(\gamma - 1)}$$

$$R = \frac{c_p(\gamma - 1)}{\gamma} = \frac{1.11 \times 10^3 \times 0.333}{1.333}$$

i.e. $R = 277.3$ N m/kg K

Then from equation (2.5)

$$v_c = \frac{RT_c}{p_c} = \frac{277.3 \times 598.4}{10^5 \times 1.89} = 0.878 \text{ m}^3/\text{kg}$$

Now the mass flow for isentropic flow, $\dot{m}_s$, is given from equation (10.16) as

$$0.99 = \frac{18}{\dot{m}_s} \quad \text{i.e.} \quad \dot{m}_s = \frac{18}{0.99} = 18.18 \text{ kg/s}$$

Then using equation (1.11)

$$A_c = \frac{\dot{m}_s v_c}{C_c} = \frac{18.18 \times 0.878}{470.3} = 0.0339 \text{ m}^2$$

i.e.   Throat area $= 0.0339 \text{ m}^2$

For an isentropic expansion from the inlet conditions down to the back pressure, the temperature exit is T_2, given by equation (3.21)

$$\frac{T_1}{T_2} = \left(\frac{p_1}{p_2}\right)^{(\gamma-1)/\gamma} \quad \text{i.e.} \quad \frac{698}{T_2} = \left(\frac{3.5}{0.97}\right)^{0.333/1.333} = 1.378$$

therefore

$$T_2 = \frac{698}{1.378} = 506.6 \text{ K}$$

The expansion is shown on a T–s diagram in Fig. 10.14, line 1–c–2s representing the isentropic expansion, and line 1–2 representing the actual expansion.

Fig. 10.14 T–s diagram for Example 10.4

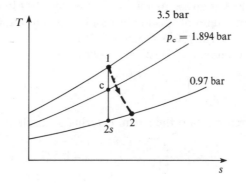

From equation (10.12)

$$\text{Nozzle efficiency} = 0.94 = \frac{T_1 - T_2}{T_1 - T_{2s}} = \frac{698 - T_2}{698 - 506.6}$$

i.e. $T_2 = 698 - 0.94(698 - 506.6) = 698 - 180 = 518.1 \text{ K}$

Then from equation (2.5)

$$v_2 = \frac{RT_2}{p_2} = \frac{277.3 \times 518.1}{10^5 \times 0.97} = 1.48 \text{ m}^3/\text{kg}$$

303

Also from equation (10.4)

$$C_2 = \sqrt{\{2(h_1 - h_2)\}} = \sqrt{\{2c_p(T_1 - T_2)\}}$$
$$= \sqrt{\{(2 \times 1.11 \times 10^3(698 - 518.1)\}} = 632 \text{ m/s}$$

Then substituting in equation (1.11),

$$A_2 = \frac{\dot{m}_2 v_2}{C_2} = \frac{18 \times 1.48}{632} = 0.0422 \text{ m}^2$$

It should be noted that in Example 10.3, the critical pressure, $p_c = 1.89$ bar, is the pressure at the throat of the nozzle when the flow is isentropic. The actual conditions at the nozzle throat are unknown, since no information is given about the proportion of the friction losses which occurs in the convergent portion. In order to allow for the friction present in the convergent portion the coefficient of discharge has been used. If the flow to the throat is isentropic the coefficient of discharge is unity.

10.6 The steam nozzle

The properties of steam can be obtained from tables or from an h–s chart, but in order to find the critical pressure ratio, and hence the critical velocity and the maximum mass flow rate, approximate formulae may be used. It is a good approximation to assume that steam follows an isentropic law $pv^k = $ constant, where k is an isentropic index for steam. (Although k is an isentropic index it is not a ratio of specific heats.) For steam initially dry saturated, $k = 1.135$; for steam initially superheated, $k = 1.3$.

Equation (10.7) can be re-written as

$$\frac{p_c}{p_1} = \left(\frac{2}{k+1}\right)^{k/k-1} \tag{10.17}$$

Therefore when the steam entering a nozzle is dry saturated

$$\frac{p_c}{p_1} = \left(\frac{2}{2.135}\right)^{1.135/0.135} = 0.577$$

and when the steam entering a nozzle is superheated

$$\frac{p_c}{p_1} = \left(\frac{2}{2.3}\right)^{1.3/0.3} = 0.546$$

The temperature at the throat, i.e. the critical temperature, can be found from steam tables at the value of p_c and $s_c = s_1$. The critical velocity can be found as before from equation (10.4)

$$C_c = \sqrt{\{2(h_1 - h_c)\}}$$

where h_c is read from tables or the h–s chart at p_c and s_c.

The case of steam initially superheated expanding into the wet region requires special treatment and is considered later.

For isentropic flow, since $v\,dp = dh$ and $vp^{1/k} = $ constant, we can write between any two states 1 and 2:

$$h_1 - h_2 = \int_1^2 v\,dp = \frac{-vp^{1/k}}{-(1/k) + 1}\left\{p_2^{-(1/k)+1} - p_1^{-(1/k)+1}\right\}$$

i.e. $$h_1 - h_2 = \frac{k}{k-1}(p_1 v_1 - p_2 v_2) = \frac{C_2^2 - C_1^2}{2} \qquad (10.18)$$

Example 10.4   Estimate the critical pressure and the throat area per unit mass flow rate of a convergent–divergent nozzle expanding steam from 10 bar, dry saturated, down to atmospheric pressure of 1 bar. Assume that the inlet velocity is negligible and that the expansion is isentropic.

Solution   Using equation (10.17)

$$p_c = 10\left(\frac{2}{2.135}\right)^{1.135/0.135} = 5.77 \text{ bar}$$

From the h–s chart at 5.77 bar and on a vertical line below point 1 at 10 bar dry saturated, we have

$$h_c = 2675 \text{ kJ/kg} \quad \text{and} \quad x_c = 0.962$$

$$v_c = 0.962 \times (v_g \text{ at } 5.77 \text{ bar})$$

Interpolating from tables at 5.77 bar we have $v_g = 0.328 \text{ m}^3/\text{kg}$, then

$$v_c = 0.962 \times 0.328 = 0.316 \text{ m}^3/\text{kg}$$

From equation (10.4)

$$C_c = \sqrt{\{2(h_1 - h_c)\}} = \sqrt{\{2(2778 - 2675) \times 10^3\}}$$
$$= 454 \text{ m/s}$$

or

$$C_c = \sqrt{(kp_c v_c)} = \sqrt{(1.135 \times 5.77 \times 10^5 \times 0.316)} = 455 \text{ m/s}$$

Then using equation (1.11)

$$A/m = v/c = 0.316 \times 10^6/454 = 696 \text{ mm}^2$$

i.e.   Throat area per kilogram per second $= 696 \text{ mm}^2$

Supersaturation

When a superheated vapour expands isentropically, condensation within the vapour begins to form when the saturated vapour line is reached. As the expansion continues below this line into the wet region, then condensation

Fig. 10.15 Superheated steam expanding into the wet region on (a) T–s and (b) h–s diagrams

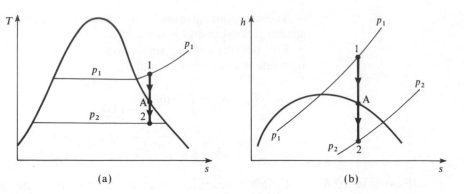

proceeds gradually and the dryness fraction of the steam becomes progressively smaller. This is illustrated on T–s and h–s diagrams in Figs 10.15(a) and (b). Point A represents the point at which condensation within the vapour just begins.

It is found that the expansion through a nozzle takes place so quickly that condensation within the vapour does not occur. The vapour expands as a superheated vapour until some point at which condensation occurs suddenly and irreversibly. The point at which condensation occurs may be within the nozzle or after the vapour leaves the nozzle.

Up to the point at which condensation occurs the state of the steam is not one of stable equilibrium, yet it is not one of unstable equilibrium, since a small disturbance will not cause condensation to commence. The steam in this condition is said to be in a *metastable state*; the introduction of a large object (e.g. a measuring instrument) will cause condensation to occur immediately.

Such an expansion is called a *supersaturated expansion*.

Assuming isentropic flow, as before, a supersaturated expansion in a nozzle is represented on a T–s and an h–s diagram in Figs 10.16(a) and (b) respectively. Line 1–2 on both diagrams represents the expansion with equilibrium throughout the expansion. Line 1–R represents supersaturated expansion. In supersaturated expansion the vapour expands as if the vapour line did not exist, so that line 1–R intersects the pressure line p_2 produced from the superheat region (shown chain-dotted). It can be seen from Fig. 10.16(a) that the temperature of the supersaturated vapour at p_2 is t_R, which is less than the saturation temperature t_2, corresponding to p_2. The vapour is said to be

Fig. 10.16 Supersaturated expansion of steam on (a) T–s and (b) h–s diagrams

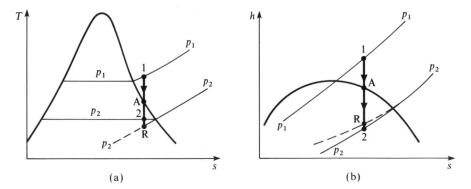

supercooled and the *degree of supercooling* is given by $(t_2 - t_R)$. Sometimes a *degree of supersaturation* is defined as the ratio of the actual pressure p_2 to the saturation pressure corresponding to the temperature t_R.

It can be seen from Fig. 10.16(b) that the enthalpy drop in supersaturated flow $(h_1 - h_R)$ is less than the enthalpy drop under equilibrium conditions. Since the velocity at exit, C_2, is given by equation (10.4), $C_2 = \sqrt{2(h_1 - h_2)}$, it follows that the exit velocity for supersaturated flow is less than that for equilibrium flow. Nevertheless, the difference in the enthalpy drop is small, and since the square root of the enthalpy drop is used in equation (10.4), then the effect on the exit velocity is small.

If the approximations for isentroic flow are applied to the equilibrium expansion, then for the process illustrated in Figs 10.16(a) and (b), the expansion from 1 to A obeys the law $pv^{1.3} = $ constant, and the expansion from A to 2 obeys the law $pv^{1.135} = $ constant. The equilibrium expansion and the supersaturated expansion are shown on a $p-v$ diagram in Fig. 10.17, using the same symbols as in Fig. 10.16. It can be seen from Fig. 10.17 that the specific volume at exit with supersaturated flow, v_R, is considerably less than the specific volume at exit with equilibrium flow, v_2. Now the mass flow through a given exit area, A_2, is given by equation (1.11), i.e. for equilibrium flow

$$\dot{m} = \frac{A_2 C_2}{v_2}$$

Fig. 10.17 Equilibrium and supersaturated expansion processes on a $p-v$ diagram

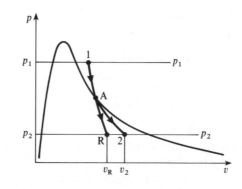

and for supersaturated flow

$$\dot{m}_s = \frac{A_2 C_R}{v_R}$$

It has been pointed out that C_2 and C_R are very nearly equal; therefore, since $v_R < v_2$, it follows that the mass flow with supersaturated flow is greater than the mass flow with equilibrium flow. It was this fact, proved experimentally, that led to the discovery of the phenomenon of supersaturation.

Example 10.5   A convergent–divergent nozzle receives steam at 7 bar and 200 °C and expands it isentropically into a space at 3 bar. Neglecting the inlet velocity, calculate the exit area required for a mass flow of 0.1 kg/s:
 (i) when the flow is in equilibrium throughout;
 (ii) when the flow is supersaturated with $pv^{1.3} = $ constant.

Fig. 10.18 Processes on the $h–s$ chart for Example 10.8

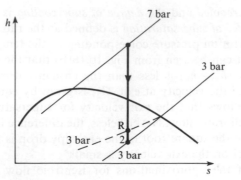

Solution A sketch of both processes (i) and (ii) is shown on an $h–s$ chart in Fig. 10.18.

Part (i) is most conveniently solved by using the $h–s$ chart.

From the chart at 7 bar, 200 °C, $h_1 = 2846$ kJ/kg. Also

$$h_2 = 2682 \text{ kJ/kg} \quad \text{and} \quad x_2 = 0.98$$

therefore

$$v_2 = x_2 v_{g_2} = 0.98 \times 0.6057 = 0.594 \text{ m}^3/\text{kg}$$

From equation (10.4)

$$C_2 = \sqrt{\{2(h_1 - h_2)\}} = \sqrt{2 \times 10^3(2846 - 2682)} = 573 \text{ m/s}$$

Then using equation (1.11)

$$A_2 = \frac{\dot{m}v_2}{C_2} = \frac{0.1 \times 0.594 \times 10^6}{573} = 103.7 \text{ mm}^2$$

i.e.   Exit area $= 103.7 \text{ mm}^2$

(ii) Although the process is represented on an $h–s$ diagram as 1–R in Fig. 10.18, nevertheless the $h–s$ chart cannot be used to find h_R, since the chain-dotted line representing 3 bar, produced from the superheat region, cannot be located easily on the chart. The enthalpy drop $(h_1 - h_R)$ and hence the velocity, C_R, can be found using equation (10.18),

i.e. $$\frac{C_R^2}{2} = \frac{k}{k-1}(p_1 v_1 - p_2 v_R)$$

From tables at 7 bar and 200 °C, $v_1 = 0.3001 \text{ m}^3/\text{kg}$. Then, since $p_1 v_1^k = p_2 v_R^k$, we have

$$\frac{v_R}{v_1} = \left(\frac{p_1}{p_2}\right)^{1/k} = \left(\frac{7}{3}\right)^{1/1.3} = 1.919$$

i.e. $v_R = 1.919 \times 0.3001 = 0.576 \text{ m}^3/\text{kg}$

Then substituting

$$\frac{C_R^2}{2} = \frac{1.3 \times 10^5}{0.3}\{(7 \times 0.3001) - (3 \times 0.576)\} = \frac{1.3 \times 10^5 \times 0.3727}{0.3}$$

$$C_R = \sqrt{\left(\frac{2 \times 1.3 \times 10^5 \times 0.3727}{0.3}\right)} = 568 \text{ m/s}$$

Then using equation (1.11)

$$A_2 = \frac{\dot{m}v_R}{C_R} = \frac{0.1 \times 0.576 \times 10^6}{568} = 101.4 \text{ mm}^2$$

10.7 Stagnation conditions

Throughout this chapter it has been assumed that the inlet velocity to the nozzle is negligible. When this is not the case the concept of stagnation conditions can be used.

Let a gas moving with velocity, C, at a temperature, T, be brought to rest adiabatically, finally reaching a temperature T_0 when at rest. Then, applying the flow equation, for a perfect gas, we have

$$c_p T + \frac{C^2}{2} = c_p T_0$$

or

$$T_0 = T + \frac{C^2}{2c_p} \tag{10.19}$$

The temperature T_0 is called the *stagnation temperature* of the moving gas.

When a thermometer is inserted in a moving gas stream, the gas around the bulb is brought to rest adiabatically and hence the thermometer measures the stagnation temperature. In order to measure the ordinary or static temperature, T, the thermometer would have to move at the gas velocity.

The term $C^2/(2c_p)$ in equation (10.19) is sometimes called the *temperature equivalent of velocity*. The error in the absolute temperature by neglecting this term is less than 1% for velocities up to about 75 m/s, for a gas at atmospheric temperature.

The stagnation pressure, p_0, of a gas stream is defined as the pressure the gas would attain if brought to rest isentropically.

From equation (3.21)

$$\frac{p_0}{p} = \left(\frac{T_0}{T}\right)^{\gamma/(\gamma-1)} \tag{10.20}$$

Using equation (10.19), we have

$$\frac{p_0}{p} = \left(1 + \frac{C^2}{2c_p T}\right)^{\gamma/(\gamma-1)}$$

Also, from equation (2.22), $c_p = \gamma R/(\gamma - 1)$, hence substituting

$$\frac{p_0}{p} = \left\{1 + \frac{(\gamma-1)C^2}{2\gamma RT}\right\}^{\gamma/(\gamma-1)}$$

309

From equation (10.10), the velocity of sound in a gas, a, is equal to $\sqrt{(\gamma RT)}$, hence

$$\frac{p_0}{p} = \left\{1 + \frac{(\gamma - 1)C^2}{2a^2}\right\}^{\gamma/(\gamma - 1)} = \left\{1 + \frac{(\gamma - 1)(Ma)^2}{2}\right\}^{\gamma/(\gamma - 1)} \quad (10.21)$$

where $Ma = C/a$ is the *Mach number*.

If the right-hand side of equation (10.21) is expanded by the binomial theorem we have

$$\frac{p_0}{p} = 1 + \frac{\gamma(\gamma - 1)(Ma)^2}{(\gamma - 1) \times 2}$$

$$+ \left(\frac{\gamma}{\gamma - 1}\right)\left\{\left(\frac{\gamma}{\gamma - 1}\right) - 1\right\}\frac{1}{2}\frac{(\gamma - 1)^2(Ma)^4}{4} + \cdots$$

i.e.
$$\frac{p_0}{p} = 1 + \frac{\gamma(Ma)^2}{2} + \frac{\gamma(Ma)^4}{8} + \cdots$$

When the velocity of the gas is low, and Ma is therefore small (say $Ma < 0.2$), then it is a good approximation to write

$$\frac{p_0}{p} = 1 + \frac{\gamma(Ma)^2}{2} = 1 + \frac{\gamma C^2}{2\gamma RT} = 1 + \frac{C^2}{2RT}$$

and
$$p_0 = p + \frac{C^2}{2}\frac{p}{RT}$$

Now the density, ρ, is the reciprocal of the specific volume,

$$\rho = \frac{1}{v} = \frac{p}{RT}$$

i.e. Stagnation pressure,
$$p_0 = p + \frac{\rho C^2}{2} \quad (10.22)$$

The term $\rho C^2/2$ in equation (10.22) is called the *velocity head*.

Applying stagnation conditions to flow through a nozzle we have at inlet

$$\frac{C_1^2}{2} + c_p T_1 = c_p T_{0_1}$$

At any other section of the nozzle where the velocity is C and the temperature is T we have

$$\frac{C^2}{2} + c_p T = c_p T_0$$

therefore

$$\frac{C_1^2}{2} + c_p T_1 = \frac{C^2}{2} + c_p T = c_p T_{0_1} = c_p T_0$$

Therefore the stagnation temperature remains constant throughout the nozzle for adiabatic flow. (The stagnation pressure remains constant throughout the nozzle for isentropic flow, but not for irreversible adiabatic flow, since there is a pressure loss due to friction.)

The nozzle inlet velocity, C_1, can be treated by imagining the nozzle extrapolated back from the inlet to a section where the velocity is zero. The conditions at this imaginary section are the stagnation conditions. This is illustrated in Fig. 10.19. At section 0 the cross-sectional area is infinite.

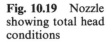

Fig. 10.19 Nozzle showing total head conditions

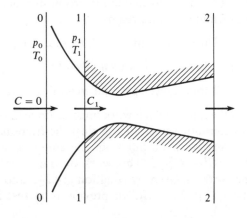

The equations derived previously can be used with p_{0_1} and T_{0_1} substituted for p_1 and T_1. Therefore, equation (10.4) can be written

$$C = \sqrt{\{2c_p(T_{0_1} - T)\}}$$

Also, from equations (10.7) and (10.8) we have

$$\frac{p_c}{p_{0_1}} = \left(\frac{2}{\gamma + 1}\right)^{\gamma/(\gamma - 1)} \quad \text{and} \quad \frac{T_c}{T_{0_1}} = \frac{2}{\gamma + 1}$$

10.8 Jet propulsion

Aircraft propulsion may be achieved by using a heat engine to drive an airscrew or propeller, or by allowing a high-energy fluid to expand and leave the aircraft in a rearward direction as a high-velocity jet. In the propeller type of aircraft engine the propeller takes a large mass flow and gives it a moderate velocity backwards relative to the aircraft. In the jet engine the aircraft induces a comparatively small airflow and gives it a high velocity backwards relative to the aircraft. In both cases the rate of change of momentum of the air provides a reactive forward thrust which propels the aircraft.

The propeller-type engine can be driven by a petrol engine or by a gas turbine unit.

If the velocity of the jet (from propeller or jet engine), backwards relative to the aircraft, is C_j, and the velocity of the aircraft is C_a, then the atmospheric air, initially at rest, is given a velocity of $(C_j - C_a)$. This is illustrated in Fig. 10.20. Assuming for the moment that the jet leaves the aircraft in the case of

Fig. 10.20 Flow through a turbojet and a turboprop

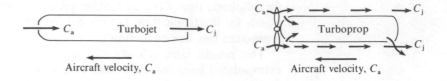

the jet engine at atmospheric pressure, then there is no thrust due to pressure forces. The thrust available for propulsion is solely due to the rate of change of momentum of the stream,

i.e. Thrust per unit mass flow rate $= C_j - C_a$ (10.23)

The propulsive power is then given by

Thrust power per unit mass flow rate $= C_a(C_j - C_a)$ (10.24)

This is the rate at which work must be done in order to keep the aircraft moving at the constant velocity C_a against the frictional resistance or drag.

The net work output from the engine is given by the increase in kinetic energy, $(C_j^2 - C_a^2)/2$. This work output is used in two ways: it provides the thrust work as given by equation (10.24), and it gives the air a kinetic energy of $(C_j - C_a)^2/2$, i.e. the air previously at rest is given an absolute velocity of $(C_j - C_a)$,

i.e. $$C_a(C_j - C_a) + \frac{(C_j - C_a)^2}{2} = C_a C_j - C_a^2 + \frac{C_j^2}{2} + \frac{C_a^2}{2} - \frac{2C_j C_a}{2}$$

i.e. Work output from engine $= \dfrac{C_j^2 - C_a^2}{2}$ (10.25)

The propulsive efficiency, η_P, is defined as the thrust work divided by the rate at which work is done on the air in the aircraft. Therefore from equations (10.24) and (10.25) we have

$$\eta_P = \frac{2C_a(C_j - C_a)}{C_j^2 - C_a^2}$$

therefore

$$\eta_P = \frac{2C_a}{C_j + C_a} \tag{10.26}$$

It can be seen from equation (10.26) that as the aircraft velocity, C_a, increases then the propulsive efficiency increases. For a propeller-driven aircraft the change of η_P is greater initially, but at speeds at which the propeller tip approaches sonic velocity the efficiency of the propeller falls off rapidly, and equation (10.26) is no longer applicable. Curves of the form shown in Fig. 10.21 are obtained. It can be seen that for aircraft speeds up to about 850 km/h the propeller is the more efficient means of propulsion, but for speeds above this the jet engine is superior.

The simplest form of jet engine is the *ramjet*. In the ramjet the air is compressed by the conversion of the kinetic energy of the atmospheric air relative to the

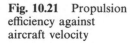

Fig. 10.21 Propulsion
efficiency against
aircraft velocity

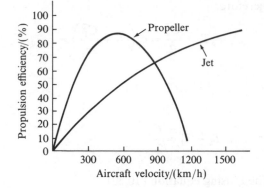

aircraft; this is known as the *ram effect*. Fuel is then burned in the compressed
air stream at approximately constant pressure, and the hot gases are allowed
to expand through a nozzle, reaching a high velocity backwards relative to the
aircraft. The ramjet is shown diagrammatically in Fig. 10.22(a), and the cycle
is represented on a T–s diagram in Fig. 10.22(b).

Fig. 10.22 Ramjet with
processes on a T–s
diagram

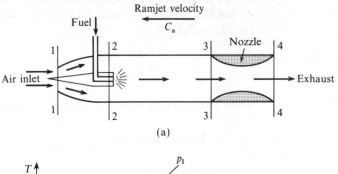

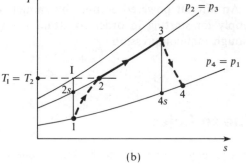

If the ramjet velocity is C_a, then the air enters the diffuser with a kinetic
energy of $C_a^2/2$ per unit mass of air. The velocity after diffusion can be allowed
for by using the stagnation temperature after diffusion, as follows:

Using the flow equation and assuming isentropic flow

$$h_1 + \frac{C_a^2}{2} = h_I + \frac{C_I^2}{2}$$

313

therefore

$$c_p T_1 + \frac{C_a^2}{2} = c_p\left(T_1 + \frac{C_1^2}{2c_p}\right) = c_p T_{0_1}$$

i.e.

$$\frac{C_a^2}{2} = c_p(T_{0_1} - T_1)$$

or

$$T_{0_1} - T_1 = \frac{C_a^2}{2c_p} \qquad (10.27)$$

Then, using equation (10.20)

$$\frac{T_{0_1}}{T_1} = \left(\frac{p_{0_1}}{p_1}\right)^{(\gamma-1)/\gamma}$$

The total pressure p_{0_1} is the pressure the air attains when the diffusion process is isentropic. When the process is irreversible, although still approximately adiabatic, then the total pressure attained is p_{0_2}, which is less than p_{0_1}, as seen from Fig. 10.22(b). Since the kinetic energy available, $C_a^2/2$, is the same whether or not the process is reversible, then the temperature change remains the same (i.e. $T_{0_2} = T_{0_1}$),

i.e.

$$T_{0_2} - T_1 = \frac{C_a^2}{2c_p} \qquad (10.28)$$

Then the intake isentropic efficiency is defined as follows:

$$\text{Intake efficiency} = \frac{T_{t_{2s}} - T_1}{T_{t_2} - T_1} \qquad (10.29)$$

An aircraft powered entirely by ramjet would require an auxiliary power supply for starting in order to attain the velocity necessary to give a large enough ram compression.

10.9 The turbojet

In a jet engine or turbojet the kinetic energy of the incoming air can be used to obtain a ram compression in the intake duct, thus raising the overall efficiency of the unit. The layout of the unit has been considered briefly in section 9.1. In aircraft gas turbine work it becomes important to use stagnation conditions, since velocity changes through the unit are no longer negligible. Also, in general, temperature-measuring instruments such as thermocouples, measure stagnation temperature and not static temperature. Using stagnation conditions, the isentropic efficiencies of the compressor and turbine can be redefined, and an intake and jet pipe efficiency can be introduced.

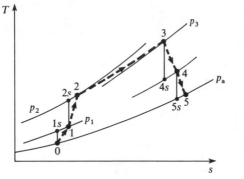

Fig. 10.23   Processes for a typical jet engine on a T–s diagram

Referring to Fig. 10.23 for a typical jet engine, we have

$$\text{Isentropic efficiency of intake duct} = \frac{T_{0_{1s}} - T_0}{T_{0_1} - T_0} \tag{10.30}$$

$$\text{Isentropic efficiency of compressor} = \frac{T_{0_{2s}} - T_{0_1}}{T_{0_2} - T_{0_1}} \tag{10.31}$$

$$\text{Isentropic efficiency of turbine} = \frac{T_{0_3} - T_{0_4}}{T_{0_3} - T_{0_{4s}}} \tag{10.32}$$

and $$\text{Jet pipe efficiency} = \frac{T_{0_4} - T_5}{T_{0_4} - T_{5s}} \tag{10.33}$$

For adiabatic flow, the total temperature remains constant, and therefore $T_0 = T_{0_1}$, and $T_{0_4} = T_{0_5}$, for the intake duct and the jet pipe respectively. Note that Fig. 10.23 is a diagram of static temperature against entropy.

In a practical unit there is a loss of pressure in the combustion chamber from 2 to 3.

Pressure thrust

It has been assumed in the foregoing analysis that the gases expand down to atmospheric pressure in the jet nozzle. In practice, particularly in the case of a convergent nozzle, the back pressure will normally be lower than the pressure of the gases at the nozzle outlet; this phenomenon is called underexpansion and is fully explained in section 10.4.

Due to the difference in pressure between the nozzle exit and the atmosphere in which the aircraft is flying there will be an additional thrust, called the pressure thrust. Also, in the case of a supersonic aircraft, the pressure at the air intake is higher than the atmospheric pressure because of compression through the shock wave formed; this causes a reduction in the net thrust calculated purely from momentum considerations.

Consider an aircraft like the turbojet in Fig. 10.24 with an air intake of area A_1, inlet air pressure p_1, and a nozzle exit area A_2, exit pressure p_2; let the atmospheric pressure be p_a. For a control volume round the working fluid in

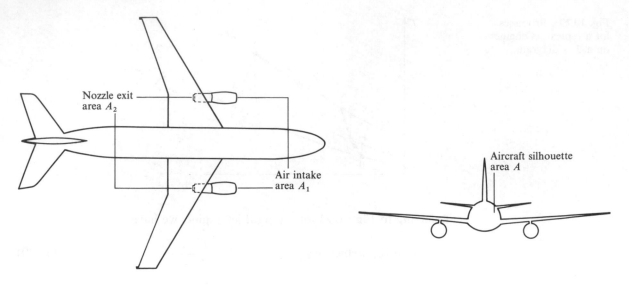

Fig. 10.24 Plan view of aircraft and aircraft silhouette

the aircraft engine we have, using Newton's second law:

$$F + p_1 A_1 - p_2 A_2 = \text{rate of change of momentum of working fluid}$$
$$\text{in the direction of motion of the fluid}$$

where F is the net force due to hydrostatic pressure and friction exerted by the inside of the aircraft on the working fluid in the direction of its motion,

i.e.
$$F + p_1 A_1 - p_2 A_2 = \dot{m}(C_j - C_a)$$

therefore

$$F = \dot{m}(C_j - C_a) - p_1 A_1 + p_2 A_2$$

There is an equal and opposite force, R, exerted by the working fluid on the inside of the aircraft engine,

i.e.
$$R = \dot{m}(C_j - C_a) - p_1 A_1 + p_2 A_2$$

in the direction of motion of the aircraft.

Consider now the forces acting on the aircraft. There is the force R, there is the total drag D, due to the air resistance, and there is a pressure force due to the atmospheric pressure acting on the projected area in the direction of flight. It may be easier to understand this force if the aircraft is imagined to be fixed to a test bed such that there is no flow of air over it. In flight there is considerable pressure variation over the aircraft surfaces, which is the cause of the lift and drag forces; the total drag force, D, incorporates all such effects and also includes form drag due to the vortices formed.

Assuming that the aircraft silhouette area in the direction of flight is A (see Fig. 10.24), then the net pressure force in the direction of flight is given by

$$p_a(A - A_2) - p_a(A - A_1) = p_a(A_1 - A_2)$$

Since the aircraft is flying at constant velocity the net force acting is zero,

i.e.
$$R - D + p_a(A_1 - A_2) = 0$$

Therefore the total thrust required to overcome the total drag force is given by

$$\text{Total thrust} = D = R + p_a(A_1 - A_2)$$

$$= \dot{m}(C_j - C_a) - p_1 A_1 + p_2 A_2 + p_a(A_1 - A_2)$$

therefore

$$\text{Total thrust} = \dot{m}(C_j - C_a) + A_2(p_2 - p_a) - A_1(p_1 - p_a)$$

For subsonic aircraft the last term is zero, since in that case $p_1 = p_a$,

i.e.
$$\text{Total thrust} = \dot{m}(C_j - C_a) + A_2(p_2 - p_a) \tag{10.34}$$

$$\underset{\text{(momentum thrust)}}{\quad\quad\quad\quad\quad} \underset{\text{(pressure thrust)}}{\quad\quad\quad}$$

Example 10.6

A turbojet aircraft is flying at 800 km/h at 10 700 m where the pressure and temperature of the atmosphere are 0.24 bar and $-50\,°\text{C}$ respectively. The compressor pressure ratio is 10/1 and the maximum cycle temperature is 820 °C. Calculate the thrust developed and the specific fuel consumption, using the following information: entry duct efficiency 0.9; isentropic efficiency of compressor 0.9; stagnation pressure loss in the combustion chamber 0.14 bar; calorific value of fuel 43 300 kJ/kg; combustion efficiency 98%; isentropic efficiency of turbine 0.92; mechanical efficiency of drive 98%; jet pipe efficiency 0.92; nozzle outlet area 0.08 m²; c_p and γ for the compression process 1.005 kJ/kg K and 1.4; c_p and γ for the combustion and expansion processes 1.15 kJ/kg K and 1.333; assume that the nozzle is convergent.

Solution

The cycle is shown on a T–s diagram in Fig. 10.25. The exhaust condition of the gases leaving the nozzle is not known until it is ascertained whether or not the nozzle is choked.

Fig. 10.25 Cycle on a T–s diagram for Example 10.6

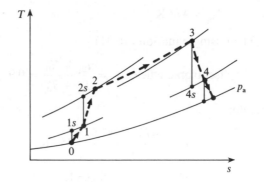

Kinetic energy of air at inlet

$$= \tfrac{1}{2} \times \left(\frac{800 \times 1000}{3600}\right)^2 = \tfrac{1}{2} \times (222.2)^2 \text{ N m/kg}$$

$$= 24.7 \text{ kJ/kg}$$

Therefore

$$T_{0_1} - T_0 = \frac{24.7}{c_p} = \frac{24.7}{1.005} = 24.6 \text{ K}$$

therefore

$$T_{0_1} = (-50 + 273) + 24.6 = 247.6 \text{ K}$$

Now from equation (10.30),

$$\text{Intake efficiency} = \frac{T_{0_{1s}} - T_0}{T_{0_1} - T_0} = 0.9$$

therefore

$$T_{0_{1s}} - T_0 = 0.9 \times 24.6 = 22.1 \text{ K}$$

i.e. $T_{0_{1s}} = (-50 + 273) + 22.1 = 245.1 \text{ K}$

$$\frac{p_{0_1}}{p_a} = \left(\frac{T_{0_{1s}}}{T_0}\right)^{\gamma/(\gamma-1)} = \left(\frac{245.1}{223}\right)^{1.4/0.4} = 1.1^{3.5} = 1.393$$

therefore

$$p_{0_1} = 1.393 \times 0.24 = 0.334 \text{ bar}$$

For the compressor, we have

$$\frac{T_{0_{2s}}}{T_{0_1}} = \left(\frac{p_{0_2}}{p_{0_1}}\right)^{(\gamma-1)/\gamma} = 10^{0.4/1.4} = 1.931$$

therefore

$$T_{0_{2s}} = 1.932 \times 247.6$$

i.e. $T_{0_{2s}} = 478 \text{ K}$

Then using equation (10.31)

$$\text{Isentropic efficiency} = \frac{T_{0_{2s}} - T_{0_1}}{T_{0_2} - T_{0_1}} = 0.9$$

therefore

$$T_{0_2} - T_{0_1} = \frac{478 - 247.6}{0.9} = 256 \text{ K}$$

i.e. $T_{0_2} = 247.6 + 256 = 503.6 \text{ K}$

Also $p_{0_2} = 10 \times p_{0_1} = 10 \times 0.334 = 3.34 \text{ bar}$

Hence

$$p_{0_3} = p_{0_2} - (\text{loss of total pressure in combustion})$$

i.e. $p_{0_3} = 3.34 - 0.14 = 3.2 \text{ bar}$

Now the turbine develops just enough work to drive the compressor and overcome mechanical losses,

i.e.
$$c_{p_g}(T_{0_3} - T_{0_4}) = \frac{c_{p_a}(T_{0_2} - T_{0_1})}{0.98}$$

Note that the work output from the turbine and the work input to the compressor are given by the product of c_p and the difference in total temperature, when the flow in each is adiabatic, e.g. for the turbine, using the flow equation we have

$$c_{p_g} T_3 + \frac{C_3^2}{2} + W = c_{p_g} T_4 + \frac{C_4^2}{2}$$

therefore

$$-W = c_{p_g}(T_{0_3} - T_{0_4})$$

Therefore

$$T_{0_3} - T_{0_4} = \frac{1.005(503.6 - 247.6)}{1.15 \times 0.98} = 228.3 \text{ K}$$

i.e.
$$T_{0_4} = (820 + 273) - 228.3 = 864.7 \text{ K}$$

Then using equation (10.32),

$$\text{Isentropic efficiency} = \frac{T_{0_3} - T_{0_4}}{T_{0_3} - T_{0_{4s}}} = 0.92$$

therefore

$$T_{0_3} - T_{0_{4s}} = \frac{228.3}{0.92} = 248.2 \text{ K}$$

i.e.
$$T_{0_{4s}} = (820 + 273) - 248.2 = 844.9 \text{ K}$$

Then
$$\frac{p_{0_3}}{p_{0_4}} = \left(\frac{T_{0_3}}{T_{0_{4s}}}\right)^{\gamma/(\gamma-1)} = \left(\frac{1093}{844.9}\right)^{1.333/0.333} = 2.803$$

therefore

$$p_{0_4} = \frac{3.24}{2.803} = 1.156 \text{ bar}$$

For choked flow in the nozzle the critical pressure ratio is given by equation (10.7),

i.e.
$$\frac{p_c}{p_{0_4}} = \left(\frac{2}{\gamma + 1}\right)^{\gamma/(\gamma-1)} = \left(\frac{2}{2.333}\right)^{1.333/0.333} = 0.54$$

therefore

$$p_c = 0.54 \times 1.156 = 0.624 \text{ bar}$$

Since the atmospheric pressure is 0.24 bar it follows that the nozzle is choking and hence the actual velocity of the gas at exit is sonic. The expansion in the nozzle is shown on a T–s diagram in Fig. 10.26.

Fig. 10.26 T–s diagram for Example 10.6 showing expansion outside nozzle

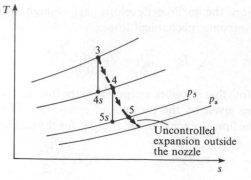

The temperature at exit, T_5, is given by equation (10.8) since the velocity at exit is sonic,

i.e. $$\frac{T_5}{T_{0_4}} = \frac{2}{\gamma + 1} = \frac{2}{1.333 + 1}$$

$$T_5 = \frac{864.7}{1.1665} = 741.3 \text{ K}$$

Using equation (10.33)

$$\text{Jet pipe efficiency} = \frac{T_{0_4} - T_5}{T_{0_4} - T_{5s}} = 0.92$$

therefore

$$T_{5s} = 864.7 - \frac{(864.7 - 741.3)}{0.92} = 730.6 \text{ K}$$

From equation (10.20)

$$\frac{p_{0_4}}{p_5} = \left(\frac{T_{0_4}}{T_{5s}}\right)^{\gamma/\gamma - 1} = \left(\frac{864.7}{730.6}\right)^{1.333/0.333} = 1.963$$

therefore

$$p_5 = \frac{1.156}{1.963} = 0.589 \text{ bar}$$

Now, from equation (2.22)

$$R = \frac{c_p(\gamma - 1)}{\gamma} = \frac{1.15 \times 0.333}{1.333}$$

i.e. $R = 0.2873 \text{ kJ/kg K}$

Hence $v_5 = \dfrac{RT_5}{p_5} = \dfrac{287.3 \times 741.3}{0.589 \times 10^5} = 3.616 \text{ m}^3/\text{kg}$

Also Jet velocity, $C_j = \sqrt{\gamma R T_5} = \sqrt{1.333 \times 287.3 \times 741.3} = 532.8 \text{ m/s}$

Then   Mass flow $= \dfrac{AC_j}{v_5} = \dfrac{0.08 \times 532.8}{3.616} = 11.788 \text{ kg/s}$

The momentum thrust is given by equation (10.23),

i.e.   Momentum thrust $= \dot{m}(C_j - C_a) = 11.788(532.8 - 222.2)$

$$= 3661 \text{ N}$$

The pressure thrust is given by equation (10.34),

i.e.   Pressure thrust $= (p_5 - p_a)A = (0.589 - 0.24) \times 0.08 \times 10^5$

$$= 2792 \text{ N}$$

therefore

Total thrust $= 3661 + 2792 = 6453 \text{ N}$

Also   Heat supplied $= \dot{m}c_{p_g}(T_{0_3} - T_{0_2})$

$$= 11.788 \times 1.15(1093 - 503.6) = 7990 \text{ kJ/s}$$

If curves of theoretical total temperature rise against fuel–air ratio were available for the fuel used, then the fuel consumption could be found in this way. In this case it is sufficient to write

Heat supplied $= \dot{m}_f \times$ calorific value $= \dfrac{7990}{0.98} \text{ kJ/s}$

where $\dot{m}_f$ is the mass of fuel supplied in kg/s,

$$\dot{m}_f = \dfrac{7990}{43\,300 \times 0.98} = 0.188 \text{ kg/s}$$

i.e.   Specific fuel consumption $= \dfrac{0.188 \times 10^3}{6453} = 0.0291 \text{ kg/kN s}$

(Note: because of the low value of fuel–air ratio it is a good approximation to assume that the mass flow rate of air is equal to the mass flow rate of gases in the turbine and nozzle.)

At sea-level conditions with the same exit area for the propulsion nozzle, the thrust produced will be much higher since the mass flow through the unit will be considerably increased. It is possible to have a variable-area nozzle which can be adjusted to give maximum thrust at any given altitude. When a convergent–divergent nozzle is used, the jet pipe efficiency is less because of the increased friction of the divergent portion, and the jet pipe is bulkier. When the pressure ratio across the jet pipe is not large a convergent nozzle is preferred, since the pressure thrust obtained due to underexpansion makes up for the loss of momentum thrust. When the pressure ratio across the jet pipe becomes large, a convergent–divergent nozzle is required to make full use of the energy available. The nozzle throat area must be made variable to avoid the losses which would occur if the nozzle were underexpanding.

Another means of obtaining a thrust boost is by using *afterburning*; this is thermodynamically equivalent to reheat. Fuel is sprayed into the gases leaving the turbine thus increasing the jet velocity leaving the nozzle. A 50% increase in thrust can be obtained in this way, but it is very wasteful on fuel. Afterburning can be used on starting and as a reserve power source for thrust augmentation over short periods. It is usually arranged to keep the same pressure ratio across the nozzle when afterburning is used. Therefore, since jet velocity changes approximately directly with the square root of the exit temperature, but specific volume changes directly with the temperature, it follows that the exit area of the nozzle must be increased proportionately with the increase of jet velocity, i.e.

$$A = \dot{m}v/C_j = \text{constant} \times T/\sqrt{T} = \text{constant} \times \sqrt{T} = \text{constant} \times C_j$$

Thus a variable-area nozzle is necessary when afterburning is used in order to keep the mass flow through the unit constant, and therefore to allow the compressor and turbine to operate efficiently. The volume of the afterburner is large compared with a normal combustion chamber, since it operates at a lower pressure. This additional bulk and weight must be set against the increase in thrust available.

10.10 The turboprop

In a turboprop aircraft the ram effect of the incoming air relative to the aircraft can be used as in the turbojet. At exit from the turboprop engine, the gases should theoretically have a velocity relative to the aircraft, just high enough to carry the exhaust clear of the aircraft. In practice the whole pressure drop available is not used by the turbine, and the gases leaving the turbine expand in the jet pipe, thus leaving with a velocity relative to the aircraft which is higher than the relative velocity of the air entering the engine. This provides a momentum thrust which complements the thrust from the propeller. For basic calculations this additional thrust will be neglected, and it will be assumed that the gases leave the turbine at the ambient pressure. However, in order to make use of the stagnation isentropic efficiency of the turbine (given by equation (10.32)), it is necessary to know the stagnation temperature at turbine exhaust. This can be found if the exhaust velocity is known.

Example 10.7 A turboprop aircraft is flying at 650 km/h at an altitude where the ambient temperature is $-18\,°\mathrm{C}$. The compressor pressure ratio is 9/1 and the maximum cycle temperature is 850 °C. The intake duct efficiency is 0.9, and the stagnation isentropic efficiencies of the compressor and turbine are 0.89 and 0.93 respectively. Calculate the specific power output and the cycle efficiency, taking a mechanical efficiency of 98% and neglecting the pressure loss in the combustion chamber. Assume that the exhaust gases leave the aircraft at 650 km/h relative to the aircraft, and take c_p and γ as in Example 10.6.

Solution The cycle is shown on a T–s diagram in Fig. 10.27.

Fig. 10.27 Cycle on a T–s diagram for Example 10.7

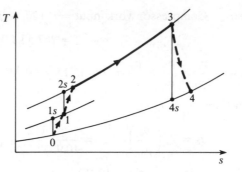

Kinetic energy of the air at inlet

$$= \frac{1}{2} \times \left(\frac{650 \times 10^3}{3600} \right)^2 = \frac{180.5^2}{2} \, \text{N m/kg}$$

therefore

$$T_{0_1} - T_0 = \frac{180.5^2}{2c_p} = \frac{180.5^2}{2 \times 10^3 \times 1.005} = 16.2 \, \text{K}$$

i.e. $T_{0_1} = (-18 + 273) + 16.2 = 271.2 \, \text{K}$

Now from equation (10.30)

$$\text{Intake efficiency} = \frac{T_{0_{1s}} - T_0}{T_{0_1} - T_0} = 0.9$$

therefore

$$T_{0_{1s}} - T_0 = 0.9 \times 16.2 = 14.6 \, \text{K}$$

i.e. $T_{0_{1s}} = 269.6 \, \text{K}$

Then $\dfrac{p_{0_1}}{p_0} = \left(\dfrac{T_{0_{1s}}}{T_0} \right)^{\gamma/(\gamma-1)} = \left(\dfrac{269.6}{255} \right)^{1.4/0.4} = 1.215$

For the compressor we have

$$T_{0_{2s}} = T_{0_1} \left(\frac{p_{0_2}}{p_{0_1}} \right)^{(\gamma-1)/\gamma} = 271.2 \times 9^{0.4/1.4} = 508 \, \text{K}$$

Then using equation (10.31)

$$\text{Isentropic efficiency} = \frac{T_{0_{2s}} - T_{0_1}}{T_{0_2} - T_{0_1}} = 0.89$$

therefore

$$T_{0_2} - T_{0_1} = \frac{508 - 271.2}{0.89} = 266 \, \text{K}$$

i.e. $T_{0_2} = 271.2 + 266 = 537.2 \, \text{K}$

Also   Compressor work input $= c_{p_a}(T_{0_2} - T_{0_1}) = 1.005 \times 266$
$$= 267.5 \text{ kJ/kg}$$

Now $\dfrac{p_{0_3}}{p_4} = \dfrac{p_{0_2}}{p_{0_1}} \times \dfrac{p_{0_1}}{p_0} = 9 \times 1.215 = 10.935$

therefore

$$T_4 = T_{0_3}\left(\frac{p_4}{p_{0_3}}\right)^{(\gamma - 1)/\gamma} = \frac{1123}{10.935^{0.333/1.333}} = 617.8$$

Then using equation (10.19)

$$T_{0_4} = T_4 + \frac{C_4^2}{2c_{p_g}} = 617.8 + \frac{180.5^2}{2 \times 10^3 \times 1.15} = 617.8 + 14.2$$

i.e. $T_{0_4} = 632 \text{ K}$

Then using equation (10.32)

$$\text{Isentropic efficiency} = \frac{T_{0_3} - T_{0_4}}{T_{0_3} - T_{0_{4s}}} = 0.93$$

therefore

$$T_{0_3} - T_{0_4} = 0.93(1123 - 632) = 456.6 \text{ K}$$

Then   Turbine work output $= c_{p_g}(T_{0_3} - T_{0_4}) = 1.15 \times 456.6$
$$= 525.1 \text{ kJ/kg}$$

therefore

Net work output $= (525.1 - 267.5)0.98 = 252.5 \text{ kJ/kg}$

i.e.   Specific power output $= 252.5 \text{ kW per kg/s}$

Also   Heat supplied $= c_{p_g}(T_{0_3} - T_{0_2}) = 1.15(1123 - 537.2)$
$$= 675 \text{ kJ/kg}$$

Then   Thermal efficiency $= \dfrac{252.5}{675} = 0.374$ or 37.4%

For aircraft flying at low speeds the amount of thrust from the jet pipe of a turboprop engine is usually very small, but as the aircraft speed is raised it becomes an advantage to use a portion of the available energy to obtain thrust in a nozzle, since the propulsion efficiency for a jet increases as shown in Fig. 10.21 (p. 313).

It was stated at the beginning of this section that the turbojet is more efficient than the turboprop at speeds of about 850 km/h and above. At lower speeds the propulsive efficiency can be increased by using a *ducted fan* engine. In the ducted fan engine a turbine drives a fan which draws air through a duct surrounding the engine, and delivers it either to the main gas stream leaving the turbine, or to atmosphere where it surrounds the main jet. Some of the energy available is used in driving the fan, but the decreased jet velocity at a higher mass flow gives a higher propulsive efficiency and a similar thrust to that of the turbojet engine. A ducted fan engine is shown diagrammatically in Fig. 10.28. Against the advantage of increased propulsive efficiency must be set the disadvantage of increased size and weight due to the ducting required to deal with the comparatively large volumes of air.

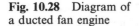

Fig. 10.28 Diagram of a ducted fan engine

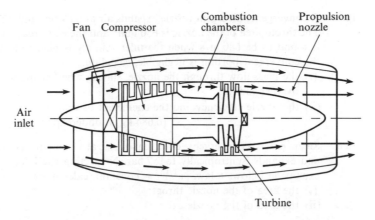

An alternative to the ducted fan engine is the *bypass* engine, in which some of the air flow is bypassed at an intermediate stage in the compression and passed directly to the main jet. In this engine and the ducted fan, fuel may be injected into the secondary airstream to give thrust boost over short periods.

Another important feature of the ducted fan and bypass engines (sometimes called turbofan engines) is the great reduction in noise level compared with the turbojet engine. This is of great importance for aircraft flying on commercial airlines, taking off and landing at airports in congested areas.

In considering the performance of gas turbine units many variables are involved. It is usual to express the variables in non-dimensional form. The non-dimensional characteristics of the compressor and turbine must then be matched (together with the propulsion nozzle, if a jet engine is considered), and a set of equilibrium running curves obtained for the unit. A good introductory treatment of equilibrium running is given in ref. 10.3.

Problems

10.1 Calculate the throat and exit areas of a nozzle to expand air at the rate of 4.5 kg/s from 8.3 bar, 327 °C into a space at 1.38 bar. Neglect the inlet velocity and assume isentropic flow.

(3290 mm^2; 4840 mm^2)

10.2 It is required to produce a stream of helium at the rate of 0.1 kg/s travelling at sonic velocity at a temperature of 15 °C. Assuming negligible inlet velocity, isentropic flow,

and a back pressure of 1.013 bar, calculate:
 (i) the required inlet pressure and temperature;
 (ii) the exit area of the nozzle.
For helium take the molar mass as 4 kg/kmol, and $\gamma = 1.66$.

(2.077 bar; 110 °C; 593 mm^2)

10.3 Recalculate Problem 10.1 assuming a coefficient of discharge of 0.96 and a nozzle efficiency of 0.92.

(3430 mm^2; 5310 mm^2)

10.4 A convergent–divergent nozzle expands air at 6.89 bar and 427 °C into a space at 1 bar. The throat area of the nozzle is 650 mm^2 and the exit area is 975 mm^2. The exit velocity is found to be 680 m/s when the inlet velocity is negligible. Assuming that friction in the convergent portion is negligible, calculate:
 (i) the mass flow through the nozzle, stating whether the nozzle is underexpanding or overexpanding;
 (ii) the nozzle efficiency and the coefficient of velocity.

(0.684 kg/s; underexpanding, $p_2 = 1.39$ bar; 0.895, 0.946)

10.5 Steam enters a convergent–divergent nozzle at 11 bar, dry saturated at a rate of 0.75 kg/s, and expands isentropically to 2.7 bar. Neglecting the inlet velocity, and assuming the expansion follows a law, $pv^{1.135} = $ constant, calculate:
 (i) the area of the nozzle throat;
 (ii) the area of the nozzle exit.

(474 mm^2; 646 mm^2)

10.6 Steam at 20 bar and 240 °C expands isentropically to a pressure of 3 bar in a convergent–divergent nozzle. Calculate the mass flow per unit exit area:
 (i) assuming equilibrium flow;
 (ii) assuming supersaturated flow.
For supersaturated flow assume that the process follows the law, $pv^{1.3} = $ constant.

(1540 kg/s; 1751 kg/s)

10.7 A thermometer inserted into an airstream flowing at 33.5 m/s records a temperature of 15 °C; the static pressure in the duct is found to be 1.01 bar. Assuming that the air round the thermometer bulb is brought to rest adiabatically, calculate:
 (i) the true static temperature of the air;
 (ii) the stagnation pressure of the air.

(14.44 °C; 1.017 bar)

10.8 A turbojet aircraft is travelling at 925 km/h in atmospheric conditions of 0.45 bar and −26 °C. The compressor pressure ratio is 8, the air mass flow rate is 45 kg/s, and the maximum allowable cycle temperature is 800 °C. The compressor, turbine, and jet pipe stagnation isentropic efficiencies are 0.85, 0.89, and 0.9 respectively, the mechanical efficiency of the drive is 0.98, and the combustion efficiency is 0.99. Assuming a convergent propulsion nozzle, a loss of stagnation pressure in the combustion chamber of 0.2 bar, and a fuel of calorific value 43 300 kJ/kg, calculate:
 (i) the required nozzle exit area;
 (ii) the net thrust developed;
 (iii) the air–fuel ratio;
 (iv) the specific fuel consumption.
For the gases in the turbine and propulsion nozzle take $\gamma = 1.333$ and $c_p = 1.15$ kJ/kg K; for the combustion process assume an equivalent c_p value of 1.15 kJ/kg K.

(0.216 m^2; 19.94 kN; 70.87; 0.0319 kg/kN s)

10.9 In a turboprop engine the compressor pressure is 6 and the maximum cycle temperature is 760 °C. The stagnation isentropic efficiencies of the compressor and turbine are 0.85 and 0.88 respectively; the mechanical efficiency is 99%. The aircraft is travelling at 725 km/h at an altitude where the ambient temperature is −7 °C. Taking an intake duct efficiency of 0.9, neglecting the pressure loss in the combustion chamber, and assuming that the gases in the turbine expand down to atmospheric pressure, leaving the aircraft at 725 km/h relative to the aircraft, calculate:

(i) the specific power output;
(ii) the cycle efficiency.

For the gases in the turbine take $\gamma = 1.333$ and $c_p = 1.15$ kJ/kg K; for the combustion process assume an equivalent c_p value of 1.15 kJ/kg K.

(170.2 kW per kg/s; 28.4%)

10.10 Afterburning is used in the aircraft of Problem 10.8 to obtain an increase in thrust. The stagnation temperature after the afterburner is 700 °C and the pressure loss in the afterburning process is 0.07 bar. Calculate the nozzle exit area now required to pass the same mass flow rate as in Problem 10.8, and the new net thrust.

(0.244 m²; 22.0 kN)

References

10.1 SHAPIRO A H 1983 *The Dynamics and Thermodynamics of Compressible Flow* vols 1 and 2 Kreiger

10.2 DOUGLAS J F, GASIOREK J M, and SWAFFIELD J A 1985 *Fluid Mechanics* 2nd edn Longman

10.3 COHEN H, ROGERS G F C, and SARAVANAMUTTOO H I H 1987 *Gas Turbine Theory* 3rd edn Longman

11

Rotodynamic Machinery

A rotodynamic machine is one in which a fluid flows freely through an impeller or rotor; the transfer of energy between the fluid and the rotor is continuous and the change of angular momentum of the fluid causes, or is the result of, a torque on the rotor. When energy is transferred from the fluid to the rotor the machine is known as a turbine; when energy is transferred to the fluid from the rotor the machine is known as a fan, pump, or compressor. Note that rotary machines such as the vane type discussed in Chapter 12 are defined as positive displacement rather than rotodynamic since the fluid does not flow freely through the rotating part of the machine but is displaced from sealed spaces.

This chapter covers the basic theory of turbines and compressors used in steam plant and gas turbine plant. The analysis of pumps, fans, and turbines using liquids or incompressible gases is not dealt with in this text; the reader is referred to books on fluid mechanics such as that by Douglas, Gasiorek and Swaffield (ref. 11.1) for the analysis of such rotodynamic machines. Compressors and turbines with air, hot gases, or steam, as the working substance can be analysed in a generalized way (refs 11.2, 11.3, 11.4), but in this text a more pragmatic approach is used; for example, steam and gas turbine plant use rotodynamic turbines, but the conditions of use and the working substance are so different that over the years designers have developed separate methods, and different design conventions.

11.1 Rotodynamic machines for steam and gas turbine plant

A rotodynamic turbine or compressor may be classified in two ways: firstly by the direction of flow of the fluid relative to the rotor; secondly by the way in which the rate of change of angular momentum of the fluid is achieved.

The flow of fluid can be either in a direction parallel to the axis of the rotor, in a radial direction, or less commonly in a mixed mode. The bulk of this chapter is concerned with axial-flow machines which are used most often in practice; the most commonly used radial machine is the centrifugal compressor considered in section 11.9; mixed mode machines are found mostly in pump applications and are not considered in this text.

Impulse turbine

The most basic turbine takes a high-pressure, high-enthalpy fluid, expands it in a fixed nozzle, and then uses the rate of change of angular momentum of the fluid in a rotating passage to provide the torque on the rotor. Such a machine is called an *impulse turbine*. A simple example of an impulse turbine is shown in Figs 11.1(a) and (b). Since the fluid flows through the wheel at a

Fig. 11.1 Simple impulse turbine (a) showing cross-section through blades and nozzles (b)

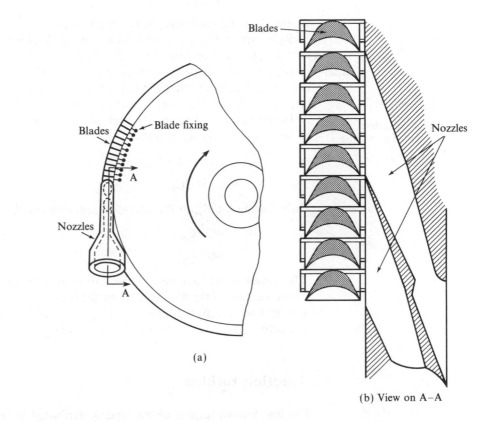

(a)

(b) View on A–A

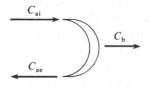

Fig. 11.2 Tangential flow on to a moving blade

fixed mean radius, then the change of linear momentum tangential to the wheel gives a tangential force that causes the wheel to rotate. Assume initially that the fluid is able to enter and leave the wheel passages in the tangential direction with an absolute velocity at inlet, C_{ai}, and an absolute velocity at exit, C_{ae}, as shown in Fig. 11.2; the blade velocity is denoted by C_b. The rate of increase of fluid momentum in the tangential direction from left to right in Fig. 11.2 gives the tangential force acting on the fluid,

i.e.   Force on the fluid from left to right $= \dot{m}(-C_{ae} - C_{ai})$

(assuming a constant mass flow rate, $\dot{m}$).

An equal and opposite force, F, must act on the blades,

i.e. $F = \dot{m}(C_{ai} + C_{ae})$ from left to right

329

The torque acting on the wheel is then given by

$$T = \dot{m}R(C_{ai} + C_{ae})$$

where R is the radius of the wheel, and the rate at which work is done for a rotational speed of N is

$$\dot{W} = 2\pi NT = 2\pi N\dot{m}R(C_{ai} + C_{ae})$$
$$= \dot{m}C_b(C_{ai} + C_{ae})$$

where C_b is the blade tangential speed $= 2\pi NR$.

Referring to Fig. 11.3, the velocity of the fluid relative to the blade at inlet is $(C_{ai} - C_b)$ and the velocity of the fluid relative to the blades at outlet in the direction of the blade movement is $(-C_{ae} - C_b)$. (Note: the relative velocity can be obtained by imagining stopping the blade by applying an equal and opposite velocity, C_b, from right to left at inlet and at outlet.)

In the absence of friction the relative velocity at inlet is equal *in magnitude* to the relative velocity at outlet,

i.e.
$$C_{ai} - C_b = -(-C_{ae} - C_b)$$
$$C_{ae} = C_{ai} - 2C_b$$

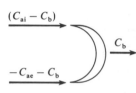

Fig. 11.3 Relative velocities for a moving blade

Substituting for C_{ae} in the previous equation for $\dot{W}$, we have

$$\dot{W} = \dot{m}C_b(C_{ai} + C_{ai} - 2C_b)$$
$$= 2\dot{m}C_b(C_{ai} - C_b)$$

This result could have been derived directly by considering the rate of change of momentum of the fluid relative to the blades. Since, in the absence of friction, the relative velocity changes from $(C_{ai} - C_b)$ at inlet to $-(C_{ai} - C_b)$ at outlet, the rate of increase of momentum is $-2\dot{m}(C_{ai} - C_b)$.

Reaction turbine

The first known turbine of this type is attributed to Hero of Alexandria about the first century AD. A turbine built on Hero's principle was demonstrated by Gustaf de Laval in 1883. Figure 11.4 shows the reaction type; the radial tubes, which are connected to the supply tube, are free to rotate about a vertical axis. The end of each tube is shaped as a nozzle and the steam from the supply tube expands through the nozzles to atmosphere in a tangential direction. There is an increase of velocity of the steam, and the rate of increase of momentum is provided by a force on the steam from the nozzle walls in the direction of the steam flow; an equal and opposite force acts on the nozzle walls causing the tubes to spin round in a direction opposite to the steam flow. De Laval's turbine was an achievement of its time and showed the possibility of high shaft speeds using steam, but it is now only of historical interest.

The reaction principle is used in modern turbines and compressors although in a very different way to that shown above. In modern reaction turbines half the expansion or compression occurs in fixed passages and the remainder of

Fig. 11.4   Simple reaction turbine

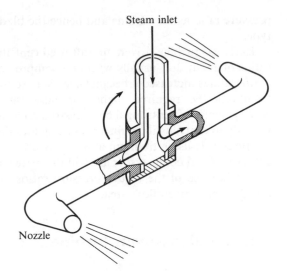

Steam inlet

Nozzle

the expansion or compression in moving passages. An expansion in the moving passages of a turbine causes the velocity of the fluid relative to the passages to increase, giving a reactive force on the blade walls. This type of blading is called 50% reaction, or Parsons blading after C A Parsons who built the first commercial steam turbine in 1885 using this principle.

Choice of rotodynamic machine

Because of the principle of using the rate of change of momentum of a flowing fluid to obtain a torque on a rotor, a rotodynamic machine has a high rotational speed. The rotating wheel and blades are subject to high stresses due to the centripetal accelerations involved and there is thus a limit to the energy that can be utilized in a single row of blades. Most rotodynamic machines are therefore multistage in operation.

There is a considerable difference in the pressure range of a typical steam turbine compared with that of a gas turbine. A steam turbine expands down to a condenser pressure governed by the temperature of the cooling water available (say about 0.035 bar), and the supply pressure may be as high as 170 bar giving overall pressure ratios of the order of 4500:1; even with a much lower supply pressure of about 40 bar the pressure ratio is still over 1000:1. The gas turbine, on the other hand, has an overall pressure range of the order of 10:1 with a supply pressure at ground level of about 10 bar.

It is common steam turbine practice to use one or more impulse stages at the HP end of the turbine so that the steam pressure is reduced rapidly over the first few stages; this prevents the high leakage losses that would otherwise occur due to the large pressure differences if reaction-type blading were used. Reaction blading is then used for the LP end of the turbine. It is rarely necessary to use an impulse stage for a gas turbine since the supply pressure is relatively low. Axial-flow compressors in a gas turbine plant have the same order of

331

pressure ratio as the turbine and hence the blading is also of the 50% reaction type.

Early gas turbine-driven aircraft used centrifugal compressors with pressure ratios of about 2:1, usually with two compressors in series. As maximum cycle temperatures increased it became necessary to introduce axial-flow compressors to obtain the required higher pressure ratios. Further research into the aerodynamics of centrifugal compressors and into blade materials has led more recently to centrifugal compressors using aluminium alloys with pressure ratios of about 4:1; using titanium alloys pressure ratios greater than 8:1 have been achieved. Axial compressors are still the better option for units developing high power because of the higher pressure ratios and the greater mass flow rates possible for a given flow area.

11.2 The impulse steam turbine

In this section it will be assumed that steam is the working substance since, for reasons given in section 11.1, most steam turbine plant use impulse steam turbines, whereas gas turbine plant seldom do. The general principles are the same whether steam or gas is the working substance.

The steam supplied to a single wheel impulse turbine expands completely in the nozzles and leaves with a high absolute velocity. This is the absolute inlet velocity to the blades as shown in Fig. 11.5(a). The steam is delivered to the wheel at an angle α_i. The selection of the angle α_i is one of compromise since an increase in α_i reduces the value of the useful component, $C_{ai} \cos \alpha_i$, and increases the value of the axial, or flow, component, $C_{ai} \sin \alpha_i$. The absolute velocity, C_{ai}, can be considered as the resultant of the blade velocity, C_b, and the velocity of the steam relative to the blade at inlet, C_{ri}. The two points of particular interest are those at the inlet and exit of the blades. The velocities of these points are as shown in Fig. 11.5(b) as C_{ri} and C_{re} respectively, and the directions are defined by the angles β_i and β_e as shown. If the steam is to enter and leave the blades smoothly without shock, then β_i is the angle of the blades at inlet, and β_e the angle of the blades at exit. The velocity triangle for the inlet conditions is drawn in Fig. 11.6(a) from the information of Fig. 11.5. The steam leaves the blade with a velocity, C_{re}, relative to the blade, and at the blade exit angle of β_e. The absolute velocity at exit C_{ae} is determined from the velocity triangle of Fig. 11.6(b), and its direction is α_e. Since both triangles have the common side OA $= C_b$, the triangles can be combined to give a single diagram as shown in Fig. 11.6(c).

Fig. 11.5 Absolute and relative velocities for a simple impulse turbine blade

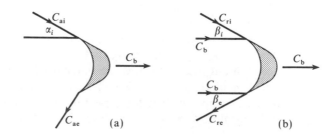

Fig. 11.6 Inlet (a) and outlet (b) blade velocity diagrams for an impulse turbine and a composite diagram (c)

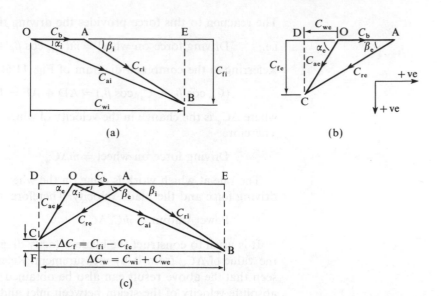

(a)

(b)

(c)

If the blade is symmetrical then $\beta_i = \beta_e$, and if the friction effects of the blade on the steam are zero, then $C_{re} = C_{ri}$. It is usual, however, that the velocity of the steam relative to the blade is reduced by friction, and this is expressed by

$$C_{re} = kC_{ri} \qquad (11.1)$$

where k is a *blade velocity coefficient*.

The velocities of flow across the blade at inlet and exit are given by C_{fi} and C_{fe} (i.e. EB and DC respectively in Fig. 11.6(c)). There may be a difference between these values which means that there is a change in velocity in the axial direction and an associated *axial thrust*. The horizontal components of the absolute velocities at inlet and exit are called the *whirl velocities*, C_{wi} and C_{we}, as shown in Fig. 11.6(c).

From Newton's second law the tangential force acting on the jet is given by $F = \dot{m} \times$ (change of velocity in the tangential direction).

The tangential velocity of the steam relative to the blade at inlet is given by

$$AE = C_{ri} \cos \beta_i$$

The tangential velocity of the steam relative to the blade at exit is given by

$$AD = -C_{re} \cos \beta_e$$

Therefore

Change in velocity in tangential direction

$$= -C_{re} \cos \beta_e - C_{ri} \cos \beta_i$$
$$= -(C_{re} \cos \beta_e + C_{ri} \cos \beta_i)$$

therefore

Tangential force $= -\dot{m}(C_{re} \cos \beta_e + C_{ri} \cos \beta_i)$

The reaction to this force provides the driving thrust on the wheel,

i.e.   Driving force on wheel $= \dot{m}(C_{re} \cos \beta_e + C_{ri} \cos \beta_i)$

Referring to the combined diagram of Fig. 11.6(c),

$$(C_{re} \cos \beta_e + C_{ri} \cos \beta_i) = AD + AE = DE = \Delta C_w$$

where ΔC_w is the change in the velocity of whirl $= C_{wi} - (-C_{we}) = C_{wi} + C_{we}$. Therefore

$$\text{Driving force on wheel} = \dot{m}\Delta C_w \qquad (11.2)$$

The rate at which work is done on the wheel is given by the product of the driving force and the blade velocity. Therefore, using equation (11.2)

$$\text{Power output} = \dot{m}C_b\Delta C_w \qquad (11.3)$$

It is usual to construct the blade velocity diagram to scale and to determine the value of ΔC_w from it by measurement. Referring to Fig. 11.6(c) it can be seen that the above result can also be obtained by considering the changes in absolute velocity of the steam between inlet and exit.

A part of this diagram is repeated in Fig. 11.7(a) as defined by the lines OB and OC. The change in absolute velocity of the jet is given by BC and the resultant force on the jet is equal to $\dot{m} \times BC$. The reaction to this is the force on the wheel (i.e. $\dot{m} \times CB$). This force can be expressed as its components $\dot{m} \times CF$ and $\dot{m} \times FB$ as in Fig. 11.7(b). From Fig. 11.6(c), $FB = DE = \Delta C_w$ and $\dot{m} \times FB$ is the tangential driving force as determined previously. The axial component of the force, $\dot{m} \times CF$, is the axial thrust on the wheel, which must be taken up by the bearings in which the shaft is mounted. From Fig. 11.6(c) it can be seen that $CF = EB - DC = C_{fi} - C_{fe} = \Delta C_f$,

i.e.   Axial thrust $= \dot{m}\Delta C_f$ (11.4)

Fig. 11.7 Absolute velocities at inlet and exit and the forces produced

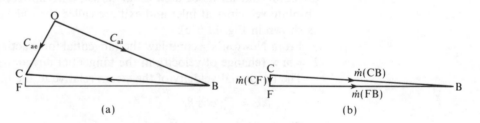

(a)   (b)

If the enthalpy of the steam at entry to the nozzle is h_0, and at the nozzle exit is h_i, then the maximum velocity of the steam impinging on the blades is given by equation (10.4) as

$$C_{ai} = \sqrt{\{2(h_0 - h_i)\}} \qquad (11.5)$$

(assuming negligible velocity at inlet to the nozzles).

The energy supplied to the blades is the kinetic energy of the jet, $C_{ai}^2/2$, and the blading efficiency or *diagram efficiency* is defined by

$$\text{Diagram efficiency, } \eta_d = \frac{\text{rate of doing work per unit mass of steam}}{\text{energy supplied per unit mass of steam}}$$

The numerator is obtained from equation (11.3) as $C_b \Delta C_w$ per unit mass flow rate of steam, therefore,

$$\eta_d = C_b \Delta C_w \times \frac{2}{C_{ai}^2} = \frac{2 C_b \Delta C_w}{C_{ai}^2} \tag{11.6}$$

For purpose of analysis, ΔC_w can be expressed as

$$\Delta C_w = C_{re} \cos \beta_e + C_{ri} \cos \beta_i \tag{11.7}$$

Also, if the blade velocity coefficient, $k = 1$ (i.e. $C_{re} = C_{ri}$), and if the blades are symmetrical (i.e. $\beta_e = \beta_i$), then we have

$$\Delta C_w = 2 C_{re} \cos \beta_i$$

But $\quad C_{re} \cos \beta_i = C_{ai} \cos \alpha_i - C_b \quad$ (see Fig. 11.6(a))

Hence

$$\Delta C_w = 2(C_{ai} \cos \alpha_i - C_b)$$

and $\quad$ Rate of doing work per unit mass $= 2(C_{ai} \cos \alpha_i - C_b)C_b \tag{11.8}$

Therefore

$$\eta_d = 2(C_{ai} \cos \alpha_i - C_b)C_b \times \frac{2}{C_{ai}^2}$$

$$= \frac{4(C_{ai} \cos \alpha_i - C_b)C_b}{C_{ai}^2}$$

i.e. $\quad$ Diagram efficiency $= \dfrac{4 C_b}{C_{ai}} \left(\cos \alpha_i - \dfrac{C_b}{C_{ai}} \right) \tag{11.9}$

where C_b / C_{ai} is called the *blade speed ratio*.

The simple impulse turbine is called the de Laval turbine, since it was invented by Dr Gustaf de Laval and patented by him in 1888. The first turbine produced by de Laval was a 11 kW marine turbine in 1892; this turbine ran at 16 000 rev/min with an output shaft speed of 330 rev/min using a double reduction in a double helical gear. It is a tribute to de Laval that so much of his work is either current practice or has received little modification up to the present day. He pioneered the use of high pressures and high speeds for steam turbines. His mechanical design from the point of view of blading shape and blade attachment compare with the best in modern practice. He also developed a form of double helical reduction gearing which is similar to that used in ships of modern construction.

Example 11.1 $\quad$ The velocity of steam leaving the nozzles of an impulse turbine is 900 m/s and the nozzle angle is 20°. The blade velocity is 300 m/s and the blade velocity coefficient is 0.7. Calculate for a mass flow of 1 kg/s, and symmetrical blading:

(i) the blade inlet angle;
(ii) the driving force on the wheel;

(iii) the axial thrust;

(iv) the diagram power;

(v) the diagram efficiency.

Solution The information given is indicated on Fig. 11.8, and the required quantities can be obtained either by construction or by calculation. In many cases calculation becomes tedious and long, and a graphical construction is recommended for all but the simplest problems.

(i) The blade inlet angle can be measured directly from Fig. 11.8 if this figure is drawn to scale, but the angle will be found by an analytical method to illustrate the procedure.

Fig. 11.8 Blade velocity diagram for Example 11.1

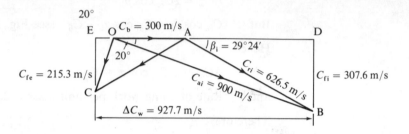

Applying the cosine rule to triangle OAB

$$C_{ri}^2 = C_{ai}^2 + C_b^2 - 2C_b C_{ai} \cos \alpha_i$$

$$= 900^2 + 300^2 - (2 \times 300 \times 900 \times \cos 20°) = 39.25 \times 10^4$$

therefore

$$C_{ri} = 626.5 \text{ m/s}$$

Then using the sine rule in triangle OAB,

$$\frac{C_{ai}}{\sin OAB} = \frac{C_{ri}}{\sin \alpha_i}$$

also $\sin OAB = \sin(180 - \beta_i) = \sin \beta_i$

therefore

$$\sin \beta_i = \frac{C_{ai} \sin \alpha_i}{C_{ri}} = \frac{900 \times \sin 20°}{626.5} = 0.491$$

i.e. $\beta_i = 29° 24' = \beta_e$

(ii) Using equation (11.1)

$$C_{re} = kC_{ri} = 626.5 \times 0.7 = 438.5 \text{ m/s}$$

$$AD = C_{ri} \cos \beta_i = 626.5 \cos 29° 24' = 545.8 \text{ m/s}$$

$$AE = C_{re} \cos \beta_e = 438.5 \cos 29° 24' = 381.9 \text{ m/s}$$

therefore

$$\Delta C_w = 545.8 + 381.9 = 927.7 \text{ m/s}$$

From equation (11.2)

Driving force on wheel $= \dot{m}\Delta C_w = 1 \times 927.7 = 927.7$ N per kg/s

(iii) $C_{fi} = C_{ri} \sin \beta_i = 626.5 \sin 29°\,24' = 307.6$ m/s

$C_{fe} = C_{re} \sin \beta_e = 438.5 \sin 29°\,24' = 215.3$ m/s

therefore

$\Delta C_f = 307.6 - 215.3 = 92.3$ m/s

Then from equation (11.4)

Axial thrust $= \dot{m}\Delta C_f = 1 \times 92.3 = 92.3$ N per kg/s

(iv) From equation (11.3)

Diagram power per unit mass flow rate $= C_b \Delta C_w$

$$= \frac{1 \times 927.7 \times 300}{10^3}\,\text{kW}$$

$$= 278.3 \text{ kW}$$

(v) From equation (11.6)

Diagram efficiency, $\eta_d = \dfrac{C_b \Delta C_w}{C_{ai}^2} = \dfrac{2 \times 927.7 \times 300}{900^2}$

$$= 0.687 \text{ or } 68.7\%$$

Optimum operating conditions from the blade velocity diagrams

From equation (11.8) the rate of doing work on the blade wheel per unit mass flow rate of steam is given by $2C_b(C_{ai}\cos \alpha_i - C_b)$. For a given steam velocity C_{ai} and a given blade velocity C_b, it can be seen that the rate of doing work is a maximum when $\cos \alpha_i = 1$ (i.e. when $\alpha_i = 0$). For this value of α_i the axial-flow component would be zero. It is essential to have an axial-flow component to allow the steam to reach the blades and to clear the blades on leaving. As α_i is increased the rate of working on the blades is reduced, but the blade annulus area required for a given mass flow is also reduced since the axial-flow component of the velocity is increased. Further, the surface area of the blades will be reduced at higher values of α_i, and this means friction losses will be less. A selection of α_i must be made based on these conflicting requirements, and the usual values of α_i lie between 15° and 30°. For a fixed value of α_i the optimum blade speed ratio for maximum diagram efficiency can be obtained by differentiating equation (11.9) and putting the result equal to zero,

i.e. Diagram efficiency, $\eta_d = 4\dfrac{C_b}{C_{ai}}\left(\cos \alpha_i - \dfrac{C_b}{C_{ai}}\right)$

337

therefore

$$\frac{\mathrm{d}(\eta_\mathrm{d})}{\mathrm{d}\left(\dfrac{C_\mathrm{b}}{C_\mathrm{ai}}\right)} = 4\cos\alpha_\mathrm{i} - 8\frac{C_\mathrm{b}}{C_\mathrm{ai}} = 0$$

therefore

$$\frac{C_\mathrm{b}}{C_\mathrm{ai}} = \frac{\cos\alpha_\mathrm{i}}{2} \tag{11.10}$$

i.e. $\qquad$ Maximum diagram efficiency $= \dfrac{4\cos\alpha_i}{2}\left(\cos\alpha_i - \dfrac{\cos\alpha_i}{2}\right)$

$$= \cos^2\alpha_i \tag{11.11}$$

The rate of doing work corresponding to the maximum diagram efficiency is then given by substituting in equation (11.8), i.e.

Power output per unit mass flow rate at maximum diagram efficiency
$$= 2C_\mathrm{b}^2 \tag{11.12}$$

The variation in η_d with $C_\mathrm{b}/C_\mathrm{ai}$ is shown in Fig. 11.9.

The single-stage impulse steam turbine is used only as a small power machine. The steam velocities may be as high as 1070 m/s, and for $\alpha_i = 20°$ the optimum blade speed ratio would be about 0.47, giving the maximum blade speed as 500 m/s. In practice the blade speed is limited to about 420 m/s. This value of velocity used in small machines would give high speeds of rotation of the order of 30 000 rev/min. Smaller-diameter rotors mean a more economic construction, but high rotational speeds mean high stresses. Further, the high turbine speeds are too high for direct use, and a reduction gear is required to give output speeds which are in the useful range.

The expansion of steam in the simple impulse turbine is carried out in a single stage, so the steam velocity at inlet to the wheel is high. There is no drop in pressure in the wheel casing. The blade velocity must be limited for mechanical reasons of strength and operating speed. From these considerations, and an inspection of the velocity diagrams, it is evident that the steam leaves the blade wheel with a high velocity. This constitutes a loss in the work available on the wheel, although a moderate velocity must be accepted in order to take the steam to the condenser. This *leaving loss* may amount to 11% of the input energy. The leaving velocity in the velocity diagram of Fig. 11.6(c) is C_ae, and the leaving loss is given by $C_\mathrm{ae}^2/2$.

Methods of improving the efficiency of the simple impulse turbine, known as compounding, will be considered in the following section.

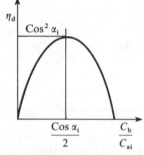

Fig. 11.9 Diagram efficiency against blade speed ratio for a single-stage impulse turbine

11.3 Pressure and velocity compounded impulse steam turbines

Pressure compounding (the Rateau turbine)

The pressure drop available to the turbine is used in a series of small increments, each increment being associated with one stage of the turbine. The physical

Fig. 11.10
Pressure-compounded impulse turbine showing pressure and velocity variations

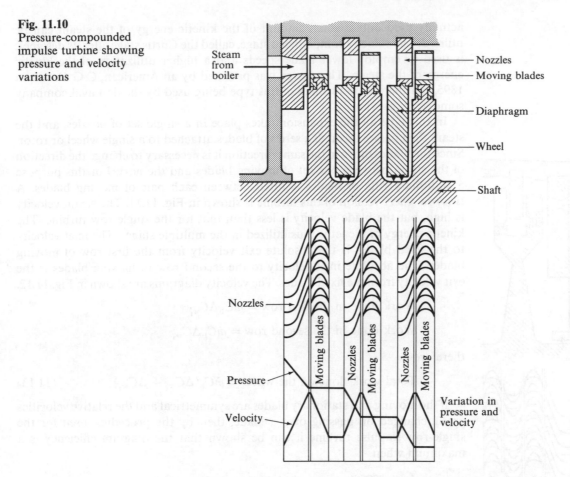

arrangement is shown in Fig. 11.10. The nozzles are carried in diaphragms which separate each stage from the next. The steam pressure in the space between each pair of diaphragms is constant, but there is a pressure drop across each diaphragm as required by the nozzles. Precaution must be taken to prevent leakage of the steam from one section to the next at the shaft and outer casing. The steam speeds, and hence the blade speeds, are low if the number of stages is high. In Fig. 11.10 the variations in pressure and velocity through the turbine are shown, the final pressure being that of the condenser, and the final velocity that required for the steam to leave the turbine. In Fig. 11.10 only one set of wheels is shown, but these may be followed by another set with a larger mean radius. Each of the stages can be analysed by the method used previously for the single stage. A turbine with a series of simple impulse stages is called a Rateau turbine.

Velocity compounding (the Curtis turbine)

From previous considerations it is seen that in the simple impulse stage the optimum condition of blade speed is hardly practicable, and with the speeds actually used only a small amount of the kinetic energy of the steam can be utilized. The velocity compounded stage, called the Curtis stage after its designer, is used to employ lower blade speeds and a higher utilization of the kinetic energy of the steam. This design was patented by an American, C G Curtis, in 1895, but there were turbines of this type being used by the de Laval company some years previously.

In this type all the expansion takes place in a single set of nozzles, and the steam then passes through a series of blades attached to a single wheel or rotor. Since the blades move in the same direction it is necessary to change the direction of the steam between one set of moving blades and the next. For this purpose a stationary ring of blades is fitted between each pair of moving blades. A *two-row wheel* version of this turbine is shown in Fig. 11.11. The steam velocity is high, but the blade velocity is less than that for the single row turbine. The kinetic energy of the jet is thus utilized in the multiple stages. The inlet velocity to the fixed blades is the absolute exit velocity from the first row of moving blades. The absolute inlet velocity to the second row of moving blades is the exit velocity from the fixed blades. The velocity diagrams are shown in Fig. 11.12.

$$\text{Work done in the first row} = \dot{m}C_b\Delta C_{w_1}$$

$$\text{Work done in the second row} = \dot{m}C_b\Delta C_{w_2}$$

therefore

$$\text{Total work done on the wheel} = \dot{m}C_b(\Delta C_{w_1} + \Delta C_{w_2}) \qquad (11.13)$$

If the moving and stationary blades are symmetrical and the relative velocities are unchanged on passing over a blade, then by the procedure used for the single-row impulse turbine it can be shown that the diagram efficiency is a maximum when

$$\frac{C_b}{C_{ai_1}} = \frac{\cos \alpha_{i_1}}{4} \qquad (11.14)$$

and for this condition the absolute velocity at exit is in the axial direction. The maximum diagram efficiency is then

$$\eta_d = \cos^2 \alpha_{i_1} \qquad (11.15)$$

The corresponding rate at which work is done is given by

$$\text{Rate of doing work per unit mass flow rate} = 8C_b^2 \qquad (11.16)$$

Comparing equations (11.16) and (11.12) it can be seen that the enthalpy drop used in the two-row stage is four times that of the single-row stage. The variation of η_d with C_b/C_{ai_1} is shown in Fig. 11.13.

The *stage efficiency* of a turbine is defined in section 11.7 by equation (11.33), and is a measure of the useful enthalpy drop for a given available enthalpy drop. The single-row, two-row, and three-row velocity compounded wheels show maximum stage efficiencies of approximately 0.8, 0.67 and 0.52 respectively, at blade-speed ratios of 0.46, 0.23, and 0.13 respectively. Hence the steam consumption increases with the increase in the number of rows of blades. The three-row wheel is used usually for small turbines employed on auxiliary work.

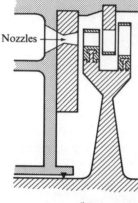

Nozzles —

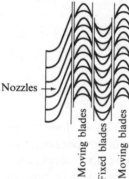

Nozzles →

Moving blades Fixed blades Moving blades

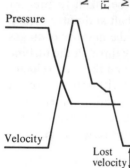

Pressure

Velocity

Lost velocity

Fig. 11.11 Two-row velocity-compounded impulse turbine showing pressure and velocity variations

Fig. 11.12 Velocity diagrams for a two-row velocity-compounded impulse turbine

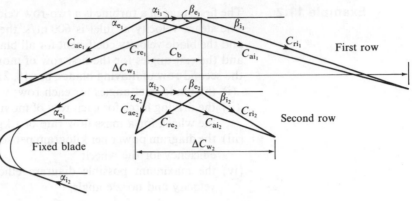

Fig. 11.13 Diagram efficiency against blade speed ratio for a two-row velocity-compounded impulse turbine

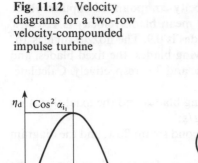

The Curtis stage permits a large expansion of the steam in a machine of compact dimensions.

For maximum economy in a two-row impulse turbine, the blade-speed ratio is about 0.23, which means blade velocities of about 275 m/s. In order to reduce the blade velocity and at the same time utilize a large enthalpy drop a fairly obvious solution is to pressure compound two or more two-row wheels in series, as shown in Fig. 11.14.

Fig. 11.14 Pressure-compounded two-row velocity-compounded impulse turbine showing pressure and velocity variations

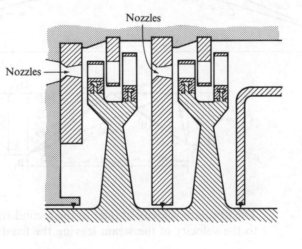

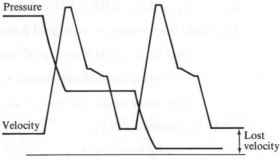

Example 11.2   The first stage of a turbine is a two-row velocity-compounded impulse wheel. The steam velocity at inlet is 600 m/s, the mean blade velocity is 120 m/s, and the blade velocity coefficient for all blades is 0.9. The nozzle angle is 16° and the exit angles for the first row of moving blades, the fixed blades, and the second row of moving blades, are 18, 21, and 35° respectively. Calculate:

(i) the blade inlet angles for each row;

(ii) the driving force for each row of moving blades and the axial thrust on the wheel, for a mass flow rate of 1 kg/s;

(iii) the diagram power per kilogram per second steam flow, and the diagram efficiency for the wheel;

(iv) the maximum possible diagram efficiency for the given steam inlet velocity and nozzle angle.

Solution   (i) The velocity diagrams are drawn to scale, as shown in Fig. 11.15, and the relative velocities, C_{re_1} and C_{re_2} are calculated using equation (11.1),

i.e. $C_{re_1} = k_1 C_{ri_1} = 0.9 \times 486 = 437.4$ m/s

$C_{re_2} = k_2 C_{ri_2} = 0.9 \times 187.5 = 169$ m/s

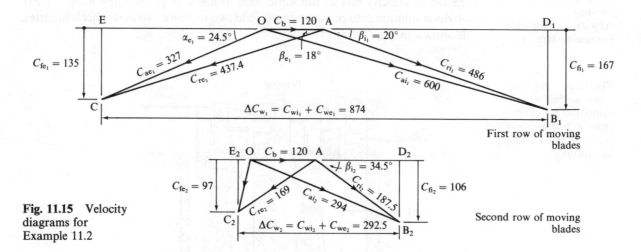

Fig. 11.15 Velocity diagrams for Example 11.2

The absolute velocity at inlet to the second row of moving blades, C_{ai_2}, is equal to the velocity of the steam leaving the fixed row of blades,

i.e. $C_{ai_2} = kC_{ae_1} = 0.9 \times 327 = 294$ m/s

The blade inlet angles are measured from the velocity diagram as

Inlet blade angle, first row of moving blades, $\beta_{i_1} = 20°$

Inlet blade angle, fixed blades, $\alpha_{e_1} = 24.5°$

Inlet blade angle, second row of moving blades, $\beta_{i_2} = 34.5°$

(ii) Using equation (11.2)

Driving force $= \dot{m}\Delta C_w$

Then   First row of moving blades, $\dot{m}\Delta C_{w_1} = 1 \times 874 = 874$ N

Second row of moving blades, $\dot{m}\Delta C_{w_2} = 1 \times 292.5 = 292.5$ N

where ΔC_{w_1} and ΔC_{w_2} are scaled from the velocity diagram.

Using equation (11.4)

Axial thrust $= \dot{m}\Delta C_f = \dot{m}(C_{fi} - C_{fe})$

Then,   First row of moving blades, $\dot{m}\Delta C_{f_1} = 1 \times (167 - 135) = 32$ N

Second row of moving blades, $\dot{m}\Delta C_{f_2} = 1 \times (106 - 97) = 9$ N

i.e.   Total axial thrust $= 32 + 9 = 41$ N per kg/s

(iii) Now

Total driving force $= 874 + 292.5 = 1166.5$ N per kg/s

and   Power output $=$ driving force $\times$ blade velocity

i.e.   Power output $= \dfrac{1166.5 \times 120}{10^3} = 140$ kW per kg/s

Energy supplied to the wheel $= \dfrac{\dot{m}C_{ai_1}^2}{2} = \dfrac{1 \times 600^2}{2 \times 10^3}$ kW

therefore

Diagram efficiency $= \dfrac{140 \times 10^3 \times 2}{600^2} = 0.779$ or 77.9%

(iv) From equation (11.15)

Maximum diagram efficiency $= \cos^2 \alpha_{i_1} = \cos^2 16°$

$= 0.923$ or 92.3%

Turbine blade height

In the impulse turbine the nozzles do not occupy the complete circumference leading into the blade annulus, and this is referred to as *partial admission*.

The total nozzle exit area must be such as to satisfy the conditions of continuous mass flow of steam. Referring to Fig. 11.16, if n is the length of the arc covered by the nozzles and the nozzle height is l, then the nozzle area in the exit plane is nl. If the specific volume of the steam at the nozzle exit condition is v_{i_1}, and the mass flow is $\dot{m}$, then the volume flow rate is $\dot{m}v_{i_1}$. The component of the steam velocity at exit from the nozzles and perpendicular to the area nl is $C_{ai_1} \sin \alpha_i$. Therefore we have

$$\dot{m}v_{i_1} = C_{ai_1} \sin \alpha_i \times nl \tag{11.17}$$

Fig. 11.16 Diagram
showing blade passage
width (a) and length (b)
for impulse blading

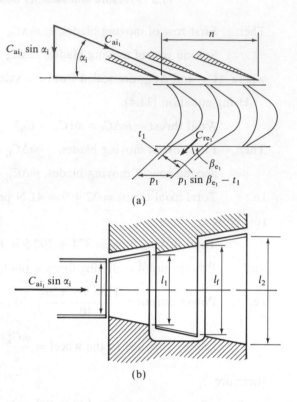

(a)

(b)

The mass flow of steam, $\dot{m}$, passes through the channels of the first moving row and, due to friction, the relative velocity of the steam at exit is C_{re_1}. If the blade pitch at exit is p_1 and the blade thickness is t_1, then each blade channel exit area is $(p_1 \sin \beta_{e_1} - t_1)l_1$, where l_1 is the height of the blades at exit. These quantities are shown in Fig. 11.16(a) and (b) where $p_1 \sin \beta_{e_1}$ is shown to be the effective width of the channel perpendicular to the direction of the relative velocity, and t_1 is the blade thickness measured in the same direction. The arc covered by the nozzles is of length n, therefore the number of blade channels accepting steam is given by n/p_1, and the total blade channel exit area by $(n/p_1)(p_1 \sin \beta_{e_1} - t_1)l_1$. As before, for the condition of continuity of mass flow of steam, we have

$$\dot{m}v_{e_1} = \frac{n}{p_1}(p_1 \sin \beta_{e_1} - t_1)l_1 C_{re_1} \tag{11.18}$$

Figure 11.16(a) shows only the nozzles and the first moving row of blades, and Fig. 11.16(b) shows the nozzles as supplying steam to a two-row velocity compounded wheel. The blade height is increased progressively, and for each row of blades, fixed and moving, an expression similar to that of equation (11.18) can be established.

Due to friction effects as the steam passes over the blades, the enthalpy at exit is higher than it would be if friction were absent. There is an associated increase in the specific volume of the steam compared with the frictionless case, but this is small enough to be negligible compared with the reduction in the relative velocity of the steam.

Example 11.3
For the nozzles and wheel of Example 11.2 the steam flow is 5 kg/s and the nozzle height is 25 mm. The specific volume of the steam leaving the nozzles is 0.375 kg/m^3. Neglecting the wall thickness between the nozzles, and assuming that all blades have a pitch of 25 mm and exit tip thickness of 0.5 mm, calculate:

 (i) the length of the nozzle arc;
 (ii) the blade height at exit from each row.

Solution   (i) Using equation (11.17)

$$\dot{m}v_{i_1} = C_{ai_1} \sin \alpha_i \times nl = 600 \times \sin 16° \times n \times \frac{25}{10^3}$$

i.e. $5 \times 0.375 = 600 \times 0.2756 \times n \times \frac{25}{10^3}$

therefore

$$n = 0.454 \text{ m}$$

i.e.   Length of nozzle arc = 0.454 m

(ii) Using equation (11.18)

$$\dot{m}v_{e_1} = \frac{n}{p}(p \sin \beta_e - t)lC_{re_1}$$

applied to each row of blades, we have, for the first row:

$$5 \times 0.375 = \frac{0.454}{0.025}(0.025 \times \sin 18° - 0.0005)l_1 \times 437.4$$

therefore

$$l_1 = \frac{5 \times 0.375 \times 0.025}{0.454 \times 0.00723 \times 437.4} = 0.0327 \text{ m}$$

i.e.   Blade height at exit = 32.7 mm

For the fixed row:

$$5 \times 0.375 = \frac{0.454}{0.025}(0.025 \times \sin 21° - 0.0005)l_f \times 294$$

therefore

$$l_f = \frac{5 \times 0.375 \times 0.025}{0.454 \times 0.00846 \times 294} = 0.0415 \text{ m}$$

i.e.   Blade height at exit = 41.5 mm

For the second row:

$$5 \times 0.375 = \frac{0.454}{0.025}(0.025 \times \sin 35° - 0.0005)l_2 \times 169$$

therefore

$$l_2 = \frac{5 \times 0.375 \times 0.025}{0.454 \times 0.01384 \times 169} = 0.0442 \text{ m}$$

i.e.   Blade height at exit = 44.2 mm

11.4 Axial-flow reaction turbines

The reaction turbine applies the principle of both the pure impulse and the pure reaction turbine. Each stage of the reaction turbine consists of a fixed row of blades over the whole of the circumferential annulus, and an equal number of blades on a wheel. Admission of fluid in the reaction turbine takes place over the complete annulus, and so there is *full admission*. The fixed blade channels are of nozzle shape and there is a comparatively small drop in pressure accompanied by an increase in velocity. The fluid then passes over the moving blades and, as in the pure impulse turbine, a force is exerted on the blades by the fluid. There is a further drop in pressure as the fluid passes through the moving blades, since the moving blade channels are also of nozzle shape, and therefore there is an increase in the fluid velocity relative to the blades. This is illustrated in the velocity diagram of Fig. 11.17(a). With a simple impulse type the value of C_{re} would be given by AD, but in the reaction turbine this velocity is increased to AC by further expansion of the fluid in the blade channels. The net change in velocity of the fluid is given by BC and the resultant force on the blades by $\dot{m}(CB)$, as shown in Fig. 11.17(b). This force can be resolved into the tangential and axial thrusts, $\dot{m}(CE)$ and $\dot{m}(EB)$ as shown in Fig. 11.17(b).

Fig. 11.17 Blade velocity (a) and thrust diagram (b) for an axial-flow reaction turbine

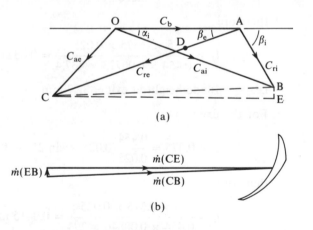

Axial-flow steam turbines

The reaction steam turbine was invented by Sir Charles Parsons who produced a 7.5 kW steam turbine running at 17 000 rev/min in 1884. Parsons was a contemporary of de Laval.

The reaction effect is one of degree since the impulse effect is always there. The degree of reaction, Λ, is defined as

$$\Lambda = \frac{\text{the enthalpy drop in the moving blades}}{\text{the enthalpy drop in the stage}} = \frac{h_1 - h_2}{h_0 - h_2} \qquad (11.19)$$

where h_0, h_1, and h_2 are the enthalpies of the steam at inlet to the fixed blades, at entry to the moving blades, and at exit from the moving blades respectively.

For the simple impulse turbine, $\Lambda = 0$; for the Parsons type of blading, which has the same section for both the fixed and moving blades, $\Lambda = 0.5$. This arrangement has the practical advantage that most of the blades used in a turbine of this type can be extruded from one set of dies. The two set of blades, fixed and moving, are mounted in the relationship shown in Fig. 11.18(a), and the h–s diagram for the expansion is shown in Fig. 11.18(b). The moving blade exit angle β_e is the same as the fixed blade exit angle α_i, and the velocity diagram for this blade arrangement is shown in Fig. 11.19. The steam leaving the moving blade must be travelling in such a direction that it enters the next row of fixed blades smoothly without shock; the fixed blades have an inlet angle equal

Fig. 11.18 Fixed and moving blades of a reaction turbine blade pair showing pressure and velocity variations (a) and the process on an h–s chart (b)

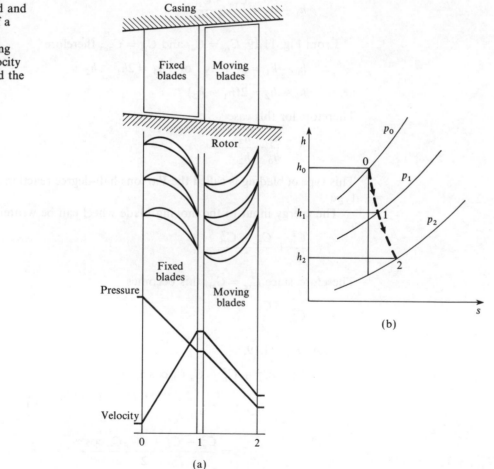

Fig. 11.19 Blade
velocity diagram for a
reaction turbine stage

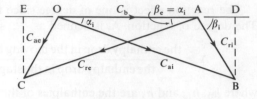

to that of the inlet angle of the moving blades. It follows therefore that when
the fixed and moving blades are geometrically similar, then the velocity diagram
must be symmetrical, as shown in Fig. 11.19. Then, applying the steady-flow
energy equation to the fixed blades

$$h_0 - h_1 = \frac{C_{ai}^2 - C_{ae}^2}{2} \tag{11.20}$$

(This assumes that the velocity of the steam entering the fixed blade is equal
to the absolute velocity of the steam leaving the previous moving row; it therefore
applies to a stage which is not the first.)

Similarly, for the moving blades,

$$h_1 - h_2 = \frac{C_{re}^2 - C_{ri}^2}{2} \tag{11.21}$$

From Fig. 11.19, $C_{re} = C_{ai}$ and $C_{ri} = C_{ae}$, therefore

$$h_0 - h_1 = h_1 - h_2 \quad \text{or} \quad h_0 = 2h_1 - h_2$$

i.e. $\qquad h_0 - h_2 = 2(h_1 - h_2)$

Therefore for this case

$$\Lambda = \frac{h_1 - h_2}{h_0 - h_2} = \frac{1}{2}$$

This type of blading is called the Parsons half-degree reaction or 50% reaction
type.

The energy input to the moving blade wheel can be written as

$$\frac{C_{ai}^2}{2} + \frac{C_{re}^2 - C_{ri}^2}{2}$$

Therefore, since $C_{re} = C_{ai}$, this becomes

$$C_{ai}^2 - \frac{C_{ri}^2}{2}$$

From Fig. 11.19,

$$C_{ri}^2 = C_{ai}^2 + C_b^2 - C_{ai} C_b \cos \alpha_i$$

i.e. $\qquad$ Energy input $= C_{ai}^2 - \left(\dfrac{C_{ai}^2 + C_b^2 - C_{ai} C_b \cos \alpha_i}{2} \right)$

$$= \frac{C_{ai}^2 - C_b^2 + 2C_{ai} C_b \cos \alpha_i}{2}$$

From equation (11.3),

Rate of doing work per unit mass flow rate $= C_b \Delta C_w$

Also $\Delta C_w = \mathrm{ED} = 2C_{ai} \cos \alpha_i - C_b$, therefore

Rate of doing work per unit mass flow rate

$$= C_b(2C_{ai} \cos \alpha_i - C_b) \tag{11.22}$$

Therefore the diagram efficiency of the 50% reaction turbine is given by

$$\eta_d = \frac{\text{rate of doing work}}{\text{energy input}}$$

i.e.
$$\eta_d = \frac{2C_b(2C_{ai} \cos \alpha_i - C_b)}{C_{ai}^2 - C_b^2 + 2C_{ai}C_b \cos \alpha_i}$$

$$= \frac{2(C_b/C_{ai})\{2 \cos \alpha_i - (C_b/C_{ai})\}}{1 - (C_b/C_{ai})^2 + 2(C_b/C_{ai}) \cos \alpha_i} \tag{11.23}$$

where (C_b/C_{ai}) is the blade speed ratio.

Example 11.4   A stage of a steam turbine with Parsons blading delivers dry saturated steam at 2.7 bar from the fixed blades at 90 m/s. The mean blade height is 40 mm, and the moving blade exit angle is 20°. The axial velocity of the steam is three quarters of the blade velocity at the mean radius. Steam is supplied to the stage at the rate of 9000 kg/h. The effect of the blade tip thickness on the annulus area can be neglected. Calculate:
  (i) the rotational speed of the wheel;
 (ii) the diagram power;
(iii) the diagram efficiency;
 (iv) the enthalpy drop of the steam in this stage.

Solution   The velocity diagram is shown in Fig. 11.20(a), and the blade wheel annulus is represented in Fig. 11.20(b).

(i) $\quad C_f = \dfrac{3}{4} C_b = C_{ai} \sin \alpha_i = 90 \times \sin 20° = 30.78$ m/s

Fig. 11.20 Blade velocity diagram (a) and turbine blading annulus (b) for Example 11.4

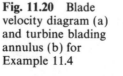
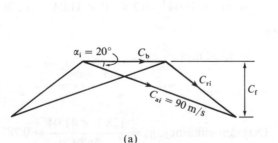
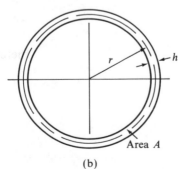

(a)

(b)

therefore

$$C_b = \frac{4 \times 30.78}{3} = 41.04 \text{ m/s}$$

The mass flow of steam is given by

$$\dot{m} = \frac{C_f A}{v}$$

where A is the annulus area, and v is the specific volume of the steam.
In this case, $v = v_g$ at 2.7 bar $= 0.6686 \text{ m}^3/\text{kg}$. Therefore

$$\dot{m} = \frac{9000}{3600} = \frac{30.78A}{0.6686} \quad \text{therefore } A = 0.054 \text{ m}^2$$

Now   Annulus area, $A = 2\pi r h$

where r is the mean radius, and h the mean blade height.

i.e. $0.054 = 2\pi r \times 0.04 \quad \text{therefore } r = 0.215 \text{ m}$

then   Wheel speed, $N = \dfrac{\text{blade speed}}{2\pi r} = \dfrac{41.04}{2\pi \times 0.215} = 30.4 \text{ rev/s}$

(ii) Using equation (11.3)

Diagram power $= \dot{m} C_b \Delta C_w$

Now $\Delta C_w = 2C_{ai} \cos \alpha_i - C_b = (2 \times 90 \times \cos 20°) - 41.04 = 128.1 \text{ m/s}$

therefore

$$\text{Diagram power} = \frac{9000 \times 128.1 \times 41.04}{3600 \times 10^3} = 13.14 \text{ kW}$$

(iii)   Rate of doing work per unit mass flow rate $= C_b \Delta C_w$

$$= 128.1 \times 41.04 \text{ N m/s}$$

Also   Energy input to the moving blades per stage $= C_{ai}^2 - (C_{ri}^2/2)$

Referring to Fig. 11.20(a),

$$C_{ri}^2 = C_{ai}^2 + C_b^2 - C_{ai} C_b \cos \alpha_i$$
$$= 90^2 + 41.04^2 - (2 \times 90 \times 41.04 \times \cos 20°)$$

therefore

$$C_{ri} = 53.32 \text{ m/s}$$

$$\text{Energy input} = 90^2 - \frac{53.32^2}{2} = 6679 \text{ N m per kg/s}$$

$$\text{Diagram efficiency, } \eta_d = \frac{128.1 \times 41.04}{6679} = 0.787 \text{ or } 78.7\%$$

(iv)   Enthalpy drop in the moving blades $= \dfrac{C_{re}^2 - C_{ri}^2}{2} = \dfrac{90^2 - 53.32^2}{2 \times 10^3}$

$$= 2.63 \text{ kJ/kg}$$

therefore

Total enthalpy drop per stage $= 2 \times 2.63 = 5.26 \text{ kJ/kg}$

Optimum operating conditions from the blade velocity diagram

The diagram efficiency for the 50% reaction wheel is given by equation (11.23) as

$$\eta_d = \frac{2(C_b/C_{ai})\{2 \cos \alpha_i - (C_b/C_{ai})\}}{1 - (C_b/C_{ai})^2 + 2(C_b/C_{ai}) \cos \alpha_i}$$

By equating $d\eta_d/d(C_b/C_{ai})$ to zero, the value of blade speed ratio for maximum diagram efficiency can be shown to be given by

$$\frac{C_b}{C_{ai}} = \cos \alpha_i \tag{11.24}$$

From equation (11.22)

Rate of doing work $= C_b(2C_{ai} \cos \alpha_i - C_b)$

therefore for maximum diagram efficiency

$$\text{Rate of doing work} = C_b\left(2C_{ai}\frac{C_b}{C_{ai}} - C_b\right) = C_b^2 \tag{11.25}$$

Also substituting from equation (11.24) in equation (11.23)

$$\text{Maximum diagram efficiency} = \frac{2 \cos^2 \alpha_i}{1 + \cos^2 \alpha_i} \tag{11.26}$$

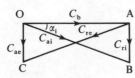

Fig. 11.21 Blade velocity diagram for the optimum blade speed ratio for a reaction turbine stage

For the optimum blade speed ratio a blade velocity diagram as shown in Fig. 11.21 is obtained (i.e. $C_b = C_{ai} \cos \alpha_i$).

The variation of η_d with blade speed ratio for the simple impulse and the reaction stage are shown in Fig. 11.22. It can be seen that for the reaction turbine the curve is reasonably flat in the region of the maximum value of diagram efficiency, so that a variation in $\cos \alpha_i$, and hence C_b/C_{ai}, can be accepted without much variation in the diagram efficiency from the maximum value.

The variation of pressure and velocity through a reaction turbine is shown in Fig. 11.23. The pressure falls continuously as the steam passes over the fixed and moving blades of each stage. The steam velocities are low compared with those of the impulse turbine, and it can be seen from the diagram that the steam velocity is increased in each set of fixed blades. It is no longer convenient to talk of 'nozzles' and 'blades', since in the reaction turbine both fixed and moving blades act as nozzles. It is usual to refer to the two sets of blades as the stator blades and the rotor blades.

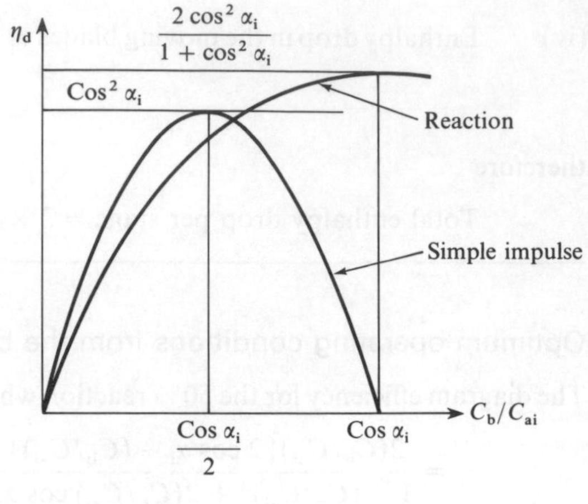

Fig. 11.22 Diagram
efficiency against blade
speed ratio for impulse
and reaction turbines

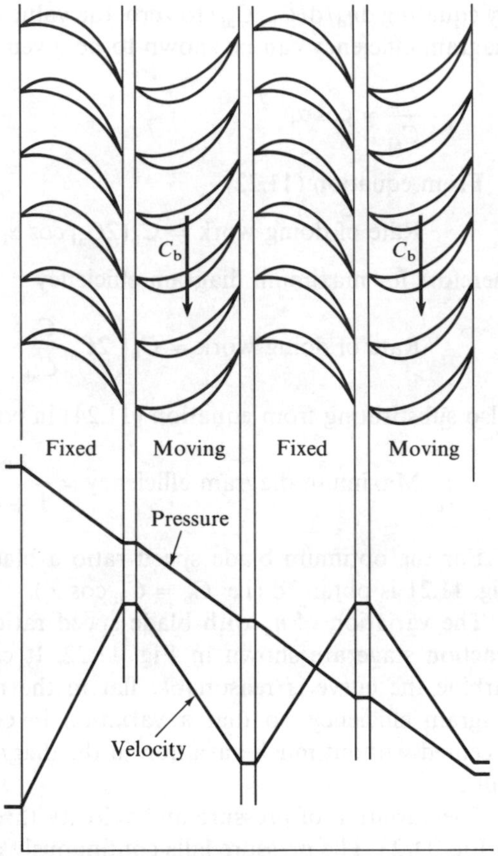

Fig. 11.23 Reaction
turbine showing
pressure and velocity
variations

The pressure drop across the rotor produces an end thrust equal to the
product of the pressure difference and the area of the annulus in contact with
the steam. For the 50% reaction turbine the thrust due to the change in axial
velocity is zero, but the side thrust is nevertheless greater than that of an
equivalent impulse turbine, and larger thrust bearings are fitted. The net end

thrust can be reduced by admitting the steam to the casing at the mid-section and allowing it to expand outwards to each end of the casing, passing over identical sets of blades. This has the additional feature of reducing the blade height at a given wheel for a given total mass flow of steam.

Steam turbine blade design

Reaction turbines are used for the lower pressure stages of most large steam turbines. A common combination is that of a Curtis stage followed by a series of reaction stages.

Since the blade velocity of any blade is greater at the tip than at the root due to the increased radius at the same angular velocity, then a turbine designed for 50% reaction at the mean radius, say, will have a degree of reaction at the root lower than 50% and a degree of reaction at the tip greater than 50%. In steam turbine design practice it has been usual to keep the blade angles the same from root to tip and to accept the consequent slight reduction in efficiency, except when the root to tip radius ratio is very low as is the case for the LP turbine of a large unit. For such units the root to tip radius may be as low as 0.4.

For the steam turbine of Example 11.4 the mean radius is found to be 0.215 m with the given blade height of 40 mm, giving a root to tip radius ratio of $0.195/0.235 = 0.83$. The blade velocity at the mean radius is calculated as 41.04 m/s, hence the blade velocity at the blade tip is given by $(41.04 \times 0.235/0.215) = 44.86$ m/s, and the blade velocity at the blade root is given by $(41.04 \times 0.195/0.215) = 37.22$ m/s. The flow velocity is calculated in the example as 30.78 m/s; assuming this is constant from root to tip, and that the steam enters the moving blade at 90 m/s at 20° at every radius along the moving blade, then the angles at which the steam flows relative to the moving blade are as shown in Fig. 11.24. The blade angle at inlet to the blade can be

Fig. 11.24 Blade velocity diagrams at the blade tip, blade root, and at the blade mean diameter

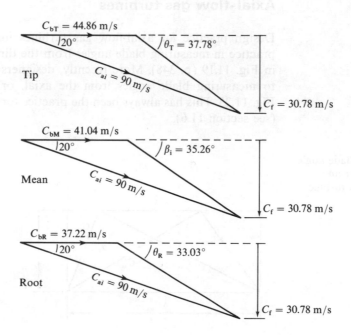

calculated for the mean radius as

$$\beta_i = \tan^{-1}\left\{\frac{30.78}{90\cos 20° - 41.04}\right\} = 35.26°$$

The angle at which the steam flows on to the blade at the tip is then given by

$$\beta_T = \tan^{-1}\left\{\frac{30.78}{90\cos 20° - 44.86}\right\} = 37.78°$$

and at the root the steam flow enters the blade at the angle

$$\beta_R = \tan^{-1}\left\{\frac{30.78}{90\cos 20° - 37.22}\right\} = 33.03°$$

It can be seen that the steam flow enters the moving blade at the wrong angle relative to the blade except at the mean radius, thus causing flow losses. For this order of root to tip radius ratio the losses incurred would be acceptable in steam turbine practice.

A three-dimensional analysis of the flow through the blades enables a twisted blade to be designed in which the steam flow enters the blade at the correct angle at all blade sections. Such a blade is much more expensive to manufacture and therefore the manufacturing costs must be set against the potential savings due to the increased efficiency of the turbine. In steam turbine practice it is found that only for LP turbines with root to tip radius ratios of the order of 0.4 is it considered feasible to design twisted blades. For axial-flow gas turbines and axial-flow air compressors the root to tip radius ratios are usually low and twisted blades are used more frequently, particularly in aircraft practice.

Axial-flow gas turbines

Design practice for axial-flow gas turbines initially followed steam turbine practice in measuring blade angles from the direction of the blade velocity as in Fig. 11.19 (p. 348). More recently, designers of gas turbines have changed to measuring blade angles from the axial, or flow, direction, as shown in Fig. 11.25. This has always been the practice for axial-flow compressor blading (see section 11.6).

Fig. 11.25 Blade angle convention for an axial-flow gas turbine

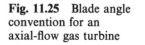

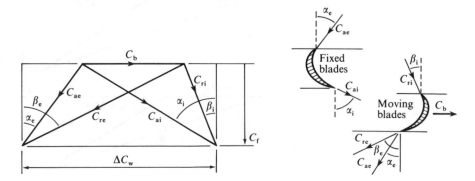

Figure 11.25 is drawn for the general case for a stage where the degree of reaction is not 50%; the diagram is asymmetric unlike the steam turbine stage of Fig. 11.19 where the degree of reaction is 50%. It is usual to assume that for any stage the absolute velocity at inlet to each stage is equal to the absolute velocity at exit from the moving blades, and that the flow velocity, C_f, is constant throughout the turbine.

The degree of reaction, Λ, is defined by equation 11.19 as the ratio of the enthalpy drop in the moving blades to the enthalpy drop in the stage. Referring to Fig. 11.25 we therefore have

$$\Lambda = \frac{(C_{re}^2 - C_{ri}^2)}{2C_b \Delta C_w}$$

$$= \frac{C_f^2(\sec^2 \beta_e - \sec^2 \beta_i)}{2C_b(C_f \tan \beta_e + C_f \tan \beta_i)}$$

Then, since $(\sec^2 = 1 + \tan^2)$, we have

$$\Lambda = \frac{C_f(\tan^2 \beta_e - \tan^2 \beta_i)}{2C_b(\tan \beta_e + \tan \beta_i)}$$

i.e.
$$= \frac{C_f}{2C_b}(\tan \beta_e - \tan \beta_i) \tag{11.27}$$

Note that putting $\Lambda = 0.5$ in equation (11.27) gives

$$C_f(\tan \beta_e - \tan \beta_i) = C_b$$

i.e.
$$C_b + C_f \tan \alpha_e - C_f \tan \beta_i = C_b$$

therefore

$$\alpha_e = \beta_i$$

From the diagram (Fig. 11.25) it follows also that $\alpha_i = \beta_e$. The fixed and moving blades have the same cross-section and the diagram is symmetrical as shown earlier (Fig. 11.19).

Vortex blading

Vortex blading is the name given to the twisted blades designed using three-dimensional flow equations in order to decrease fluid flow losses. A radial equilibrium equation can be derived (see for example reference 11.5), and it can be shown that one set of conditions which satisfies this equation is as follows:

(i) constant axial velocity along the blades, i.e. $C_f = $ constant;
(ii) constant specific work over the annulus, i.e. $C_b \Delta C_w = $ constant;
(iii) free vortex at entry to the moving blades, i.e. $C_{wi}r = $ constant, where r is the blade radius at any point.

Since the specific work output is constant over the annulus it can be calculated at the mean radius, say, and multiplied by the mass flow rate to obtain the power for the stage. On the other hand, the fluid density varies along the blade and hence the continuity equation should be integrated across the annulus to give the mass flow rate; it is usually sufficiently accurate for an initial design to use the density at the mean radius, ρ_M, then $\dot{m} = \rho_M C_f A$, where A is the annulus area.

The free vortex condition is not the only condition that satisfies the radial equilibrium equation. Two other possible methods are the *constant nozzle angle* method and the *constant specific mass flow* method. For further discussion on blade design consult reference 11.5.

Example 11.5
At the mean diameter of a gas turbine stage the blade velocity is 350 m/s. The blade angles at inlet and exit are 20 and 54° respectively, and the blades at this section are designed to have a degree of reaction of 50%. The mean radius of the blades is 0.216 m and the mean blade height is 0.07 m. Assuming the blades are designed according to vortex theory calculate:

   (i) the flow velocity;
 (ii) the angles of blades at the tip and at the root;
(iii) the degree of reaction at the tip and at the root of the blades.

Solution
(i) Figure 11.26 shows the blade velocity diagram at the mean radius. The flow velocity can be calculated from

$$C_f \tan 54° - C_f \tan 20° = 350$$

Fig. 11.26 Blade velocity diagram at the blade mean diameter for Example 11.5

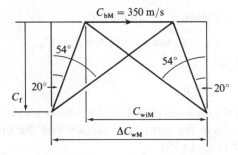

therefore

$$C_f = 345.7 \text{ m/s}$$

(ii) The flow velocity at the tip and root is also 345.7 m/s. Considering the tip of the blades, the blade velocity at the tip is given by $C_{bT} = 350 \times 0.251/0.216 = 406.7$ m/s. The whirl velocity at blade inlet at the mean radius is $C_{wiM} = C_f \tan 54° = 475.8$ m/s. Therefore using the free vortex condition

$$C_{wiT} = 475.8 \times 0.216/0.251 = 409.5 \text{ m/s}$$

Then using the condition for constant specific work:

$$\begin{aligned}
\Delta C_{wT} &= \Delta C_{wM} C_{bM}/C_{bT} \\
&= \{350 + (2 \times 345.7 \tan 20°)\} \times 350/406.7 \\
&= 517.8 \text{ m/s}
\end{aligned}$$

Fig. 11.27 Blade velocity diagram at the blade tip for Example 11.5

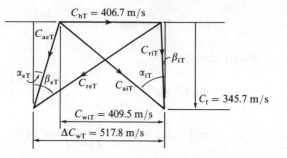

$C_{bT} = 406.7$ m/s

$C_f = 345.7$ m/s

$C_{wiT} = 409.5$ m/s

$\Delta C_{wT} = 517.8$ m/s

The blade velocity diagram for the tip is therefore as shown in Fig. 11.27. For the blade angles we then have the following:

Moving blades:

$$\text{inlet} \quad \beta_{iT} = \tan^{-1}\left\{\frac{409.5 - 406.7}{345.7}\right\} = 0.5°$$

$$\text{exit} \quad \beta_{eT} = \tan^{-1}\left\{\frac{517.8 - 409.5 + 406.7}{345.7}\right\} = 56.1°$$

Fixed blades:

$$\text{inlet} \quad \alpha_{eT} = \tan^{-1}\left\{\frac{517.8 - 409.5}{345.7}\right\} = 17.4°$$

$$\text{exit} \quad \alpha_{iT} = \tan^{-1}\left\{\frac{409.5}{345.7}\right\} = 49.8°$$

Using the same method for the root:

$$C_{bR} = 350 \times 0.181/0.216 = 293.3 \text{ m/s}$$

$$C_{wR} = 475.8 \times 0.216/0.181 = 567.8 \text{ m/s}$$

$$\Delta C_{wR} = 589.4 \times 350/293.3 = 703.3 \text{ m/s}$$

Then, referring to Fig. 11.28:

Moving blades:

$$\text{inlet} \quad \beta_{iR} = \tan^{-1}\left\{\frac{567.8 - 293.3}{345.7}\right\} = 38.5°$$

Fig. 11.28 Blade velocity diagram at the blade root for Example 11.5

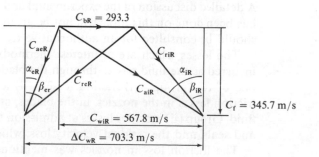

$C_{bR} = 293.3$

$C_f = 345.7$ m/s

$C_{wiR} = 567.8$ m/s

$\Delta C_{wR} = 703.3$ m/s

$$\text{exit} \quad \beta_{eR} = \tan^{-1} 703.3 - \frac{567.8}{345.7} + 293.3 = 51.1°$$

Fixed blades:

$$\text{inlet} \quad \alpha_{eR} = \tan^{-1}\left\{\frac{703.3 - 567.8}{345.7}\right\} = 21.4°$$

$$\text{exit} \quad \alpha_{iR} = \tan^{-1}\left\{\frac{567.8}{345.7}\right\} = 58.7°$$

(iii) The degree of reaction of the blades at the tip and at the root can be found using equation (11.22),

$$\text{i.e.} \quad \Lambda_T = \frac{345.7}{2 \times 406.7}(\tan 56.1° - \tan 0.5°) = 0.63$$

$$\text{and} \quad \Lambda_R = \frac{345.7}{2 \times 293.3}(\tan 51.1° - \tan 38.5°) = 0.26$$

Table 11.1 summarizes the blade angles.

Table 11.1 Summary of blade angles for Example 11.5

	Inlet angles			Exit angles		
	Root	Mean	Tip	Root	Mean	Tip
Fixed blades	21.4°	20.0°	17.4°	58.7°	54.0°	49.8°
Moving blades	38.5°	20.0°	0.5°	51.1°	54.0°	56.1°

11.5 Losses in turbines

In the previous sections it has been shown that the work done per stage at optimum speed conditions for the simple impulse, two-row impulse, and the 50% reaction steam turbines, is given by $2\dot{m}C_b^2$, $8\dot{m}C_b^2$, and $\dot{m}C_b^2$ respectively. The arrangement of a particular turbine to suit a specified purpose requires a consideration of other factors. The most important of these is the losses involved. A detailed discussion of the experimental and associated theoretical work which has been done on this topic will not be given here, and references 11.5 and 11.6 should be consulted for an authoritative treatment.

The losses which are of interest thermodynamically are the internal losses incurred as the fluid passes through the blades. The losses may be classified in one of two groups: (i) friction losses; (ii) leakage losses. Group (i) indicates friction losses in the nozzles, in the blades, and at the discs which rotate in the fluid. Group (ii) includes losses at admission to the stages and leakage at glands and seals, and the residual velocity loss, which has already been mentioned.

The friction loss in nozzles was mentioned in section 10.5, and the effect of friction was allowed for by means of a nozzle efficiency defined by equation (10.11) as

$$\text{Nozzle efficiency} = \frac{h_1 - h_2}{h_1 - h_{2s}}$$

where $(h_1 - h_2)$ is the actual enthalpy drop in the nozzle, and $(h_1 - h_{2s})$ the ideal isentropic enthalpy drop. Since the actual enthalpy drop is less than in the isentropic case, then the actual velocity of the fluid leaving the nozzle is less than that obtained with isentropic expansion. The effect of friction in nozzle and blade passages is to cause losses which increase with the mean relative velocity through them, and with the surface area exposed to the fluid. The losses are influenced by the nature of the flow, whether it is laminar or turbulent. The friction effects occur in the boundary layer on the surface and losses are higher in a turbulent boundary layer than in a laminar one. It is found that with curved blades the boundary layer is usually turbulent at the concave surface, and initially laminar at the convex surface. This laminar condition persists for some distance along the blade and then gives way to turbulence. In reaction stages, where a continuous pressure drop, and hence an acceleration of the fluid exists, the laminar condition persists over a greater length of the passage, so the friction loss is less than in the impulse stage. For a reaction turbine the enthalpy drop per stage is low and a large number of stages is required. This increases the blade surface area required, which increases the friction loss, but the average velocities are low and this helps to reduce friction losses.

In Fig. 11.10 (p. 339), a blade wheel and a diaphragm arrangement for a compounded impulse steam turbine is shown. It is evident that the wheel is rotating in a space full of steam. There is a loss at the surfaces of the wheel due to viscous friction of the steam on the wheel, and there is an admission loss as the steam passes from the nozzles to the wheel. Also, in impulse turbines the nozzles occupy only a part of the area opposite the blade annulus, and the blades pass areas in which they are not being served by nozzles. This tends to create eddies in the blade channels. The effects of this partial admission are referred to as blade windage losses.

The leakage loss in the impulse steam turbine between one stage and the next through the clearance space between the diaphragm and the shaft has been mentioned previously. Another leakage loss occurs at the external glands where the turbine shaft passes through the casing. At one end the tendency is for high pressure steam to escape into the atmosphere, and at the condenser end for air to leak in from the atmosphere. In diaphragm and external glands it is usual to use a form of labyrinth packing, examples of which are shown in Figs 11.29(a) and (b). Figure 11.29(a) shows a possible application to a diaphragm, and Fig. 11.29(b) shows a section through part of an external gland. At the LP gland it is usual to feed a supply of steam at low pressure to the centre of the gland. By this means there is a leakage of steam into the turbine, but the leakage of air into the turbine is prevented. For the diaphragm gland the labyrinth packing fits across the clearance space, but there is a small clearance at the tip through which some steam is throttled. For a more detailed description and discussion of the operation of this type of packing reference 11.6 should be consulted.

Fig. 11.29 Examples of labyrinth packing for a steam turbine

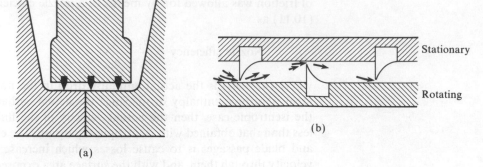

(a)

(b)

Stationary

Rotating

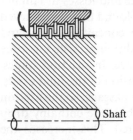

Shaft

Fig. 11.30 Drum construction for a reaction turbine

There is a leakage loss between the blade tips and the casing in all turbines, and this is greater in the reaction turbine than in the impulse type due to the pressure difference across the clearance passage.

In reaction turbines the drum construction as shown in Fig. 11.30 is usually preferred to the diaphragm and wheel construction of Fig. 11.10.

The effect of leakages is more important in the smaller turbines since, although the wheels, etc. can be scaled down, the working clearance cannot be reduced on the same scale. Leakages are highest where the pressure is high and are predominant over the friction losses in that case. It is advantageous to use impulse stages at the HP end of the turbine and reaction stages thereafter. At the lower pressures the friction losses become more important than the leakage losses.

11.6 Axial-flow compressors

In the earliest gas turbine units for aircraft the centrifugal type of compressor was used. For low pressure ratios (no greater than about 4/1) the centrifugal compressor is lighter, and is able to operate effectively over a wider range of mass flows at any one speed than its axial flow counterpart. Using titanium alloys pressure ratios of above 8 have now been achieved.

For larger units with higher pressure ratios the axial-flow compressor is more efficient and is usually preferred. For industrial and large marine gas turbine plants axial compressors are usually used, although some units may employ two or more centrifugal compressors with intercooling between stages. For aircraft the trend has been to higher pressure ratios, and the compressor is usually of the axial-flow type. In aircraft units the advantage of the smaller diameter axial-flow compressor can offset the disadvantage of the increased length and weight compared with an equivalent centrifugal compressor. However, centrifugal compressors are cheaper to produce, more robust, less prone to icing troubles at high altitudes, and have a wider operating range than the axial-flow types.

All design is a compromise, and it may be that under certain circumstances two centrifugal compressors in series may be preferred to an axial compressor. Perhaps the best example of this was in the series of Rolls-Royce Dart turboprop engines which used pressure ratios from 5.4/1 to 6.35/1 with two centrifugal compressor stages.

The design of the blading of a rotary compressor is more the province of the aerodynamicist, particularly in the case of axial-flow compressors for high-speed aircraft. For a more extensive theoretical treatment references 11.2, 11.4, and 11.5 should be consulted.

Axial compressor blading

An axial-flow compression stage is similar to an axial-flow turbine stage; it consists of a row of moving blades arranged round the circumference of a rotor, and a row of fixed blades arranged round the circumference of a stator. The air flows axially through the moving and fixed blades in turn; stationary guide vanes are provided at entry to the first row of moving blades. The work input to the rotor shaft is transferred by the moving blades to the air, thus accelerating it. The blades are arranged so that the spaces between blades form diffuser passages, and hence the velocity of the air relative to the blades is decreased as the air passes through them, and there is a rise in pressure. The air is then further diffused in the stator blades, which are also arranged to form diffuser passages. In the fixed stator blades the air is changed in direction so that it can pass to a second row of moving rotor blades. It is usual to have a relatively large number of stages and to maintain a constant work input per stage (e.g. from 5 to 14 stages have been used).

The necessary reduction in volume may be allowed for by flaring the stator or by flaring the rotor. It is more common to use a flared rotor, and this type is shown diagrammatically in Fig. 11.31. The rotor is built up of discs of steel or light alloy and the blades are fitted into tee-shaped, or dove-tailed, slots in the periphery of the disc. The stator blades are normally spot welded on to a ring at one end of the blade, and loosely fitted to a ring at the other end, to allow for expansion of the blade. This annulus of blades is then fixed to the casing by set screws. There are many possible alternative methods of blade fixing and rotor and stator design; the above brief description has been included to give a general impression only.

Fig. 11.31 Axial compressor with flared rotor

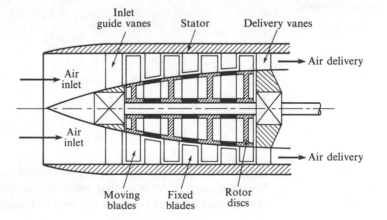

It is usually arranged to have an equal temperature rise in the moving and the fixed blades, and to keep the axial velocity of the air constant throughout the compressor. Thus each stage of the compression is exactly similar with regard to air velocity and blade inlet and outlet angles.

A diffusing flow is less stable than a converging flow, and for this reason the blade shape and profile is more important for a compressor than for a reaction turbine.

Typical blade sections are shown in Fig. 11.32; note that the convention is to measure the blade angles from the axial direction as for axial-flow gas turbines,

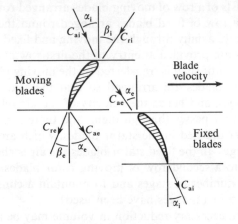

Fig. 11.32 Typical axial-flow compressor blade sections

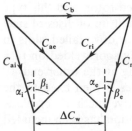

Fig. 11.33 Blade velocity diagram for an axial-flow compressor

and not the tangential direction as in the case of steam turbines. The corresponding blade velocity diagram is shown in Fig. 11.33.

$$\text{Power input} = \dot{m} C_b \Delta C_w$$

From the geometry of the diagram

$$\Delta C_w = C_{ri} \sin \beta_i - C_{re} \sin \beta_e = C_f(\tan \beta_i - \tan \beta_e)$$

As with the axial flow turbine a degree of reaction is defined:

$$\text{Degree of reaction, } \Lambda = \frac{\text{enthalpy rise in rotor}}{\text{enthalpy rise in the stage}}$$

$$= \frac{h_1 - h_0}{h_2 - h_0} = \frac{C_{ri}^2 - C_{re}^2}{2 C_b \Delta C_w}$$

i.e.
$$\Lambda = \frac{C_f^2(\sec^2 \beta_i - \sec^2 \beta_e)}{2 C_b C_f(\tan \beta_i - \tan \beta_e)}$$

Then, since $\sec^2 = 1 + \tan^2$, we have

$$\Lambda = \frac{C_f}{2 C_b}(\tan \beta_e + \tan \beta_i) \tag{11.28}$$

In practice the blades are usually of twisted section designed according to free vortex theory or by a similar method (see ref. 11.5). This not only improves the efficiency but also delays the onset of surging.

Due to the non-uniformity of the velocity profile in the blade passages the work that can be put into a given blade passage is less than that given by the ideal diagram. A work done factor, Y, is introduced, defined by

$$\text{Work done factor, } Y = \frac{\text{actual power input}}{\dot{m}C_b \Delta C_w} \tag{11.29}$$

This is usually about 0.85 for a compressor stage; for a turbine stage the velocity profile is much more uniform and this effect can be neglected.

Example 11.6   In an axial-flow air compressor producing a pressure ratio of 6/1 with air entering at 20 °C the mean velocity of the rotor blades is 200 m/s, and the inlet and exit angles of both the moving and the fixed blades at the mean radius are 45 and 15° respectively. The degree of reaction at the mean radius is 50%, the work done factor is 0.86 throughout, there are 12 stages, and the axial velocity may be taken as constant through the compressor.
Calculate the isentropic efficiency of the compressor.

Solution   By drawing the blade diagram to scale (see Fig. 11.33). the value of ΔC_w may be found,

i.e. $\Delta C_w = 115 \text{ m/s}$

Specific power input per stage $= C_b \Delta C_w \times$ work done factor

$$= 200 \times 115 \times 0.86 = 19\,780 \text{ J}$$

$$= 19.78 \text{ kJ}$$

i.e.   Compressor specific power input $= 12 \times 19.78 = 237.4 \text{ kJ}$

For isentropic compression,

Exit temperature $= (20 + 273) \times 6^{0.4/1.4} = 488.9 \text{ K}$

i.e.   Isentropic specific power input $= 1.005(488.9 - 293) = 196.9 \text{ kJ}$

therefore

$$\text{Compressor isentropic efficiency} = \frac{196.9}{237.4} = 0.83 \text{ or } 83\%$$

11.7 Overall efficiency, stage efficiency, and reheat factor

Overall efficiency

It has been shown that as a fluid expands through a turbine or compressor there are friction effects between the fluid and the enclosing boundary surfaces of the nozzles and blade passages. Further losses are produced by leakage. Both of these are irreversibilities in the expansion or compression process and there is a reduction in the useful enthalpy drop in the case of a turbine and an increase in the enthalpy rise required in the case of a compressor (see Fig. 11.34(a) and (b)).

Fig. 11.34 Actual and
isentropic enthalpy
changes for a
turbine (a) and a
compressor (b)

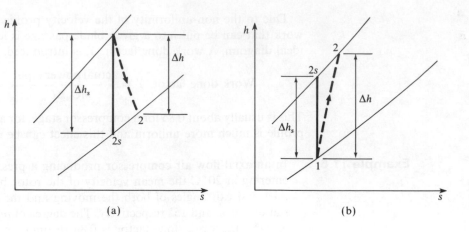

Fig. 11.34 Actual and isentropic enthalpy changes for a turbine (a) and a compressor (b)

The overall isentropic efficiency of a turbine is defined as

$$\text{Overall efficiency, } \eta_0 = \frac{h_1 - h_2}{h_1 - h_{2s}} = \frac{\Delta h}{\Delta h_s} \qquad (11.30)$$

For a gas turbine

$$\eta_0 = \frac{\Delta T}{\Delta T_s} \qquad (11.31)$$

and for an air compressor

$$\eta_0 = \frac{\Delta T_s}{\Delta T} \qquad (11.32)$$

The overall efficiency so defined depends only on the change of properties of the fluid during the expansion or compression.

Stage efficiency and reheat factor

The expansion of the fluid through the successive stages of a reaction turbine can be represented on an h–s chart as shown in Fig. 11.35. The procedure followed above for the whole turbine can be applied to each stage separately, and the dotted line joins the points representing the state of the steam between each stage. The dotted line is called the condition curve, although it does not give a continuous state path since in between the known points the processes are irreversible.

Considering any one stage, the available enthalpy drop of the stage can be represented by Δh_i, where subscript i refers to any stage from 1 to n, and the isentropic enthalpy drop between the same pressures can be represented by Δh_{si}. Then a stage efficiency can be defined as

$$\text{Stage efficiency, } \eta_s = \frac{\Delta h_i}{\Delta h_{si}} \qquad (11.33)$$

Fig. 11.35 Multistage expansion for a reaction turbine on an h–s chart

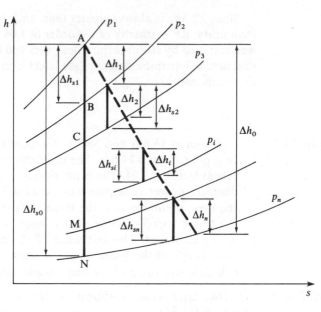

From an inspection of Fig. 11.35 it is seen that BC $< \Delta h_{s2}$, etc. since the lines of constant pressure diverge from left to right on the diagram,

$$\sum_1^n \Delta h_{si} > AB + BC + \ldots + MN$$

i.e. $$\sum_1^n \Delta h_{si} > \Delta h_{s0}$$

From equation (11.30), $\Delta h_i = \eta_s \Delta h_{si}$, and if it can be assumed that the stage efficiency is the same for each stage, then

$$\sum_1^n \Delta h_i = \eta_s \sum_1^n \Delta h_{si}$$

therefore

$$\Delta h_0 = \eta_s \sum_1^n \Delta h_{si}$$

Dividing by Δh_{s0} we have

$$\frac{\Delta h_0}{\Delta h_{s0}} = \eta_s \frac{\sum_1^n \Delta h_{si}}{\Delta h_{s0}}$$

or $$\eta_0 = \eta_s \times (RF)$$

where RF is known as the *reheat factor*

i.e. $$RF = \frac{\sum_1^n \Delta h_{si}}{\Delta h_{s0}} = \frac{\eta_0}{\eta_s} \tag{11.34}$$

365

Since $\sum_{1}^{n} \Delta h_{si}$ is always greater than Δh_{s0}, it follows that RF is always greater than unity; RF is usually of the order of 1.04 for a steam turbine. Reheat factor was first used by steam turbine designers and is not normally used in gas turbine practice; polytropic efficiency (see later) is used for axial-flow gas turbines and air compressors.

Example 11.7
(i) Steam at 15 bar and 350 °C is expanded through a 50% reaction turbine to a pressure of 0.14 bar. The stage efficiency is 75% for each stage, and the reheat factor is 1.04. The expansion is to be carried out in 20 stages and the diagram power is required to be 12 000 kW. Calculate the flow of steam required, assuming that the stages all develop equal work.

(ii) In the turbine above at one stage the pressure is 1 bar and the steam is dry saturated. The exit angle of the blades is 20°, and the blade speed ratio is 0.7. If the blade height is one-twelfth of the blade mean diameter, calculate the value of the mean blade diameter and the rotor speed.

Solution
(i) The expansion is shown on an h–s chart in Fig. 11.36(a). Using equation (11.34)

$$\eta_0 = \eta_s \times \text{RF} = 0.75 \times 1.04 = 0.78$$

Fig. 11.36
h–s chart (a) and blade velocity diagram (b) for Example 11.5

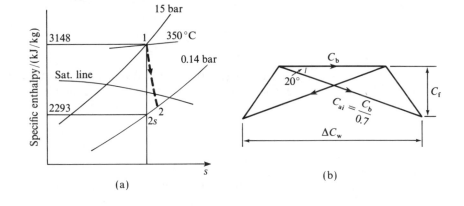

From the chart, $h_1 = 3148$ kJ/kg, and state 1 is fixed.

Now $s_1 = s_2$, therefore state 2 is fixed, and $h_{2s} = 2293$ kJ/kg,

i.e.   Overall isentropic enthalpy drop $= 3148 - 2293 = 855$ kJ/kg

From equation (11.30)

$$\Delta h_0 = \eta_0 \times \Delta h_{s0} = 0.78 \times 855 = 667 \text{ kJ/kg}$$

i.e.   Enthalpy drop per stage $= \dfrac{667}{20} = 33.35$ kJ/kg

Also   Total diagram power $= \dot{m} \Delta h_0$

where $\dot{m}$ is the mass flow in kg/s, therefore

$$\dot{m} = \frac{12\,000}{667} = 17.99 \text{ kg/s}$$

i.e.   Steam mass flow $= 17.99 \times 3600 = 64\,770 \text{ kg/h}$

(ii) The blade velocity diagram for any one stage is shown in Fig. 11.36(b). From the diagram we have

$$\text{Work done per unit mass flow rate of steam} = C_b \Delta C_w$$

$$= C_b(2C_{ai}\cos 20° - C_b)$$

In this case $C_{ai} = C_b/0.7$, therefore

$$\text{Work done per unit mass flow} = \frac{C_b}{10^3}\left\{\frac{(2C_b\cos 20°)}{0.7} - C_b\right\}$$

$$= 1.685 C_b^2$$

Now   Work done per unit mass flow per stage = enthalpy drop per stage

$$= 33.35 \text{ kJ/kg}$$

therefore

$$1.685 C_b^2 = 33.35 \times 10^3 \quad \text{i.e. } C_b = 140.7 \text{ m/s}$$

Also, from the diagram,

$$C_f = C_{ai}\sin 20° = C_b \sin 20°/0.7 = 140.7 \sin 20°/0.7$$

i.e. $C_f = 68.8 \text{ m/s}$

therefore

$$\text{Volume flow per second at 1 bar} = C_f \times \pi D \times h$$

where D is the blade mean diameter and h the blade height at this stage,

i.e. $$\text{Volume flow} = 68.8 \times \pi D \times \frac{D}{12} = 18 D^2 \text{ m}^3/\text{s}$$

At 1 bar, $v_g = 1.694 \text{ m}^3/\text{kg}$, therefore

$$\text{Mass flow} = \frac{18 D^2}{1.694} = 17.99 \text{ kg/s}$$

i.e. $$D^2 = \frac{17.99 \times 1.694}{18} \quad \text{therefore } D = 1.3 \text{ m}$$

i.e.   Blade mean diameter $= 1.3 \text{ m}$

Then   Blade speed, $C_b = \dfrac{\pi D N}{60}$

where N is the rotor speed in revolutions per minute.

Therefore

$$140.7 = \frac{\pi \times 1.3 \times N}{60}$$

i.e.   Rotor speed = 2067 rev/min

11.8 Polytropic efficiency

The polytropic efficiency of an expansion or compression process is the isentropic efficiency of an infinitely small stage. It is a useful concept since, unlike isentropic efficiency, it is independent of the pressure ratio. For an expansion polytropic efficiency is defined as

$$\eta_{\infty e} = \frac{dh}{dh_s} \tag{11.35}$$

For a compression:

$$\eta_{\infty c} = \frac{dh_s}{dh} \tag{11.36}$$

Taking an expansion process 1–2 as shown in Fig. 11.37, and substituting for $dh_s = v\,dp$, see p. 60, we have:

$$\eta_{\infty e} = \frac{dh}{dh_s} = \frac{dh}{v\,dp}$$

Fig. 11.37 Expansion process on a T–s diagram

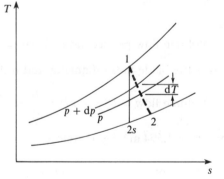

Also, for a perfect gas,

$$dh = c_p\,dT \quad \text{and} \quad v = \frac{RT}{p}$$

therefore

$$\eta_{\infty e} = \frac{c_p\,dT\,p}{RT\,dp}$$

i.e.

$$\eta_{\infty e} \int_1^2 \frac{dp}{p} = \frac{c_p}{R} \int_1^2 \frac{dT}{T}$$

$$\eta_{\infty e} \ln\left(\frac{p_1}{p_2}\right) = \frac{\gamma}{\gamma - 1} \ln\left(\frac{T_1}{T_2}\right)$$

or

$$\frac{T_1}{T_2} = \left(\frac{p_1}{p_2}\right)^{(\gamma - 1)\eta_{\infty e}/\gamma} \tag{11.37}$$

For a compression process it can be shown in a similar way that

$$\frac{T_2}{T_1} = \left(\frac{p_2}{p_1}\right)^{(\gamma - 1)/\gamma \eta_{\infty c}} \tag{11.38}$$

For a gas turbine with pressure ratio, r, and polytropic efficiency, $\eta_{\infty e}$, we have

$$\text{Isentropic efficiency, } \eta_T = \frac{T_1 - T_2}{T_1 - T_{2s}} = \frac{1 - \dfrac{1}{r^{(\gamma - 1)\eta_{\infty e}/\gamma}}}{1 - \dfrac{1}{r^{(\gamma - 1)/\gamma}}} \tag{11.39}$$

Similarly for a compressor

$$\text{Isentropic efficiency, } \eta_c = \frac{r^{\gamma - 1/\gamma} - 1}{r^{\gamma - 1/\gamma \eta_{\infty c}} - 1} \tag{11.40}$$

Example 11.8

A gas turbine unit operates in ambient conditions of 1.012 bar, 17 °C, and the maximum cycle temperature is limited to 1000 K. The compressor, which has a polytropic efficiency of 88%, is driven by the HP turbine, and a separate LP turbine is geared to the power output on a separate shaft; both turbines have polytropic efficiencies of 90%. There is a pressure loss of 0.2 bar between the compressor and the HP turbine inlet. Neglecting all other losses, and assuming negligible kinetic energy changes, calculate:

(i) the compressor pressure ratio which will give maximum specific power output;
(ii) the isentropic efficiency of the power turbine.

For the gases in both turbines, take $c_p = 1.15$ kJ/kg K and $\gamma = 4/3$.
For air take $c_p = 1.005$ kJ/kg K and $\gamma = 1.4$

Solution

(i) The cycle is shown on a T–s diagram in Fig. 11.38. Let $p_2/p_1 = r$. From equation (11.38),

$$T_2 = T_1 r^{\gamma - 1/\gamma \eta_{\infty c}} = (17 + 273) \times r^{0.4/1.4 \times 0.88} = 290 r^{0.325}$$

Now

$$p_3 = p_2 - 0.2 = (p_1 \times r) - 0.2 = 1.012r - 0.2$$

$$p_5 = p_1 = 1.012 \text{ bar}$$

i.e.

$$\frac{p_3}{p_5} = \frac{1.012r - 0.2}{1.012} = r - 0.198$$

369

Fig. 11.38 Gas turbine cycle on a $T–s$ diagram for Example 11.8

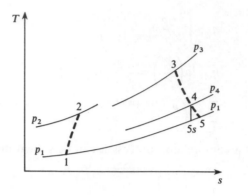

Since the polytropic efficiency of both turbines is the same then, using equation (11.37),

$$\frac{T_3}{T_5} = \frac{(p_3)^{(\gamma-1)\eta\infty e/\gamma}}{(p_1)} = (r - 0.198)^{0.9/4}$$

i.e.
$$T_5 = \frac{1000}{(r - 0.198)^{0.225}}$$

Turbine specific power output $= c_p(T_3 - T_5)$

$$= 1.15\left(1000 - \frac{1000}{(r - 0.198)^{0.225}}\right)$$

$$= 1150[1 - (r - 0.198)^{-0.225}]$$

Compressor specific power output $= c_p(T_2 - T_1)$

$$= 1.005(290r^{0.325} - 290)$$

$$= 291.5(r^{0.325} - 1)$$

Net specific power output,

$$\dot{W} = 1150[1 - (r - 0.198)^{-0.225}] - 291.5(r^{0.325} - 1)$$

This is a maximum when $d\dot{W}/dr = 0$, i.e. when

$$0.225 \times 1150 \times (r - 0.198)^{-1.225} = 0.325 \times 291.5 \times r^{-0.675}$$

Trial and error, or graphical, solution gives, $r = 6.65$.

i.e.   Compressor pressure ratio for maximum specific power output $= 6.65$

(ii) $T_2 = 290r^{0.325} = 290(6.65)^{0.325} = 536.8$ K

Now

HP turbine power output $=$ compressor power input

therefore

$$1.15(1000 - T_4) = 1.005(536.8 - 290)$$

i.e. $T_4 = 784.3$ K

Then

$$\frac{p_3}{p_4} = \frac{(T_3)^{\gamma/(\gamma-1)\eta\infty e}}{(T_4)} = \frac{(1000)^{4/0.9}}{(784.3)} = 2.944$$

Also

$$p_3 = 6.65p_1 - 0.2 = (6.65 \times 1.012) - 0.2 = 4.73 \text{ bar}$$

therefore

$$p_4 = \frac{4.73}{2.944} = 1.607 \text{ bar}$$

Then

$$\frac{T_4}{T_5} = \frac{(p_4)^{(\gamma-1)/\eta\infty e/\gamma}}{(p_5)} = \frac{(1.607)^{0.9/4}}{(1.012)} = 1.110$$

$$\frac{T_4}{T_{5s}} = \frac{(p_4)^{\gamma-1/\gamma}}{(p_5)} = \frac{(1.607)^{0.25}}{(1.012)} = 1.123$$

Using equation (9.2)

$$\text{Turbine isentropic efficiency, } \eta_T = \frac{T_4 - T_5}{T_4 - T_{5s}} = \frac{1 - (T_5/T_4)}{1 - (T_{5s}/T_4)}$$

i.e.

$$\eta_T = \frac{1 - (1/1.11)}{1 - (1/1.123)} = 0.905 \text{ or } 90.5\%$$

Example 11.9   Derive an expression for the pressure ratio for the yth stage of an axial-flow air compressor in terms of the stage efficiency, the blade velocity, the change of the velocity of whirl across the stage, the inlet temperature to the compressor, the work done factor, and the specific heat at constant pressure of air. Explain why the pressure ratio for any stage of an axial-flow air compressor is not equal to the overall pressure ratio to the power of $1/n$, where n is the number of stages.

Assume (a) the compressor is designed such that each stage requires the same work input; (b) the stage efficiency of each stage is the same.

Solution   Each stage has the same work input and hence the same temperature rise. Then, using equation (11.29) for unit mass flow rate,

$$\text{Actual work done per stage} = YC_b\Delta C_w$$

$$\text{Actual temperature rise} = \frac{YC_b\Delta C_w}{c_p}$$

i.e.

$$\text{Isentropic temperature rise for any one stage} = \frac{\eta_s YC_b\Delta C_w}{c_p}$$

where η_s is the stage efficiency and Y is the work done factor.

371

Let the temperature at inlet to the compressor be T_1. Then at the yth stage of the compressor

$$\text{Actual temperature at exit} = T_1 + \frac{y Y C_b \Delta C_w}{c_p}$$

$$\text{Actual temperature at inlet} = T_1 + \frac{(y-1) Y C_b \Delta C_w}{c_p}$$

$$\frac{\text{Isentropic temperature rise}}{\text{Temperature at inlet}} = \frac{\eta_s C_b \Delta C_w Y / c_p}{T_1 + (y-1)(C_b \Delta C_w / c_p) Y}$$

$$\frac{\text{Isentropic temperature at exit}}{\text{Temperature at inlet}} = \frac{\eta_s C_b \Delta C_w Y}{c_p T_1 + (y-1) C_b \Delta C_w Y} + 1$$

$$\text{Pressure ratio for the } y\text{th stage} = \left\{ \frac{\eta_s C_b \Delta C_w Y}{c_p T_1 + (y-1) C_b \Delta C_w Y} + 1 \right\}^{\gamma/\gamma - 1}$$

Since η_s, C_b, Y and ΔC_w are constant throughout the compressor, and T_1 is fixed, it follows that the pressure ratio for any stage varies according to the number of the stage, y. As y increases the pressure ratio decreases. The overall pressure ratio for the compressor must be equal to the product of the stage pressure ratios, but since these are not equal then the overall pressure ratio is not equal to the pressure ratio for a stage raised to the power of n, where n is the number of stages.

11.9 Centrifugal compressors

A centrifugal compressor consists of an impeller with a series of curved radial vanes as shown in Fig. 11.39. Air is drawn in near the hub, called the impeller eye, and is whirled round at high speed by the vanes on the impeller as the impeller rotates at high rotational speed. The static pressure of the air increases from the eye to the tip of the impeller in order to provide the centripetal force on the air. As the air leaves the impeller tip it is passed through diffuser passages which convert most of the kinetic energy of the air into an increase in enthalpy,

Fig. 11.39 Centrifugal compressor impeller

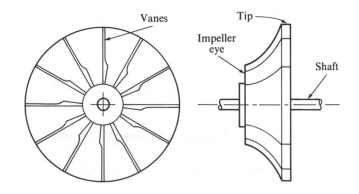

and hence the pressure of the air is further increased. The complete compressor is shown diagrammatically in Fig. 11.40. In a gas turbine plant the air from the discharge scroll passes to the combustion chamber.

Fig. 11.40 Centrifugal compressor showing discharge scroll

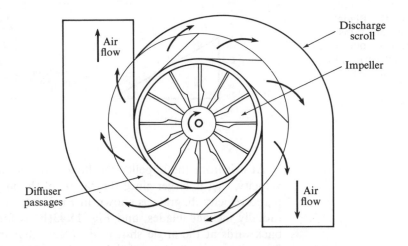

The impeller may be double-sided, having an eye on either side of the compressor, so that air is drawn in on both sides, as shown diagrammatically in Fig. 11.41. The advantage of this type is that the impeller is subjected to approximately equal forces in an axial direction.

Fig. 11.41 Double-sided impeller for a centrifugal compressor

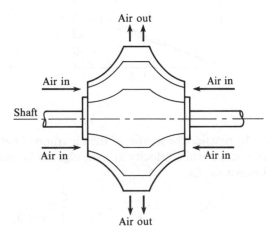

In practice about half the pressure rise occurs in the impeller vanes and half in the diffuser passages.

Centrifugal compressors or blowers are used for a wide variety of purposes in engineering in addition to their use in gas turbine units (e.g. blowers and superchargers for IC engines), and there is no basic difference in the design for any of the different applications.

Referring to Fig. 11.39, if the airflow into the impeller eye is in the axial direction the blade velocity diagram at inlet is as shown in Fig. 11.42(a). (Note that in this figure the plane of the figure is a horizontal plane through the axis of the compressor.) In Fig. 11.42(b) the inlet velocity to the impeller eye is inclined at an angle by using fixed guide vanes; this is known as pre-whirl.

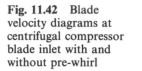

Fig. 11.42 Blade velocity diagrams at centrifugal compressor blade inlet with and without pre-whirl

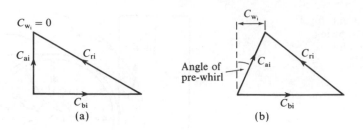

At exit from the impeller the flow is in the radial direction and the blade velocity, C_{be}, is larger since the radius of the impeller is larger at outlet. The blade velocity diagram is shown in Fig. 11.43; Fig. 11.43(a) is for the case of radially inclined blades, and Fig. 11.43(b) is for the case of blades inclined backwards at the angle shown, β_e. (Note that in this figure the plane of the figure is a vertical plane at right angles to the axis of the compressor.)

Fig. 11.43 Blade velocity diagrams at centrifugal compressor blade outlet for radially inclined (a) and backward inclined (b) blading

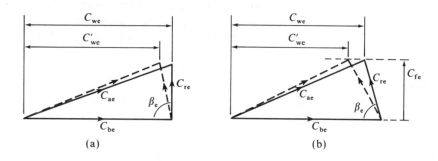

In practice the inertia of the air trapped between the impeller blades causes the actual whirl velocity at exit, C'_{we}, to be less than C_{we}; this phenomenon is known as *slip*.

$$\text{Slip factor} = \frac{C'_{we}}{C_{we}} = \frac{C'_{we}}{C_{be} - C_{fe} \cot \beta_e}$$

Also, using Newton's second law,

$$\text{Power input} = \dot{m}(C_{be}C'_{we} - C_{bi}C_{wi})$$

Example 11.10 A centrifugal compressor has a pressure ratio of 4/1 with an isentropic efficiency of 80% when running at 15 000 rev/min and inducing air at 20 °C. Guide vanes at inlet give the air a pre-whirl of 25° to the axial direction at all radii and the mean diameter of the eye is 250 mm; the absolute air velocity at inlet is 150 m/s. At exit the blades are radially inclined and the impeller tip diameter is 590 mm. Calculate the slip factor of the compressor.

Solution   Temperature after isentropic compression

$$= (20 + 273) \times 4^{0.4/1.4} = 435.4 \text{ K}$$

i.e.   Isentropic temperature rise $= 435.4 - 293 = 142.4 \text{ K}$

therefore

$$\text{Actual temperature rise} = \frac{142.4}{0.8} = 178 \text{ K}$$

i.e.   Power input per unit mass flow rate

$$= c_p \times \text{actual temperature rise}$$

$$= 1.005 \times 178 = 178.9 \text{ kJ}$$

Referring to Fig. 11.42(b) $C_{ai} = 150 \text{ m/s}$ (given), and the angle of pre-whirl is given as $25°$:

$$C_{bi} = \frac{15\,000 \times \pi \times 250}{60 \times 10^3} = 196.4 \text{ m/s}$$

and $C_{wi} = C_{ai} \sin 25 = 150 \times \sin 25 = 63.4 \text{ m/s}$

At exit, referring to Fig. 11.43(a):

$$C_{be} = \frac{15\,000 \times \pi \times 590}{60 \times 10^3} = 463.4 \text{ m/s}$$

i.e. $C_{we} = 463.4 \text{ m/s}$, since the blades are radial.

Also,

$$\text{Power input per unit mass flow rate} = C_{be} C'_{we} - C_{bi} C_{wi}$$

$$= 178.9 \text{ kJ (see above)}$$

i.e. $178.9 \times 10^3 = 463.4 C'_{we} - (196.4 \times 63.4)$

$C'_{we} = 412.9 \text{ m/s}$

Hence

$$\text{Slip factor} = \frac{C'_{we}}{C_{we}} = \frac{412.9}{463.4} = 0.89$$

11.10 Radial-flow turbines

The turbines considered in the previous sections were of the axial-flow type. In the radial flow turbine the fluid is supplied near the axis and expands in concentric rings. This is called an outward-flow radial turbine, but inward-flow turbines are also used. The concentric rings of blades may be alternatively fixed and moving or all moving with alternate rings moving in opposite directions. The blade velocities are effectively doubled by this relative velocity in the latter, and higher fluid speeds are used. One turbine using this arrangement is the Ljungström steam turbine which has impulse-reaction blading. The turbine is shown diagrammatically in Fig. 11.44. The blade wheels A and B rotate in opposite directions and their output shafts may be connected to different loads which operate at the same speed. The steam is supplied to the centre of the turbine and then expands radially outwards through the blades of wheels A and B. The blades are of the 50% reaction type. For a more detailed description and a theoretical analysis of this type of turbine, reference 11.6 should be consulted.

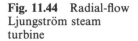

Fig. 11.44 Radial-flow Ljungström steam turbine

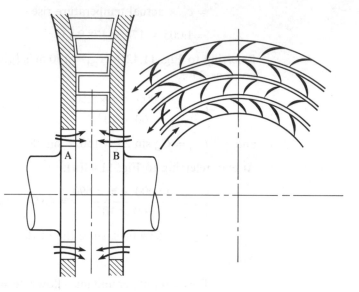

Problems

11.1 The velocity of steam at inlet to a simple impulse turbine is 1000 m/s, and the nozzle angle is 20°. The blade speed is 400 m/s and the blades are symmetrical. Determine the blade angles if the steam is to enter the blades without shock. If the friction effects on the blade are negligible, calculate the tangential force on the blades and the diagram power for a mass flow of 0.75 kg/s. What is the axial thrust and the diagram efficiency?

If the relative velocity at exit is reduced by friction to 80% of that at inlet, what is then the diagram power and the axial thrust? Calculate also the diagram efficiency in this case.

(32.36°; 810 N; 324 kW; 0; 86.4%; 291.5 kW; 51.3 N; 77.7%)

11.2 The steam from the nozzles of a single-wheel impulse turbine discharges with a velocity of 600 m/s and at 20° to the plane of the wheel. The blade wheel rotates at 3000 rev/min and the mean blade radius is 590 mm. The axial velocity of the steam at exit from the blades is 164 m/s and the blades are symmetrical. Calculate:

(i) the blade angles;
(ii) the diagram work per unit mass flow rate of steam;
(iii) the diagram efficiency;
(iv) the blade velocity coefficient.

(28° 47′; 126.2 kJ per kg/s; 70%; 0.799)

11.3 The nozzles of the impulse stage of a turbine receive steam at 15 bar and 300 °C and discharge it at 10 bar. The nozzle efficiency is 95% and the nozzle angle is 20°. The blade speed is that required for maximum work, and the inlet angle of the blades is that required for entry of the steam without shock. The blade exit angle is 5° less than the inlet angle. The blade velocity coefficient is 0.9. Calculate for a steam flow of 1350 kg/h:
   (i) the diagram power;
   (ii) the diagram efficiency.

(30.3 kW; 86.3%)

11.4 The following particulars apply to a two-row velocity compounded impulse stage of a turbine: nozzle angle 17°; blade speed 125 m/s; exit angles of the first row moving blades, the fixed blades, and the second row moving blades, 22, 26, and 30° respectively. Take the blade velocity coefficient for each row of blades as 0.9, and assume that the absolute velocity of the steam leaving the stage is in the axial direction. Draw the velocity diagram for the stage and obtain:
   (i) the absolute velocity of the steam leaving the stage;
   (ii) the diagram efficiency.

(72.2 m/s; 80%)

11.5 The first stage of a turbine is a two-row velocity compounded wheel. Steam at 40 bar and 400 °C is expanded in the nozzles to 15 bar, and has a velocity at discharge of 700 m/s. The inlet velocity to the stage is negligible. The relevant exit angles are: nozzle 18°; first row blades 21°; fixed blades 26.5°; second row blades 35°. Take the blade velocity coefficient for all blades as 0.9. The mean diameter of the blading is 750 mm and the turbine shaft speed is 3000 rev/min. Draw the velocity diagram for this wheel and calculate:
   (i) the diagram efficiency;
   (ii) the stage efficiency.

(70.8%; 67.4%)

11.6 For the turbine of Problem 11.5 the mass flow of steam for each set of nozzles is 4.5 kg/s. Calculate the length of arc occupied by the nozzles if the nozzle height is 25 mm and the wall thickness between them is negligible. If the blades of the wheel have a pitch of 25 mm and the blade tip thickness at exit is 0.5 mm, calculate the blade exit height for each row.

(132.3 mm; 30.2 mm; 33.4 mm; 38.9 mm)

11.7 In a reaction stage of a steam turbine the nozzle angle is 20° and the absolute velocity of the steam at inlet to the moving blades is 240 m/s. The blade velocity is 210 m/s. If the blading is designed for 50% reaction, determine:
   (i) the blade angle at inlet and exit;
   (ii) the enthalpy drop per unit mass of steam in the moving blades and in the complete stage;
   (iii) the diagram power for a steam flow of 1 kg/s;
   (iv) the diagram efficiency.

(79.3°, 20°; 25.3 kJ/kg; 50.6 kJ/kg; 50.6 kW; 93.5%)

11.8 The speed of rotation of a blade group of a reaction turbine is 3000 rev/min. The mean blade velocity is 100 m/s. The blade speed ratio is 0.56 and the exit angle of the blades is 20°. If the mean specific volume of the steam is 0.65 m³/kg, and the mean height of the blades is 25 mm, calculate the mass flow of steam through the turbine. Neglect the effect of blade thickness on the annulus area, and assume 50% reaction blading.

(16 900 kg/h)

377

11.9 In the blade group of Problem 11.8 there are five pairs of blades. Calculate the useful enthalpy drop required for the group and the diagram power.

(117.8 kJ/kg; 553.7 kW)

11.10 Calculate the optimum diagram efficiency for the reaction stage of Problems 11.8 and 11.9, assuming the nozzle angle and the axial velocity remain unchanged, the blade speed being adjusted. Calculate also the blade speed at optimum efficiency.

(93.8%; 167.8 m/s)

11.11 Ten stages of an ideal reaction turbine develop 3000 kW when the mass flow of steam is 18 000 kg/h. The mean value of the blade velocity is 0.8 of the steam velocity from the fixed blades. The exit angle of each blade is constant at 20° and the axial velocity is constant throughout the turbine. Calculate the inlet angle of the blades and the enthalpy drop in each moving row.

(67.8°; 30 kJ/kg)

11.12 At a particular stage of an axial-flow gas turbine the radius at the blade root is 0.225 m and the blade tip radius is 0.375 m. The blades are to be designed according to vortex theory with 50% reaction at the mean radius; the angle of the blade inlet at the tip is fixed at 0°. (Blade angles are measured from the axial direction.)

The turbine rotates at 6000 rev/min, the flow velocity is to be 0.8 of the blade velocity at the mean radius, and the mean density in the blade annulus may be taken as 0.7 kg/m^3. Assuming the conditions required for radial equilibrium using free vortex design, calculate:
  (i) the inlet and exit angles for the moving and fixed blades at the mean radius;
 (ii) the inlet and exit angles for the moving and fixed blades at the tip;
(iii) the inlet and exit angles for the moving and fixed blades at the root;
 (iv) the degree of reaction at the tip and at the root;
  (v) the power output of the stage.

(35.1°, 62.9°; 0°, 64.8°, 29.4°, 57.4°; 59.0°, 61.9°, 43.2°, 69.0°; 0.68, 0.11; 2253.5 kW)

11.13 Show that for an axial-flow turbine stage designed with free vortex flow twisted blading with 50% reaction at the mean radius, r_m, that the degree of reaction, Λ, at any other radius, r, is given by

$$1 - \frac{1}{2}\left\{\frac{r_m}{r}\right\}^2$$

Hence show that for zero reaction at the root, the ratio of the mean radius to the radius at the root is $\sqrt{2}$, and that the degree of reaction at the blade tip is 0.7.

11.14 At a particular stage of an axial-flow compressor the required temperature rise is 20 K. The blade velocity at the mean radius is 200 m/s and the flow velocity is 150 m/s. Assuming that the degree of reaction is to be 50% at the mean radius and that the work done factor is 0.9, calculate the required blade angles at the mean radius.

(16.4°, 46.1°)

11.15 For the compressor stage of Problem 11.14 the root to tip radius ratio is 0.6. Assuming that twisted blades are used based on free vortex flow design, calculate:
  (i) the required blade angles at the root;
 (ii) the required blade angles at the tip;
(iii) the degree of reaction at the root and at the tip.

(31.3°, −21.1°, 21.4°, 54.2°; 55.1°, 39.9°, 13.3°, 39.7°; 0.11, 0.68)

11.16 A reaction turbine is supplied with steam at 60 bar and 600 °C. The condenser pressure is 0.07 bar. If the reheat factor can be assumed to be 1.04 and the stage efficiency is constant throughout at 80%, calculate the steam flow required for a diagram power of 25 000 kW.

(75 600 kg/h)

11.17 A reaction turbine expands 34 000 kg/h of steam from 20 bar, 400 °C to a pressure of 0.2 bar. The turbine is designed such that the steam leaving is just dry saturated. The reheat factor is 1.05 and the isentropic efficiency of each stage is the same throughout. There are 14 stages and the enthalpy drop is the same in each. All the blades have an exit angle of 22° and the mean value of the blade speed ratio is 0.82. Calculate the stage efficiency, the diagram power, the drum diameter, and the blade height for the last row of moving blades. The turbine speed is 2400 rev/min. Calculate also the pressure at the entry to the last stage and make a sketch on the h–s diagram showing the last stage expansion.

(67.7%; 6035 kW; 1.34 m; 175 mm; 0.31 bar)

11.18 The gases enter an axial-flow gas turbine at 8 bar, 850 °C and leave at 2.5 bar. There are 10 stages, each developing the same specific work with the same stage efficiency; the flow velocity is constant throughout at 110 m/s, and the polytropic efficiency is 0.85. At one particular stage the mean blade velocity is 140 m/s, the stage is designed for 50% reaction at the mean blade height, and the specific work output is constant across the stage at all radii.

Assuming that the gas velocities entering and leaving the turbine are approximately equal, and taking $\gamma = 4/3$ and $c_p = 1.15$ kJ/kg K for the gases, calculate:
  (i) the blade angles at the mean radius for the stage;
  (ii) the overall isentropic efficiency of the turbine;
  (iii) the stage efficiency of each turbine stage.

(15.7°, 57.3°; 86.8%; 85.2%)

11.19 (a) Show that for an axial compressor of 50% reaction design with blade velocity C_b, axial flow component of airflow, C_f, and inlet blade angle, β_i, that

$$\text{Specific power input per stage} = C_b^2 \left[2\frac{C_f}{C_b} \tan \beta_i - 1 \right] \times (\text{work done factor})$$

(b) A 10-stage axial-flow compressor of 50% reaction design has a mean blade velocity of 250 m/s and the blade inlet angle for each row is 45°. The ratio of flow velocity to blade velocity is 0.75, the work done factor for each stage is 0.87 and the isentropic efficiency of the compressor is 0.85. Assuming an air inlet temperature of 20 °C, calculate:
  (i) the exit angle of the blades;
  (ii) the pressure ratio of the compressor;
  (iii) the pressure ratio of the first stage.
(Hint: For part (iii) as a first approximation take the stage efficiency to be equal to the compressor polytropic efficiency.)

(18.4°; 7.6/1; 1.317/1)

11.20 A centrifugal compressor running at 16 000 rev/min takes in air at 17 °C and compresses it through a pressure ratio of 4 with an isentropic efficiency of 82%. The blades are radially inclined and the slip factor is 0.85. Guide vanes at inlet give the air an angle of pre-whirl of 20° to the axial direction; take the mean diameter of the impeller eye as 200 mm and the absolute air velocity at inlet as 120 m/s. Calculate the impeller tip diameter.

(549 mm)

References

11.1 DOUGLAS J F, GASIOREK J M and SWAFFIELD J A 1986 *Fluid Mechanics* 2nd edn Longman

11.2 DIXON S L 1978 *Fluid Mechanics, Thermodynamics of Turbomachinery* 3rd edn Pergamon

11.3 SHEPHERD D G 1969 *Principles of Turbomachinery* Macmillan

11.4 TURTON R K 1984 *Principles of Turbomachinery* E & F Spon

11.5 COHEN H, ROGERS G F C and SARAVANAMUTTOO H I H 1987 *Gas Turbine Theory* 3rd edn Longman

11.6 KEARTON W J 1960 *Steam Turbine Theory and Practice* Pitman

12

Positive Displacement Machines

The function of a compressor is to take a definite quantity of fluid (usually a gas, and most often air) and deliver it at a required pressure. The most efficient machine is one which will accomplish this with the minimum input of mechanical work. Both reciprocating and rotary positive displacement machines are used for a variety of purposes. On the basis of performance a general distinction can be made between the two types by defining the reciprocating type as having the characteristics of a low mass rate of flow and high-pressure ratios, and the rotary type as having a high mass rate of flow and low-pressure ratios. The pressure range of atmospheric to about 9 bar is common to both types.

Some rotary machines are suitable only for low-pressure ratio work, and are applied to the scavenging and supercharging of engines, and the various applications of exhausting and vacuum pumping. For pressures above 9 bar the vane-type rotary machine can be used to supply boost pressures, but for sustained high-pressure work up to 500 bar and above, for special purposes, the reciprocating type is used.

Both basic types exist in many different forms each having its own characteristics. They may be single or multistage, and have either air or water cooling. The reciprocating machine is pulsating in action which limits the rate at which fluid can be delivered, but the rotary machine is continuous in action and does not have this disadvantage. The rotary machines are smaller in size for a given flow, lighter in weight and mechanically simpler than their reciprocating counterparts. The treatment and scope of the following sections is fundamental and is not exhaustive. Many compressors are designed to overcome the deficiencies of the basic machines and to satisfy special requirements. For descriptions of these machines the excellent literature supplied by the manufacturers concerned should be consulted.

For a compressor which operates in a cyclic or pulsating manner, such as a reciprocating compressor, the properties at inlet and outlet are the average values taken over the cycle. Alternatively the boundary of the control volume is chosen such that states 1 and 2 are constant with time, the positions selected being remote from the pulsating disturbance.

12.1   Reciprocating compressors

Typical reciprocating compressor cylinder arrangements are shown in Fig. 12.1(a) and (b). The mechanism involved is the basic piston, connecting-rod,

Fig. 12.1
Single-acting (a) and double-acting (b) reciprocating air compressors

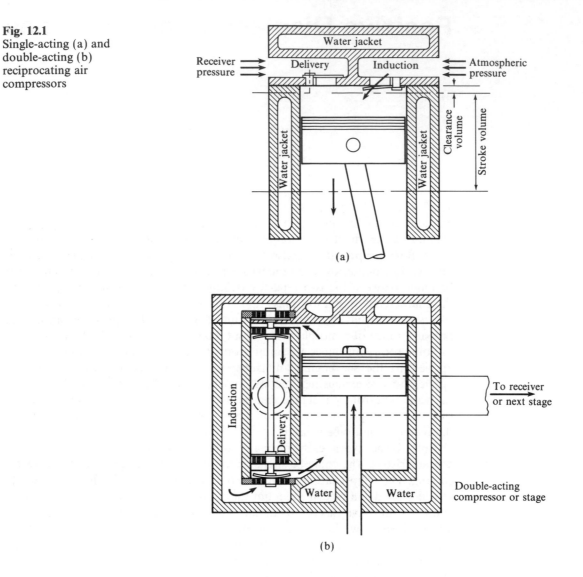

crank, and cylinder arrangement. Initially the clearance volume in the cylinder will be considered negligible. Also the working fluid will be assumed to be a perfect gas. The cycle takes one revolution of the crankshaft for completion and the basic indicator diagram is shown in Fig. 12.2.

The valves employed in most air compressors are designed to give automatic action. They are of the spring-loaded type operated by a small difference in pressure across them, the light spring pressure giving a rapid closing action.

Fig. 12.2 Pressure–volume diagram for a reciprocating compressor with clearance neglected

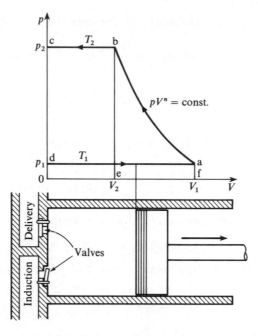

The lift of the valve to give the required airflow should be as small as possible and should operate without shock.

In Fig. 12.2 the line d–a represents the induction stroke. The mass in the cylinder increases from zero at d to that required to fill the cylinder at a. In the ideal case the temperature is constant at T_1 for this process and there is no heat exchange with the surroundings. Induction commences when the pressure difference across the valve is sufficient to open it. Line abc represents the compression and delivery stroke. As the piston begins its return stroke the pressure in the cylinder rises and closes the inlet valve. The pressure rise continues with the returning piston as shown by line ab until the pressure is reached at which the delivery valve opens (a value decided by the valve and the pressure in the receiver). The delivery takes place as shown by the line bc, which is a process at constant temperature T_2, constant pressure p_2, zero heat exchange, and decreasing mass. At the end of this stroke the cycle is repeated. The value of the delivery temperature T_2 depends upon the law of compression between a and b, which in turn depends upon the heat exchange with the surroundings during this process. It may be assumed that the general form of compression is the reversible polytropic (i.e. $pV^n = $ constant).

The net work done in the cycle is given by the area of the p–V diagram and is the work done on the gas.

Indicated work done on the gas per cycle

= area abcd

= area abef + area bc0e − area ad0f

Using equation (3.24) for area abef,

$$\text{Work input} = \frac{(p_2 V_b - p_1 V_a)}{n-1} + p_2 V_b - p_1 V_a$$

$$= (p_2 V_b - p_1 V_a)\left(\frac{1}{n-1} + 1\right)$$

i.e. $\quad\text{Work input} = (p_2 V_b - p_1 V_a)\dfrac{1 + n - 1}{n-1}$

$$= \frac{n}{n-1}(p_2 V_b - p_1 V_a) \tag{12.1}$$

From equation (2.6) we can write

$$p_1 V_a = mRT_1 \quad \text{and} \quad p_2 V_b = mRT_2$$

where m is the mass induced and delivered per cycle. Then

$$\text{Work input per cycle} = \frac{n}{n-1}mR(T_2 - T_1) \tag{12.2}$$

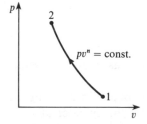

Fig. 12.3 Compression process on a $p\text{--}v$ diagram

Work done on the air per unit time is equal to the work done per cycle times the number of cycles per unit time. The rate of mass flow is more often used than the mass per cycle; if the rate of mass flow is given the symbol $\dot{m}$, and replaces m in equation (12.2), then the equation gives the rate at which work is done on the air, or the indicated power.

The working fluid changes state between a and b in Fig. 12.2, from p_1 and T_1 to p_2 and T_2, the change being shown in Fig. 12.3, which is a diagram of properties (i.e. p against v).

The delivery temperature is given by the equation (3.29),

i.e. $\quad T_2 = T_1\left(\dfrac{p_2}{p_1}\right)^{(n-1)/n}$

Example 12.1 A single-stage reciprocating compressor takes $1\,\text{m}^3$ of air per minute at 1.013 bar and $15\,^\circ\text{C}$ and delivers it at 7 bar. Assuming that the law of compression is $pV^{1.35} = \text{constant}$, and that clearance is negligible, calculate the indicated power.

Solution $\quad$ Mass delivered per min, $\dot{m} = \dfrac{p_1 V_1}{R T_1}$

$$= \frac{1.013 \times 1 \times 10^5}{287 \times 288} = 1.226\,\text{kg/min}$$

where $T_1 = 15 + 273 = 288$ K.

Delivery temp., $T_2 = T_1\left(\dfrac{p_2}{p_1}\right)^{(n-1)/n} = 288\left(\dfrac{7}{1.013}\right)^{(1.35-1)/1.35}$

$$= 475.4\,\text{K}$$

From equation (12.2)

$$\text{Indicated power} = \frac{n}{n-1} \dot{m}R(T_2 - T_1)$$

where $\dot{m}$ is the mass flow rate,

i.e. $\quad$ Indicated power $= \dfrac{1.35 \times 1.226 \times 287 \times (475.4 - 288)}{10^3 \times (1.35 - 1) \times 60}$

$$= 4.238 \text{ kW}$$

The actual power input to the compressor is larger than the indicated power, due to the work necessary to overcome the losses due to friction, etc.

i.e. $\quad$ Shaft power = indicated power + friction power $\qquad$ (12.3)

The mechanical efficiency of the machine is given by

$$\text{Compressor mechanical efficiency} = \frac{\text{indicated power}}{\text{shaft power}} \qquad (12.4)$$

To determine the power input required the efficiency of the driving motor must be taken into account, in addition to the mechanical efficiency. Then

$$\text{Input power} = \frac{\text{shaft power}}{\text{efficiency of motor and drive}} \qquad (12.5)$$

Example 12.2 $\quad$ If the compressor of Example 12.1 is to be driven at 300 rev/min and is a single-acting, single-cylinder machine, calculate the cylinder bore required, assuming a stroke to bore ratio of 1.5/1. Calculate the power of the motor required to drive the compressor if the mechanical efficiency of the compressor is 85% and that of the motor transmission is 90%.

Solution $\quad$ Volume dealt with per unit time at inlet $= 1 \text{ m}^3/\text{min}$

therefore

$$\text{Volume drawn in per cycle} = \frac{1}{300} = 0.003\,33 \text{ m}^3/\text{cycle}$$

i.e. $\quad$ Cylinder volume $= 0.003\,33 \text{ m}^3$

therefore

$$\frac{\pi}{4}d^2 L = 0.003\,33$$

where d is the bore and L the stroke,

i.e. $\quad \dfrac{\pi}{4}d^2(1.5 \times d) = 0.003\,33$

therefore

$$d^3 = 0.002\,83 \text{ m}^3$$

i.e.   Cylinder bore $= 141.4 \text{ mm}$

Power input to the compressor $= \dfrac{4.238}{0.85} = 4.99 \text{ kW}$

therefore

Motor power $= \dfrac{4.99}{0.9} = 5.54 \text{ kW}$

Proceeding from equation (12.2), other expressions for the indicated work can be derived, i.e.

$$\text{Indicated power} = \frac{n}{n-1} \dot{m} R (T_2 - T_1) = \frac{n}{n-1} \dot{m} R T_1 \left(\frac{T_2}{T_1} - 1 \right)$$

Also from equation (3.29)

$$\frac{T_2}{T_1} = \left(\frac{p_2}{p_1} \right)^{(n-1)/n}$$

Therefore

$$\text{Indicated power} = \frac{n}{n-1} \dot{m} R T_1 \left\{ \left(\frac{p_2}{p_1} \right)^{(n-1)/n} - 1 \right\} \tag{12.6}$$

or $$\text{Indicated power} = \frac{n}{n-1} p_1 \dot{V} \left\{ \left(\frac{p_2}{p_1} \right)^{(n-1)/n} - 1 \right\} \tag{12.7}$$

where $\dot{V}$ is the volume induced per unit time.

The condition for minimum work

The work done on the gas is given by the area of the indicator diagram, and the work done will be a minimum when the area of the diagram is a minimum. The height of the diagram is fixed by the required pressure ratio (when p_1 is fixed), and the length of the line da is fixed by the cylinder volume, which is itself fixed by the required induction of gas. The only process which can influence the area of the diagram is the line ab. The position taken by this line is decided by the value of the index n; Fig. 12.4 shows the limits of the possible processes.

Line ab_1 is according to the law $pV = \text{constant}$ (i.e. isothermal)

Line ab_2 is according to the law $pV^\gamma = \text{constant}$ (i.e. isentropic)

Both processes are reversible.

Isothermal compression is the most desirable process between a and b, giving the minimum work to be done on the gas. This means that in an actual

Fig. 12.4 Possible compression processes on a p–v diagram

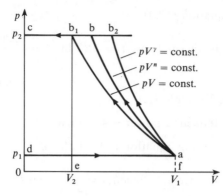

compressor the gas temperature must be kept as close as possible to its initial value, and a means of cooling the gas is always provided, either by air or by water.

The indicated work done when the gas is compressed isothermally is given by the area ab_1cd.

$$\text{Area } ab_1cd = \text{area } ab_1ef + \text{area } b_1c0e - \text{area } ad0f$$

$$\text{Area } ab_1ef = p_2 V_{b_1} \ln \frac{p_2}{p_1} \text{ (from equation (3.9))}$$

i.e. $\text{indicated work per cycle} = p_2 V_{b_1} \ln \dfrac{p_2}{p_1} + p_1 V_{b_1} - p_1 V_a$

Also $p_1 V_a = p_2 V_{b_1}$, since the process ab_1 is isothermal, therefore

$$\text{indicated work per cycle} = p_2 V_{b_1} \ln \frac{p_2}{p_1}$$

$$= p_1 V_a \ln \frac{p_2}{p_1} \qquad (12.8)$$

$$= mRT \ln \frac{p_2}{p_1} \qquad (12.9)$$

When m and V_a in equations (12.8) and (12.9) are the mass and volume induced per unit time, then these equations give the isothermal power.

Isothermal efficiency

By definition, based on the indicator diagram

$$\text{Isothermal efficiency} = \frac{\text{isothermal work}}{\text{indicated work}} \qquad (12.10)$$

Example 12.3   Using the data of Example 12.1 calculate the isothermal efficiency of the compressor.

Solution　From equation (12.9)

$$\text{Isothermal power} = \dot{m}RT\ln\frac{p_2}{p_1} = \frac{1.226 \times 0.287 \times 288}{60} \times \ln\frac{7}{1.013}$$

$$= 3.265 \text{ kW}$$

From Example 12.1,

Indicated power $= 4.238$ kW

Therefore using equation (12.10) above

$$\text{Isothermal efficiency} = \frac{3.265}{4.238} = 0.77 \text{ or } 77\%$$

The least desirable form of compression in reciprocating compressors is that given by the isentropic process (see Fig. 12.4). The actual form of compression will usually be one between these two limits. The three processes are shown represented on a T–s diagram in Fig. 12.5:

1–2' represents isothermal compression
1–2" represents isentropic compression
1–2 represents compression according to a law $pv^n = $ constant

Fig. 12.5 Isothermal, polytropic, and isentropic compression processes on a T–s diagram

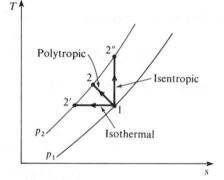

The value of n is usually between 1.2 and 1.3 for a reciprocating air compressor. The main method used for cooling the air is by surrounding the cylinder by a water jacket and designing for the best ratio of surface area to volume of the cylinder.

12.2 Reciprocating compressors including clearance

Clearance is necessary in a compressor to give mechanical freedom to the working parts and allow the necessary space for valve operations.

Figure 12.6 shows the ideal indicator diagram with the clearance volume included. For good-quality machines the clearance volume is about 6% of the swept volume, and with a sleeve-valve machine it can be as low as 2%, but machines with clearances of 30–35% are also common.

Fig. 12.6 Ideal
indicator diagram for a
reciprocating
compressor with
clearance

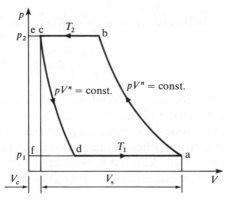

When the delivery stroke bc is completed the clearance volume V_c is full of gas at pressure p_2 and temperature T_2. As the piston proceeds on the next induction stroke the air expands behind it until the pressure p_1 is reached. Ideally as soon as the pressure reaches p_1, the induction of fresh gas will begin and continue to the end of this stroke at a. The gas is then compressed according to the law $pV^n = C$, and delivery begins at b as controlled by the valves. The effect of clearance is to reduce the induced volume at p_1 and T_1 from V_s to $(V_a - V_d)$. The masses of gas at the four principal points are such that $\dot{m}_a = \dot{m}_b$ and $\dot{m}_c = \dot{m}_d$. The mass delivered per unit time is given by $(\dot{m}_b - \dot{m}_c)$, which is equal to that induced, given by $(\dot{m}_a - \dot{m}_d)$. The properties of the working fluid change in processes a–b and c–d as shown in Fig. 12.7.

Fig. 12.7 Compression
and re-expansion of
masses of gas in a
reciprocating
compressor

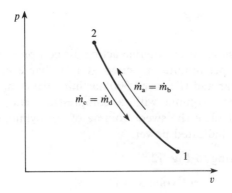

Referring to Fig. 12.6 the indicated work done is given by the area of the p–V diagram.

$$\text{Indicated work} = \text{area abcd}$$

$$= \text{area abef} - \text{area cefd}$$

Then, using equation (12.2)

$$\text{Indicated power} = \frac{n}{n-1}\dot{m}_a R(T_2 - T_1) - \frac{n}{n-1}\dot{m}_d R(T_2 - T_1)$$

i.e.   Indicated power $= \dfrac{n}{n-1} R(\dot{m}_a - \dot{m}_d)(T_2 - T_1)$

$$= \dfrac{n}{n-1} R\dot{m}(T_2 - T_1) \tag{12.11}$$

where $\dot{m}$ is the mass induced per unit time $= (\dot{m}_a - \dot{m}_d)$.

A comparison of equations (12.11) and (12.2) shows that they are identical. The work done on compressing the mass of gas $\dot{m}_c$ (or $\dot{m}_d$) on compression, a–b, is returned when the gas expands from c to d. Hence the work done per unit mass of air delivered is unaffected by the size of the clearance volume.

Other expressions can be derived as before. From equation (12.7)

$$\text{Indicated power} = \dfrac{n}{n-1} p_1 \dot{V} \left\{ \left(\dfrac{p_2}{p_1}\right)^{(n-1)/n} - 1 \right\}$$

Also, if there are f cycles per unit time, then we have:

$$\dot{V} = f(V_a - V_d) \tag{12.12}$$

therefore

$$\text{Indicated power} = \dfrac{n}{n-1} p_1 f(V_a - V_d) \left\{ \left(\dfrac{p_2}{p_1}\right)^{(n-1)/n} - 1 \right\} \tag{12.13}$$

The mass delivered per unit time can be increased by designing the machine to be double acting, i.e. gas is dealt with on both sides of the piston, the induction stroke for one side being the compression stroke for the other (see Fig. 12.1).

Example 12.4   A single-stage, double-acting air compressor is required to deliver 14 m³ of air per minute measured at 1.013 bar and 15 °C. The delivery pressure is 7 bar and the speed 300 rev/min. Take the clearance volume as 5% of the swept volume with a compression and re-expansion index of $n = 1.3$. Calculate the swept volume of the cylinder, the delivery temperature, and the indicated power.

Solution   Referring to Fig. 12.8

Swept volume $= (V_a - V_c) = V_s$

and   Clearance volume, $V_c = 0.05 V_s$

i.e. $V_a = 1.05 V_s$

Using equation (12.12) for a double-acting machine

$$\text{Volume induced per cycle}, (V_a - V_d) = \dfrac{14}{300 \times 2}$$

$$= 0.0233 \text{ m}^3/\text{cycle}$$

(cycles per minute = revolutions per minute × cycles per revolution).

Fig. 12.8 Pressure–volume diagram for Example 12.4

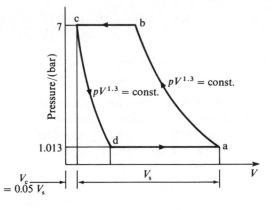

Now

$$V_d = V_c \left(\frac{p_2}{p_1}\right)^{1/n} = 0.05 V_s \left(\frac{7}{1.013}\right)^{1/1.3}$$

i.e. $V_d = 0.221 V_s$

therefore

$$(V_a - V_d) = 1.05 V_s - 0.221 V_s = 0.0233 \text{ m}^3/\text{cycle}$$

therefore

$$V_s = \frac{0.0233}{0.829} = 0.0281 \text{ m}^3/\text{cycle}$$

i.e.   Swept volume of compressor $= 0.0281 \text{ m}^3$

Delivery temp., $T_2 = T_1 \left(\frac{p_2}{p_1}\right)^{(n-1)/n}$ from equation (3.29)

and $T_1 = 15 + 273 = 288 \text{ K}$

i.e. $T_2 = 288 \left(\frac{7}{1.013}\right)^{(1.3-1)/1.3}$

$$= 450 \text{ K}$$

therefore

Delivery temp. $= 177\,^\circ\text{C}$

Using equation (12.7)

Indicated power

$$= \frac{n}{n-1} p_1 \dot{V} \left\{ \left(\frac{p_2}{p_1}\right)^{(n-1)/n} - 1 \right\}$$

$$= \frac{1.3}{0.3} \times \frac{1.013 \times 10^5 \times 14}{10^3 \times 60} \left\{ \left(\frac{7}{1.013} \right)^{(1.3-1)/1.3} - 1 \right\} \text{kW}$$

i.e.   Indicated power $= 57.6 \text{ kW}$

The approach used for a particular problem depends on how the data are stated and the quantities evaluated during the solution. In some problems it is better to evaluate $\dot{m}$ and T_2 and then use equation (12.11) for the indicated power; e.g. in Example 12.4 above, T_2 has been calculated, and the mass induced is given by

$$\dot{m} = \frac{1.013 \times 14 \times 10^5}{0.287 \times 288 \times 10^3} = 17.16 \text{ kg/min}$$

Then, using equation (12.11)

$$\text{Indicated power} = \frac{n}{n-1} \dot{m} R (T_2 - T_1)$$

$$= \frac{1.3 \times 17.16 \times 0.287 (450 - 288)}{0.3 \times 60}$$

$$= 57.6 \text{ kW} \quad \text{(as before)}$$

The diagrams previously shown (e.g. Fig. 12.8) are ideal diagrams. An actual indicator diagram is similar to the ideal except for the induction and delivery processes which are modified by a valve action. This is shown in Fig. 12.9. The waviness of the lines d–a and b–c is due to valve bounce. Automatic valves are in general use (see Fig. 12.1), and these are less definite in action than cam-operated valves; they also give more throttling of the gas. The induction stroke d–a is a mixing process, the induced air mixing with that in the cylinder.

Fig. 12.9 Actual indicator diagram for a reciprocating compressor

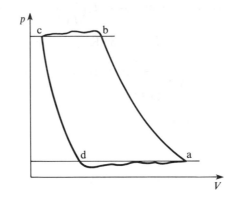

Volumetric efficiency, η_v

It has been shown that one of the effects of clearance is to reduce the induced volume to a value less than that of the swept volume. This means that for a required induction the cylinder size must be increased over that calculated on the assumption of zero clearance. The volumetric efficiency is defined as follows:

η_v = the mass of gas delivered, divided by the mass of gas which would fill the swept volume at the free air conditions of pressure and temperature (12.14)

or

η_v = the volume of gas delivered measured at the free air pressure and temperature, divided by the swept volume of the cylinder (12.15)

The volume of air dealt with per unit time by an air compressor is quoted as the free air delivery (FAD), and is the rate of volume flow delivered, measured at the pressure and temperature of the atmosphere in which the machine is situated.

Equations (12.14) and (12.15) can be shown to be identical, i.e. if the FAD per cycle is V, at p and T, then the mass delivered per cycle is

$$m = \frac{pV}{RT}$$

The mass required to fill the swept volume, V_s, at p and T is given by

$$m_s = \frac{pV_s}{RT}$$

Therefore by equation (12.14),

$$\eta_v = \frac{m}{m_s} = \frac{pV}{RT} \times \frac{RT}{pV_s} = \frac{V}{V_s}.$$

The volumetric efficiency can be obtained from the indicator diagram. Referring to Fig. 12.10

$$\text{Volume induced} = V_a - V_d = V_s + V_c - V_d$$

Fig. 12.10 Indicator diagram for a reciprocating compressor

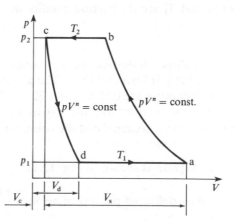

and using equation (3.25)

$$\frac{V_d}{V_c} = \left(\frac{p_2}{p_1}\right)^{1/n} \quad \text{i.e. } V_d = V_c\left(\frac{p_2}{p_1}\right)^{1/n}$$

therefore

$$\text{Volume induced} = V_s + V_c - V_c\left(\frac{p_2}{p_1}\right)^{1/n}$$

$$= V_s - V_c\left\{\left(\frac{p_2}{p_1}\right)^{1/n} - 1\right\}$$

Hence using equation (12.15),

$$\eta_v = \frac{V_a - V_d}{V_s} = \frac{V_s - V_c\{(p_2/p_1)^{1/n} - 1\}}{V_s}$$

i.e.

$$\eta_v = 1 - \frac{V_c}{V_s}\left\{\left(\frac{p_2}{p_1}\right)^{1/n} - 1\right\} \tag{12.16}$$

It is important to note that this definition of volumetric efficiency is only consistent with that of equations (12.14) and (12.15) if the conditions of pressure and temperature in the cylinder during the induction stroke are identical with those of the free air. In fact the gas will be heated by the cylinder walls, and there will be a reduction in pressure due to the pressure drop required to induce the gas into the cylinder against the resistance to flow. These modifications to the ideal case require a more careful application of the formulae previously derived.

For example, when the FAD per cycle is denoted by V at p and T, then

$$m = \frac{pV}{RT} = \frac{p_1(V_a - V_d)}{RT_1}$$

i.e.

$$\text{FAD/cycle, } V = (V_a - V_d)\frac{T}{T_1}\frac{p_1}{p} \tag{12.17}$$

where p_1 and T_1 are the suction conditions.

Example 12.5 A single-stage, double-acting air compressor has a FAD of $14\,\text{m}^3/\text{min}$ measured at 1.013 bar and 15 °C. The pressure and temperature in the cylinder during induction are 0.95 bar and 32 °C. The delivery pressure is 7 bar and the index of compression and expansion, n, is equal to 1.3. Calculate the indicated power required and the volumetric efficiency. The clearance volume is 5% of the swept volume.

Solution The p–V diagram is shown in Fig. 12.11

$$\text{Mass delivered per unit time, } \dot{m} = \frac{p\dot{V}}{RT}$$

Fig. 12.11 Pressure–volume diagram for Example 12.5

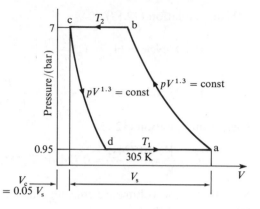

where the FAD is $\dot{V}$ at p and T,

i.e.

$$\dot{m} = \frac{1.013 \times 14 \times 10^5}{0.287 \times 288 \times 10^3} = 17.16 \text{ kg/min}$$

where $T = 15 + 273 = 288$ K.

$$T_2 = T_1 \left(\frac{p_2}{p_1}\right)^{(n-1)/n} \quad \text{from equation (3.29)}$$

i.e.

$$T_2 = 305 \times \left(\frac{7}{0.95}\right)^{(1.3-1)/1.3} = 483.6 \text{ K}$$

where $T_1 = 32 + 273 = 305$ K.

From equation (12.11)

$$\text{Indicated power} = \frac{n}{n-1} \dot{m} R (T_2 - T_1)$$

$$= \frac{1.3 \times 17.16 \times 0.287(483.6 - 305)}{0.3 \times 60}$$

$$= 63.5 \text{ kW}$$

As before

$$V_d = V_c \left(\frac{p_2}{p_1}\right)^{1/n}$$

i.e.

$$V_d = 0.05 V_s \left(\frac{7}{0.95}\right)^{1/1.3} = 0.05 V_s \times 7.368^{0.769}$$

$$= 0.232 V_s$$

therefore

$$V_a - V_d = V_a - 0.232 V_s = 1.05 V_s - 0.232 V_s = 0.818 V_s$$

Using equation (12.17)

$$\text{FAD/cycle} = (V_a - V_d)\frac{T}{T_1}\frac{p_1}{p}$$

i.e. $\text{FAD/cycle} = 0.818V_s \times \dfrac{288}{305} \times \dfrac{0.95}{1.013} = 0.724V_s$

Then from equation (12.15)

$$\eta_v = \frac{V}{V_s} = \frac{0.724V_s}{V_s} = 0.724 \text{ or } 72.4\%$$

Note that if the volumetric efficiency in the above example is evaluated using equation 12.16 then

$$\eta_v = 1 - \frac{V_c}{V_s}\left\{\left(\frac{p_2}{p_1}\right)^{1/n} - 1\right\} = 1 - \frac{0.05V_s}{V_s}\left\{\left(\frac{7}{0.95}\right)^{1/1.3} - 1\right\}$$

$$= 0.818 \text{ or } 81.8\%$$

There is a considerable difference between the two values, since the latter answer ignores the difference in temperature and pressure between the free air conditions and the suction conditions.

12.3 Multistage compression

It is shown in section 12.1 that the condition for minimum work is that the compression process should be isothermal. In general the temperature after compression is given by equation (3.29), $T_2 = T_1(p_2/p_1)^{(n-1)/n}$. The delivery temperature increases with the pressure ratio. Further, from equation (12.16)

$$\eta_v = 1 - \frac{V_c}{V_s}\left\{\left(\frac{p_2}{p_1}\right)^{1/n} - 1\right\}$$

it can be seen that as the pressure ratio increases the volumetric efficiency decreases. This is illustrated in Fig. 12.12.

For compression from p_1 to p_2 the cycle is abcd and the FAD per cycle is $V_a - V_d$; for compression from p_1 to p_3 the cycle is ab'c'd' and the FAD per cycle is $V_a - V_{d'}$; for compression from p_1 to p_4 the cycle is ab''c''d'' and the FAD per cycle is $V_a - V_{d''}$. Therefore for a required FAD the cylinder size would have to increase as the pressure ratio increases.

The volumetric efficiency can be improved by carrying out the compression in two stages. After the first stage of compression the fluid is passed into a smaller cylinder in which the gas is compressed to the required final pressure. If the machine has two stages, the gas will be delivered at the end of this stage, but it could be delivered to a third cylinder for higher pressure ratios. The cylinders of the successive stages are proportioned to take the volume of gas delivered from the previous stage.

Fig. 12.12 Effect on
the volumetric efficiency
of increasing the
delivery pressure

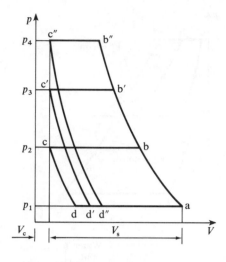

The indicator diagram for a two-stage machine is shown in Fig. 12.13. In this diagram it is assumed that the delivery process from the first or LP stage and the induction process of the second or HP stage are at the same pressure.

Fig. 12.13 Pressure–
volume diagram for
two-stage compression

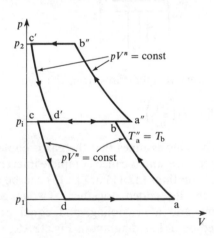

The ideal isothermal compression can only be obtained if ideal cooling is continuous. This is difficult to obtain during normal compression. With multistage compression the opportunity presents itself for the gas to be cooled as it is being transferred from one cylinder to the next, by passing it through an intercooler. If intercooling is complete, the gas will enter the second stage at the same temperature at which it entered the first stage. The saving in work obtained by intercooling is shown by the shaded area in Fig. 12.14 and the diagram of the plant is shown in Fig. 12.15. The two indicator diagrams abcd and a'b'c'd' are shown with a common pressure, p_i. This does not occur in a real machine as there is a small pressure drop between the cylinders. An after-cooler can be fitted after the delivery process to cool the gas.

Fig. 12.14 Effect of intercooling on the compression work

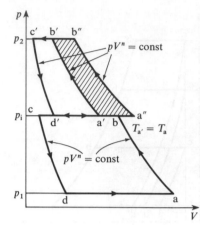

Fig. 12.15 Plan showing intercooling between compressor stages

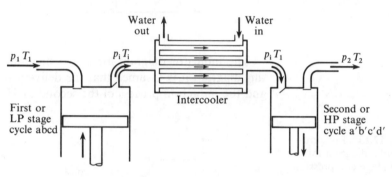

The delivery temperatures from the two stages are given by

$$T_i = T_1 \left(\frac{p_i}{p_1}\right)^{(n-1)/n} \quad \text{and} \quad T_2 = T_1 \left(\frac{p_2}{p_1}\right)^{(n-1)/n}$$

respectively. This assumes that the gas is cooled in the intercooler back to the inlet temperature, and is called complete intercooling. To calculate the indicated power the equations (12.11) or (12.13) can be applied to each stage separately and the results added together. Two-stage compression with complete intercooling and after-cooling, and equal pressure ratios in each stage, is represented on a T–s diagram in Fig. 12.16.

Fig. 12.16 T–s diagram showing intercooling and aftercooling

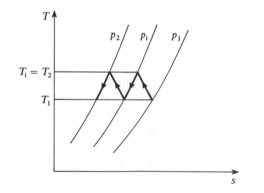

Example 12.6

In a single-acting, two-stage reciprocating air compressor 4.5 kg of air per minute are compressed from 1.013 bar and 15 °C through a pressure ratio of 9 to 1. Both stages have the same pressure ratio, and the law of compression and expansion in both stages is $pV^{1.3} = $ constant. If intercooling is complete, calculate the indicated power and the cylinder swept volumes required. Assume that the clearance volumes of both stages are 5% of their respective swept volumes and that the compressor runs at 300 rev/min.

Solution

The two indicator diagrams are shown superimposed in Fig. 12.17. The LP stage cycle is abcd and the HP cycle is a'b'c'd'.

Fig. 12.17 Pressure–volume diagram showing both stages for Example 12.7

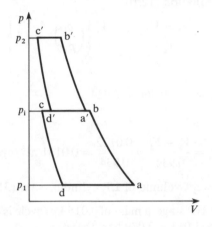

Now $p_2 = 9p_1$, also $p_i/p_1 = p_2/p_i$, therefore

$$p_i^2 = p_1 p_2 = 9p_1^2$$

therefore

$$p_i/p_1 = \sqrt{9} = 3$$

Using equation (3.29)

$$\frac{T_i}{T_1} = \left(\frac{p_i}{p_1}\right)^{(n-1)/n} \qquad \text{therefore } \frac{T_i}{288} = 3^{(1.3-1)/1.3}$$

where $T_1 = 15 + 273 = 288$ K, and T_i is the temperature of the air entering the intercooler,

i.e. $\qquad T_i = 288 \times 1.289 = 371$ K

Now as n, $\dot{m}$, and the temperature difference are the same for both stages, then the work done in each stage is the same. Therefore using equation (12.11)

$$\text{Total indicated power} = 2 \times \frac{n}{n-1} \dot{m} R(T_i - T_1)$$

$$= \frac{2 \times 1.3 \times 4.5 \times 0.287(371 - 288)}{0.3 \times 60}$$

$$= 15.5 \text{ kW}$$

399

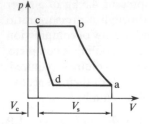

Fig. 12.18 Pressure–volume diagram for LP stage for Example 12.6

The mass induced per cycle is

$$\frac{4.5}{300} = 0.015 \text{ kg/cycle}$$

This mass is passed through each stage in turn.
For the LP cylinder, referring to Fig. 12.18,

$$V_a - V_d = \frac{mRT_1}{p_1} = \frac{0.015 \times 287 \times 288}{1.013 \times 10^5} = 0.0122 \text{ m}^3/\text{cycle}$$

Using equation (12.16)

$$\eta_v = \frac{V_a - V_c}{V_s} = 1 - \frac{V_c}{V_s}\left\{\left(\frac{p_i}{p_1}\right)^{1/n} - 1\right\} = 1 - 0.05(3^{1/1.3} - 1)$$

therefore

$$\eta_v = 1 - 0.066 = 0.934$$

Then

$$V_s = \frac{V_a - V_d}{0.934} = \frac{0.0122}{0.934} = 0.0131 \text{ m}^3/\text{cycle}$$

i.e.   Swept volume of LP cylinder = 0.0131 m^3

For the HP stage, a mass of 0.015 kg/cycle is drawn in at 15 °C and a pressure of $p_i = 3 \times 1.013 = 3.039$ bar, therefore

$$\text{Volume drawn in} = \frac{0.015 \times 287 \times 288}{3.039 \times 10^5}$$

$$= 0.004\,06 \text{ m}^3/\text{cycle}$$

Using equation (12.16) for the HP stage

$$\eta_v = 1 - \frac{V_c}{V_s}\left\{\left(\frac{p_2}{p_i}\right)^{1/n} - 1\right\}$$

and since V_c/V_s is the same as for the LP stage and also $p_2/p_i = p_i/p_1$ then η_v is 0.934 as above. Therefore

$$\text{Swept volume of HP stage} = \frac{0.004\,06}{0.934} = 0.004\,36 \text{ m}^3$$

Note that the clearance ratio is the same in each cylinder, and the suction temperatures are the same since intercooling is complete, therefore the swept volumes are in the ratio of the suction pressures,

i.e. $$V_{s_H} = \frac{V_{s_L}}{3} = \frac{0.0131}{3} = 0.004\,36 \text{ m}^3 \quad \text{(as above)}$$

The ideal intermediate pressure

The value chosen for the intermediate pressure p_i influences the work to be done on the gas and its distribution between the stages. The condition for the work done to be a minimum will be proved for two-stage compression but can be extended to any number of stages.

Total work = LP stage work + HP stage work.

Therefore using equation (12.6)

$$\text{Total power} = \frac{n}{n-1} \dot{m} R T_1 \left\{ \left(\frac{p_i}{p_1} \right)^{(n-1)/n} - 1 \right\}$$

$$+ \frac{n}{n-1} \dot{m} R T_1 \left\{ \left(\frac{p_2}{p_i} \right)^{(n-1)/n} - 1 \right\} \tag{12.18}$$

It is assumed that intercooling is complete and therefore the temperature at the start of each stage is T_1.

i.e. $\qquad \text{Total power} = \frac{n}{n-1} \dot{m} R T_1 \left\{ \left(\frac{p_i}{p_1} \right)^{(n-1)/n} - 1 + \left(\frac{p_2}{p_i} \right)^{(n-1)/n} - 1 \right\} \tag{12.19}$

If p_1, T_1, and p_2 are fixed, then the optimum value of p_i which makes the power a minimum can be obtained by equating d (power)/(dp_i) to zero, i.e. optimum value of p_i when

$$\frac{d}{dp_i} \left\{ \left(\frac{p_i}{p_1} \right)^{(n-1)/n} + \left(\frac{p_2}{p_i} \right)^{(n-1)/n} - 2 \right\} = 0$$

i.e. when

$$\frac{d}{dp_i} \left\{ \left(\frac{1}{p_1} \right)^{(n-1)/n} p_i^{(n-1)/n} + p_2^{(n-1)/n} \left(\frac{1}{p_i} \right)^{(n-1)/n} - 2 \right\} = 0$$

therefore

$$p_1^{-(n-1)/n} \left(\frac{n-1}{n} \right) p_i^{\{(n-1)/n\}-1} + p_2^{(n-1)/n} \left(\frac{1-n}{n} \right) p_i^{\{(1-n)/n\}-1} = 0$$

therefore

$$p_1^{-(n-1)/n} \left(\frac{n-1}{n} \right) p_i^{-1/n} = p_2^{(n-1)/n} \left(\frac{n-1}{n} \right) p_i^{(1-2n)/n}$$

therefore

$$p_i^{\{2(n-1)\}/n} = (p_1 p_2)^{(n-1)/n}$$

therefore

$$p_i^2 = p_1 p_2 \tag{12.20}$$

or $\qquad \dfrac{p_i}{p_1} = \dfrac{p_2}{p_i} \tag{12.21}$

i.e. the pressure ratio is the same for each stage.

$$\text{Total minimum power} = 2 \times (\text{power required for one stage})$$

$$= 2 \times \frac{n\dot{m}RT_1}{n-1}\left\{\left(\frac{p_i}{p_1}\right)^{(n-1)/n} - 1\right\}$$

Or in terms of the overall pressure ratio p_2/p_1, we have, using equation (12.20),

$$\frac{p_i}{p_1} = \frac{\sqrt{p_1 p_2}}{p_1} = \sqrt{\frac{p_2}{p_1}}$$

therefore

$$\text{Total minimum power} = 2 \times \frac{n\dot{m}RT_1}{n-1}\left\{\left(\frac{p_2}{p_1}\right)^{(n-1)/2n} - 1\right\}$$

This can be shown to extend to z stages giving in general,

$$\text{Total minimum power} = z\frac{n}{n-1}\dot{m}RT_1\left\{\left(\frac{p_2}{p_1}\right)^{(n-1)/zn} - 1\right\} \qquad (12.22)$$

Also

$$\text{Pressure ratio for each stage} = \left(\frac{p_2}{p_1}\right)^{1/z} \qquad (12.23)$$

Hence the condition for minimum work is that the pressure ratio in each stage is the same and that intercooling is complete. (Note that in Example 12.6 the information given implies minimum work.)

Example 12.7   A three-stage, single-acting air compressor running in an atmosphere at 1.013 bar and 15 °C has a free air delivery of 2.83 m³/min. The suction pressure and temperature are 0.98 bar and 32 °C respectively. Calculate the indicated power required, assuming complete intercooling, $n = 1.3$, and that the machine is designed for minimum work. The delivery pressure is to be 70 bar.

Solution $$\text{Mass of air delivered} = \frac{pV}{RT} = \frac{1.013 \times 10^5 \times 2.83}{287 \times 288} = 3.47 \text{ kg/min}$$

where $T = 15 + 273 = 288$ K.
Then using equation (12.22)

Total indicated power

$$= z\frac{n}{n-1}\dot{m}RT_1\left\{\left(\frac{p_2}{p_1}\right)^{(n-1)zn} - 1\right\}$$

$$= 3 \times \frac{1.3}{0.3} \times \frac{3.47 \times 0.287 \times 288}{60}\left\{\left(\frac{70}{0.98}\right)^{(1.3-1)/(3 \times 1.3)} - 1\right\}$$

$$= 24.2 \text{ kW}$$

Besides the benefits of multistage compression already dealt with there are also mechanical advantages. The higher pressures are confined to the smaller cylinders and a multicylinder machine has less variation in rotational speed and requires a smaller flywheel.

Energy balance for a two-stage machine with intercooler

Referring to Fig. 12.19, the steady-flow energy equation (1.10) can be applied to the LP stage, the intercooler, and the HP stage, in turn. Changes in kinetic energy and height can be neglected, i.e. from equation (1.10)

$$h_1 + \frac{C_1^2}{2} + Q + W = h_2 + \frac{C_2^2}{2}$$

Fig. 12.19 Steady flow through a two-stage reciprocating compressor with intercooler

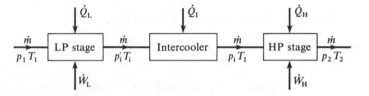

for the LP stage, for unit mass flow rate,

$$h_1 + Q_L + W_L = h_i$$

or for mass flow rate, $\dot{m}$

$$\dot{m}c_p T_1 + \dot{Q}_L + \dot{W}_L = \dot{m}c_p T_i$$

therefore

$$\dot{Q}_L = -\{\dot{W}_L - \dot{m}c_p(T_i - T_1)\}$$

i.e.   Heat rejected in LP stage $= \dot{W}_L - \dot{m}c_p(T_i - T_1)$ (12.24)

for the intercooler, for unit mass flow rate

$$h_i + Q_I = h_1$$

or for mass flow rate, $\dot{m}$

$$\dot{m}c_p T_i + \dot{Q}_I = \dot{m}c_p T_1$$

therefore

$$\dot{Q}_I = -\dot{m}c_p(T_i - T_1)$$

i.e.   Heat rejected in intercooler $= \dot{m}c_p(T_i - T_1)$ (12.25)

for the HP stage, for unit mass flow rate,

$$h_1 + Q_H + W_H = h_2$$

or for mass flow rate, $\dot{m}$

$$\dot{m}c_p T_1 + \dot{Q}_H + \dot{W}_H = \dot{m}c_p T_2$$

403

therefore

$$\dot{Q}_\mathrm{H} = -\{\dot{W}_\mathrm{H} - \dot{m}c_p(T_2 - T_1)\}$$

i.e.   Heat rejected in HP stage = $\dot{W}_\mathrm{H} - \dot{m}c_p(T_2 - T_1)$ (12.26)

With complete intercooling, as assumed in Fig. 12.19, and the compressor designed for minimum work, then, from equation (12.11),

$$\dot{W}_\mathrm{L} = \dot{W}_\mathrm{H} = \frac{n}{n-1}\dot{m}R(T_2 - T_1)$$

Example 12.8   Using the data of Example 12.6 determine the rate of heat loss to the cylinder jacket cooling water and the rate of heat loss to the intercooler circulating water.

Solution   From Example 12.6 we have

$$\dot{W}_\mathrm{L} = \dot{W}_\mathrm{H} = \frac{15.5}{2}\,\mathrm{kW}$$

and $T_2 = T_\mathrm{i} = 371\,\mathrm{K}$

Then, from equation (12.24)

$$-\dot{Q}_\mathrm{L} = \dot{W}_\mathrm{L} - \dot{m}c_p(T_2 - T_1)$$

therefore

$$-\dot{Q}_\mathrm{L} = \frac{15.5}{2} - \frac{4.5 \times 1.005}{60}(371 - 288)$$

i.e. $-\dot{Q}_\mathrm{L} = 7.75 - 6.26 = 1.49\,\mathrm{kW}$

From equation (12.26)

$$-\dot{Q}_\mathrm{H} = \dot{W}_\mathrm{H} - \dot{m}c_p(T_2 - T_1)$$

and $\dot{W}_\mathrm{H} = \dot{W}_\mathrm{L}$ and $T_2 = T_\mathrm{i}$

therefore

$$\dot{Q}_\mathrm{H} = \dot{Q}_\mathrm{L} = -1.49\,\mathrm{kW}$$

i.e.   Heat loss from the cylinder in each stage = $1.49\,\mathrm{kW}$

From equation (12.25)

$$-\dot{Q}_\mathrm{I} = \dot{m}c_p(T_\mathrm{i} - T_1) = \frac{4.5 \times 1.005}{60} \times (371 - 288)$$

i.e.   Heat to intercooler circulating water = $6.26\,\mathrm{kW}$

The quantities $\dot{W}_\mathrm{L}$ and $\dot{W}_\mathrm{H}$, as defined by Fig. 12.19, are the rates of work done on the air. The actual power inputs exceed this by the amounts necessary to overcome frictional resistance to the moving parts of the machine. It can be assumed that about 50% of the friction power goes to increasing the energy transferred to the cooling water, in addition to the heat transferred to the cooling water from the air in the cylinder.

12.4 Steady-flow analysis

In section 12.2 an expression was obtained (equation (12.11)) for the indicated power required to take a mass flow rate of gas, $\dot{m}$, in state 1 and deliver it at a higher pressure in state 2. This was done by analysing the internal processes of the machine. Another approach is to consider the compression process as one of steady flow, as shown in Fig. 12.20, with the change of state from 1 to 2 being achieved by a non-flow process of polytropic compression, as indicated in the property diagram of Fig. 12.21.

Fig. 12.20 Steady flow through a reciprocating compressor

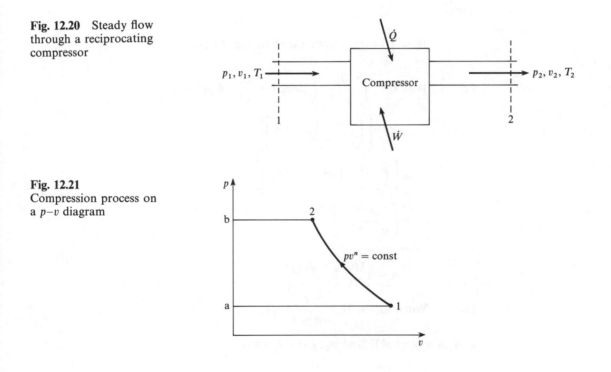

Fig. 12.21 Compression process on a p–v diagram

The steady-flow energy equation for the system shown in Fig. 12.20, neglecting changes in potential and kinetic energy and for unit mass flow rate, is

$$h_1 + Q + W = h_2$$

therefore

$$Q + W = h_2 - h_1$$

or for an elemental process

$$dQ + dW = dh \tag{a}$$

405

Assuming that no heat is transferred on induction or delivery the heat transferred, to or from the system, takes place during the polytropic non-flow compression process. The non-flow equation for a reversible process states

$$dQ = du + p\,dv \tag{b}$$

Combining (a) and (b) gives

$$dh = du + p\,dv + dW$$

and, by definition, $h = u + pv$, hence $dh = du + p\,dv + v\,dp$, therefore substituting

$$du + p\,dv + v\,dp = du + p\,dv + dW$$

therefore

$$dW = v\,dp$$

Then

$$W = \int_1^2 v\,dp = \text{area 12ba1 in Fig. 12.21,}$$

i.e.

$$W = C^{1/n} \int_1^2 \frac{dp}{p^{1/n}} \quad \left(\text{since } v = \frac{C^{1/n}}{p^{1/n}} \text{ if } pv^n = C \right)$$

$$= C^{1/n} \left[\left(\frac{n}{n-1} \right) p^{(n-1)/n} \right]_1^2$$

$$= \left[\left(\frac{n}{n-1} \right) p^{(n-1)/n} p^{1/n} v \right]_1^2$$

$$= \left[\frac{n}{n-1} pv \right]_1^2$$

$$= \frac{n}{n-1} (p_2 v_2 - p_1 v_1)$$

i.e. Work input, $W = \dfrac{n}{n-1}(p_2 v_2 - p_1 v_1)$

and as $p_1 v_1 = RT_1$ and $p_2 v_2 = RT_2$ then

$$W = \frac{nR}{n-1}(T_2 - T_1)$$

12.5 Rotary machines

Because of the continuous rotary action, the rotary positive displacement machine is smaller for a given flow than its reciprocating counterpart. The machines in this category are generally uncooled and as the compression is carried out at a high rate the conditions are approximately adiabatic. Examples of this type are: (i) the Roots blower; (ii) vane type.

Roots blower

The two-lobe type is shown in Fig. 12.22, but three- and four-lobe versions are in use for higher pressure ratios. One of the rotors is connected to the drive and the second rotor is gear driven from the first. In this way the rotors rotate in phase and the profile of the lobes is of cycloidal or in volute form giving correct mating of the lobes to seal the delivery side from the inlet side. This sealing continues until delivery commences. There must be some clearance between the lobes and between the casing and the lobes to reduce wear; this clearance forms a leakage path which has an increasingly adverse effect on efficiency as the pressure ratio increases.

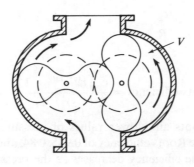

Fig. 12.22 Roots blower with a two-lobe rotor

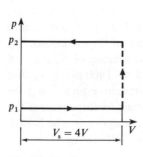

Fig. 12.23 Pressure–volume diagram for a Roots blower

As each side of each lobe faces its side of the casing a volume of gas V, at pressure p_1, is displaced towards the delivery side at constant pressure. A further rotation of the rotor opens this volume to the receiver, and the gas flows back from the receiver, since this gas is at a higher pressure. The gas induced is compressed irreversibly by that from the receiver to the pressure p_2, and then delivery begins. This process is carried out four times per revolution of the driving shaft.

The p–V diagram for this machine is shown in Fig. 12.23, in which the pressure rise from p_1 to p_2 is shown as an irreversible process at constant volume.

$$\text{Work done per cycle} = (p_2 - p_1)V$$

therefore

$$\text{Work done per revolution} = 4(p_2 - p_1)V \tag{12.27}$$

If $\dot{V}_s$ is the volume dealt with per unit time at p_1 and T_1, then

$$\text{Power input} = (p_2 - p_1)\dot{V}_s \tag{12.28}$$

The ideal compression process from p_1 to p_2 is a reversible adiabatic (i.e. isentropic) process. The work done per minute ideally is thus given by equation (12.7) with $n = \gamma$,

i.e. $\quad \text{Power input} = \dfrac{\gamma}{\gamma - 1} p_1 \dot{V}_s \left\{ \left(\dfrac{p_2}{p_1} \right)^{(\gamma - 1)/\gamma} - 1 \right\}$

407

Then a comparison may be made on the basis of a Roots efficiency,

i.e. $\text{Roots efficiency} = \dfrac{\text{work done isentropically}}{\text{actual work done}}$

i.e. $\text{Roots efficiency} = \dfrac{\{\gamma/(\gamma-1)\}p_1 \dot{V}_s\{(p_2/p_1)^{(\gamma-1)/\gamma}-1\}}{\dot{V}_s(p_2-p_1)}$

$= \dfrac{\gamma\{r^{(\gamma-1)/\gamma}-1\}}{(\gamma-1)(r-1)}$

where r = pressure ratio, p_2/p_1.

From equation (2.22), we can write

$$\frac{\gamma}{\gamma-1} = \frac{c_p}{R}$$

therefore

$$\text{Roots efficiency} = \frac{c_p}{R}\left\{\frac{r^{(\gamma-1)/\gamma}-1}{(r-1)}\right\} \tag{12.29}$$

For a Roots air blower values of pressure ratio, r, of 1.2, 1.6, and 2 give values for the Roots efficiency of 0.945, 0.84, and 0.765 respectively. These values show that the efficiency decreases as the pressure ratio increases.

The actual compression process is not quite as simple as that described. When the displacement volume V is opened to the delivery space a pressure wave enters which increases with the opening and moves at the velocity of sound. This wave is reflected from the approaching lobe to the delivery space. The pressure oscillations set up unsteady conditions in the delivery space which vary considerably from one design to another. The actual torque and loading on the rotors are higher than is suggested by the p–V diagram, and fluctuate with high frequency. This fluctuation is transmitted to the drive and creates difficulties due to vibrations. This machine has a number of imperfections, but is well suited to such tasks as the scavenging and supercharging of IC engines.

Roots blowers are built for capacities of from 0.14 to 1400 m³/min, and pressure ratios of the order of 2 to 1 for a single-stage machine and 3 to 1 for a two-stage machine. Other designs have been produced to improve on the Roots blower, one of these being the Bicera compressor, designed by the British Internal Combustion Engineering Research Association (BICERA).

Vane type

The simple vane type is shown in Fig. 12.24 and consists of a rotor mounted eccentrically in the body, and supported by ball- and roller-bearings in the end covers of the body. The rotor is slotted to take the blades which are of a non-metallic material, usually fibre or carbon. As each blade moves past the inlet passage, compression begins due to decreasing volume between the rotor and casing. Delivery begins with the arrival of each blade at the delivery passage.

Fig. 12.24 Vane-type positive displacement compressor

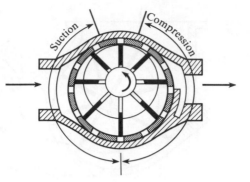

This type of compression differs from that of the Roots blower in that some or all of the compression is obtained before the trapped volume is opened to delivery. Further compression can be obtained by the back-flow of air from the receiver which occurs in an irreversible manner.

The $p-V$ diagram is shown in Fig. 12.25. V_s is the induced volume at pressure p_1 and temperature T_1. Compression occurs to the pressure p_i, the ideal form for an uncooled machine being isentropic. At this pressure the displaced gas is opened to the receiver and gas flowing back from the receiver raises the pressure irreversibly to p_2. The work input is given by the sum of the areas A and B, referring to Fig. 12.25. Comparing the areas of Figs 12.23 and 12.25 it can be seen that for a given airflow and given pressure ratio the vane type requires less work input than the Roots blower.

Fig. 12.25 Pressure–volume diagram for a vane-type compressor

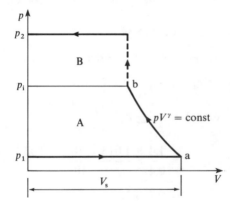

A rotary sliding vane two-stage machine is shown in Fig. 12.26; in this type the vanes are in contact with the cylinder walls.

Example 12.9

Compare the work inputs required for a Roots blower and a vane-type compressor having the same induced volume of 0.03 m³/rev, the inlet pressure being 1.013 bar and the pressure ratio 1.5 to 1. For the vane type assume that internal compression takes place through half the pressure range.

Solution $p_1 = 1.013$ bar

Fig. 12.26 Rotary
sliding vane two-stage
positive displacement
compressor

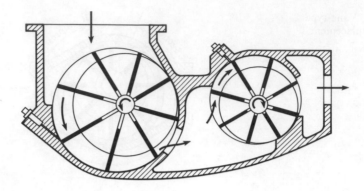

therefore

$$p_2 = 1.013 \times 1.5 = 1.520 \text{ bar}$$

For the Roots blower, referring to Fig. 12.23

$$\text{Work done per revolution} = (p_2 - p_1)V_s$$

$$= (1.520 - 1.013) \times \frac{10^5 \times 0.03}{10^3}$$

$$= 1.52 \text{ kJ/rev}$$

For the vane type

$$p_i = \frac{(1.5 \times 1.013) + 1.013}{2} = 1.266 \text{ bar}$$

Referring to Fig. 12.25

$$\text{Work required} = (\text{area A} + \text{area B})$$

Now using equation (12.7) with $n = \gamma$

$$\text{area A} = \frac{\gamma}{\gamma - 1} p_1 V_s \left\{ \left(\frac{p_i}{p_1} \right)^{(\gamma - 1)/\gamma} - 1 \right\}$$

$$= \frac{1.4}{0.4} \times \frac{1.013 \times 10^5 \times 0.03}{10^3} \left\{ \left(\frac{1.266}{1.013} \right)^{0.4/1.4} - 1 \right\} \text{kJ/rev}$$

$$= 0.70 \text{ kJ/rev}$$

$$\text{area B} = (p_2 - p_i)V_b$$

where V_b is given by equation (3.19),

i.e. $\quad V_b = V_s \left(\frac{p_1}{p_i} \right)^{1/\gamma} = 0.03 \times \left(\frac{1.013}{1.266} \right)^{1/1.4}$

$$= 0.0256 \text{ m}^3$$

i.e. $\quad$ area B $= (1.520 - 1.266) \times 10^2 \times 0.0256 \text{ kJ/rev}$

$$= 0.65 \text{ kJ/rev}$$

410

therefore

Work required $= 0.70 + 0.65 = 1.35 \, \text{kJ/rev}$

(compared with the work required for the Roots machine of 1.52 kJ/rev).

Rotary sliding vane compressors are used with free air deliveries of up to 150 m³/min and pressure ratios up to 8.5 to 1. For special applications and boosting, pressure ratios of the order of 20 to 1 have been obtained from this type. The larger machines are usually water-cooled.

Lubrication is important with vane-type machines and is accomplished by injecting oil to the vane tips in contact with the casing. Some machines, having carbon vanes, require no lubrication. Another version is designed to reduce the friction between vane and casing. This employs a floating drum which rotates between the rotor and casing, and does not allow the vanes to make contact with the casing. The only movement of the blades relative to the floating drum is along the slots. See Fig. 12.26.

12.6 Vacuum pumps

Rotary positive displacement pumps are used to produce a vacuum or to scavenge a vessel. An example of this type of pump is shown in Fig. 12.27. The rotor is eccentrically mounted in the stator and carries two blades which sweep the space between the rotor and stator. The gas being exhausted enters through

Fig. 12.27 Rotary positive displacement vacuum pump

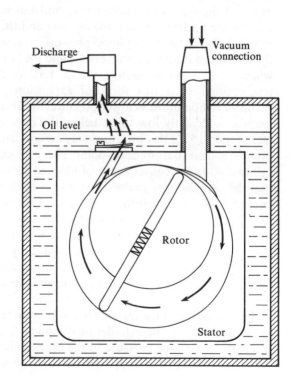

the vacuum connection and is compressed before delivery through the discharge valve. The efficiency of such pumps is impaired by the presence of condensable vapours, and means must be provided to deal with these if necessary. The vapours tend to condense before delivery through the discharge valve and mix with the sealing oil. The liquid eventually evaporates into the vacuum system and lowers the vacuum obtainable, as well as impairing the sealing and lubricating properties of the oil.

12.7 Air motors

Compressed air is used in a wide variety of applications in industry. For some purposes air-operated motors are the most suitable forms of power, especially where there are safety requirements to be met as in mining applications.

Pneumatic breakers, picks, spades, rammers, vibrators, riveters, etc. form a range of hand tools which have wide applications in constructional work. They are light in construction and suitable for operation in remote situations for which other forms of power tools may not be suitable. The action required of such tools, with the associated simplicity and robustness of construction, is obtained with air-operated design.

Basically the cycle in the reciprocating expander is the reverse of that in the reciprocating compressor. Air is supplied to the air motor from an air receiver in which the air is at approximately ambient temperature. There is a pressure drop in the air line between the receiver and the motor. The air expands in the motor cylinder to atmospheric pressure in a manner which is polytropic (i.e. the expansion is internally reversible and the law of expansion is $pv^n = $ constant, where $n < \gamma$, and is usually about 1.3). If the air is initially at ambient temperature, then this form of expansion will bring about a reduction in the air temperature as lower pressures are reached. The temperatures reached may be sufficiently low to be below the dew-point of the moisture in the air (see section 15.2); this moisture may be condensed, and the water formed may even be cooled to its freezing-point. This may lead to the formation of ice in the cylinder with the consequence of blocked valves. To prevent this condition it may be necessary to preheat the air to an initial temperature which is high enough to prevent the formation of ice. This heating of the air causes an increase in volume at the supply pressure and reduces the demand from the compressor. Further, the temperature at which the heat transfer is required is low, and a low-grade supply of heat or 'waste heat' may be utilized for the purpose.

A hypothetical indicator diagram for an air motor is shown in Fig. 12.28. In this case the air expands from 1 to the pressure p_2 at the end of the stroke. There is then a *blow-down* of air from 2 to 3. Air is exhausted from 3 to 4, and at 4 compression of the trapped or *cushion air* begins. Air at the supply pressure, p_6, is admitted to the cylinder at the point 5 where it mixes irreversibly with the cushion air. The pressure in the cylinder is rapidly brought up to the inlet value, p_6. The further supply of air is made at constant pressure behind the

Fig. 12.28 Pressure–volume diagram for an air motor

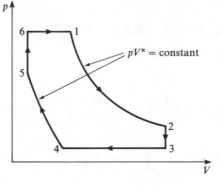

moving piston to the point of cut-off at 1. The cut-off ratio is given by

$$\text{Cut-off ratio} = \frac{V_1 - V_6}{V_3 - V_6}$$

The effect of the cushion air is to give a smoother-running motor. The position of the point 5 depends on the point of initial compression 4, and on the law of compression $pV^n = $ constant. The conditions may be such that the points 5 and 6 coincide.

The analysis of such a diagram is best carried out from basic principles, as illustrated in the following example.

Example 12.10

The cylinder of an air motor has a bore of 63.5 mm and a stroke of 114 mm. The supply pressure is 6.3 bar, the supply temperature 24 °C, and the exhaust pressure is 1.013 bar. The clearance volume is 5% of the swept volume and the cut-off ratio is 0.5. The air is compressed by the returning piston after it has travelled through 0.95 of its stroke. The law of compression and expansion is $pV^{1.3} = $ constant. Calculate:

(i) the temperature at the end of expansion;

(ii) the indicated power of the motor which runs at 300 rev/min;

(iii) the air supplied per minute.

Solution   (i) Referring to the cycle of Fig. 12.28

$$\text{Clearance volume} = V_6 = V_5 = 0.05V_s$$

Since the cut-off ratio, $(V_1 - V_6)/(V_3 - V_6)$, is 0.5, therefore

$$V_1 = 0.5V_s + 0.05V_s = 0.55V_s$$
$$V_2 = V_s + V_6 = 1.05V_s$$

Now

$$V_3 - V_4 = 0.95(V_3 - V_5) \quad \text{(given)}$$

or
$$V_4 - V_5 = 0.05V_s$$

413

therefore

$$V_4 = 0.05V_s + 0.05V_s = 0.1V_s$$

$$p_1 V_1^n = p_2 V^n$$

therefore

$$p_2 = p_1 \left(\frac{V_1}{V_2}\right)^n = 6.3\left(\frac{0.55}{1.05}\right)^{1.3} = 2.718 \text{ bar}$$

also $$T_2 = T_1 \left(\frac{V_1}{V_2}\right)^{n-1} = 297\left(\frac{0.55}{1.05}\right)^{0.3} = 244.6 \text{ K}$$

i.e.   Temperature after expansion $= 244.6 - 273 = -28.4\,^{\circ}\text{C}$

(ii) Now

$$p_5 = p_4 \left(\frac{V_4}{V_5}\right)^n = 1.013\left(\frac{0.1}{0.05}\right)^{1.3} = 2.494 \text{ bar}$$

Work output per cycle = area 1 234 561

$$\text{Work output} = p_1(V_1 - V_6) + \left(\frac{p_1 V_1 - p_2 V_2}{n-1}\right)$$

$$- p_3(V_3 - V_4) - \left(\frac{p_5 V_5 - p_4 V_4}{n-1}\right)$$

and   Swept volume $= \dfrac{\pi \times 63.5^2 \times 114}{4 \times 10^9} = 0.361 \times 10^{-3} \text{ m}^3$

therefore

Work output per cycle

$$= (6.3 \times 10^5 \times 0.361 \times 10^{-3} \times 0.5)$$

$$+ \frac{10^5 \times 0.361 \times 10^{-3}}{0.3}\{(6.3 \times 0.55) - (2.718 \times 1.05)\}$$

$$- (1.013 \times 10^5 \times 0.361 \times 10^{-3} \times 0.95)$$

$$- \frac{10^5 \times 0.361 \times 10^{-3}}{0.3}\{(2.494 \times 0.05) - (1.013 \times 0.1)\}$$

i.e.   Work output per cycle $= 113.7 + 73.5 - 34.7 - 2.8$

$$= 149.7 \text{ J/cycle}$$

Power developed $= \dfrac{149.7 \times 300}{60 \times 10^3} = 0.749 \text{ kW}$

(iii) The mass induced per cycle is given by $(m_1 - m_4)$. It is necessary to determine the temperature of the air at 4, which can be taken as equal to that at 3. It is assumed that the air in the cylinder at the point 2 expands isentropically to the exhaust pressure. Therefore

$$T_3 = T_2\left(\frac{p_3}{p_2}\right)^{(\gamma-1)/\gamma} = 244.6\left(\frac{1.013}{2.718}\right)^{0.4/1.4} = 184.5 \text{ K}$$

i.e.

$$m_4 = \frac{p_4 V_4}{R T_4} = \frac{1.013 \times 10^5 \times 0.0361 \times 10^{-3}}{287 \times 184.5} = 0.0691 \times 10^{-3} \text{ kg}$$

$$m_1 = \frac{p_1 V_1}{R T_1} = \frac{6.3 \times 10^5 \times 0.55 \times 0.361 \times 10^{-3}}{287 \times 297} = 1.4675 \times 10^{-3} \text{ kg}$$

therefore

$$\text{Induced mass per cycle} = (1.4675 \times 10^{-3}) - (0.0691 \times 10^{-3})$$
$$= 1.398 \times 10^{-3} \text{ kg}$$

i.e.   Mass flow rate of air supplied $= 1.398 \times 10^{-3} \times 300 = 0.42 \text{ kg/min}$

Air motors can be rotary in action and are then similar in form to their compressor counterparts, see section 12.5. Figure 12.29(a), (b), and (c) show the forms of the performance characteristics of a small, 0.3 kW, vane type air motor in terms of power/speed, torque/speed, and air consumption/speed. An air motor which receives air from a constant pressure supply can be controlled to meet the load requirements by fitting a restrictor either before or after the motor. It can be shown by a consideration of a simplified p–V diagram, neglecting clearance, that fitting the restrictor before the motor requires a lower airflow than fitting it after. The reader should establish this for himself and also show that the airflow rate required is approximately proportional to the supply pressure to the motor for a given duty. Figure 12.30 shows the results of a test on a small air motor which gives a 25% reduction in air requirement if the restrictor is on the inlet side to the motor.

Fig. 12.29
Characteristics of a small vane-type air motor: (a) power–speed, (b) torque–speed, (c) air consumption–speed

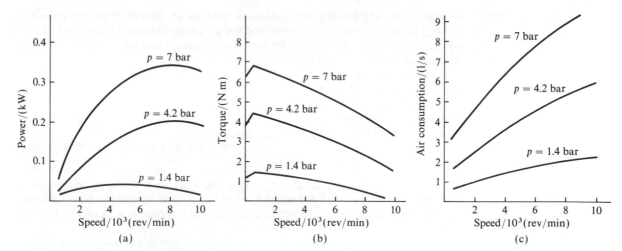

Fig. 12.30 Test results
for a vane-type air
motor with restrictor
control (no load)

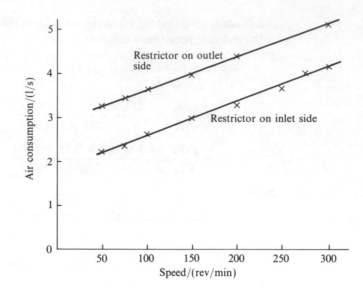

Problems

12.1 Air is to be compressed in a single-stage reciprocating compressor from 1.013 bar and 15 °C to 7 bar. Calculate the indicated power required for a free air delivery of 0.3 m³/min, when the compression process is as follows:
  (i) isentropic;
 (ii) reversible isothermal;
(iii) polytropic, with $n = 1.25$.
What will be the delivery temperature in each case?
(1.31 kW; 0.98 kW; 1.20 kW; 227.3 °C; 15 °C; 150.9 °C)

12.2 The compressor of Problem 12.1 is to run at 1000 rev/min. If the compressor is single-acting and has a stroke/bore ratio of 1.2/1, calculate the bore size required.
(68.3 mm)

12.3 A single-stage, single-acting air compressor running at 1000 rev/min delivers air at 25 bar. For this purpose the induction and free air conditions can be taken as 1.013 bar and 15 °C, and the FAD as 0.25 m³/min. The clearance volume is 3% of the swept volume and the stroke/bore ratio is 1.2/1. Calculate:
  (i) the bore and stroke;
 (ii) the volumetric efficiency;
(iii) the indicated power;
(iv) the isothermal efficiency.
Take the index of compression and re-expansion as 1.3.
(73.2 mm; 87.8 mm; 67.7%; 2 kW; 67.7%)

12.4 The compressor of Problem 12.3 has actual induction conditions of 1 bar and 40 °C, and the delivery pressure is 25 bar. Taking the bore and stroke as calculated in Problem 12.3, calculate the FAD referred to 1.013 bar and 15 °C and the indicated power required. Calculate also the volumetric efficiency and compare it with that of Problem 12.3.
(0.226 m³/min; 1.98 kW; 61.2%; 67.7%)

12.5 A single-acting compressor is required to deliver air at 70 bar from an induction pressure of 1 bar, at the rate of 2.4 m^3/min measured at free air conditions of 1.013 bar and 15 °C. The compression is carried out in two stages with an ideal intermediate pressure and complete intercooling. The clearance volume is 3% of the swept volume in each cylinder and the compressor speed is 750 rev/min. The index of compression and re-expansion is 1.25 for both cylinders and the temperature at the end of the induction stroke in each cylinder is 32 °C. The mechanical efficiency of the compressor is 85%. Calculate:
  (i) the indicated power required;
  (ii) the saving in power over single-stage compression between the same pressures;
  (iii) the swept volume of each cylinder;
  (iv) the required power output of the drive motor.
  (22.74 kW; 5.98 kW; 0.00396 m^3, 0.000474 m^3; 26.75 kW)

12.6 For the compressor of Problem 12.5 calculate the heat rejected per minute to the jacket cooling water of each stage, and the heat rejected per minute to the intercooler. Assume that 50% of the friction power in each stage is transferred to the jacket cooling water.
  (264 kJ/min; 478 kJ/min)

12.7 A single-cylinder, single-acting air compressor of 200 mm bore by 250 mm stroke is constructed so that its clearance can be altered by moving the cylinder head, the stroke being unaffected.
  (a) Using the data below calculate:
  (i) the free air delivery;
  (ii) the power required from the drive motor.

Data Clearance volume set at 700 cm^3; rotational speed, 300 rev/min; delivery pressure, 5 bar; suction pressure and temperature, 1 bar and 32 °C; free air conditions, 1.013 bar and 15 °C; index of compression and re-expansion, 1.25; mechanical efficiency, 80%.
  (b) To what minimum value can the clearance volume be reduced when the delivery pressure is 4.2 bar, assuming that the same driving power is available and that the suction conditions, speed, value of index, and mechanical efficiency, remain unaltered?
  (1.68 m^3/min; 7.1 kW; 458 cm^3)

12.8 A single-acting, single-cylinder air compressor running at 300 rev/min is driven by an electric motor. Using the data given below, and assuming that the bore is equal to the stroke, calculate:
  (i) the free air delivery;
  (ii) the volumetric efficiency;
  (iii) the bore and stroke.

Data Air inlet conditions, 1.013 bar and 15 °C; delivery pressure, 8 bar; clearance volume, 7% of swept volume; index of compression and re-expansion, 1.3; mechanical efficiency of the drive between motor and compressor, 87%; motor power output, 23 kW.
  (4.47 m^3/min; 72.7%; 297 mm)

12.9 A two-stage air compressor consists of three cylinders having the same bore and stroke. The delivery pressure is 7 bar and the FAD is 4.2 m^3/min. Air is drawn in at 1.013 bar, 15 °C and an intercooler cools the air to 38 °C. The index of compression is 1.3 for all three cylinders. Neglecting clearance calculate:
  (i) the intermediate pressure;
  (ii) the power required to drive the compressor;
  (iii) the isothermal efficiency.

  (2.19 bar; 16.2 kW; 84.5%)

12.10 A four-stage compressor works between limits of 1 bar and 112 bar. The index of compression in each stage is 1.28, the temperature at the start of compression in each stage is 32 °C, and the intermediate pressures are so chosen that the work is divided equally among the stages. Neglecting clearance, calculate:

(i) the temperature at delivery from each stage;

(ii) the volume of free air delivered per kilowatt-hour at 1.013 bar and 15 °C;

(iii) the isothermal efficiency.

$$(122 °C; 6.23 m^3/kW h; 87.6\%)$$

12.11 A single-cylinder, single-acting reciprocating air compressor supplies a water-cooled receiver from which the air is drawn off for process work. Taking the polytropic index of compression and re-expansion as 1.3, and using the data below, calculate:

(i) the pressure in the receiver;

(ii) the rate of heat rejection from the receiver;

(iii) the volumetric efficiency of the compressor;

(iv) the required power input to the compressor.

Data   Cylinder bore, 200 mm; stroke, 250 mm; rotational speed, 440 rev/min; clearance volume, 5% of swept volume; ambient pressure and temperature in compressor house, 1.01 bar and 10 °C; average pressure and temperature during the induction stroke, 1 bar and 20 °C; volume flow rate of air drawn off for process work, 0.6 m³/min at 17 °C.

Note: Use a trial-and-error method for part (i).

$$(5 bar; 8.14 kW; 83.9\%; 9.85 kW)$$

12.12 Air at 1.013 bar and 15 °C is to be compressed at the rate of 5.6 m³/min to 1.75 bar. Two machines are considered: (a) the Roots blower; and (b) a sliding vane rotary compressor. Compare the powers required, assuming for the vane type that internal compression takes place through 75% of the pressure rise before delivery takes place, and that the compressor is an ideal uncooled machine.

$$(6.88 kW; 5.71 kW)$$

12.13 Air is compressed in a two-stage vane-type compressor from 1.013 bar to 8.75 bar. Using the data below, and assuming equal pressure ratios in each stage, that compression is complete in each stage, that the machine operates in an ideal manner, and is uncooled apart from the intercooler, calculate:

(i) the power required;

(ii) the volume flow rate measured at the delivery pressure.

Data   Free air delivery, 42 m³/min at 1.013 bar and 15 °C; intercooling between stages is 75% complete.

$$(187 kW; 7.21 m^3/min)$$

12.14 The following particulars refer to a single-acting air motor: cylinder diameter 380 mm; stroke 610 mm; speed 200 rev/min; supply pressure and temperature 6.2 bar and 150 °C; back pressure 1.03 bar; index of expansion and compression 1.35; cut-off ratio 0.46; clearance volume 20% of swept volume; mechanical efficiency 95%.

Assuming that the temperature and pressure of the air in the clearance space at the beginning of admission are 6.2 bar and 150 °C, calculate:

(i) the air consumption;

(ii) the air temperature after blow-down;

(iii) the fraction of stroke travelled by the piston before recompression begins;

(iv) the shaft power developed.

$$(0.54 kg/s; -14.3 °C; 0.463; 72.9 kW)$$

Reference

12.1 BS 1571 *Testing of Positive Displacement Compressors and Exhausters* Part I (1987), Part II (1984).

13

Reciprocating Internal-combustion Engines

Theoretical power cycles are considered in Chapter 5, and the $p-v$ diagrams analysed are similar to those obtained from actual reciprocating engines. There are, however, fundamental mechanical and thermodynamic differences between the cycles, which make comparison less valuable than might be expected.

In the theoretical cycles there is no chemical change in the working fluid, which is assumed to be air, and the heat exchanges in the cycle are made externally to the working fluid. In the practical cycle the heat supply is obtained from the combustion of a fuel in air and thus the air charge is consumed during combustion and the combustion products must be exhausted from the cylinder before a fresh charge of air can be induced for the next cycle. The practical cycle consists of the exhaust and induction processes together with the compression and expansion processes as in the theoretical cycle. Further differences between the ideal and the actual cycles are discussed in section 13.8.

The reciprocating engine mechanism consists of a piston which moves in a cylinder and forms a movable gas-tight plug, a connecting-rod and a crankshaft (see Fig. 13.1). If the engine has more than one cylinder then the cylinders, pistons, etc. are identical, and all the connecting-rods are fastened to a common crankshaft. The angular positions of the crank-pins are such that the cylinders contribute their power strokes in a selected and regular sequence. By means of this arrangement the reciprocating motion of the piston is converted to rotary motion at the crankshaft.

There are many types and arrangements of engines, and some classification is necessary to describe a particular engine adequately. The methods of classification are as follows:

(i) By the fuel used and the way in which the combustion is initiated. Petrol engines and gas engines have spark ignition (SI). Diesel engines or oil engines have compression ignition (CI). In the SI engine the air and the fuel are mixed before compression. In the CI engine the air only is compressed, and the fuel is injected into the air which is then at a sufficiently high temperature to initiate combustion.

(ii) By the way in which the cycle of processes is arranged. This is defined by the number of complete strokes of the piston required for one complete

Fig. 13.1 Reciprocating
IC engine

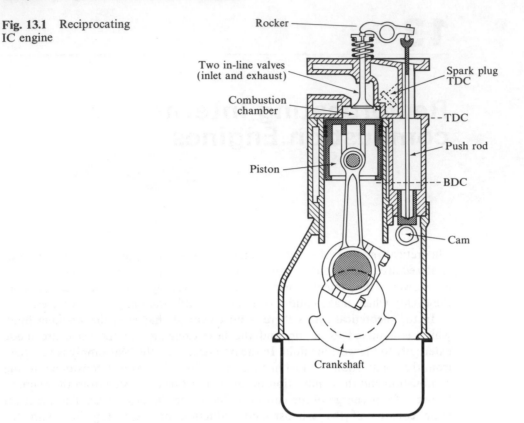

Rocker

Two in-line valves
(inlet and exhaust)

Combustion
chamber

Piston

Spark plug
TDC

TDC

Push rod

BDC

Cam

Crankshaft

cycle. The stroke of the piston is the distance it moves from the position most extreme from the crankshaft to that nearest it. This takes place over half a revolution of the crankshaft. In petrol engine practice the extreme positions of the piston are referred to as *top dead centre* (TDC), and *bottom dead centre* (BDC) (see Fig. 13.1). In oil-engine practice they are referred to as *outer dead centre* and *inner dead centre* respectively. An engine which requires four strokes of the piston (i.e. two revolutions of the crankshaft) to complete its cycle is called a *four-stroke cycle engine*. An engine which requires only two strokes of the piston (i.e. one crankshaft revolution) is called a *two-stroke cycle engine*.

In all reciprocating internal-combustion (IC) engines the gases are induced into and exhausted from the cylinder through ports, the opening and closing of which are related to the piston position. In a two-stroke engine the ports can be opened or closed by the piston itself, but in the four-stroke engine a separate shaft, called the camshaft, is required; this is driven from the crankshaft through a 2 to 1 speed reduction. The cams on this shaft operate valves, called poppet valves, either directly or by means of push rods. Modern high-speed petrol engines have two camshafts, one operating the exhaust valves, and the other operating the inlet valves. The timing of the valves and the point of ignition are fundamental to the engine performance, and the specified timing

IC   Inlet valve closes. This occurs 20–40° after BDC to take advantage of the momentum of the rapidly moving gas.

S   Spark occurs. This is 20–40° before TDC when the ignition is fully advanced, and is at TDC when the ignition is fully retarded.

EO   Exhaust valve opens. The average value of this position is about 50° before BDC, but it is greater than this in racing-car engines.

EC   Exhaust valve closes. This occurs 0° to 10° after TDC.

There may be an overlap between IO and EC such that both valves are open at the same instant.

It should be appreciated that the values of angular position quoted are average ones, and considerable differences occur between different engines. The points shown are the normal opening and closing positions for each valve. The time required to open and close the valves means that each valve will be fully open for a crank angle movement much less than that indicated by the timing diagram. Another point to be borne in mind is that at the dead centre positions a considerable amount of crank movement produces a small corresponding movement of the piston; this is shown by a consideration of crank-connecting rod geometry.

Crank angle displacements can be translated into time values if the engine rotational speed is known and can be assumed to be constant.

The p–V diagram for a CI engine with mechanical or solid injection of the fuel is shown in Fig. 13.4. The ideal cycle for the engine is the dual combustion cycle (see section 5.8) in which the heat input is partly at approximately constant volume, and partly at approximately constant pressure. With modern high-speed engines the fuel injection is well advanced and the greater proportion of fuel is burnt at approximately constant volume. A diagram from the same engine with a later point of injection would be flatter at the top, as indicated by the dotted line in Fig. 13.4. The shape of the diagram is also influenced by the design of the combustion chamber.

Fig. 13.4 Pressure–volume diagram for a CI engine

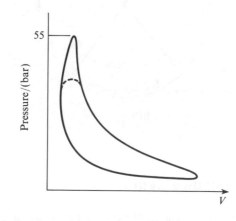

With engines having the fuel injected by means of an air blast, the p–V diagram is similar to that shown in Fig. 13.5. In this cycle the air which enters the cylinder with the fuel on injection helps to maintain a constant pressure over the early part of the return stroke, and this increases the area of the

Fig. 13.5 Pressure–
volume diagram for a
blast-injection CI engine

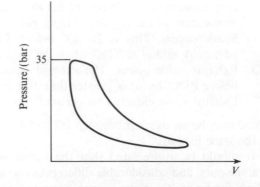

diagram. This method of injection is now practically obsolete due to the
difficulties and cost of supplying the high-pressure air required.

Figure 13.6 shows a typical timing diagram for a four-stroke oil engine which
has average values for the valve positions as follows:

IO   Up to 30° before TDC.
IC   Up to 50° after BDC.
EO   About 45° before BDC.
EC   About 30° after TDC.
Injection   About 15° before TDC.

Fig. 13.6 Timing
diagram for a
four-stroke CI engine

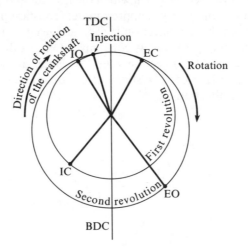

13.2   Two-stroke cycle

Figure 13.7 represents the cylinder of a two-stroke petrol engine with crankcase
compression. As the piston ascends on the compression stroke the next charge
is drawn into the crankcase, C, through the spring-loaded automatic valve, S.
Ignition occurs before TDC, and at TDC the working stroke begins. As the
piston descends through about 80% of the working stroke, the exhaust port,

Fig. 13.7 Two-stroke SI engine with crankcase compression

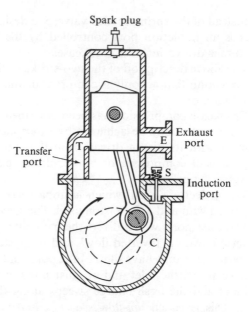

E, is uncovered by the piston and exhaust begins. The transfer port, T, is uncovered later in the stroke due to the shape of the piston or the position of the port in relation to the port E, and the charge in the crankcase, C, which has been compressed by the descending piston, enters the cylinder through the port T.

The piston can be shaped to deflect the fresh gas across the cylinder to assist the scavenging of the cylinder; this is called *cross-flow scavenge*. As the piston rises, the transfer port, T, is closed slightly before the exhaust port E, and after E is closed compression of the charge in the cylinder begins. The $p-V$ diagram and the timing diagram for a two-stroke petrol engine are shown in Fig. 13.8(a) and (b).

Fig. 13.8 Pressure–volume (a) and timing (b) diagrams for a two-stroke SI engine. (I) inlet angle (80° approx.); (E) exhaust angle (120° approx.); (T) transfer angle (100° approx.)

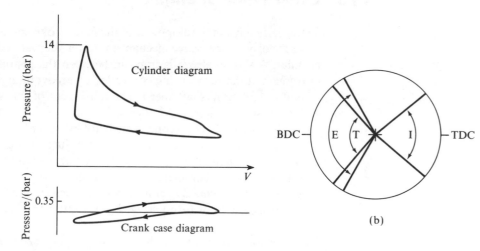

Instead of the spring-loaded valve, S, a design with a third port may be used. This is an induction port controlled by the piston, and through which the mixture is drawn into the crankcase.

The above description of the two-stroke cycle applies also to CI engines with the exception that air only is compressed, and the sparking plug is replaced by a fuel injector.

Crankcase compression has been described but the scavenging and charging of the cylinder may be achieved by other means. A separately phased pump cylinder with its piston driven from the crankshaft may be used. A positive displacement compressor or blower driven from the engine is a third way of charging the cylinder.

The deflector piston, which is unbalanced and can cause 'rattle', may be dispensed with and a flat piston used. The scavenging is then obtained by using two transfer ports which divert the incoming air up the cylinder. This is called 'reverse flow', or 'inverted flow', and the system is called *loop scavenge*.

In engines which have simple inlet ports and poppet or sleeve valve controlled exhaust ports, the inlet and exhaust ports are placed at opposite ends of the cylinder and the fresh charge sweeps along the cylinder towards the exhaust port. This is called *uni-flow scavenge* and is applied with great mechanical simplicity in opposed piston engines.

For several reasons the two-stroke cycle has more application in the CI field than in the SI field, especially for stationary constant-speed engines. In such engines a number of ingenious arrangements have been patented in an attempt to dispense with the scavenge blower. Some designs have used the Kadenacy effect, which employs the high vacuum created by suddenly releasing the exhaust gas through large-area, sharp-edged ports. With constant-speed engines it is possible to 'tune' the exhaust system such that the HP exhaust gas leaving one cylinder can be used to 'pack' another cylinder which is on the early part of its induction stroke.

13.3 Other types of engine

In the early days of development of the four-stroke engine one of the difficulties was the noisy poppet valve mechanism. As an alternative the *sleeve valve* became popular. A sliding sleeve is fitted in between the piston and the cylinder, the movement of the sleeve being controlled by an overhung crank-pin driven from a shaft at half crankshaft speed. The movement of this valve controls the inlet and exhaust ports in the cylinder.

Engines have been developed which are called *multi-fuel engines*; such engines will run on any petroleum fuel from diesel fuel to premium petrol. The main application of such engines is for military purposes and it is unlikely that they will have extensive commercial application.

The *dual-fuel engine* is of considerable industrial interest. Some diesel engines, naturally aspirated or turbocharged (see section 13.12) can be used as dual-fuel engines. They are started on diesel oil and then run on an available gaseous fuel such as methane, natural gas, sewage gas, coal gas, etc. The combustion process requires a pilot injection of oil which amounts to 7–10% of the full power supply when running as a diesel engine. The changeover from diesel fuel to gas can be done automatically or manually. To supply the pilot injection of oil a second set of pumps is required which delivers fuel to the standard injectors.

A considerable amount of effort has been put into the development of the *free piston engine*. In this type of engine the crankshaft and connecting-rod are dispensed with, and two opposed but connected pistons operating in the same cylinder are used. The free piston engine is described more fully in section 13.14. One of the main applications of the engine is that instead of having a separate air compressor driven by an IC engine the compressor and the engine can be combined with the result that the intermediate rotating shafts are eliminated and a more compact unit is obtained. This is especially important for portable air compressors which are used extensively. Another use for the free piston engine is as a 'gasifier' from which the gaseous products of combustion are exhausted at a suitable pressure and allowed to expand through a gas turbine. No power output is taken from the free piston engine, the power output of the unit being that obtained from the turbine. The potential field for the smaller size free piston engine is regarded as being with road, rail, and tracked vehicles, earth-moving equipment, high-speed marine craft, cargo ships, and in generating stations.

Some classification of IC engines has been given, but this is not exhaustive. The applications for such engines are wide, both in the type of duty to be performed and in the power required. Modern developments promise the application of IC engines not only as individual units but as part of a complete plant to suit some specialized purpose.

13.4 Criteria of performance

An engine is selected to suit a particular application, the main consideration being its power/speed characteristics. Important additional factors are initial capital cost and running cost. In order that different types of engines or different engines of the same type may be compared, certain performance criteria must be defined. These are obtained by measurement of the quantities concerned during bench tests, and calculation is by standard procedures. The results are plotted graphically in the form of performance curves.

Indicated power (ip)

This is defined as the rate of work done by the gas on the piston as evaluated from an indicator diagram obtained from the engine. An indicator diagram has the form shown in Fig. 13.9. Figure 13.9 shows both the power and the pumping loops.

The mean effective pressure has been defined in section 5.9 and may be applied here.

Net work done per cycle $\propto$ (area of power loop − area of pumping loop)

427

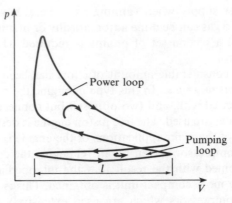

Fig. 13.9 Pressure–volume diagram for a reciprocating engine

Therefore indicated mean effective pressure, p_i, is given by

$$p_i = \frac{\text{net area of diagram}}{\text{length of diagram}} \times \text{constant} \tag{13.1}$$

(the constant depends on the scales of the recorder).

Considering one engine cylinder

$$\text{Work done per cycle} = p_i \times A \times L$$

where A is the area of piston and L the length of stroke.

Power output/unit time = work done per cycle × cycles per minute

or $ip = p_i AL \times (\text{cycles/unit time})$

The number of cycles per unit time depends on the type of engine; for four-stroke engines the number of cycles per unit time is $N/2$, and for two strokes the number of cycles per unit time is N, where N is the engine speed. The formula for ip then becomes for four-stroke engines

$$ip = \frac{p_i ALNn}{2} \tag{13.2}$$

for two-stroke engines

$$ip = p_i ALNn \tag{13.3}$$

where n is the number of cylinders.

The main recorder in use is the electronic engine indicator, which uses a cathode-ray oscilloscope (CRO) to display the pressure trace obtained (see Fig. 13.10).

Modern methods of instrumentation, data acquisition, and analysis, based on recent advances in electronics, digital methods, and the application of microprocessors, have become applicable to engine testing, development, and control. Engine test and control systems have become increasingly 'computerized' and it is regretted that space requirements do not allow a comprehensive treatment of this very important aspect of engineering in this book. Readers are advised to obtain copies of the excellent publications by engine and equipment supplies which will give a state-of-the-art description of the latest equipment and techniques.

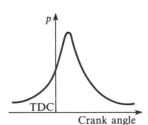

Fig. 13.10 Pressure against crank angle for a typical reciprocating IC engine

Brake power (bp)

This is the measured output of the engine. The engine is connected to a brake or dynamometer which can be loaded in such a way that the torque exerted by the engine can be measured. The dynamometer may be of the absorption or the transmission type. Absorption dynamometers are the more usual and can be classified as: (i) friction type, used for the smaller powered, lower-speed engines; (ii) hydraulic; (iii) electrical; (iv) air-fan type. With types (i), (ii), and (iii), the torque is obtained by reading off a net load, W, at a known radius, R, from the axis of rotation, and hence the torque, T, is given by

$$T = WR \tag{13.4}$$

The brake power is then given by

$$\text{bp} = 2\pi NT \tag{13.5}$$

In the transmission type of dynamometer the torque, T, transmitted by the driving shaft is measured directly, and the bp is obtained by substitution in equation (13.5). With air fans the torque is obtained from a calibration curve for the fan.

Friction power (fp) and mechanical efficiency, η_M

The difference between the ip and the bp is the friction power (fp), and is that power required to overcome the frictional resistance of the engine parts,

i.e. $$\text{fp} = \text{ip} - \text{bp} \tag{13.6}$$

The mechanical efficiency of the engine is defined as

$$\text{Mechanical efficiency, } \eta_M = \frac{\text{bp}}{\text{ip}} \tag{13.7}$$

where η_M usually lies between 80 and 90%.

The fp is very nearly constant at a given engine speed; if the load is decreased giving lower values of bp, then the variation in η_M with bp is as shown in Fig. 13.11. At zero bp at the same speed the engine is developing just sufficient power to overcome the frictional resistance.

Mechanical efficiency depends on the ip and bp, and is therefore found by evaluating these experimentally. A considerable amount of literature has been published dealing with mechanical efficiency and the analysis of engine power losses. A useful report on this subject is that of ref. 13.1, and the conclusion is that no single method is satisfactory in every respect for the evaluation of mechanical efficiency. The four main methods are as follows:

(i) Measurement of the ip and the bp by the means already described in the preceding subheadings.
(ii) Measurement of the bp at a given speed followed by 'motoring' of the engine with the fuel supply cut off. This method can only be used on engines

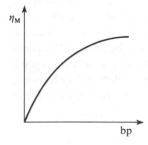

Fig. 13.11 Variation of mechanical efficiency with brake power

429

with an electrical dynamometer, the dynamometer being used as a motor instead of as a generator in order to motor the engine at the firing speed. The torque can be measured under firing and under motoring conditions and the mechanical efficiency evaluated. The fact that an electrical dynamometer can be used to find the mechanical efficiency in this way is one of the main advantages of this type of dynamometer.

(iii) The Morse test: this is only applicable to multicylinder engines. The engine is run at the required speed and the torque is measured. One cylinder is cut out, by shorting the plug if an SI engine is under test, or by disconnecting an injector if a CI engine is under test. The speed falls because of the loss of power with one cylinder cut out, but is restored by reducing the load. The torque is measured again when the speed has reached its original value. If the values of ip of the cylinders are denoted by I_1, I_2, I_3, and I_4 (considering a four-cylinder engine), and the power losses in each cylinder are denoted by L_1, L_2, L_3, and L_4, then the value of bp, B, at the test speed with all cylinders firing is given by

$$B = (I_1 - L_1) + (I_2 - L_2) + (I_3 - L_3) + (I_4 - L_4)$$

If number 1 cylinder is cut out, then the contribution I_1 is lost; and if the losses due to that cylinder remain the same as when it is firing, then the bp, B_1, now obtained at the same speed is

$$B_1 = (0 - L_1) + (I_2 - L_2) + (I_3 - L_3) + (I_4 - L_4)$$

Subtracting the second equation from the first gives

$$B - B_1 = I_1 \qquad (13.8)$$

By cutting out each cylinder in turn the values I_2, I_3, and I_4 can be obtained from equations similar to (13.8). Then for the engine

$$I = I_1 + I_2 + I_3 + I_4 \qquad (13.9)$$

(iv) 'Willan's line': this method is applicable to CI engines only. At a constant engine speed the load is reduced in increments and the corresponding bp and gross fuel consumption readings are taken. A graph is then drawn of fuel consumption against bp, as in Fig. 13.12. The graph drawn is called the 'Willan's line', and is extrapolated back to cut the bp axis at the point A. The reading OA is taken as the power loss of the engine at that speed. The fuel consumption at zero bp is given by OB; if the relationship between fuel consumption and bp is assumed to be linear, then a fuel consumption OB is equivalent to a power loss of OA.

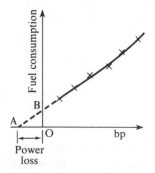

Fig. 13.12 Fuel consumption against brake power for a CI engine

Brake mean effective pressure (bmep), thermal efficiency, and fuel consumption

The bp of an engine can be obtained accurately and conveniently using a dynamometer.

From equation (13.7)

$$\text{bp} = \eta_M \times \text{ip}$$

Therefore, substituting for ip from equation (13.2) for a four-stroke engine, we have

$$\text{bp} = \frac{\eta_M \times p_i ALNn}{2}$$

Since η_M and p_i are difficult to obtain they may be combined and replaced by a brake mean effective pressure, p_b,

i.e. $\qquad \text{bp} = \frac{p_b ALNn}{2}$ \hfill (13.10)

where $p_b = \eta_M \times p_i$.

The bmep may be thought of as that mean effective pressure acting on the pistons which would give the measured bp if the engine were frictionless. The bmep is a useful criterion for comparing engine performance. Taking the two equations for bp, equations (13.5) and (13.10), and putting them together we have

$$\frac{p_b ALNn}{2} = 2\pi NT$$

therefore

$$p_b = K \times T$$

where K is a constant.

Therefore bmep is directly proportional to the engine torque and is independent of the engine speed.

The power output of the engine is obtained from the chemical energy of the fuel supplied. The overall efficiency of the engine is given by the *brake thermal efficiency*, η_{BT},

i.e. $\qquad \eta_{BT} = \dfrac{\text{brake work}}{\text{energy supplied}}$

therefore

$$\eta_{BT} = \frac{\text{bp}}{\dot{m}_f \times Q_{net,v}}$$ \hfill (13.11)

where $\dot{m}_f$ is the mass of fuel consumed per unit time, and $Q_{net,v}$ the net calorific value of the fuel.

The *specific fuel consumption* (sfc) is the mass flow rate of fuel consumed per unit power output, and is a criterion of economical power production,

i.e. $\qquad \text{sfc} = \dfrac{\dot{m}_f}{\text{bp}}$ \hfill (13.12)

The *indicated thermal efficiency*, η_{IT}, is defined in a similar way to η_{BT},

i.e. $\qquad \eta_{IT} = \dfrac{\text{ip}}{\dot{m}_f \times Q_{net,v}}$ \hfill (13.13)

Dividing equation (13.11) by equation (13.13) gives

$$\frac{\eta_{BT}}{\eta_{IT}} = \frac{bp}{ip} = \eta_M$$

therefore

$$\eta_{BT} = \eta_M \times \eta_{IT} \qquad (13.14)$$

Example 13.1   A four-cylinder petrol engine has a bore of 57 mm and a stroke of 90 mm. Its rated speed is 2800 rev/min and it is tested at this speed against a brake which has a torque arm of 0.356 m. The net brake load is 155 N and the fuel consumption is 6.74 l/h. The specific gravity of the petrol used is 0.735 and it has a lower calorific value, $Q_{net,v}$ of 44 200 kJ/kg. A Morse test is carried out and the cylinders are cut out in the order 1, 2, 3, 4, with corresponding brake loads of 111, 106.5, 104.2, and 111 N, respectively. Calculate for this speed, the engine torque, the bmep, the brake thermal efficiency, the specific fuel consumption, the mechanical efficiency and the imep.

Solution   Using equation (13.4)

$$\text{Torque, } T = WR = 155 \times 0.356 = 55.2 \text{ N m}$$

Using equation (13.5),

$$bp = 2\pi NT = \frac{2\pi \times 2800 \times 55.2}{60 \times 10^3} \text{ kN m/s} = 16.2 \text{ kW}$$

From equation (13.10),

$$bmep = \frac{bp \times 2}{ALNn} = \frac{16.2 \times 2 \times 4 \times 60 \times 10^3}{\pi \times 0.057^2 \times 0.09 \times 2800 \times 4 \times 10^5}$$

$$= 7.55 \text{ bar}$$

Using equation (13.11)

$$\eta_{BT} = \frac{bp}{\dot{m}_f \times Q_{net,v}} = \frac{16.2}{0.001\,376 \times 44\,200} = 0.266 \text{ or } 26.6\%$$

where $\dot{m}_f = (6.74/3600) \times 1 \times 0.735 = 0.001\,376 \text{ kg/s}$.

Using equation (13.12),

$$sfc = \frac{\dot{m}_f}{bp} = \frac{0.001\,376}{16.2} = 0.000\,085 \text{ kg/kJ}$$

It is more convenient to express sfc in terms of fuel consumption rate per unit power and to express the fuel consumption rate in kilograms per hour rather than kg/s,

i.e. $sfc = 0.000\,085 \times 3600 = 0.306 \text{ kg/kW h}$

Using equation (13.8) for each cylinder in turn, and substituting brake loads instead of the values of bp since the speed is constant, we have

$$I_1 = B - B_1 = 155 - 111 = 44\text{ N}$$

$$I_2 = B - B_2 = 155 - 106.5 = 48.5\text{ N}$$

$$I_3 = B - B_3 = 155 - 104.2 = 50.8\text{ N}$$

$$I_4 = B - B_4 = 155 - 111 = 44\text{ N}$$

Hence for the engine, the indicated load, I, is given by

$$I = I_1 + I_2 + I_3 + I_4 = 44 + 48.5 + 50.8 + 44 = 187.3\text{ N}$$

Therefore from equation (13.7)

$$\eta_M = \frac{\text{bp}}{\text{ip}} = \frac{155}{187.3} = 0.828 \text{ or } 82.8\%$$

From the definition of bmep given by equation 13.10, we have

$$\text{bmep} = \eta_M \times \text{imep}$$

i.e. $\quad \text{imep} = \dfrac{7.55}{0.828} = 9.12\text{ bar}$

Volumetric efficiency, η_v

The power output of an IC engine depends directly upon the amount of charge which can be induced into the cylinder. This is referred to as the *breathing capacity* of the engine and is expressed quantitatively by the *volumetric efficiency*, which is defined as for reciprocating compressors by equations (12.14) and (12.15). For IC engines, the volumetric efficiency is the ratio of the volume of air induced, measured at the free air conditions to the swept volume of the cylinder,

i.e. $\quad \eta_v = \dfrac{V}{V_s}$ $\hfill$ (13.15)

The power output of an engine depends on its capacity to breathe, and if a particular engine had a constant thermal efficiency then its output would be in proportion to the amount of air induced. The volumetric efficiency with normal aspiration is seldom above 80%, and to improve on this figure, *supercharging* is used. Air is forced into the cylinder by a blower or fan which is driven by the engine. More will be said about supercharging in section 13.12.

The volumetric efficiency of an engine is affected by many variables such as compression ratio, valve timing, induction and port design, mixture strength, specific enthalpy of vaporization of the fuel, heating of the induced charge, cylinder temperature, and the atmospheric conditions.

Example 13.2 In Example 13.1 an analysis of the dry exhaust showed no oxygen and negligible carbon monoxide. The engine was tested in an atmosphere at 1.013 bar and 15 °C. Estimate the volumetric efficiency of the engine.

Solution The condition of the exhaust implies a stoichiometric air–fuel ratio which for petrols can be taken to be 14.5/1.

From Example 13.1

$$\dot{m}_f = 0.001\,376 \text{ kg/s}$$

therefore

$$\text{Air mass flow rate} = 14.5 \times 0.001\,376 = 0.019\,95 \text{ kg/s}$$

therefore

$$\text{Volume drawn in per unit time, } \dot{V} = \frac{0.019\,95 \times 287 \times 288}{10^5 \times 1.013}$$

$$= 0.0163 \text{ m}^3/\text{s}$$

Now

$$\text{Swept volume of engine} = ALn \text{ m}^3/\text{cycle} = \frac{ALnN}{2} \text{ m}^3/\text{min}$$

i.e.

$$\dot{V}_s = \frac{\pi \times 0.057^2 \times 0.09 \times 4 \times 2800}{4 \times 2 \times 60} = 0.0214 \text{ m}^3/\text{s}$$

Then using equation (13.15)

$$\eta_v = \frac{\dot{V}}{\dot{V}_s} = \frac{0.0163}{0.0214} = 0.76 \text{ or } 76\%$$

13.5 Engine output and efficiency

The power output of an engine depends on the conditions under which it is tested. In order to make reported performances acceptable and comparable, standard procedures are established which define the quantities to be measured, the methods of measurement to be used, and the procedure to be adopted for reporting. The standards which apply in the UK are defined in British Standard (BS) 5514 (ref. 13.2) which is equivalent to the corresponding standards ISO 3046 of the International Organization for Standardization (ISO). In the United States the standards are those of the Society of Automotive Engineers (SAE) and are described in the *SAE Handbook, 1984*, volume 3 or its replacement. In Germany the standards are those of the Deutsche Industrie Norm (DIN) and in the UK also the power and torque outputs of engines are often quoted to the DIN procedures. Standards are withdrawn or updated from time to time and the different standards may become fewer as a single international standard emerges.

The power output changes with the atmospheric conditions of pressure, temperature, and humidity and if the test conditions are not those of the relevant standard the readings of power and fuel consumption taken must be corrected to the defined conditions and procedures of the standard. The power output also depends on the form in which the engine is tested with regard to the auxiliaries with which it is fitted, i.e. fully or partly equipped. Diesel engine power, torque and fuel consumption values or complete characteristics are readily available; for petrol engines maximum values of power (kilowatts) and torque (Newton metres) are quoted, but fuel consumption figures are not so readily come by. For private vehicle use the overall vehicle performance figures are quoted as power and torque to the specified standard, usually DIN, and the fuel consumption figures to official tests under the Passenger Car Fuel Consumption Order 1983 in litres/100 km at the specified test speeds of 90 and 120 km/h (56 and 75 mph).

For example, for one particular engine the power (in kilowatts) and torque (in Newton metres) outputs quoted to different test standards were respectively 69 and 359 to BS 5514, 72 and 377 to DIN 70020, 79 and 369 to BS AU 1141a, and 72.6 and 366 to BS 649, showing a considerable variation. This range of variation of the main performance parameters indicates the necessity for a rationalization of engine test standards.

Table 13.1 details the power and torque performances of engines (mainly petrol) to DIN ratings and the fuel consumptions of vehicles in which they are fitted. Some basic engines appear in different forms with different carburation or fuel injection systems, have alternative forms of ignition and may even be supercharged by means of a turbocharger (see section 13.12). Similarly, the same engine may be available in different forms of vehicle, e.g. saloon, estate or cabriolet which have different aerodynamic characteristics with corresponding vehicle performance figures. The comparison of engine power outputs by the criterion power/volume should be done with the understanding that for a given total engine capacity the small capacity multicylinder engine will give a higher

Table 13.1 Some private vehicle engine and road performances

Engine size/(litres) [cylinders]	Fuel type/ [compression ratio]	Maximum output			Vehicle fuel consumption (litre/100 m) [mpg]		
		Power/(kW) [speed/(rev/min)]	Torque/(N m) [speed/(rev/min)]	Power–capacity ratio/(kW/litre)	56 mph (90 km/h)	75 mph (120 km/h)	Simulated urban drive
0.96 [4]	Petrol [8.5]	33 [5750]	68 [3700]	34.5	5.2 [54.3]	7.1 [39.8]	7.0 [40.4]
1.60 [4]	Petrol [8.5]	58 [5800]	125 [3300]	36.3	5.6 [50]	7.6 [37]	8.9 [32]
1.60 [4]	Petrol FI [8.5]	77 [6000]	138 [4800]	48.1	6.2 [45.6]	7.9 [37.5]	10.3 [27.7]
1.76 [4] (16 valve)	Petrol [10.0]	99 [6500]	158 [4250]	56.1	6.3 [44.8]	7.6 [37.2]	10.6 [26.6]
2.30 [V6]	Petrol [9.0]	84 [5300]	176 [3000]	36.5	6.8 [41.5]	8.7 [32.5]	12.6 [22.4]
2.30 [4]	Diesel [22.2]	49 [4200]	139 [2000]	21.2	5.05 [56]	6.7 [42.2]	8.35 [33.8]
2.80 [V6]	Petrol FI [9.2]	110 [5700]	216 [4000]	39.3	7.75 [36.5]	10.2 [27.7]	14.9 [19]
2.80 [V6]	Petrol TC [9.2]	151 [5000]	353 [3500]	54.0	—	—	—

FI—fuel injection; TC—turbocharged.

value than an engine with fewer, larger cylinders because the volume is swept more frequently for equal piston speeds.

The last two engines listed are different versions of the same engine and are the power ratings for intermittent running. The continuous running characteristics for the same engines are shown in Fig. 13.21 (p. 443).

The specific fuel consumptions at maximum power vary between 0.20 and 0.25 kg/kWh with somewhat lower values (by 0.005 to 0.010 kg/kWh at the maximum torque condition).

The details quoted in Table 13.2 are for the smaller size diesel engines for industrial and commercial vehicle use. Engines in the higher power range, 1000–3000 kW, are in service in fast patrol boats, for electrical power generation on warships, on offshore oil rigs, in rail traction including high-speed trains and in submarine applications.

Table 13.2 Some diesel engine performances to BS 5514/DIN 6270

Capacity (litre)	Cylinders	Compression ratio	Breathing	Max. power/(kW) at speed/(rev/min)	Max. torque/(N m) at speed/(rev/min)	Power (kW/litre)
2.9	3	17.3	Normal	29 at 2000	158 at 1600	9.9
3.3	3	16.3	Normal	43 at 2200	211 at 1400	13.0
4.4	4	16.3	Normal	57.5 at 2100	284 at 1600	13.0
4.4	4	15.6	Turbocharged	68.7 at 2100	359 at 1600	15.6
6.6	6	16.3	Normal	90 at 2300	436 at 1400	13.6
6.6	6	15.6	Turbocharged	103 at 2200	500 at 1600	15.6
6.6	6	15.6	Turbo + intercooling	130 at 2200	665 at 1600	19.7

Diesel engine competition with the petrol engine for the private car market is an interesting study exercise. The principles of the diesel engine were well known at the start of the century and prior to the Second World War a few diesel-engined cars had been built as conversions from petrol engines. The advantages of the diesel of better fuel consumption and longer engine life were well known, particularly for use in taxis which are used for short journeys and have high annual mileages. For such applications the savings in fuel cost favour the diesel engine but its relative progress has been slow. In the late 1970s only about 69 000 diesel private cars and vans were in use in the UK, but the number rose to about 519 000 in 1986, and currently is in excess of 1 300 000 of which over 650 000 are private cars. The performance of an early (1974) diesel-engined vehicle compared favourably with that of a smaller petrol engine, e.g. a 1.8 litre diesel compared with a 1.3 litre petrol engine gave similar journey times and overall fuel costs of about half those of the petrol engine. The noise level of the diesel engine was not acceptable and the diesel was regarded as being sluggish in comparison with the petrol engine. Modern diesel engine design has narrowed the gap between the two with engines becoming lighter, quieter, and more responsive. The better fuel consumption of the diesel makes it an attractive power unit for the taxi and vehicle fleet owner, provided the price of diesel fuel remains sufficiently lower than that of petrol. Another important factor is the frequency and cost of service work required by the diesel and petrol equivalents. Diesel engines required more frequent servicing, but the intervals between servicing are approaching equality.

13.6 Performance characteristics

The testing of IC engines consists of running them at different loads and speeds and taking sufficient measurements for the performance criteria to be calculated. As well as the measurements required for the criteria of section 13.4 the airflow is required to give the air–fuel ratio and the combustion products can be analysed (see section 7.6).

An energy balance is sometimes presented for an engine, and the heat taken by the cooling water is obtained by measuring the rate of flow of the water and its temperature rise. The outlet temperature of the cooling water is usually limited to about 80 °C to prevent the formation of steam pockets. To estimate the energy of the exhaust gas, an exhaust calorimeter can be fitted; this is simply a heat exchanger in which the exhaust gas is cooled by circulating water, the rate of flow and temperature rise of which are measured. In order to avoid condensation of the steam in the gas, the gas is not usually cooled below about 50 °C.

The items usually included in an energy balance and expressed as percentages of the energy supplied by the fuel (i.e. $\dot{m}_f \times Q_{net,v}$) are as follows: (a) bp; (b) the heat to cooling water; (c) the energy of the exhaust referred to inlet conditions, or as obtained by an exhaust calorimeter; (d) unaccounted losses obtained by difference and which include radiation and convection losses, etc.

The energy balance usually presented is not an accurate account of the energy distribution, but it is a useful one. The bp is conveniently and accurately measured and the percentage of the input energy to the bp is the most important item in the balance. The heat transferred to the cooling water may be used as an indication of how much heat could be usefully obtained if this water were used for other heating purposes in a combined plant.

The energy to exhaust is best obtained by means of the exhaust calorimeter as described. Ideally the exhaust gas should be cooled to the temperature of the inlet air, and the heat taken by the cooling water in the calorimeter per minute would give item (c) of the balance. The temperature at which the exhaust gas enters the calorimeter is most likely not that at which it passes through the exhaust valve, and some of the energy to the exhaust will have been taken by the cooling water or lost to the atmosphere. To obtain item (c) by calculation is more speculative since the gas is chemically different from the inlet air and the mass flow has increased due to the addition of the fuel. The error involved in choosing the datum is likely to be less than that produced by using an inaccurate value for the exhaust temperature, which is not easy to obtain with accuracy. It is sufficient to write

$$\text{Energy to exhaust} = (\dot{m}_a + \dot{m}_f)h_e - \dot{m}_a h_a$$

where $\dot{m}_a$ and $\dot{m}_f$ are the air and fuel mass flow rates, h_e is the specific enthalpy of the exhaust gas (dry exhaust + steam), reckoned from $0\,°C$, and h_a is the specific enthalpy of the air at inlet reckoned from $0\,°C$. A suitable value for c_p for the dry exhaust gas must be calculated or assumed.

For a diesel engine at full load typical values would be: to bp 35%; to cooling water 20%; to exhaust 35%; to radiation, etc. 10%. The heat to the jacket cooling water and exhaust can be utilized in industries which have heating loads such as space and heating and hot-water systems, and which require either steam or hot water for process work. The heat to the jacket water is recoverable and about 18% of the total energy supplied can be recovered from the exhaust gas.

The most elementary power test is that which gives the power–speed and the torque–speed characteristics, as shown in Fig. 13.13. The test is carried out at constant throttle setting in the petrol engine, and at constant fuel pump setting in the CI engine.

Fig. 13.13 Engine characteristics of power, imep, bmep, and mechanical efficiency against engine speed

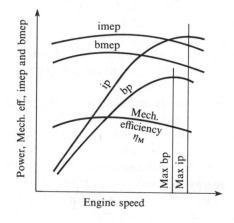

In Fig. 13.13 are shown the engine power characteristics for both ip and bp. As the speed increases from the lower values the two curves are similar, the difference between the ip and the bp at any speed being the fp, which increases with speed. Both curves show maximum values, but they occur at different speeds. The ip falls after the maximum because of a reduction in volumetric efficiency with increased speed. This is influenced by gas temperatures, valve timing, valve mechanism dynamics, and the pressure pulsation patterns in the induction and exhaust manifolds. This fall in volumetric efficiency affects also the bp curve, but this is further decreased by an increase in the fp. This latter effect is predominant since the bp reaches its maximum at a lower speed than the ip.

The variation of volumetric efficiency with speed is indicated in Fig. 13.14, and that of mechanical efficiency with speed in Fig. 13.13. The various methods used for the determination of mechanical efficiency are discussed in section 13.4.

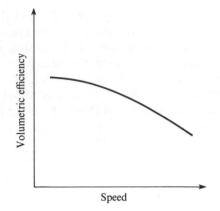

Fig. 13.14 Variation of volumetric efficiency with engine speed

SI engines

These engines are *quantity governed* by the opening or closing of a throttle valve which regulates the mass flow of charge to the cylinders. Some gas engines are throttled by alteration of the lift of the admission valve, and this can be controlled from the engine governor. The governed speed can be adjusted to select any value in its range.

The petrol engine will operate on air–fuel ratios in the range 10/1–22/1, but not necessarily satisfactorily at the extremes. There is some variation between engines. An important test is to run the engine with the air–fuel ratio as the only variable. This is carried out at constant speed, constant throttle opening, and constant ignition setting. The specific fuel consumption is plotted to a base of bmep and a 'hook curve' or consumption 'loop' is obtained. For a single-cylinder engine at full throttle the curve is sharply defined as in Fig. 13.15. The air–fuel ratio is a minimum at A (i.e. the richest mixture). As the air–fuel ratio is increased the bmep increases until a maximum is reached at B (usually for an air–fuel ratio between 10/1 and 13/1). Further increase in the air–fuel ratio produces a decrease in bmep with increasing economy until the position of maximum economy is reached at D. Beyond D, for increasing air–fuel ratios, both bmep and consumption values are adversely affected. Near the point A

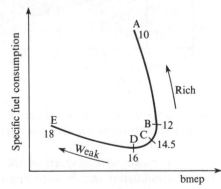

Fig. 13.15 Specific fuel consumption against bmep for a spark-ignition engine

the engine could be running unsteadily and there may be combustion of the mixture in the exhaust system. At E, with the weakest mixture, running will be unsteady and the combustion may be so slow that the gases continue burning in the clearance volume until the next induction stroke begins; this causes *popping back* through the carburettor. Point C is the point of chemically correct or stoichiometric air–fuel ratio, and is about 14.5/1. The mixture strengths range between those at B and D, which are for maximum power and maximum economy respectively. The indicator diagrams corresponding to mixtures B, C, and D are shown in Fig. 13.16.

Fig. 13.16 Pressure–volume and timing diagrams for rich, weak, and stoichiometric mixtures for a SI engine

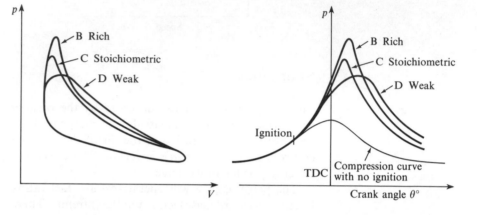

For multicylinder engines the consumption loops are less distinct, but are generally similar in shape to that for the single-cylinder engine. This is also true for tests made at part throttle openings. A series of readings obtained at different throttle positions at constant speed is shown in Fig. 13.17.

Fig. 13.17 SI engine consumption loops at different throttle settings

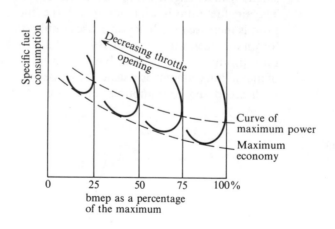

In the above tests the ignition has been assumed to be constant, but other tests can be included to show the effect of ignition timing on the consumption loop. Alternatively the ignition setting can be adjusted at each mixture strength to give maximum power at the speed of the test; by this means the rate of pressure rise on combustion can be kept approximately constant.

Bmep and sfc may be plotted against air–fuel ratio as shown in Fig. 13.18. To the same base of air–fuel ratio the variation of carbon dioxide, oxygen, and carbon monoxide contents of the dry exhaust can be plotted as shown in Fig. 13.19.

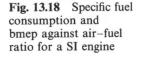

Fig. 13.18 Specific fuel consumption and bmep against air–fuel ratio for a SI engine

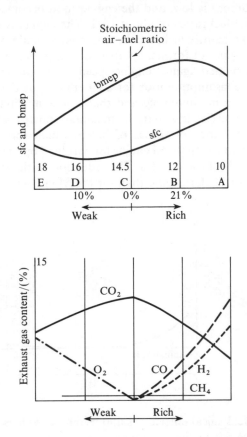

Fig. 13.19 Exhaust gas analysis against air–fuel ratio for a SI engine

Energy balances can be drawn up for the principal points taken from the consumption loop.

Testing the SI engine at part load shows the deficiency of the method of governing by throttling the charge since the efficiency falls with decreasing load. The induction pressure is reduced and the pumping losses increase. The dilution of the fresh charge by exhaust gas increases at lower loads, the clearance volume containing practically the same amount of exhaust gas since the back pressure in the exhaust process remains constant. For this type of mixture a greater amount of fuel is required for combustion to be possible.

CI engines

In the main CI engines are not controlled by throttling but by adjusting the amount of fuel supplied to the engine, and hence are *quality governed*. When

441

adjusting the fuel supplied to a CI engine the limiting condition is given by the *smoke limit*, which is the appearance of black smoke in the exhaust. Engines should not be operated with mixtures rich enough to produce smoke, although such a mixture may give a greater power output. The efficiency under these conditions is low, and the engine soon becomes dirty. The smoke limit occurs at air–fuel ratios of about 16/1. The engine is tested at different speeds to the smoke limit, which can be observed visually or measured by a smoke meter. The values of torque, bp, fuel consumption, and specific fuel consumption are then plotted against engine speed in revolutions per minute.

A consumption loop for a CI engine has the form shown in Fig. 13.20. This shows a minimum sfc and therefore a maximum brake thermal efficiency at part load (i.e. less than the maximum bmep). The curve is reasonably flat over a wide range of values of bmep, which shows the virtue of the CI engine compared with the SI engine for part load operation, a condition often prevailing in road vehicle engines. The reduction in the thermal efficiency at part load is less for the CI engine than for the SI engine.

Fig. 13.20 Specific fuel consumption against bmep for a CI engine

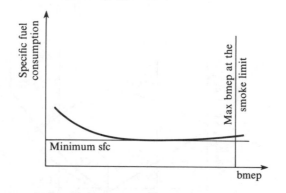

Mechanical efficiencies and minimum values of sfc are plotted against speed to provide further characteristics.

For a given engine the individual characteristics described in this section are usually sufficient for normal use, but if comparisons are to be made with other engines, perhaps of different types, then the more comprehensive presentation by means of a performance map is an advantage. Figure 13.21 shows the form of such a map for a petrol engine and Fig. 13.22 shows the form for a diesel engine. In the maps bmep is plotted against engine (or piston) speed with curves of constant specific fuel consumption and power per unit piston area. If the engine is to be used in a vehicle the road requirement can be included as a bmep–speed characteristic.

13.7 Factors influencing performance

SI engines

The analysis of the Otto cycle given in section 5.6 shows the dependence of thermal efficiency on compression ratio. A graph of air standard thermal

Fig. 13.21 Performance map for a SI engine

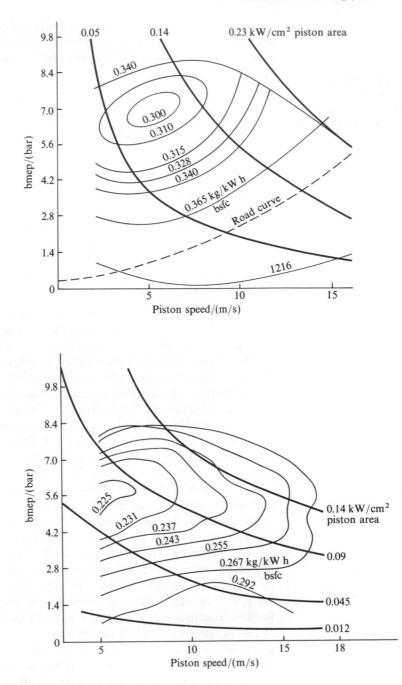

Fig. 13.22 Performance map for a CI engine

efficiency against compression ratio is shown in Fig. 13.23. This graph indicates the form engine development should take, and over the early years increases in compression ratio were made. However, since 1960 the compression ratios have not increased greatly and are in the range 9–10 for production vehicles in the UK. The ability to use higher ratios has depended on the provision of better-quality fuels and of improved designs of combustion chamber. The main features of the combustion chamber are the distances to be travelled by the flame after initiation of combustion, and the gas flow pattern established.

Fig. 13.23 Thermal efficiency against compression ratio for an ideal cycle for a SI engine

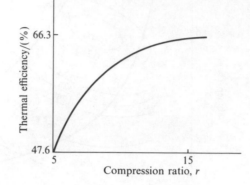

It is evident that if a petrol–air mixture is compressed sufficiently it will ignite spontaneously. This suggests one limit to compression ratio if controlled combustion is to be obtained from spark ignition. However, before this limit is reached for the whole charge, spontaneous ignition can occur in the unburnt charge after combustion has commenced normally. The unburnt gas, compressed by the advancing flame front, is raised in temperature and may reach the point of *self-ignition*. This produces an uncontrolled combustion and its occurrence may be heard as a knocking sound. A critical condition can be reached which is called *detonation*, or 'heavy knock'. The advancing flame front is suddenly accelerated by the occurrence of a high-pressure wave and the flame front and shock wave traverse the cylinder together. The *detonation wave* suffers successive reflections, and a high-frequency noise is created. This is an extreme condition which has been produced in test rigs, but such intense conditions are less likely to be produced in a normal engine. These combustion phenomena are usually referred to collectively as 'knock'. One of the results of knock is that local hot spots can be created which remain at a sufficiently high temperature to ignite the next charge before the spark occurs. This is called *pre-ignition*, and can help to promote further knocking. The result is a noisy, overheated, and inefficient engine, and perhaps eventual mechanical failure. The chemical behaviour during this type of combustion is still not fully understood, although a considerable amount of empirical data are available. The pressure/crank angle diagram for normal combustion is shown in Fig. 13.24 with maximum pressure occurring at 10–12° after TDC, and a rate of pressure rise of 1.38 bar per degree of crank angle, with a compression ratio of 8/1. The spark occurs at the point S on the normal compression curve, but there is a *delay period* between the occurrence of the spark and a noticeable departure of the pressure curve from that of normal compression. This is a time delay which is independent of engine speed so that as the engine speed is increased the point S must occur earlier in the cycle to obtain the best position of the peak pressure. This *ignition advance* can be accomplished manually, but can also be controlled automatically by a mechanism in the distributor which is sensitive to engine speed; an additional control is obtained at small throttle openings by a pressure connection from

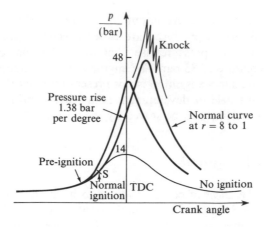

Fig. 13.24 Pressure against crank angle for various conditions for a SI engine

the distributor to the induction manifold. Modern petrol engines now have electronic ignition systems.

The compression ratio which can be utilized depends on the fuel to be used and a scale has been developed against which the knock tendency of a fuel can be rated. The rating is given as an *octane number*. The fuel under test is compared with a mixture of iso-octane (high rating) and normal heptane (low rating), by volume. The octane number of the fuel is the percentage of octane in the reference mixture which knocks under the same conditions as the fuel. The number obtained depends on the conditions of the test and the two main methods in use (the *research* and the *motor* methods; see ref. 13.3) give different ratings for the same fuel. The motor test is carried out at the higher temperature and gives the lower rating. The difference between the two is taken as a measure of the temperature sensitivity of the fuel. High-octane fuels (up to 100) can be produced by refining techniques, but it is done more cheaply and more frequently by the use of anti-knock *additives*, such as *tetraethyl lead*. (An addition of 1.1 cm^3 of tetraethyl lead to 1 litre of 80-octane petrol increases the octane number to 90.) Unfortunately it has been shown that the levels of lead now in the atmosphere due to the SI engine are harmful to health. Public opinion has gradually caused governments to bring in financial incentives for the use of unleaded fuel and car manufacturers have adjusted engines for the use of unleaded fuel (see section 13.13).

Fuels have been developed which have a higher anti-knock rating than iso-octane and this has led to an extension of the octane scale. Aviation conditions of operation lead to another scale which gives a better indication of the detonation characteristics; this is the *performance number* (PN). The relationship between octane number (ON), above 100 and performance number is given by

$$\text{ON above } 100 = 100 + \frac{\text{PN} - 100}{3}$$

With higher compression ratio engines other phenomena are observed. From compression ratios of 9.5/1 upwards there are high rates of pressure rise which have their origin in the additional flame fronts started from surface deposits in the cylinder. At about 9.5/1 compression ratio the low-frequency engine vibrations produced are called *rumble* or *pounding*. At compression ratios of 12/1 the pressure rise is about 8.3 bar per degree crank angle with a peak pressure of 83 bar. The engine noises produced are known as *thud* or *pressure rap*; surface ignition is not present and fuel characteristics have little influence. This field of development is one which brings many new problems which are more likely to be solved by the chemist than the engineer.

CI engines

The effect of compression ratio in the CI engine is somewhat simpler than in the SI engine. For combustion to occur at the temperature produced by the compression of the air a compression ratio of 12/1 is required. The efficiency of the cycle increases with higher values of compression ratio and the limit is a mechanical one imposed by the high pressures developed in the cylinder, a factor which adversely affects the power–weight ratio. The normal range of compression ratios is 13/1 to 17/1, but may be anything up to 25/1.

The combustible mixture in the SI engine is formed before compression, but with the CI engine this mixture has to be formed after compression and after injection begins. This leads to delay periods in the CI engine which are greater than those in the SI engine (see Fig. 13.25). The fuel droplets injected have to evaporate and mix with oxygen to give a combustible mixture. The delay period forms the first phase of the combustion process, and is dependent on the nature of the fuel. The second phase consists of the spread of flame from the initial nucleus to the main body of the charge. There is a rapid increase in pressure during this phase and the rate of pressure rise depends to some extent on the availability of oxygen to the fuel spray, which in turn depends on the turbulence in the cylinder. The main factor, however, is that of the delay period. A long delay period means more combustible mixture has had time to form, and so more charge will be involved in the initial combustion. As the speed increases the rate of pressure rise in this phase also increases. This is because the delay period is a function of time if surrounding conditions remain constant, and at the higher engine speeds more mixture will be formed in the delay period. The initial rapid combustion can give rise to rough running and a characteristic

Fig. 13.25   Pressure against crank angle for a CI engine

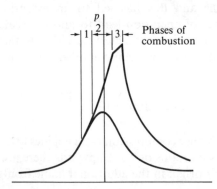

noise called *diesel knock*. During the third phase of combustion the fuel burns as it is injected into the cylinder, and this phase gives more controlled combustion than that of phase two. One of the main factors in a controlled combustion is the swirl which is induced by the design of the combustion chamber. It has been stated that the delay period depends on the nature of the fuel, and a fuel with a short delay period, or high ignitability, is required. The *ignitability* of a fuel oil is indicated by its *cetane number*, and the procedure for obtaining it is similar to that for obtaining the octane number of petrols. Reference mixtures of cetane ($C_{16}H_{34}$) (high ignitability), and α-methyl-naphthalene ($C_{11}H_{10}$) (low ignitability), are used. The mixture is made by volume and the ignitability of the test fuel is quoted as the percentage of cetane in the reference mixture which has the same ignitability. For higher-speed engines the cetane number required is about 50, for medium-speed engines about 40, and for slow-speed engines about 30. For details of the standard tests see reference 13.3.

The air–fuel ratios used in CI engines lie between 20/1 and 25/1. As these mixtures are much weaker than the stoichiometric proportion then the imep will be limited, and this also means that for a given fuel consumption the swept volume of the engine will be greater than that of the equivalent SI engine. For further reading on the combustion process in IC engines see refs. 13.4 and 13.9.

Engines are affected in performance by the atmosphere in which they operate and some allowance must be made in performance figures quoted for variations in pressure, temperature, and relative humidity. The variations in performance can be represented graphically, but the normal values quoted apply up to 30 °C and 150 m altitude from sea-level, for normally aspirated engines. The reduction in output per 300 m of altitude above 150 m is about 3%, and for every 5 K above 30 °C the reduction is also about 3%. The reduction in output for changes in relative humidity can be up to 6% depending on the conditions.

13.8   Real cycles and the air standard cycle

Some of the differences between practical and ideal cycles were discussed as an introduction to this chapter. Other differences are evident from a comparison of the indicator diagram, such as the rounding off of the corners of the diagram due to valve throttling and the fact that the combustion process is not truly at constant volume or at constant pressure. The working fluid is not air throughout; it starts off the combustion process as a mixture of air and fuel vapour, and this mixture changes as combustion proceeds. One of the properties of a perfect gas is that its specific heats remain constant, but in the IC engine there is a considerable change in these values due to the high temperatures reached. At the higher temperatures dissociation takes place (see section 7.7), and the work output is reduced. The maximum pressure and temperature reached in the practical cycle are much less than those obtained by calculation based on the theoretical cycle, assuming the same heat input. Heat losses from a heat engine must be avoided, but with a real engine some form of cooling is essential so that temperatures will not be reached such that there will be failure of the engine materials. The compression and expansion processes are thus not adiabatic but are of the form $pv^n = $ constant; the assumption of internal reversibility is nevertheless a good approximation. A comparison of the real cycle and the ideal cycle is shown in Fig. 13.26 for a petrol engine; the pumping loop of the practical cycle has been omitted.

Fig. 13.26 Comparison of real and ideal cycles for a SI engine

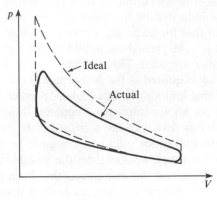

A most important criterion is the temperature (and hence pressure) which can be attained by the 'constant volume' combustion phase. Figure 13.27 shows the temperatures reached on the combustion of mixtures of fuel and air of different strengths between 50% weak and 50% rich, for a compression ratio of 5 to 1. The maximum temperature should theoretically be produced by a stoichiometric mixture strength, but in fact it occurs at about 20% rich. The shape of the temperature curve is due to the dissociation of CO_2 and H_2O into CO, H_2, and O_2 which is slight at a temperature of 2000 °C, but increases rapidly above that. In engine combustion the dissociation of CO_2 affects engine behaviour more than that of H_2O which is relatively slight. Combustion theory, as described in Chapter 7, allows the dissociated proportions of CO, H_2, and O_2 to be calculated and represented in the combustion equations. Hence the energy released on combustion can be calculated. If the formation of nitric oxide which occurs at the higher temperatures is taken into account the temperatures attained are even lower, due to the absorption of energy to form NO as shown in Fig. 13.27. At the higher compression ratios of modern engines higher temperatures and pressures are attained and dissociation occurs, not only to a greater extent for NO, but also to some common radicals mainly OH, H, and O, e.g. at 3000 °C a stoichiometric mixture of octane and air at equilibrium will contain OH (1.4%), H (0.3%), O (0.3%), and NO (0.3%).

Mixture strength, particularly for a spark-ignited engine, can vary considerably due to the way in which the mixture is created and distributed. It is not a controlled process as it is in the diesel engine. One of the results of this is a considerable variation in the cycle to cycle and cylinder to cylinder performance of petrol engines.

An important parameter which is open to selection is that of the spark timing as described in sections 13.6 and 13.7. Figure 13.28 shows the form of the variation of the maximum pressure reached in the cylinder with the advance of the ignition before the TDC position.

Fig. 13.27 Combustion temperature against mixture strength for a SI engine with and without dissociation

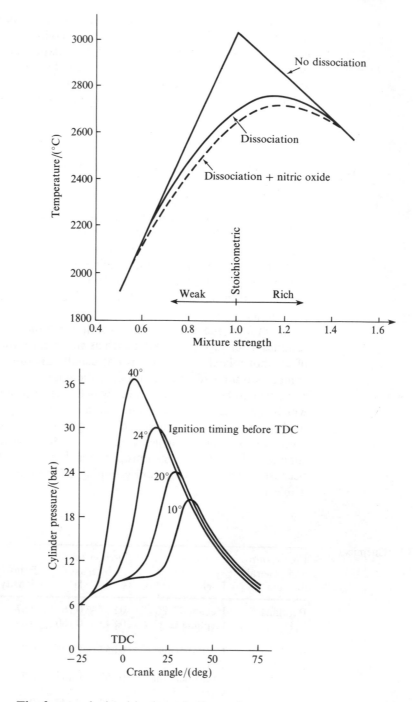

Fig. 13.28 Effect of ignition advance on the pressure–crank angle diagram

The factors dealt with above indicate the complexity of the processes of the IC engine that give the engine a character which has fascinated many distinguished engineers over lifetimes of service, and continues to do so. Calculation becomes more difficult as the information required is increased more and more, the relationship between data is complex, and the time taken for computation increases rapidly. To this end charts have been prepared from established information to describe the behaviour of different fuels at different mixture ratios including the effects of dissociation and different proportions of recycling exhaust gas etc. (see ref. 13.5). In serious work this information will be included as data stored in a computer ready for use in the analysis or computation of different engine cycles and running conditions.

13.9   Properties of fuels for IC engines

Some of the requirements of the fuels used in IC engines have been indicated by the earlier sections of this chapter, and these will be added to in this section.

Fuels suitable for petrol engines are obtained from several sources. The majority are obtained from petroleum; these consist of straight petrol distilled from crude petroleum, natural gasoline (which is a light spirit), and cracked or re-formed spirits from certain petroleum fractions. Petrol can be obtained from the distillation of oil shale, but the product is more expensive than that from petroleum.

Alcohol in the main is ethyl alcohol obtained from the fermentation of residues from vegetable matter such as sugar-cane and sugar-beet. The amount of alcohol mixed with petrol is not usually greater than 20%, since a higher content would lead to carburation difficulties.

Fuels can be classified as those which belong to a chemical series and those which are manufactured. The main chemical groups are the paraffins, naphthenes, and aromatics, each group consisting of carbon and hydrogen atoms to a characteristic general formula. Members from two groups can have similar carbon to hydrogen proportions, but different chemical characteristics because of the different atom grouping in the molecule. This is illustrated in Table 13.3.

Table 13.3   Chemical fuels

Fuel group and general formula	Fuel	Proportions by mass C	H	Relative density	Air/fuel ratio	NCV (kJ/kg)
Paraffins C_nH_{2n+2}	Hexane C_6H_{14}	0.837	0.163	0.67	15.3	45 124
	Heptane C_7H_{16}	0.840	0.160	0.69	15.2	44 660
Naphthenes C_nH_{2n}	Cyclo-hexane C_6H_{12}	0.857	0.143	0.79	14.7	43 030
	Cyclo-heptane C_7H_{14}	0.857	0.143	0.78	14.7	43 960
	Cyclo-octane C_8H_{16}	0.857	0.143	0.74	14.7	43 960
Aromatics C_nH_{2n-6}	Benzene C_6H_6	0.923	0.077	0.88	13.2	40 700
	Toluene C_7H_8	0.913	0.087	0.87	13.4	41 054
	Xylene C_8H_{10}	0.905	0.095	0.95	13.6	41 868

The common manufactured fuels are as follows:

Petrol: a refined distillate of petroleum; the boiling-point is not usually greater than 200 °C.

Kerosene: a refined distillate of petroleum which boils between 150 and 200 °C.

Benzole: a refined distillate of coal tar which consists mainly of benzene and toluene.

Gas oil: an unrefined distillate after kerosene fractions have been removed.

Diesel fuel: mixtures of gas oils from various crude oils or blends with fuel oils.

Blends of alcohol: petrol with about 15% alcohol and 15% benzole.

Kerosene is commonly called 'paraffin', but as this is the name of a chemical series as mentioned earlier, and as the commercial paraffin contains members of other series, the name kerosene is to be preferred. Petrol contains members from several different series and special additivies. The actual content of the straight petrols depends on the geographical origin of the petroleum.

The specific enthalpy of vaporization of fuels is an important quantity in both CI and SI engines. With SI engines this factor together with the volatility of the fuel determines the density of the charge in the cylinder; the density is inversely proportional to the absolute temperature in the cylinder at the closing of the inlet valve. This temperature depends on the specific enthalpy of vaporization of the fuel and the amount of heating which takes place during induction. Most liquid fuels have similar specific enthalpies of vaporization, the exception being alcohol which has a high specific enthalpy of vaporization, a higher density of charge, and a marked increase in power compared with other fuels. In CI engines the specific enthalpy of vaporization has no bearing on volumetric efficiency due to the fact that the fuel is injected after compression; this is detrimental when the calorific value and delay period are considered since the energy necessary for evaporation of the fuel must come from the compressed air.

Fuel *volatility* is of importance in SI engines from the point of view of ease of starting and the heating required to provide an even distribution of mixture to different cylinders.

In CI engines the volatility of the fuel influences the time taken for a combustible envelope to form round the fuel droplets, and hence influences the delay period.

The calorific value of a fuel is important in determining the amount of fuel required, but does not indicate how much power can be obtained from it. Engines using gaseous fuels are affected more by variation in calorific value than petrol and diesel engines. The power output depends on the heating value of the mixture of fuel and air and the stoichiometric mixtures of all hydrocarbon fuels give the same heating value per unit volume at the same reference, temperature, and pressure, within close limits.

Tetraethyl lead has been mentioned as an anti-knock additive in petrols. Other additives are oxidation inhibitors, corrosion inhibitors, combustion control compounds, and anti-icing additives to counteract freezing in the carburettor.

The properties of fuels are determined by standard tests, the procedure for which may be obtained from ref. 13.3.

13.10 Fuel systems

The purpose of an engine fuel system is to provide the cylinder with a mixture of air and fuel in the correct proportions for the engine requirements at any particular instant. There are basically two methods available, one is called *carburation* and is used for petrol engines and the other is a type of *fuel injection* which is a characteristic method for diesel engines. There are many different designs of each and only the basic principles will be dealt with in this book.

The petrol engine for automotive purposes has been developed on the basis of the carburettor although petrol injection is becoming more common, the latter having been used for a long time for aircraft and special engines, such as military vehicles and racing cars. The carburettor is a simple, cheap device which has served its purpose for many years, but the trend to higher powered, multicylinder engines has shown the single carburettor system to be inadequate. As a consequence multichoke carburettors and twin or triple carburettor layouts have been used to meet increasingly sophisticated engine requirements, but several designs of petrol injection systems have also been introduced.

The fundamentally different methods of charge ignition by spark and by compression in the petrol and diesel engines respectively have dictated different means of fuel supply for the two engines. However, the features of the oil injection system from the point of view of control, accurate fuel metering, and good fuel consumption characteristics in comparison with the carburated petrol engine with its poor fuel consumption, particularly at part load, have created the belief that ultimately petrol injection would be the preferred method if a sufficiently simple system could be produced at a cost which would make it competitive with the carburettor. In recent years the added requirement for engines to meet exhaust gas emission regulations (see section 13.13) has increased the demand for accurate fuel metering and engine control. These factors have increased the interest in petrol injection although the earlier regulations were more satisfactorily met by carburated engines than by those with the first petrol injection systems. In general the fundamental problem is to measure, or compute accurately, the mass flow rate of air into the engine at any instant and to mix the correct amount of petrol into it in such a way that the air and fuel mixture produced is right for the engine running condition. If the airflow were steady in an engine and at constant temperature and pressure the problem would be relatively straightforward, but airflow rates, pressures and temperatures change; also the engine is in dynamic operation with phases of acceleration, deceleration, and overrun (when the throttle is closed and the vehicle wheels are turning the engine). Thus a fuel supply system, to be successful, must be able to cope with a wide range of running conditions and demands.

Carburation

The term 'carburation' covers the whole process of supplying continuously to a petrol engine a mixture of vaporized fuel and air which is suitable to each engine condition of load, speed, and temperature. In section 13.5 it is seen that the air–fuel ratio varies between maximum economy and maximum power conditions, and the stoichiometric air–fuel ratio is not adequate to all demands. The function of the carburettor is to measure out the correct proportions of liquid fuel and air for the particular engine condition. The liquid fuel must be 'atomized' at the carburettor (i.e. broken up into a fine spray to assist in the evaporation of the fuel, so that the mixture entering the cylinders is homogeneous). The necessary energy for the evaporation of the fuel must be supplied to the mixture and this is not the function of the carburettor but of the whole induction tract. The metering process is carried out at the carburettor, but the actual mixture ratio, its condition and distribution between cylinders, depends also on the design of the complete induction system and the temperatures therein.

The perfect carburettor would supply the air–fuel ratio required at all speeds and throttle openings no matter what the climatic conditions or the rate at which the demand was changing. A consideration of all the factors which influence the final mixture burned in the cylinder (i.e. engine condition required, mechanical characteristics of the engine, the physical differences between the constituents of the mixture, the rapid fluctuation in demand, and the temperature and humidity variation) shows the difficulty in obtaining the ideal carburettor. The carburettor is a highly developed component which is produced in a number of different designs each having its own refinements, with the object of supplying to the engine the air–fuel mixture it needs. The elementary metering process will be considered, since this of fundamental importance to all types.

In Fig. 13.29 a simplified carburation system is shown. The petrol pump, either electrically driven or mechanical driven by the crankshaft, pumps petrol from the tank to the float chamber of the carburettor. The function of the float chamber is to maintain a constant level of petrol in the chamber by shutting off the supply from the pump when this level is about to be exceeded. The float chamber is vented to atmosphere through a small hole in the cover, hence the pressure on the surface of the petrol is constant and equal to that of the atmosphere. The air is induced by the depression created by the piston moving downwards in the engine cylinder, and, after passing through a filter, enters the carburettor at about atmospheric pressure. The petrol engine is quantity governed, which means that when less power is required at a particular speed the amount of charge delivered to the cylinders is reduced. This is achieved by means of a throttle valve of the butterfly type which is situated in the air inlet.

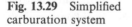

Fig. 13.29 Simplified carburation system

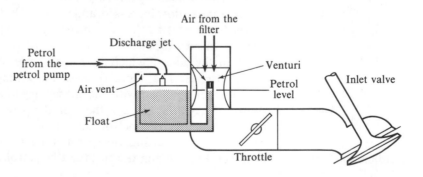

The air on induction enters the venturi or choke tube. This is a tube of decreasing cross-section which reaches a minimum at the throat or choke of the venturi, which is shaped to give the minimum resistance to the airflow. The petrol discharge jet is situated at the throat and is subject to the air pressure there. The pressure at the throat is below atmospheric since the air velocity has been increased from that at inlet to the carburettor to a maximum at the throat. Thus the two petrol surfaces, that in the float chamber and that at the discharge jet, are subject to different pressures. This pressure difference acting on the petrol column causes the petrol to flow into the airstream, and the rate of flow is controlled or metered by the size of the smallest section in the petrol passage. This is provided by the main jet and the size of this jet is chosen to give the required engine performance, and is an empirical selection. The pressure at the throat at the fully open throttle condition lies usually between 38 and 50 mm Hg below atmospheric, and seldom exceeds 76 mm Hg below atmospheric.

In the carburettor described the choke or throat has a constant area and the pressure changes with throttle opening and engine speed. It is referred to as the *fixed choke* type. Another type, of which the SU carburettor is the main example, is the *constant vacuum* type. In this type the pressure at the throat, and thus the air velocity, remains constant, but the area of the throat is varied. Similarly the area of the petrol orifice or jet must also vary, and this is achieved by means of a tapered needle, attached to a piston. The needle moves in the orifice thus forming a discharge annulus for the petrol (see Fig. 13.30).

Fig. 13.30 Constant vacuum carburettor

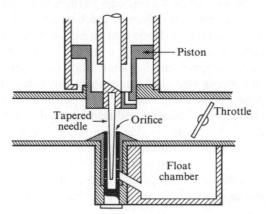

With the elementary form of carburettor described, the mixture would become richer as the airflow increased, and this must be 'corrected'. In some carburettors this is done by fitting the main metering jet about 2.5 cm below the petrol level, and it is called a submerged jet (see Fig. 13.31). The jet is situated at the bottom of a well, the sides of which have small holes which are in communication with the atmosphere. The second object of this arrangement is to achieve atomization which is not obtained with the elementary system described previously. Air is drawn through the holes in the well, the petrol is emulsified, and the pressure difference across the petrol column is not as great as in the elementary carburettor. On starting, the petrol in the well is at the level of that in the float chamber. On opening the throttle this petrol, being subject to the low throat

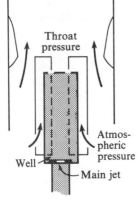

Fig. 13.31 Submerged jet carburettor

pressure, is drawn into the air. This continues with decreasing mixture richness as the holes in the central tube are progressively uncovered. Normal flow then takes place from the main jet.

A petrol engine operates in service mainly at part throttle and a carburettor which gives a correct full throttle delivery would not be able to meet part throttle requirements. As the throttle is closed the point of lowest pressure in the induction system moves from the venturi, at full throttle, to the engine side of the throttle when the engine is idling with the throttle practically closed. The position of minimum pressure changes between these limits. At the idling condition the pressure in the induction trace is 400–460 mm Hg below atmospheric. The main jet will not supply petrol when the engine is idling because there is no longer sufficient depression at the throat. Another petrol supply is provided which delivers to the engine side of the throttle. This is indicated diagrammatically in Fig. 13.32.

A rich mixture is required for starting and idling, but as the throttle is opened the demand is for a weaker mixture. There is a merging of the idling and main jet deliveries, but in order to avoid flat spots and be able to deliver the required mixture throughout the range of throttle positions, it is usual to provide a third 'progression' system. On acceleration the throttle is opened suddenly and it is necessary to provide a rich mixture. If the part throttle characteristics of the

Fig. 13.32 Injector pump and injector for a CI engine

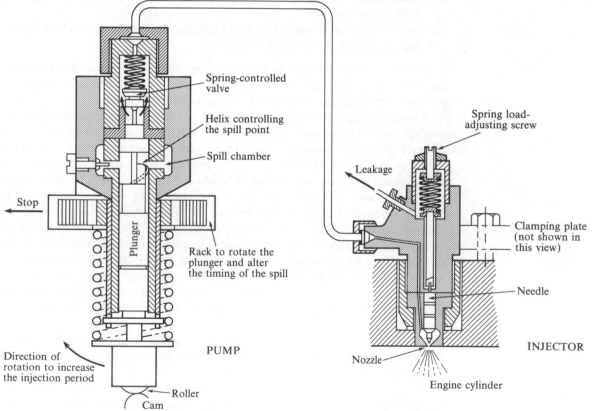

carburettor are satisfactory the fuel for acceleration is best provided, in the interests of economy, by a separate accelerating device. The means of doing this vary and the types may be classified as using an *accelerating pump*, or an *accelerating well*.

Fuel injection: diesel engines

The function of a fuel injection system is to meter the fuel accurately and uniformly to the engine cylinders under all operating conditions from idling to full load. The timing of the injection should be accurate enough to give the required combustion characteristics. The fuel from the tank is filtered before passing to the pump, and the metered fuel is then passed to the injection which is fitted in the engine cylinder. The 'jerk pump' system is the one which is used almost universally over the whole range of oil engines, and will be the one described here. The jerk pump is a piece of precision equipment, and consists of a barrel with plunger; the close fit required to prevent leakage at the high pressures reached is obtained by lapping. The plunger is driven from the camshaft and the control of the amount of fuel to be delivered can be made by the following:

(i) using a plunger with a variable stroke; or
(ii) measuring the quantity at the beginning of the plunger stroke and spilling back the excess; or
(iii) using a constant stroke plunger and bringing delivery to an end by suddenly spilling off the fuel from the cylinder.

The spill of the fuel may be controlled by a cam-controlled spill-valve (see ref. 13.6), or a port control. Pumps with port control are produced by a number of manufacturers and their action will be described.

A simplified injection system is shown in Fig. 13.32. The spring-loaded injector needle is set to lift at a predetermined pressure in the delivery line which contains oil at a high residual pressure. There are complex pressure and velocity variations set up in the system, the changes in the pressure being propagated in the oil at the speed of sound in the oil. The effect of such variations will not be considered here (see ref. 13.6 for a full treatment of the subject). The plunger is provided with a helical groove, the upper edge of which controls the uncovering of the spill port. The timing of the spill is thus decided by the shape of the helix, and this is most important. The part of the helix presented to the port is varied by rotating the plunger in the barrel, and a means must be provided for this to be done automatically or manually while the engine is running. An extreme position of rotation of the plunger in the barrel gives the position at which the pump will not deliver fuel to the engine, and this is the 'stop' position.

Figure 13.33 indicates the way in which the pressure in the fuel line changes during the injection cycle. At point 1 the pressure begins to rise above the residual pressure due to delivery from the pump. At 2 the injector needle begins to lift and the pressure variation from 2 to 3 depends on the delivery from the pump and the injector characteristics. At 3 the spill port is opened and the

Fig. 13.33 Pressure variation in the fuel line during injection

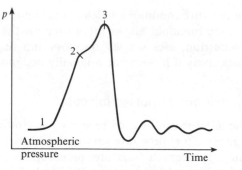

pressure falls with a characteristic depending on the spill port action and that of the closing of the injector needle. The pressure variation after this is due to the reflection of pressure waves in the line which are damped out unless the next cycle commences before the damping is complete. The pressure in the line after spill may reach such a value that the needle opens for a second time in the cycle. This is undesirable and is known as *secondary injection* (see Fig. 13.34).

Fig. 13.34 Needle lift against time showing undesirable secondary injection

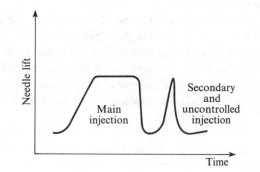

Fuel injection: petrol engines

At first sight it may appear that the application of diesel engine experience with injection systems to the petrol engine would produce a satisfactory mechanical petrol injection system. However, the different requirements of the two types of engine and the increased knowledge and use of electrical and electronically operated valves and transducers have made it evident that a petrol injection system would be wholly or mainly an electronic system. The development of electronic devices continues with increased component reliability, greater design sophistication, more miniaturization, and reduced cost giving flexible systems which are simple to locate, robust in operation, and easy to replace.

There are certain basic parameters to be decided upon for each system such as those given under the headings below.

Manifold or in-cylinder injection

Direct injection into the cylinder is attractive from the point of view of efficient fuel distribution, but the injector used would be subject to high pressure and temperature conditions which it would not experience in the manifold. Injection into the manifold allows more time for the fuel and air to mix, giving better combustion, uses simpler injectors and requires easier access to the engine, particularly if it were not originally designed for injection.

Continuous or pulsed injection

The injector can be used to spray fuel continuously into the manifold and this requires a fuel flow rate varying by 50 to 1 between idle and maximum speed running; accurate mixture preparation requires good control on airflow measuring and fuel metering. With pulsed injection the fuel is injected near each cylinder inlet valve in each cycle by an injector which is solenoid operated for a measured period of time. The range of pulse durations required is about 5 to 1 and 'time' is a simple quantity, conveniently and accurately measured electronically; the demands of several engine parameters can be computed and expressed as a single controlling value. The very small pulse durations required at idling speed, of the order of 1 ms, set the criterion of quality of design for the injector.

It would appear that the ability to time the injection pulse anywhere in the induction stroke would be a powerful facility, but experience shows that the effect of such timing on power output and hydrocarbon emissions is not great. This allows some relaxation on the injector design as it enables injectors to be grouped to receive the same pulse, over a longer duration, and so injection occurs into different cylinders at different points in the cycle or by 'non-time injection' procedure.

The layout of a typical injection system is illustrated in Fig. 13.35 and includes provision for the various engine parameters which it is believed it is necessary to take into account. Alternative, and possibly simpler, designs could employ fewer parameters and so eliminate the need for some of the transducers, etc. described in this illustration. For instance, in this case the steady airflow requirement is computed from measurements of manifold pressure, inlet air temperature, and engine speed; a single reading of air mass flow rate would simplify considerably the requirements of the system. The system shown takes into account in addition the parameters of cooling water temperature, starter motor action, a measurement of change of manifold pressure by electronic differentiation of the pressure signal, and a throttle-actuated switch which cuts off the fuel supply when the throttle is closed. Information from each of the transducers or switches involved is 'computed' electronically to give a single pulse width, or a main and a second pulse, and is communicated to the pulse generators.

Fuel is supplied to the injectors at a pressure of 1.72 bar above the manifold pressure and fuel not injected is returned through a pressure control valve to the supply tank. This circulation of fuel cools the injectors and purges them of air and vapour. The quantity of fuel injected depends on the pulse duration as decided by the electronic control unit, the orifice area and the pressure difference across it and the last two of these are constant, giving only the pulse duration as a variable. It would appear to be essential that the nozzles be matched very closely with respect to area and coefficient of discharge as small variations in

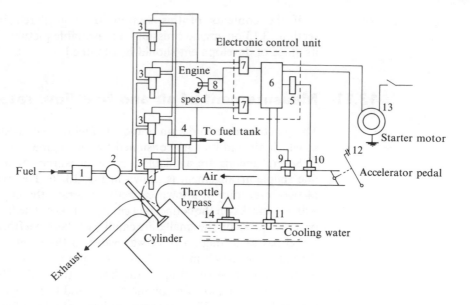

Fig. 13.35 Typical direct fuel injection system for a SI engine. (1) fuel filter; (2) fuel pump (electrically driven); (3) injectors; (4) fuel pressure control valve; (5) pressure transducer; (6) computer; (7) pulse generator; (8) trigger unit (engine driven). Transducers: (9) manifold air temperature; (10) manifold air temperature; (11) cooling water temperature; (12) throttle switch; (13) starter motor switch; (14) idling air control valve (throttle bypass)

the fuel supply could affect the exhaust emissions of the engine quite considerably. A reduction in air–fuel ratio promotes the carbon monoxide content and an increase usually adversely affects the hydrocarbon content depending on the air–fuel ratio required, with a value of 16–16.5 to 1 being good average optimum values for both types of emission.

Information from the manifold temperature and pressure transducers and the engine speed is computed to give the basic pulse duration. If the engine is starting from cold a signal from the starter motor produces fuel enrichment, and a temperature signal from the cooling water controls the additional fuel during warm-up. Additional fuel for acceleration is provided by a control signal arising from the rate of change of pressure in the manifold as computed in the control unit. When the vehicle is in the overrun condition the throttle closes and switches off the fuel, thus eliminating the high output of hydrocarbons in the exhaust gas obtained with a carburated engine during this phase of operation. A throttle bypass valve allows an increase in the air supply during idling when the throttle is in the closed position and the opening of this is controlled by cooling water temperature.

Charge stratification

One of the developments that can utilize petrol injection is known as *charge stratification*. The objective of this is to initiate combustion in a region where the mixture is rich and to produce a flame that can then travel easily through the rest of the charge which is, on the whole, weak. One such system employs two injectors, one supplying the main combustion chamber and the other a small directly connected auxiliary chamber in which the spark plug is fitted. In the main chamber the air–fuel ratio is 2.18 to 1, in the auxiliary it is 0.66 to 1 and overall it is 1.64 to 1 relative to the stoichiometric ratio.

If the contents of this section are considered in relation to that of section 13.13 on engine emissions a reasonable picture of the petrol engine and its stage of development should be obtained.

13.11 Measurement of air and fuel flow rates

The practical determination of the air–fuel ratio consists of measuring the rate at which air and fuel are consumed by the engine.

Several means are available for the measurement of fuel consumption and one simple arrangement is shown in Fig. 13.36. The measuring vessel consists of two reservoirs of known capacity in series, the capacities being measured between marks on the connecting capillary tubes. The fuel level thus falls quickly past the marks on the small bore capillary tube. Fuel from the tank or measuring vessel passes through a three-way valve to the engine. The three positions of the valve are shown in Fig. 13.36. Provision must be made to allow the fuel to fill the measuring vessel up to the level in the tank. The time is taken for the consumption of a known volume of fuel, and thus the rate of fuel consumption can be determined. This arrangement has a number of possible variations which are suitable for particular purposes.

Fig. 13.36 Simple test measurement of fuel consumption

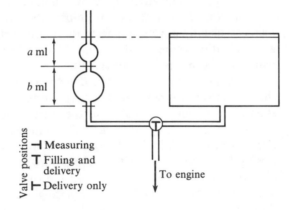

a ml

b ml

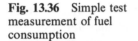

Valve positions

⊣ Measuring
T Filling and delivery
⊢ Delivery only

To engine

There are types of instruments available called flow meters, which give the rate of flow directly and the calibration can be made in litres per hour or kilograms per hour as required. The Amal flow meter is used widely in industry and is illustrated diagrammatically in Fig. 13.37. A float-controlled constant-level tank A is connected to a vertical glass tube B. The connecting tube carries an orifice at C, and a tap D is fitted between the instrument and the engine. When no fuel is being delivered to the engine the fuel is at the same level in A and B. As fuel is supplied to the engine the level in B falls until the fuel taken by the engine is equal to that passing through the orifice, and the level in B reaches an equilibrium position. The gauge reading is proportional to the fuel consumption and the calibration is an empirical one. Each instrument has two tubes such as B, each with its own range of flow.

Such instruments as those described which depend on volumetric measurement are satisfactory for carburettor engines, but may not be satisfactory for CI

Fig. 13.37 Amal flow meter for engine testing

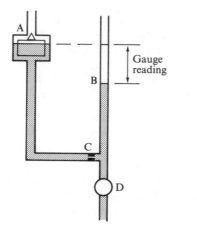

engines with fuel pumps. There may be variation in the volume of fuel between the inlet port of the pump and the measuring instrument, due to entrapped air. If the head of fuel under which the engine operates varies during the fuel measurement (as in the type in Fig. 13.36), the entrapped air in the system expands thus causing an error in the volume of fuel measured.

An instrument which overcomes this difficulty and gives an accurate measurement of the weight of fuel consumed is the BICERA flow meter. In this instrument the fuel during the measuring operation is drawn from a vessel which moves upwards as its weight decreases, thus keeping the fuel level constant. The vessel is attached to an arm pivoted on a block, and a counterweight on a rod also pivoted on the block provides the necessary balancing torque. The weight of fuel consumed is measured by observing the upward movement of the vessel; this is done by an optical system which magnifies the movement of the vessel approximately 30 times.

The satisfactory measurement of air consumption is a more difficult task. It is essentially the measurement of the rate of flow of a compressible fluid, complicated by the fact that the flow is pulsating due to the cyclic nature of the engine. This prevents the use of an orifice in the induction pipe since a steady and reliable reading would not be obtained. The usual method of damping out the pulsations is to fit an air box of suitable volume to the engine with an orifice in the side of the box remote from the engine. The pressure difference across the orifice is steady if the system is correctly designed, and is measured by means of a suitable manometer (see Fig. 13.38).

The criteria for the selection of the size of air-box have been established and

Fig. 13.38 Air box for air flow rate measurement

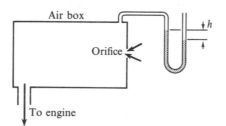

are described in ref. 13.7. It is a conclusion of this reference that the dimensions of a suitable air-box meter can be determined by the use of a non-dimensional factor U, defined by

$$U = \frac{1}{40.94 \times 10^5}\left(\frac{\eta_v C V N^2 n^2}{T d^4 p^2}\right)$$

where η_v is the volumetric efficiency, V the engine swept volume (m^3), C the airbox volume (m^3), N the engine speed (rev/min), n the number of cylinders, T the air temperature (K), d the orifice diameter (m), and p the strokes of the piston per induction stroke.

For reasonable accuracy, U should not be less than 2.5, and the depression across the orifice should not exceed 100–150 mm of water.

Another design of air meter is available for the measurement of airflow, this is the Alcock viscous-flow air meter and it is not subject to the errors of the simpler types of flow meters. With the air-box the flow is proportional to the square root of the pressure difference across the orifice. With the Alcock meter the air flows through a form of honeycomb so that the flow is viscous. The resistance of the element is directly proportional to the air velocity and is measured by means of an inclined manometer. Felt pads are fitted in the manometer connections to damp out fluctuations. The meter is shown diagrammatically in Fig. 13.39. The accuracy is improved by fitting a damping vessel between the meter and the engine to reduce the effect of pulsations.

Fig. 13.39 Alcock viscous flow air meter

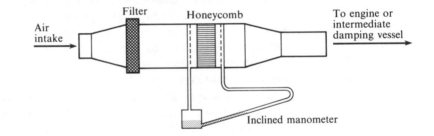

An entirely different principle on which the velocity of an airflow can be measured is that of vortex shedding. As air flows over a bluff body, vortices or local regions of low pressure are created in the flow behind the body as shown in Fig. 13.40. A sensor of some kind, e.g. a pressure transducer, optical or ultrasonic sensor, built into the bluff body, registers the asymmetric conditions in the flow during vortex shedding and the frequency of the disturbance is measured. An alternative method is to record the frequency by a hot-wire anemometer placed in the airstream. The flow velocity C and the frequency f

Fig. 13.40 Air flow measurement using vortex shedding

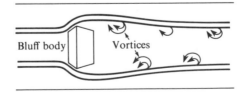

are calculated by the Reynolds number (Cd/v) and the Strouhal number (fd/C). The lower limit of Re is 1000 for vortex shedding to occur after which the Strouhal number is constant for a wide range of Reynolds numbers. Hence the volumetric flow rate is given by $\dot{V} = K \times f$, where K is the calibration coefficient for the instrument which can be refined in value by more elaborate calibration techniques.

The need to be able to measure accurately the instantaneous mass flow rate of air into an engine, or possibly into each engine cylinder, is an important requirement of modern transducer and instrument design and engine development work. One method of development is to use the pressure drop across an orifice or throttle plate, accurately measure and record the pressure difference, the air pressure and temperature, and then compute the air mass flow rate using modern microcomputing techniques. Another development is to seek new types of transducers or sensors which will respond, preferably in a linear manner, to the mass flow rate of air.

Such rapid response devices are particularly desirable when engine intakes suffer reversals in the direction of the airflow, and one principle on which a suitable unit may be devised is that of *corona* discharge, see Fig. 13.41. A thin wire maintained at a potential of 10 kV is fitted across the direction of the airflow and the discharge between the wire and the electrode collectors is disturbed by the airflow such that the discharge current to the electrodes is proportional to the mass flow rate of the air.

Fig. 13.41
Measurement of air flow rate using a corona discharge method

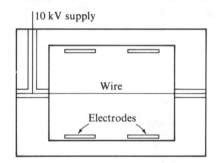

13.12 Supercharging

Figure 13.14 (p. 439) shows how the power output of an engine is affected by the reduction in volumetric efficiency at increased engine speed. The purpose of supercharging is to raise the volumetric efficiency above that obtained with normal aspiration. The main attraction of supercharging engines is to obtain a high power output from a small engine hence giving a good power–weight ratio with a corresponding saving in space which is important in some applications such as road and rail vehicles. In this case the engine is designed as a supercharged engine to withstand the higher loads and temperatures reached in supercharging compared with normal aspiration.

Greater benefits are to be expected from supercharging the diesel engine than from the petrol engine because of the different methods of charging the cylinders and the quite different combustion characteristics of the two types of engine. The diesel induces air only and the fuel is injected under pressure into the cylinder with self-ignition of the fuel in the air; the petrol engine induces a mixture of air and fuel which is spark ignited and burns as described in sections 13.6 and 13.7 giving fundamental combustion problems which do not occur with the diesel engine. To avoid charge detonation or 'knocking' in the petrol engine, giving uncontrolled combustion, the compression ratio may have to be reduced, an action which adversely affects the thermal efficiency of the engine. Alternatively a fuel of higher octane rating may be necessary. With the diesel engine higher boost pressures can give more satisfactory combustion conditions with a wider range of usable fuels, reduced delay periods, controlled pressure rise and an engine which is smoother and quieter in operation.

The main features of supercharging are illustrated in the p–V diagrams for the idealized constant-volume four-stroke cycle in Fig. 13.42 and the plant line diagrams in Fig. 13.43. Figure 13.42(a) shows the normally aspirated cycle with line 1–5 representing both the induction and exhaust strokes at about the ambient air pressure p_a. The early applications of supercharging were for piston-engined aircraft in which the 'blower' was driven mechanically from the engine as shown in Fig. 13.43(a). The power output of the engine was increased by the higher flow of air, and hence the fuel consumed, but part of this increase in power was required to drive the blower. The effects on the p–V diagram, as shown in Fig. 13.42(b), are to increase the pressures (and temperatures) reached during the cycle and to give a positive pumping loop, 15671, to add to the main working loop 12341.

Fig. 13.42 Pressure–volume diagram for a four-stroke CI engine with supercharging (a) and without supercharging (b)

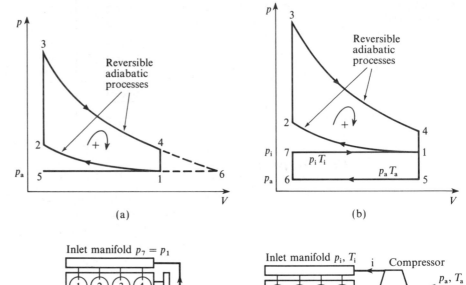

(a)   (b)

Fig. 13.43 Diagrams of a four-stroke, four-cylinder CI engine with mechanical supercharging (a) and with turbocharging (b) (see p. 466)

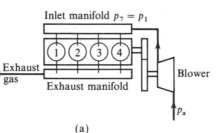

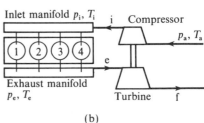

(a)   (b)

The power required to drive a blower mechanically connected to the engine must be subtracted from the engine output to obtain the net bp of the supercharged engine. Then

$$\text{imep} = \left(\frac{\text{area } 12341 + \text{area } 15671}{\text{length of diagram}} \right) \times \text{constant} \qquad (13.16)$$

and $\quad \text{bp} = (\eta_M \times \text{ip}) - (\text{power to drive blower}) \qquad (13.17)$

(for mechanically driven blowers only).

Example 13.3   The average ip developed in a CI engine is 12.9 kW/m³ of free air induced per minute. The engine is a 3-litre four-stroke engine running at 3500 rev/min, and has a volumetric efficiency of 80%, referred to free air conditions of 1.013 bar and 15 °C. It is proposed to fit a blower, driven mechanically from the engine. The blower has an isentropic efficiency of 75% and works through a pressure ratio of 1.7. Assume that at the end of induction the cylinders contain a volume of charge equal to the swept volume, at the pressure and temperature of the delivery from the blower. Calculate the increase in bp to be expected from the engine. Take all mechanical efficiencies as 80%.

Solution   Engine capacity = 3 litres = 0.003 m³

Swept volume $= \dfrac{3500}{2} \times 0.003 = 5.25 \text{ m}^3/\text{min}$

Unsupercharged induced volume $= 0.8 \times 5.25 = 4.2 \text{ m}^3/\text{min}$

Blower delivery pressure $= 1.7 \times 1.013 = 1.72 \text{ bar}$

Temperature after isentropic compression $= 288 \times 1.7^{(1.4-1)/1.4}$

$$= 335.2 \text{ K}$$

Therefore

$$\text{Blower delivery temperature} = 288 + \left(\frac{335.2 - 288}{0.75} \right) = 351 \text{ K}$$

The blower delivers 5.25 m³/min at 1.72 bar and 351 K.

$$\text{Equivalent volume at } 1.013 \text{ bar and } 15 \,^\circ\text{C} = \frac{5.25 \times 1.72 \times 288}{1.013 \times 351}$$

$$= 7.32 \text{ m}^3/\text{min}$$

therefore

Increase in induced volume $= 7.32 - 4.2 = 3.12 \text{ m}^3/\text{min}$

Increase in ip due to the increased induction pressure

$$= \frac{(1.72 - 1.013) \times 10^5 \times 5.25}{10^3 \times 60} = 6.2 \text{ kW}$$

i.e.   Total increase in ip $= 40.2 + 6.2 = 46.4 \text{ kW}$

therefore

$$\text{Increase in engine bp} = 0.8 \times 46.4 = 37.1 \text{ kW}$$

From this must be deduced the power required to drive the blower.

$$\text{Mass of air delivered by blower} = \frac{1.72 \times 10^5 \times 5.25}{60 \times 287 \times 351} = 0.149 \text{ kg/s}$$

$$\text{Work input to blower} = \dot{m}c_p(351 - 288) = 0.149 \times 1.005 \times 63$$

therefore

$$\text{Power required} = \frac{0.149 \times 1.005 \times 63}{0.8} = 11.8 \text{ kW}$$

therefore

$$\text{Net increase in bp} = 37.1 - 11.8 = 25.3 \text{ kW}$$

Figure 13.42(a) shows the start of the exhaust process at 4 at a pressure substantially greater than the ambient, p_a. This means that over 60% of the cylinder charge is suddenly exhausted by a free expansion which constitutes a considerable loss of the energy released on combustion; of about 30–40% between diesel and petrol engines. The attraction of 'turbocharging' is evident as the energy lost in this way is used to drive a turbine wheel integral with a compressor wheel which delivers compressed air or charge to the cylinder. The additional work available from the gas is indicated, after continuing the reversible adiabatic expansion line 3–4 down to the pressure p_a at 6, by area 4614. The physical arrangement is shown in Fig. 13.43(b) and there is no mechanical connection to the engine. The turbocharger combination is a free-running unit with approximately equal mass flow rates over the turbine and compressor wheels reaching an equilibrium speed in the range 20 000–80 000 rev/min.

The simplest form of the supercharged cycle shows constant pressures created in the inlet manifold, p_i, and in the exhaust manifold, p_e, and it is essential that $p_i > p_e$, see Fig. 13.44(a). This pressure difference, $p_i - p_e$, can be utilized to scavenge residual gas from the combustion chamber if there is some overlap between the exhaust and inlet valve operation and particularly so for the diesel engine. This is called *constant pressure* supercharging and requires a large enough exhaust manifold to create a constant pressure supply to the turbine from a highly pulsating delivery from the engine cylinders. The T–s diagram for the turbocharger is shown in Fig. 13.44(b) and, using the methods of Chapter 9, the energy balance for the unit is obtained as follows:

The compressor power input

$$\dot{W}_c = \dot{m}_a c_{pa} T_a \left[\left(\frac{p_i}{p_a} \right)^{(\gamma_a - 1)/\gamma_a} - 1 \right] \bigg/ \eta_c$$

The turbine power output,

$$-\dot{W}_T = \dot{m}_e c_{pe} T_e \left[1 - \left(\frac{p_a}{p_e} \right)^{(\gamma_g - 1)/\gamma_g} \right] \times \eta_T$$

Fig. 13.44 Pressure–volume diagram (a) and T–s diagram (b) for the engine and turbocharger for 'constant-pressure' supercharging

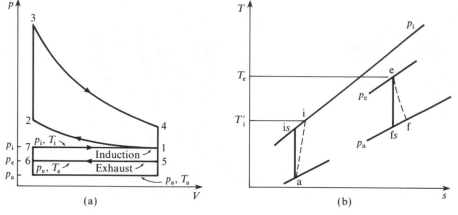

where η_c and η_T are the isentropic efficiencies of the compressor and turbine respectively.

For a balance of mass flow rates

$$\dot{m}_e = \dot{m}_a + \dot{m}_f, \qquad \frac{\dot{m}_e}{\dot{m}_a} = 1 + \frac{\dot{m}_f}{\dot{m}_a} = 1 + F/A$$

where F/A = fuel to air ratio = $\dot{m}_f/\dot{m}_a$.

Also $\dot{W}_c = -\dot{W}_T \times \eta_M$, where η_M = mechanical efficiency of the drive, therefore

$$\left[\left(\frac{p_i}{p_a}\right)^{(\gamma_a - 1)/\gamma_a} - 1\right] = \left[1 - \left(\frac{p_a}{p_e}\right)^{(\gamma_g - 1)/\gamma_g}\right]\frac{c_{pe}}{c_{pa}}\left(\frac{T_e}{T_a}\right)(1 + F/A) \times \eta_0$$

where $\eta_0 = \eta_M \times \eta_T \times \eta_C$ = the overall efficiency of the turbocharger.

This expression shows how the manifold pressure p_i depends mainly upon η_0 and T_e as the effect of the F/A ratio is small. A set of characteristics can be drawn of p_i/p_a against η_0 for different values of T_e. A set is obtained for each value of p_e and the minimum requirement to sustain the limit is $p_e = p_i$ as shown in Fig. 13.45.

For example for $p_i/p_a = 2$, and $T_3 = 773$ K (500 °C) $\eta_0 = 42\%$. The overall efficiency is higher for lower values of p_e and higher values of p_i and T_3. The study of turbocharging should continue into the design of the turbocharger unit to meet its service requirements, but this is outside the scope of this book and specialist references such as 13.8 should be consulted.

The usual arrangement for a turbocharger is a single-stage centrifugal compressor driven by a single-stage axial-flow turbine for the medium and large-size engines for industrial, rail, and marine applications, and by a radial-flow turbine for the smaller engines used in automotive applications, transport vehicles and cars. It is somewhat against earlier expectations that most car manufacturers now include supercharged petrol-engined cars in their product range with apparent overlap with their normally aspirated engines of different capacities. This is in spite of the fact that cars spend a great deal of their time at part throttle and that additional control is necessary to restrict the boost pressure and prevent the onset of knock by retarding the ignition.

Fig. 13.45
Turbocharger
characteristics of
pressure ratio against
turbocharger overall
efficiency for various
exhaust gas temperatures

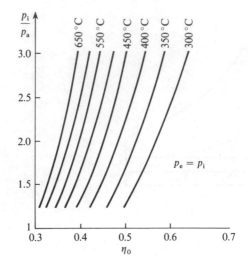

The above description has been confined to the constant-pressure charging of four-stroke cycle engines. The blowing of the two-stroke engine is attractive as the cycle does not include a separate exhaust stroke, and a means of improving the scavenging process would improve the breathing and hence the power output. The two-stroke is particularly sensitive to exhaust back pressure which is increased by turbocharging and its use would require additional care in port timing and the matching of characteristics.

In any cycle the gas leaves the engine cylinders at high speed through the opening valve and possesses high kinetic energy which is dissipated in the large manifold of the constant-pressure system. To utilize the kinetic energy more fully a *pulsed* system of charging can be used provided the turbocharger is designed to cope with the conditions created. For multicylinder engines the cylinders are grouped, taking cylinders alternately from their firing order for entry into the turbine and this also groups them into the front and rear cylinders, as shown in Fig. 13.46. To improve the charging of the cylinders the compressed air can be cooled by passing it through a water- or air-cooled intercooler thereby increasing its density.

Fig. 13.46 Grouping of
cylinders in a
turbocharged
multicylinder CI engine

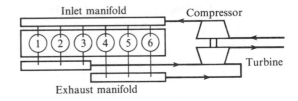

The turbine and compressor are complex devices and further study will show that the complexities of design, matching of the units to each other and to the engine, etc. are such that the performance may not be satisfactory over the whole speed range. Compound engines have been proposed, and some have been built which are turbocharged and have a mechanical connection between the engine crankshaft and the turbocharger as shown in Fig. 13.47(a). A fixed

Fig. 13.47 Two compound arrangements of CI engine and gas turbine

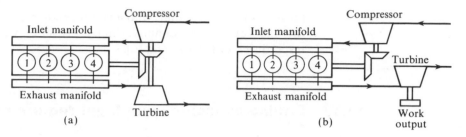

(a)

(b)

drive improves the low-speed performance but restrains the turbine at high speeds. A variable-speed drive is called for but this is an expensive addition. An extreme arrangement for a compound engine is shown in Fig. 13.47(b) in which the engine drives the compressor only and the exhaust gas drives an independent turbine from which the power output of the combination is taken.

Figure 13.48 shows the continuous running performance characteristics of a diesel engine in three modes: normally aspirated (a), turbocharged (b), and turbocharged with intercooling (c). The engines are of 6.6 l capacity

Fig. 13.48 Continuous running performance characteristics of a CI engine in three modes: (a) normally aspirated; (b) turbocharged; (c) turbocharged with intercooling

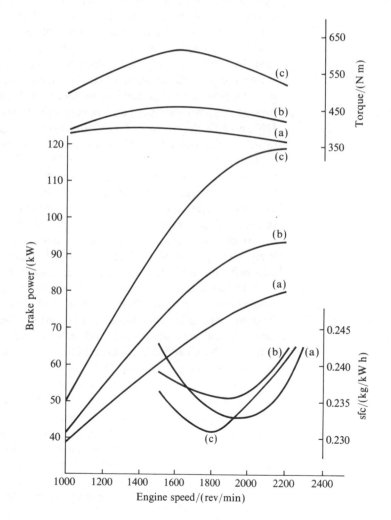

(bore = stroke = 112 mm) with a compression ratio of 16.3 for (a) and 15.6 for (b) and (c). The maximum cylinder pressures were 90, 121, and 138 bar respectively for (a), (b), and (c).

13.13   Engine emissions and legal requirements

An atmospheric phenomenon called 'smog' was experienced on an increasing scale in the State of California, USA, around the city of Los Angeles until it became a severe social problem with political and subsequently commercial and engineering consequences which have had a fundamental effect upon the design and operation of the IC engine.

Smog is a type of light fog which is unpleasant and a cause of irritation to the eyes and nasal passages, and although not affecting visibility greatly does affect vegetation and has caused serious economic losses in horticulture and agriculture. Smog is created by the action of sunlight on hydrocarbons (HC) in the atmosphere and the main source of HC is the exhaust gases of motor vehicles, the rapid increase in volume of town traffic causing the increase in smog. It was against a background of increasing public nuisance and distress that, in 1960, the State of California legislated against the motor vehicle and stated the procedures against which vehicles would be tested and the levels of pollutants which would have to be met by vehicles to be sold in that State. The procedures were extensive and covered the types of emissions to be controlled, HC and CO, the sources of emissions (exhaust, crankcase, carburettor, and fuel tank), the test procedures, the instrumentation and the test equipment to be used.

The initial aim of the Californian Motor Vehicle Pollution Control Board was to reduce the total exhaust emissions to pre-1940 levels by 1980. This was to be done by legislation against new vehicles, as no requirement was made for existing vehicles to be modified, and included the effects of a continuous increase in the number of vehicles on the road. Although mainly American manufacturers of vehicles were affected so were the manufacturers of imported vehicles and these were penalized by having to meet prescribed rates of emission output, and not total amounts, as their engines were generally of small capacity.

The legislation was introduced before the process of smog creation was understood and it was believed that CO and unburned HC were the cause. Mixture strengths were weakened to reduce the CO and HC output, but the smog levels were unaffected and it was later established that the oxides of nitrogen (NO_x) and the unburned HC combined photochemically in the strong sunlight to create the smog. The Californian legislation was followed by that of the US Federal Government through the Environmental Protection Agency (EPA) although smog was not the primary concern in other parts of the USA where the atmospheric conditions for its creation did not occur. The Californian and federal authorities each have emission standards.

The introduction of emission controls for motor vehicles has spread in the wake of the legislation introduced in the USA to other countries including Japan, the EC, Canada, Sweden, the UK, Australia, and Finland. In Europe the most significant effort has been to reduce the use of leaded fuels as a means of raising the knock rating of petrol. The reduction of the lead content in fuels is a response to the demand to reduce the quantity of airborne lead particles in the atmosphere, 90% of which are attributed to petrol consumption. Inhaled lead is a smaller contributor to the consumed lead in the human body than food or water but it is not negligible. The lead content of fuel can be reduced if either a lower knock rating or an increase in the cost of fuel processing is accepted. Premium petrol has a research octane number of 98.5 with a lead concentration of 0.4 g/litre. To reduce this to 0.15 g/litre would require an increase in crude oil consumption of 0.5% by mass at the refinery – a much more expensive operation than the proportion suggests – but which can be avoided by accepting the octane rating of 95.5 as a consequence of the reduction in the lead content to 0.15 g/litre. This in turn could lead to a reduction in usable compression ratio, and hence thermal efficiency of the engine, and an increase in fuel consumption. In Japan and North America lead has been removed from petrol because it poisoned the catalyst of the chemical converters fitted to engines to reduce the exhaust emissions and not for environmental reasons related to the lead itself.

The legislation of each country requires progressive reductions to be made in the HC, CO and NO_x content of the exhaust gas from vehicles and this anticipates that the technology will be available to achieve the levels defined by the time stated. The requirement in the USA in 1970 was to obtain a 90% reduction in emissions from the 1970 levels within a stated time-scale. The aims of the different countries are not directly comparable as there are differences in the sampling and measuring techniques over test cycles which are selected to represent the driving patterns of the country concerned. The diesel engine is low on HC and CO emissions but is higher on NO_x and particulate emission, the latter being 10 times that of a comparable petrol engine; the technical objective is to reduce the levels without losing on fuel economy.

Over the years during which emission control has applied the fuel consumption of the smaller-engined vehicles has improved, but the larger engines have given better emission figures at the expense of fuel economy as the engine has had to be 'detuned' by adjusting mixture strength and ignition timing to meet the emission requirements. The optimum settings were restored when catalytic reactors were fitted to the engines to reduce the emission levels chemically by units fitted external to the engine combustion systems.

Some approaches to emission control such as increasing the air–fuels ratio and recirculating part of the exhaust gas through the cylinders reduces the CO and HC output, controls the NO_x, and does not increase the fuel consumption, but is comparatively cheap in the additional facilities it requires on the engine. The modification to basic engine parameters such as those affecting mixture preparation and spark ignition involve small costs, but others which require thermal reactors or catalytic converters to be added as additional units can involve large cost increases. Advanced aspects of engine design involving stratified charge, with torch ignition or open chamber systems, are expensive to produce but have considerable effects on emission levels, increasing the possibilities if taken together with exhaust gas recirculation and a catalytic converter.

A vehicle without emission control contributes to the pollution from several sources with unburned HC coming from the fuel tank vent and the carburettor bowl vent to the atmosphere; from piston blow-by into the crankcase and leaks from fuel lines as well as from the exhaust gas itself which puts, in addition, CO, CO_2, NO_x, and particulates into the atmosphere. Crankcase emissions are dealt with by venting all crankcase fumes directly into the engine intake system to be burned in the engine cylinders.

The exhaust gas emissions are affected by many of the engine variables which also control the completeness of the combustion process, an appreciation of which is essential to an understanding of emission control. The way in which engines are used in the vehicles has a complex effect on emissions, and so it was necessary to devise representative test methods to reproduce on the road running as closely as possible on the test rig. As mentioned earlier, different countries have different test procedures and the following comments can only be regarded as being general to the problem.

A chassis dynamometer is used to load a vehicle to reproduce the road condition with regard to wind and inertia loads, the latter being experienced during acceleration and deceleration. This is not easy to achieve as the tyre/roller contact is quite different from that between tyre and road, and the engine operating parameters may not be the same as those experienced on the road for apparently corresponding conditions.

The drive cycle includes periods of acceleration, deceleration, steady-state cruising, and idling and the exhaust gases are continuously analysed for CO, CO_2, NO, and HC (as a hexane equivalent). The analyses are recorded and may be interpreted by a weighting procedure to give an estimate of the total mass emitted over the test cycle. Acceleration periods have high weight factors and idle periods have low. More recent techniques described as 'constant-volume sampling' include the collection of quantities of gas in bags either during or over all of the test cycle for subsequent analysis to a mass basis.

As the test procedures have developed and experience has been gained, the test techniques have become established in relation to specified instrumentation (see section 7.6).

Emission control is a complex matter which was brought into existence by legislation and has had to develop as a science since. Different countries have emission-control programmes to meet their own requirements albeit related to the experience and procedures pioneered in the USA. The EC countries have their set of tests which have been adopted by member countries to their own requirements to suit their road systems, vehicle population, and atmospheric conditions.

The number of official tests for all of the countries involved is high and the cover is comprehensive including the measurements of HC, CO, CO_2, NO_x and smoke (for diesel engines) with different classifications of tests for vehicles of different size, light- and heavy-duty passenger vehicles and trucks, petrol and diesel engines, the tests including different procedures and modes of testing.

Emissions vary with the engine parameters and tests can be carried out on research engines with the facilities of variable parameters to investigate the characteristics of emissions. Some of these are illustrated in Figs 13.49–53 inclusive, for petrol engines, and serve to show the complexity of the processes involved. It is probably true to say that the legal emission requirements have

Fig. 13.49 Emission characteristics of a SI engine showing HC and NO_x against mixture strength

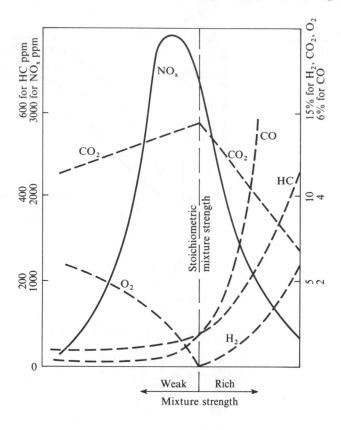

Fig. 13.50 HC emissions for a SI engine against air–fuel ratio

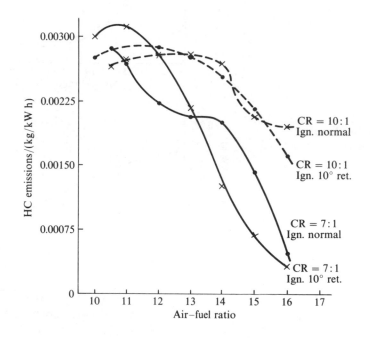

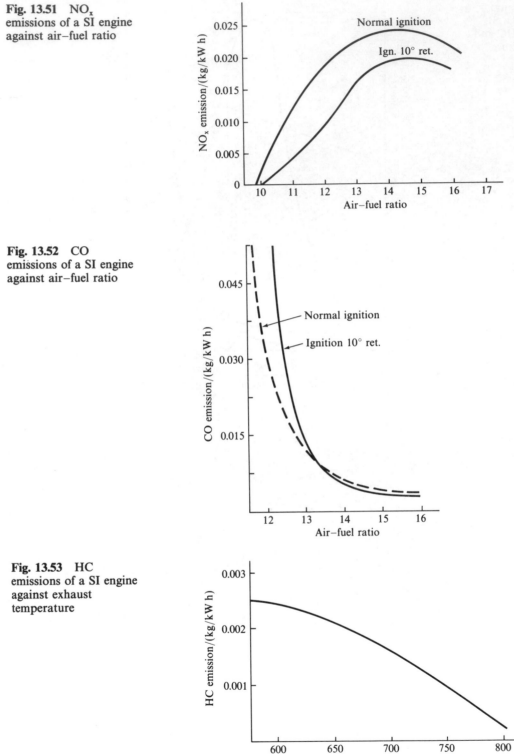

Fig. 13.51 NOₓ
emissions of a SI engine
against air–fuel ratio

Fig. 13.52 CO
emissions of a SI engine
against air–fuel ratio

Fig. 13.53 HC
emissions of a SI engine
against exhaust
temperature

added a new dimension to the development of the IC engine and have probably led to more fundamental research into the process of internal combustion since the onset of legislation than was done before, (see ref. 13.9).

The pollution problem is closely linked to that of fuel economy with the increase in cost of fuel and oil and the need to conserve natural resources. It has therefore become essential to rationalize the experience of generations of engine builders and integrate modern, analytical, computational, and experimental methods into the science of IC engineering. Section 13.15 describes some of the modern developments in IC engines which are a result of advanced understanding of the basic processes and modern techniques. Further progress can be made by studying research papers on individual topics.

13.14 Alternative forms of IC engines

In recent years there have been several attempts to produce power units which would be superior to the IC engine in its conventional forms. Some attempts have been merely to improve the breathing of the reciprocating engine by alternative designs of the valve mechanism. The more ambitious projects have had the object of basic improvements, and have included engines which have a fundamentally different geometry.

The Wankel engine

One engine with this object is the Wankel rotary engine, the most successful of many proposed rotary engine designs. For many years inventors have worked to produce an engine which would fulfil all the promises of the rotary engine concept, with improved power-to-weight ratio, and compete successfully with the reciprocating engine. A simplified representation of the Wankel engine is shown in Fig. 13.54 and the engine is briefly described below.

The rotor, which has a profile defined by three circular arcs PQ, QS, and SP, is attached through a roller-bearing to an eccentric which is an integral part of the engine main shaft. The bearing (and rotor) centre is at an eccentricity e to the mainshaft centre. The rotor radius R to the apex points P, Q and S is the generator radius for the enclosing 'cylinder' or housing. The profile of the cylinder is of epitrochoidal form and is followed by the rotor by means of the epicyclic gear formed by the gear-wheel which is fixed to the casing and the internal gear of the rotor. The pitch circle diameter of the fixed gear-wheel is two-thirds that of the rotor gear-wheel. With this mechanism the rotor turns at one-third of the speed of the mainshaft (or eccentric). As each of the three faces of the rotor is concerned in turn with one power cycle, for each revolution of the rotor the mainshaft receives one working 'stroke' per revolution. Thus the single-rotor Wankel compares with a single-cylinder two-stroke cycle engine.

The rotor is shown in Fig. 13.54 in two positions, one in dotted outline and the other in full line; for the former position the eccentric OA' ($= e$) and the generator radius A'P' ($= P$) are in line to show OP' equal to half the length of the major axis ($= R + e$). The rotor turns clockwise and to give the position

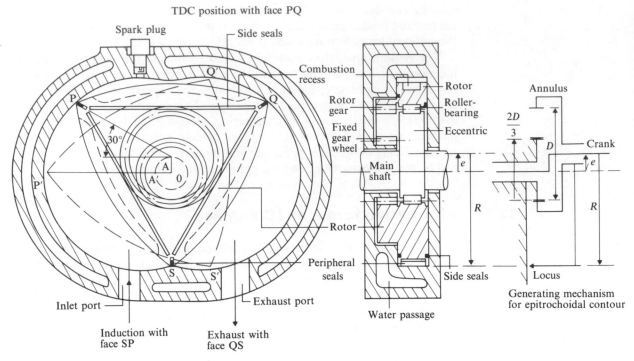

Fig. 13.54 Simplified representation of a Wankel engine

P the mainshaft (or eccentric) has turned through 90° from OA′ and the rotor through 30° (i.e. 90/3). By continuing this construction for other mainshaft positions the epitrochoidal profile can be generated and is repeated for apex S which is displaced 120° with regard to P and similarly for Q. For the full line position OS forms half of the minor axis length ($= R - e$). In the diagram the pitch circle of the bearing coincides with that of the rotor gear for simplicity of drawing.

The three parts of the cylinder are sealed by peripheral seals at P, Q, and S and side seals which are held in the flanks of the rotor on both sides and these constitute a sealing system corresponding to that of the piston rings and valves in the reciprocating engine. The peripheral seals take gas loads only and do not influence the movement of the rotor.

In the full line position shown PQ is at TDC and defines the minimum or clearance volume of the engine. This volume can be varied by forming a recess in the rotor. The maximum cylinder volume will be shown when S takes up a position just on completion of its movement across the inlet port. The difference between the two volumes gives the swept volume of the engine per rotor face. As PQ moves towards the position occupied by QS the working stroke is performed and exhaust begins when the cylinder space is uncovered by the seal as it passes over the exhaust port. Continued movement of the rotor opens the space to the inlet and the induction of the fresh charge and so there is a considerable overlap between exhaust and inlet phases and this can be considered to be a basic limitation of the design. When the apex seal passes over the inlet port, compression of the charge begins up to the TDC position after which the cycle is repeated. To the right of Fig. 13.54 a 'skeleton' figure shows the essentials of the 'generating' mechanism.

The basic design is attractive and offers advantages in comparison with a reciprocating four-stroke engine with regard to compactness for a given power output, fewer working parts, lower mass, lack of vibrations and elimination of the poppet valve. There have been many development problems which are not yet solved to the extent that the automotive market shows a significant conversion to the rotary engine.

Some of the criticisms are basic and include the belief that the combustion space is the wrong shape (long and narrow) for good combustion and the limitations on porting are a disadvantage for good breathing, particularly at low speeds. These are fundamental to the epitrochoidal geometry which also limits the attainable compression ratio, thereby making a diesel version impossible without increased complication such as two cylinders working in series. Another outcome of the geometry is the re-entrant shape or 'cusp' on the minor axis which the peripheral seal finds difficult to follow at high speed because of the reversal of acceleration experienced. The effectiveness of the sealing is held to be in doubt in spite of the considerable development work which has taken place on these. It is also believed that the rotor bearing is subject to adverse conditions, being close to the hot rotor surface without adequate cooling. Solutions to these problems have been applied such as dual, phased ignition and side porting as well as peripheral ports and the use of different seal designs and materials. The assessment and development of the Wankel engine has been undertaken by companies in many countries of the world and for many applications other than for motor vehicle purposes, for much smaller engines and much larger power, lower speed units.

The SI petrol engine is predominant in the light power and high-speed field, and for automobile purposes, but is receiving some competition from the high-speed diesel engine. They have geometries which are basically identical and both are subject to continuous development; it is not likely that they will be replaced to any great extent in their particular field by any other unit for some time. Other power units may appear for particular purposes and may be modifications to the conventional layouts, or they may be combined units.

Free piston engine

The *free piston* engine is a unit of interest and is referred to in section 13.3 in an introductory manner. Two versions are referred to: the free piston air compressor and the free piston gasifier. The free piston air compressor cycle will be described and the gasifier action can be deduced from it. The free piston engine is usually constructed as an opposed-piston, two-stroke diesel cycle with a conventional fuel injection system, but it could be a gas engine or have spark ignition. In Fig. 13.55 the engine/compressor unit is shown on its power stroke. The two diesel pistons are connected mechanically and externally by lightly loaded links through a centrally situated rocking lever. Alternatively the connection may be by a twin rack and pinion arrangement. On the power stroke the air in the compressor cylinder is compressed and then delivered to

Fig. 13.55 Free piston engine operating on a two-stroke CI cycle

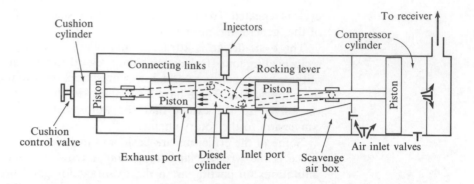

the receiver. At the same time scavenge air enters the compressor cylinder on the underside of the piston, and the air in the cushion cylinder is compressed. Towards the end of this stroke the exhaust port is uncovered and exhaust begins. Slightly later in the stroke air enters from the scavenge air box to complete the exhaust process and to charge the cylinder. The piston then returns on the compression stroke in the diesel cylinder due to the pressure of the air in the cushion cylinder assisted by the air pressure in the compressor cylinder. The air from the underside of the compressor piston is forced into the scavenging air box during this stroke and is ready for the next scavenge process. At the inner dead centre, ignition occurs and the cycle is then repeated.

When used as a gasifier the two pistons have identical cushion cylinder arrangements and no power is taken from the engine. The HP exhaust gas, which is at a moderate temperature (<480 °C), is taken from the engine to an expansion turbine from which the whole of the mechanical power output is taken. This combined unit gives an overall thermal efficiency which is higher than that obtainable with a complete turbine unit fitted with a heat exchanger. The output of the gasifier is given as a *gas power* (gp), which is calculated on the assumption of isentropic expansion of the gas leaving the cylinder, and is the potential power.

The free piston engine can utilize a wider range of fuel oils than the conventional diesel engine and has a higher compression ratio, which indicates a higher thermal efficiency. The compression pressures in the diesel cylinder are in the order of 69 bar, and the thermal efficiency is 40–45%, which gives an overall thermal efficiency of about 35%.

Gas turbines for vehicles

The gas turbine has been a competitor to the piston engine for use in vehicles apart from its natural field of application in the higher power range for aircraft, industrial turbines, power stations, and small ships. For vehicle use the gas turbine appeared as the power unit in a few prototype cars, but it is for the heavier vehicles, trains, buses, and the larger trucks that the gas turbine has more to offer. The small power single-shaft turbines cannot give competitive fuel consumptions to the petrol engine without a highly effective heat exchanger and even then, although the full load figures are good, the part load fuel consumption is poor. The advantages offered by the gas turbine are small volume and mass for a given power output giving a good power-to-weight ratio, a light and relatively simple and robust construction, a potential for burning a wide range of fuels giving a running cost advantage if cheaper fuels are available, and low emission output. Power units can be scaled down to give the required power output, but the inherent source of losses in the engine, e.g. essential running clearances, cannot be scaled down and so some losses are proportionately increased.

In the search for an alternative power unit to the petrol and diesel engines interest in the Stirling engine, see section 5.10, has had some revival. It has external continuous combustion, high thermal efficiency, quiet operation, good torque at low speed, and low emissions, but requires a highly effective heat exchanger, as for the gas turbine, which must be cheap in construction and reliable in operation at high temperatures ($>1000\,^{\circ}\text{C}$) over the life of the engine.

13.15 Developments in IC engines

There is still no serious replacement for the SI and CI engines for the mass-produced power units required for cars and other road vehicles and this is likely to remain so as long as the existing fuels are available. Petrol and diesel engines have been subject to continuous development to improve their performance particularly with respect to fuel consumption and specific power output. The constant volume cycle analysis in section 5.6 shows the fundamental dependence of thermal efficiency on compression ratio. This study can be extended to deal with air–fuel cycles by the selection of appropriate values of γ for the mixture. As the fuel content is increased the value of γ falls and a plot of the thermal efficiency against compression ratio for mixture strengths 0.8, 1.0 and 1.2 is drawn in Fig. 13.56, showing the increase in efficiency obtained with the weaker mixtures. The development problem is to incorporate these principles into practical engine design; prototype engines, radically designed and shown to give improved performances, have appeared only very slowly as production

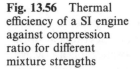

Fig. 13.56 Thermal efficiency of a SI engine against compression ratio for different mixture strengths

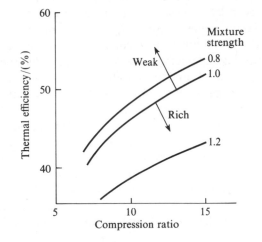

479

engines. The simple reciprocating piston-in-cylinder arrangement offers scope for development mainly in the combustion chamber, the induction and exhaust system, the preparation and movement of the combustible charge, and in the timing and quality of the ignition. The use of multiple valves, e.g. two inlet and two exhaust or two inlet and one exhaust, improves the charging and exhausting of the cylinder if the added complexity and cost are acceptable; the former arrangement requires two overhead camshafts and the latter can be done with one. In addition to the quantity of gas induced, the movement of the charge is important with the gas swirling as the piston descends on induction and through compression until the gas is forced by an increasing amount of 'squish' into the combustion chamber as it is squeezed between the rising piston and the flat face of the exposed cylinder head. The purpose of combustion chamber design is to obtain the necessary characteristics and the various shapes are described as bathtub, hemispherical, pentroof, etc. The fireball design is an advanced form of combustion chamber in which the main part of the volume is concentrated around the exhaust valve and charge movement is encouraged by the shape of path from the inlet valve side to the main part of the chamber round the exhaust valve. This design has shown good improvements in the average fuel consumption figures, with prospects of 20% being achieved with production engines, because of its ability to burn lean mixtures due to the high temperatures obtained with the high compression ratio and the highly turbulent movement of the charge produced by the geometry of the combustion chamber. There are other forms of stratified charge/combustion chambers, such as the divided chamber of the Honda CVCC engine and the Ford PROCO open chamber, which have shown improvements in mixture control and performance.

Modern engine development starts with an understanding of the characteristics of the combustion process arising from the theoretical modelling of this highly complex two-phase reaction which describes the initiation of the ignition and the propagation of the flame through a mixture of the fuel–air charge induced and the proportion of residual exhaust gas in the cylinder. This must describe the nature of the flame front, the rate of burn, the charge turbulence and combustion kinetics. The equations established are complex and require a great deal of computing time. The prediction by such programmes has to be compared with engine results and then the models are further refined to improve the correlation between theory and practice.

Such complex theoretical investigations have only become possible because of the availability of adequate computing facilities, but the impact of the computer and microprocessor have not ended there. The computer is a powerful support to all analytical and experimental work and can be involved directly in the experimental work for data collection, analysis, and control of the experiment. Once the characteristics are known, from theory or experiment, the combustion process must be controlled and increasingly the microprocessor is in evidence as the brain of the engine management system. In the early 1980s engines appeared in production vehicles with programmed ignition and electronic fuel control and the inclusion of knock sensors to warn the ignition control unit of the onset of 'knock' which then retards the ignition of individual cylinders to avoid detonation. Similarly, during warm-up the output of an electrically heated hot spot in the inlet manifold is automatically reduced as the engine warms up and is switched off at the normal running temperature. Fuel supply is controlled by microchip logic as an automatic choke, low speed control, and overrun cut-off. Apart from the control of the basic engine parameters computer electronic control also leads to simpler ignition devices of more robust construction than the traditional contact-breaker type, requiring less maintenance and giving longer life.

The experimental investigation of engine performance has made increasing demands for refined techniques, improved transducers and instrumentation for the measurement of gas flow, turbulence, pressure, temperature, and charge composition throughout the process. It is necessary to be able to take the measurements in production engines under test-bed conditions and also, although it is more difficult, in the vehicle in which the engine is used if that is its application. A modern technique to allow such measurements to be made on the test bed is that which employs lasers such that no intrusion is made into the combustion process, a fault of the earlier techniques which by the presence of transducers or other measuring devices changed the combustion system.

One of the disadvantages of the SI engine is the cycle-to-cycle and cylinder-to-cylinder variation in performance which is readily illustrated by taking a continuous sample of pressure/crank angle diagrams. Investigation has shown that a significant factor in the cycle-to-cycle variation is the size of the eddies created by the induction system. The smaller the eddy size variation about a critical mean size the more consistent the combustion process, but the grids used to control the generation of eddies, although improving the efficiency of combustion, cause restrictions to gas flow and hence reduce the full throttle volumetric efficiency and power output. Greater knowledge of the induction and combustion processes supported by reliable experimental measurements should lead to further improvements in engine power output and fuel efficiency. For further reading see references 13.4 and 13.10.

Problems

13.1 A quality governed four-stroke, single-cylinder gas engine has a bore of 146 mm and a stroke of 280 mm. At 475 rev/min and full load the net load on the friction brake is 433 N, and the torque arm is 0.45 m. The indicator diagram gives a net area of 578 mm^2 and a length of 70 mm with a spring rating of 0.815 bar per mm. Calculate the ip, bp, and mechanical efficiency.

(12.49 kW; 9.69 kW; 77.6%)

13.2 A two-cylinder, four-stroke gas engine has a bore of 380 mm and a stroke of 585 mm. At 240 rev/min the torque developed is 11.86 kN m. Calculate: (i) the bp; (ii) the mean piston speed in m/s; (iii) the bmep.

(298.1 kW; 4.68 m/s; 11.23 bar)

13.3 The engine of Problem 13.2 is supplied with a mixture of gas and air in the proportion of 1 to 7 by volume. The estimated volumetric efficiency is 85% and the $Q_{net,p}$ of the gas is 38.6 MJ/m^3. Calculate the brake thermal efficiency of the engine.

(27.4%)

13.4 A four-cylinder racing engine of capacity 2.495 litres has a bore of 94 mm and a compression ratio of 12/1. When tested against a dynamometer with a torque arm of 0.461 m a maximum load of 622 N was obtained at 5000 rev/min, and at the peak speed of 6750 rev/min the load was 547 N. The minimum fuel consumption was 17.2 ml/s at a speed of 5000 rev/min, the specific gravity of the fuel being 0.735, and $Q_{net,v} = 44\,200$ kJ/kg. Calculate the maximum bmep, the maximum bp, the minimum specific fuel consumption, and the maximum brake thermal efficiency at maximum torque, and compare this latter answer with the air standard efficiency.

(14.44 bar; 178 kW; 0.303 kg/kW h; 26.9%; 63%)

13.5 A three-cylinder, direct-injection, water-cooled, two-stroke oil engine with two horizontally opposed pistons per cylinder has a bore of 82.6 mm and each piston has a stroke of 102 mm. The engine was tested against a brake with a torque arm of 0.381 m. The results taken on a variable speed test are as in Table 13.4. Plot curves of torque, bp, and specific fuel consumption against speed. Convert the torque curve to a bmep curve by calculation of the appropriate scale factor.

Table 13.4 Data for Problem 13.5

Speed	Brake load	Fuel
(rev/min)	(N)	(kg/min)
1000	607.8	0.146
1100	614.6	0.157
1200	621.4	0.172
1300	621.4	0.185
1400	621.4	0.201
1500	621.4	0.216
1600	616.0	0.229
1700	609.0	0.241
1800	596.5	0.252

13.6 A four-cylinder, four-stroke diesel engine has a bore of 212 mm and a stroke of 292 mm. At full load at 720 rev/min the bmep is 5.93 bar and the specific fuel consumption is 0.226 kg/kW h. The air–fuel ratio as determined by exhaust gas analysis is 25/1. Calculate the brake thermal efficiency and the volumetric efficiency of the engine. Atmospheric conditions are 1.01 bar and 15 °C, and $Q_{net,v}$, for the fuel may be taken as 44 200 kJ/kg.

(36%; 76.1%)

13.7 The engine of Problem 13.6 is to be used as a dual-fuel engine. It is to burn methane (calorific value 33 480 kJ/m^3 at 1.013 bar and 15 °C), and has a pilot injection of oil of 10% of the input when running as a diesel engine. The air–fuel ratio for the oil is 25/1 as before, and for the methane 8.5/1. If the volumetric efficiency and the power output remain the same, what is the brake thermal efficiency of the engine when running on the dual fuel?

(23%)

13.8 A four-cylinder petrol engine with a bore of 63 mm and a stroke of 76 mm was tested at full throttle at 3000 rev/min over a range of mixture strengths. The following readings were taken during the test:

Brake load/(N)	162	165.5	169	170	169	162	159
Fuel consumption/(ml/s)	2.08	2.04	2.17	2.50	2.84	3.40	3.56

The relative density of the fuel is 0.724. Calculate the corresponding values of bmep and specific fuel consumption. Plot a consumption loop and obtain from it the corresponding values for maximum power and maximum economy. The bp in kilowatts is given by $WN/26\,830$, where W is the brake load in newtons and N is the engine speed in revolutions per minute.

(8.02 bar; 0.34 kg/kW h; 7.84 bar; 0.28 kg/kW h)

13.9 For the test outlined in Problem 13.8 an air box was fitted which had an orifice of 41.65 mm diameter with a discharge coefficient of 0.6. The corresponding manometer readings in millimetres of water were as follows:

Manometer/(mm water)   33.50   33.50   33.50   33.80   33.80   34.25   34.80

Take the density of the air at inlet as $1.215\ kg/m^3$. Plot to a base of air–fuel ratio graphs of bp and specific fuel consumption. What are the air–fuel ratios at maximum power and maximum economy?

(12.8/1; 15.7/1)

13.10 A four-cylinder petrol engine has an output of 52 kW at 2000 rev/min. A Morse test is carried out and the brake torque readings are 177, 170, 168, and 174 N m respectively. For normal running at this speed the specific fuel consumption is 0.364 kg/kW h. The $Q_{net,v}$ of the fuel is 44 200 kJ/kg. Calculate the mechanical and brake thermal efficiencies of the engine.

(81.6%; 22.4%)

13.11 A V-8 four-stroke petrol engine is required to give 186.5 kW at 4400 rev/min. The brake thermal efficiency can be assumed to be 32% at the compression ratio of 9/1. The air–fuel ratio is 12/1 and the volumetric efficiency at this speed is 69%. If the stroke to bore ratio is 0.8, determine the engine displacement required and the dimensions of the bore and stroke. The $Q_{net,v}$, of the fuel is 44 200 kJ/kg, and the free air conditions are 1.013 bar and 15 °C.

(5.1 l; 100.5 mm; 80.4 mm)

13.12 A four-cylinder, four-stroke diesel engine develops 83.5 kW at 1800 rev/min with a specific fuel consumption of 0.231 kg/kW h, and air–fuel ratio of 23/1. The analysis of the fuel is 87% carbon and 13% hydrogen, and the $Q_{net,v}$, is 43 500 kJ/kg. The jacket cooling water flows at 0.246 kg/s and its temperature rise is 50 K. The exhaust temperature is 316 °C. Draw up an energy balance for the engine. Take $R = 0.302$ kJ/kg K and $c_p = 1.09$ kJ/kg K for the dry exhaust gas, and $c_p = 1.86$ kJ/kg K for superheated steam. The temperature in the test house is 17.8 °C, and the exhaust gas pressure is 1.013 bar.

(bhp 35.8%; cooling water 22.1%; exhaust 25.3%; radiation and unaccounted 16.9%)

13.13 An eight-cylinder, four-stroke diesel engine of 229 mm bore, 304 mm stroke, and compression ratio 14/1, has an output of 375 kW at 750 rev/min. The volumetric efficiency is 78%, the mechanical efficiency is 90%, and the air–fuel ratio is 25/1. If the imep for the pumping loop is 0.345 bar, calculate the imep for the working loop. The engine is now fitted with an exhaust-driven turbo-blower which delivers air to the cylinders at 1.43 bar. The compression ratio is reduced to 13/1 and the measured volumetric efficiency is 102%. It can be assumed that the exhaust pressure remains constant and equal to 1.013 bar, and that the imep of the main loop is directly proportional to the mass of air induced. Calculate the bp which can be expected if the speed and mechanical efficiency remain the same. Compare the specific fuel consumption for the two cases, if the air–fuel ratio for the supercharged engine is 26.8/1. The free air conditions are 1.013 bar and 15 °C.

(7.0 bar; 514 kW; 0.23 kg/kW h; 0.205 kg/kW h)

13.14 A six-cylinder, four-stroke CI engine of 75 mm bore and 100 mm stroke has a brake power output of 110 kW at 3750 rev/min. The volumetric efficiency at this operating condition referred to ambient conditions of 1.013 bar and 20 °C is 80%.

The engine is now fitted with a mechanically driven supercharger which has an isentropic efficiency of 0.7 and a pressure ratio of 1.6. The supercharged version has a volumetric efficiency of 100% referred to the supercharger delivery pressure and temperature. If it is assumed that the indicated power developed per unit volume flow rate of induced air at ambient conditions is the same for normal aspiration and supercharging, calculate the net increase in brake power to be expected from the supercharged engine. Take the mechanical efficiency of the engine as 80% in both cases and the mechanical efficiency of the drive from engine to supercharger as 95%.

(64.1 kW)

References

13.1 *The Mechanical Efficiency of IC Engines* MIRA report no 1958/5

13.2 BS 5514 *Reciprocating Internal Combustion Engines: Performance.* Part 1 1987 *Specification for Standard Reference Conditions and Declarations of Power, Fuel Consumption and Lubricating Oil Consumption*; Part 2 1988 *Test Methods*; Part 3 1979 *Test Measurements*

13.3 INSTITUTE OF PETROLEUM STANDARDS FOR PETROLEUM AND ITS PRODUCTS 1984 Part 1 *Methods for Analysis and Testing*; Part 2 *Methods for Rating Fuels; Engine Tests* Wiley

13.4 HEYWOOD J B 1988 *Internal Combustion Engines Fundamentals* McGraw-Hill

13.5 TAYLOR C F 1977 *The Internal Combustion Engine in Theory and Practice* vols 1 and 2 MIT Press

13.6 WASSENAAR H 1955 Injection phenomena in high-speed diesel engines *Proc. Inst. Mech. Eng. Automobile Division*

13.7 KASTNER L J 1947 Investigation of the air-box method of measuring the air consumption of IC engines *Proc. Inst. Mech. Eng.* **157**

13.8 WATSON N and JANOTA M S 1984 *Turbocharging the IC engine* Macmillan

13.9 *Combustion in engines: technology, applications and the environment* International Conference Dec 1992 IMechE, London

13.10 ARCOUMANIS C (ed) 1988 *Internal Combustion Engines* Academic Press

14

Refrigeration and Heat Pumps

The purpose of a refrigerator is to transfer heat from a cold chamber which is at a temperature lower than that of its surroundings. A temperature gradient is thus established from the surroundings to the chamber and heat will flow naturally in this direction. The heat flow can be resisted by insulating the chamber from the surroundings by the use of suitable insulating materials, but practical requirements and conditions make necessary a continuous means of transfer of heat from the chamber.

Elementary refrigerators have been used which utilize the melting of ice or the sublimation of solid carbon dioxide at atmospheric pressure to provide the cooling effect. The continuous consumption of the refrigerating substance with the means of replenishment required from another source of supply make these methods inconvenient. The temperatures attainable by these methods are limited, but although inefficient for continuous refrigeration, these methods are sometimes convenient forms of cooling in the laboratory and workshop.

The nature of the problem suggests a means of refrigeration which consists of a cycle of processes with the same quantity of working fluid, called the refrigerant, in continuous circulation. If the refrigerant receives energy in the cold chamber at a temperature below that of the surroundings, then this energy must be rejected before the refrigerant can return to the cold chamber in its initial state. This energy rejection must be carried out at a temperature above that of the surroundings. The energy at rejection is of a higher quality, because of its higher temperature, than that received in the cold chamber. This energy can be used for heating purposes and refrigerating plants designed entirely for this purpose are called heat pumps. The term 'heat pump' is appropriate to the action of the plant since energy is transferred against the natural temperature gradient from a low-temperature to a higher one. It is analogous to the pumping of water from a low level to a higher one against the natural gradient of gravitational force. Both actions require an input of energy for their accomplishment. There is no difference in operation between a refrigerator and a heat pump. With the refrigerator the important quantity is the energy removed from the cold chamber called the *refrigerating effect*, and with the heat pump it is the energy to be rejected by the refrigerant for heating purposes. The machine can be used for both purposes and one particular domestic unit provides for the cooling of a larder and the heating of water.

The refrigerating plant chosen depends on the particular purpose since each application has to meet specific requirements. A number of substances are utilized as refrigerants and most methods use the refrigerants in the liquid–vapour states. The reasons for this will be discussed later (see section 14.2).

The choice of a suitable refrigerant depends not only on the thermodynamic, heat transfer, and chemical properties, but also on whether the refrigerant is flammable and/or toxic. Leaks can occur during service and personnel who are involved in the manufacture of the refrigerant or in the commissioning or decommissioning of the plant must not be subjected to poisonous or dangerously inflammable gases. Unfortunately it has been discovered that one group of refrigerants (called CFCs), although non-flammable and non-toxic, are partly responsible for a thinning of the ozone layer in the earth's stratosphere leading to an increase in the ultraviolet radiation reaching the earth's surface from the sun. This important topic is considered in more detail in section 14.11.

For a more thorough treatment of refrigeration the reader is recommended to reference 14.1; a concise analytical treatment of refrigeration plant used for air-conditioning is given in reference 14.2.

14.1 Reversed heat engine cycles

In section 2.1 the First Law of Thermodynamics was stated as follows:

When a system undergoes a thermodynamic cycle then the net heat supplied to the system from its surroundings plus the net work input to the system from its surroundings is equal to zero.

Figure 14.1 shows a reversed heat engine cycle. The effect of the reversed heat engine is to transfer a quantity of heat, Q_1, from a cold source at temperature, T_1.

The reversed heat engine fulfils the requirements of a refrigerator and the First Law of Thermodynamics applied to the system of Fig. 14.1 gives

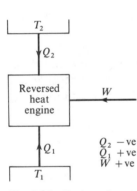

$$\sum dQ + \sum dW = 0$$

$$Q_2 \ -ve$$
$$Q_1 \ +ve$$
$$W \ +ve$$

or $\quad Q_1 + Q_2 + W = 0$

therefore

$$W + Q_1 = -Q_2 \tag{14.1}$$

Fig. 14.1 Reversed heat engine

For a refrigerator the important quantity is the heat supplied to the system from the surroundings, Q_1, and for a heat pump it is the heat rejected from the system, $-Q_2$. The power input, W, is important because it is the quantity which has to be paid for and constitutes the main item of the running cost.

Refrigerator and heat-pump performances are defined by means of the coefficient of performance, (COP), which is given by

$$\mathrm{COP_{ref}} = \frac{Q_1}{\sum W} \tag{14.2}$$

$$\mathrm{COP_{hp}} = \frac{-Q_2}{\sum W} \qquad (14.3)$$

($\mathrm{COP_{hp}}$ is sometimes called the performance ratio.)

The best COP will be given by a cycle which is a Carnot cycle operating between the given temperature conditions. Such a cycle using a wet vapour as the working substance is shown diagrammatically in Fig. 14.2(a). Wet vapour is used as the example, since the processes of constant-pressure heat supply and heat rejection are made at constant temperature, a necessary requirement of the Carnot cycle and one which is not fulfilled by using a superheated vapour.

Fig. 14.2 Reversed heat engine system operating on the Carnot cycle

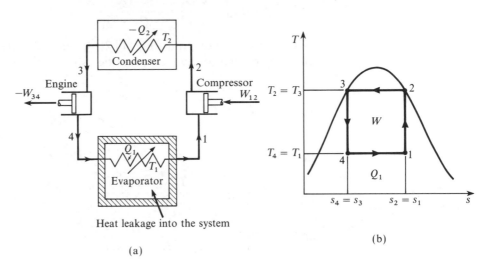

(a)

(b)

The changes in the thermodynamic properties of the refrigerant throughout the cycle are indicated on the T–s diagram of Fig. 14.2(b). The cycle events are as follows:

1–2. Wet vapour at state 1 enters the compressor and is compressed isentropically to state 2. The work input for this process is represented by W_{1-2}.

2–3. The vapour enters the condenser at state 2 and is condensed at constant pressure and temperature to state 3 when it is completely liquid. The heat rejected by the refrigerant is $-Q_2$.

3–4. The liquid expands isentropically behind the piston of the engine doing work of amount $-W_{3-4}$.

4–1. At the lower pressure and temperature of state 4 the refrigerant enters the evaporator where the heat necessary for evaporation, Q_1, is supplied from the cold source.

The boundaries of the system are as shown in Fig. 14.2(a) and therefore

The net work input to the system $\sum W = W_{1-2} + W_{3-4}$

The net heat supplied to the system $\sum Q = Q_2 + Q_1$

and

$$\sum Q + \sum W = 0$$

therefore

$$\sum W = W_{1-2} + W_{3-4} = -Q_2 - Q_1$$

and from equations (14.2) and (14.4)

$$COP_{ref} = \frac{Q_1}{\sum W} = \frac{Q_1}{-Q_2 - Q_1}$$

and

$$COP_{hp} = \frac{-Q_2}{\sum W} = \frac{-Q_2}{-Q_2 - Q_1}$$

From the T–s diagram (Fig. 14.2(b)), since the areas on the T–s diagram are proportional to the heat quantities, then

$$Q_1 = T_1(s_1 - s_4)$$

and

$$-Q_2 = T_2(s_2 - s_3)$$
$$= T_2(s_1 - s_4)$$

therefore

$$COP_{ref} = \frac{T_1(s_1 - s_4)}{(T_2 - T_1)(s_1 - s_4)}$$

and

$$COP_{hp} = \frac{T_2(s_1 - s_4)}{(T_2 - T_1)(s_1 - s_4)}$$

i.e.

$$COP_{ref} = \frac{T_1}{T_2 - T_1} \qquad (14.4)$$

and

$$COP_{hp} = \frac{T_2}{T_2 - T_1} \qquad (14.5)$$

Equations (14.4) and (14.5) give the maximum possible values of COP_{ref} and COP_{hp} between given values of T_1 and T_2, the temperatures of the refrigerant in the evaporator and condenser coils respectively. For steady operation of any system there is a constant mass flow rate, $\dot{m}$, and the power input is $\dot{W} = \dot{m}W$; similarly the rates of heat transfer become $\dot{Q}_1 = \dot{m}Q_1$ and $\dot{Q}_2 = \dot{m}Q_2$.

Example 14.1   A refrigerator has working temperatures in the evaporator and condenser coils of -30 and $32\,°C$ respectively. What is the maximum COP possible? If the actual refrigerator has a COP of 0.75 of the maximum calculate the required power input for a refrigerating effect of 5 kW.

Solution $T_1 = -30 + 273 = 243$ K   and $T_2 = 32 + 273 = 305$ K

From equation (14.4)

$$COP_{ref} = \frac{T_1}{T_2 - T_1} = \frac{243}{305 - 243} = 3.92$$

$$\text{Actual COP}_{ref} = 0.75 \times 3.92 = 2.94$$

Using equation (14.2), where $\dot{Q}_1$ is 5 kW, we have

$$\text{COP}_{ref} = \frac{Q_1}{\sum W} = 2.94$$

therefore

$$\dot{W} = \frac{5}{2.94} = 1.7\,\text{kW}$$

i.e.　　Required power input $= 1.7\,\text{kW}$

The areas representing the quantities Q_1 and W are shown in Fig. 14.2(b). A consideration of these areas and equations (14.2) and (14.3) shows the relationship between COP_{ref} and COP_{hp}.

From equation (14.1)

$$-Q_2 = Q_1 + W$$

Dividing through by W gives

$$\frac{-Q_2}{W} = \frac{Q_1}{W} + 1$$

Then using the definitions of equations (14.2) and (14.3)

$$\text{COP}_{hp} = \text{COP}_{ref} + 1 \tag{14.6}$$

Equation (14.6) indicates that ideally the heating effect of a heat pump is greater than the work input, and this suggests that it would provide an effective heater. Since a heating effect can be obtained in a number of ways which have a much lower capital cost than a heat-pump system, a careful economic analysis is necessary before deciding whether a heat pump is financially viable.

It should also be noted that a heat pump used for heating is much less effective when using the atmosphere as its low-temperature heat source. This is illustrated in Fig. 14.3. The performance of a heat pump varies with the temperature of the source of heat supply as shown in the figure; the heat loss

Fig. 14.3 Heat pump balance point

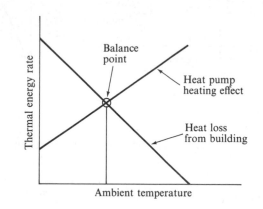

489

from a building varies linearly with ambient temperature as shown. There is therefore only one point, called the *balance point*, at which the heat pump will satisfy the demand. At ambient temperatures above the balance point the heat pump provides too much heat, and at temperatures below the balance point supplementary heating will be necessary to maintain the building at the required internal temperature. A heat pump used for heating is much more effective if the temperature of the low-temperature source is constant or approximately constant throughout the heating season; this is the case if ground water, or a river, or a lake, or the sea, is used as the source.

Heat pumps are being used in increasing numbers for energy recovery (see Ch. 17), and also in plants where both heating and refrigeration are required (e.g. supermarkets and hypermarkets): reference 14.3 gives a concise analytical treatment. For a full treatment of heat pumps consult reference 14.4.

The definitions of equations (14.4) and (14.5) show that the values of COP_{ref} and COP_{hp} decrease with increased temperature difference $(T_2 - T_1)$. In a practical unit this temperature difference is increased above that between the source and receiver because of the temperature differences required for the purpose of heat transfer. Figure 14.4(a) shows a heat pump which takes low-grade energy from a large river and utilizes it to heat a building. The temperature differences required are indicated on the $T–s$ diagram in Fig. 14.4(b). It can be seen that $(T_2 - T_1) > (T_b - T_a)$. A secondary working fluid is required to take the heat rejected by the refrigerant in the condenser and reject it in the space to be heated. If the temperature limits of the refrigerant are 0 and 40 °C, as indicated in Fig. 14.4(b), then, using equation (14.5), the ideal coefficient of performance of the heat pump is

$$COP_{hp} = \frac{313}{313 - 273} = \frac{313}{40} = 7.825$$

(without using a secondary fluid).

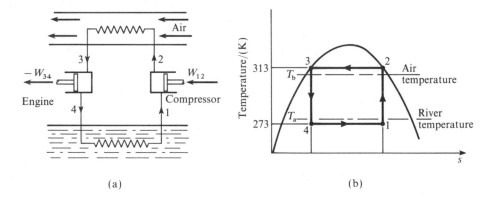

Fig. 14.4 Heat pump working on a reversed Carnot cycle

If it was proposed to use hot water as a secondary fluid at a temperature of 60 °C and heat the space using radiators, then the COP_{hp} becomes $333/60 = 5.55$.

Practical values of the COP_{hp} for these temperature limits might be as low as 3.5. The values calculated using equation (14.5) are ideal values and the difference between these and those for the actual plant are due to the modifications made to the ideal cycle, and to irreversibilities.

14.2 Vapour–compression cycles

The most widely used refrigerators and heat pumps are those which use a liquefiable vapour as the refrigerant. The evaporation and condensation processes take place when the fluid is receiving and rejecting the specific enthalpy of vaporization, and these are constant-temperature and constant-pressure processes. The cycle is one in which these two processes correspond to those of the reversed Carnot cycle for a vapour, and this enables the temperature range for a given duty to be kept low. The resistance to heat transfer during the change of state from liquid to vapour, or from vapour to liquid, is less than that for the refrigerant in the liquid or gaseous states. For a required rate of heat transfer the area of the surfaces required is less if this fact is utilized.

The properties of the various refrigerants must be considered when a selection is to be made for a particular purpose. A high specific enthalpy of vaporization at the evaporator temperature means a low mass flow rate for a given refrigerating effect.

Practical considerations have led to several modifications to the ideal cycle of section 14.1, using a vapour as the working fluid. These will be considered in turn.

Replacement of the expansion engine by a throttle valve

The plant is simplified by replacing the expansion cylinder with a simple throttle valve. Throttling was discussed in section 3.5 and the process was shown to occur such that the initial enthalpy equals the final enthalpy. The process is highly irreversible so that the whole cycle becomes irreversible. The process is represented by the dotted line 3–4 on Fig. 14.5(a). A comparison of Figs 14.5(a) and 14.2(b) shows that the refrigerating effect, $Q_1 = T_1(s_1 - s_4)$, is reduced by using a throttle valve instead of the expansion cylinder.

Fig. 14.5 Reversed cycle using a throttle valve

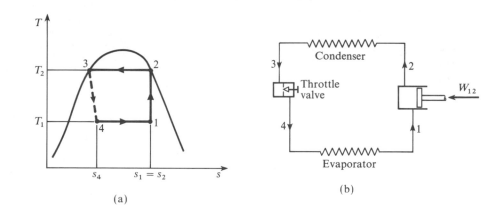

491

Condition at the compressor inlet

To make complete use of the specific enthalpy of vaporization of the refrigerant in the evaporator it is desirable to continue the process until the vapour is dry saturated. In a practical unit this process is extended to give the vapour a definite amount of superheat as it leaves the evaporator. This is really undesirable, since the work to be done by the compressor is increased, as will be shown later. It is a practical necessity to allow the refrigerant to become superheated in this way in order to prevent the carry-over of liquid refrigerant into the compressor, where it interferes with the lubrication. The amount of superheat should be kept to a minimum. The compression process under these conditions is shown in Fig. 14.6 and it is seen that the isentropic compression takes the refrigerant well into the superheat region. The rejection of heat in the condenser cannot now be carried out at constant temperature, and this represents another departure from the ideal reversed Carnot cycle.

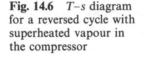

Fig. 14.6 *T–s* diagram for a reversed cycle with superheated vapour in the compressor

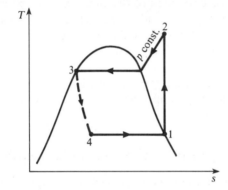

Undercooling of the condensed vapour

The condensed vapour can be cooled at constant pressure to a temperature below that of the saturation temperature corresponding to the condenser pressure. This effect is shown in Fig. 14.7, in which the constant-pressure line is shown further from the liquid line than it would actually appear, in order to illustrate the point. The effect of undercooling is to move the line 3–4,

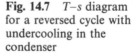

Fig. 14.7 *T–s* diagram for a reversed cycle with undercooling in the condenser

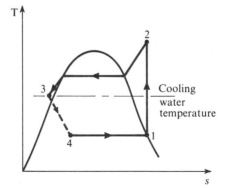

representing the throttling process, to the left on the diagram. The result of this is that the refrigerating effect in process 4–1 is increased. The amount of undercooling is limited by the temperature of the cooling water and the essential temperature difference required for the transfer of heat.

Figure 14.7 includes all the modifications of this section, and this can be taken as showing the practical cycle. The term 'ideal' in this context means that all the individual processes behave exactly as specified; the compression process is isentropic, the refrigerant passes through the evaporator and condenser coils at constant pressure, there are no pressure losses in the interconnecting pipework, and no heat transfers except in the evaporator and condenser, requirements which are not met with in practice.

The cycle consists of a number of flow processes and can be analysed by the application of the steady-flow energy equation, (1.10),

$$h_1 + \frac{C_1^2}{2} + Q + W = h_2 + \frac{C_2^2}{2}$$

If the changes in kinetic energy are neglected, then

$$h_1 + Q + W = h_2$$

Applying this equation to each of the processes in Fig. 14.7, we have for process 4–1:

$$h_4 + Q_1 + 0 = h_1$$

therefore

$$Q_1 = (h_1 - h_4)$$

i.e.   Refrigerating effect $Q_1 = (h_1 - h_4)$ (14.7)

For process 1–2:

$$h_1 + 0 + W = h_2$$

therefore

$$W = h_2 - h_1$$

i.e.   Work done on the refrigerant $= h_2 - h_1$ (14.8)

Equation (14.8) applies equally well to irreversible and reversible compression between states 1 and 2, the only condition being that the process is adiabatic.

If the process is reversible and adiabatic, then it is isentropic (i.e. $s_1 = s_2$).

For process 2–3:

$$h_2 + Q_2 + 0 = h_3$$

therefore

$$Q_2 = h_3 - h_2 = -(h_2 - h_3)$$

i.e.   Heat rejected by the refrigerant, $-Q_2 = h_2 - h_3$ (14.9)

493

For process 3–4:

$$h_3 + 0 + 0 = h_4$$

i.e. $h_3 = h_4$ in throttling process

The solution to numerical problems depends on the means of obtaining the enthalpies h_1, h_2, and h_3. Two methods are available:

1. Using tabulated values of the thermodynamic properties of the refrigerant.
2. Using a chart which gives the properties of the refrigerant, the most useful chart for this purpose being the pressure–enthalpy, p–h, chart. This method is dealt with in section 14.4; charts are available for most common refrigerants.

Example 14.2   The pressure in the evaporator of an ammonia refrigerator is 1.902 bar and the pressure in the condenser is 12.37 bar. Calculate the refrigerating effect per unit mass of refrigerant and the COP_{ref} for the following cycles:

(i) the ideal reversed Carnot cycle;
(ii) dry saturated vapour delivered to the condenser after isentropic compression, and no undercooling of the condensed liquid;
(iii) dry saturated vapour delivered to the compressor where it is compressed isentropically, and no undercooling of the condensed liquid;
(iv) dry saturated vapour delivered to the compressor, and the liquid after condensation undercooled by 10 K.

Solution   (i) The COP_{ref} of a reversed Carnot cycle is given by equation (14.4). The cycle is shown in Fig. 14.8.

$$COP_{ref} = \frac{T_1}{T_2 - T_1} = \frac{-20 + 273}{32 - (-20)} = \frac{253}{52} = 4.86$$

where $t_1 = -20\,°C$ is the saturation temperature corresponding to the evaporator pressure of 1.902 bar and $t_2 = 32\,°C$ the saturation temperature corresponding to the condenser pressure of 12.37 bar; a table of properties of ammonia is given in reference 14.5. Note that $(t_2 - t_1) = (T_2 - T_1)$.

Fig. 14.8 Reversed Carnot cycle for Example 14.2(i)

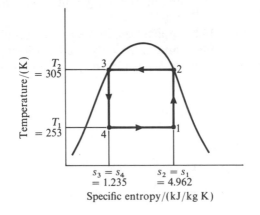

Specific entropy/(kJ/kg K)

The refrigerating effect, Q_1, is given by

$$Q_1 = T_1(s_1 - s_4) = T_1(s_2 - s_3)$$

From tables

$$s_2 = s_g \text{ at } 12.37 \text{ bar} = 4.962 \text{ kJ/kg K}$$

and $\quad s_3 = s_f$ at 12.37 bar $= 1.235$ kJ/kg K

therefore

$$Q_1 = 253(4.962 - 1.235) = 942.8 \text{ kJ/kg}$$

i.e.   Ideal refrigerating effect $= 942.8$ kJ/kg

(ii) The cycle is shown in Fig. 14.9. At 12.37 bar,

$$h_2 = h_g = 1469.9 \text{ kJ/kg} \quad \text{and} \quad h_3 = h_f = 332.8 \text{ kJ/kg}$$

Fig. 14.9 T–s diagram for Example 14.2(ii)

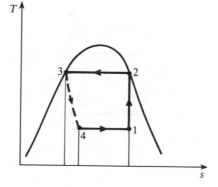

At 1.902 bar,

$$h_4 = h_3 = 332.8 \text{ kJ/kg}$$

To find h_1 use the fact that process 1–2 is isentropic,

i.e. $s_1 = s_2 = s_g$ at 12.37 bar $= 4.962$ kJ/kg K

therefore

$$0.368 + x_1(5.623 - 0.368) = 4.962$$

therefore

$$x_1 = \frac{4.594}{5.255} = 0.874$$

Then from equation (2.2)

$$h_1 = h_{f_1} + x_1 h_{f_{g1}}$$

i.e. $h_1 = 89.8 + \{0.874 \times (1420 - 89.8)\} = 1251.8$ kJ/kg

From equation (14.7)

Refrigerating effect $= h_1 - h_4 = 1251.8 - 332.8 = 919 \text{ kJ/kg}$

From equation (14.8)

Work done on refrigerant $= h_2 - h_1$

i.e. Work done on refrigerant $= 1469.9 - 1251.8 = 218.1 \text{ kJ/kg}$

Then from equation (14.2)

$$\text{COP}_{\text{ref}} = \frac{Q_1}{\sum W} = \frac{919}{218.1} = 4.2$$

(iii) The cycle is shown in Fig. 14.10. At 1.902 bar

$$h_1 = h_g = 1420 \text{ kJ/kg}$$

Fig. 14.10 *T–s* diagram for Example 14.2(iii)

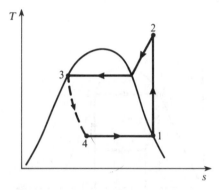

as before

$$h_4 = h_3 = 332.8 \text{ kJ/kg}$$

Also at 1.902 bar,

$$s_1 = s_g = 5.623 \text{ kJ/kg K} = s_2$$

At 12.37 bar, $s_g = 4.962 \text{ kJ/kg K}$, hence the refrigerant is superheated at state 2. Interpolating

$$h_2 = 1613 + \left(\frac{5.623 - 5.397}{5.731 - 5.397}\right) \times (1739.3 - 1613) = 1698.5 \text{ kJ/kg}$$

From equation (14.7)

Refrigerating effect, $Q_1 = (h_1 - h_4) = 1420 - 332.8$

$$= 1087.2 \text{ kJ/kg}$$

From equation (14.8)

Work done on refrigerant $= (h_2 - h_1) = (1698.5 - 1420)$

$$= 278.5 \text{ kJ/kg}$$

Then from equation (14.2)

$$COP_{ref} = \frac{Q_1}{\sum W} = \frac{1077.2}{278.5} = 3.9$$

(iv) The cycle is shown in Fig. 14.11. The values of h_1 and h_2 are as determined for part (iii). The value of $h_3 = h_4$ can be found by assuming that the undercooling takes place along the saturated liquid line, and therefore $h_3 = h_f$ at t_3. This is a good approximation for most refrigerants. Another way of obtaining h_3 is by assuming a constant specific heat, c, for the ammonia liquid, and then

$$h_3 = (h_f \text{ at } t_a) - c(t_a - t_3)$$

Fig. 14.11 *T–s* diagram for Example 14.2(iv)

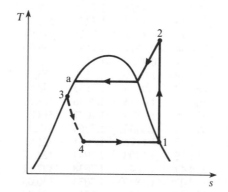

It is usually more convenient to use the first approximation, i.e.

$$h_3 = h_f \text{ at } t_3 = 284.6 \text{ kJ/kg}$$

where $t_3 = 32 - 10 = 22\,°C$. Then, from equation (14.7),

$$\text{Refrigerating effect}, Q_1 = (h_1 - h_4) = 1420 - 284.6$$
$$= 1135.4 \text{ kJ/kg}$$

Also from equation (14.2)

$$COP_{ref} = \frac{Q_1}{\sum W} = \frac{1135.4}{278.5} = 4.08$$

where $\sum W$ is the same as in part (ii).

Example 14.3　Recalculate Example 14.2(iv) for a cycle between the same saturation temperature using Refrigerant (R) 134a instead of ammonia.

Solution　Some properties of R134a are given in Table 14.1; these are derived from properties of the ICI refrigerant KLEA 134a by permission of ICI.

From tables, interpolating for undercooled liquid at 22 °C,

$$h_3 = h_{f3} \text{ at } 22\,°C = h_4 = 126.92 + 0.4(133.89 - 126.92)$$
$$= 129.71 \text{ kJ/kg}$$

497

Table 14.1 Some
properties of R134a

	Saturation values						Superheat			
							10 K		20 K	
t_g	p_g	v_g	h_f	h_g	s_f	s_g	h	s	h	s
(°C)	(bar)	(m³/kg)	(kJ/kg)		(kJ/kg K)		(kJ/kg)	(kJ/kg K)	(kJ/kg)	(kJ/kg K)
−45	0.3908	0.458	42.67	268.49	0.7716	1.7613	276.73	1.7973	284.89	1.8288
−40	0.5188	0.356	48.94	271.40	0.7987	1.7528	279.77	1.7895	288.12	1.8231
−35	0.6612	0.279	55.22	274.30	0.8253	1.7452	283.02	1.7818	291.59	1.8156
−30	0.8436	0.222	61.52	277.21	0.8514	1.7385	286.23	1.7752	295.02	1.8092
−25	1.0638	0.178	67.83	280.12	0.8771	1.7325	289.44	1.7694	298.44	1.8036
−20	1.3271	0.145	74.18	283.03	0.9023	1.7273	292.62	1.7645	301.84	1.7989
−15	1.6389	0.118	80.56	285.94	0.9272	1.7228	295.81	1.7603	305.24	1.7948
−10	2.0051	0.098	86.98	288.86	0.9517	1.7189	299.00	1.7567	308.64	1.7914
−5	2.4371	0.081	93.46	291.77	0.9760	1.7155	302.18	1.7536	312.05	1.7884
0	2.9252	0.068	100.00	294.69	1.0000	1.7128	305.36	1.7511	315.44	1.7861
5	3.4920	0.058	106.61	297.60	1.0238	1.7105	308.55	1.7491	318.84	1.7842
10	4.1390	0.049	113.29	300.50	1.0475	1.7086	311.71	1.7475	322.23	1.7828
15	4.8734	0.042	120.06	303.38	1.0709	1.7071	314.86	1.7463	325.59	1.7817
20	5.7024	0.036	126.92	306.22	1.0943	1.7060	317.97	1.7454	328.93	1.7809
25	6.6337	0.031	133.89	309.03	1.1176	1.7051	321.08	1.7448	332.25	1.7804
30	7.6752	0.027	140.96	311.79	1.1408	1.7044	324.12	1.7444	335.52	1.7802
35	8.8351	0.023	148.15	314.47	1.1641	1.7038	327.11	1.7442	338.75	1.7801
40	10.1219	0.020	155.47	317.07	1.1873	1.7033	330.03	1.7440	341.91	1.7802
45	11.5447	0.017	162.93	319.54	1.2105	1.7028	332.87	1.7440	345.04	1.7804
55	14.8368	0.013	178.33	324.04	1.2572	1.7012	338.01	1.7432	350.74	1.7803
65	18.7960	0.010	194.53	327.62	1.3047	1.6983	342.73	1.7424	356.19	1.7805

The saturation pressure in the evaporator is that corresponding to $-20\,°\text{C}$, i.e. 1.3271 bar,

i.e. $\qquad h_1 = h_{g1} = 283.03 \text{ kJ/kg}$

Also

$$s_1 = s_{g1} = s_2 = 1.7273 \text{ kJ/kg K}$$

The saturation temperature in the condenser is $32\,°\text{C}$, hence using a double interpolating to find h_2:

at $30\,°\text{C}$,

$$h = 311.79 + \frac{(1.7273 - 1.7044)}{(1.7444 - 1.7044)}(324.12 - 311.79) = 318.85 \text{ kJ/kg}$$

at $35\,°\text{C}$,

$$h = 314.47 + \frac{(1.7273 - 1.7038)}{(1.7442 - 1.7038)}(327.11 - 314.47) = 321.82 \text{ kJ/kg}$$

Therefore

at 32 °C,

$$h_2 = 318.85 + 0.4(321.82 - 318.85) = 320.04 \text{ kJ/kg}$$

Then

$$\text{Refrigerating effect} = h_1 - h_4 = 283.03 - 129.71$$
$$= 153.32 \text{ kJ/kg}$$

(compared with 1087.2 kJ/kg for the ammonia cycle).

Also

$$\text{COP}_{\text{ref}} = \frac{h_1 - h_4}{h_2 - h_1} = \frac{153.32}{320.04 - 283.03} = \frac{153.32}{37.01}$$
$$= 4.14$$

(compared with 3.90 for the ammonia cycle).

In most systems the space to be cooled is not cooled directly by the refrigerant in the evaporator. The space is encircled by pipes carrying a secondary fluid, (e.g. a solution of sodium, or calcium, chloride in water). The secondary fluid is cooled in the evaporator of the refrigerator before being pumped through pipes passing through the cold chamber. This introduces a further complication to the system, but when large distances have to be covered by the pipes carrying the cold fluid, the refrigerant leakage problem is reduced by using a secondary fluid. Also, in some cases (e.g. in storage warehouses) there could be damage to goods due to leakage if direct refrigerant cooling were used. The cost of the plant is increased by introducing a secondary fluid, and the cost of pumping it may be an important factor in the running costs. A calcium chloride brine solution is used more often than a sodium chloride solution since it can be applied to cases where the temperature is to be below $-18\,°\text{C}$. For air-conditioning applications the secondary fluid is usually water.

14.3 Refrigerating load

The most important quantity in the application of a refrigerator is the amount of heat which must be transferred per unit time from the cold chamber and is known as the *refrigeration capacity*. The American unit of refrigeration is called the *ton* and is defined as a rate of heat transfer of 200 Btu/min, based on the cooling rate required to produce 2000 lb of ice at 32 °F from water at 32 °F in a time of 24 h, i.e. 1 ton = 200 Btu/min = 3.516 kW. The refrigeration capacity decides the mass flow rate of a given refrigerant when working under specified conditions, i.e.

Mass flow rate of refrigerant, $\dot{m}$

$$= \frac{\text{refrigeration capacity}}{\text{refrigerating effect per unit mass}} \qquad (14.10)$$

Example 14.4   Calculate for the data of Example 14.2(iv) and Example 14.3 the mass flow rates required of ammonia and R134a, and the indicated power of the compressor per kilowatt of refrigeration capacity in each case.

Solution   From Example 14.2(iv), we have for ammonia

Refrigerating effect = 1087.2 kJ/kg

For Example 14.3, we have for R134a

Refrigerating effect = 153.32 kJ/kg

Therefore from equation (14.10)

$$\text{Mass flow rate of ammonia} = \frac{1}{1087.2} = 0.000\,919\,8 \text{ kg/s}$$

$$\text{Mass flow rate of R134a} = \frac{1}{153.32} = 0.006\,522 \text{ kg/s}$$

Also, for Example 14.2(iv)

Work input on ammonia = 273.8 kJ/kg

i.e.   Power input on ammonia = 278.5 × 0.000 919 8 = 0.256 kW

For Example 14.3

Work input on R134a = 37.01 kJ/kg

i.e.   Power input on R134a = 37.01 × 0.006 522 = 0.241 kW

It can be seen from the above example, and from Example 14.3, that the COP_{ref} is greater for the cycle using R134a, and therefore the power input required is less for the same refrigerating capacity. The mass flow rate required for R134a is very much greater than for ammonia (a factor of just over 7), but at the compressor intake the specific volume of saturated ammonia vapour at the suction pressure is $0.6237 \text{ m}^3/\text{kg}$ compared with a value of $0.145 \text{ m}^3/\text{kg}$ for R134a. Hence the volume flow rate of vapour into the compressor per kilowatt of refrigeration capacity is for ammonia

$$0.6237 \times 0.000\,919\,8 = 0.000\,574 \text{ m}^3/\text{s}$$

for R134a

$$0.145 \times 0.006\,522 = 0.000\,946 \text{ m}^3/\text{s}$$

so that the required volume flow rate for R134a is a factor of only about 1.6 greater than for ammonia although the mass flow rate is about a factor of 7 times greater. It is the volume flow rate that determines the size of the compressor and the other cycle components.

14.4 The pressure–enthalpy diagram

Up to this point refrigeration cycles have been represented on a T–s diagram. The pressure–enthalpy diagram is more convenient for refrigeration cycles since the enthalpies required for the calculation can be read off direct. The essential features of the diagram are given in Fig. 14.12, and a typical refrigeration cycle is shown as a p–h diagram in Fig. 14.13. The points 1, 2, 3, and 4 represent the same positions in the cycle as they did in the previous sections; a cycle with undercooling is shown. For the rest of this chapter all refrigeration cycles will be shown on p–h diagrams.

Fig. 14.12 Sketch of a pressure–enthalpy chart for a refrigerant

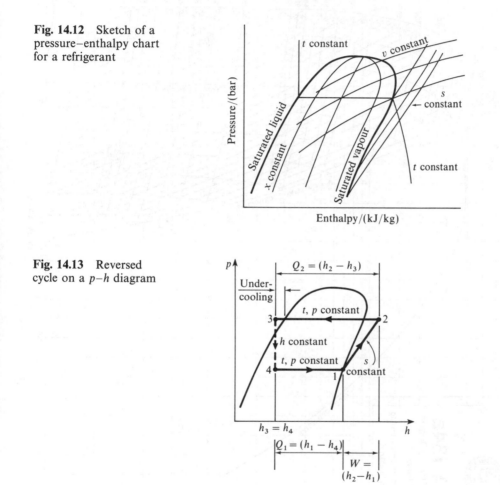

Fig. 14.13 Reversed cycle on a p–h diagram

Example 14.5 A vapour–compression plant using R134a operates with an evaporator pressure of 2 bar and a condenser saturation pressure of 8 bar. The vapour entering the compressor is saturated at 2 bar and the liquid leaving the condenser is saturated at 8 bar. Assuming the compression process is isentropic, calculate the coefficient of performance. Use the p–h chart for R134a.

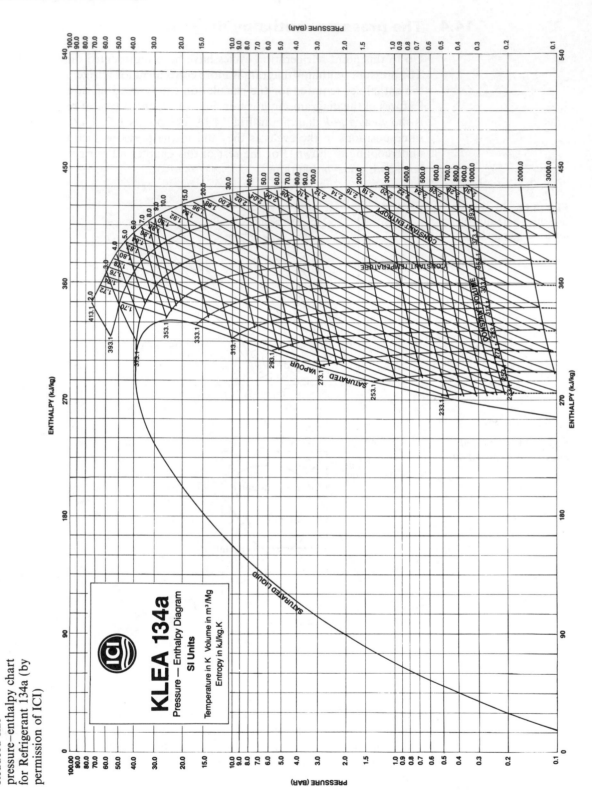

Fig. 14.14
Reduced-size
pressure–enthalpy chart
for Refrigerant 134a (by
permission of ICI)

502

Solution A reduced size pressure–enthalpy chart for R134a is given by permission of ICI as Fig. 14.14. A sketch of the processes on a *p–h* chart is given as Fig. 14.15.

From the chart: at $p_1 = 2$ bar and $T_1 = 273.1$ K the vapour is superheated and $h_1 = 301$ kJ/kg; at $p_3 = 8$ bar, $h_3 = h_{f3} = 145.5$ kJ/kg $= h_4$; at $p_2 = 8$ bar and $s_1 = s_2 = 1.759$ kJ/kg K, $h_2 = 330$ kJ/kg. Then

$$\text{COP}_{\text{ref}} = \frac{h_1 - h_4}{h_2 - h_1} = \frac{301 - 145.5}{330 - 301} = 5.36$$

Fig. 14.15 Pressure–enthalpy diagram for Example 14.5

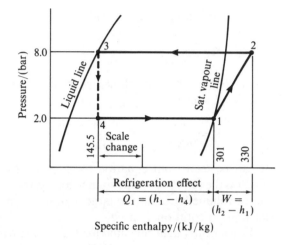

14.5 Compressor type

Both centrifugal and positive displacement compressors are used in refrigeration plant and the general theory is covered in Chapters 11 and 12 with air as the working fluid; the basic theory is the same but with a vapour as the working substance in place of a perfect gas.

Reciprocating compressors are in common use up to a power input of about 600 kW; at the lower end of the power scale, particularly for domestic plant and small air-conditioning units, vane-type compressors are used. Centrifugal compressors are used in the range from about 300 kW up to 15 MW with screw-type compressors also used from 300 kW up to about 3 MW. Centrifugal compressors are ideally suited to high volume flow machines; they run at speeds from 3000 rev/min up to about 20 000 rev/min; several stages are used for high pressure ratios so that the diameter of the impeller can be kept to a reasonable size, thus avoiding destructively high centripetal stresses.

Reciprocating machines run at much lower speeds, usually in the range from about 200 to 600 rev/min, and increased capacity is obtained by using multicylinder machines. The required mass flow of refrigerant is given by equation 14.10. The volume flow rate of the refrigerant drawn into the compressor is given by

$$\dot{V} = \dot{m}v \tag{14.11}$$

where v is the specific volume of the refrigerant at entry to the compressor.

If the compressor is of the reciprocating type and has a volumetric efficiency, η_v, (usually between about 65 and 85%), then the swept volume is given by

$$V_s = \frac{\dot{V}}{nN\eta_v}$$

for a single-acting machine and

$$V_s = \frac{\dot{V}}{2nN\eta_v}$$

for a double-acting machine, where n is the number of cylinders and N the rotational speed,

or $\qquad V_s = \dfrac{\dot{m}v}{nN\eta_v} \quad$ and $\quad \dfrac{\dot{m}v}{2nN\eta_v} \tag{14.12}$

It can be seen from equation (14.12) that the larger the specific volume the larger is the required swept volume for a given mass flow rate. When the required swept volume becomes too large then the number of cylinders can be increased for a given rotational speed.

For a centrifugal machine for the simple case where the refrigerant vapour enters the impeller in an axial direction and leaves in a radial direction with a slip factor of unity, we have from section 11.9

$$\text{Power input} = \dot{m}(C_{be})^2$$

where C_{be} is the blade velocity in the tangential direction at the impeller exit.

For an impeller diameter, D, we have $C_{be} = \pi ND$, hence

$$\text{Power input} = \dot{m}(\pi ND)^2 = \dot{m}(h_2 - h_1) \tag{14.13}$$

Example 14.6   For the refrigeration plant of Example 14.5 the compressor is a double-acting reciprocating compressor with a bore of 250 mm, a stroke of 300 mm, running at a speed of 200 rev/min, with a volumetric efficiency of 85%. Calculate:

(i) the mass flow rate of refrigerant;
(ii) the refrigeration capacity;
(iii) the required power input to the electric motor when the mechanical efficiency is 90%.

Solution   (i) At the compressor inlet the specific volume is found from the chart to be $0.106 \text{ m}^3/\text{kg}$.

The swept volume of the compressor is

$$V_s = \frac{\pi}{4}D^2L = \frac{\pi}{4}(0.25)^2 \times 0.3 = 0.014\,726 \text{ m}^3/\text{cycle}$$

Then using equation (14.12)

$$\dot{m} = \frac{2 \times 1 \times 200 \times 0.85 \times 0.014\,726}{0.106 \times 60} = 0.7873 \text{ kg/s}$$

i.e.   Mass flow of refrigerant $= 0.7873$ kg/s

(ii) Taking the enthalpy values from Example 14.5

Refrigerating capacity $= \dot{m}(h_1 - h_4)$

$$= 0.7873(301 - 145.5)$$

$$= 122.4 \text{ kW}$$

(iii)   Power input from motor $= \dot{m}(h_2 - h_1)$

$$= 0.7873(330 - 301)$$

$$= 22.8 \text{ kW}$$

Example 14.7   A plant using R22 has an evaporator saturation temperature of $-1\,°\text{C}$ and a condenser saturation temperature of $45\,°\text{C}$. The vapour is dry saturated at entry to the compressor and is at a temperature of $75\,°\text{C}$ after compression to the condenser pressure. The compressor is a two-stage centrifugal compressor, each stage having the same pressure ratio and enthalpy rise. Assuming no undercooling in the condenser, a slip factor of unity, axial flow of refrigerant into the compressor, radial flow of refrigerant at the impeller exit, and using the properties of R22 given in Table 14.2, calculate:

(i) the coefficient of performance;

Table 14.2 Some properties of Refrigerant 22

Saturated values

Temp.	Press.	Specific enthalpy		Specific entropy		Specific volume
		h_f	h_g	s_f	s_g	v_g
(°C)	(bar)	(kJ/kg)		(kJ/kg K)		(m³/kg)
-1	4.816	198.83	404.99	0.996	1.753	0.0487
45	17.290	256.40	417.31	1.187	1.693	0.0133

Superheat values at 17.290 bar

Temp.	Specific enthalpy	Specific entropy
(°C)	(kJ/kg)	(kJ/kg K)
65	436.27	1.751
70	440.77	1.764
75	445.21	1.777

    (ii) the power input required for a refrigeration capacity of 2 MW;

    (iii) the diameter of the impeller in each stage when the rotational speed is 300 rev/min.

Solution   (i) A sketch of the processes on a p–h chart is given in Fig. 14.16. Lines 1–i and i–2 represent the actual compression process in each stage and lines 1–is and i–2s represent isentropic compression between the stage pressures. (Note: a chart for R22 is given in reference 14.6.)

Fig. 14.16 Pressure–enthalpy diagram for Example 14.7

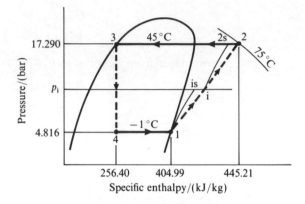

From equation (14.8)

$$\text{Work input} = h_2 - h_1 = 445.21 - 404.99$$

$$= 40.22 \text{ kJ/kg}$$

From equation (14.7)

$$\text{Refrigerating effect} = h_1 - h_4 = 404.99 - 256.40$$

$$= 148.59 \text{ kJ/kg}$$

Then, from equation (14.2)

$$\text{COP}_{\text{ref}} = \frac{148.59}{40.22} = 3.69$$

(ii) For a refrigeration capacity of 2 MW we have

$$\dot{m} = \frac{2 \times 10^3}{148.59} = 13.46 \text{ kg/s}$$

therefore

$$\text{Power input} = 13.46 \times 40.22 = 541.4 \text{ kW}$$

(iii) The work input to each stage is equal to $40.22/2 = 20.11$ kJ/kg, therefore using equation (14.13)

$$20.11 \times 10^3 = \left\{ \frac{\pi \times 3000 \times D}{60} \right\}^2$$

$$D = \frac{60}{\pi \times 3000}(20.11 \times 10^3)^{1/2} = 0.9028 \text{ m}$$

i.e.   Diameter of impeller in each stage = 903 mm

14.6 The use of the flash chamber

Figure 14.17 shows the cycle which was discussed in section 14.2(c). Consider the throttling process 3–4 which shows the refrigerant as an undercooled liquid at state 3, and a wet vapour at 4. Vapour begins to form at the pressure at which the line 3–4 crosses the liquid line, and as expansion proceeds more liquid becomes vapour. In an actual process the change from liquid to vapour is not gradual as might be suggested by the p–h diagram, but the liquid refrigerant is immediately exposed to the evaporator pressure in passing through the valve and some of it 'flashes' into vapour. The vapour is a wet mixture (i.e. a mixture of dry saturated vapour and saturated liquid), and the proportions of dry vapour to liquid are given by the ratio ab/bc at any pressure as shown on Fig. 14.17.

Fig. 14.17 Proportions of dry saturated vapour and saturated liquid on a p–h diagram

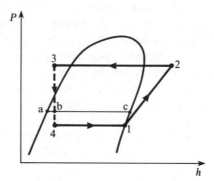

The basic practical cycle of section 14.2 can be improved if an increase in the complexity of the plant is acceptable. It was shown that undercooling of the condensate improved the refrigerating effect per unit mass of refrigerant, and hence increased the COP$_{\text{ref}}$. It would be an advantage if the dry saturated vapour formed by flashing as the liquid expands through the throttle valve did not pass through the evaporator, since it can make no contribution to the refrigerating effect.

Using a flash chamber at some intermediate pressure, the flash vapour at this pressure can be bled off and fed back to the compression process. The throttling process is then carried out in two stages, each one starting with a liquid. Suitable valving is required and the compression process is best done in two separate compressor stages, the flash vapour at the intermediate pressure mixing with the refrigerant after the first stage of compression. Figure 14.18(a) shows a diagrammatic arrangement of the plant, and Fig. 14.18(b) shows the cycle plotted on a p–h diagram. Some care is necessary in making the calculations since the mass flow is not the same in all parts of the circuit.

Fig. 14.18 Two-stage refrigeration cycle using a flash chamber

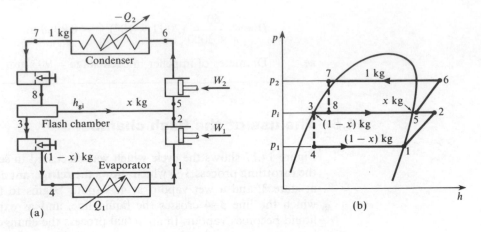

Consider 1 kg of refrigerant flowing through the condenser, and use the subscript i to denote the interstage and flash chamber conditions. At the flash chamber, x kg of dry saturated vapour at pressure p_i and enthalpy h_{g_i} are bled off to the interstage of the compressor. The remaining mass of $(1 - x)$ kg of liquid of enthalpy h_{f_i} passes through the second throttle valve to the evaporator. In the first stage of compression $(1 - x)$ kg of fluid are compressed from p_1 to p_i (i.e. from state 1 to state 2). At the interstage pressure, $(1 - x)$ kg of vapour at state 2 are mixed adiabatically with x kg of flash vapour of enthalpy h_{g_i}. The resultant mixture at some state 5 is compressed in the second-stage compressor from p_i to p_6 at which state it is delivered to the condenser.

The amount of dry saturated vapour bled off is given by the dryness fraction at state 8. The mixture at 8 consists of x kg of dry saturated vapour and $(1 - x)$ kg of liquid for every kilogram of refrigerant in the condenser. The value of x can be determined by equating the enthalpy before throttling at 7 to the enthalpy after throttling at 8, therefore

$$h_{f_i} + x h_{fg_i} = h_7 \quad \text{or} \quad x = \frac{h_7 - h_{f_i}}{h_{fg_i}} \tag{14.14}$$

Using equation (14.8)

$$\text{Total work input} = W_1 + W_2$$
$$= (h_2 - h_1)(1 - x) + (h_6 - h_5) \tag{14.15}$$

Using equation (14.7)

$$\text{Refrigerating effect} = (h_1 - h_4)(1 - x) \tag{14.16}$$

The heat rejected in the condenser is given by equation (14.10),

i.e. $\text{Heat rejected in condenser} = (h_6 - h_7)$

Note that the work and heat quantities given by equations (14.15) and (14.16) are for 1 kg of refrigerant in the condenser.

A consideration of Fig. 14.18 shows that if the compression process were carried out in an infinite number of stages then the compression line would be coincident with the saturated vapour line. The throttling process would show a decreasing amount of liquid in a succession of states from p_6 to p_1 along the liquid line. This cycle is the reverse of that used in steam power plant for regenerative feed heating (see section 8.5).

Example 14.8   A vapour compression plant uses R134a and has a suction saturation temperature of $-5\,°C$ and a condenser saturation temperature of $45\,°C$. The vapour is dry saturated on entering the compressor and there is no undercooling of the condensate. The compression is carried out isentropically in two stages and a flash chamber is employed at an interstage saturation temperature of $15\,°C$. Calculate:

(i) the amount of vapour bled off at the flash chamber;
(ii) the state of the vapour at the inlet to the second stage of compression;
(iii) the refrigerating effect per unit mass of refrigerant in the condenser;
(iv) the work done per unit mass of refrigerant in the condenser;
(v) the coefficient of performance.

Use the extract of properties for R134a given as Table 14.1 (p. 498).

Solution   The cycle is shown on Fig. 14.19.

(i) From equation (14.14)

$$x = \frac{h_3 - h_{f_i}}{h_{fg_i}} = \frac{162.93 - 120.06}{303.38 - 120.06} = \frac{42.87}{183.32} = 0.234$$

i.e.   Vapour bled off $= 0.234$ kg per kg in the condenser.

Fig. 14.19 Pressure–enthalpy diagram for Example 14.8

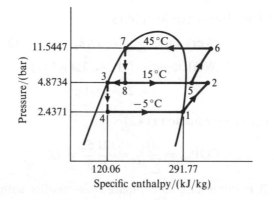

(ii) Adiabatic mixing was dealt with in section 3.4. In this case x kg of dry saturated vapour at $15\,°C$ are mixed with $(1 - x)$ kg of superheated vapour at 11.5447 bar. Since $s_1 = s_2 = 1.7155$ kJ/kg K then the value of h_2 can be found by interpolating from superheat tables,

$$h_2 = 303.38 + \left(\frac{1.7155 - 1.7071}{1.7463 - 1.7071}\right) \times (314.86 - 303.38)$$

$$= 305.84 \text{ kJ/kg}$$

509

From equation (3.33) for adiabatic mixing

$$1 \times h_5 = \{(1 - x) \times h_2\} + x h_{g_i}$$
$$= \{(1 - 0.234) \times 305.84\} + (0.234 \times 303.38) = 305.26 \text{ kJ/kg}$$

Therefore the vapour at inlet to the second stage compressor is still superheated.

(iii) From equation (14.16)

$$\text{Refrigerating effect} = (1 - x)(h_1 - h_4)$$
$$= (1 - 0.234) \times (291.77 - 120.06) = 131.53 \text{ kJ/kg}$$

(iv) The entropy s_5 is found by interpolation at 4.8734 bar and $h_5 = 305.26$ kJ/kg, between the saturated state and state 2,

i.e. $\quad s_5 = 1.7071 + \left(\dfrac{305.26 - 303.38}{305.84 - 303.38}\right) \times (1.7155 - 1.7071)$

therefore

$$s_5 = 1.7071 + 0.00645 = 1.7135 \text{ kJ/kg K}$$

The process 5 to 6 is isentropic, hence interpolating at p_6

$$h_6 = 319.54 + \left(\dfrac{1.7135 - 1.7028}{1.7440 - 1.7028}\right) \times (332.87 - 319.54)$$

therefore

$$h_6 = 323.08 \text{ kJ/kg}$$

Then from equation (14.15),

$$\text{Work input} = (1 - x)(h_2 - h_1) + (h_6 - h_5)$$

i.e. $\quad$ Work input $= \{(1 - 0.234)(305.84 - 291.77) + (323.08 - 305.26)\}$
$$= 28.6 \text{ kJ/kg}$$

(v) From equation (14.2)

$$\text{COP}_{\text{ref}} = \frac{Q_1}{\sum W} = \frac{131.53}{28.6} = 4.6$$

It is interesting to compare these results with those obtained with the basic cycle as shown in Fig. 14.20.

As before, $s_1 = 1.7155$ kJ/kg K, therefore, since $s_1 = s_2$, interpolating

$$h_2 = 319.54 + \left(\dfrac{1.7155 - 1.7028}{1.7440 - 1.7028}\right) \times (332.87 - 319.54)$$

$$= 323.65 \text{ kJ/kg}$$

Then from equation (14.8)

$$W = (h_2 - h_1) = (323.65 - 291.77) = 31.88 \text{ kJ/kg}$$

Fig. 14.20 Basic cycle for Example 14.8

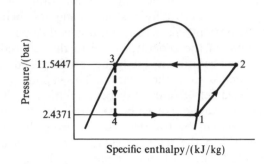

From equation (14.7)

Refrigerating effect, $Q_1 = (h_1 - h_4) = 291.77 - 162.93 = 128.84 \, \text{kJ/kg}$

Also from equation (14.2),

$$\text{COP}_{\text{ref}} = \frac{Q_1}{\sum W} = \frac{128.84}{31.88} = 4.04$$

14.7 Vapour-absorption cycles

Vapour-absorption cycles have been in use for some time, but recently they have become of more interest because of their potential use as part of energy-saving plant; also, they use more environmentally friendly refrigerants than current vapour-compression cycles.

A vapour-absorption system operates with a condenser, a throttle valve, and an evaporator in the same way as a vapour–compression system, but the compressor is replaced by an absorber and generator as shown in Fig. 14.21.

Fig. 14.21 Diagrammatic arrangement of a vapour-absorption system

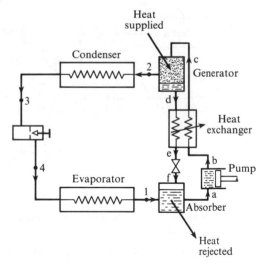

The refrigerant on leaving the evaporator is readily absorbed in a low-temperature absorbing medium, some heat being rejected during the process. The refrigerant–absorbent solution is then pumped to the higher pressure and is heated in the generator. Due to the reduced solubility of the refrigerant–absorbent solution at the higher pressure and temperature, refrigerant vapour is separated from the solution. The vapour passes to the condenser and the weakened refrigerant–absorbent solution is throttled back to the absorber. A heat exchanger placed between the absorber and generator makes the system more efficient by transferring heat from the weak solution coming from the generator to the stronger solution pumped from the absorber, as shown in Fig. 14.21. The work done in pumping the liquid solution is much less than that required to compress the vapour in the compressor of an equivalent vapour–compression cycle. The main energy input to the system is the heat supplied in the generator; this may be supplied in any convenient form such as a fuel-burning device, direct electrical heating, steam if already available, solar energy, or waste heat. A refrigerant and absorbent must be found which have suitable solubility properties. Two combinations are in general use: in one, ammonia is used as the refrigerant with water as the absorbent; in the other, water is used as the refrigerant with lithium bromide as the absorbent. The principle of operation is the same for both types.

One absorption system, known as the Electrolux, dispenses with the pump and uses natural circulation. The system is shown diagrammatically in Fig. 14.22; the refrigerant is ammonia and the absorbent is water, with hydrogen used to enable circulation to take place by natural means due to density differences. The total pressure is constant throughout the system.

Fig. 14.22 Absorption Electrolux refrigeration system

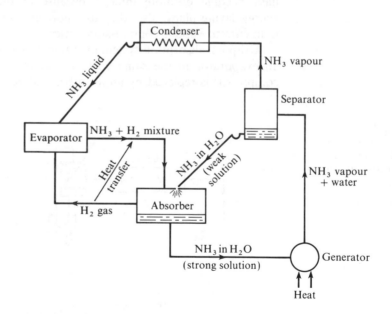

The ammonia liquid leaving the condenser enters the evaporator and evaporates into the hydrogen at the low temperature corresponding to its low partial pressure. The ammonia–hydrogen gas mixture passes to the absorber into which is also admitted a weak ammonia–water solution from the separator. The water absorbs the ammonia vapour, and the hydrogen returns to the evaporator. In the absorber the ammonia vapour passes from the ammonia circuit into the water circuit as a strong ammonia–water solution. This strong solution passes to the generator where it is heated and the vapour given off rises to the separator. The water with the vapour is separated out and a weak ammonia–water solution passes back to the absorber, thus completing the water circuit. The ammonia vapour rises from the separator to the condenser where it is condensed and then returned to the evaporator. An actual plant includes refinements and practical modifications which are not included here. The main advantage of the system is that all the energy supplied can be in the form of heat; no pump is necessary and hence a supply of electricity is not essential to run the plant. The heat could be in the form of solar energy or waste heat.

Example 14.9
A water-chilling plant is to have a refrigeration capacity of 100 kW when operating between condenser and evaporator saturation temperatures of 34 and 3 °C. A vapour-absorption cycle is used with water as refrigerant and lithium bromide as absorbent. The generator is maintained at 80 °C using a supply of waste heat from hot gases, and the absorber is at 25 °C. Assume that the solution leaving the absorber is saturated at the evaporator pressure, and the solutions entering and leaving the generator are saturated at the evaporator pressure. There is no undercooling in the condenser and dry saturated vapour leaves the evaporator. There is a heat exchanger between the absorber and heat exchanger as shown in Fig. 14.21 (p. 511). Assuming that the enthalpy of superheated steam at low pressures may be taken as approximately equal to the saturated value at the same temperature, and neglecting pump work and all pressure and heat losses, calculate:

(i) the mass flow rate of water through the evaporator;
(ii) the heat supplied in the generator;
(iii) the heat rejected in the absorber;
(iv) the COP_{ref};
(v) the specific enthalpy of the solution entering the absorber.

Solution
To solve the problem it is necessary to have information about the properties of lithium bromide–water solutions. A chart of specific enthalpy against concentration is given in Fig. 14.23. When a solution of a particular concentration is saturated at any pressure the temperature is fixed at that concentration. If a saturated solution is heated at constant pressure, water evaporates to maintain saturation at the higher temperature. For this example the pressure in the condenser and generator is the pressure corresponding to the condenser saturation pressure of 34 °C, i.e. from steam tables $p_2 = 0.0538$ bar. The weak solution leaving the generator is saturated at 80 °C and 0.0538 bar and hence the point d can be fixed on the specific enthalpy-concentration chart where the pressure line 5.38 kN/m² cuts the 80 °C temperature line. The concentration and specific enthalpy at point d can then be read off the chart,

i.e. $x_d = 0.6$ and $h_d = -88$ kJ/kg

Fig. 14.23
Restricted specific
enthalpy-concentration
chart for a lithium
bromide–water solution

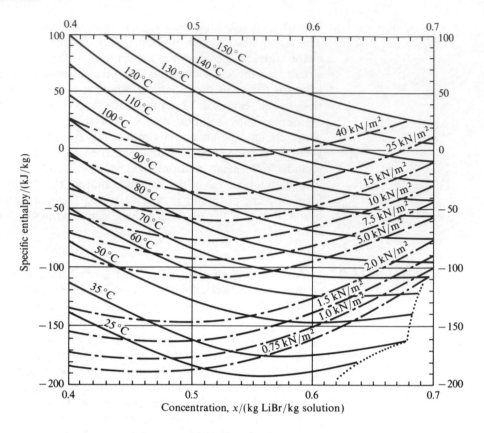

where x represents the concentration of lithium bromide in water at any point.

The pressure in the evaporator and absorber is that corresponding to the evaporator temperature of $3\,°C$, i.e. from steam tables, $p_1 = 0.7575\ kN/m^2$. The strong solution is saturated leaving the absorber at $25\,°C$, hence from the chart

$$x_a = 0.51 \quad \text{and} \quad h_a = -185\ kJ/kg$$

The solution entering the generator at c must have the same concentration as that leaving the absorber at a, i.e. $x_c = x_a$. Hence at the generator pressure of $5.38\ kN/m^2$ and a concentration of 0.51 from the chart we have

$$h_c = -104\ kJ/kg$$

Properties of steam are taken from the tables of reference 14.5.

(i) At $34\,°C$,

$$h_3 = h_{f3} = 142.4\ kJ/kg = h_4$$

(The numbers and letters throughout the solution refer to those of Fig. 14.21 (p. 511).)

For dry saturated steam at $3\,°C$,

$$h_1 = 2506.2\ kJ/kg$$

Then

$$\text{Refrigerating effect} = h_1 - h_4 = 2506.2 - 142.4$$
$$= 2363.8 \text{ kJ/kg}$$

and since the refrigeration capacity is given as 100 kW then from equation (14.10),

$$\text{Mass flow rate of water} = \frac{100}{2363.8} = 0.0423 \text{ kg/s}$$

(ii) Applying a total mass balance to the absorber:

$$\dot{m}_1 = \dot{m}_a - \dot{m}_f$$

i.e. $\dot{m}_a - \dot{m}_f = 0.0423$ [1]

From a lithium bromide mass balance

$$x_1 \dot{m}_1 = x_a \dot{m}_a - x_f \dot{m}_f$$

Now, x_1 is zero, $x_f = x_e = x_d = 0.6$, and $x_a = 0.51$, therefore

$$0.51 \dot{m}_a = 0.6 \dot{m}_f$$ [2]

From [1] and [2],

$$0.51(0.0423 + \dot{m}_f) = 0.6 \dot{m}_f$$

therefore

$$\dot{m}_f = 0.2397 \text{ kg/s} = \dot{m}_d$$

and $\dot{m}_a = 0.2876 \text{ kg/s}$

For the generator:

$$\dot{m}_2 = \dot{m}_1 = 0.0423 \text{ kg/s}$$

and $\dot{m}_c = \dot{m}_b = \dot{m}_a = 0.2876 \text{ kg/s}$

From steam tables, for superheated steam at 80 °C

$$h_2 = h_g \text{ at } 80\,°C = 2643.2 \text{ kJ/kg}$$

Also $h_d = -88 \text{ kJ/kg}, \quad h_a = -185 \text{ kJ/kg}, \quad \text{and} \quad h_c = -104 \text{ kJ/kg}$

Then applying an energy balance to the generator

$$\dot{Q}_G + \dot{m}_c h_c = \dot{m}_2 h_2 + \dot{m}_d h_d$$

i.e. $\dot{Q}_G = (0.0423 \times 2643.2) + (0.2397 \times -88)$

$$- (0.2876 \times -104)$$
$$= 120.60 \text{ kW}$$

(iii) Now for the condenser, taking the heat rejected as $-\dot{Q}_C$,

$$\dot{Q}_C = -m_1(h_2 - h_3) = -0.0423(2643.2 - 142.4)$$
$$= -105.78 \text{ kW}$$

Then applying an energy balance to the complete system, neglecting pump work

$$\dot{Q}_G + \dot{Q}_E + \dot{Q}_C + \dot{Q}_A = 0$$

where $\dot{Q}_E$ is the heat supplied in the evaporator,

i.e. $120.60 + 100 - 105.78 + \dot{Q}_A = 0$

$$\dot{Q}_A = -114.82 \text{ kW}$$

i.e.   Heat rejected in the absorber $= 114.82 \text{ kW}$

(iv) The COP_{ref} can be defined as the heat supplied in the generator divided into the refrigeration capacity since pump work is negligible,

i.e. $$\text{COP}_{\text{ref}} = \frac{100}{120.6} = 0.83$$

(v) From an energy balance on the absorber

$$\dot{Q}_A + \dot{m}_1 h_1 + \dot{m}_f h_f = \dot{m}_a h_a$$

$$-114.82 + (0.0423 \times 2506.2) + 0.2397 h_f = -0.2876 \times 185$$

therefore

$$h_f = -185.2 \text{ kJ/kg}$$

i.e.   Specific enthalpy of solution entering the absorber $= -185.2 \text{ kJ/kg}$

The answer could also be obtained from an energy balance on the heat exchanger,

i.e. $\dot{m}_d(h_d - h_e) = \dot{m}_a(h_c - h_b)$

Also, for throttling $h_e = h_f$, and neglecting pump work, $h_b = h_a$, therefore

$$0.2397(-88 - h_f) = 0.2876(-104 + 185)$$

therefore

$$h_f = -185.2 \text{ kJ/kg (as before)}$$

The solution entering the generator in the above example is stated to be saturated; this implies that the heat transferred to the solution in the heat exchanger is just enough to change its state to the saturated value at the generator temperature. This is the ideal case. If the solution is not raised to the generator temperature it is not saturated and its state cannot be located on the specific enthalpy-concentration chart; more heat must then be supplied in the generator to maintain the given temperature. The worst case would be when there is no heat exchanger; in that case the solution leaving the absorber would be pumped to the generator pressure with negligible increase in specific enthalpy.

Example 14.10   Recalculate the heat required to be supplied in the generator for Example 14.9 when a heat exchanger is not used.

Solution   Referring to Fig. 14.21 (p. 511), when there is no heat exchanger then $h_a = h_c$ when pump work is neglected. For the generator all other factors are as before, therefore

$$\dot{Q}_G + \dot{m}_c h_c = \dot{m}_2 h_2 + \dot{m}_d h_d$$

i.e. $\dot{Q}_G = (0.0423 \times 2643.2) + (0.2397 \times -88)$

$\qquad\qquad - (0.2876 \times -185)$

$\qquad\quad = 143.92 \text{ kW}$

This compares with the value found in Example 14.9 of 120.6 kW. Since the heat supplied in the evaporator and the heat rejected in the condenser remain the same, then it follows that the heat rejected in the absorber is increased by the same amount as the heat supplied in the generator.

14.8   Gas cycles

Gases can be used as refrigerants and the gas cycles are basically similar to those described previously for vapours. The exception made here is that of the means of throttling. It has been shown in section 3.5 that when a perfect gas is throttled then its temperature is unchanged. It has also been shown in section 3.2 that if a gas expands adiabatically and does external work then there is a reduction in temperature. If the expansion process is also reversible and therefore isentropic then the relationship between the initial and final temperature is given by equation (3.21)

$$\frac{T_1}{T_2} = \left(\frac{p_1}{p_2}\right)^{(\gamma-1)/\gamma}$$

It is necessary, in the case of a gaseous refrigerant, to provide an expansion cylinder to replace the throttling process. The plant arrangement is shown in Fig. 14.24(a), and the cycle is represented on a T–s diagram in Fig. 14.24(b). It will be noticed that the cycle is a reversed constant pressure or Joule cycle.

Fig. 14.24   Reversed constant-pressure gas refrigeration system

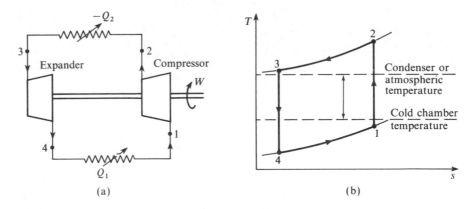

The gas does not receive and reject heat at constant temperature, and this shows a deficiency of a gas as a refrigerant when compared with a liquefiable vapour. The temperature range for the wet vapour is seen from Fig. 14.24(b) to be t_1 to t_3, and the temperature range for the gas is seen to be t_4 to t_2. The significance of this is that the gas cycle will be less efficient than the vapour cycle for given evaporator and condenser temperatures. The volume of refrigerant to be dealt with is much greater for a gas and larger surfaces are required for heat transfer.

The use of a gas as a refrigerant becomes more attractive when a double purpose is to be met. This is so in the case of air-conditioning when the air can be both the refrigerating and the conditioning medium. Another advantage of using air is that it is safe as a refrigerant. The reversed constant-pressure cycle was used in the early days of refrigeration for this reason, the refrigerator using this cycle being known as the Bell–Coleman refrigerator. In modern air refrigeration cycles the large displacement volumes can be handled best by the rotary-type compressor and expander, as these are more compact for a given flow than the reciprocating machines.

Figure 14.24(b) shows the ideal cycle for a plant using a rotary compressor and turbine. The work done by the air expanding is used to help to drive the compressor. The net input to the plant is given by $\sum W = W_{1-2} + W_{3-4}$. Then applying the steady-flow equation to the cycle and neglecting changes in kinetic energy

$$W_{1-2} = (h_2 - h_1) \quad \text{and} \quad W_{3-4} = -(h_3 - h_4)$$

Therefore for a perfect gas

$$W_{1-2} = c_p(T_2 - T_1) \quad \text{and} \quad W_{3-4} = -c_p(T_3 - T_4)$$

The refrigerating effect is obtained from equation (14.7)

$$Q_1 = (h_1 - h_4) = c_p(T_1 - T_4) \quad \text{for a perfect gas}$$

The actual cycle would be as shown in Fig. 14.25 which includes the effects of irreversibilities in the compression and expansion processes. For this cycle we have

$$W_{1-2} = c_p(T_2 - T_1) \quad \text{and} \quad W_{3-4} = -c_p(T_3 - T_4)$$

Fig. 14.25
T–s diagram for an actual reversed constant-pressure cycle

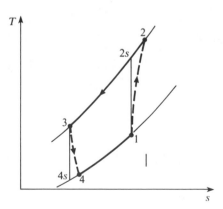

and from the definition of isentropic efficiency in section 9.1, and equations (9.1) and (9.2), we have

$$T_2 = T_1 + \frac{T_{2s} - T_1}{\eta_C} \quad \text{and} \quad T_4 = T_3 - \eta_T(T_3 - T_{4s})$$

where η_C and η_T are the isentropic efficiencies of the compressor and the turbine respectively.

The refrigerating effect in the actual cycle is then reduced to

$$Q_1 = c_p(T_1 - T_4)$$

One application of this system is that met with in modern aircraft practice in which the air delivered to the cabin must be conditioned. Air is bled from the compressor of the engine. The proportion of this air to be used for air-conditioning is passed through a heat exchanger and is cooled by ram air at atmospheric temperature, which passes over the outside of the heat exchanger. This cooling takes place ideally at constant pressure and the cooled air then passes to the refrigerator turbine through which it expands and gives a work output. The air, after expanding approximately adiabatically, is passed to the cabin at a low temperature. The work output of the turbine is used to drive essential auxiliaries such as pumps, and perhaps a fan to draw air over the heat exchanger. As this turbine would not develop sufficient power to drive all the auxiliaries its effort is joined to that of another turbine which uses the rest of the bleed air from the main compressor. The arrangement is shown in Fig. 14.26(a), and the cycle is shown on a T–s diagram in Fig. 14.26(b). The process follows an open cycle, starting from the atmosphere and exhausting to the atmosphere.

Fig. 14.26 Example of an aircraft air-conditioning system (a) with the processes on a T–s diagram (b)

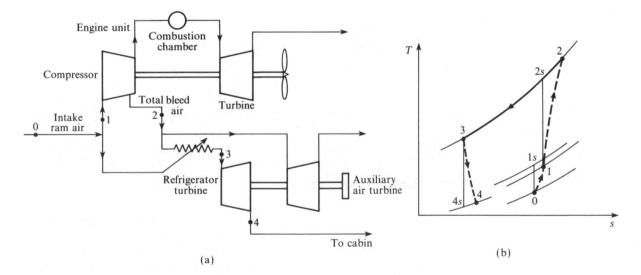

(a)

(b)

Example 14.11   In the air-cooling system of a jet aircraft, air is bled from the engine compressor at 3 bar, and is cooled in a heat exchanger to 105 °C. It is expanded to 0.69 bar in an air turbine, the isentropic efficiency of the process being 85%.

The air is then delivered to the cockpit and leaves the aircraft at 27 °C. Calculate the temperature at which the air enters the cockpit and the mass flow of air required for a refrigerating effect of 4 kW. If the air turbine is used to help to drive the auxiliaries, calculate its contribution in power.

Solution The expansion process is shown by the line 3–4 on the T–s diagram of Fig. 14.26(b).

Using equation (3.21)

$$T_4 = T_3\left(\frac{p_4}{p_3}\right)^{(\gamma - 1)/\gamma} = 378 \times \frac{1}{(3/0.69)^{0.4/1.4}} = 248.4 \text{ K}$$

where $T_3 = 105 + 273 = 378$ K.
Using equation (9.2)

$$T_4 = T_3 - 0.85(T_3 - T_4)$$

where the isentropic efficiency is 0.85,

i.e. $T_4 = 378 - 0.85(378 - 248.4) = 267.8$ K $= -5.2\,°$C

therefore

The air enters the cockpit at $-5.2\,°$C

Refrigerating effect per unit time $= \dot{m}c_p(t_C - t_4)$

where t_C is the temperature of the air leaving the cockpit.
Now the refrigerating effect per unit time is given as 4 kW,

i.e. $4 = \dot{m} \times 1.005(27 - (-5.2))$

therefore

$\dot{m} = 0.124$ kg/s

Power developed by turbine $= \dot{m}c_p(T_3 - T_4)$

$$= 0.124 \times 1.005 \times (378 - 267.8) \text{ kJ/s}$$

$$= 13.7 \text{ kW}$$

14.9 Liquefaction of gases

If a gas is to be liquefied, its temperature must be reduced to a value below its critical temperature (e.g. for nitrogen the critical temperature is $-147\,°$C). In order to reach such low temperatures various means are used. The arrangement described in section 14.8 can be extended to employ four refrigerating systems in series, the refrigerants being ammonia, NH_3, ethylene, C_2H_4, methane, CH_4, and nitrogen, N_2, for the production of liquid nitrogen. The plant required for this would be complex and a simpler method is available for the liquefaction of some other gases. The arrangement is shown in Fig. 14.27(a), and the

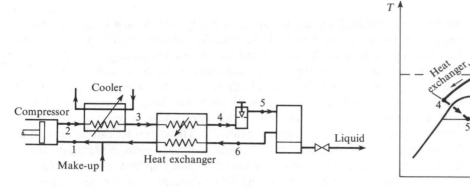

Fig. 14.27 The Linde system for gas liquefaction (a) with the processes on a *T–s* diagram (b)

corresponding *T–s* diagram is shown in Fig. 14.27(b). This is called the *Linde process*. The corresponding state points are indicated on both diagrams. Substances which solidify at a temperature above that of the required liquid temperature must be removed from the gas before admission to the plant. The gas is compressed to a pressure of 100–200 atm before delivery to the cooler. The gas is cooled in the cooler to a temperature which depends on the temperature of the cooling water available. The gas passes through a heat exchanger where heat is transferred from it to the returning low-temperature vapour. It is cooled at 3 to a temperature which is in the region of its critical value and is then throttled to atmospheric pressure, at which pressure it exists as a wet vapour. The liquid is drawn off and the vapour is returned to the compressor. The quantity is made up from an external supply before induction into the compressor.

Lower temperatures could be obtained by replacing the throttling operation with an expansion machine of the turbine type.

From what has been said previously about the throttling of a gas it would appear that the process described in this section is impossible, as there would be no change in the temperature of the gas in throttling and therefore no cooling effect. With real gases there is a small change in temperature on throttling and this may be either an increase or a decrease. At any particular pressure there is a temperature above which the gas will not be reduced in temperature by a throttling process. This temperature is called the *temperature of inversion*.

At the commencement of the Linde process there is no cold gas returning through the heat exchanger and so there will be no cooling effect from 3 to 4. If, however, the temperature at 3 is below the temperature of inversion there will be some cooling as the gas is throttled from 3 to 7. As the process continues the amount of cooling due to returning cold vapour will increase and the line 3–7 will gradually move down to position 4–5. If conventional cooling is not able to cool the gas to below the temperature of inversion then a refrigeration process must first take place to do so.

14.10 Steam-jet refrigeration

Water is a refrigerant which, like air, is perfectly safe. At low temperatures the saturation pressures are low and the specific volumes high (e.g. at 4 °C the saturation pressure is 0.008 129 bar and the specific volume, v_g, is 157.3 m³/kg). The temperatures which can be attained using water as a refrigerant are not low enough for most applications of refrigeration, but are in the range which may satisfy the conditions of air-conditioning. The large volumes can be dealt with by means of a steam jet, and so this application is more attractive if a supply of medium-pressure steam is available. The steam jet can be applied to high-temperature cooling and can be adjusted quickly to meet variations in the load. In the summer, when steam for heating is not required, a supply of steam is readily available for refrigeration, and the equipment of a steam-jet refrigerator can be installed for a low capital expenditure. Another essential is a liberal supply of cooling water for the steam condenser. Steam jets are not used when temperatures below 5 °C are required.

The layout of a steam-jet plant is as shown in Fig. 14.28. The secondary ejectors are for the removal of air and take approximately 10% of the total amount of steam required to produce the refrigerating effect. Steam expands through a nozzle to form a high-speed, low-pressure jet which entrains the vapour to be extracted from the vacuum chamber. The combined flow is diffused in the diverging part of the venturi to the exhaust pressure. After condensation some of the water can be returned to the evaporator, the rest is returned to the boiler feed.

Fig. 14.28
Diagrammatic
arrangement of a
steam-jet refrigeration
system

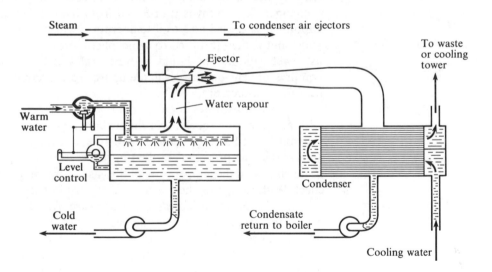

14.11 Refrigerants

Until recently the suitability of a refrigerant for a given refrigeration temperature and capacity was judged by its physical and chemical properties alone; safety was considered but only as it related to the risks of fire, explosion, or poisonous leaks. This has changed with the discovery of the damage being done to the earth's ozone layer (see below). The refrigeration industry is now in turmoil, and will remain so for some years ahead as chemists try to find replacement refrigerants for those causing the damage to the ozone layer.

From an engineering standpoint the important parameters in the choice of a refrigerant are as follows (not necessarily in order of importance):

(i) the saturation pressure at the desired low temperature should be above atmospheric, but not too high, otherwise leakage is harder to prevent and components must be designed to withstand the higher pressures; the pressure at the condenser must not be excessively high for the same reasons;

(ii) the specific enthalpy of vaporization at the low temperature should be as high as possible to give a reasonably low mass flow rate for a given refrigeration capacity;

(iii) the specific volume at compressor suction should not be excessively high to avoid over-large compressors for the required mass flow rate;

(iv) the refrigerant should not react with the lubricating oil of a reciprocating compressor and should be miscible with the oil (i.e. form a homogeneous solution which remains as such throughout the cycle); in a plant using an immiscible refrigerant with a reciprocating compressor an oil separator may have to be fitted after the compressor to prevent oil fouling the condenser and evaporator surfaces and hence reducing the heat transfer;

(v) there should be an acceptably low fire or explosion risk;

(vi) the refrigerant should be non-toxic, or at least toxic to a limit below an acceptable level in the time required to evacuate personnel from the danger area in the event of a leak; harmless irritants may need to be added so that leaks are detectable.

In the early days of refrigeration the commonly used refrigerants included ammonia (still widely used), methyl chloride, sulphur dioxide, and carbon dioxide. Of these, only carbon dioxide is non-toxic; ammonia and sulphur dioxide are irritants, but methyl chloride is odourless and hence more dangerous. Ammonia and methyl chloride also create a fire and explosion risk. (Other possible refrigerants such as ethyl chloride, propane, or butane, are usually ruled out completely because of the very high risk of explosion.) Refrigeration plants using refrigerants which are toxic, or which are a fire or explosion risk, should always use a secondary refrigerant such as brine so that the risk is confined to a smaller area. Carbon dioxide, although non-toxic and with a zero risk of fire or explosion, is not a very suitable refrigerant because of its high saturation pressures at normal working temperatures (e.g. at $-15\,°C$ the saturation pressure is 22.9 bar and at $30\,°C$ is 71.9 bar); this adds considerably to the cost because of the need to design the cycle components to withstand the high pressures. Leakage of carbon dioxide in a confined space is dangerous in spite of the fact that the gas is strictly non-toxic since oxygen is displaced upwards and suffocation can be caused. Since carbon dioxide is also the main greenhouse gas its reintroduction into the refrigeration industry is extremely unlikely.

CFCs and HCFCs

The apparently ideal refrigerant, the halocarbon compound, dichlorodifluoromethane (now known as Refrigerant 12), was invented in the 1930s by Thomas Midgley; it is non-toxic with a zero fire and explosion risk and has ideal thermodynamic properties. Chemical companies such as Du Pont and ICI developed a whole range of similar compounds, under the respective trade names of Freon and Arcton. These are now familiarly called CFCs and are known to be damaging the ozone layer. (It is ironic that the other great legacy to science of Thomas Midgley is tetraethyl lead to inhibit knock in IC engines; the build-up of lead in the atmosphere due to car exhaust fumes is now known to be a possible cause of brain damage, particularly in children, hence the move towards the use of unleaded fuel.)

The designation of the CFC family of refrigerants is arrived at by assigning each refrigerant a notional three-digit number. The first digit from the left represents one less than the number of carbon atoms; if there is only one carbon atom then this digit is left blank. The second digit represents one more than the number of hydrogen atoms; if there are no hydrogen atoms this digit is unity; if there is at least one hydrogen atom then the refrigerant is familiarly called an HCFC. The last digit represents the number of fluorine atoms. For example, some typical CFCs are as follows:

dichlorodifluoromethane, CCl_2F_2 is R12;
trichlorofluoromethane, CCl_3F is R11;
trichlorotrifluoromethane, CCl_2FCClF_2 is R113

and typical HCFCs:

chlorodifluoromethane, $CHClF_2$ is R22;
dichlorotrifluoroethane, $CHCl_2CF_3$ is R123;
tetrafluoroethane, CH_2FCF_3 is R134

It will be seen from the above that the number of chlorine atoms plays no part in the designation; this is unfortunate since the ozone layer depletion potential of the refrigerant depends on the chlorine content. Refrigerants other than halocarbon compounds are sometimes designated by putting the number seven before the relative molar mass of the substance. For example, ammonia, NH_3 is sometimes designated as R717, and water as 718.

Miscibility with oil was previously identified as one of the desirable properties of a refrigerant. Miscible refrigerants include R11, R12, R21, and R113; immiscible refrigerants include R13, R14, and ammonia, sulphur dioxide and carbon dioxide. Azeotropic mixtures of refrigerants are used to produce a refrigerant combining the good properties of two refrigerants, making up a miscible refrigerant. (An azeotropic mixture is one which cannot be separated into its component parts by distillation.) One commonly used example of an azeotropic mixture is approximately 50% R22 and 50% R115 and is designated as R502.

The ozone layer

Ozone gas consists of three atoms of oxygen per molecule, chemical formula, O_3. It is a vigorous oxidizing agent and for this reason was at one time used to 'clean' the air by producing it in atmospheres such as kitchens. Ozone can be produced by an electrical discharge in oxygen and its presence can be detected in the vicinity of some electrical machines. At one time it was thought to have health-giving properties; the air at the seaside was believed to contain ozone. This theory is now discredited; the characteristic smell of the sea is probably due to seaweed and perhaps partly to ionization of the air. It is now known that ozone is positively harmful at ground level; it is one of the byproducts of photochemical smog, the action of sunlight on a trapped layer of car and chimney exhaust fumes producing the chemical reaction to convert oxygen to ozone.

Ozone is present in the layer in the earth's atmosphere known as the stratosphere, about $11-50$ km above the surface of the earth. At these altitudes it is formed by the action of sunlight on oxygen and decomposes back to oxygen plus a free oxygen atom also due to sunlight; the free oxygen atom combines with ozone to form two molecules of oxygen gas,

i.e. $\qquad O_3 + \text{sunlight} \rightarrow O_2 + O \rightarrow O_3$

At any one temperature stability will eventually be reached, but the process is slow. Ozone is produced more easily above the equator, and natural circulation causes the ozone to move towards the two poles where the lower temperatures encourage higher concentrations. Similar cyclic changes occur in the seasons of the year, causing the ozone layer at the poles to be thicker in winter for example. Due to this continuous reaction in the stratosphere, ultraviolet light from the sun is being absorbed in the process thus partially shielding the earth from the sun's harmful radiation.

A second way in which ozone is destroyed is in combination with certain reactive atoms, or free radicals, such as H, OH, NO, and Cl; such reactions are catalytic which means that the reactive atoms are not consumed. The CFCs and other chlorine-based compounds in the stratosphere are decomposed by the intense solar radiation; the chlorine atoms then enter into a catalytic chain reaction, possibly with OH and NO radicals to form HCl, NO_2, and eventually $ClNO_3$, consuming ozone in the process. The HCl acts as a temporary sink for the chlorine atoms before it descends to the troposphere (the layer between the earth's surface and the stratosphere), and then as rain. By this chain reaction one atom of chlorine is able to destroy thousands of ozone molecules.

Sources of chlorine in the atmosphere include CFCs from refrigeration plant, from aerosols where they are used as a propellant, and from foam blowing for insulation materials. Other major sources are carbon tetrachloride used for making CFCs, rubber, and paints, and for cleaning and fire-fighting; halons used for fire-fighting; methyl chloroform and other solvents used for cleaning; methyl chloride generated by seaweed and burning wood. All of these compounds are stable under normal temperatures and therefore eventually rise into the stratosphere.

In the Antarctic the ozone depletion is greater than in the Arctic. This is because of the unique winter conditions made considerably worse by the strong winds (known as the roaring forties), which tend to isolate the air over the polar region. The intense cold leads to the formation of ice crystal clouds providing a surface which helps to accelerate the chemical reactions. An additional reaction occurs in which HCl and $ClNO_3$ react to form HNO_3 and chlorine atoms, the latter then re-entering the ozone destruction cycle.

The breakdown of CFCs and other chlorine-based compounds in the stratosphere occurs more readily in the low winter temperatures at the poles (down to $-80\,°C$), creating free chlorine atoms. In the spring, when the sun returns, the chain reaction accelerates and the destruction of ozone begins again.

There is therefore a natural cycle which is as old as the earth itself; ozone, produced more readily at the equator where the sunlight is more intense, circulates naturally to the poles where it eventually breaks down as described above. It is estimated that 300 million tonnes of ozone are created and destroyed each *day*.

In the mid-1980s instruments for measuring ozone concentrations were becoming more accurate and an Ozone Trends Panel was formed to co-ordinate measurements; instruments in satellites were used to supplement measurements taken by ground instruments such as those used by the British Antarctic Survey. It was found that the natural annual cyclic thinning of the ozone layer over the Antarctic appeared to be increasing and a 'hole' had been created. This extreme thinning of the ozone layer has now moved over parts of Australia causing fears of skin cancer due to increased intensities of ultraviolet-B from the sun's rays. An account of the problem of the measurement of the ozone layer and its depletion is given in reference 14.7.

The Montreal protocol

The United Nations through its Environment Programme persuaded many nations to sign what has become known as the Vienna Convention, a treaty specifically intended to control the production of substances known to be depleting the ozone layer. The Montreal protocol to this treaty in 1987 outlines the means for achieving certain limits to production of particular substances. At this stage individual compounds were allocated an *ozone depletion potential* (ODP); this is a measure of the possible effect on the ozone layer of the release of chlorine from the compound. Table 14.3 lists some common compounds with their ODP, and also their greenhouse potential (see Ch. 13), relative to R11.

The aim of the protocol is to achieve a phased reduction of compounds, but the rate of reduction is subject to continuous review. In 1990 it was agreed that CFCs should be phased out by the year 2000 with interim cuts of 50% of the 1986 production taking place by 1995, and 87% of the 1986 production taking place by 1997. At the same time the ban was extended to halons and carbon tetrachloride by the same date of 2000, although CFCs and halon would continue to be allowed in essential medical applications, and aircraft fire-fighting. In 1992 it was discovered that the thinning of the ozone layer over the Arctic had become much more significant than expected, presenting a threat to the Northern Hemisphere of increased cancers of the eye and skin. The USA immediately

Table 14.3 Ozone depletion potential (ODP), and greenhouse potential for certain refrigerants

Refrigerant name	Chemical formula	Designation	ODP	Greenhouse potential
Trichlorofluoro-methane	CCl_3F	R11	1.00	1.00
Dichlorodifluoro-methane	CCl_2F_2	R12	1.00	3.10
Chlorodifluoro-methane	$CHClF_2$	R22	0.05	0.37
Dichlorotrifluoro-ethane	$CHCl_2F_3$	R123	0.02	0.02
Tetrafluoro-ethane	CH_2FCF_3	R134a	0	0.37
Ammonia	NH_3	R717	0	0

brought forward the phasing out of CFCs to the end of 1995; Germany, Denmark, and The Netherlands had previously brought forward the phasing out to the beginning of 1995. At the same time, chemical manufacturer ICI announced that it would cease all production of CFCs in 1995 and would stop production in the UK in 1993; this was followed by an announcement by the UK Government of a phasing out by the end of 1995. The European Commission then called upon member nations to stop all CFC production by the end of 1995.

Also under the Montreal protocol, methyl chloroform is to be banned by 2005 with a cut of 70% by the year 2000. HCFCs, such as R22, are not covered by the protocol at present, although they may be phased out possibly by 2020 or at the latest by 2040. Chemical manufacturer Du Pont has announced that it will cease production of R22 for new machines by 2005 although it will continue to produce it for existing equipment until 2020. Therefore HCFCs are at best a short-term replacement for CFCs. The problem of developing countries has been recognized by the creation of a special fund to help them adapt to the changes needed; this has enabled India and China to sign the protocol.

The position on the use of all types of CFCs is continually changing as scientific measurements of the ozone layer are improved and updated. More and more pressure is being put on the international community through internal conferences on the environment; agreements to phase out various substances are being modified with dates being continually brought forward as the worry over the effects of holes in the ozone layer increases.

The air-conditioning industry has now phased out the use of R11 and R12 and has moved over to the use of R22 as a short-term solution. A new HCFC refrigerant, R123, which was expected to replace R11 in centrifugal chillers, has recently been discovered to be a carcinogen when present in the atmosphere above certain limits, causing extreme doubts about its future use. The refrigerant R134a, tetrafluoroethane, CH_2FCF_3, produced by ICI as KLEA 134a, is described by ICI as a hydrofluoroalkane, HFA; it contains no chlorine atoms and hence has a zero ODP. Suitable lubricants have been found for R134a and at present it appears to be the best replacement refrigerant for CFCs and, in the long term, HCFCs. Ammonia, which is an ideal refrigerant for most applications, is argued by many as a good long-term solution to the replacement of CFCs and HCFCs; its toxicity is the main disadvantage, but secondary, or even tertiary, coolant systems can overcome this problem by confining the risk to a well-ventilated plant room, preferably situated on the roof of the building; modern instrumentation can provide efficient detection and warning systems.

In the manufacture, commissioning, maintenance and decommissioning of refrigeration plant it is estimated that approximately 320 tonnes of refrigerant are lost to the atmosphere each year in the UK. This is a very high figure which could be substantially reduced by better handling practices. The Heating and Ventilation Contractors Association (HVCA) has called for mandatory rules and licences for the recovery and recycling of refrigerants, but to date (1992) the Government has not responded by introducing the necessary legislation. Other countries have not been so slow to react to the problem; for example, in Sweden a code has been introduced which lays down maintenance procedures and stipulates that special permits are required for refrigerant retrieval; permit holders must also ensure that their operatives are trained to a specified high level in refrigeration engineering. Refrigerants, unlike solvents, fire-fighting fluid, or aerosol gases, are only a threat to the ozone layer when they are not contained safely within the refrigeration plant; skilled maintenance fitters should be able to keep leakage down to a minimum level, and obselete plant should be returned to approved centres so that the refrigerant can be recycled.

14.12 Control of refrigerating capacity

The refrigerating load is seldom constant and the refrigerator must be controlled to meet the demand. It is desirable that automatic controls should be available and small refrigerating plants are operated on the on–off principle. The control in larger plants is obtained by regulating the mass flow of the refrigerant. This can be done by means of a manual speed control on the compressor, but this is uneconomical and inconvenient.

With the automatic on–off device a metal bellows, charged with a volatile liquid, is connected to a temperature-sensitive element which is located at the evaporator coil. An increasing temperature at the evaporator causes the temperature and therefore the pressure in the bellows to increase. The bellows expand and the end of the bellows operates a switch which closes the compressor motor circuit. A decreasing temperature in the evaporator produces the reverse effect.

In a multicylinder reciprocating compressor the capacity can be controlled by regulating the number of cylinders which are effective at any time. When a reduction in refrigerant flow is required the suction valves of one or more cylinders are held open by a hydraulically operated mechanism. The control also applies for starting conditions when all suction valves are held open. The capacity of centrifugal compressors can be controlled by means of a speed control, but this involves expensive equipment. There is a limit to the reduction in mass flow rate, which is determined by the surging characteristics of the compressor. With a constant-speed machine, control can be achieved by the following:

(a) Reducing the cooling water flow to the condenser. The temperature and pressure in the condenser increase and this reduces the effective capacity of the machine. The compressor characteristics should be carefully studied.

(b) Throttling the inlet to the compressor. This reduces the inlet pressure and hence the density of the incoming charge. The delivery pressure is reduced because of the lower inlet pressure.

A simple analytical treatment of plant operation is given in reference 14.2.

Problems

Use Table 14.1 or the chart of Fig. 14.14 for properties of R134a, and use the chart of Fig. 14.23 for properties of saturated lithium bromide–water solutions. All other properties can be found in the tables of reference 14.5.

14.1   The temperature in a refrigerator evaporator coil is $-6\,°C$ and that in the condenser coil is $22\,°C$. Assuming that the machine operates on the reversed Carnot cycle, calculate the COP_{ref}, the refrigerating effect per kilowatt of input work, and the heat rejected to the condenser.

(9.54; 9.54 kW; 10.54 kW)

14.2   An ammonia vapour-compression refrigerator operates between an evaporator pressure of 2.077 bar and a condenser pressure of 12.37 bar. The following cycles are to be compared; in each case there is no undercooling in the condenser, and isentropic compression may be assumed:

  (i) the vapour has a dryness fraction of 0.9 at entry to the compressor;

 (ii) the vapour is dry saturated at entry to the compressor;

(iii) the vapour has 5 K of superheat at entry to the compressor.

In each case calculate the COP_{ref} and the refrigerating effect per unit mass. What would be the COP_{ref} of a reversed Carnot cycle operating between the same saturation temperatures?

(4.5; 957.5 kJ/kg; 4.13; 1089.9 kJ/kg; 4.1; 1101.5 kJ/kg; 5.1)

14.3   A heat pump using ammonia as the refrigerant operates between saturation temperatures of 6 and $38\,°C$. The refrigerant is compressed isentropically from dry saturation and there is 6 K of undercooling in the condenser. Calculate:

  (i) the COP_{hp};

 (ii) the mass flow of refrigerant per kilowatt power input;

(iii) the heat available per kilowatt power input.

(8.77; 25.03 kg/h; 8.77 kW)

14.4   (a) In a refrigerating plant using R12 the vapour leaves the evaporator dry saturated at 1.826 bar and is compressed to 7.449 bar. The temperature of the vapour leaving the compressor is $45\,°C$. The liquid leaves the condenser at $25\,°C$ and is throttled to the evaporator pressure. Calculate:

  (i) the refrigerating effect;

 (ii) the specific work input;

(iii) the COP_{ref}.

(b) Compare the results found in (a) with a plant using R134a between the same saturation temperatures, with the same maximum cycle temperature and the same degree of undercooling of the condensate.

(121.27 kJ/kg; 29.66 kJ/kg; 4.09; 152.05 kJ/kg; 43.65 kJ/kg; 3.48)

14.5   An ammonia vapour-compression refrigerating plant has a single-stage, single-acting reciprocating compressor which has a bore of 127 mm, a stroke of 152 mm, and a speed of 240 rev/min. The pressure in the evaporator is 1.588 bar and that in the condenser is 13.89 bar. The volumetric efficiency of the compressor is 80% and its mechanical efficiency is 90%. The vapour is dry saturated on leaving the evaporator and the liquid leaves the condenser at 32 °C. Calculate:

  (i) the mass flow of refrigerant;

 (ii) the refrigerating effect;

(iii) the power required to drive the compressor.

(0.5 kg/min; 9.01 kW; 2.74 kW)

14.6   A cold storage plant is used to cool 9000 litres of milk per hour from 27 to 4 °C, and the heat leakage into the plant is estimated to be 60 kW. The refrigerant used is ammonia and the temperature required in the evaporator is −6 °C. The compressor delivery pressure is 10.34 bar and the condenser liquid is undercooled to 24 °C before throttling. The plant has a brine circulating system and the rise in temperature of the brine is to be limited to 3 K. Assuming that the vapour is dry saturated on leaving the evaporator and that the compression process is isentropic, calculate:

  (i) the power input required taking the mechanical efficiency of the compressor as 90%;

 (ii) the swept volume of each cylinder of the twin-cylinder, single-acting compressor, for which the volumetric efficiency can be taken as 85% at a rotational speed of 200 rev/min;

(iii) the rate at which the brine must be circulated in litres per second.

For milk: specific heat capacity, 3.77 kJ/kg K; density, 1030 kg/m^3.

For brine: specific heat capacity, 2.93 kJ/kg K; density, 1190 kg/m^3.

(42.7 kW; 0.0157 m^3; 27.1 litres/s)

14.7   A refrigeration plant running on R134a operates between saturation temperatures of 0 and 45 °C and has a refrigeration capacity of 1 MW. The vapour is dry saturated at entry to the compressor and there is no undercooling in the condenser. Compression takes place in a two-stage centrifugal compressor, each stage having the same pressure ratio and enthalpy rise. Taking a slip factor of unity and an overall isentropic efficiency for the compressor of 0.8, calculate:

  (i) the coefficient of performance;

 (ii) the required power input;

(iii) the diameter of the impeller in each stage when the rotational speed is 3000 rev/min;

(iv) the flow area required at compressor inlet when the axial velocity of the refrigerant at inlet is 60 m/s.

(3.75; 266.5 kW; 0.843 m; 0.0086 m^2)

14.8   A combined domestic unit serves the dual purpose of cooling the kitchen larder and providing hot water. The motor driving the compressor operates for approximately one-third of the day, and has an electrical input of 0.225 kW. The heat leakage into the larder from the kitchen is 0.29 kW. The mechanical efficiency of the compressor is 85% and the compression process can be taken to be adiabatic. All the heat rejected by the refrigerant is taken by the water in the domestic hot-water tank, which is heated from 10 to 60 °C. Calculate the amount of hot water which can be supplied by this plant in litres per hour.

(6.08 l/h)

14.9   It is proposed to use a heat pump working on the ideal vapour-compression cycle for the purpose of heating the air supply to a building. The supply of heat is taken from a river at 7 °C. Air is required to be delivered into the building at 1.013 bar and 32 °C at a rate of $0.5 \text{ m}^3/\text{s}$. The air is heated at constant pressure from $10\,^{\circ}\text{C}$ as it passes over the condenser coils of the heat pump. The refrigerant is R134a which is dry saturated leaving the evaporator; there is no undercooling in the condenser. A temperature difference of 17 K is necessary for the transfer of heat from the river to the refrigerant in the evaporator. The delivery pressure of the compressor is 11.545 bar. Calculate:

 (i) the mass flow of refrigerant;

 (ii) the motor power required to drive the compressor if the mechanical efficiency is 87%;

 (iii) the COP_{hp};

 (iv) the swept volume of the compressor which is single-acting and which runs at 240 rev/min with a volumetric efficiency of 85%.

$$(0.079 \text{ kg/s}; \ 3.26 \text{ kW}; \ 3.92; \ 2277 \text{ cm}^3)$$

14.10 (a) An ammonia refrigerator operates between evaporating and condensing temperatures of -16 and $50\,^{\circ}\text{C}$. The vapour is dry saturated at the compressor inlet, and there is no undercooling of the condensate. Calculate:

 (i) the refrigerating effect;

 (ii) the mass flow rate per kilowatt of refrigeration capacity;

 (iii) the power input per kilowatt of refrigeration capacity;

 (iv) the COP_{ref}.

 (b) In the plant of part (a) a flash chamber is introduced with an interstage pressure of 5.346 bar. Recalculate parts (iii) and (iv) for this arrangement.

$$(1003.4 \text{ kJ/kg}; \ 3.59 \text{ kg/h}; \ 0.338 \text{ kW}; \ 2.96; \ 0.313 \text{ kW}; \ 3.19)$$

14.11 A vapour-compression refrigeration plant using R134a operates with a compressor suction pressure and temperature of 2.005 bar and $-10\,^{\circ}\text{C}$. The condenser pressure is 7.675 bar and there is no undercooling of the condensate. Compression takes place in two stages, and the condensate is throttled into a flash chamber at 4.139 bar from which dry saturated vapour is drawn off to mix with the refrigerant from the LP compressor before entry to the HP compressor. The liquid from the flash chamber is throttled into the evaporator. Assuming isentropic compression in both compressors and neglecting all losses, calculate:

 (i) the coefficient of performance of the plant;

 (ii) the mass flow rate of refrigerant in the evaporator when the power input to the plant is 100 kW.

$$(5.93; \ 3.38 \text{ kg/s})$$

14.12 A vapour-absorption cycle using lithium bromide–water solution operates between pressures of 0.01 and 0.05 bar. The refrigeration capacity is 150 kW, the temperature in the generator is $70\,^{\circ}\text{C}$, and the temperature in the absorber is $25\,^{\circ}\text{C}$. The steam entering the absorber is dry saturated and there is no undercooling in the condenser. The solution may be taken to be saturated at exit from both the generator and absorber; a heat exchanger transfers heat from the weak solution leaving the generator to the strong solution pumped from the absorber such that the solution entering the generator is saturated at the generator temperature. Assuming that the specific enthalpy of superheated steam is approximately equal to the specific enthalpy of saturated steam at the same temperature, and neglecting pressure losses, heat losses, and pump work, calculate:

 (i) the mass flow rate of refrigerant through the evaporator and condenser;

 (ii) the heat supplied in the generator;

 (iii) the heat rejected in the absorber;

 (iv) the specific enthalpy of the solution entering the absorber;

 (v) the heat to be supplied in the generator if the heat exchanger is not used.

$$(0.06313 \text{ kg/s}; \ 161.07 \text{ kW}; \ 153.98 \text{ kW}; \ -216.1 \text{ kJ/kg}; \ 204.84 \text{ kW})$$

14.13 An air refrigeration plant operates with a centrifugal compressor and an air turbine mounted coaxially such that the power output of the turbine contributes to the work required to drive the compressor. The temperature of the air at the compressor inlet is 15 °C and the pressure ratio is 2.5. The air during its passage from the compressor to the turbine passes through an intercooler and enters the turbine at 40 °C. The cold space temperature is required to be maintained at 15 °C. Taking the isentropic efficiency of both compressor and turbine to be 84%, and the mechanical efficiency of the turbine–compressor drive as 90%, calculate:

  (i) the refrigerating effect;

 (ii) the mass flow rate per kilowatt of refrigeration capacity;

(iii) the driving power required per kilowatt of refrigeration capacity.

           (35.74 kJ/kg; 1.679 kg/min; 1.312 kW)

14.14 In an aircraft refrigeration unit air is bled from the engine compressor at 3.5 bar and 270 °C and is passed through an air-cooled heat exchanger. The refrigerant air bleed leaves the exchanger at 3.5 bar and 75 °C and is expanded through a turbine to 0.76 bar. The isentropic efficiency of the turbine is 85%. The air is then delivered to the aircraft cabin and leaves the aircraft at 16 °C. Calculate:

 (i) the refrigerating effect;

(ii) the power developed by the air turbine per unit mass flow rate.

           (45.8 kJ/kg; 105.1 kW per kg/s)

References

14.1 DOSSAT R J 1990 *Principles of Refrigeration* 2nd edn Wiley

14.2 EASTOP T D and WATSON W E 1992 *Mechanical Services for Buildings* Longman

14.3 EASTOP T D and CROFT D R 1990 *Energy Efficiency* Longman

14.4 REAY D A and MACMICHAEL D B A 1987 *Heat Pumps* 2nd edn Pergamon

14.5 ROGERS G F C and MAYHEW Y R 1987 *Thermodynamic and Transport Properties of Fluids* 4th edn Basil Blackwell

14.6 CIBSE 1986 *Guide to Current Practice* volume B14.16

14.7 KEMP D D 1990 *Global Environmental Issues* Routledge

15

Psychrometry and Air-conditioning

Mixtures of air and water vapour are considered in Chapter 6; in this chapter moist atmospheric air (i.e. a mixture of dry air and water vapour) is introduced as a separate topic.

It is often necessary to provide a controlled atmosphere in buildings where industrial processes are to be carried out, or to provide air-conditioning in private and public buildings. The properties of atmospheric air have to be considered in these problems, and this is a subject which is receiving an increasing amount of attention and application. Another topic which will be considered is that of the cooling tower by means of which large quantities of cooling water are cooled for recirculation. These topics come under the title of psychrometry (sometimes called hygrometry).

15.1 Psychrometric mixtures

In section 6.6 the evaporation of water into an evacuated space or into a volume occupied by a gas is described. Before the saturated condition is reached the vapour exists in the mixture as a superheated vapour. At the saturation condition the partial pressure of the vapour can be obtained from steam tables as that pressure corresponding to the temperature of the mixture. If the space or gas is not saturated at a particular temperature, then the partial pressure of the vapour will be less than the saturation pressure corresponding to that temperature.

Consider atmospheric air at 1.013 bar and 15 °C. The saturation pressure of water vapour corresponding to 15 °C is 0.017 04 bar. Unless the water vapour is in contact with its liquid it will not be saturated, and its pressure will be below the saturation value of 0.017 04 bar. In normal applications the atmosphere is well removed from the saturated condition. At such low vapour pressures the vapour can be considered to act as a perfect gas, and the properties of the mixture can be found using the Gibbs–Dalton law. The properties of the mixture depend on its pressure and temperature, and are determined for a particular state with reference to the properties of saturated vapour.

Assume that in a quantity of atmospheric air the vapour pressure is 0.01 bar at 15 °C and the total pressure is 1.013 bar. From equation (6.2)

$$p = p_a + p_s$$

where p_a is the partial pressure of the dry air, and p_s the partial pressure of the superheated vapour,

i.e. $\qquad p_a = p - p_s = 1.013 - 0.01 = 1.003 \text{ bar}$

The saturation temperature corresponding to 0.01 bar is 7 °C, hence the vapour in atmospheric air under these conditions has a degree of superheat of $(15 - 7) = 8$ K. This state is indicated by point 1 on a T–s diagram in Fig. 15.1. Suppose a metal beaker containing water is placed in this atmosphere, and the water is progressively cooled by adding ice. At a particular temperature of the water it will be noticed that condensation begins to appear on the outside surface of the beaker. The vapour in contact with the beaker has been cooled at constant pressure to 7 °C, as indicated by point 2 in Fig. 15.1. This is the condition of saturation, and further cooling causes condensation of the water vapour. This temperature is called the *dew point* of the mixture, and it is the temperature to which an unsaturated mixture must be cooled in order to become just saturated. The dew point temperature is denoted by t_d.

Fig. 15.1 T–s diagram for superheated water vapour in atmospheric air

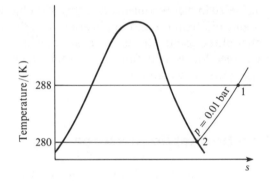

If a room is warm and the outside atmosphere is cold, then a window may produce condensation on its inside surface. A person wearing spectacles entering a warm room after a time spent in a cold atmosphere very often finds the vapour in the air condensing on the lenses as the vapour is cooled to its dew point. Condensation is noticed on cold-water pipes which are situated in an atmosphere which is at a higher temperature and which is sufficiently humid.

15.2 Specific humidity, relative humidity, and percentage saturation

The *specific humidity* (or *moisture content*) is the ratio of the mass of water vapour to the mass of dry air in a given volume of the mixture, and is denoted by the symbol ω,

i.e. $\qquad \omega = \dfrac{m_s}{m_a}$ $\hfill (15.1)$

where the subscript s denotes the vapour, and the subscript a denotes the dry air.
Since both masses occupy the volume V then

$$\omega = \frac{m_s/V}{m_a/V} = \frac{1/v_s}{1/v_a} = \frac{v_a}{v_s} \qquad (15.2)$$

where v_a and v_s are the specific volumes of the dry air and vapour respectively.
Since both the vapour and the dry air are considered as perfect gases then

$$m_s = \frac{p_s V}{R_s T} \quad \text{and} \quad m_a = \frac{p_a V}{R_a T}$$

Also $\quad R_s = \dfrac{\tilde{R}}{\tilde{m}_s} \quad$ and $\quad R_a = \dfrac{\tilde{R}}{\tilde{m}_a}$

Therefore

$$m_s = \frac{p_s V \tilde{m}_s}{\tilde{R} T} \quad \text{and} \quad m_a = \frac{p_a V \tilde{m}_a}{\tilde{R} T}$$

Then, substituting in equation (15.1)

$$\omega = \frac{p_s V \tilde{m}_s}{\tilde{R} T} \times \frac{\tilde{R} T}{p_a V \tilde{m}_a}$$

i.e. $\quad \omega = \dfrac{\tilde{m}_s}{\tilde{m}_a} \times \dfrac{p_s}{p_a}$

therefore

$$\omega = \frac{18}{28.96} \times \frac{p_s}{p_a} = 0.622 \frac{p_s}{p_a}$$

If the total pressure is p, then from equation (6.2), $p = p_a + p_s$, therefore

$$\omega = 0.622 \times \left(\frac{p_s}{p - p_s}\right) \qquad (15.3)$$

(the total pressure p is usually the barometric pressure).

The *relative humidity* of the atmosphere is the ratio of the actual mass of the
water vapour in a given volume to that which it would have if it were saturated
at the same temperature,

$$\text{Relative humidity, } \phi = \frac{m_s}{(m_s)_{sat}}$$

Relative humidity is frequently expressed as a percentage.

$$m_s = \frac{p_s V}{R_s T} \quad \text{and} \quad (m_s)_{sat} = \frac{p_g V}{R_s T}$$

where p_g is the saturation pressure at the temperature of the mixture,

i.e. $\quad \phi = \dfrac{p_s}{p_g}$ \hfill (15.4)

The term *percentage saturation* is also used, defined as the ratio of the specific humidity of a mixture to the specific humidity of the mixture when saturated at the same temperature, expressed as a percentage,

i.e.   Percentage saturation, $\psi = \dfrac{\omega}{\omega_g}$ (15.5)

Note: 'percentage saturation' is the name used by CIBSE in its *Guide to Current Practice* (reference 15.1), but it would be more logical to call the ratio ω/ω_g, 'relative saturation', or 'degree of saturation' (the name used by Threlkeld (reference 15.2)); to conform with common practice in the UK the term 'percentage saturation' will be used in this book.

From equations (15.3), (15.4) and (15.5) it can be seen that

$$\text{Percentage saturation, } \psi = 100\phi \times \frac{(p - p_g)}{(p - p_s)}$$ (15.6)

In air-conditioning practice the percentage difference between ψ and ϕ is in the range 0.5–2%, approximately.

Example 15.1   The air supplied to a room of a building in winter is to be at 17 °C and have a percentage relative humidity of 60%. If the barometric pressure is 1.013 26 bar, calculate the specific humidity. What would be the dew point under these conditions?

Solution   At 17 °C, $p_g = 0.019\,36$ bar, hence using equation (15.4)

$$0.6 = \frac{p_s}{0.019\,36} \quad \text{therefore } p_s = 0.6 \times 0.010\,39 = 0.011\,616 \text{ bar}$$

Using equation (15.3), $\omega = 0.622 \times p_s/(p - p_s)$, we have

$$\omega = 0.622 \times \frac{0.011\,616}{1.013\,25 - 0.011\,616} = 0.007\,213$$

i.e. the atmosphere contains 0.007 213 kg of vapour per kilogram of dry air.

If the air is cooled at constant pressure the vapour will begin to condense at the saturation temperature corresponding to 0.011 616 bar. By interpolation from tables, the dew point temperature t_d is then

$$t_d = 9 + (10 - 9) \times \left(\frac{0.011\,616 - 0.011\,47}{0.012\,27 - 0.011\,47}\right) = 9.18\,°C$$

Example 15.2   If air at the condition of Example 15.1 is passed at the rate of 0.5 m³/s over a cooling coil which is at a temperature of 6 °C, calculate the amount of vapour which will be condensed. Assume that the barometric pressure is the same as in Example 15.1, and that the air leaving the coil is saturated.

Solution   The system is shown in Fig. 15.2. The mass flow rate of dry air, $\dot{m}_a$, is given by

$$\dot{m}_a = \frac{p_a \dot{V}}{R_a T}$$

Fig. 15.2 Cooling coil for Example 15.2

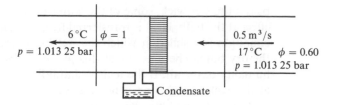

From equation 6.2, $p_a = p - p_s$, therefore

$$p_a = 1.013\,25 - 0.011\,616 = 1.001\,63 \text{ bar}$$

therefore

$$\dot{m}_a = \frac{10^5 \times 1.001\,63 \times 0.5}{10^3 \times 0.287 \times 290} = 0.6017 \text{ kg/s}$$

The mass flow rate of air is constant throughout the process.

From equation (15.1), $\omega = m_s/m_a$, and ω has been determined as 0.007 213, therefore

$$\dot{m}_{s_1} = 0.007\,213 \times \dot{m}_a$$

After passing the cooling coil, $\phi = 1$, since the air is saturated. From equation (15.4), $p_s = p_g$ for this condition, and at 6 °C, $p_g = 0.009\,346$ bar, therefore, from equation (15.3)

$$\omega_2 = 0.622 \times \left(\frac{0.009\,346}{1.013\,25 - 0.009\,346} \right) = 0.005\,79$$

therefore

$$\dot{m}_{s_2} = 0.005\,79 \times \dot{m}_a$$

Hence

$$\text{Mass of condensate} = \dot{m}_{s_1} - \dot{m}_{s_2} = (0.007\,213 - 0.005\,79) \times \dot{m}_a$$
$$= 0.001\,423 \times 0.6017 \times 3600 = 3.082 \text{ kg/h}$$

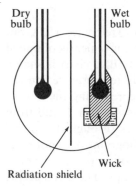

Radiation shield

Fig. 15.3 Wet and dry bulb psychrometer

Measurement of relative humidity

An instrument used to measure relative humidity is called a psychrometer, or a hygrometer. A simple psychrometer has been referred to in section 15.1 (i.e. by determining the dew point using a metal beaker of water which is cooled).

Another method is by the determination of wet and dry bulb temperatures. The principle is illustrated in Fig. 15.3. Two thermometers situated in a stream of unsaturated air are separated by a radiation screen. One of them indicates the air temperature and is called the *dry bulb* thermometer. The bulb of the second is surrounded by a wick which dips into a small reservoir of water and the temperature indicated is called the *wet bulb* temperature. As the air stream

537

passes the wet wick, some of the water evaporates and this produces a cooling effect at the bulb. Heat is transferred from the air to the wick and an equilibrium condition is reached at which the wet bulb indicates a lower temperature than the dry bulb. The amount of this *wet bulb depression* depends on the relative humidity of the air. If the relative humidity is low, then the rate of evaporation at the wick is high, and hence the wet bulb depression is high.

The instrument may be made for use in stationary air, but satisfactory results are obtained only when the air velocity past the bulbs is between 1.85 and 40 m/s. Over this range the results are fairly constant and enable the relative humidity to be calculated from the temperatures obtained. The air current can be produced by a small fan driving the air over the thermometer bulbs, or by mounting the thermometers on a frame which is whirled round by hand. This latter instrument is called a *sling psychrometer*. Another portable instrument has a fan which has a battery or clockwork drive. The wet and dry bulb temperatures are measured by thermocouples and read off an indicator. The advantages claimed are compactness and rapid response. The specifications for hygrometers are given in reference 15.3.

Instruments are available which will give a continuous reading of humidity in the form of an electrical signal which may then be used as part of a control system. A common form of sensor is a thin polymer film which absorbs and desorbs moisture thus changing the dielectric constant and hence the capacitance. By measuring temperature and humidity simultaneously the enthalpy can be obtained and hence used as the control. This type of approach has been made possible using microprocessors which convert the readings from the sensors directly into relative humidity and/or enthalpy; calibration of the sensors against wet and dry bulb, or dew point, standard instruments can be programmed into the instrument.

A more accurate type of instrument uses an opto-electronic detection of vapour condensation on an electrically chilled solid gold mirror; the chilling is done using a thermoelectric solid-state device.

Psychrometric chart

The properties of moist air can be obtained from tables (reference 15.1), but the specific humidity and percentage saturation are most conveniently obtained from a psychrometric chart. A reduced size copy of the CIBSE chart is shown in Fig. 15.4. An ordinate is erected at the known dry bulb temperature and the point of intersection between it and the diagonal line representing the known wet bulb temperature is found. The percentage saturation is then found from the curve of constant percentage saturation which passes through this point. The specific humidity is read off the ordinate scale in kilograms of vapour per kilogram of dry air. The enthalpy of the mixture in kilojoules per kilogram of dry air can be read off the diagonal scale of specific enthalpy. The zero specific enthalpy for the vapour is always taken at $0\,°C$. For the dry air the zero for specific enthalpy is also taken at $0\,°C$.

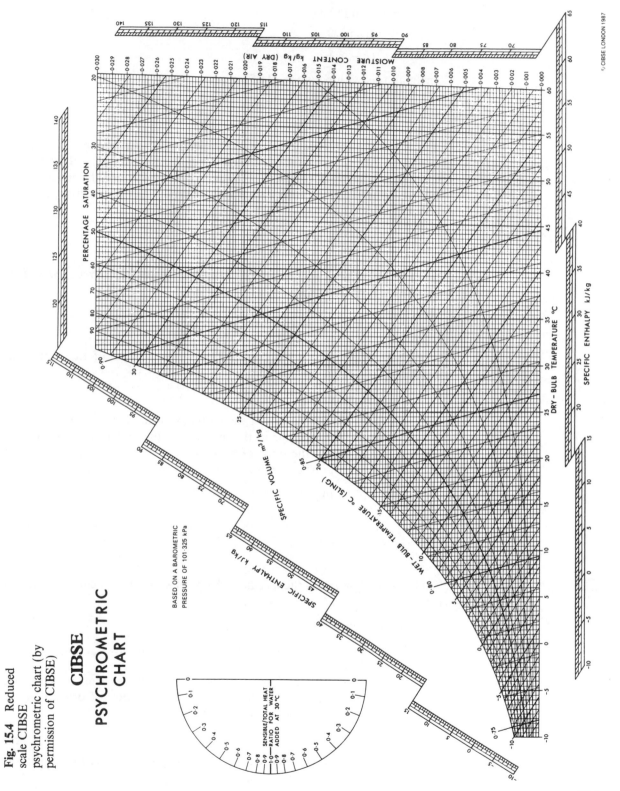

Fig. 15.4 Reduced scale CIBSE psychrometric chart (by permission of CIBSE)

CIBSE PSYCHROMETRIC CHART

BASED ON A BAROMETRIC PRESSURE OF 101·325 kPa

© CIBSE LONDON 1987

From equation (15.3),

$$\omega_g = \frac{0.622 p_g}{(p - p_g)}$$

Combining this with equation (15.5), we have

$$\psi = \frac{100\omega}{\omega_g} = \frac{100\omega(p - p_g)}{0.622 p_g} \tag{15.7}$$

For a given barometric pressure, p, the percentage saturation is a function of ω and p_g. Also p_g corresponds to the dry bulb temperature, t. The chart is prepared for a given barometric pressure and ω and h are the independent variables; values are accurate for all practical purposes for barometric pressures in the range 0.95–1.05 bar.

15.3 Specific enthalpy, specific heat capacity, and specific volume of moist air

Specific enthalpy of moist air

The enthalpy of a mixture is the sum of the enthalpies of the individual constituents (see equation (6.6)),

i.e. $mh = m_a h_a + m_s h_s$

where m is the mass of mixture, h the enthalpy of mixture per unit mass of mixture m_a the mass of dry air in the mixture, h_a the enthalpy of dry air per unit mass of dry air, m_s the mass of water vapour in the mixture, and h_s the enthalpy of water vapour per unit mass of water vapour. Therefore

Enthalpy of mixture per unit mass of dry air $= mh/m_a$

$$= h_a + \frac{m_s h_s}{m_a}$$

$$= h_a + \omega h_s$$

At low partial pressures the specific enthalpy of water vapour can be expressed as

$$h_s = (h_g \text{ at } p_s) + c_{ps}(t - t_g \text{ at } p_s) \tag{15.8}$$

where the mean specific heat of superheated water vapour, c_{ps}, may be taken as approximately 1.88 kJ/kg K.

Since the specific enthalpy of a vapour from steam tables is expressed above a datum of approximately $0\,^{\circ}\text{C}$ then the specific enthalpy of dry air in the mixture is also expressed above the same datum,

i.e. $h_a = c_{pa} t$ $\tag{15.9}$

where the specific heat capacity of dry air, c_{pa}, may be taken as 1.005 kJ/kg K.

Then

Enthalpy of mixture per unit mass of dry air

$$= c_{pa}t + \{(h_g \text{ at } p_s) + c_{ps}(t - t_g \text{ at } p_s)\}\omega$$

It can be shown that over the temperature range encountered in air-conditioning the term $\{(h_g \text{ at } p_s) - c_{ps}(t_g \text{ at } p_s)\}$ may be taken as a constant, C. Therefore we can write

Enthalpy of mixture per unit mass of dry air

$$= c_{pa}t + \omega(C + c_{ps}t) \tag{15.10}$$

where $C = 2500 \text{ kJ/kg}$.

Alternatively, since for low pressures the enthalpy of superheated vapour is approximately equal to the saturation value at the same *temperature*, then we have

Enthalpy of mixture per unit mass of dry air

$$= c_{pa}t + \omega(h_g \text{ at } t) \tag{15.11}$$

where $c_{pa} = 1.005 \text{ kJ/kg K}$, as before.

Accurate values of the enthalpy of a mixture per unit mass of dry air are given by CIBSE (reference 15.1). For example at $5\,°C$ and $\omega = 0.002\,820$, $h = 12.11 \text{ kJ/kg dry air}$.

From equation (15.10)

$$h = 1.005 \times 5 + 0.002\,82\{2500 + (1.88 \times 5)\} = 12.10 \text{ kJ/kg dry air}$$

From equation (15.11)

$$h = 1.005 \times 5 + (0.002\,82 \times 2509.9) = 12.10 \text{ kJ/kg dry air}$$

Similarly, at $30\,°C$ and $\omega = 0.01420$, $h = 66.48 \text{ kJ/kg dry air}$.

From equation (15.10)

$$h = 1.005 \times 30 + 0.0142\{2500 + (1.88 \times 30)\}$$

$$= 66.45 \text{ kJ/kg dry air}.$$

From equation (15.11)

$$h = 1.005 \times 30 + (0.0142 \times 2555.7)$$

$$= 66.44 \text{ kJ/kg dry air}.$$

It can be seen that the error is negligible over a wide temperature range for both equations (15.10) and (15.11). Equation (15.11) is easier to use than equation (15.10) except for cases where the temperature, t, is the unknown term.

Specific heat capacity of moist air

Assuming that the superheated water vapour acts as a perfect gas, then using equation (6.19),

541

Specific heat capacity of mixture per unit mass of mixture:

$$c_p = \frac{m_a c_{pa}}{m} + \frac{m_s c_{ps}}{m}$$

Then

Specific heat capacity of mixture per unit mass of dry air, c_{pma}, is given by

$$c_{pma} = c_{pa} + \frac{m_s c_{ps}}{m_a}$$

i.e. $\qquad c_{pma} = c_{pa} + \omega c_{ps}$ $\hfill$ (15.12)

where $c_{pa} = 1.005 \text{ kJ/kg K}$, and $c_{ps} = 1.88 \text{ kJ/kg K}$, as before.

Specific volume

Since the enthalpy of the mixture is expressed per unit mass of dry air it is convenient to use the specific volume of the dry air. Therefore when the volume flow of the mixture is known the rate of mass flow of dry air can be found directly,

i.e. $\qquad$ Specific volume of dry air, $v_a = \dfrac{R_a T}{p_a}$

and

$$\dot{m}_a = \frac{\dot{V}}{v_a} \hfill (15.13)$$

The specific volume of dry air is plotted on the psychrometric chart. By reference to the chart it may be noted that over the normal range of room temperatures and humidities the density expressed as kilograms dry air per cubic metre of mixture is approximately 1.2 (i.e. $v_a = 1/1.2 = 0.833 \text{ m}^3/\text{kg dry air}$); this is a useful approximation for many practical problems.

15.4 Air-conditioning systems

In the UK air-conditioning is used mainly for industrial purposes and to supply a controlled atmosphere to public buildings such as offices, cinemas, halls, etc. In tropical and subtropical countries cooling by means of air-conditioning is a necessary feature of modern development.

The following classification of air-conditioning systems may be made:

(a) *Conventional:* the air is processed in a central plant and is distributed to the conditioned spaces via ducts.
(b) *Terminal reheat:* air supply to the units in the room is from ducting as in (a) and provides the cooling and dehumidification load; the units in the room, provided with water coils, supply the necessary reheat.

(c) *Induction:* similar to (b) but only a small quantity of primary conditioned air is supplied to each unit in the rooms where it expands through nozzles thus inducing a large volume of secondary room air into the unit; the secondary air passes over a coil before mixing with the primary air, the mixture then being delivered to the room.

(d) *Fan-coil:* air is drawn from the room and from outside the building into the room units and passed over coils as necessary, the coils carrying chilled or hot water supplied from a central plant.

(e) *Dual-duct:* twin ducts of high-velocity air, one with hot air the other with cold air, are supplied to room units from a central plant as in (a); mixing at the units gives the required condition of the room air.

(f) *Variable air volume* (VAV): a high-velocity flow of cooled air is supplied in a single duct and the change of load is met by varying the air volume while maintaining the same supply temperature; winter conditions can be catered for using a terminal reheat unit as (b) or by adding a second duct carrying hot air with a mixing box as in (e) or by providing a perimeter system of LP hot-water heating.

(g) *Panel air systems:* chilled water is circulated through radiant ceiling panels at a temperature above the room dew point.

(h) *Integrated environmental design:* this is an all-embracing term to cover complex systems incorporating some of the features described above, but with the emphasis on energy recovery, incorporating airflow through luminaires, the use of thermal wheels, and air-to-air or air-to-water heat pumps.

References 15.4 and 15.5 should be consulted for a more detailed discussion.

Summer air-conditioning

The air-conditioning load on a room or space may be considered in two parts: the sensible heat load which is defined as the energy added per unit time which increases the dry bulb temperature; and the latent heat load which is defined as the energy added per unit time due to the enthalpy of the moisture added plus the heat required to evaporate the moisture added.

The sensible heat gains are due to heat transfer through the fabric, including solar radiation, plus internal gains from people, lighting, machinery, etc. The latent heat gains are mainly due to the occupants of the room.

Figure 15.5 shows a typical room condition line on the psychrometric chart; point 1 represents the moist air from the air-conditioning plant entering the room; point 2 represents the moist air leaving the room. It may be assumed that the air at state 2 is at the design conditions for the room.

Point X is such that

$$\omega_1 = \omega_X \quad \text{and} \quad t_2 = t_X$$

Sensible heat load $= \dot{m}_a(h_X - h_1) = \dot{m}_a c_{pma}(t_X - t_1)$

Or using equation (15.12),

Sensible heat load $= \dot{m}_a(c_{pa} + \omega c_{ps})(t_X - t_1)$ (15.14)

Fig. 15.5 Typical room
condition line on a
psychrometric chart

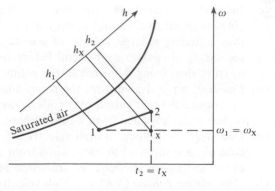

Also

$$\text{Latent head load} = \dot{m}_a(h_2 - h_x)$$

Or using equation (15.11) with $t_2 = t_x$

$$\text{Latent heat load} = \dot{m}_a(\omega_2 - \omega_1)(h_g \text{ at } t_2) \qquad (15.15)$$

$$\text{Total heat load} = \text{sensible heat load} + \text{latent heat load}$$

$$= \dot{m}_a(h_2 - h_1) \qquad (15.16)$$

The *room ratio line* 1–2 is given by

$$\frac{h_x - h_1}{h_2 - h_1} = \frac{\text{sensible heat load}}{\text{total heat load}}$$

where, for zero latent heat load the ratio is unity and the line on the chart is horizontal, and for zero sensible heat load the ratio is zero and the line is vertical. The ratio of sensible heat load to total heat load, and hence the slope of the room condition line on the chart, is given by a protractor in the top left-hand corner of the chart (see Fig. 15.4). A more detailed description of the chart construction and the room ratio line is given in reference 15.5.

Note: the use of the term 'heat gain' is contrary to the accepted thermodynamic definition of heat as a transitory form of energy; also, the term 'latent heat' has now been replaced by 'specific enthalpy of vaporization'. The terms 'sensible heat gain' and 'latent heat gain' are still used extensively in the building services industry in the UK and therefore will be used in this book.

A typical conventional air-conditioning system is shown diagrammatically in Fig. 15.6(a). Some of the air is recirculated and mixed with a quantity of fresh air. Assuming adiabatic mixing, we have

$$rh_2 + (1 - r)h_3 = h_4$$

where r is the mass flow of dry recirculated air per unit mass flow of dry air supplied to the room. Therefore

$$r = \frac{h_3 - h_4}{h_3 - h_2}$$

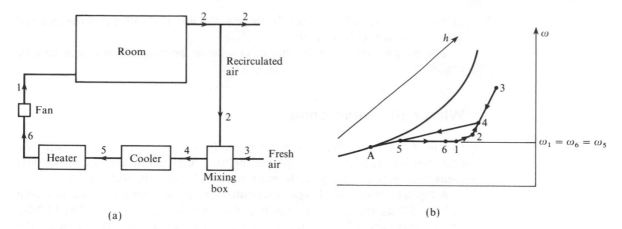

(a)

(b)

Fig. 15.6 Summer
air-conditioning
system (a) with the
processes on a
psychrometric chart (b)

A mass balance of the moisture gives

$$r\omega_2 + (1 - r)\omega_3 = \omega_4$$

therefore

$$r = \frac{\omega_3 - \omega_4}{\omega_3 - \omega_2}$$

i.e.

$$r = \frac{h_3 - h_4}{h_3 - h_2} = \frac{\omega_3 - \omega_4}{\omega_3 - \omega_2} = \frac{\text{line } 3 - 4}{\text{line } 3 - 2} \tag{15.17}$$

Hence point 4 can be fixed by proportion along the line 3–2 on Fig. 15.6(b) when r is known.

In the cooling coil the air undergoes sensible cooling and dehumidification. Point A in Fig. 15.6(b) is called the *apparatus dew point*. The moist air leaving the coil is at some intermediate state 5, and points 5 and A would only coincide if the coil surface were infinitely large.

To define the efficiency of the cooler a term is introduced as follows:

$$\text{Coil bypass factor} = \frac{\text{line } 5 - A}{\text{line } 4 - A} \tag{15.18}$$

This is sometimes defined in terms of a contact factor

$$\text{Contact factor} = \frac{\text{line } 4 - 5}{\text{line } 4 - A} \tag{15.19}$$

Dehumidification may also be achieved by passing the air through a spray cooler supplied with chilled water. The apparatus dew point is then the water temperature. In this case the contact factor given by equation (15.19) is usually renamed the spray cooler, or washer, efficiency and is expressed as a percentage.

The actual condition line of the moist air in both a coil type and a spray-type cooler is not straight on the psychrometric chart; the exact path can be plotted using the theory of combined heat and mass transfer (see reference 15.5).

The reheat coil provides sensible heating of the air to bring the air intake to the room to state 1 such that the slope of the line 1–2 will match the required sensible and latent heat loads.

The fan provides a small amount of sensible heating which can usually be neglected.

Winter air-conditioning

In winter the fabric heat losses are partially compensated by solar radiation and internal heat gains from people, lighting, machinery, etc. the latent heat gains from people are again the main source of moisture addition.

A typical conventional type air-conditioning system for winter use is shown in Fig. 15.7(a) and the corresponding state points are shown on Fig. 15.7(b). The various parts of the system are similar to those of Fig. 15.6 except for the humidifier. The humidification process 5–6 in the case shown is assumed to be adiabatic and takes place at constant wet bulb temperature if pumped recirculation of the water is used as shown in Fig. 15.7(a).

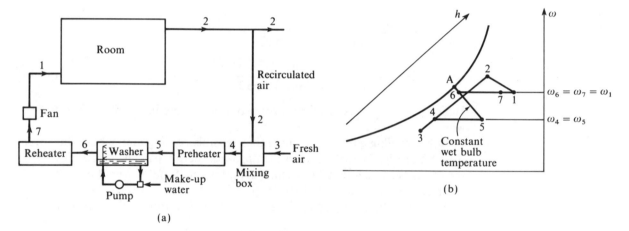

Fig. 15.7 Winter air-conditioning system (a) with the processes on a psychrometric chart (b)

As before, a washer efficiency is defined as (line 5–6)/(line 5–A).

In general, direct-contact air washers and humidifiers may be classified as follows:

(a) pumped recirculation;
(b) (i) no recirculation, with a water spray which is continuously evaporated;
    (ii) no recirculation, with steam blown into the air stream.

It can be shown for (a), assuming an adiabatic process, that the process occurs at a constant thermodynamic wet bulb temperature (see reference 15.5).

For cases (b)(i) and (b)(ii), assuming that the process changes from state 5 to state 6, we have

$$\text{Mass of water or steam added, } \dot{m}_\text{s} = \dot{m}_\text{a}(\omega_6 - \omega_5)$$

Also

$$\dot{m}_a(h_6 - h_5) = \dot{m}_s h_s$$

i.e. $$\dot{m}_a(h_6 - h_5) = \dot{m}_a(\omega_6 - \omega_5)h_s$$

or

$$h_s = \frac{h_6 - h_5}{\omega_6 - \omega_5} \tag{15.20}$$

It can be seen from equation (15.20) that the slope of the process line 5–6 depends on the enthalpy of the water or steam added. When water is added it can be shown that over the range of possible water temperatures the process line can be approximated to a line of constant wet bulb temperature as in (a). For example for a change in moisture content of 0.01, say, and water at $100\,°C$ then the enthalpy change, $(h_6 - h_5)$, is only $0.01 \times 419.1 = 4.2 \text{ kJ/kg}$; for water at $20\,°C$ the enthalpy change is only 0.8 kJ/kg; it can be seen from the chart that these enthalpy changes are of the order found in following a line of constant wet bulb over this sort of moisture content range.

When steam is injected the enthalpy, h_s, is much larger than for water, and heating and humidification can be obtained. For example, from equations (15.15), (15.16) and (15.20) it can be seen that for a value of $h_s = (h_g \text{ at } t_5)$ then there will be no sensible heating or cooling and the process will be such that $t_6 = t_5$.

Example 15.3

An air-conditioned room is to be maintained at $18\,°C$, percentage saturation 40%. The fabric heat gains are 3000 W and there are a maximum of 20 people in the room at any time. Neglecting all other heat gains or losses calculate the required volume flow rate of air to be supplied to the room and its percentage saturation when the air supply temperature is $10\,°C$.

Data

Sensible heat gains per person = 100 W; latent heat gains per person = 30 W; barometric pressure = 1.013 25 bar.

Solution

Sensible heat gain = $3000 + (20 \times 100) = 5000$ W

Latent heat gain = $20 \times 30 = 600$ W

The process is shown on a sketch of the psychrometric chart in Fig. 15.5, (p. 544), where point 2 represents the design state and point 1 the state of the supply air to the room.

At $18\,°C$, from tables $p_{g2} = 0.020\,63$ bar. Using equation (15.7)

$$\omega_2 = \frac{0.4 \times 0.622 \times 0.020\,63}{1.013\,25 - 0.020\,63} = 0.005\,17$$

From equation (15.15)

Latent head load = $\dot{m}_a(\omega_2 - \omega_1)(h_g \text{ at } t_2)$

therefore

$$\omega_2 - \omega_1 = \frac{600}{\dot{m}_a \times 2533.9}$$

and $\qquad \omega_1 = 0.005\,17 - \dfrac{0.2368}{\dot{m}_\mathrm{a}}$ $\qquad\qquad$ (a)

Also, using equation (15.14)

$$\text{Sensible heat load} = \dot{m}_\mathrm{a}(c_{pa} + \omega_1 c_{ps})(t_2 - t_1)$$

therefore

$$5000 = \dot{m}_\mathrm{a}(1.005 + 1.88\omega_1)(18 - 10)$$

and $\qquad \dot{m}_\mathrm{a} = \dfrac{5000}{8.04 + 15.04\omega_1}$ $\qquad\qquad$ (b)

Substituting from (b) into (a)

$$\omega_1 = 0.005\,17 - \frac{0.2368}{5000}(8.04 + 15.04\omega_1)$$

therefore

$$\omega_1 = 0.004\,79$$

Substituting in (b)

$$\dot{m}_\mathrm{a} = 616.4\ \mathrm{kg/s}$$

From equation (15.3)

$$\frac{p_{s1}}{p_{a1}} = \frac{0.004\,79}{0.622} = 0.0077$$

i.e. $\qquad \dfrac{p - p_{a1}}{p_{a1}} = 0.0077 \quad$ therefore $p_{a1} = 1.005\,51$

Then $\qquad \rho_{a1} = \dfrac{1.005\,51 \times 10^5}{287 \times 283} = 1.238\ \mathrm{kg/m^3}$

i.e. $\qquad \text{Volume flow rate of supply air} = \dfrac{616.4}{1.238} = 498\ \mathrm{m^3/s}$

Using equation (15.7),

$$\psi_1 = \frac{100 \times 0.004\,79(1.013\,25 - 0.012\,27)}{0.622 \times 0.012\,27} = 62.8\%$$

i.e. $\qquad \text{Percentage saturation of supply air} = 62.8\%$

This problem can be solved very quickly, if less accurately, using the psychrometric chart. Point 2 can be located on the chart from the information given, then we have

$$\frac{\text{Sensible heat load}}{\text{Total heat load}} = \frac{5000}{5600} = 0.893$$

Using the chart protractor and drawing a line of slope 0.893 from point 2 to where it cuts the 10 °C dry bulb line fixes point 1. Hence the percentage saturation at point 1 and h_1 can be read from the chart. Also

$$\text{Total heat load} = 5600 = \dot{m}_a(h_2 - h_1)$$

Hence $\dot{m}_a$ can be found; the specific volume at state 1 can also be read from the chart and hence the volume flow rate may be calculated.

Example 15.4   Air at 1 °C dry bulb and 80% percentage saturation mixes adiabatically with air at 18 °C dry bulb, 40% percentage saturation, in the ratio of 1 to 3 by volume. Calculate the temperature and percentage saturation of the mixture. Take the barometric pressure as 1.013 25 bar.

Solution   The process is shown on a sketch of the chart in Fig. 15.8, where point 3 represents the condition of the mixture.

Fig. 15.8 Mixing process for Example 15.4

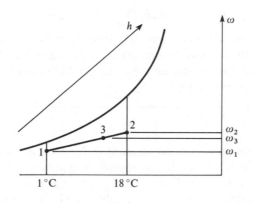

The problem can be solved very easily using the psychrometric chart. From equation (15.17)

$$r = \frac{\text{line } 3\text{--}2}{\text{line } 1\text{--}2} = \frac{\dot{m}_{a1}}{\dot{m}_{a1} + \dot{m}_{a2}}$$

From the chart

$$v_{a1} = 0.78 \text{ m}^3/\text{kg} \quad \text{and} \quad v_{a2} = 0.831 \text{ m}^3/\text{kg}$$

And from the information given

$$\frac{\dot{V}_2}{\dot{V}_1} = 3$$

Then using equation (15.13)

$$\frac{\dot{m}_{a2}}{\dot{m}_{a1}} = \frac{\dot{V}_2}{\dot{V}_1} \times \frac{v_{a1}}{v_{a2}} = \frac{3 \times 0.78}{0.831} = 2.82$$

$$r = \frac{1}{1 + 2.82} = 0.262$$

Then by measurement from the chart

$$\text{Line } 1-2 = 7.9 \text{ mm}$$

$$\text{Line } 3-2 = 0.262 \times 7.9 = 2.07 \text{ mm}$$

Hence point 3 can be located on the line 1–2.
    From the chart

$$t_3 = 13.6\,°\text{C} \quad \text{and} \quad \text{percentage saturation} = 48\%$$

It can be seen from equation (15.17) that the enthalpies and specific humidities are in proportion to the lengths on the line 1–2. This is approximately true also for the dry bulb temperature,

i.e. $\qquad \dfrac{t_2 - t_3}{t_2 - t_1} \simeq \dfrac{\text{line } 3-2}{\text{line } 1-2} = 0.262$

therefore

$$t_3 \simeq 18 - 0.262(18 - 1) = 13.6\,°\text{C}$$

This is only approximately true since the dry bulb temperature lines are not exactly vertical. In constructing the chart the dry bulb line at 30 °C is made vertical and hence all the other dry bulb lines have a slight slope to the vertical (see reference 15.5).

Example 15.5   An air-conditioning plant is designed to maintain a room at a condition of 20 °C dry bulb and specific humidity 0.0079 when the outside condition is 30 °C dry bulb and 40% percentage saturation and the corresponding heat gains are 18 000 W (sensible), and 3600 W (latent). The supply air contains one-third outside air by mass and the supply temperature is to be 15 °C dry bulb.
    The plant consists of a mixing chamber for fresh and recirculated air, an air washer with chilled spray water with an efficiency of 80%, an after heater battery and supply fan.
    Neglecting temperature changes in fan and ducting, calculate:

  (i) the mass flow rate of supply air necessary;
 (ii) the specific humidity of the supply air;
(iii) the cooling duty of the washer;
(iv) the heating duty of the after heater.

Use the psychrometric chart assuming the barometric pressure is 1.013 25 bar.

Solution   The plant is shown in Fig. 15.9(a) and the processes are shown on the chart in Fig. 15.9(b). Points 2 and 3 can be fixed since the conditions are known. Fresh air is to be one-third by mass of the total air to the room, hence point 4 is fixed one-third of the way from 2 to 3.

$$\frac{\text{Sensible heat load}}{\text{Total heat load}} = \frac{18\,000}{18\,000 + 3600} = 0.833$$

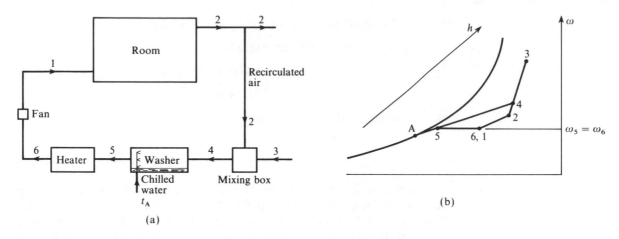

Fig. 15.9 Summer air-conditioning plant (a) and psychrometric chart for Example 15.5 (b)

Using the chart protractor a line of slope 0.833 is drawn from point 2 and where it cuts the dry bulb line of 15 °C gives point 1. Neglecting the fan work then points 6 and 1 are coincident.

The washer efficiency is 80% and point 5 must lie on the horizontal line through point 1 since there is no change in moisture content across the heater,

i.e. $\quad \dfrac{\text{Line } 4\text{--}5}{\text{Line } 4\text{--}A} = 0.8$

or $\quad \dfrac{\omega_4 - \omega_5}{\omega_4 - \omega_A} = 0.8 \quad$ therefore $\omega_A = \omega_4 - \dfrac{(\omega_4 - \omega_1)}{0.8}$

i.e. $\quad \omega_A = 0.0089 - \dfrac{(0.0089 - 0.0075)}{0.8} = 0.007\,15$

Point 5 is fixed by joining points 4 and A; where this line cuts the horizontal line through 1 fixes point 5 at $t_5 = 12\,°C$ dry bulb.

(i) From the chart $h_1 = 33.9 \text{ kJ/kg}$, $h_2 = 40.2 \text{ kJ/kg}$.

$$\text{Total heat load} = 18\,000 + 3600 = 21.6 \text{ kW}$$

therefore

$$\dot{m}_{a1} = \frac{21.6}{(40.2 - 33.9)} = 3.43 \text{ kg/s}$$

Also $\omega_1 = 0.0075$, therefore

$$\text{Mass flow rate of supply air} = 3.43(1 + 0.0075)$$
$$= 3.45 \text{ kg/s}$$

(ii) From the chart

$$\text{Specific humidity of supply air} = 0.007\,45$$

551

(iii) From the chart $h_4 = 46.2 \text{ kJ/kg}$, $h_5 = 31.1 \text{ kJ/kg}$.

$$\text{Cooling load on washer} = \dot{m}_{a1}(h_4 - h_5)$$
$$= 3.43(46.2 - 31.1)$$
$$= 51.8 \text{ kW}$$

(iv) $\text{Heating load} = \dot{m}_{a1}(h_6 - h_5) = 3.43(33.9 - 31.1)$
$$= 9.6 \text{ kW}$$

Example 15.6

The plant in Example 15.5 is modified for winter use by using pumped circulation in the washer as in Fig. 15.10(a). The summer room design conditions of 20 °C and specific humidity 0.0079 are to be maintained when the outside conditions are 0 °C and 100% saturation; the ratio of fresh air to recirculated air is to be the same as in summer, and the mass flow rate of air is also the same. The sensible heat loss is 14 400 W. Taking the washer efficiency as 80% as before, and neglecting temperature changes in the fan and ducting, calculate:

Fig. 15.10 Winter air-conditioning plant (a) and psychrometric chart (b) for Example 15.6

(i) the temperature of the air supplied to the room;
(ii) the heat supplied in the heater.

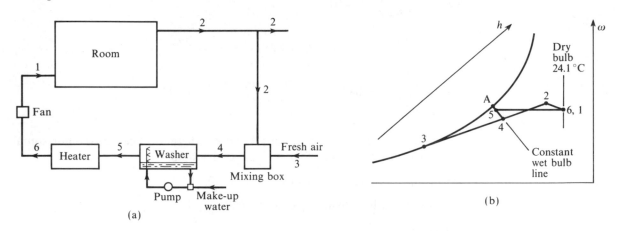

(a)

(b)

Solution (i) The system is shown diagrammatically in Fig. 15.10(a) with the processes on the psychrometric chart in Fig. 15.10(b). The system is slightly different from that of Fig. 15.7 (p. 546); in this case the pre-heater has been dispensed with by using recirculated air to raise the mixed air to a high enough temperature before entry to the washer as shown.

The recirculated air leaving the room at state 2 is at the same condition as in summer, therefore point 2 is fixed on the chart. Point 3 is fixed from the information given, and point 4 is fixed by dividing the line 2–3 in the ratio one to two since there is one-third outside air and two-thirds recirculated air by mass. From the chart, $h_2 = 40.2 \text{ kJ/kg}$, $h_4 = 29.8 \text{ kJ/kg}$, and $\omega_4 = 0.00655$. The best method in this case is to calculate a value of specific heat capacity per unit mass of dry air, and to write

$$\text{Sensible heat loss} = \dot{m}_a c_{pma}(t_1 - t_2)$$

From equation (15.12)

$$c_{pma} = 1.005 + (1.880\omega)$$

and taking the specific humidity at point 2 we have

$$c_{pma} = 1.005 + (1.88 \times 0.0079) = 1.02 \, \text{kJ/kg K}$$

Then, since the mass flow rate, $\dot{m}_a = 3.43$ kg/s as in Example 15.5, we have

$$14.4 = 3.43 \times 1.02(t_1 - 20)$$

therefore

$$t_1 = 24.1 \,°\text{C}$$

(ii) Drawing a line of constant wet bulb temperature through point 4 until it cuts the saturation line fixes point A. Since the washer efficiency is 80%, point 5 is fixed by proportioning line 4–A. Then drawing a horizontal line through point 5 to cut the dry bulb temperature line at 24.1 °C fixes point 6, which is coincident with point 1 if fan work and duct losses are neglected.

From the chart, $h_1 = 43.5$ kJ/kg, and $h_5 = 29.9$ kJ/kg, then

$$\text{Heater load} = \dot{m}_a(h_1 - h_5)$$

$$= 3.43(43.5 - 29.9)$$

$$= 46.7 \, \text{kW}$$

15.5 Cooling towers

Some industrial processes require large quantities of cooling water. The position of the plant may be such that a convenient supply of water (e.g. from the sea or a river) is not available and a recirculatory system is necessary. A necessary part of this system is a cooler which cools down the cooling water. A convenient cooling medium is necessary and this is inevitably the atmosphere. It would be possible to produce some cooling by means of a heat exchanger, the cooling water passing through it and the air passing over it. A more satisfactory method employs the cooling effect produced when water evaporates. This is done by spraying the water into the air over a pond, or into the air passing through a cooling tower. A current of air rises, in the latter method, by means of a natural or forced draught, through a cooling tower and the hot water enters at some point and is sprayed into the air. The cooling effect is greater with a forced draught due to the increased flow of air.

As the water falls, some of it evaporates and to assist this process the tower contains packing which breaks up the stream. The warm water is cooled, the temperature of the air is raised, and it becomes almost completely saturated with water vapour. The cooling water can theoretically be cooled to the wet bulb temperature of the incoming air, but a compromise is reached between the amount of cooling obtained and the size of the tower, and the figure used in design for the cooling water leaving the tower is about 8 K above the wet bulb temperature. Induced and natural draught cooling towers are shown in Figs 15.11 and 15.12, respectively. The packing of the tower is usually formed from wooden slats. A modern design of tower employs a plastic impregnated cellulose material which has high water-absorption qualities and a long working life. For a given duty the size of tower using this type of packing is about one-fifth of that using wooden packing, and it is of much lighter construction. The compact design means that the tower could possibly be situated on the top of a building without a special structure being required. This design, which is of the induced-draught type, delivers the warm water over the packing from a rotating header.

Some of the cooling water is lost to the atmosphere in the evaporation process in all cooling towers, and so a small amount of make-up water is required.

Fig. 15.11 Induced-draught cooling tower

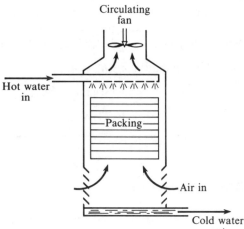

Fig. 15.12 Natural-draught cooling tower

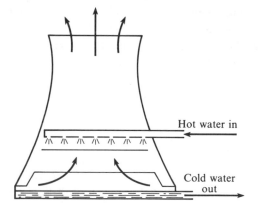

Example 15.7

A small-size cooling tower is designed to cool 5.5 litres of water per second, the inlet temperature of which is 44 °C. The motor-driven fan induces 9 m³/s of air through the tower and the power absorbed is 4.75 kW. The air entering the tower is at 18 °C, and has a relative humidity of 60%. The air leaving the tower can be assumed to be saturated and its temperature is 26 °C. Assuming that the pressure throughout the tower is constant at 1.013 bar, and make-up water is added outside the tower, calculate:

(i) the mass flow rate of make-up water required;
(ii) the final temperature of the water leaving the tower.

Solution

(i) The cooling tower is shown diagrammatically in Fig. 15.13. At inlet, using equation (15.4)

$$\phi = p_s/p_g \quad \text{and} \quad p_g \text{ at } 18\,°\text{C} = 0.020\,63 \text{ bar}$$

Fig. 15.13 Induced-draught cooling tower for Example 15.7

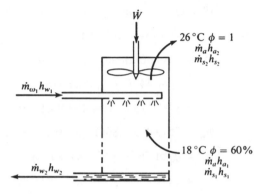

therefore

$$p_{s_1} = 0.6 \times 0.020\,63 = 0.012\,38 \text{ bar}$$

From equation (6.2),

$$p_{a_1} = 1.013 - 0.012\,38 = 1.0006 \text{ bar}$$

Then
$$\dot{m}_a = \frac{10^5 \times 1.0006 \times 9}{10^3 \times 0.287 \times 291} = 10.78 \text{ kg/s}$$

and
$$\dot{m}_{s_1} = \frac{10^5 \times 0.012\,38 \times 9}{10^3 \times 0.4618 \times 291} = 0.0829 \text{ kg/s}$$

At exit at 26 °C

$$p_g = 0.033\,60 \text{ bar} \quad \text{and} \quad \phi = 1$$

therefore

$$p_{s_2} = 0.033\,60 \text{ bar}$$

Using equation (15.3),

$$\omega_2 = 0.622\left(\frac{p_{s_2}}{p - p_{s_2}}\right) = \left(\frac{0.622 \times 0.0336}{1.013 - 0.0336}\right) = 0.02133$$

Then using equation (15.1),

$$\dot{m}_{s_2} = 10.78 \times 0.021\,33 = 0.23 \text{ kg/s}$$

Hence

Make-up water required $= 0.23 - 0.0829 = 0.1471 \text{ kg/s}$

(ii) Also

$$\dot{m}_{w_1} = 5.5 \times 1 = 5.5 \text{ kg/s}$$

and

$$\dot{m}_{w_2} = \dot{m}_{w_1} - (\text{make-up water}) = 5.5 - 0.1471 = 5.353 \text{ kg/s}$$

Applying the steady-flow energy equation and neglecting changes in kinetic energy and potential energy, and all heat losses, we have

$$\dot{W} + \dot{m}_{w_1} h_{w_1} + \dot{m}_{a_1} h_{a_1} + \dot{m}_{s_1} h_{s_1} = \dot{m}_{a_2} h_{a_2} + \dot{m}_{s_2} h_{s_2} + \dot{m}_{w_2} h_{w_2} \qquad [1]$$

Now $\quad \dot{W} = 4.75 \text{ kW} = 4.75 \text{ kJ/s}$

Evaluating the enthalpies from a datum of $0\,°C$, we have

$$h_{w_1} = h_f \text{ at } 44\,°C = 184.2 \text{ kJ/kg}$$

$$h_{a_1} = 1.005(18 - 0) = 18.09 \text{ kJ/kg}$$

$$h_{s_1} = 2519.4 + 1.86(18 - 10.13) = 2534 \text{ kJ/kg}$$

$$h_{s_2} = h_g \text{ at } 26\,°C = 2548.4 \text{ kJ/kg}$$

$$h_{a_2} = 1.005(26 - 0) = 26.13 \text{ kJ/kg}$$

The vapour is superheated at 1, being above $10.13\,°C$, the saturation temperature corresponding to $0.012\,38$ bar. Then substituting in [1]

$$4.75 + (5.5 \times 184.2) + (10.78 \times 18.09) + (0.0829 \times 2534)$$

$$= (10.78 \times 26.13) + (0.23 \times 2548.4) + 5.353 h_{w_2}$$

i.e. $\quad 5.353 h_{w_2} = 556.3 \quad$ therefore $h_{w_2} = 104 \text{ kJ/kg}$

By interpolation, $h_f = 104 \text{ kJ/kg}$ at $24.8\,°C$

i.e. $\quad$ Temperature of water leaving tower $= 24.8\,°C$

Problems

15.1 Air at $32\,°C$ is saturated with water vapour at a barometric pressure of 1.013 bar. Calculate:
 (i) the partial pressures of the vapour and dry air;
 (ii) the volume of the mixture per kilogram of vapour;
 (iii) the mass of dry air per kilogram of vapour;
 (iv) the specific humidity of the mixture;
 (v) the relative humidity of the mixture.

$\qquad$ (0.04754 bar; 0.9655 bar; 29.57 m^3; 32.6 kg; 0.031; 1)

15.2 (a) The pressure of the water vapour in an atmosphere which is at 32 °C and 1.013 bar is 0.02063 bar. Calculate:
   (i) the degree of superheat of the water vapour;
   (ii) the specific humidity;
   (iii) the relative humidity;
   (iv) the temperature to which the air must be cooled for it to become just saturated.
   (b) If the air in (a) is cooled to 10 °C from its original condition, calculate the mass of condensate formed per kilogram of dry air.

   (14 K; 0.01292; 43.4%; 18 °C; 0.0053 kg)

15.3 An air and water vapour mixture at 1 bar and 26.7 °C has a specific humidity of 0.0085. Calculate the percentage saturation.

   (37.7%)

15.4 A mixture of air and water vapour at 1.013 bar and 16 °C has a dew point of 5 °C. Calculate the relative and specific humidities.

   (48%; 0.0054)

15.5 Atmospheric air at a pressure of 760 mm Hg has a temperature of 32 °C and a percentage saturation as determined from a psychrometric chart of 52%. Calculate:
   (i) the partial pressures of the vapour and dry air;
   (ii) the specific humidity;
   (iii) the dew point;
   (iv) the density of the mixture.

   (0.0252 bar; 0.988 bar; 0.0159; 21.2 °C; 1.146 kg/m^3)

15.6 Compare the gas constant for dry air with the value for air saturated with water vapour at 16 °C and 1.013 bar.

   (0.2871 kJ/kg K; 0.2891 kJ/kg K)

15.7 The temperature in a room of volume 38 m^3 is 25 °C and the pressure is 1.013 bar; the dew point of the air in the room is 14 °C. If a vessel containing water is placed in the room calculate the maximum that can be lost by evaporation. Assume that the pressure in the room remains constant.

   (0.433 kg)

15.8 (a) An air-conditioned room is maintained at a temperature of 21 °C and a relative humidity of 55% when the barometric pressure is 740 mm Hg. Calculate:
   (i) the specific humidity;
   (ii) the temperature of the inside of the windows in the room if moisture is just beginning to form on them.
   (b) For the room in part (a) what mass of water vapour per kilogram of dry air must be removed from the mixture in order to prevent condensation on the windows when their temperature drops to 4 °C, and what is the initial relative humidity to satisfy this condition when the room temperature remains at 21 °C?

   (0.00874; 11.62 °C; 0.0035 kg; 32.7%)

15.9 A mixture of air and water vapour at 50 °C and 1.013 bar has an analysis by mass of 4% water vapour to 96% dry air. Calculate:
   (i) the analysis by volume;
   (ii) the partial pressures of the water vapour and dry air;
   (iii) the relative humidity.

   (6.28%, 93.72%; 0.0636 bar, 0.9494 bar; 51.6%)

15.10 For the mixture of Problem 15.9 calculate:
  (i) the enthalpy per kilogram of the mixture reckoned from 0 °C;
  (ii) the heat to be rejected at constant pressure of 1.013 bar for condensation to begin.
  (151.9 kJ/kg; 13.3 kJ/kg)

15.11 For the mixture of Problem 15.8 calculate:
  (i) the specific volume of the vapour;
  (ii) the heat rejected per kilogram of mixture when the mixture is cooled at constant volume until condensation just begins;
  (iii) the dew point for the mixture at this condition;
  (iv) the pressure in the room after cooling.
  (99.3 m^3/kg; 7 kJ; 11.35 °C; 0.954 bar)

15.12 The readings taken in a room from a sling psychrometer give a dry and a wet bulb temperature of 25 °C and 19.9 °C. Using a psychrometric chart, taking the atmospheric pressure as 1.01325 bar, calculate:
  (i) the specific humidity;
  (ii) the percentage saturation;
  (iii) the dew point;
  (iv) the specific volume of the mixture;
  (v) the specific enthalpy of the mixture.
  (0.0125; 62%; 17.4 °C; 0.861 m^3/kg; 57 kJ/kg)

15.13 The atmosphere of Problem 15.12 is cooled until the air is saturated at 5 °C and then heated at constant pressure until the dry bulb temperature is 17.5 °C, both processes occurring at constant pressure. Using the psychrometric chart, calculate:
  (i) the final percentage saturation;
  (ii) the final specific humidity;
  (iii) the final wet bulb temperature;
  (iv) the mass of condensate collected at the cooler per kilogram of dry air;
  (v) the heat rejected in the cooling process per kilogram of dry air;
  (vi) the heat supplied in the heating process per kilogram of dry air.
  (43%; 0.0054; 11 °C; 0.0071 kg; 38.4 kJ; 12.9 kJ)

15.14 It is required to maintain the air in a building at 20 °C and 40% saturation when the outside conditions are 28 °C, 50% saturation with a barometric pressure of 1.01325 bar. The total heat gains to the room (sensible plus latent) are 15 kW, and the latent heat gains are 3 kW. There is no recirculation and the fresh air passes over a cooling coil to dehumidify it, and then a heater, before entering the room. The cooling coil by pass factor is 0.2; the volume flow of fresh air is 5 m^3/s. Calculate:
  (i) the temperature of the air leaving the cooling coil;
  (ii) the refrigeration load for the coil;
  (iii) the heat supplied in the heater.
  (6.6 °C; 219 kW; 66 kW)

15.15 (a) The air-conditioning system in Problem 15.14 is to be adapted for use in winter with the same room conditions of 20 °C and 40% saturation. The mass flow rate of dry air from the heater is to be kept the same as in summer. The outside conditions are −5 °C, 100% saturation, and the sensible heat loss from the conditioned space is 34.5 kW with the same latent heat gain of 3 kW. A saving in energy is obtained by recirculating a proportion of the room air using a mixing box and dispensing with the washer. Calculate:
  (i) the ratio of room air to fresh air that must be used for the system to operate satisfactorily;
  (ii) the heat supplied in the heater under these conditions.

(b) The fresh air requirements for the building in part (a) are such that the ratio of recirculated air to fresh air must be no greater than 3. A steam humidifier is introduced between the mixing box and the heater; the air in the humidifier is heated at approximately constant dry bulb temperature. Calculate the heat now required in the heater.

(16.5; 43 kW; 71 kW)

15.16 A room in summer is to be maintained at 18 °C, 50% saturation when the outside conditions are 30 °C, 80% saturation. The sensible heat gains and latent heat gains are 4.4 kW and 1.89 kW respectively.

The conditioned air is supplied through ducts from a central station consisting of a cooler battery, a reheat battery, and a fan. Fresh air is supplied to a mixing unit where it mixes with a certain percentage of air recirculated from the room, the remainder of the room air being expelled to atmosphere.

The air entering the room is at 12.5 °C, the air temperature rise in the fan and duct work is 1 K, the air leaving the cooler battery and entering the reheat battery is at 7 °C, and the apparatus dew point of the cooler is 1.5 °C.

Draw a sketch of the plant, numbering the relevant points, and calculate:
 (i) the ratio of the mass rate of flow of recirculated air to the mass rate of air supplied to the room;
 (ii) the cooler battery load;
 (iii) the reheater battery load;
 (iv) the cooler battery bypass factor.
Use the psychrometric chart, taking the barometric pressure as 1.01325 bar.

(0.88; 15.4 kW; 3.5 kW; 0.3)

15.17 Air enters a natural draught cooling tower at 1.013 bar and 13 °C and relative humidity 50%. Water at 60 °C from turbine condensers is sprayed into the tower at the rate of 22.5 kg/s and leaves at 27 °C. The air leaves the tower at 38 °C, 1.013 bar and is saturated. Calculate:
 (i) the airflow required in cubic metres per second;
 (ii) the make-up water required in kilograms per second.

($21 \, \text{m}^3/\text{s}$; 1 kg/s)

15.18 In a forced-draught cooling tower hot water enters at a rate of 15 kg/s at 27 °C and leaves the tower at 21 °C. The ambient air drawn into the tower is at 1.01325 bar, 23 °C dry bulb and 17 °C wet bulb, and the air leaving the tower is saturated at 25 °C. The fan power input is 5 kW. Assuming that the specific enthalpy of superheated vapour is approximately equal to the specific enthalpy of saturated vapour at the same temperature, and that the pressure in the tower remains constant throughout, calculate:
 (i) the required mass flow rate of air;
 (ii) the required mass flow rate of make-up water to be added exterior to the tower.
Take the partial pressure of water vapour in air as

$$p = p_{\text{g}} - \{6.748 \times 10^{-4} (t_{\text{DB}} - t_{\text{WB}})\}$$

where the pressures are in bar, and the dry bulb temperature, t_{DB}, and the wet bulb temperature, t_{WB}, are in degrees Centigrade.

(13.67 kg/s; 0.144 kg/s)

References

15.1 CIBSE 1986 *Guide to Current Practice* volume C
15.2 THRELKELD J L 1970 *Thermal Environmental Engineering* 2nd edn Prentice-Hall

15.3   BS 2842 1982 *Specification for Whirling Hygrometers*
          BS 5248 1982 *Specification for Aspirated Hygrometers*
          BS 3292 1988 *Specification for Direct Reading Hygrometers*
15.4   JONES W P 1985 *Air Conditioning Engineering* 3rd edn Edward Arnold
15.5   EASTOP T D and WATSON W E 1992 *Mechanical Services for Buildings* Longman

16

Heat Transfer

The transfer of heat across the boundaries of a system, either to or from the system, is considered in previous chapters for non-flow and flow processes; the definition of heat used throughout this book, and given in section 1.1, simply states that heat is a form of energy which is transferred from one body to another body at a lower temperature by virtue of the temperature difference between the bodies. When the mechanism of the transfer of heat is considered a slightly different approach is necessary compared with the approach of fundamental thermodynamics. For instance it becomes difficult to define a system. In order to illustrate this point consider a bar of metal being heated at one end and cooled at the other. Now a boundary may be put round the source of heat or round the sink for the rejection of heat, but a boundary encircling the metal bar encloses a body the temperature of which varies throughout its length. In order to apply the laws of thermodynamics to the system consisting of the metal bar, a mean temperature must be assumed.

In previous chapters many problems have been considered in which a certain quantity of heat has been transferred from one system to another. In this chapter we shall be concerned with the rate at which heat is transferred. The rate of heat transfer may be constant or variable, depending on whether conditions are such that the temperatures remain the same or change continually with time. Most problems in practice are concerned with steady-state heat transfer, in which heat flows continuously at a uniform rate, but there are many cases of transient heat transfer and some of these will also be considered.

In general there are three ways in which heat may be transferred, given under the headings below.

By conduction

Conduction is the transfer of heat from one part of a substance to another part of the same substance, or from one substance to another in physical contact with it, without appreciable displacement of the molecules forming the substance. For example, the heat transfer in the metal bar mentioned previously is by conduction.

By convection

Convection is the transfer of heat within a fluid by the mixing of one portion of the fluid with another. The movement of the fluid may be caused by differences in density resulting from the temperature differences as in *natural convection* (or *free convection*), or the motion may be produced by mechanical means, as in *forced convection*. For example, the heat transferred from a hot-plate to the atmosphere is by natural convection, whereas the heat transferred by a domestic fan-heater, in which a fan blows air across an electric element, is by forced convection.

The transfer of heat through solid bodies is by conduction alone, whereas the heat transfer from a solid surface to a liquid or gas takes place partly by conduction and partly by convection. Whenever there is an appreciable movement of the gas or liquid, the heat transfer by conduction in the gas or liquid becomes negligibly small compared with the heat transfer by convection. However, there is always a thin boundary layer of fluid on a surface, and through this thin film the heat is transferred by conduction.

By radiation

All matter continuously emits electromagnetic radiation unless its temperature is absolute zero. It is found that the higher the temperature then the greater is the amount of energy radiated. If, therefore, two bodies at different temperatures are so placed that the radiation from each body is intercepted by the other, then the body at the lower temperature will receive more energy than it is radiating, and hence its internal energy will increase; similarly the internal energy of the body at the higher temperature will decrease. Thus there is a net transfer of energy from the high-temperature body to the low-temperature body by virtue of the temperature difference between the bodies. This form of energy transfer satisfies the definition of heat given in section 1.1, and hence we may say that heat is transferred by radiation.

Radiant energy, being electromagnetic radiation, requires no medium for its propagation, and will pass through a vacuum. Heat transfer by radiation is most frequent between solid surfaces, although radiation from gases also occurs. Certain gases emit and absorb radiation on certain wavelengths only, whereas most solids radiate over a wide range of wavelengths.

In any particular example in practice heat may be transferred by a combination of conduction, convection, and radiation, and it is usually possible to assess the effects of each mode of heat transfer separately and then to sum up the results. There are two main groups of problems; first, the desirable transfer of heat to or from a fluid as in a heat exchanger, boiler, or condenser, and second, the prevention of heat losses from a fluid to its surroundings.

16.1 Fourier's law of conduction

Fourier's law states that the rate of flow of heat through a single homogeneous solid is directly proportional to the area A of the section at right angles to the direction of heat flow, and to the change of temperature with respect to the length of the path of the heat flow, dt/dx. This is an empirical law based on observation.

The law is illustrated in Fig. 16.1(a) in which a thin slab of material of thickness dx and surface area A has one face at a temperature t and the other at a temperature $(t + dt)$. Then applying Fourier's law we have for the rate of heat flow in the direction x,

$$\text{Rate of heat flow, } \dot{Q} \propto A\frac{dt}{dx}$$

or $\qquad \dot{Q} = -\lambda A\dfrac{dt}{dx}$ $\qquad\qquad\qquad\qquad\qquad$ (16.1)

Fig. 16.1 Heat flow through a thin slab of material

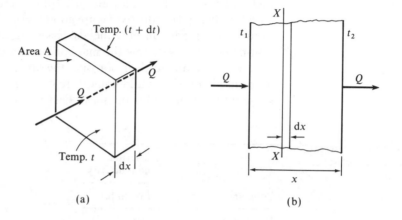

(a) $\qquad\qquad\qquad\qquad\qquad$ (b)

The rate of heat flow in the direction x is taken as positive, hence the negative sign in equation (16.1) since dt is always negative. The term λ is called the *thermal conductivity* of the material. The thermal conductivity of a substance can be defined as the heat flow per unit area per unit time when the temperature decreases by one degree in unit distance.

The units of λ are usually written as W/m K or kW/m K.

Consider the transfer of heat through a slab of material as shown in Fig. 16.1(b). At section X–X, using equation (16.1)

$$\dot{Q} = -\lambda A\frac{dt}{dx} \quad \text{or} \quad \dot{Q}\,dx = -\lambda A\,dt$$

Integrating

$$\int_0^x \dot{Q}\,dx = -\int_{t_1}^{t_2} \lambda A\,dt$$

or $\qquad \dot{Q}x = -A\displaystyle\int_{t_1}^{t_2} \lambda\,dt$

This equation can be solved when the variation of thermal conductivity, λ, with temperature, t, is known. Now for most solids the value of the thermal

563

conductivity is approximately constant over a wide range of temperatures, and therefore λ can be taken as constant,

i.e. $\quad \dot{Q}x = -A\lambda \int_{t_1}^{t_2} \mathrm{d}t$

or $\quad \dot{Q} = -\dfrac{\lambda A}{x}(t_2 - t_1) = \dfrac{\lambda A}{x}(t_1 - t_2)$ \hfill (16.2)

Note that in this case the area in the direction at right angles to the heat flow remains constant through the slab. Cases will be considered later in which the area varies.

The thermal conductivities of some materials encountered in engineering are shown in Table 16.1. It follows from equation (16.1) that materials with high thermal conductivities are good conductors of heat, whereas materials with low thermal conductivities are good thermal insulators. Conduction of heat occurs most readily in pure metals, less so in alloys, and much less readily in non-metals. The very low thermal conductivities of certain thermal insulators (e.g. cork) are due to their porosity, the air trapped within the material acting as an insulator. Gases and liquids are good insulators, but unless a completely stagnant layer of fluid is obtained, heat is transferred by convection currents.

Table 16.1 Thermal conductivity of some substances

Substance	Thermal conductivity (W/m K)
Pure copper	386
Pure aluminium	229
Duralumin	164
Cast iron	52
Mild steel	48.5
Lead	34.6
Concrete	0.85–1.4
Building brick	0.35–0.7
Wood (oak)	0.15–0.2
Rubber	0.15
Cork board	0.043

Example 16.1 The inner surface of a plane brick wall is at $40\,°C$ and the outer surface is at $20\,°C$. Calculate the rate of heat transfer per unit area of wall surface; the wall is 250 mm thick and the thermal conductivity of the brick is 0.52 W/m K.

Solution From equation (16.2)

$$\dot{Q} = \frac{\lambda A}{x}(t_1 - t_2)$$

therefore

$$\frac{\dot{Q}}{A} = \dot{q} = \frac{10^3 \times 0.52}{250} \times (40 - 20) = 41.6 \text{ W/m}^2$$

Note that the symbol $\dot{q}$ is used for the rate of heat transfer per unit area.

16.2 Newton's law of cooling

In order to consider the rate at which heat is transferred from one fluid to another through a plane wall it is necessary to know something of the way in which heat is transferred from a solid surface to a fluid and vice versa.

Newton's law of cooling states that the heat transfer from a solid surface of area A, at a temperature t_w, to a fluid of temperature t, is given by

$$\dot{Q} = \alpha A(t_w - t) \tag{16.3}$$

where α is called the *heat transfer coefficient*.

The units of α are seen to be $\text{W/m}^2\,\text{K}$, or $\text{kW/m}^2\,\text{K}$. The heat transfer coefficient, α, depends on the properties of the fluid and on the fluid velocity; it is usually necessary to evaluate it by experiment. This will be discussed more fully in section 16.9.

Equation (16.3) does not include the heat loss from the surface by radiation. This effect can be calculated separately (see section 16.18), and in many cases is negligible compared with the heat transferred by conduction and convection from the surface to the fluid. When the surface temperature is high, or when the surface loses heat by natural convection, then the heat transfer due to radiation is of a similar magnitude to that lost by convection.

Consider the transfer of heat from a fluid A to a fluid B through a dividing wall of thickness x, and thermal conductivity λ, as shown in Fig. 16.2. The variation of temperature in the direction of the heat transfer is also shown. In fluid A the temperature decreases rapidly from t_A to t_1 in the region of the wall, and similarly in fluid B the temperature decreases rapidly from t_2 to t_B in the

Fig. 16.2 Temperature variation for heat transfer from one fluid to another through a dividing wall

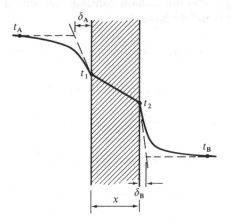

region of the wall. In most practical cases the fluid temperature is approximately constant throughout its bulk, apart from a thin film near the solid surface bounding the fluid. The dotted lines drawn on Fig. 16.2 show that the thickness of this film of fluid is given by δ_A for fluid A and δ_B for fluid B. The heat transfer in these films is by conduction only, hence applying equation (16.2) we have, considering unit surface area, from fluid A to the wall

$$\dot{q} = \frac{\lambda_A}{\delta_A}(t_A - t_1) \tag{a}$$

from the wall to fluid B

$$\dot{q} = \frac{\lambda_B}{\delta_B}(t_2 - t_B) \tag{b}$$

Also from equation (16.3), from fluid A to the wall

$$\dot{q} = \alpha_A(t_A - t_1) \tag{c}$$

from the wall to fluid B

$$\dot{q} = \alpha_B(t_2 - t_B) \tag{d}$$

Comparing equations (a) and (c), and equations (b) and (d), it can be seen that

$$\alpha_A = \frac{\lambda_A}{\delta_A} \quad \text{and} \quad \alpha_B = \frac{\lambda_B}{\delta_B}$$

In general, $\alpha = \lambda/\delta$, where δ is the thickness of the stagnant film of fluid on the surface.

The heat flow through the wall in Fig. 16.2 is given by equation (16.2).
For unit surface area

$$\dot{q} = \frac{\lambda}{x}(t_1 - t_2)$$

For steady-state heat transfer, the heat flowing from fluid A to the wall is equal to the heat flowing through the wall, which is also equal to the heat flowing from the wall to fluid B. If this were not so, then the temperatures t_A, t_1, t_2, and t_B would not remain constant but would change with time.
We therefore have

$$\dot{q} = \alpha_A(t_A - t_1) = \frac{\lambda}{x}(t_1 - t_2) = \alpha_B(t_2 - t_B)$$

Rewriting these equations in terms of the temperatures, then

$$(t_A - t_1) = \frac{\dot{q}}{\alpha_A}; \quad (t_1 - t_2) = \frac{\dot{q}x}{\lambda}; \quad (t_2 - t_B) = \frac{\dot{q}}{\alpha_B}$$

Hence adding the corresponding sides of the three equations

$$(t_A - t_1) + (t_1 - t_2) + (t_2 - t_B) = \frac{\dot{q}}{\alpha_A} + \frac{\dot{q}x}{\lambda} + \frac{\dot{q}}{\alpha_B}$$

therefore

$$(t_A - t_B) = \dot{q}\left(\frac{1}{\alpha_A} + \frac{x}{\lambda} + \frac{1}{\alpha_B}\right)$$

i.e. $$\dot{q} = \frac{(t_A - t_B)}{(1/\alpha_A + x/\lambda + 1/\alpha_B)}$$

By analogy with equation (16.3) this can be written as

$$\dot{q} = U(t_A - t_B) \tag{16.4}$$

or $$\dot{Q} = UA(t_A - t_B) \tag{16.5}$$

where $$\frac{1}{U} = \left(\frac{1}{\alpha_A} + \frac{x}{\lambda} + \frac{1}{\alpha_B}\right) \tag{16.6}$$

U is called the *overall heat transfer coefficient*, and it has the same units as α.

Example 16.2   A mild steel tank of wall thickness 10 mm contains water at 90 °C when the atmospheric temperature is 15 °C. The thermal conductivity of mild steel is 50 W/m K, and the heat transfer coefficients for the inside and outside of the tank are 2800 and 11 W/m² K respectively. Calculate:

(i) the rate of heat loss per unit area of tank surface;
(ii) the temperature of the outside surface of the tank.

Solution   (i) The wall of the tank is shown diagrammatically in Fig. 16.3. From equation (16.6)

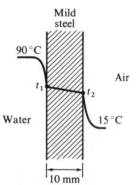

Mild steel

90 °C

t_1

Water

Air

t_2

15 °C

10 mm

Fig. 16.3 Tank wall for Example 16.2

$$\frac{1}{U} = \frac{1}{\alpha_A} + \frac{x}{\lambda} + \frac{1}{\alpha_B} = \frac{1}{2800} + \frac{10}{10^3 \times 50} + \frac{1}{11}$$

$$= 0.000\,357 + 0.0002 + 0.0909$$

i.e. $$\frac{1}{U} = 0.0915$$

Then substituting in equation (16.4), $\dot{q} = U(t_A - t_B)$, we have

$$\dot{q} = \left(\frac{90 - 15}{0.0915}\right) = 820 \text{ W/m}^2$$

i.e.   Rate of heat loss per square metre of surface area = 0.82 kW

(ii) From equation (16.3)

$$\dot{q} = \alpha_B(t_2 - t_B) \quad \text{or} \quad 820 = 11 \times (t_2 - 15)$$

where t_2 is the temperature of the outside surface of the tank as shown in Fig. 16.3.

$$t_2 = \frac{820}{11} + 15 = 89.6 \text{ °C}$$

i.e.   Temperature of outside surface of tank = 89.6 °C

16.3   The composite wall and the electrical analogy

There are many cases in practice when different materials are constructed in layers to form a composite wall. An example of this is the wall of a building, which usually consists of a layer of plaster, a row of bricks, an air gap, a second row of bricks, and perhaps a cement rendering on the outside surface.

Consider the general case of a composite wall as shown in Fig. 16.4. There are n layers of material of thickness x_1, x_2, x_3, etc. and of thermal conductivity $\lambda_1, \lambda_2, \lambda_3$, etc. On one side of the composite wall there is a fluid A at temperature t_A, and the heat transfer coefficient from fluid to wall is α_A; on the other side of the composite wall there is a fluid B, and the heat transfer coefficient from wall to fluid is α_B. Let the temperature of the wall in contact with fluid A be t_0 and the temperature of the wall in contact with fluid B be t_n; the interface temperatures are then t_1, t_2, t_3, etc. as shown. The most convenient method of solving such a problem is by making use of an electrical analogy. The flow of heat can be thought of as analogous to an electric current. The heat flow is caused by a temperature difference whereas the current flow is caused by a potential difference, hence it is possible to postulate a *thermal resistance* analogous to an electrical resistance. From Ohm's law we have

$$V = IR \quad \text{or} \quad I = \frac{V}{R}$$

where V is the potential difference, I the current, and R the resistance.

Fig. 16.4 Heat transfer through a composite wall

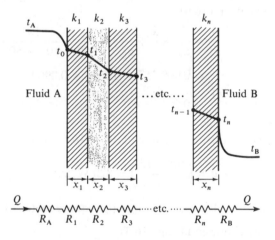

Comparing this equation with equation (16.2), $\dot{Q} = \{\lambda A/x\}(t_1 - t_2)$, we have

$$\text{Thermal resistance, } R = \frac{x}{\lambda A} \tag{16.7}$$

where $\dot{Q}$ is analogous to I, and $(t_1 - t_2)$ is analogous to V.

The composite wall is analogous to a series of resistances, as shown in Fig. 16.4, and resistances in series can be added to give the total resistance. To find the resistance of a fluid film it is necessary to compare Ohm's law with equation (16.3), $\dot{Q} = \alpha A(t_{\text{w}} - t)$,

i.e.   Thermal resistance of a fluid film, $R = \dfrac{1}{\alpha A}$ (16.8)

where $\dot{Q}$ is analogous to I and $(t_{\text{w}} - t)$ is analogous to V. Note that the units of thermal resistance are K/W or K/kW.

Referring to Fig. 16.4 we therefore have

$$R_A = \frac{1}{\lambda_A A}, \qquad R_1 = \frac{x_1}{\lambda_1 A}, \qquad R_2 = \frac{x_2}{\lambda_2 A}, \text{ etc.}$$

$$R_n = \frac{x_n}{\lambda_n A} \quad \text{and} \quad R_B = \frac{1}{\alpha_B A}$$

The total resistance to heat flow is then

$$R_{\text{T}} = R_A + R_1 + R_2 + \ldots + R_n + R_B = \frac{1}{\alpha_A A} + \frac{x_1}{\lambda_1 A} + \text{etc.} \frac{x_n}{\lambda_A A} + \frac{1}{\alpha_B A}$$

Or for any number of layers of material

$$\text{Total resistance, } R_{\text{T}} = \frac{1}{\alpha_A A} + \sum \frac{x}{\lambda A} + \frac{1}{\alpha_B A}$$ (16.9)

It can be seen from equation (16.9) that in this case the surface area, A, remains constant through the wall, and it is usual to calculate the total resistance for unit surface area in such problems. Cases in which the area varies through the various layers are considered in section 16.4.

Using the electrical analogy for the overall heat transfer we have

$$\dot{Q} = \frac{t_A - t_B}{R_{\text{T}}}$$ (16.10)

(analogous to $I = V/R$).

In equation (16.6) the overall heat transfer coefficient, U, is defined as

$$\frac{1}{U} = \frac{1}{\alpha_A} + \frac{x}{\lambda} + \frac{1}{\alpha_B}$$

For any number of walls we have

$$\frac{1}{U} = \frac{1}{\alpha_A} + \sum \frac{x}{\lambda} + \frac{1}{\alpha_B}$$

It can be seen that the reciprocal of U is simply the thermal resistance for unit area,

i.e. $\dfrac{1}{U} = R_{\text{T}} A \quad \text{or} \quad U = \dfrac{1}{R_{\text{T}} A}$ (16.11)

If the inner and outer wall surface temperatures are known then the heat transfer can be found by calculating the thermal resistance of the composite wall only,

569

i.e. $R = \sum \dfrac{x}{\lambda A}$

The overall heat transfer coefficient from one wall surface to the other is given by

$$\frac{1}{U} = \sum \frac{x}{\lambda}$$

It should be noted that there may be an additional thermal resistance at the various interfaces of a composite wall, due to the small pockets of air trapped between the surfaces.

Example 16.3   A furnace wall consists of 125 mm wide refractory brick and 125 mm wide insulating firebrick separated by an air gap. The outside wall is covered with a 12 mm thickness of plaster. The inner surface of the wall is at 1100 °C and the room temperature is 25 °C. The heat transfer coefficient from the outside wall surface to the air in the room is 17 W/m² K, and the resistance to heat flow of the air gap is 0.16 K/W. The thermal conductivities of refractory brick, insulating firebrick, and plaster are 1.6, 0.3, and 0.14 W/m K, respectively. Calculate:

  (i) the rate of heat loss per unit area of wall surface;
  (ii) the temperature at each interface throughout the wall;
  (iii) the temperature at the outside surface of the wall.

Solution (i) The wall is shown in Fig. 16.5. Consider 1 m² of surface area. Then using equation (16.7), $R = x/\lambda A$, we have

$$\text{Resistance of refractory brick} = \frac{125}{10^3 \times 1.6} = 0.0781 \text{ K/W}$$

$$\text{Resistance of insulating firebrick} = \frac{125}{10^3 \times 0.3} = 0.417 \text{ K/W}$$

$$\text{Resistance of plaster} = \frac{12}{10^3 \times 0.14} = 0.0857 \text{ K/W}$$

Fig. 16.5 Composite wall for Example 16.3

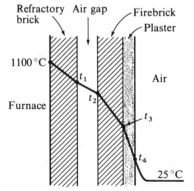

Also, using equation (16.8) for a fluid film, $R = 1/\alpha A$, we have

$$\text{Resistance of air film on outside surface} = \frac{1}{17}\,\text{K/W}$$

Hence

$$\text{Total resistance, } R_T = 0.0781 + 0.417 + 0.0857 + \frac{1}{17} + 0.16$$

where the resistance of the air gap is 0.16 K/W,

i.e. $R_T = 0.8$ K/W

Then using equation (16.10)

$$\dot{Q} = \frac{t_A - t_B}{R_T} = \frac{1100 - 25}{0.8} = 1344\text{ W}$$

i.e.   Rate of heat loss per square metre of surface area = 1.344 kW

(ii) Referring to Fig. 16.5, the interface temperatures are t_1, t_2, and t_3; the outside surface is at t_4. Applying the electrical analogy to each layer and using the values of thermal resistance calculated above, we have

$$\dot{Q} = 1344 = \frac{1100 - t_1}{0.0781}$$

i.e. $t_1 = 1100 - (1344 \times 0.0781) = 995\,°C$

Also $\dot{Q} = 1344 = \dfrac{t_1 - t_2}{0.16}$

i.e. $t_2 = 995 - (0.16 \times 1344) = 780\,°C$

$$\dot{Q} = 1344 = \frac{t_2 - t_3}{0.417}$$

i.e. $t_3 = 780 - (1344 \times 0.417) = 220\,°C$

And $\dot{Q} = 1344 = \dfrac{t_3 - t_4}{0.0857}$

i.e. $t_4 = 220 - (1344 \times 0.0857) = 104\,°C$

(iii) The temperature t_4 can also be found by considering the air film,

i.e. $\dot{Q} = 1344 = \dfrac{t_4 - 25}{1/17}$

i.e. $t_4 = \left(1344 \times \dfrac{1}{17}\right) + 25$

therefore

$$t_4 = 104.1\,°C$$

i.e.   Temperature at outside surface of wall = 104.1 °C

16.4 Heat flow through a cylinder and a sphere

One of the most commonly occurring problems in practice is the case of heat being transferred through a pipe or cylinder. Less common is the case of heat being transferred through a spherical wall, but both cases will now be considered.

The cylinder

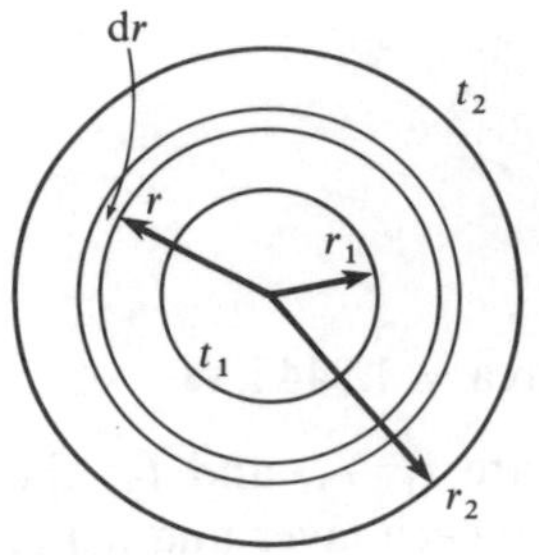

Fig. 16.6 Cross-section through a cylinder

Consider a cylinder of internal radius r_1, and external radius r_2 as shown in Fig. 16.6. Let the inside and outside surface temperatures be t_1 and t_2, respectively. Consider the heat flow through a small element, thickness dr, at any radius r, where the temperature is t. Let the conductivity of the material be k. Then applying equation (16.1), for unit length in the axial direction, we have

$$\dot{Q} = -\lambda A \frac{dt}{dx} = -\lambda (2\pi r \times 1) \frac{dt}{dr}$$

i.e.

$$\dot{Q} \frac{dr}{r} = -2\pi\lambda \, dt$$

Integrating between the inside and outside surfaces

$$\dot{Q} \int_{r_1}^{r_2} \frac{dr}{r} = -2\pi\lambda \int_{t_1}^{t_2} dt$$

where $\dot{Q}$ and λ are both constant

$$\dot{Q} \ln \frac{r_2}{r_1} = -2\pi\lambda(t_2 - t_1) = 2\pi\lambda(t_1 - t_2)$$

i.e.

$$\dot{Q} = \frac{2\pi\lambda(t_1 - t_2)}{\ln(r_2/r_1)} \qquad (16.12)$$

Now from equation (16.2),

$$\dot{Q} = \frac{\lambda A}{x}(t_1 - t_2)$$

If we substitute a mean area A_m in this equation, and substitute also for the thickness $x = (r_2 - r_1)$, we have

$$\dot{Q} = \frac{\lambda A_m(t_1 - t_2)}{(r_2 - r_1)}$$

Comparing this equation with equation (16.12), then

$$\dot{Q} = \frac{\lambda A_m(t_1 - t_2)}{(r_2 - r_1)} = \frac{2\pi\lambda(t_1 - t_2)}{\ln(r_2/r_1)}$$

therefore

$$\frac{A_m}{r_2 - r_1} = \frac{2\pi}{\ln(r_2/r_1)}$$

$$A_m = \frac{2\pi(r_2 - r_1)}{\ln(r_2/r_1)} = \frac{A_2 - A_1}{\ln(r_2/r_1)}$$

Here A_m is called the logarithmic mean area, and using this area in equation (16.2) an exact solution is obtained. It can be seen from the above that there is also a logarithmic mean radius given by

$$r_m = \frac{r_2 - r_1}{\ln(r_2/r_1)}$$

In the case of a composite cylinder (e.g. a metal pipe with several layers of lagging) the most convenient approach is again that of the electrical analogy; by using equation (16.7)

$$R = \frac{x}{\lambda A_m}$$

where x is the thickness of a layer, and A_m is the logarithmic mean area for that layer.

From equation (16.12), applying the electrical analogy ($I = V/R$), it can be seen that

$$R = \frac{\ln(r_2/r_1)}{2\pi\lambda} \qquad (16.13)$$

The film of fluid on the inside and outside surfaces can be treated as before using equation (16.8),

i.e. $\qquad R_{outside} = \dfrac{1}{\alpha_0 A_0}$

where A_0 is the outside surface area, $2\pi r_2$, referring to Fig. 16.6, and α_0 is the heat transfer coefficient for the outside surface.

$$R_{inside} = \frac{1}{\alpha_i A_i}$$

where A_i is the inside surface area, $2\pi r_1$ and α_i is the heat transfer coefficient for the inside surface.

It can be seen from equation (16.12),

$$\dot{Q} = \frac{2\pi\lambda(t_1 - t_2)}{\ln(r_2/r_1)}$$

that the heat transfer rate depends on the ratio of the radii, r_2/r_1, and not on the difference $(r_2 - r_1)$. The smaller the ratio, r_2/r_1, then the higher is the heat flow for the same temperature difference. In many practical problems the ratio, r_2/r_1, tends towards unity since the pipe-wall thickness or lagging thickness is

573

usually small compared with the mean radius. In these cases it is a sufficiently close approximation to use the arithmetic mean radius,

i.e.   Arithmetic mean radius $= \dfrac{r_2 + r_1}{2}$

The error in the rate of heat transfer in using the arithmetic mean instead of the logarithmic mean is just over 4% for a ratio $r_2/r_1 = 2$. Most heat transfer experiments in practice cannot give better accuracy than about 4 or 5%, hence it is a good approximation to use the arithmetic mean area when $r_2/r_1 < 2$.

Example 16.4   A steel pipe of 100 mm bore and 7 mm wall thickness, carrying steam at 260 °C, is insulated with 40 mm of a moulded high-temperature diatomaceous earth covering, this covering in turn insulated with 60 mm of asbestos felt. The atmospheric temperature is 15 °C. The heat transfer coefficients for the inside and outside surfaces are 550 and 15 W/m² K respectively, and the thermal conductivities of steel, diatomaceous earth, and asbestos felt are 50, 0.09, and 0.07 W/m K respectively. Calculate:

  (i) the rate of heat loss by the steam per unit length of pipe;
  (ii) the temperature of the outside surface.

Solution   (i) A cross-section of the pipe is shown in Fig. 16.7. Consider 1 m length of the pipe. From equation (16.8)

$$R = \frac{1}{\alpha A}$$

i.e.   Resistance of steam film $= \dfrac{10^3}{550 \times 2\pi \times 50 \times 1} = 0.00579 \text{ K/W}$

From equation (16.13), for the steel pipe

$$R = \frac{\ln(r_2/r_1)}{2\pi\lambda}$$

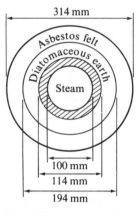

314 mm

Asbestos felt

Diatomaceous earth

Steam

100 mm
114 mm
194 mm

Fig. 16.7 Cross-section through an insulated cylinder for Example 16.4

therefore

Resistance of pipe $= \dfrac{\ln(57/50)}{2\pi \times 50} = 0.000417 \text{ K/W}$

Similarly

Resistance of diatomaceous earth $= \dfrac{\ln(97/57)}{2\pi \times 0.09}$

$$= 0.94 \text{ K/W}$$

and   Resistance of asbestos felt $= \dfrac{\ln(157/97)}{2\pi \times 0.07}$

$$= 1.095 \text{ K/W}$$

From equation (16.8), for the air film on the outside surface

$$\text{Resistance of air film} = \frac{1}{\alpha A} = \frac{10^3}{15 \times 2\pi \times 157 \times 1} = 0.0675 \text{ K/W}$$

Hence

$$\text{Total resistance, } R_\text{T} = 0.005\,79 + 0.000\,417 + 0.94 + 1.095 + 0.0675$$

i.e. $R_\text{T} = 2.1087 \text{ K/W}$

Note that the resistance to heat flow of the pipe metal is very small; also in this case the resistance of the film on the inside surface is very small because the heat transfer coefficient for steam is high.

Then, using equation (16.10)

$$\dot{Q} = \frac{t_\text{A} - t_\text{B}}{R_\text{T}} = \frac{260 - 15}{2.1087} = 116 \text{ W}$$

i.e.   Rate of heat loss per metre length of pipe = 116 W

(ii) Using the electrical analogy for the air film we have

$$\dot{Q} = 116 = \frac{t - 15}{0.0675}$$

where t is the temperature of the outside surface

$$t = (116 \times 0.0675) + 15 = 22.8\,°\text{C}$$

i.e.   Temperature of outside surface = 22.8 °C

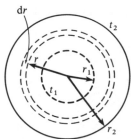

Fig. 16.8 A hollow sphere

The sphere

Consider a hollow sphere of internal radius r_1 and external radius r_2, as shown in Fig. 16.8. Let the inside and outside surface temperatures be t_1 and t_2, and let the thermal conductivity be λ. Consider a small element of thickness dr at any radius r. It can be shown that the surface area of this spherical element is given by $4\pi r^2$. Then, using equation (16.1)

$$\dot{Q} = -\lambda A \frac{\text{d}t}{\text{d}r} = -\lambda 4\pi r^2 \frac{\text{d}t}{\text{d}r}$$

Integrating

$$\dot{Q} \int_{r_1}^{r_2} \frac{\text{d}r}{r^2} = -4\pi\lambda \int_{t_1}^{t_2} \text{d}t$$

therefore

$$-\dot{Q}\left(\frac{1}{r_2} - \frac{1}{r_1}\right) = -4\pi\lambda(t_2 - t_1)$$

575

$$\frac{\dot{Q}(r_2 - r_1)}{r_1 r_2} = 4\pi\lambda(t_1 - t_2)$$

i.e.
$$\dot{Q} = \frac{4\pi\lambda r_1 r_2(t_1 - t_2)}{(r_2 - r_1)} \qquad \text{(a)}$$

Hence applying the electrical analogy, $(I = V/R)$, we have

$$R = \frac{(r_2 - r_1)}{4\pi\lambda r_1 r_2} \qquad \text{(16.14)}$$

If a mean area, A_m, is introduced, then from equation (16.2)

$$\dot{Q} = \frac{\lambda A_m}{x}(t_1 - t_2) = \frac{\lambda A_m(t_1 - t_2)}{(r_2 - r_1)} \qquad \text{(b)}$$

Comparing the equations (a) and (b) above, we have

$$A_m = 4\pi r_1 r_2$$

A mean radius, r_m, can be defined,

i.e.
$$A_m = 4\pi r_m^2 = 4\pi r_1 r_2$$

therefore

$$\text{Mean radius, } r_m = \sqrt{(r_1 r_2)}$$

It can be seen that r_m is a geometric mean radius.

Example 16.5   A small hemispherical oven is built of an inner layer of insulating firebrick 125 mm thick, and an outer covering of 85% magnesia 40 mm thick. The inner surface of the oven is at 800 °C and the heat transfer coefficient for the outside surface is 10 W/m² K; the room temperature is 20 °C. Calculate the rate of heat loss through the hemisphere if the inside radius is 0.6 m. Take the thermal conductivities of firebrick and 85% magnesia as 0.31 and 0.05 W/m K respectively.

Solution   For the insulating firebrick: from equation (16.14), for a hemisphere

$$\text{Resistance of firebrick} = \frac{0.125}{2\pi \times 0.31 \times 0.6 \times 0.725}$$

$$= 0.1478 \text{ K/W}$$

For the 85% magnesia

$$\text{Resistance of 85\% magnesia} = \frac{0.04}{2\pi \times 0.05 \times 0.725 \times 0.765}$$

$$= 0.2295 \text{ K/W}$$

For the outside surface: from equation (16.8)

$$\text{Resistance of outside air film} = \frac{1}{\alpha A} = \frac{1}{10 \times 2\pi \times 0.765^2}$$

$$= 0.0272 \text{ K/W}$$

Hence

$$\text{Total resistance}, R_T = 0.1478 + 0.2295 + 0.0272$$

$$= 0.4045 \text{ K/W}$$

Then using equation (16.10)

$$\dot{Q} = \frac{t_A - t_B}{R_T} = \frac{800 - 20}{0.4045} = 1930 \text{ W}$$

i.e.   Rate of heat loss from the oven $= 1.93 \text{ kW}$

16.5 General conduction equation

A general equation may be derived for a three-dimensional solid in which there is uniform internal heat generation (due to ohmic heating for example), and a change of temperature with time.

Consider an element at a temperature, t, at any instant of time, τ, within a homogeneous solid as shown in Fig. 16.9. Let the rate of internal heat generation

Fig. 16.9 A small element within a homogeneous solid

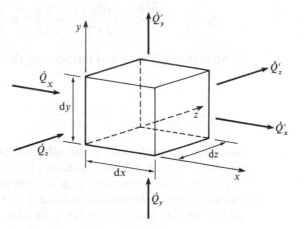

per unit volume be $\dot{q}_g$. Let the density of the material be ρ, the specific heat c, and the thermal conductivity λ; assume that these properties are uniform and constant with time. By Fourier's law:

$$\dot{Q}_x = -\lambda(\mathrm{d}y\,\mathrm{d}z)\frac{\partial t}{\partial x}$$

$$\dot{Q}_y = -\lambda(\mathrm{d}x\,\mathrm{d}z)\frac{\partial t}{\partial y}$$

$$\dot{Q}_z = -\lambda(\mathrm{d}x\,\mathrm{d}y)\frac{\partial t}{\partial z}$$

577

For the x-direction

$$\dot{Q}'_x - \dot{Q}_x = \frac{\partial \dot{Q}}{\partial x}\,dx = -\lambda\frac{\partial^2 t}{\partial x^2}\,dy\,dz\,dx$$

Similarly, for the y- and z-directions.

Also   Rate of heat generation $= \dot{q}_g(dx\,dy\,dz)$

and   Rate of increase of energy of the element

      $=$ mass $\times$ specific heat $\times$ rate of change of temperature with time

$$= \rho(dx\,dy\,dz)c\frac{\partial t}{\partial \tau}$$

Hence an energy balance on the element gives

$$\dot{q}_g(dx\,dy\,dz) - \{(\dot{Q}'_x - \dot{Q}_x) + (\dot{Q}'_y - \dot{Q}_y) + (\dot{Q}'_z - \dot{Q}_z)\}$$

$$= \rho c(dx\,dy\,dz)\frac{\partial t}{\partial \tau}$$

therefore

$$\dot{q}_g - \left(-\lambda\frac{\partial^2 t}{\partial x^2} - \lambda\frac{\partial^2 t}{\partial y^2} - \lambda\frac{\partial^2 t}{\partial z^2}\right) = \rho c\frac{\partial t}{\partial \tau}$$

Dividing through by λ and introducing the thermal diffusivity, $\kappa = \lambda/\rho c$, we have

$$\frac{\partial^2 t}{\partial x^2} + \frac{\partial^2 t}{\partial y^2} + \frac{\partial^2 t}{\partial z^2} + \frac{q_g}{\lambda} = \frac{1}{\kappa}\frac{\partial t}{\partial \tau} \qquad (16.15)$$

Equations using cylindrical or spherical coordinates may be derived in a similar way, or obtained from equation (16.15) by transforming the coordinates. For one-dimensional problems (e.g. an infinitely long cylinder or a sphere), it is simpler to derive the equations directly as in section 16.4, and as below.

(a) Infinite slab (see Fig. 16.1(b)). From equation (16.15) we have

$$\frac{\partial^2 t}{\partial x^2} + \frac{\dot{q}_g}{\lambda} = \frac{1}{\kappa}\frac{\partial t}{\partial \tau} \qquad (16.16)$$

(b) Infinitely long cylinder (see Fig. 16.6). Applying an energy balance to an element of thickness, dr, we have

$$\dot{q}_g 2\pi r\,dr - \frac{\partial \dot{Q}}{\partial r}\,dr = \rho c 2\pi r\,dr\frac{\partial t}{\partial \tau}$$

therefore

$$\dot{q}_g 2\pi r\,dr - \frac{\partial}{\partial r}\left(-\lambda 2\pi r\frac{\partial t}{\partial r}\right)dr = \rho c 2\pi r\,dr\frac{\partial t}{\partial \tau}$$

therefore

$$\dot{q}_g r + \left(\lambda r \frac{\partial^2 t}{\partial r^2} + \lambda \frac{\partial t}{\partial r} \right) = \rho c r \frac{\partial t}{\partial \tau}$$

i.e.

$$\frac{\partial^2 t}{\partial r^2} + \frac{1}{r}\frac{\partial t}{\partial r} + \frac{\dot{q}_g}{\lambda} = \frac{1}{\kappa}\frac{\partial t}{\partial \tau} \qquad (16.17)$$

(c) Sphere (see Fig. 16.8). Applying an energy balance

$$\dot{q}_g 4\pi r^2 \, dr - \frac{\partial \dot{Q}}{\partial r} \, dr = \rho c 4\pi r^2 \, dr \frac{\partial t}{\partial \tau}$$

therefore

$$\dot{q}_g 4\pi r^2 \, dr - \frac{\partial}{\partial r}\left(-\lambda 4\pi r^2 \frac{\partial t}{\partial r} \right) dr = \rho c 4\pi r^2 \, dr \frac{\partial t}{\partial \tau}$$

i.e.

$$\frac{\partial^2 t}{\partial r^2} + \frac{2}{r}\frac{\partial t}{\partial r} + \frac{\dot{q}_g}{\lambda} = \frac{1}{\kappa}\frac{\partial t}{\partial \tau} \qquad (16.18)$$

For steady-state cases the right-hand side of equations (16.16), (16.17), and (16.18) becomes zero, and the equations become ordinary differential equations.

Example 16.6   A hollow cylindrical copper conductor of 30 mm outside diameter and 14 mm inside diameter has a current density of 40 A/mm^2. The external surface is covered with a uniform layer of insulation of thickness 10 mm, and the ambient temperature is 10 °C. Neglecting axial conduction and assuming that the temperature of the insulation must not exceed 135 °C at any point, calculate:

(i) the heat required to be removed per unit time by forced cooling from the inside of the conductor;
(ii) the temperature at the inside surface of the conductor.

Data   Thermal conductivity of copper = 380 W/m K; thermal conductivity of insulating material = 0.3 W/m K; heat transfer coefficient at outside surface = 40 W/m^2 K; electrical resistivity of copper = 2 × 10^{-5} Ω mm.

Solution   From equation (16.17), for the steady state

$$\frac{d^2 t}{dr^2} + \frac{1}{r}\frac{dt}{dr} + \frac{\dot{q}_g}{\lambda} = 0$$

or

$$\frac{1}{r}\frac{d}{dr}\left(r\frac{dt}{dr} \right) = -\frac{\dot{q}_g}{\lambda}$$

Hence integrating

$$r\frac{dt}{dr} = -\frac{\dot{q}_g r^2}{2\lambda} + C_1$$

therefore

$$\frac{dt}{dr} = -\frac{\dot{q}_g r}{2\lambda} + \frac{C_1}{r} \tag{a}$$

Integrating

$$t = -\frac{\dot{q}_g r^2}{4\lambda} + C_1 \ln r + C_2 \tag{b}$$

where C_1 and C_2 are integration constants.

The heat generated per unit volume due to the current flowing is given by

$$\dot{q}_g = \frac{I^2 R}{AL}$$

where I is the current, R the electrical resistance, A the cross-sectional area, and L the length.

Current density, $J = I/A$ and resistance, $R = sL/A$

where s is the electrical resistivity of the conductor material,

i.e. $\qquad \dot{q}_g = \frac{J^2 A^2 sL}{AL \ A} = J^2 s \tag{16.19}$

$$\dot{q}_g = 40^2 \times 2 \times 10^{-5} \text{ W/mm}^3 = 32 \times 10^6 \text{ W/m}^3$$

(i) The maximum temperature of the insulation ($= 135\,°C$) occurs at the interface between the insulation and the copper tube. Hence for the insulation, using equation (16.12)

$$\text{Heat transfer to the outside, } \dot{Q}_0 = \frac{2\pi \times 0.3(135 - t)}{\ln(50/30)}$$

where t is the temperature of the outside surface of the insulation,

i.e. $\qquad 135 - t = 0.271\dot{Q}_0 \tag{c}$

For the heat transferred from the outside surface of the insulation by convection, from equation (16.3)

$$\dot{Q}_0 = \alpha A(t - t_{\text{fluid}})$$

i.e. $\qquad \dot{Q}_0 = 40 \times 2\pi \times 0.025(t - 10)$

therefore

$$t - 10 = 0.159\dot{Q}_0 \tag{d}$$

Adding equations (c) and (d)

$$135 - 10 = 0.43\dot{Q}_0 \quad \text{therefore } \dot{Q}_0 = 290.7 \text{ W}$$

Total heat generated internally $= \dot{q}_g \times$ volume

$$= 32 \times 10^6 \times \frac{\pi}{4}(0.03^2 - 0.014^2)$$

$$= 17\,693.5 \text{ W}$$

Hence

$$\text{Heat removed from inside of conductor} = (17\,693.5 - 290.7)\ \text{W}$$

$$= 17.4\ \text{kW}$$

(ii) Two boundary conditions are required to find the constants C_1 and C_2 and hence to obtain the solution of equation (b).

At the inside surface of the conductor

$$\text{Heat supplied to conductor} = -\lambda A\left(\frac{dt}{dr}\right)_{r=0.007} = -17\,400\ \text{W}$$

therefore

$$\left(\frac{dt}{dr}\right)_{r=0.007} = \frac{17\,400}{380 \times \pi \times 0.014} = 1041.1\ \text{K/m}$$

Substituting in equation (a)

$$1041.1 = -\left\{\frac{32 \times 10^6 \times 0.007}{2 \times 380}\right\} + \frac{C_1}{0.007}$$

therefore

$$C_1 = 9.351$$

At the outside surface of the conductor, $t = 135\,°\text{C}$, hence in equation (b)

$$135 = -\left\{\frac{32 \times 10^6 \times 0.015^2}{4 \times 380}\right\} + 9.351\ \ln(0.015) + C_2$$

therefore

$$C_2 = 179$$

Therefore the complete solution for the temperature distribution in the conductor is

$$t = -\left(\frac{32 \times 10^6}{4 \times 380}\right)r^2 + (9.351\ \ln r) + 179$$

Hence, at the inside surface, when $r = 0.007$

$$t = -\left\{\frac{32 \times 10^6 \times 0.007^2}{4 \times 380}\right\} + (9.351\ \ln 0.007) + 179$$

$$= 131.6\,°\text{C}$$

Transient conduction in one dimension

The equations for one-dimensional transient conduction, (16.16), (16.17), and (16.18), can be solved using the separation of variables method. For example, from equation (16.16) for the case when there is no internal heat generation

$$\frac{\partial^2 t}{\partial x^2} = \frac{1}{\alpha} \frac{\partial t}{\partial \tau} \tag{16.20}$$

it can be shown that

$$t = e^{-\alpha p^2 \tau} \{ C_1 \sin(px) + C_2 \cos(px) \}$$

where p, C_1 and C_2 are determined by the boundary conditions.

For an infinite slab of half-thickness, L, initially at a uniform temperature, t_i, throughout, which is suddenly exposed to a fluid at a constant temperature, t_F, the temperature at any point, x, at time, τ, is given by

$$\frac{t - t_F}{t_i - t_F} = 2 \sum_{n=1}^{\infty} e^{-(p_n L)^2 Fo} \left\{ \frac{\sin(p_n L) \cos(p_n x)}{(p_n L) + \sin(p_n L) \cos(p_n L)} \right\} \tag{16.21}$$

and $\qquad (p_n L) \tan(p_n L) = Bi \tag{16.22}$

where Bi is the Biot number, $\alpha L / \lambda$ and Fo the Fourier number, $\lambda \tau / \rho c L^2$ or $\kappa \tau / L^2$.

Similar equations can be derived from the cases of the infinitely long cylinder and the sphere, and graphs of non-dimensional temperature against Fourier number for various values of $1/Bi$ have been drawn (see for example reference 16.1).

Newtonian heating or cooling

This approach, which is sometimes known as *lumped capacity*, may be used when the temperature within a body does not vary appreciably as the body's average temperature changes with time due to exposure of the body to a fluid at a different temperature. This is the case when the surface thermal resistance is very much greater than the internal thermal resistance, and hence the heat transfer from the surface is the controlling factor.

For a body of surface area, A, volume, V, specific heat, c, and density, ρ, with an average temperature, $\bar{t}$, at any time, τ, we have

$$\alpha A (\bar{t} - t_F) = -\rho V c \frac{d\bar{t}}{d\tau}$$

where α is the surface heat transfer coefficient and t_F the temperature of the fluid surrounding the body, assumed constant with time. Therefore

$$\int_{t_i}^{\bar{t}} \frac{d\bar{t}}{t - t_F} = - \int_0^{\tau} \frac{\alpha A}{\rho V c} d\tau$$

where t_i is the initial temperature of the body,

i.e. $\qquad \ln \left(\frac{\bar{t} - t_F}{t_i - t_F} \right) = - \frac{\alpha A \tau}{\rho V c}$

or $\qquad \dfrac{\bar{t} - t_F}{t_i - t_F} = e^{-(\alpha A \tau / \rho V c)} \tag{16.23}$

Equation (16.23) can also be written as

$$\frac{\bar{t} - t_F}{t_i - t_F} = e^{-(AL/V)BiFo} \tag{16.24}$$

The dimension of length, L, may be the half-thickness of an infinite slab, or the radius of an infinite cylinder, or the radius of a sphere. The term AL/V for an infinite slab, cylinder or sphere, may be shown to be 1, 2, or 3 respectively, e.g. for a sphere

$$\frac{AL}{V} = \frac{(4\pi L^2)L}{(4\pi L^3/3)} = 3$$

In the previously considered exact solution, equation (16.21), it can be shown that when $Fo > 0.2$ then only the first term of the summation need be considered within engineering accurancy. Also, when Bi is small, then in equation (16.22) $\tan(p_1 L)$ approximates to $(p_1 L)$, and hence Bi approximates to $(p_1 L^2)$. Similarly, $\sin(p_1 L)$ approximates to $(p_1 L)$ and $\cos(p_1 L)$ approaches unity. Therefore, substituting these approximations into equation (16.21), for the centre where $x = 0$, and hence $\cos(p_1 x)$ is 1, we have

$$\frac{t - t_F}{t_i - t_F} = 2e^{-BiFo}\left(\frac{p_1 L}{p_1 L + p_1 L}\right) = e^{-BiFo}$$

Comparing this with equation (16.24) it can be seen to be equivalent to Newtonian cooling of an infinite slab, i.e. when $Fo > 0.2$ and Bi is very small the problem approximates to Newtonian cooling.

Example 16.7

For transient conduction in a sphere when $Fo > 0.2$ it can be shown that the solution of equation (16.18) for the temperature at the centre of the sphere, t_c, when initially at t_i, and plunged into a fluid at t_F, is given by

$$\frac{t_c - t_F}{t_i - t_F} = \frac{\{\sin(p_1 L) - p_1 L\cos(p_1 L)\}}{\{p_1 L - \sin(p_1 L)\cos(p_1 L)\}} 2e^{-(p_1 L)^2 Fo}$$

and $\qquad 1 - p_1 L\cot(p_1 L) = Bi$

Using the data below determine the temperature at the centre of a sphere, initially at a uniform temperature of 500 °C, twenty minutes after it is plunged into a large bath of liquid at a temperature of 20 °C:
(i) from the above equation; (ii) assuming Newtonian cooling.

Data

Radius of sphere = 50 mm; density of sphere = 7600 kg/m³; thermal conductivity of sphere = 40 W/m K; specific heat of sphere = 0.5 kJ/kg K; heat transfer coefficient from sphere surface to liquid = 88.8 W/m² K. It may be assumed that the heat transfer coefficient and the temperature of the liquid remain constant over the time period.

Solution (i) $\qquad Bi = \alpha L/\lambda = (88.8 \times 0.05)/40 = 0.111$

therefore

$$1 - p_1 L\cot(p_1 L) = 0.111$$

or $\qquad p_1 L\cot(p_1 L) = 0.889$

This equation may be solved by trial and error,

i.e.

$p_1 L$	0.7	0.6	0.5
$p_1 L \cot(p_1 L)$	0.831	0.877	0.915

By further trial and error, or by drawing a graph, it can be shown that $p_1 L = 0.57$. Also

$$Fo = \kappa\tau/R^2 = \lambda\tau/\rho c R^2 = (40 \times 20 \times 60)/(7600 \times 0.5 \times 10^3 \times 0.05^2)$$

$$= 5.053$$

therefore

$$\frac{t_c - t_F}{t_i - t_F} = \frac{(\sin 0.57 - 0.57 \cos 0.57)}{(0.57 - \sin 0.57 \cos 0.57)} 2e^{-(0.57)^2 \times 5.053}$$

i.e.

$$\frac{t_c - t_F}{t_i - t_F} = 0.5161 \times 2 \times 0.1936 = 0.2$$

therefore

$$t_c = 20 + 0.2(500 - 20) = 116\,°C$$

(ii) For Newtonian cooling of a sphere, from equation (16.24),

$$\frac{\bar{t} - t_F}{t_i - t_F} = e^{-3BiFo} = e^{-3 \times 0.111 \times 5.053}$$

therefore

$$\bar{t} = 20 + 0.1859(500 - 20) = 109.2\,°C$$

16.6 Numerical methods for conduction

The most commonly used numerical method is the finite difference method in which a differential equation is replaced by an approximate algebraic expression. The set of equations thus produced can be solved using a computer. The reader is recommended to consult references 16.2 and 16.3 for a fuller treatment of numerical methods and their application in heat transfer.

A different method known as the finite element method is increasingly being used for heat transfer applications, but is not considered in this book.

In specialized texts the derivation of finite difference expressions is given in detail, using for example the Taylor series, but in this book only the following brief illustration will be given. Referring to a graph of t against x (Fig. 16.10), three approximations to the true tangent to the curve dt/dx are illustrated,

i.e.

$$\frac{dt}{dx} = \frac{(t_x - t_{x-\delta x})}{\delta x} \quad \text{backward difference} \tag{16.25}$$

$$\frac{dt}{dx} = \frac{(t_{x+\delta x} - t_x)}{\delta x} \quad \text{forward difference} \tag{16.26}$$

Fig. 16.10
Diagrammatic
definition of backward,
forward, and central
difference
approximations

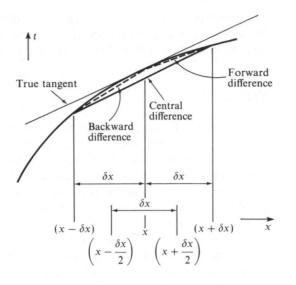

$$\frac{dt}{dx} = \frac{(t_{x+\delta x} - t_{x-\delta x})}{2\delta x} \quad \text{central difference} \tag{16.27}$$

It can be seen from the Fig. 16.10 that the central difference approximation, equation (16.27), is a more accurate approximation to the true slope.

The second derivative, d^2t/dx^2, is the rate of change of slope at the point x. This may be approximated as the change of dt/dx over the distance δx. From Fig. 16.10 it can be seen that the slope at $x + (\delta x/2)$ is approximately $(t_{x+\delta x} - t_x)/\delta x$, and the slope at $x - (\delta x/2)$ is approximately $(t_x - t_{x-\delta x})/\delta x$. Hence the rate of change of slope over the distance δx is given by

$$\frac{d^2t}{dx^2} = \left\{ \frac{(t_{x+\delta x} - t_x)}{\delta x} - \frac{(t_x - t_{x-\delta x})}{\delta x} \right\} \frac{1}{\delta x}$$

i.e.
$$\frac{d^2t}{dx^2} = \frac{t_{x+\delta x} + t_{x-\delta x} - 2t_x}{\delta x^2} \tag{16.28}$$

To solve a conduction problem by the finite difference method the relevant partial differential equation is replaced using expressions such as the above. The space and time dimensions are divided into a number of increments of finite size and the approximate expression which replaces the differential equation applies to every point in the grid of points, or nodes; separate equations are derived for the boundary conditions.

Hence the relevant differential equation is effectively replaced by a large number of identical algebraic expressions for the temperature at each point in the space at any time.

A set of simultaneous equations can be put in the form of a matrix and solved by *matrix inversion* methods. However, in the case of conduction the matrix of temperature coefficients has a small number of non-zero terms and hence matrix inversion is not recommended. It is better to solve such a matrix by a direct method such as Gaussian elimination (see p. 598).

585

In the case of steady conduction in two dimensions the initial temperatures are unknown and hence the set of equations is more conveniently solved by a relaxation method such as Gauss–Siedel iteration (see p. 590).

Errors

Using a finite difference method the answer obtained converges towards the exact solution as the size of the increments chosen approaches zero. Finite difference expressions must be chosen such that the computer solution converges towards the exact solution; in certain cases the solution will become unstable because errors generated are increasing in size as the solution proceeds, or are growing at a faster rate than the rate of convergence.

There are basically two types of error: round-off error and discretization error. Round-off error occurs when the answer is taken to a specific number of significant figures, and is cumulative; fortunately in modern computers this error is not usually important. Discretization error is mainly due to the inaccuracy of the finite difference expression, see Fig. 16.10, and can be reduced by reducing the size of the increments.

Notation

Referring to Fig. 16.11, a two-dimensional space may be divided into a grid of nodes as shown. The temperature at any point may then be designated as $t_{i,j}$. Note that i increases from left to right, and j from bottom to top, of the grid, following the normal x-direction and y-direction respectively.

For a problem in transient conduction the temperature at any time will be denoted by $t_{i,j}^{\tau}$; the next time is therefore $\tau + 1$ and the temperature at that instant is $t_{i,j}^{\tau+1}$. For transient problems in one-dimension the j-direction will be omitted.

Fig. 16.11 Grid definition for two-dimensional steady conduction

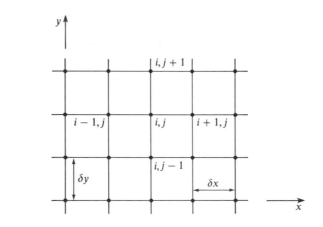

16.7 Two-dimensional steady conduction

From equation (16.15), for zero internal heat generation and for steady conduction in two dimensions the equation reduces to the Laplace equation:

$$\frac{\partial^2 t}{\partial x^2} + \frac{\partial^2 t}{\partial y^2} = 0 \tag{16.29}$$

This equation may be put into finite difference form using the central difference expression, equation (16.28). Using the notation outlined in section 16.6 (see Fig. 16.11), we have

$$\frac{(t_{i+1,j} + t_{i-1,j} - 2t_{i,j})}{\delta x^2} + \frac{(t_{i,j+1} + t_{i,j-1} - 2t_{i,j})}{\delta y^2} = 0$$

The grid may be chosen such that $\delta x = \delta y$, then

$$t_{i,j} = (t_{i,j-1} + t_{i+1,j} + t_{i,j+1} + t_{i-1,j})/4 \tag{16.30}$$

All the internal points within the boundaries of the two-dimensional space are represented by equation (16.30).

Conducting rod analogy

Equation (16.30) may be derived using the basic Fourier equation and the concept of heat flow paths. In Fig. 16.12 conducting paths of width, δx, from each point towards the centre point, are shown cross-hatched. Fourier's law can be applied to each conducting path; for example, the heat transferred from point $(i + 1, j)$ to point (i, j) is given by

Thermal conductivity × area × temperature gradient

$$= \lambda(\delta x)\frac{(t_{i+1,j} - t_{i,j})}{\delta x}$$

Fig. 16.12 Conducting rod analogy for two-dimensional steady conduction

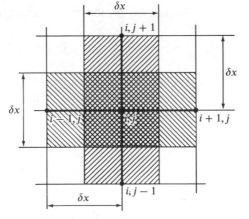

Then a simple energy balance gives

$$\lambda(\delta x)\frac{(t_{i+1,j} - t_{i,j})}{\delta x} + \lambda(\delta x)\frac{(t_{i,j+1} - t_{i,j})}{\delta x} + \lambda(\delta x)\frac{t_{i-1,j} - t_{i,j}}{\delta x}$$

$$+ \lambda(\delta x)\frac{(t_{i,j-1} - t_{i,j})}{\delta x} = 0$$

This equation reduces to the same expression as in equation (16.30).

The conducting rod analogy can be used in more complex cases, including the case with internal heat generation or at a boundary convecting to a fluid; it may be found easier to apply since it relates to a simple physical model.

Boundary conditions

Surface convecting to a fluid

For a point (i, j) on the surface (see Fig. 16.13):

$$-\lambda\frac{dt}{dx} = \alpha(t_F - t_{i,j})$$

Fig. 16.13 Grid for a left-hand surface convecting to a fluid in two-dimensional steady conduction

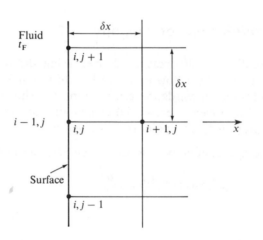

Using a central difference expression for dt/dx, equation (16.27),

$$-\lambda\frac{(t_{+1,j} - t_{i-1,j})}{2\delta x} = \alpha(t_F - t_{i,j}) \tag{16.31}$$

The point $(i - 1, j)$ is fictitious and can be eliminated from the equation by assuming that it lies on the extrapolated temperature distribution line (see Fig. 16.14), i.e. from equation (16.30):

$$t_{i-1,j} = 4t_{i,j} - t_{i+1,j} - t_{i,j+1} - t_{i,j-1}$$

Substituting in equation (16.31)

$$t_{i+1,j} - 4t_{i,j} + t_{i+1,j} + t_{i,j+1} + t_{i,j-1} = \frac{2\alpha\,\delta x}{\lambda}(t_{i,j} - t_F)$$

Fig. 16.14 Fictitious point for a left-hand surface in two-dimensional steady conduction

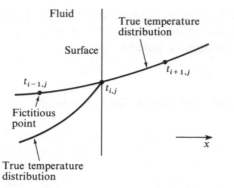

therefore

$$t_{i,j} = \frac{2t_{i+1,j} + t_{i,j+1} + t_{i,j-1} + (2\alpha\,\delta x t_F/\lambda)}{4 + (2\alpha\,\delta x/\lambda)} \qquad (16.32)$$

Similar expressions may be obtained for a surface with the fluid on the right, or at the top or bottom.

Equation (16.32) can be derived using the conducting rod analogy with a rod of half-width, $\delta x/2$, running from point $(i, j+1)$ to point (i, j), and from point $(i, j-1)$ to (i, j), i.e.

$$\lambda(\delta x)\frac{(t_{i+1,j} - t_{i,j})}{\delta x} + \lambda\frac{(\delta x)}{2}\frac{(t_{i,j+1} - t_{i,j})}{\delta x} + \lambda\frac{(\delta x)}{2}\frac{(t_{i,j-1} - t_{i,j})}{\delta x}$$

$$= \alpha\,\delta x(t_{i,j} - t_F)$$

Simplifying this equation the same expression as in equation (16.32) is obtained.

Insulated surface

At an insulated surface, $-\lambda\,dt/dx = 0$, or $\alpha = 0$, and hence for a left-hand surface which is insulated equation (16.32) reduces to

$$t_{i,j} = \frac{2t_{i+1,j} + t_{i,j+1} + t_{i,j-1}}{4}$$

It should be noted that a line of thermal symmetry within a two-dimensional space will act as an insulated surface (see Example 16.8).

Corners

Expressions can be derived for the temperatures at outside corners (top left, bottom right, etc.) and at inside corners. For example, for a top left outside corner

$$t_{i,j} = \frac{t_{i+1,j} + t_{i,j-1} + (2\alpha\,\delta x t_F/\lambda)}{2 + (2\alpha\,\delta x/\lambda)}$$

For a bottom right inside corner

$$t_{i,j} = \frac{2t_{i+1,j} + 2t_{i,j-1} + t_{i,j+1} + t_{i-1,j} + (2\alpha\,\delta x t_F/\lambda)}{6 + (2\alpha\,\delta x/\lambda)}$$

589

The derivation of expressions for corner points such as the above is left as an exercise for the reader; the conducting rod analogy is the best method, particularly for inside corners.

Choice of numerical method

The Laplace equation is an elliptic equation and as stated in section 16.6 the recommended numerical method in this case is *Gauss–Siedel iteration*.

In the Gauss–Siedel method the grid points are assigned initial values at every point, then, from the set of equations for the internal nodes and all the surface points, new values for each point are calculated point by point. The new value at a point is compared with the previous value and the process continued until the difference between successive values is small compared with the actual temperature, according to the accuracy required. Provided the solution converges then the complete temperature field is obtained.

This method is called a relaxation method because the values of temperature are modified, or relaxed, at each point in turn to satisfy the equation, which then alters the temperatures at adjacent points. In general, at any point (i, j) the value after one iteration $t_{i,j}^{(2)}$ in terms of the initial values is given by

$$t_{i,j}^{(2)} = \omega \frac{(t_{i,j-1}^{(1)} + t_{i+1,j}^{(1)} + t_{i,j+1}^{(1)} + t_{i-1,j}^{(1)})}{4} + (1 - \omega)t_{i,j}^{(1)}$$

where ω is a relaxation factor.

For $\omega = 0$, $t_{i,j}^{(2)} = t_{i,j}^{(1)}$, and no relaxation occurs. For $\omega = 1$, equation (16.30) applies, which is Gauss–Siedel iteration.

Values of ω between 0 and 1 will slow down the relaxation process, whereas for $\omega > 1$ the solution will converge more rapidly. It can be shown that the solution will not converge for $\omega > 2$ and hence a value of ω between 1 and 2 is chosen, determined largely by experience. This method is known as *successive over relaxation* (SOR).

Example 16.8 A long duct of square cross-section 0.5 m × 0.5 m is buried in deep soil with one of its sides parallel to the surface of the soil as shown in Fig. 16.15; the centre-line of the duct is at a depth of 1.25 m. The surface of the soil is at an equilibrium temperature of 0 °C, and at a soil depth of 2.5 m it may be assumed that the uniform equilibrium temperature is −10 °C across the horizontal cross-section. The temperature of each side of the duct is 50 °C and it may be assumed that at a vertical cross-section a horizontal distance of 2.25 m from the duct centre-line, the heat transfer vertically downwards from the surface is simple one-dimensional conduction.

Taking a square mesh of side 0.5 m, use a numerical method to obtain an approximation for the temperatures within the soil to the nearest degree, and estimate the heat loss per metre length of duct.

The thermal conductivity of the soil is 1 W/m K.

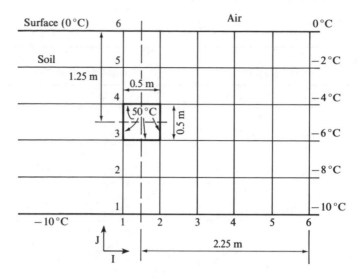

Fig. 16.15 Duct buried in soil for Example 16.8

Solution   At the vertical cross-section where the conduction is one-dimensional the temperature must vary linearly, hence the temperatures can be written in as shown on Fig. 16.15.

At the centre-line of the duct there is thermal symmetry, i.e. no heat can flow across the centre-line and hence the temperatures on either side of this line are equal.

Number the grid as shown with columns, I, from 1 to 6 horizontally, and rows, J, from 1 to 6 vertically.

The temperatures which are known are as follows: I = 1 to 6, J = 6, $t = 0\,°C$; I = 1, J = 3 and 4, $t = 50\,°C$; I = 2, J = 3 and 4, $t = 50\,°C$; I = 1, 2, 3, 4, 5, and 6, J = 1, $t = -10\,°C$; I = 6, J = 2, $t = -8\,°C$; I = 6, J = 3, $t = -6\,°C$; I = 6, J = 4, $t = -4\,°C$; I = 6, J = 5, $t = -2\,°C$.

For all other points the temperature is given by equation (16.30)

$$t_{i,j} = (t_{i,j-1} + t_{i+1,j} + t_{i,j+1} + t_{i-1,j})/4$$

Also, from the condition of symmetry as stated above, the temperature at I = 1, J = 2 is equal to the temperature at I = 2, J = 2; similarly the temperature at I = 1, J = 5 is equal to the temperature at I = 2, J = 5.

To solve the problem we assign initial values to the temperatures at the internal points and then proceed by iteration, using equation (16.30) for internal points, and the values given above for the other points. This is best done by computer when the size of mesh and the required accuracy can be greater than those specified in the problem. To illustrate the method a procedure is outlined in Table 16.2 using mental arithmetic only; the first column on the left-hand side is the set of initial guesses, the second column is the first iteration, and so on. The temperature values in brackets for each iteration are taken in the same order as in equation (16.30) at each point. The temperatures at each point after the third iteration are within one degree of the previous value which is within the required accuracy. Note that in this approximate solution the temperatures at each point have been rounded up to the nearest degree and this error will be accumulative.

Table 16.2 Solution for
Example 16.8

Temperature at (I, J)	First iteration	Second iteration	Third iteration
$t(2, 2) = 10$	$\dfrac{(-10 - 5 + 50 + 10)}{4} = 11$	$\dfrac{(-10 + 3 + 50 + 11)}{4} = 14$	$\dfrac{(-10 + 5 + 50 + 14)}{4} = 15$
$t(3, 2) = -5$	$\dfrac{(10 - 8 + 20 + 11)}{4} = 3$	$\dfrac{(-10 - 4 + 21 + 14)}{4} = 5$	$\dfrac{(-10 - 1 + 21 + 15)}{4} = 6$
$t(4, 2) = -8$	$\dfrac{(-10 - 8 + 0 + 3)}{4} = -4$	$\dfrac{(-10 - 4 + 4 + 5)}{4} = -1$	$\dfrac{(-10 - 5 + 6 + 6)}{4} = -1$
$t(5, 2) = -8$	$\dfrac{(-10 - 8 + 0 - 4)}{4} = -6$	$\dfrac{(-10 - 8 - 2 - 1)}{4} = -5$	$\dfrac{(-10 - 8 - 1 - 1)}{4} = -5$
$t(3, 3) = 20$	$\dfrac{(3 + 0 + 30 + 50)}{4} = 21$	$\dfrac{(5 + 4 + 25 + 50)}{4} = 21$	$\dfrac{(6 + 16 + 23 + 50)}{4} = 21$
$t(4, 3) = 0$	$\dfrac{(-4 + 0 + 0 + 21)}{4} = 4$	$\dfrac{(-1 - 2 + 7 + 21)}{4} = 6$	$\dfrac{(-1 - 1 + 9 + 21)}{4} = 7$
$t(5, 3) = 0$	$\dfrac{(-6 - 6 + 10 + 4)}{4} = -2$	$\dfrac{(-5 - 6 + 0 + 6)}{4} = -1$	$\dfrac{(-5 - 6 + 1 + 7)}{4} = -1$
$t(3, 4) = 30$	$\dfrac{(21 + 0 + 30 + 50)}{4} = 25$	$\dfrac{(21 + 7 + 13 + 50)}{4} = 23$	$\dfrac{(21 + 9 + 13 + 50)}{4} = 24$
$t(4, 4) = 0$	$\dfrac{(4 + 0 + 0 + 25)}{4} = 7$	$\dfrac{(6 + 0 + 5 + 23)}{4} = 9$	$\dfrac{(7 + 1 + 6 + 24)}{4} = 10$
$t(5, 4) = 0$	$\dfrac{(-2 - 4 + 0 + 7)}{4} = 0$	$\dfrac{(-1 - 4 + 1 + 9)}{4} = 1$	$\dfrac{(-1 - 4 + 1 + 10)}{4} = 2$
$t(2, 5) = 30$	$\dfrac{(50 + 30 + 0 + 30)}{4} = 28$	$\dfrac{(50 + 13 + 0 + 28)}{4} = 23$	$\dfrac{(50 + 13 + 0 + 23)}{4} = 22$
$t(3, 5) = 30$	$\dfrac{(25 + 0 + 0 + 28)}{4} = 13$	$\dfrac{(23 + 5 + 0 + 23)}{4} = 13$	$\dfrac{(24 + 6 + 0 + 22)}{4} = 13$
$t(4, 5) = 0$	$\dfrac{(7 + 0 + 0 + 13)}{4} = 5$	$\dfrac{(9 + 1 + 0 + 13)}{4} = 6$	$\dfrac{(10 + 1 + 0 + 13)}{4} = 6$
$t(5, 5) = 0$	$\dfrac{(0 - 2 + 0 + 5)}{4} = 1$	$\dfrac{(1 - 2 + 0 + 6)}{4} = 1$	$\dfrac{(2 - 2 + 0 + 6)}{4} = 2$

From the conducting rod analogy the heat transfer from the duct per unit length is given by

$$\dot{q} = \left\{ \lambda \frac{(0.5)}{2} \frac{(50 - 22)}{0.5} \times 2 \right\} + \left\{ \lambda \frac{(0.5)}{2} \frac{(50 - 24)}{0.5} \times 2 \right\}$$

$$+ \left\{ \lambda \frac{(0.5)}{2} \frac{(50 - 21)}{0.5} \times 2 \right\} + \left\{ \lambda \frac{(0.5)}{2} \frac{(50 - 15)}{0.5} \times 2 \right\}$$

i.e. $\quad \dot{q} = 118 \text{ W/m}$

It is left to the reader to write a simple program in BASIC or FORTRAN to obain a more accurate solution using the same method.

16.8 One-dimensional transient conduction by finite difference

The case of the infinite slab will be considered as given in section 16.5, equation (16.20). This equation is parabolic and therefore the solution starts from an initial boundary value and proceeds step by step in time.

In the finite difference method of solution a central difference expression is used as before for $\partial^2 t/\partial x^2$, but it is not possible to use a central difference expression for $\partial t/\partial \tau$ because this leads to instability. Here $\partial t/\partial \tau$ can either be replaced by a forward difference expression, equation (16.26) (leading to an *explicit solution*), or by a backward difference expression, equation (16.25) (leading to an *implicit solution*).

Explicit solution (or Euler solution)

For an infinite slab, equation (16.20)

$$\frac{\partial^2 t}{\partial x^2} = \frac{1}{\kappa}\frac{\partial t}{\partial \tau}$$

is replaced by

$$\frac{t_{i+1}^{\tau} + t_{i-1}^{\tau} - 2t_i^{\tau}}{\delta x^2} = \frac{1}{\kappa}\frac{(t_i^{\tau+1} - t_i^{\tau})}{\delta \tau}$$

i.e.
$$t_i^{\tau+1} = Fo(t_{i+1}^{\tau} + t_{i-1}^{\tau} - 2t_i^{\tau}) + t_i^{\tau} \tag{16.33}$$

where $Fo = \kappa\,\delta\tau/\delta x^2$

or
$$t_i^{\tau+1} = Fo\left\{t_{i+1}^{\tau} + t_{i-1}^{\tau} + \left(\frac{1}{Fo} - 2\right)t_i^{\tau}\right\} \tag{16.34}$$

When the term $(1/Fo - 2)$ becomes negative the solution becomes unstable; hence the condition for stability is

$$Fo \leqslant 1/2 \tag{16.35}$$

The size of the increments, δx, and $\delta\tau$, must be chosen such that Fo is less than or equal to 0.5, and then a set of equations such as equation (16.33) can be solved at each time step.

Note that when Fo is chosen to be equal to 0.5 then from equation (16.34) it can be seen that

$$t_i^{\tau+1} = (t_{i+1}^{\tau} + t_{i-1}^{\tau})/2 \tag{16.36}$$

593

In this type of problem the initial temperatures are known at all the nodes, and the values at the next time step can be found one by one using equation (16.33). This process is repeated for the next time step and so on until the required time is reached, or until the temperature within the slab reaches a certain value.

Boundary condition (surface convecting to a fluid)

At any instant of time the equation for heat transfer at the surface is the same as that previously considered in section 16.7 for two-dimensional conduction. Referring to Fig. 16.13 (page 588) and equation (16.31), in this case the equation may be written

$$-\lambda \frac{(t_{i+1}^{\tau} - t_{i-1}^{\tau})}{2\delta x} = \alpha(t_{\mathrm{F}}^{\tau} - t_{i}^{\tau}) \tag{16.37}$$

As before, the point $(i-1)$ is fictitious and can be eliminated from equation (16.37) using, in this case, equation (16.34), i.e. substituting for t_{i-1}^{τ} from equation (16.37) in equation (16.34) we have

$$t_{i}^{\tau+1} = Fo\left\{2t_{i+1}^{\tau} + \left(\frac{1}{Fo} - 2 - \frac{2\alpha\,\delta x}{\lambda}\right)t_{i}^{\tau} + \frac{2\alpha\,\delta x}{\lambda}t_{\mathrm{F}}^{\tau}\right\} \tag{16.38}$$

It can be seen from equation (16.38) that the condition for stability is now more restrictive,

i.e. $\qquad Fo \leqslant \dfrac{1}{2 + (2\alpha\,\delta x)/\lambda}$ \hfill (16.39)

By sacrificing some accuracy but retaining the condition for stability given by equation (16.35), the term $\partial t/\partial x$ can be replaced by a forward difference expression,

i.e. $\qquad \lambda\dfrac{(t_{i+1}^{\tau} - t_{i}^{\tau})}{\delta x} = \alpha(t_{i}^{\tau} - t_{\mathrm{F}}^{\tau})$

therefore

$$t_{i}^{\tau} = \frac{t_{i+1}^{\tau} + (\alpha\,\delta x t_{\mathrm{F}}^{\tau}/\lambda)}{1 + (\alpha\,\delta x/\lambda)} \tag{16.40}$$

Boundary condition (insulated face or line of thermal symmetry)

As considered previously in section 16.7, at an insulated face $\partial t/\partial x$ is zero and hence the temperature at an insulated surface is obtained from equation (16.33) by putting $t_{i+1}^{\tau} = t_{i-1}^{\tau}$,

$$t_{i}^{\tau+1} = 2Fo(t_{i+1}^{\tau} - t_{i}^{\tau}) + t_{i}^{\tau}$$

for a left-hand insulated surface. Note that when $Fo = 0.5$ then $t_{i}^{\tau+1} = t_{i+1}^{\tau}$.

In an important class of problem called *quenching*, a slab is suddenly subjected to surroundings at a different temperature. In such a case the centre-line of the slab is a line of thermal symmetry and hence only one half of the slab need be considered, and is treated as a slab with one face insulated.

Boundary condition (sudden increase in temperature)

When the fluid in contact with a surface, or the surface itself, is suddenly changed in temperature, a decision is required as to which temperature to use for the initial time-step calculations. Provided the time taken for the change of temperature is very small compared with the time step $\delta\tau$, a suitable value to take is the arithmetic mean of the initial and final values.

For example, if the surface of a slab is suddenly raised from 20 to 100 °C, then for the first set of calculations use a surface temperature of 60 °C, then for the next time step use a surface temperature of 100 °C.

Example 16.9   A large steel plate, 300 mm thick, is heated to 800 °C in a furnace and then allowed to cool in air at 25 °C. Find the temperature distribution across the plate thickness 14 min after the start of cooling.

For the plate, $\lambda = 18.75$ W/m K and $\kappa = 7.5 \times 10^{-6}$ m^2/s. The heat transfer coefficient from the plate surfaces to the air may be taken as 125 W/m^2 K throughout the cooling process.

Solution   An explicit solution will be chosen with $Fo = 0.5$, hence equation (16.30) applies for the internal points. The less accurate expression for the surface is chosen to ensure stability with the chosen value of Fo, i.e. for the surface, equation (16.40) applies.

The plate half-thickness of 150 mm is divided into five increments of thickness, $x = 0.03$ m.

Then

$$Fo = \kappa \, \delta\tau/\delta x^2 = 0.5$$

therefore

$$\delta\tau = \frac{0.5 \times 0.03^2}{7.5 \times 10^{-6}} = 60 \text{ s} = 1 \text{ min}$$

For the surface, from equation (16.40)

$$t_i^{\tau} = \frac{t_i^{\tau} + (125 \times 0.03 \times 25/18.75)}{1 + (125 \times 0.03/18.75)} = \frac{t_{i+1}^{\tau} + 5}{1.2}$$

For internal points, from equation (16.36)

$$t_i^{\tau+1} = \frac{t_{i+1}^{\tau} + t_{i-1}^{\tau}}{2}$$

The problem can now be solved by calculating all the temperatures, one by one, at each time step until the time period of 14 min has elapsed. For a simple problem like this one the calculation can easily be done with sufficient accuracy using a hand calculator as in Table 16.3.

Table 16.3 Solution for Example 16.9

$\tau/(\text{min})$	Surface 0	1	2	3	4	Centre-line 5
0	800	800	800	800	800	800
1	670.8	800	800	800	800	800
2	617.0	735.4	800	800	800	800
3	594.6	708.5	767.7	800	800	800
4	571.8	681.2	754.3	783.9	800	800
5	556.8	663.1	732.6	777.2	792.0	800
6	541.4	644.7	720.2	762.3	788.6	792.0
7	529.8	630.8	703.5	754.4	777.2	788.6
8	518.1	616.7	692.6	740.4	771.5	777.2
9	508.7	605.4	678.6	732.1	758.8	771.5
10	498.9	593.7	668.8	718.7	751.8	758.8
11	490.8	583.9	656.2	710.3	738.8	751.8
12	482.1	573.5	647.1	697.5	731.1	738.8
13	474.7	564.6	635.5	689.1	718.2	731.1
14	466.8	555.1	626.9	676.9	710.1	718.2

To illustrate how the figures are calculated: at time 6 min from the start, at the surface $t_0 = (644.7 + 5)/1.2 = 541.4\,°C$; at internal section 2 at the same time, $t_2 = (663.1 + 777.2)/2 = 720.2\,°C$; at the centre-line at the same time, $t_5^{\tau+1} = t_4^{\tau}$, i.e. $t_5 = 792.0\,°C$.

Implicit solution

In equation (16.20) a backward difference approximation is used by expressing the $\partial^2 t/\partial x^2$ term at the next time interval while retaining the same expression for $\partial t/\partial \tau$,

i.e.
$$\frac{t_{i+1}^{\tau+1} + t_{i-1}^{\tau+1} - 2t_i^{\tau+1}}{\delta x^2} = \frac{1}{\kappa}\frac{(t_i^{\tau+1} - t_i^{\tau})}{\delta \tau} \tag{16.41}$$

All the temperatures at $(\tau + 1)$ in equation (16.41) are unknown, hence a set of simultaneous equations such as the above must be solved at each time interval. This makes the method longer, with greater computer storage requirements than the explicit method, but there is not the limitations on the size of Fourier number because the solution is stable for all values of *Fo*. Also, the accuracy is not strongly dependent on the size of the time step and therefore some of the increased computer time and storage can be offset by using a larger time interval.

From equation (16.41):

$$t_i^{\tau+1} - t_i^{\tau} = Fo(t_{i+1}^{\tau+1} + t_{i-1}^{\tau+1} - 2t_i^{\tau+1}) \tag{16.42}$$

This is the fully implicit method. In general we can write

$$\left(\frac{\delta^2 t}{\delta x^2}\right) = \theta\left(\frac{\delta^2 t}{\delta x^2}\right)^{\tau+1} + (1-\theta)\left(\frac{\partial^2 t}{\partial x^2}\right)^{\tau}$$

When $\theta = 0$ the equation reduces to that for the explicit method; when $\theta = 1$ the equation reduces to the fully implicit method.

When θ is put equal to 0.5 the method is known as the *Crank–Nicholson* method,

i.e. $\quad t_i^{\tau+1} - t_i^{\tau} = Fo\frac{1}{2}(t_{i+1}^{\tau+1} + t_{i-1}^{\tau+1} - 2t_i^{\tau+1})$

$$+ (1 - \tfrac{1}{2})Fo(t_{i+1}^{\tau} + t_{i-1}^{\tau} - 2t_i^{\tau})$$

Since there is no stability restriction on the value of Fo it can be put equal to unity. Then we have

$$4t_i^{\tau+1} - t_{i+1}^{\tau+1} - t_{i-1}^{\tau+1} = t_{i+1}^{\tau} + t_{i-1}^{\tau} \tag{16.43}$$

For a surface convecting to a fluid a central difference expression can be used with a fictitious point which is then eliminated by substitution from equation (16.42).

For example, for a left-hand boundary

$$t_{i+1}^{\tau} - t_{i-1}^{\tau} = \frac{2\alpha\,\delta x}{\lambda}(t_i^{\tau} - t_F^{\tau})$$

and $\quad t_{i+1}^{\tau+1} - t_{i-1}^{\tau+1} = \dfrac{2\alpha\,\delta x}{\lambda}(t_i^{\tau+1} - t_F^{\tau+1})$

Adding these equations

$$t_{i+1}^{\tau} + t_{i+1}^{\tau+1} - (t_{i-1}^{\tau} + t_{i-1}^{\tau+1}) = \frac{2\alpha\,\delta x}{\lambda}(t_i^{\tau} + t_i^{\tau+1} - t_F^{\tau} - t_F^{\tau+1}) \tag{16.44}$$

From equation (16.43)

$$t_{i-1}^{\tau} + t_{i-1}^{\tau+1} = 4t_i^{\tau+1} - t_{i+1}^{\tau+1} - t_{i+1}^{\tau}$$

Substituting in equation (16.44) and rearranging

$$-t_i^{\tau+1}\left(2 + \frac{\alpha\,\delta x}{\lambda}\right) + t_{i+1}^{\tau+1} + \frac{\alpha\,\delta x}{\lambda}t_F^{\tau+1}$$

$$= -\frac{\alpha\,\delta x}{\lambda}t_F^{\tau} - t_{i+1}^{\tau} + \frac{\alpha\,\delta x}{\lambda}t_i^{\tau} \tag{16.45}$$

Equation (16.45), with equation (16.43) for the internal points, form a set of equations which must be solved at each time step.

As an illustration of the Crank–Nicholson method consider the previous Example, 16.9. Taking $\delta x = 0.03$ m as before, we then have $\delta\tau = 2$ min since $Fo = 1$ instead of 0.5 as previously. Numbering the points across the slab as in the previous table (p. 596), then from equation (16.45) for the convecting surface

$$-t_0^{\tau+1}\left(2 + \frac{125 \times 0.03}{18.75}\right) + t_1^{\tau+1}$$

$$= \frac{125 \times 0.03}{18.75}t_0^{\tau} - \frac{2 \times 125 \times 0.03 \times 25}{18.75} - t_1^{\tau}$$

therefore

$$-2.2t_0^{\tau+1} + t_1^{\tau+1} = 0.2t_0^{\tau} - 10 - t_1^{\tau}$$

i.e. for the first time step

$$-2.2t_0^{\tau+1} + t_1^{\tau+1} = (0.2 \times 800) - 10 - 800 = -650 \qquad (a)$$

For the internal points, using equation (16.43):

$$4t_1^{\tau+1} - t_2^{\tau+1} - t_0^{\tau+1} = t_2^{\tau} + t_0^{\tau} = 800 + 800 = 1600 \qquad (b)$$

$$4t_2^{\tau+1} - t_3^{\tau+1} - t_1^{\tau+1} = t_3^{\tau} + t_1^{\tau} = 1600 \qquad (c)$$

$$4t_3^{\tau+1} - t_4^{\tau+1} - t_2^{\tau+1} = t_4^{\tau} + t_2^{\tau} = 1600 \qquad (d)$$

$$4t_4^{\tau+1} - t_5^{\tau+1} - t_3^{\tau+1} = t_5^{\tau} + t_3^{\tau} = 1600 \qquad (e)$$

and $\quad 4t_5^{\tau+1} - t_4^{\tau+1} - t_4^{\tau+1} = t_4^{\tau} + t_4^{\tau}$

(using the insulated surface condition),

i.e. $\quad 4t_5^{\tau+1} - 2t_4^{\tau+1} = 2t_4^{\tau} = 1600 \qquad (f)$

In equations (a)–(f) the left-hand side remains the same for each time step, and only the right-hand side must be updated after each set of calculations. These equations form the following matrix for the time step, $\tau = 2$ min.

$$\begin{vmatrix} -2.2 & +1 & 0 & 0 & 0 & 0 \\ -1 & +4 & -1 & 0 & 0 & 0 \\ 0 & -1 & +4 & -1 & 0 & 0 \\ 0 & 0 & -1 & +4 & -1 & 0 \\ 0 & 0 & 0 & -1 & +4 & -1 \\ 0 & 0 & 0 & 0 & -2 & +4 \end{vmatrix} \begin{vmatrix} t_0 \\ t_1 \\ t_2 \\ t_3 \\ t_4 \\ t_5 \end{vmatrix} = \begin{vmatrix} -650 \\ +1600 \\ +1600 \\ +1600 \\ +1600 \\ +1600 \end{vmatrix}$$

This matrix, which has a large number of zero values, is called tridiagonal and may be solved particularly easily on a computer by *Gaussian elimination*. The temperatures $t_0, t_1, t_2, t_3, t_4,$ and t_5 are thus obtained at time, $\tau = 2$ min. These values are then substituted in the right-hand sides of equations (a)–(f) and a new matrix obtained for the temperatures at time, $\tau = 4$ min, which is solved in the same way. The solution proceeds thus until the time, $\tau = 14$ min, as specified in the problem.

Note: The Gaussian elimination method of solving a set of simultaneous equations is the traditional method taught in school for solving two or three simultaneous equations; unlike iterative methods it leads to an exact solution. In the above matrix the coefficient of t_0 in the first row is put equal to -1 by dividing the equation by 2.2; this row is then subtracted from the second row thus eliminating t_0 from the second equation. In the modified matrix t_1 is then eliminated from the third row in the same way, and so on until an equation for t_5 is obtained in the last row. The value of t_5 is then back substituted into the preceding row to give t_4, and t_4 then back substituted into the preceding row to give t_3, and so on until all the temperatures are found. Subroutines are available for solving a tridiagonal matrix by Gaussian elimination as described above.

16.9 Forced convection

The study of forced convection is concerned with the transfer of heat between a moving fluid and a solid surface. In order to apply Newton's law of cooling, given by equation (16.3), it is necessary to find a value for the heat transfer coefficient, α. It is stated in section 16.2 that α is given by λ/δ, where λ is the thermal conductivity of the fluid and δ is the thickness of the fluid film on the surface. The problem is then to find a value for δ in terms of the fluid properties and the fluid velocity; δ depends on the type of fluid flow across the surface and this is governed by the Reynolds number.

The Reynolds number is a dimensionless group given by

$$Re = \frac{\rho Cl}{\eta} \quad \text{or} \quad \frac{Cl}{v}$$

where ρ is the fluid density, C the fluid mean velocity, l the characteristic linear dimension, η the dynamic viscosity of the fluid, and v the kinematic viscosity of the fluid, η/ρ.

The various kinds of forced convection, such as flow in a tube, flow across a tube, flow across a flat plate, etc. can be solved mathematically when certain assumptions are made with regard to the boundary conditions. It is exceedingly difficult to obtain an exact mathematical solution to such problems, particularly in the case of turbulent flow, but approximate solutions can be obtained by making suitable assumptions.

It is not within the scope of this book to approach the subject of forced convection fundamentally. However, many of the results used in heat transfer are derived from experiment, and in fact for many problems no mathematical solution is available and empirical values are essential. These empirical values can be generalized using dimensional analysis, which will now be considered.

Dimensional analysis

In order to apply dimensional analysis it is necessary to know from experience all the variables upon which the desired function depends. The results must apply to geometrically similar bodies, therefore one of the variables must always be a characteristic linear dimension.

Consider the dimensional analysis for forced convection, assuming that the effects of free convection, due to differences in density, may be neglected. It is

599

found that the heat transfer coefficient, α, depends on the fluid viscosity, η, the fluid density, ρ, the thermal conductivity of the fluid, λ, the specific heat capacity of the fluid, c, the temperature difference between the surface and the fluid, Δt, and the fluid velocity, C. Therefore we have

$$\alpha = f(\eta, \rho, \lambda, c, \Delta t, C, l) \tag{16.46}$$

where l is a characteristic linear dimension and f is some function.

Equation (16.46) can be written as follows:

$$\alpha = A\eta^{a_1}\rho^{b_1}\lambda^{\gamma_1}c^{d_1}\Delta t^{e_1}C^{f_1}l^{g_1} + B\eta^{a_2}\rho^{b_2}\lambda^{c_2}c^{d_2}\Delta t^{e_2}C^{f_2}l^{g_2} + \text{etc.} \tag{16.47}$$

where A and B are constants, and a_1, b_1, c_1, etc. are arbitrary indices.

Each term on the right-hand side of the equation must have the same dimensions as the dimensions of α. Considering the first term only, we can write

$$\text{Dimensions of } \alpha = \text{dimensions of } (\eta^{a_1}\rho^{b_1}\lambda^{c_1}c^{d_1}\Delta t^{e_1}C^{f_1}l^{g_1})$$

Each of the properties in the equation can be expressed in terms of five fundamental dimensions; these are mass, M, length, L, time, T, temperature, t, and heat, H.

For α the units are $\dfrac{W}{m^2\,K}$, i.e. $\dfrac{H}{L^2\,Tt}$

For η the units are $\dfrac{kg}{m\,s}$, i.e. $\dfrac{M}{LT}$

For λ the units are $\dfrac{W}{m\,K}$, i.e. $\dfrac{H}{LTt}$

For ρ the units are $\dfrac{kg}{m^3}$, i.e. $\dfrac{M}{L^3}$

For c the units are $\dfrac{kJ}{kg\,K}$, i.e. $\dfrac{H}{Mt}$

For Δt the units are K, i.e. t

For C the units are $\dfrac{m}{s}$, i.e. $\dfrac{L}{T}$

For l the units are m, i.e. L

Hence, substituting

$$\frac{H}{L^2\,Tt} = \left(\frac{M}{LT}\right)^a \times \left(\frac{M}{L^3}\right)^b \times \left(\frac{H}{LTt}\right)^c \times \left(\frac{H}{Mt}\right)^d \times (t)^e \times \left(\frac{L}{T}\right)^f \times (L)^g$$

i.e. $\quad \dfrac{H}{L^2\,Tt} = (M)^{a+b-d} \times (L)^{f+g-a-3b-c} \times (T)^{-a-c-f} \times (t)^{e-c-d} \times (H)^{c+d}$

For the dimensions of each side of the equation to be the same, the power to which each fundamental dimension is raised must be the same on both sides of the equation. Therefore, equating indices we have

For H: $1 = c + d$

For L: $-2 = f + g - a - 3b - c$

For T: $-1 = -a - c - f$

For t: $-1 = e - c - d$

For M: $0 = a + b - d$

We have five equations and seven unknowns, therefore a solution can only be obtained in terms of two of the indices. It is most useful to express a, b, c, e, and g in terms of d and f. Then it can be shown that

$$a = (d - f); \qquad b = f; \qquad c = (1 - d); \qquad e = 0; \qquad g = (f - 1)$$

Substituting these values in equation (16.47), we have

$$\alpha = A\eta^{(d_1 - f_1)}\rho^{f_1}\lambda^{(1-d_1)}c^{d_1}\Delta t^0 C^{f_1}l^{(f_1 - 1)}$$

$$+ B\eta^{(d_2 - f_2)}\rho^{f_2}\lambda^{(1-d_2)}c^{d_2}\Delta t^0 C^{f_2}l^{(f_2 - 1)} + \text{etc.}$$

i.e. $$\alpha = A\frac{\lambda}{l}\left(\frac{c\eta}{\lambda}\right)^{d_1}\left(\frac{\rho Cl}{\eta}\right)^{f_1} + B\frac{\lambda}{l}\left(\frac{c\eta}{\lambda}\right)^{d_2}\left(\frac{\rho Cl}{\eta}\right)^{f_2} + \text{etc.}$$

Hence it can be seen that

$$\frac{\alpha l}{\lambda} = K\text{F}\left\{\left(\frac{c\eta}{\lambda}\right), \left(\frac{\rho Cl}{\eta}\right)\right\}$$

where K is a constant and F is some function.

The dimensionless group, $\alpha l/\lambda$, is called the Nusselt number, Nu; the dimensionless group, $c\eta/\lambda$, is called the Prandtl number, Pr; and the dimensionless group, $\rho Cl/\eta$, is the Reynolds number, Re,

i.e. $Nu = K\text{F}\{(Pr), (Re)\}$ (16.48)

Experiments can be performed in order to evaluate K, and to determine the nature of the function F.

When evaluating Nu, Pr, and Re it is necessary to take the fluid properties at a suitable mean temperature, since the properties vary with temperature. For cases in which the temperature of the bulk of the fluid is not very different from the temperature of the solid surface, then fluid properties are evaluated at the *mean bulk temperature* of the fluid. When the temperature difference is large, errors may be caused by using a mean bulk temperature, and a *mean film temperature* is sometimes used, defined by

$$t_f = \frac{t_b + t_w}{2}$$ (16.49)

where t_b is the mean bulk temperature and t_w the surface temperature.

When using an empirical equation it is essential to know at what reference temperature the properties have been evaluated by the experimenter. It should be noted that the Prandtl number, $Pr = c\eta/\lambda$, consists entirely of fluid properties and therefore is itself a property. Some values of η, c, λ, Pr, and ρ for various fluids are tabulated in reference 16.4.

Example 16.10

Calculate the heat transfer coefficient for water flowing through a 25 mm diameter tube at the rate of 1.5 kg/s, when the mean bulk temperature is 40 °C. For turbulent flow of a liquid take

$$Nu = 0.0243 \, Re^{0.8} \times Pr^{0.4}$$

where the characteristic dimension of length is the tube diameter and all properties are evaluated at mean bulk temperature.

Solution

First it is necessary to ascertain whether the flow is turbulent or laminar. For flow through a tube it can be assumed that the flow is turbulent when $Re > 2100$ approximately. The properties of water can be taken from reference 16.4.
Then

$$\text{Volume flow} = 1.5 \times v_f = 1.5 \times 0.001 = 0.0015 \text{ m}^3/\text{s}$$

i.e.

$$\text{Velocity in tube, } C = \frac{0.0015 \times 4}{\pi \times 0.025^2} = 3.06 \text{ m/s}$$

$$Re = \frac{\rho C d}{\eta} = \frac{Cd}{v_f \eta} = \frac{3.06 \times 0.025}{0.001 \times 651 \times 10^{-6}} = 117\,500$$

The flow is therefore well into the turbulent region and the formula given for turbulent flow can be applied.
From tables, $Pr = 4.3$, hence substituting

$$Nu = 0.0243 \times (117\,500)^{0.8} \times (4.3)^{0.4}$$

$$= 0.0243 \times 11\,377 \times 1.792 = 495.5$$

i.e.

$$Nu = \frac{\alpha d}{\lambda} = 495.5$$

therefore

$$\alpha = \frac{495.5 \times 632 \times 10^{-6}}{0.025} = 12.53 \text{ kW/m}^2 \text{ K}$$

i.e.

Heat transfer coefficient $= 12.53 \text{ kW/m}^2 \text{ K}$

For laminar flow in a tube an exact mathematical solution has been found; this gives $Nu = 3.65$. It can be seen that, since $Nu = \alpha d/\lambda = 3.65$, the heat transfer coefficient, α, for any one tube, depends only on the thermal conductivity of the fluid.

In the foregoing dimensional analysis five fundamental dimensions, heat H, length L, time T, temperature t, and mass M, were chosen. The units of work, or energy in general, are given by

$$(\text{Force} \times \text{distance}) = (\text{mass} \times \text{acceleration} \times \text{distance})$$

$$= M\frac{L}{T^2}L = \frac{ML^2}{T^2}$$

Since heat is a form of energy it can be seen that there is no need to choose heat as one of the fundamental dimensions. If the dimension, H, is omitted, and the units of heat are replaced by ML^2/T^2 whenever they occur, then four dimensionless groups are obtained,

i.e. $\quad Nu = KF\left\{(Pr), (Re), \left(\dfrac{C^2}{c\Delta t}\right)\right\}$

Now if the group $C^2/c\Delta t$ is divided by $(\gamma - 1)$, which is a constant for any one gas, and if Δt is replaced by the absolute bulk temperature of the gas, T, then we have

$$\frac{C^2}{cT(\gamma - 1)} = \frac{C^2}{\gamma RT} = \frac{C^2}{a^2} = (Ma)^2$$

where a is the velocity of sound in the gas and Ma the Mach number, see section 10.7.

Hence, $\quad Nu = K'F\{(Pr), (Re), (Ma)^2\}$ \hfill (16.50)

where K' is a constant.

The influence of the Mach number, Ma, on the heat transfer is negligible for most problems. For high-speed flow however, large amounts of kinetic energy are dissipated by friction in the boundary layer near the surface, and the Mach number becomes an important parameter.

Reynolds analogy

Reynolds postulated that the heat transfer from a solid surface is similar to the transfer of fluid momentum from the surface, and hence that it is possible to express the heat transfer in terms of the frictional resistance to the flow.

Consider turbulent flow. It can be assumed that particles of mass, m, transport heat and momentum to and from the surface, moving perpendicular to the surface. Then on the average

Heat transferred per unit area, $\dot{q} = \dot{m}c\Delta t$

where c is the specific heat capacity of the fluid and Δt the temperature difference between the surface and the bulk of the fluid. Also, the rate of change of momentum across the stream is given by

$$\dot{m}(C - C_w) = \dot{m}C$$

where C is the velocity of the bulk of the fluid and C_w the fluid velocity at the surface $= 0$. Then

$$\text{Force per unit area} = \tau_w = \dot{m}C$$

where τ_w is the shear stress in the fluid at the wall.

Combining the equations for heat flow and momentum transfer, then

$$\frac{\dot{q}}{c\Delta t} = \frac{\tau_w}{C}$$

or
$$\dot{q} = \frac{\tau_w c \Delta t}{C} \qquad (16.51)$$

For turbulent flow in practice there is always a thin layer of fluid on the surface in which viscous effects predominate. This film is known as the laminar sublayer. In this layer heat is transferred purely by conduction.

Therefore, from Fourier's law, for unit area

$$\dot{q} = -\lambda \left\{ \frac{dt}{dy} \right\}_{y=0}$$

where λ is the thermal conductivity of the fluid and y the distance from the surface perpendicular to the surface. Also, for viscous flow

$$\text{Shear stress}, \ \tau = \eta \times (\text{velocity gradient})$$

Hence the shear stress at the wall is given by

$$\tau_w = \eta \left(\frac{dC}{dy} \right)_{y=0}$$

where η is the fluid viscosity and C the fluid velocity.

Now since the laminar sublayer is very thin it may be assumed that the temperature and velocity vary linearly with the distance from the wall, y,

i.e.
$$\dot{q} = -\frac{\lambda \Delta t}{\delta_b} \quad \text{and} \quad \tau_w = \frac{\lambda C}{\delta_b}$$

where δ_b is the thickness of the laminar sublayer.

Then eliminating δ_b, and neglecting the minus sign, we have

$$\frac{\dot{q}}{\lambda \Delta t} = \frac{\tau_w}{\eta C}$$

i.e.
$$\dot{q} = \frac{\tau_w \Delta t}{\eta C}$$

It can be seen that this equation is identical with equation (16.51) when

$$c = \frac{\lambda}{\eta}$$

i.e. when

$$\frac{c\eta}{\lambda} = 1 \quad \text{or} \quad Pr = 1$$

Therefore for fluids whose Prandtl number is approximately unity the simple Reynolds analogy can be applied, since the heat transferred across the laminar sublayer can be considered in a similar way to the heat transferred from the sublayer to the bulk of the fluid. For most gases, dry vapours, and superheated vapours Pr lies between about 0.65 and 1.2.

For unit surface area, $\dot{q} = \alpha \Delta t$, therefore substituting in equation (16.51), we have

$$\frac{\alpha}{c} = \frac{\tau_w}{C}$$

Dividing through by ρC, where ρ is the mean density of the fluid, we have

$$\frac{\alpha}{\rho c C} = \frac{\tau_w}{\rho C^2}$$

Both sides of this equation are dimensionless. The term on the left-hand side is called the Stanton number, St,

i.e. $$St = \frac{\alpha}{\rho C c} \tag{16.52}$$

A dimensionless friction factor, f, is defined by

$$f = \frac{\tau_w}{(\rho C^2 / 2)} \tag{16.53}$$

Therefore we have for the Reynolds analogy

$$St = \frac{f}{2} \tag{16.54}$$

The Stanton number, St, can be written as

$$St = \frac{\alpha}{\rho C c} = \frac{\alpha l}{\lambda} \times \frac{\eta}{\rho C l} \times \frac{\lambda}{c \eta} = \frac{Nu}{Re Pr}$$

i.e. $$St = \frac{Nu}{Re Pr} \tag{16.55}$$

The friction factor, f, can be derived mathematically for some cases, but in other cases a practical determination is necessary. For turbulent flow in a pipe a simple measurement of the pressure drop gives f, and then, using equation (16.51) or equation (16.54), the approximate heat flow can be found.

For flow in a pipe of diameter, d, the resistance to flow over unit length is given by

$$\text{Resistance} = \tau_w \pi d = \Delta p \frac{\pi}{4} d^2$$

where Δp is the pressure drop in unit length.

i.e. $$\tau_w = \frac{\Delta p d}{4} \tag{16.56}$$

605

An important factor in heat exchanger design is the pumping power required. The pumping power is the rate at which work is done in overcoming the frictional resistance, i.e. for flow in a pipe

Pumping power per unit length, $\dot{W} = \tau_w \pi d C$

Also, from equation (16.51),

Heat flow per unit area, $\dot{q} = \dfrac{\tau_w c \Delta t}{C}$

i.e.   Heat flow per unit length, $\dot{Q} = \dfrac{\tau_w c \Delta t \pi d}{C}$

Then the ratio of the pumping power, $\dot{W}$, to the rate of heat flow, $\dot{Q}$, can be expressed as

$$\frac{\dot{W}}{\dot{Q}} = \frac{\tau_w \pi d C C}{\tau_w c \Delta t \pi d} = \frac{C^2}{c \Delta t} \tag{16.57}$$

(for a heat exchanger, Δt is the log mean temperature difference, $\Delta \bar{t}_{ln}$, see p. 616).

It can be seen from equation (16.57) that the power required for a given heat transfer rate can be reduced by decreasing the velocity of flow, C. However, a reduction in fluid velocity means that the required surface area must be increased, and hence a compromise must be made.

Example 16.11   Air is heated by passing it through a 25 mm bore copper tube which is maintained at 280 °C. The air enters at 15 °C and leaves at 270 °C at a mean velocity of 30 m/s. Using the Reynolds analogy, calculate the length of the tube and the pumping power required. For turbulent flow in a tube take $f = 0.0791(Re)^{-1/4}$, and all properties at mean film temperature. Take the mean temperature difference as $\Delta \bar{t}_{ln} = (\Delta t_1 - \Delta t_2)/\ln(\Delta t_1/\Delta t_2)$.

Solution   The mean film temperature can be found from equation (16.49),

i.e. $t_f = \dfrac{t_b + t_w}{2} = \dfrac{1}{2}\left(\dfrac{15 + 270}{2} + 280\right) = \dfrac{142.5 + 280}{2} = 211.25\,°C$

From Tables at $t_f = 211.25\,°C = 484.4\,K$ the properties of air can be found. Then

$$Re = \frac{Cd}{v} = \frac{30 \times 0.025}{3.591 \times 10^{-5}} = 20\,900$$

therefore

$$f = \frac{0.0791}{(20\,900)^{1/4}} = \frac{0.0791}{12.01} = 0.006\,58$$

From equations (16.54) and (16.55),

$$St = \frac{Nu}{RePr} = f/2 = \frac{0.006\,58}{2} = 0.003\,29$$

i.e. $Nu = 0.003\,29 \times 20\,900 \times 0.681 = 46.8$

therefore

$$\frac{\alpha d}{\lambda} = 46.8$$

i.e. $$\alpha = \frac{46.8 \times 3.938 \times 10^{-5}}{0.025} = 0.0737 \text{ kW/m}^2 \text{ K}$$

$$\text{Mass flow air} = \frac{\pi}{4} \times 0.025^2 \times 30 \times 0.73 = 0.010\,75 \text{ kg/s}$$

Hence

$$\text{Heat received by the air} = \dot{m}c(t_{a_2} - t_{a_1})$$

$$= 0.010\,75 \times 1.027 \times (270 - 15)$$

$$= 2.815 \text{ kW}$$

Also, from equation (16.3),

$$\dot{Q} = \alpha A \Delta t = 2.815 \text{ kW}$$

and $$\Delta \bar{t}_{\text{ln}} = \frac{\Delta t_1 - \Delta t_2}{\ln(\Delta t_1/\Delta t_2)} = \frac{(280 - 15) - (280 - 270)}{\ln\{(280 - 15)/(280 - 270)\}} = 77.9 \text{ K}$$

Then $$\dot{Q} = 2.815 = 0.0737 \times 77.9 \times A$$

therefore

$$A = \frac{2.815}{0.0737 \times 77.9} = 0.49 \text{ m}^2$$

Therefore

$$\text{Tube length} = \frac{0.49}{\pi \times 0.025} = 6.24 \text{ m}$$

From equation (16.57),

$$\frac{\dot{W}}{\dot{Q}} = \frac{C^2}{c\Delta t}$$

therefore

$$\dot{W} = \frac{2.815 \times 30 \times 30}{1.027 \times 77.9} = 31.7 \text{ W}$$

i.e.   Pumping power $= 31.7$ W

Example 16.12   In a 25 mm diameter tube the pressure drop per metre length is 0.0002 bar at a section where the mean velocity is 24 m/s, and the mean specific heat capacity of the gas is 1.13 kJ/kg K. Calculate the heat transfer coefficient.

Solution   For a 1 m length

$$\Delta p = 0.0002 \text{ bar}$$

From equation (16.56)

$$\tau_{\text{w}} = \frac{\Delta p d}{4} = \frac{10^5 \times 0.0002 \times 25}{4 \times 10^3} = 0.125 \text{ N/m}^2$$

Then from equation (16.53)

$$f = \frac{\tau_{\text{w}}}{(\rho C^2 / 2)} = \frac{2 \times 0.125}{\rho C^2}$$

Also, from equation (16.54)

$$St = \frac{f}{2} \quad \text{i.e.} \quad \frac{\alpha}{\rho c C} = \frac{2 \times 0.125}{2\rho C^2}$$

therefore

$$\alpha = \frac{0.125 \rho c C}{\rho C^2} = \frac{0.125 \times 1.13}{24} \text{ kW/m}^2 \text{ K}$$

i.e.   Heat transfer coefficient $= 0.005\,88$ kW/m^2 K

$$= 5.88 \text{ W/m}^2 \text{ K}$$

Various modifications have been made to the simple Reynolds analogy in an attempt to obtain an equation which will give a solution for turbulent heat transfer over a wide range of Prandtl numbers. (For very viscous oil the Prandtl number is of the order of thousands, whereas for liquid metals it may be as low as 0.01.) Equations based on modern theories of turbulent flow give the Stanton number as a function of the Reynolds number, the Prandtl number, and the friction factor, and in general these equations reduce to $St = f/2$, when the Prandtl number is put equal to unity (see for example references 16.1, 16.3, and 16.5). Colburn found experimentally that for a wide range of Prandtl numbers

$$St Pr^{2/3} = f/2$$

The term, $St Pr^{2/3}$, is known as the Colburn *j*-factor.

Large temperature differences

When the temperature difference between the surface and the bulk of the fluid is very high, then the property variations become large enough to be taken into consideration. It is then no longer sufficient to use a mean film temperature to evaluate the properties, as given by equation (16.49). The variation of each property with temperature across the stream must be known; sometimes it is sufficiently accurate to use an equation of the form

$$Nu = K\phi\left\{(Pr), (Re), \left(\frac{T_s}{T_w}\right)\right\}$$

where T_s and T_w are the absolute temperatures at the axis of the pipe and at the pipe wall respectively, and fluid properties are taken at the mean film temperature.

Entry length

The equations for flow in a pipe do not usually allow for the effects of the *entry length*. At the entry to a heated tube the hydrodynamic and thermal boundary layers start to build up on the wall, gradually thickening until the flow becomes *fully developed*. In this initial region of the tube the heat transfer coefficient is much larger since the resistance to heat flow of the boundary layer is less, and hence an equation which neglects this effect will give a low value for the calculated heat transfer. The effect is more marked for laminar flow than for turbulent flow, and is much more important for fluids with high Prandtl numbers. In most heat exchange processes the flow is turbulent and the tube length is sufficiently long to make the entry length effect negligibly small. In the case of oil coolers the flow is laminar, the Prandtl number is high, and hence the entry effect may be appreciable.

When flow across a flat plate is considered, the characteristic dimension of length is taken as the distance from the leading edge, and the heat transfer coefficient obtained is then the local value at that section of the plate. The average value of the heat transfer coefficient over the whole plate is the value to be used in calculating the heat transfer to or from the plate. It can be shown that the average heat transfer coefficient for a heated plate over a length l is twice the local heat transfer coefficient at a distance l from the leading edge.

Example 16.13 Air at 20 °C, flowing at 25 m/s, passes over a flat plate, the surface of which is maintained at 270 °C. Calculate the rate at which heat is transferred per metre width from both sides of the plate over a distance of 0.25 m from the leading edge. For heat transfer from a flat plate

$$Nu = 0.332(Pr)^{1/3} \times (Re)^{1/2}$$

where the characteristic linear dimension is the distance from the leading edge, and all properties are evaluated at mean film temperature

Solution Mean film temperature $= \dfrac{20 + 270}{2} = 145\,°C = 418\,K$

Taking the values from tables of properties of air, we have

$$Pr = 0.687 \quad \text{and} \quad Re = \frac{Cl}{v} = \frac{25 \times 0.25 \times 10^5}{2.8} = 223\,214$$

Then $\quad Nu = 0.332 \times (0.687)^{1/3} \times (223\,214)^{1/2}$

$$= 0.332 \times 0.882 \times 472.5 = 138.4$$

and $\quad Nu = \dfrac{\alpha l}{\lambda}$

therefore

$$\alpha = \frac{138.4 \times 3.49}{0.25 \times 10^5} = 0.0193 \text{ kW/m}^2 \text{ K}$$

Hence the average heat transfer coefficient is

$$0.0193 \times 10^3 \times 2 = 38.6 \text{ W/m}^2 \text{ K}$$

Then the rate of heat transfer from both sides of the plate over the length of 0.25 m for 1 m width is given by

$$\dot{Q} = \alpha A \Delta t = 38.6 \times 0.25 \times 1 \times 2 \times (270 - 20) = 4825 \text{ W}$$

i.e. $\quad$ Rate of heat transfer $= 4.825$ kW

The friction loss for the initial length of a flat plate where the boundary layer is still laminar is given by

$$f = 0.664(Re)^{-1/2}$$

Hence it can be seen that the simple Reynolds analogy, given by equation (16.54), $St = f/2$, gives for the initial length of a flat plate

$$St = 0.332(Re)^{-1/2}$$

or $\quad \dfrac{Nu}{PrRe} = 0.332(Re)^{-1/2}$

i.e. $\quad Nu = 0.332(Pr)(Re)^{1/2}$

This is the same as the equation given in Example 16.13 if the Prandtl number is unity.

16.10 Natural convection

As stated previously, heat transfer by free or natural convection is due to differences in density in the fluid causing a natural circulation, and hence a transfer of heat. For the majority of problems in which a fluid flows across a surface, the superimposed effect of natural convection is small enough to be neglected. When there is no forced velocity of the fluid then the heat is transferred entirely by natural convection (when radiation is negligible). The heat transfer in this case depends on the coefficient of cubical expansion, β, which is given by

$$\rho_1 = \rho_2(1 + \beta\Delta t) \quad \text{or} \quad (\rho_1 - \rho_2) = \rho_2 \beta \Delta t$$

where Δt is the temperature difference between the two parts of the fluid of density ρ_1 and ρ_2.

The upthrust per unit volume of fluid is $(\rho_1 - \rho_2)g$, and the velocity of the convection current is dependent on the upthrust, i.e. natural convection depends on

$$(\rho_1 - \rho_2)g = \rho_2 \beta \Delta t g$$

The heat transfer also depends on the fluid viscosity, the thermal conductivity of the fluid, and a characteristic dimension of length. Since the coefficient of cubical expansion, β, and the local acceleration due to gravity, g, do not have a separate effect on the heat transfer, then only the product, βg, need be considered. For a dimensional analysis we then have

$$\alpha = A\eta^{a_1} \rho^{b_1} \lambda^{c_1} c^{d_1} \Delta t^{e_1} (\beta g)^{f_1} l^{g_1} + B\eta^{a_2} \rho^{b_2} \lambda^{c_2} c^{d_2} \Delta t^{e_2} (\beta g)^{f_2} l^{g_2} + \text{etc.}$$

Then by the same procedure as in section 16.9 it can be shown that

$$Nu = KF\left\{\left(\frac{c\eta}{\lambda}\right), \left(\frac{\beta g \rho^2 l^3 \Delta t}{\eta^2}\right)\right\}$$

or $\qquad Nu = KF\{(Pr), (Gr)\}$

where $\qquad Gr = \dfrac{\beta g \rho^2 l^3 \Delta t}{\eta^2} = \dfrac{\beta g l^3 \Delta t}{\nu^2}$

Gr is the Grashof number.

In many cases of natural convection it is possible to use an approximate equation to evaluate the heat transfer coefficient, α. For example, for natural convection from a horizontal pipe

$$\alpha = 1.32\left(\frac{\Delta t}{d}\right)^{1/4} \quad \text{when} \quad 10^4 < Gr < 10^9$$

and $\qquad \alpha = 1.25\Delta t^{1/3} \quad \text{when} \quad 10^9 < Gr < 10^{12}$

where α is in W/m^2 K, Δt is in K, and d is in m.

Example 16.14

Calculate the rate of heat loss by natural convection per unit length from a horizontal pipe of 150 mm diameter, the surface of which is at 277 °C. The room temperature is 17 °C. It has been shown that for a horizontal cylinder (see reference 16.10)

$$Nu = 0.53(GrPr)^{1/4}$$

where the properties are evaluated at the mean film temperature. Take the coefficient of cubical expansion, β, as $1/T$, where T is the absolute temperature of the air.

Solution   From tables, at a mean film temperature of $0.5(550 + 290) = 420$ K, we have, $Pr = 0.686$ and

$$Gr = \frac{\beta g \Delta t d^3}{\nu^2} = \frac{9.81 \times (277 - 17) \times 0.15^3}{290 \times (2.822 \times 10^{-5})^2} = 37.27 \times 10^6$$

611

Note that $g = 9.81 \text{ m/s}^2$, and

$$\beta = \frac{1}{17 + 273} = \frac{1}{290} \text{K}^{-1}$$

Substituting

$$Nu = 0.53(37.27 \times 10^6 \times 0.686)^{1/4}$$

$$= 37.7$$

i.e. $\quad Nu = \dfrac{\alpha d}{\lambda} = 37.7$

$$\alpha = \frac{37.7 \times 3.635 \times 10^{-5}}{0.15} = 0.009\,13 \text{ kW/m}^2 \text{ K}$$

Then from equation (16.3)

$$\dot{Q} = \alpha A(t_w - t) = 0.009\,13 \times \pi \times 0.15 \times 1 \times (277 - 17) = 1.119 \text{ kW}$$

i.e. $\quad$ Heat loss per metre length = 1.119 W

Example 16.15 $\qquad$ Recalculate the heat transfer coefficient for Example 16.14 using the approximate equation

$$\alpha = 1.32\left(\frac{\Delta t}{d}\right)^{1/4} \quad \text{for} \quad 10^4 < Gr < 10^9$$

where α is in W/m^2 K, Δt is in K, and d is in m.

Solution $\qquad$ $\alpha = 1.32 \times \left(\dfrac{277 - 17}{0.15}\right)^{1/4} = 1.32 \times (1733)^{1/4} = 1.32 \times 6.45$

i.e. $\quad$ Heat transfer coefficient = 8.52 W/m^2 K

(compared with the more accurate value, 9.13 W/m^2 K).

For natural convection from a vertical wall the air in rising due to the convection currents builds up a boundary layer, starting from the bottom and thickening gradually up the wall. The heat transfer coefficient varies up the wall, and the formulae for heat transfer from a vertical wall give the local heat transfer coefficient at a distance, l, from the bottom of the wall, where the characteristic linear dimension to be used in the Grashof number is the length, l.

It can be shown that the average heat transfer coefficient for the wall from the bottom up to the distance, l, is given by

$$\alpha_{av} = \frac{4}{3}\alpha$$

where α_{av} is the average heat transfer coefficient, and α the heat transfer coefficient at the section distance, l, from the bottom of the wall.

Example 16.16   A vertical surface 1 m high is at a temperature of 627 °C, and the atmospheric temperature is 27 °C. Calculate the rate at which heat is lost by convection from the surface per metre width. For natural convection from a vertical surface take

$$\alpha = 1.42\left(\frac{\Delta t}{l}\right)^{1/4} \quad \text{for} \quad 10^4 < Gr < 10^9$$

or $\alpha = 1.31\Delta t^{1/3}$ for $10^9 < Gr < 10^{12}$

where all properties are at the mean film temperature, and $\beta = 1/T$, where T is the absolute air temperature, α is in $W/m^2\,K$, Δt in K, and l in m.

Solution   The Grashof number in such problems has the same limiting function as the Reynolds number in fluid flow. For the lower range of Grashof numbers the flow of air due to natural convection remains laminar on the wall surface, whereas for the larger Grashof numbers the boundary layer on the wall is turbulent. It can be seen from the second equation above, $\alpha = 1.31\Delta t^{1/3}$, that when the boundary layer is turbulent the heat transfer coefficient is assumed to be the same at all parts of the wall, since α no longer depends on the distance, l.

Mean film temperature $= 0.5(900 + 300) = 600$ K

Taking properties from tables, we have

$$Gr = \frac{\beta g l^3 \Delta t}{v^2} = \frac{1 \times 9.81 \times l^3 \times (327 - 30)}{303 \times (5.128 \times 10^{-5})^2} = 3.65 \times 10^9$$

where $\beta = \dfrac{1}{30 + 273} = \dfrac{1}{303}$

Hence $\alpha = 1.31\Delta t^{1/3} = 1.31(327 - 30)^{1/3}$

$= 1.31 \times 6.67 = 8.75 \ W/m^2\,K$

and $\dot{Q} = \alpha A \Delta t = 8.75 \times 1 \times 1 \times (627 - 27) = 5250$ W

i.e.   Rate of heat loss per metre width $= 5.25$ kW

Note: expressions for α as above give average values.

16.11   Heat exchangers

One of the most important processes in engineering is the heat exchange between flowing fluids. In heat exchangers the temperature of each fluid changes as it passes through the exchanger, and hence the temperature of the dividing wall between the fluids also changes along the length of the exchanger. Examples in practice in which flowing fluids exchange heat are air intercoolers and preheaters, condensers and boilers in steam plant, condensers and evaporators in refrigeration units, and many other industrial processes in which a liquid or gas is required to be either cooled or heated.

There are three main types of heat exchanger: the most important type is the *recuperator* in which the flowing fluids exchanging heat are on either side of a dividing wall; the second type is the *regenerator* in which the hot and cold fluids pass alternately through a space containing a matrix of material that provides alternately a sink and a source for heat flow; the third type is the *evaporative type* in which a liquid is cooled evaporatively and continuously in the same space as the coolant. An example of the latter type is the cooling tower (see section 15.5). Very often when the term 'heat exchanger' is used it refers to the recuperative type, which is by far the most commonly used in engineering practice. This section will deal almost entirely with the recuperative type.

Parallel-flow and counter-flow recuperators

Consider the simple case of a fluid flowing through a pipe and exchanging heat with a second fluid flowing through an annulus surrounding the pipe. When the fluids flow in the same direction along the pipe the system is known as *parallel-flow*, and when the fluids flow in opposite directions to each other the system is known as *counter-flow*. Parallel-flow is shown in Fig. 16.16(a) and counter-flow is shown in Fig. 16.16(b). Let the mean inlet and outlet temperatures of fluid A be t_{A_1} and t_{A_2} respectively, and let the mean temperatures of fluid B at sections 1 and 2 be t_{B_1} and t_{B_2} respectively. Let the mass flow rates of fluid A and fluid B be $\dot{m}_A$ and $\dot{m}_B$ respectively, and let the specific heats of fluid A and fluid B be c_A and c_B respectively. The temperature difference at section 1 is $(t_{A_1} - t_{B_1}) = \Delta t_1$, and the temperature difference at section 2 is $(t_{A_2} - t_{B_2}) = \Delta t_2$.

Fig. 16.16 Parallel-flow and counter-flow heat exchangers and the temperature distributions with length

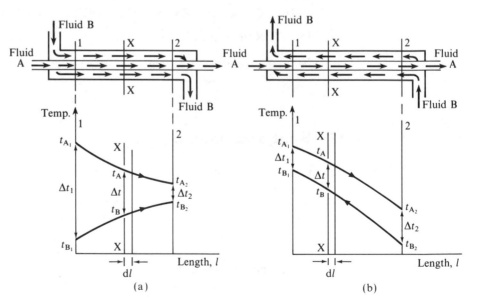

Since the tube wall separating the fluids is thin it is possible to use an overall heat transfer coefficient, U, based on equation (16.11),

i.e. $\qquad U = \dfrac{1}{R_T A}$

where A is the mean surface area of the tube.

Since the resistance of the tube wall is negligibly small we can write

$$\frac{1}{U} = \frac{1}{\alpha_A} + \frac{1}{\alpha_B} \qquad (16.58)$$

In practice the values of α_A and α_B will vary along the length of the tube, but suitable mean values can be found. A mean value of U along the tube will be assumed.

Consider any section X–X where fluid A is at t_A and fluid B is at t_B. The temperature difference at this section is $(t_A - t_B) = \Delta t$, and a small amount of heat, $\mathrm{d}\dot{Q}$, is transferred across an element of length $\mathrm{d}l$. Using equation (16.5), $\dot{Q} = UA(t_A - t_B)$, we have

$$\mathrm{d}\dot{Q} = \pi D \, \mathrm{d}l \, U\Delta t \qquad (16.59)$$

where D is the mean diameter of the tube.

Fluid A increases in temperature by $\mathrm{d}t_A$ along element $\mathrm{d}l$, and fluid B increases in temperature by $\mathrm{d}t_B$ along element $\mathrm{d}l$. Also since $\Delta t = t_A - t_B$, we have

$$\mathrm{d}(\Delta t) = \mathrm{d}t_A - \mathrm{d}t_B \qquad (16.60)$$

In the case of parallel-flow (see Fig. 16.16), temperature t_A decreases with the length l, while temperature t_B increases with the length l. The heat given up by fluid A must equal the heat received by fluid B, i.e. for parallel flow

$$\mathrm{d}\dot{Q} = -\dot{m}_A c_A \, \mathrm{d}t_A = \dot{m}_B c_B \, \mathrm{d}t_B \qquad (16.61)$$

therefore

$$\mathrm{d}t_A = \frac{-\mathrm{d}\dot{Q}}{\dot{m}_A c_A} \quad \text{and} \quad \mathrm{d}t_B = \frac{\mathrm{d}\dot{Q}}{\dot{m}_B c_B}$$

Substituting in equation (16.60)

$$\mathrm{d}(\Delta t) = \frac{-\mathrm{d}\dot{Q}}{\dot{m}_A c_A} - \frac{\mathrm{d}\dot{Q}}{\dot{m}_B c_B} = -\mathrm{d}\dot{Q}\left(\frac{1}{\dot{m}_A c_A} + \frac{1}{\dot{m}_B c_B}\right) \qquad (16.62)$$

Integrating equation (16.62) between sections 1 and 2

$$\Delta t_2 - \Delta t_1 = -\dot{Q}\left(\frac{1}{\dot{m}_A c_A} + \frac{1}{\dot{m}_B c_B}\right)$$

or $\qquad \Delta t_1 - \Delta t_2 = \dot{Q}\left(\dfrac{1}{\dot{m}_A c_A} + \dfrac{1}{\dot{m}_B c_B}\right) \qquad (16.63)$

Also, from equation (16.62),

$$\mathrm{d}\dot{Q} = \frac{-\mathrm{d}(\Delta t)}{(1/\dot{m}_A c_A) + (1/\dot{m}_B c_B)}$$

Substituting in equation (16.59)

$$\frac{-d(\Delta t)}{(1/\dot{m}_A c_A) + (1/\dot{m}_B c_B)} = \pi D \, dl \, U \Delta t$$

therefore

$$\frac{-d(\Delta t)}{\Delta t} = \pi D U \left(\frac{1}{\dot{m}_A c_A} + \frac{1}{\dot{m}_B c_B} \right) dl$$

Integrating between sections 1 and 2

$$-\ln\left(\frac{\Delta t_2}{\Delta t_1}\right) = \pi D l U \left(\frac{1}{\dot{m}_A c_A} + \frac{1}{\dot{m}_B c_B} \right) \tag{16.64}$$

where l is the total length of the tube.

Now from equation (16.63)

$$\left(\frac{1}{\dot{m}_A c_A} + \frac{1}{\dot{m}_B c_B} \right) = \frac{\Delta t_1 - \Delta t_2}{\dot{Q}}$$

Hence substituting in equation (16.64)

$$-\ln\left(\frac{\Delta t_2}{\Delta t_1}\right) = \frac{\pi D l U (\Delta t_1 - \Delta t_2)}{\dot{Q}}$$

or

$$\dot{Q} = \frac{\pi D l U (\Delta t_1 - \Delta t_2)}{\ln(\Delta t_1 / \Delta t_2)} \tag{16.65}$$

In the case of counter-flow (see Fig. 16.16), both temperature t_A and temperature t_B decrease in the direction of the length l. In place of equation (16.61) we therefore have

$$d\dot{Q} = -\dot{m}_A c_A \, dt_A = -\dot{m}_B c_B \, dt_B$$

When the same procedure as for parallel-flow is carried out equation (16.65) is again obtained; this procedure is left as an exercise for the reader.

Comparing equation (16.65) with equation (16.5), $\dot{Q} = UA(t_A - t_B)$, it can be seen that the mean temperature difference, $\Delta \bar{t}$, is given by

$$\Delta \bar{t}_{ln} = \frac{\Delta t_1 - \Delta t_2}{\ln(\Delta t_1 / \Delta t_2)} \tag{16.66}$$

$\Delta \bar{t}_{ln}$ is known as the *logarithmic mean temperature difference*.

Then we have

$$Q = UA \Delta \bar{t}_{ln} \tag{16.67}$$

where A is the mean surface area of the tube, $\pi D l$.

There are several important points that should be mentioned here:

(i) When one of the fluids is a wet vapour or a boiling liquid then its temperature remains constant. Assuming fluid A to be a wet vapour, then the temperature variations are as shown in Fig. 16.17. It follows that under these circumstances the variation in temperature of fluid B is the same whether the flow is parallel-flow or counter-flow.

Fig. 16.17
Temperature variations
with length with one
fluid condensing or
boiling

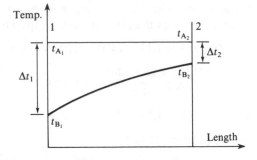

(ii) In counter-flow the temperature range possible is greater, since, in theory, the fluid being heated can be raised to a higher temperature than that of the heating fluid at exit. In parallel-flow the final temperatures of the fluids must be somewhere between the initial values of each fluid. This should be clear from Figs 16.16(a) and 16.16(b) (p. 614).

(iii) When the product $\dot{m}_A c_A$ is equal to $\dot{m}_B c_B$ then the temperature difference in counter-flow is the same all along the length of the tube. This must be the case since the heat given up by fluid A is equal to the heat received by fluid B. Referring to Fig. 16.16(b), we have

$$\dot{m}_A c_A(t_{A_1} - t_{A_2}) = \dot{m}_B c_B(t_{B_1} - t_{B_2})$$

therefore

$$(t_{A_1} - t_{A_2}) = (t_{B_1} - t_{B_2}) \quad \text{or} \quad (t_{A_1} - t_{B_1}) = (t_{A_2} - t_{B_2})$$

i.e. $\quad \Delta \bar{t} = \Delta t_1 = \Delta t_2$

Note that if we attempt to substitute in equation (16.66) under these circumstances, then the result is indeterminate,

i.e. $\quad \Delta \bar{t}_{ln} = \dfrac{\Delta t_1 - \Delta t_2}{\ln(\Delta t_1/\Delta t_2)} = \dfrac{0}{\ln 1} = \dfrac{0}{0}$

The proof of the logarithmic mean temperature difference given previously is not valid when Δt_1 is equal to Δt_2, since $d(\Delta t)$ is then zero.

(iv) From equation (16.67), $\dot{Q} = U A \Delta \bar{t}_{ln}$, it can be seen that, for a given surface area, A, and a given mean value of the overall heat transfer coefficient, U, then the logarithmic mean temperature difference, $\Delta \bar{t}_{ln}$, must be made as large as possible. It is found that for given temperature changes, $\Delta \bar{t}_{ln}$ is always greater for counter-flow than it is for parallel-flow. The initial temperature difference, Δt_1, is greater for parallel-flow, but the value of $\Delta \bar{t}_{ln}$ is always less. It follows that for given rates of mass flow of the two fluids, and for given temperature changes, the surface area required is less for counter-flow.

Example 16.17

Exhaust gases flowing through a tubular heat exchanger at the rate of 0.3 kg/s are cooled from 400 to 120 °C by water initially at 10 °C. The specific heat capacities of exhaust gases and water may be taken as 1.13 and 4.19 kJ/kg K respectively, and the overall heat transfer coefficient from gases to water is 140 W/m² K. Calculate the surface area required when the cooling water flow is 0.4 kg/s, (i) for parallel-flow, (ii) for counter-flow.

617

Solution (i) Parallel-flow. The heat exchanger is shown diagrammatically in Fig. 16.18. The heat given up by the exhaust gases is equal to the heat taken up by the water,

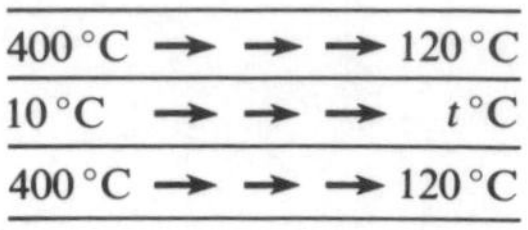

$$\dot{Q} = 0.3 \times 1.13 \times (400 - 120) = 0.4 \times 4.19 \times (t - 10)$$

i.e.

$$t = \frac{0.3 \times 1.13 \times 280}{0.4 \times 4.19} + 10 = 66.6\,°C$$

400 °C → → → 120 °C
10 °C → → → t °C
400 °C → → → 120 °C

Fig. 16.18 Parallel-flow heat exchanger for Example 16.17(a)

Also, $\dot{Q} = 0.3 \times 1.13 \times 280 = 95\,\text{kW}$

From equation (16.66)

$$\Delta \bar{t}_{\text{ln}} = \frac{\Delta t_1 - \Delta t_2}{\ln(\Delta t_1/\Delta t_2)} = \frac{(400 - 10) - (120 - 66.6)}{\ln\{(400 - 10)/(120 - 66.6)\}} = \frac{336.6}{1.99} = 169\,\text{K}$$

From equation (16.67)

$$\dot{Q} = UA\Delta \bar{t}_{\text{ln}}$$

i.e. $95 \times 10^3 = 140 \times A \times 169$

i.e.

$$A = \frac{95 \times 10^3}{140 \times 169} = 4.01\,\text{m}^2$$

i.e. Surface area required $= 4.01\,\text{m}^2$

120 °C ← ← ← 400 °C
10 °C → → → 66.6 °C
120 °C ← ← ← 400 °C

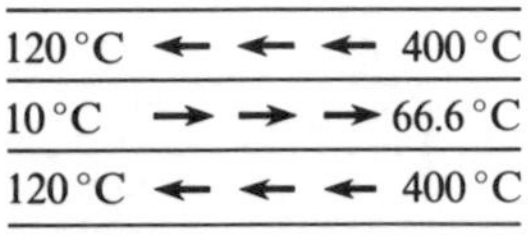

Fig. 16.19 Counter-flow heat exchanger for Example 16.17(b)

(ii) Counter-flow. The heat exchanger is shown diagrammatically in Fig. 16.19. The water temperature at outlet is 66.6 °C and $\dot{Q} = 95\,\text{kW}$ as calculated in part (i).

From equation (16.66)

$$\Delta \bar{t}_{\text{ln}} = \frac{\Delta t_1 - \Delta t_2}{\ln(\Delta t_1/\Delta t_2)} = \frac{(120 - 10) - (400 - 66.6)}{\ln\{(120 - 10)/(400 - 66.6)\}} = \frac{223.4}{1.11} = 201\,\text{K}$$

From equation (16.67)

$$\dot{Q} = UA\Delta \bar{t}_{\text{ln}} \quad \text{therefore } 95 \times 10^3 = 140 \times A \times 201$$

$$A = \frac{95 \times 10^3}{140 \times 201} = 3.37\,\text{m}^2$$

i.e. Surface area required $= 3.37\,\text{m}^2$

Cross-flow recuperator

A simple cross-flow recuperator is shown in Fig. 16.20. The calculation of the mean temperature difference is much more difficult in this case. The true mean temperature difference depends on the ratio of the product of the mass flow and specific heat capacities of fluids A and B, as well as on the ratio of the temperature difference between the fluids at inlet and outlet. Tables are available of a correction factor for various values of the ratios

$$\frac{t_{B_2} - t_{A_2}}{t_{B_1} - t_{A_1}} \quad \text{and} \quad \frac{\dot{m}_A c_A}{\dot{m}_B c_B}$$

Fig. 16.20 Simple cross-flow heat exchanger

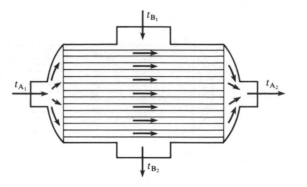

The correction factor is multiplied by the arithmetic mean temperature difference to give the true value of $\Delta \bar{t}$. When the temperature differences at inlet and outlet are not substantially different, it is a sufficiently good approximation to use the arithmetic mean temperature difference, i.e.

$$\Delta \bar{t}_a = \left(\frac{t_{A_1} + t_{A_2}}{2}\right) - \left(\frac{t_{B_1} + t_{B_2}}{2}\right) \tag{16.68}$$

It has been shown in Example 16.17 that the surface area required for a given heat flow is smaller with counter-flow than with parallel-flow. For cross-flow the required surface area is between that for parallel-flow and counter-flow. As with counter-flow, the outlet temperature of the heated fluid in cross-flow can be raised to a higher temperature than the outlet temperature of the cooled fluid (e.g. in Fig. 16.20, t_{A_2} can be higher than t_{B_2}); this is not possible in parallel-flow.

Multipass and mixed-flow recuperators

The simple parallel-flow and counter-flow heat exchangers discussed above occur very rarely in practice. To obtain the necessary surface area with a simple tube and annulus arrangement the length of the tube may be too large for practical purposes. For instance, in Example 16.17(ii), if the tube diameter were 150 mm, the length required would be

$$l = \frac{A}{\pi D} = \frac{3.37}{\pi \times 0.15} = 7.15 \text{ m}$$

In order to make the heat exchanger more compact, which is desirable from space considerations, and also to reduce the heat loss from the outside surface, it is necessary to have several tubes and perhaps several passes or bundles of tubes. The flow can be either cross-flow, or a mixture of parallel-flow, counter-flow, and cross-flow. The latter case is called mixed-flow. A typical example of a shell-type mixed-flow heat exchanger is shown in Fig. 16.21. The analysis of a mixed-flow heat exchanger is complex and correction factors have been plotted, which must be used to evaluate the mean temperature difference. The logarithmic mean temperature difference in counter-flow is evaluated and

Fig. 16.21 Shell-and-tube heat exchanger

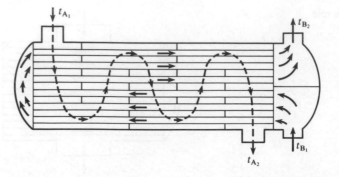

then multiplied by the correction factor. Correction factors for most types of mixed-flow heat exchangers are given in ref. 16.6. Note that when one of the fluids is a condensing vapour or a boiling liquid then the mean temperature difference is the same whether the heat exchanger is parallel-flow, counter-flow, cross-flow, or mixed-flow.

In certain heat exchangers of the multipass type, the mean temperature difference for counter-flow or parallel-flow can still be used as a reasonable approximation. For example, the heat exchanger in Fig. 16.22 is essentially a counter-flow type. The larger the number of passes made by fluid B then the nearer the heat exchanger is to pure counter-flow.

Fig. 16.22 Multipass shell-and-tube heat exchanger

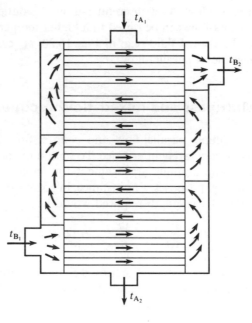

Fouling resistance

In most heat exchangers the fluid flowing is not completely free from dirt, oil, grease, and chemical deposits, and a coating tends to collect on all metal surfaces. This increases the resistance to heat transfer and must be allowed for in design calculations. It is usual to allow for the effect of this coating of dirt by adding a *fouling resistance* to the total thermal resistance. Typical values of fouling resistance for 1 m² of surface area are: 1.8 K/kW for fuel oil; 0.6 K/kW for river water; and 0.2 K/kW for boiler feedwater which has been treated. Facility must be provided for easy periodic cleaning of the tubes. For a comprehensive treatment of process heat exchangers see references 16.6 and 16.7.

Extended surface recuperators

Another form of recuperator which should be mentioned briefly is the *extended surface* type. The metal wall containing a fluid to be cooled can be extended on the outside in the form of fins, studs, or ribs. Examples of this type are the finned hot-water space-heater (sometimes misnamed 'radiator') and the air-cooled cylinders of small air compressors and IC engines. The fins on the surface give a larger outside surface area for the same internal surface area, and hence increase the cooling effect for a given volume. Details of compact heat exchangers are given in reference 16.8, two examples are shown in Figs 16.23 and 16.24. A simple analysis of run-around coils used for energy recovery (see section 17.5) is given in reference 16.9.

Fig. 16.23 Compact plate-fin heat exchanger

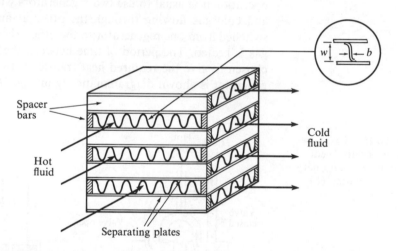

Regenerators

In the various types of recuperator described above, the hot and cold fluids are separated at all times by a metal wall. The characteristic feature of a regenerator is that the fluids occupy the same space in turn or are in contact with the same matrix in turn. The fluids used in regenerators are nearly always gaseous. When the hot gas occupies the space, it gives up heat to the walls, or to solid matter distributed throughout the space, called a matrix. The hot gas is then withdrawn, the cold gas enters the space and is heated by the walls and the matrix; the

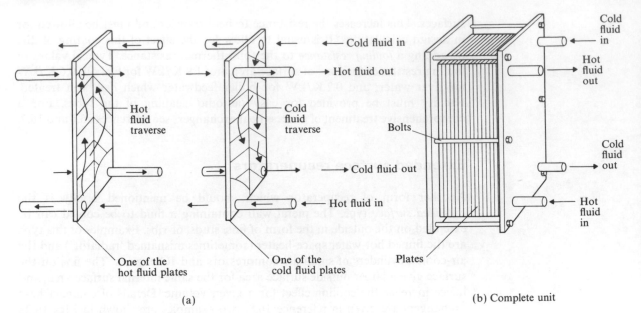

Fig. 16.24 Compact plate heat exchanger

process is a cyclic one and analysis is complex. In order to have continuous operation it is usual to use two regenerators with hot gas flowing through one, and cold gas flowing through the other at any instant. When the flows are switched from one regenerator to the other, the hot gas is cooled while the cold gas is heated. The period of time between the switching of the flows must be chosen to give the required heat transfer between the two gases. This type of generator is shown diagrammatically in Fig. 16.25(a).

Fig. 16.25 Two types of regenerative heat exchangers: stationary (a) and rotating (b) matrices

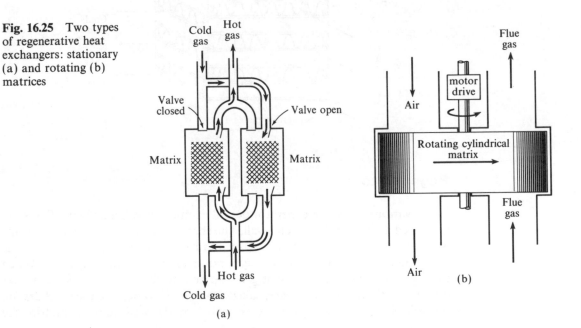

Another method used is the *rotating matrix* type, in which a cylinder containing solid inserts is rotated so that it passes alternately through cold and hot gas streams which are sealed from each other. An example of this type of regenerator is the Ljungström air pre-heater for boiler furnaces, shown diagrammatically in Fig. 16.25(b).

There are many applications for regenerators, from air pre-heaters in blast furnaces to heat exchangers in plants for gas liquefaction. One application of the rotating matrix type which is becoming important is in energy conservation (see section 17.5).

16.12   Heat exchanger effectiveness

In heat exchanger design the efficiency of the heat transfer process is very important. A method due to Nusselt and developed by Kays and London for compact heat exchangers (see reference 16.8) is described in this section.

The *effectiveness*, ε, of a heat exchanger is defined as the ratio of the actual heat transferred to the maximum possible heat transfer.

For any heat exchanger with mass flow rates of hot and cold fluids, $\dot{m}_H$ and $\dot{m}_C$, with specific heat capacities, c_H and c_C, let the overall temperature changes of each fluid be $(t_{Hi} - t_{He})$ and $(t_{Ce} - t_{Ci})$ where subscripts i and e refer to inlet and exit.

Neglecting heat losses to the surroundings:

$$\dot{Q} = \dot{m}_H c_H (t_{Hi} - t_{He}) = \dot{m}_C c_C (t_{Ce} - t_{Ci})$$

or $$\dot{Q} = C_H (t_{Hi} - t_{He}) = C_C (t_{Ce} - t_{Ci}) \qquad (16.69)$$

where $C_H = \dot{m}_H c_H$ and $C_C = \dot{m}_C c_C$ are the *thermal capacities* of the hot and cold fluids.

From equation (16.69) it can be seen that the fluid with the smaller thermal capacity, C, has the greater temperature change. The maximum possible temperature change of one of the fluids is $(t_{H_{max}} - t_{C_{min}})$, and this ideal temperature change can only be aspired to by the fluid with the minimum thermal capacity,

i.e. $$\varepsilon = \frac{\dot{Q}}{C_{min}(t_{H_{max}} - t_{C_{min}})} \qquad (16.70)$$

The object of a well-designed heat exchanger is to obtain the maximum possible change of temperature of a fluid for a given driving force, that is for a given logarithmic mean temperature difference, $\Delta \bar{t}_{ln}$. Hence another useful measure of the efficiency of a heat exchanger is the *number of transfer units*, N_{tu}, defined as

$$N_{tu} = \frac{\text{Temperature change of one fluid}}{\Delta \bar{t}_{ln}}$$

Hence we can write for the hot fluid

$$N_{tu_H} = \frac{t_{Hi} - t_{He}}{\Delta t_{ln}}$$

and for the cold fluid

$$N_{tu_C} = \frac{t_{Ce} - t_{Ci}}{\Delta \bar{t}_{ln}}$$

Now $\quad \dot{Q} = UA\Delta \bar{t}_{ln} = C_H(t_{Hi} - t_{He}) = C_C(t_{Ce} - t_{Ci})$

i.e. $\quad N_{tu_H} = \dfrac{UA}{C_H} \quad$ and $\quad N_{tu_C} = \dfrac{UA}{C_C}$

A more general definition of N_{tu} is as follows:

$$N_{tu} = \frac{\text{Greater of the two fluid temperature differences}}{\Delta \bar{t}_{ln}}$$

i.e. $\quad N_{tu} = \dfrac{UA}{C_{min}}$ \hfill (16.71)

The greater the number of transfer units the more effective is the heat exchanger.

The ratio of the minimum to the maximum thermal capacity is usually given the symbol R,

i.e. $\quad R = C_{min}/C_{max}$ \hfill (16.72)

Note that R may vary between 1 (when both fluids have the same thermal capacity) and 0 (when one of the fluids has an infinite thermal capacity, e.g. a condensing vapour or a boiling liquid).

Figure 16.26 shows a typical example of a graph of effectiveness, ε, against N_{tu} for various values of the thermal capacity ratio, R.

Fig. 16.26 Effectiveness against number of transfer units for various values of thermal capacity ratio

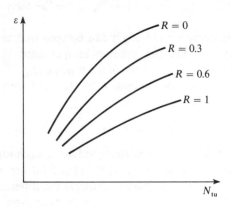

Consider a counter-flow heat exchanger as shown in Fig. 16.27. From the figure it can be seen that $C_C = C_{min}$, since $\Delta t_C > \Delta t_H$,

i.e. $\quad R = C_C/C_H$

or, using equation (16.69),

$$R = \frac{t_{H1} - t_{H2}}{t_{C1} - t_{C2}}$$ \hfill (16.73)

Fig. 16.27
Temperature variations
in a counter-flow heat
exchanger

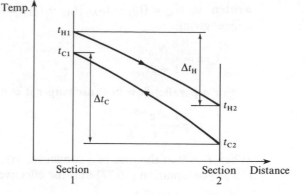

From equation (16.70)

$$\varepsilon = \frac{\dot{Q}}{C_{min}(t_{H_{max}} - t_{C_{min}})} = \frac{C_{min}(t_{C1} - t_{C2})}{C_{min}(t_{H1} - t_{C2})} = \frac{t_{C1} - t_{C2}}{t_{H1} - t_{C2}} \tag{16.74}$$

From equation (16.71)

$$N_{tu} = \frac{UA}{C_{min}} = \frac{t_{C1} - t_{C2}}{\Delta \bar{t}_{ln}}$$

From equation (16.66)

$$\Delta \bar{t}_{ln} = \frac{(t_{H1} - t_{C1}) - (t_{H2} - t_{C2})}{\ln\{(t_{H1} - t_{C1})/(t_{H2} - t_{C2})\}}$$

i.e. $$N_{tu} = \frac{(t_{C1} - t_{C2})}{(t_{H1} - t_{C1}) - (t_{H2} - t_{C2})} \ln\left\{\frac{t_{H1} - t_{C1}}{t_{H2} - t_{C2}}\right\}$$

or $$N_{tu} \frac{(t_{H1} - t_{H2}) - (t_{C1} - t_{C2})}{(t_{C1} - t_{C2})} = \ln\left\{\frac{(t_{H1} - t_{C2}) - (t_{C1} - t_{C2})}{(t_{H1} - t_{C2}) - (t_{H1} - t_{H2})}\right\}$$

therefore using equations (16.73) and (16.74),

$$N_{tu}(R - 1) = \ln\left\{\frac{[(t_{C1} - t_{C2})/\varepsilon] - (t_{C1} - t_{C2})}{[(t_{C1} - t_{C2})/\varepsilon] - R(t_{C1} - t_{C2})}\right\}$$

i.e. $$N_{tu}(R - 1) = \ln\left\{\frac{(1 - \varepsilon)}{(1 - R\varepsilon)}\right\}$$

or $$\varepsilon = \frac{1 - e^{-N_{tu}(1-R)}}{1 - R\,e^{-N_{tu}(1-R)}} \tag{16.75}$$

Note that for a counter-flow heat exchanger when $C_H = C_C$, i.e. $R = 1$ (say for a gas turbine heat exchanger), then the expression for effectiveness cannot be obtained by substituting $R = 1$ in equation (16.75). For this case the temperature change of each fluid is the same, since $C_H = C_C$, and hence $\Delta \bar{t}_{ln}$ is equal to the temperature difference between the hot and cold fluids which remains constant throughout the heat exchanger. Equation (16.71) is therefore

625

written as, $N_{tu} = (t_{C1} - t_{C2})/(t_{H1} - t_{C1})$ and the derivation then proceeds as above, giving

$$\varepsilon = \frac{N_{tu}}{1 + N_{tu}} \qquad (16.76)$$

For a parallel-flow heat exchanger it can be shown that

$$\varepsilon = \frac{1 - e^{-N_{tu}(1 + R)}}{1 + R} \qquad (16.77)$$

When $R = 0$ in the case of a condenser say, then it can be seen from equation (16.75) or equation (16.77) that the effectiveness is

$$\varepsilon = 1 - e^{-N_{tu}} \qquad (16.78)$$

Example 16.18   A single-pass shell and tube counter-flow heat exchanger uses waste gas on the shell side to heat a liquid in the tubes. The waste gas enters at a temperature of $400\,°C$ at a mass flow rate of $40\,kg/s$; the liquid enters at $100\,°C$ at a mass flow rate of $3\,kg/s$.

Assuming that the velocity of the liquid is not to exceed $1\,m/s$, using the data below calculate:

(i) the required number of tubes;
(ii) the effectiveness of the heat exchanger;
(iii) the exit temperature of the liquid.

Neglect fouling factors and the thermal resistance of the tube wall.

Data   Tube inside diameter $= 10\,mm$; tube outside diameter $= 12.7\,mm$; tube length $= 4\,m$; specific heat capacity of waste gas $= 1.04\,kJ/kg\,K$; specific heat capacity of liquid $= 1.5\,kJ/kg\,K$; density of liquid $= 500\,kg/m^3$; heat transfer coefficient on the shell side $= 260\,W/m^2\,K$; heat transfer coefficient on the tube side $= 580\,W/m^2\,K$.

Solution   (i)   Volume flow rate of liquid $= \dfrac{\dot{m}}{\rho} = \dfrac{3}{500} = 0.006\,m^3/s$

therefore

Total cross-sectional area for a velocity of $1\,m/s = 0.006\,m^2$

i.e.   Number of tubes $= \dfrac{0.006 \times 4}{\pi \times 0.01^2} = 76.39$, say 77

Note: the velocity in the tubes is then less than $1\,m/s$ as required.
(ii) From equation (16.58), taking into account the area difference,

$$\frac{1}{UA_0} = \frac{1}{\alpha_0 A_0} + \frac{1}{\alpha_i A_i}$$

where subscripts 0 and i refer to the outside and inside of the tube.

The overall heat transfer coefficient, U, is referred to the outside area which is the usual practice in heat exchanger design,

$$\frac{1}{U} = \frac{1}{\alpha_0} + \frac{A_0}{\alpha_i A_i} = \frac{1}{260} + \frac{12.7}{580 \times 10} = 0.006\,04 \text{ m}^2 \text{ K/W}$$

since $A_o/A_i = D_o/D_i$

i.e. $\qquad U = 165.68 \text{ W/m}^2 \text{ K}$

Then from equation (16.71)

$$N_{tu} = \frac{165.68 \times \pi \times 0.0127 \times 4 \times 77}{3 \times 1.5 \times 1000} = 0.452$$

Also $\qquad R = \dfrac{3 \times 1.5}{40 \times 1.04} = 0.1082$

Then from equation (16.75)

$$\varepsilon = \frac{1 - e^{-N_{tu}(1-R)}}{1 - R e^{-N_{tu}(1-R)}} = \frac{1 - e^{-0.452 \times 0.8918}}{1 - 0.1082\,e^{-0.452 \times 0.8918}}$$

i.e. $\qquad \varepsilon = 0.358$

(iii) From equation (16.74),

$$\varepsilon = 0.358 = \frac{t_{Le} - 100}{400 - 100}$$

where t_{Le} is the exit temperature of the liquid.

i.e. $\qquad t_{Le} = (300 \times 0.358) + 100 = 207.4 \,°\text{C}$

16.13 Extended surfaces

From equation (16.67) it can be seen that for a given heat transfer coefficient and given fluid temperatures the heat transfer can be increased by increasing the heat transfer area. One way of doing this is to increase the area on one side of the heat exchanger by adding fins or studs which project into the fluid; the effective heat transfer area is thus increased.

The thermal resistance on either side of a heat exchanger is $1/\alpha A$, therefore when α is very large the resistance to heat transfer is low and hence there is no advantage in increasing the area. One of the fluids usually has a much lower value of α than the other and hence the resistance on this side of the heat exchanger controls the heat transfer. It is therefore on this side that the area can be extended with advantage. (The resistance of the separating wall is usually small compared with the resistance of the fluid film on either side.)

Heat transfer coefficents are very high for condensing and boiling fluids; they are generally higher for liquids than for gases, and generally higher for forced convection than for natural convection. A typical application for an extended surface would therefore be natural convection to air.

A simplified analysis now follows for an extended one-dimensional surface in the steady state; for a fuller treatment the reader is referred to references 16.1 and 16.3.

Consider an extended surface which has a constant cross-sectional area, A, which is small compared to its length, L, so that heat transfer along it is one-dimensional (see Fig. 16.28).

Fig. 16.28 Extended surface of small cross-section in a fluid of constant temperature

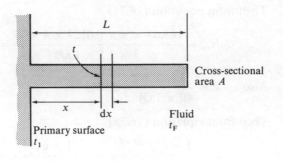

The rate of heat transfer at any section, distance x from the primary surface, where the temperature is t, is given by

$$\dot{Q} = -\lambda A \frac{dt}{dx}$$

and at a section $(x + dx)$ from the primary surface is

$$\dot{Q}' = \dot{Q} + \frac{\partial \dot{Q}}{\partial x} dx = \dot{Q} - \lambda A \frac{d^2 t}{dx^2} dx \tag{16.79}$$

For a perimeter, P, and a heat transfer coefficient, α, assumed constant with temperature and uniform over the surface, then

$$\text{Heat loss from surface} = \alpha P \, dx(t - t_F)$$

where t_F is the temperature of the surrounding fluid, assumed uniform and constant. Therefore, for the steady state

$$\dot{Q} - \dot{Q}' = \alpha P \, dx(t - t_F)$$

Substituting from equation (16.79),

$$\lambda A \frac{d^2 t}{dx^2} = \alpha P(t - t_F)$$

or $\qquad \dfrac{d^2 t}{dx^2} - \dfrac{\alpha P}{\lambda A}(t - t_F) = 0 \tag{16.80}$

The solution is found by putting $(t - t_F) = e^{mx}$,

i.e. $\qquad \dfrac{d^2(t - t_F)}{dx^2} = m^2 e^{mx} \quad$ therefore $\quad m^2 e^{mx} - \dfrac{\alpha P}{\lambda A} e^{mx} = 0$

therefore

$$m = \pm \left(\frac{\alpha P}{\lambda A}\right)^{1/2}$$

The solution is therefore

$$t - t_F = C_1 e^{mx} + C_2 e^{-mx} \qquad (16.81)$$

where C_1 and C_2 are constants determined from the boundary conditions:

(i) at $x = 0$, $t - t_F = t_1 - t_F$; therefore in equation (16.81),

$$t_1 - t_F = C_1 + C_2 \qquad (16.82)$$

(ii) at the end of the extended surface the heat convected from the end is equal to the heat conducted at the section $x = L$.

Assuming the same heat transfer coefficient, α, for the end then

$$-\lambda A \left(\frac{d(t - t_F)}{dx}\right)_{x=L} = \alpha A(t - t_F)_{x=L} \qquad (16.83)$$

From equation (16.81)

$$\left(\frac{d(t - t_F)}{dx}\right)_{x=L} = m C_1 e^{mL} - m C_2 e^{-mL}$$

and $\quad (t - t_F)_{x=L} = C_1 e^{mL} + C_2 e^{-mL}$

Substituting in equation (16.83)

$$-\lambda m C_1 e^{mL} + \lambda m C_2 e^{-mL} = \alpha C_1 e^{mL} + \alpha C_2 e^{-mL} \qquad (16.84)$$

From equations (16.82) and (16.84) the values of C_1 and C_2 can be found, and hence the solution to equation (16.80) is

$$\frac{t - t_F}{t_1 - t_F} = \frac{(e^{m(L-x)} + e^{-m(L-x)}) + \dfrac{\alpha}{m\lambda}(e^{m(L-x)} - e^{-m(L-x)})}{(e^{mL} + e^{-mL}) + \dfrac{\alpha}{m\lambda}(e^{mL} - e^{-mL})}$$

or $\quad \dfrac{t - t_F}{t_1 - t_F} = \dfrac{\cosh m(L - x) + \dfrac{\alpha}{m\lambda} \sinh m(L - x)}{\cosh mL + \dfrac{\alpha}{m\lambda} \sinh mL} \qquad (16.85)$

Note: $\cosh mL = (e^{mL} + e^{-mL})/2$ and $\sinh mL = (e^{mL} - e^{-mL})/2$.

At the end of the extended surface, at $x = L$

$$\frac{t_2 - t_F}{t_1 - t_F} = \frac{1}{\cosh mL + (\alpha/m\lambda) \sinh mL} \qquad (16.86)$$

The heat transfer from the extended surface, which is the same as the heat leaving the primary surface to the fin, is given by

$$\dot{Q}_1 = -\lambda A \left(\frac{\mathrm{d}(t - t_F)}{\mathrm{d}x} \right)_{x=0}$$

By differentiating equation (16.85) and putting $x = 0$, it can be shown that

$$\dot{Q}_1 = \alpha A(t_1 - t_F) \frac{\left(1 + \dfrac{\tanh mL}{\alpha/m\lambda}\right)}{\{1 + (\alpha/m\lambda)\tanh mL\}} \tag{16.87}$$

From equation (16.87) it can be seen that when $\alpha/m\lambda = 1$ then, $\dot{Q}_1 = \alpha A(t_1 - t_F)$, which is the heat loss from the primary surface with no extended surface, i.e. when $\alpha = m\lambda$ an extended surface of whatever length, L, will not increase the heat transfer from the primary surface.

For $\alpha/m\lambda > 1$ then $\dot{Q}_1 < \alpha A(t_1 - t_F)$ and hence adding a secondary surface reduces the heat transfer; the surface added acts as an insulation. For $\alpha/m\lambda < 1$ then $\dot{Q}_1 > \alpha A(t_1 - t_F)$ and the extended surface will increase the heat transfer. This is illustrated in Fig. 16.29.

Fig. 16.29 Heat transfer against length for various values of $\alpha/m\lambda$

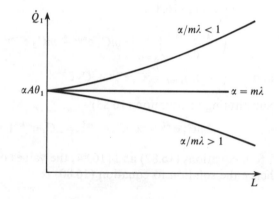

Note $\quad \dfrac{\alpha}{m\lambda} = \left(\dfrac{\alpha^2}{\lambda^2} \dfrac{\lambda A}{\alpha P} \right)^{1/2} = \left(\dfrac{\alpha A}{\lambda P} \right)^{1/2}$

Hence assuming $\alpha/m\lambda < 1$ the heat transfer becomes more effective when α/λ is low for a given geometry.

Approximate end condition

A simplified approach is possible by making the approximation that the loss of heat from the end of the extended surface is negligible, i.e. at $x = L$

$$-\lambda A \left(\frac{\mathrm{d}(t - t_F)}{\mathrm{d}x} \right)_{x=L} = 0$$

Therefore in equation (16.81)

$$-\lambda A m (C_1 \mathrm{e}^{mL} - C_2 \mathrm{e}^{-mL}) = 0 \tag{16.88}$$

Solving for C_1 and C_2 from equations (16.88) and (16.82) we have

$$\frac{t - t_F}{t_1 - t_F} = \frac{\cosh m(L - x)}{\cosh mL} \tag{16.89}$$

Then at $x = L$,

$$\frac{t_2 - t_F}{t_1 - t_F} = \frac{1}{\cosh mL} \tag{16.90}$$

The heat transfer from the extended surface is then

$$\dot{Q}_1 = m\lambda A(t_1 - t_F) \tanh mL \tag{16.91}$$

In most cases in practice the approximate expressions given by equations (16.89)–(16.91) can be used instead of the more accurate expressions given by equations (16.85)–(16.87).

In the important case of compact plate-fin heat exchangers where corrugated plates are sandwiched between flat plates (see for example Fig. 16.23, p. 621), then the expressions given in equations (16.89)–(16.91) are the accurate ones. In this case the fin bridges the two hot surfaces which may be assumed to be at the same temperature; at the mid-point of the fin the change of temperature with fin length is zero which corresponds to the condition of zero heat transfer. In the equations (16.89)–(16.91) the half-width, $w/2$, is substituted for the length, L.

Rectangular section fins

For a fin of rectangular cross-section on a plane surface as shown in Fig. 16.30, the perimeter, P, is given by $(2 + 2b)$ per unit length in the z-direction, and the cross-sectional area, A, is given by b per unit length in the z-direction,

i.e. $\qquad m = \left(\frac{\alpha P}{\lambda A}\right)^{1/2} = \left(\frac{\alpha(2 + 2b)}{\lambda b}\right)^{1/2} = \left[\frac{2\alpha}{\lambda b}(1 + b)\right]^{1/2} \simeq \left(\frac{2\alpha}{\lambda b}\right)^{1/2}$ (16.92)

Fig. 16.30 Rectangular cross-section fin on a plane surface

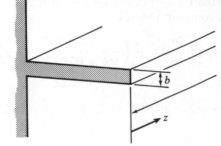

Also $\qquad \dfrac{\alpha}{m\lambda} = \left(\dfrac{\alpha^2}{\lambda^2}\dfrac{\lambda b}{2\alpha}\right)^{1/2} = \left(\dfrac{\alpha b/2}{\lambda}\right)^{1/2} = (Bi)^{1/2}$

where the Biot number, Bi, is based on the half-thickness of the fin, $b/2$.

631

Fin efficiency, η_F, is defined as the ratio of the heat loss from the fin surface to the heat loss from the fin surface if it were everywhere at the temperature of the primary surface.

Using the approximate expression, equation (16.91), for $\dot{Q}_1$, we have

$$\eta_F = \frac{m\lambda(b \times 1)(t_1 - t_F)\tanh mL}{\alpha(2L)(t_1 - t_F)} = \frac{m\lambda b \tanh mL}{2\alpha L} = \frac{\tanh mL}{mL} \quad (16.93)$$

For a finned surface with unfinned area, A_b, and total fin surface area, A_F, then for unit length in the z-direction

$$\text{Heat loss} = h(t_1 - t_F)(A_b + \eta_F A_F) \quad (16.94)$$

Example 16.19

A typical plate-fin cross-flow heat exchanger is shown in Fig. 16.23 (p. 621). For a point in one of the hot fluid channels where the fluid mean temperature within the channel is 200 °C and the separating plates on either side are at 100 °C, calculate:

(i) the mean temperature of a fin at that point in the heat exchanger;
(ii) the fin efficiency.

Data

Height of flow channel = 11.78 mm; thickness of fin = 0.203 mm; heat transfer coefficient between the hot fluid and all surfaces = 137 W/m² K; thermal conductivity of fin material = 168 W/m K.

Solution

From equation (16.89)

$$\frac{t - t_F}{t_1 - t_F} = \frac{\cosh m(w/2 - x)}{\cosh(mw/2)}$$

Then

$$\frac{t_m - t_F}{t_1 - t_F} = \frac{\int_0^{w/2}(t - t_F)\,\mathrm{d}x}{(t_1 - t_F)w/2} = \frac{[-\sinh m(w/2 - x)]_0^{w/2}}{(mw/2)\cosh(mw/2)}$$

$$= \frac{\sinh(mw/2)}{(mw/2)\cosh(mw/2)} = \frac{\tanh(mw/2)}{(mw/2)}$$

It can be seen by reference to equation (16.93) that the ratio, $(t_m - t_F)/(t_1 - t_F)$ is an alternative expression for fin efficiency.

(i) From equation (16.92),

$$m = \left(\frac{2\alpha}{\lambda b}\right)^{1/2} = \left(\frac{2 \times 137}{168 \times 0.203 \times 10^{-3}}\right)^{1/2} = 89.63$$

therefore

$$mw/2 = 89.63 \times 11.78/2 \times 10^3 = 0.5279$$

therefore

$$\frac{t_m - t_F}{t_1 - t_F} = \frac{\tanh 0.5279}{0.5279} = 0.916$$

$$t_m = 200 - 0.916 \times (200 - 100) = 108.4\,°C$$

i.e.   Mean temperature of fin = 108.4 °C

(ii)   Fin efficiency $= (t_m - t_F)/(t_1 - t_F) = (200 - 108.4)/(200 - 100)$

$$= 0.916 \quad \text{or} \quad 91.6\%$$

16.14 Black-body radiation

Thermal radiation consists of electromagnetic waves emitted due to the agitation of the molecules of a substance. The waves are similar to light waves in that they are propagated in straight lines at the speed of light and they require no medium for propagation. Radiation striking a body can be absorbed by the body, reflected from the body, or transmitted through the body. The fractions of the radiation absorbed, reflected, and transmitted are called the absorptivity, α, the reflectivity, ρ, and the transmissivity, τ, respectively. Then we have

$$\alpha + \rho + \tau = 1$$

For most solids and liquids encountered in engineering the amount of radiation transmitted through the substance is negligible, and it is possible to write

$$\alpha + \rho = 1 \tag{16.95}$$

It is useful to define an ideal body which absorbs all the radiation which falls upon it; such a body is called a *black body*. For a black body, $\alpha = 1$ and $\rho = 0$. It should be noted that the term 'black' in this context does not necessarily imply black to the eye. A surface which is black to the eye is one which absorbs all the light incident upon it, but a surface can absorb all the thermal radiation incident upon it without necessarily absorbing all the light (e.g. snow is almost 'black' to thermal radiation, $\alpha = 0.985$). Although no totally black body exists in practice, many surfaces approximate to the definition. For example, consider a small object radiating energy in a large space, as shown in Fig. 16.31. The energy striking the surface surrounding the body is reflected and absorbed many times by the surface, and the fraction of energy reflected back and intercepted by the body is exceedingly small. Therefore, when a body is placed in large surroundings the surroundings are approximately black to thermal radiation. As a better example of a black body, consider a small hole in the surface of a wall, as shown in Fig. 16.32. The hole leads into a small chamber as shown. Rays of thermal radiation entering the hole are successively absorbed by the walls of the chamber such that only a negligible amount of radiation is emitted from the hole. Thus the hole acts as a black body. This is the closest approximation to a black body which can be devised in practice; the inside surfaces of the chamber can be made of a material with a high absorptivity (e.g. lampblack).

The energy which is radiated from a body per unit area per unit time is called the *emissive power*, $\dot{E}$. It can be shown that a black body, as well as being the best possible absorber of radiation, is also the best possible emitter. Consider an enclosure at a uniform temperature, and let a black body be placed in the enclosure as shown in Fig. 16.33. If the body is at the same temperature as the enclosure then it follows that all the energy radiated by the body and absorbed

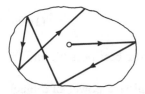

Fig. 16.31 Radiation from a small body in large surroundings

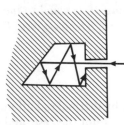

Fig. 16.32 Radiation entering a hole leading to a chamber

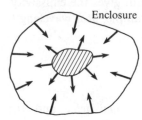

Fig. 16.33 Radiation from a body to a surrounding enclosure

by the walls of the enclosure, must exactly equal the energy radiated by the enclosure and absorbed by the body. If this were not so then the body would gain or lose energy, and this is not possible in an isolated system, by the laws of thermodynamics. Let the emissive power of the black body be $\dot{E}_B$. Therefore the rate at which energy impinges on unit surface of the black body is also $\dot{E}_B$. Now replace the black body by any other body at the same temperature, and of the same shape and size. This body must receive exactly the same amount of energy from the enclosure as the black body received when it was in the same position in the enclosure. However, this body is not black and hence will only absorb a fraction of the energy it receives,

i.e.   Rate of energy absprotion $= \alpha\dot{E}_B$

where α is the absorptivity of the body.

Now as before the energy absorbed must be equal to the energy emitted, therefore if the body has an emissive power of E, we have

$$\dot{E} = \alpha\dot{E}_B$$

or $$\alpha = \frac{\dot{E}}{\dot{E}_B} \qquad\qquad (16.96)$$

Since $\alpha < 1$ then $\dot{E} < \dot{E}_B$, and hence the black body is the best possible emitter of radiation.

The ratio of the emissive power of a body to the emissive power of a black body is called the *emissivity*, ε. From equation (16.96) it can be seen that when two bodies are at the same temperature, then the absorptivity, α, equals the emissivity, ε. This is known as *Kirchhoff's law*, which may be stated as follows:

The emissivity of a body radiating energy at a temperature, T, is equal to the absorptivity of the body when receiving energy from a source at a temperature, T.

16.15   The grey body

In section 16.14 it has been assumed that the energy emitted by thermal radiation is the same for all wavelengths of the radiation. In fact this is not the case, and Fig. 16.34 shows the emissive power per unit wavelength plotted against wavelength, λ, in micrometres for a black body at any one temperature. A corresponding curve at the same temperature is shown for a non-black body. The ratio of an ordinate of each curve at any wavelength gives the emissivity, and hence the absorptivity, at that wavelength. For example, at a wavelength of 4.5 micrometres (μm), we have

$$\varepsilon_\lambda = \alpha_\lambda = \frac{AB}{AC}$$

The terms ε_λ and α_λ are called the *monochromatic emissivity* and the *monochromatic absorptivity* respectively. It can be seen from Fig. 16.34 that the monochromatic emissivity varies with wavelength. The variation is greater for some materials than for others, and there are certain materials for which the

Fig. 16.34 Emissive power against wavelength for a black body and a non-black body

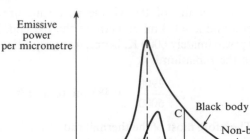
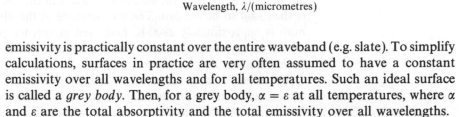

emissivity is practically constant over the entire waveband (e.g. slate). To simplify calculations, surfaces in practice are very often assumed to have a constant emissivity over all wavelengths and for all temperatures. Such an ideal surface is called a *grey body*. Then, for a grey body, $\alpha = \varepsilon$ at all temperatures, where α and ε are the total absorptivity and the total emissivity over all wavelengths.

It is an experimental fact that the emissive power of a body increases as the temperature of the body is increased. This is illustrated in Fig. 16.35 in which the emissive power of a black body per unit wavelength is plotted against the wavelength in micrometres, for several temperatures. It can be seen that the wavelength which gives maximum emissive power becomes smaller as the temperature is increased, and hence more and more of the energy emitted is radiated over the shorter wavelengths as the temperature increases. The value of the wavelength for maximum emissive power is given by Wien's law

$$\lambda_{max} = \frac{2900}{T} \tag{16.97}$$

where λ_{max} is in micrometres and T is in K.

Fig. 16.35 Emissive power against wavelength for a black body at various surface temperatures

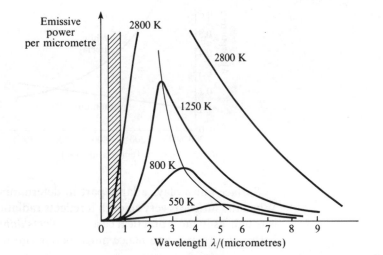

The limits of the visible spectrum are $\lambda = 0.4$ micrometre at the blue end and $\lambda = 0.8$ micrometre at the red end. Now the sun has a temperature of approximately 6000 K, hence, using equation (16.97), the maximum wavelength of the radiation is

$$\lambda_{\max} = \frac{2900}{6000} = 0.483 \text{ micrometres}$$

Therefore most of the thermal radiation from the sun is in the visible wavebend. The waveband for light is shown shaded in Fig. 16.35. At a temperature of 800 K a very small amount of the energy emitted is just within the red end of the visible spectrum. A surface at 800 K will appear as a dull red colour. At about 1250 K more of the energy emitted is in the visible range and the surface is then said to be red-hot. The temperature of the filament of an electric light bulb is approximately 2800 K, and even at this temperature only about 10% of the energy emitted is in the visible region, which shows the inefficiency of such a bulb as a light source.

For a grey body a set of curves exactly similar to those of Fig. 16.35 can be drawn, with each ordinate only a fraction, ε, of the corresponding ordinate of the curves of Fig. 16.35. In practice, although a suitable total value of the absorptivity may be taken for a large number of industrial surfaces over a wide range of wavelengths, nevertheless there is still a variation of total absorptivity with temperature. This is illustrated in Fig. 16.36. When the temperature range is small the approximation that $\alpha = \varepsilon = $ constant, for a grey body, is still sufficiently accurate for most calculations. Materials or surfaces for which the emissivity varies considerably and irregularly with wavelengths and temperature are called *selective emitters*. Some values of total emissivity over all wavelengths but for different temperatures are shown in Table 16.4.

Fig. 16.36 Absorptivity against surface temperature for various materials

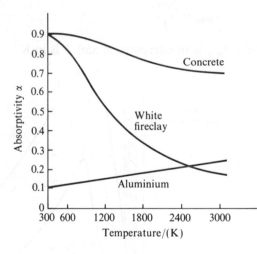

Surface finish plays a large part in determining the emissivity of a material. When the surface is very smooth it reflects radiation *specularly*; when the surface is rough, as in most practical cases, it reflects *diffusely*. Rough surfaces are much better absorbers – and hence much better emitters – of radiation than smooth

Table 16.4 Emissivities of some surfaces at various temperatures

	Emissivity			
Surface	0–40 °C	120 °C	260 °C	540 °C
White paint	0.95	0.94	0.88	0.70
Black glossy paint	0.95	0.94	0.90	0.85
Lampblack	0.97	0.97	0.97	0.97
Building brick	0.93	0.93	0.79	0.74
Concrete	0.85	0.84	0.69	0.69
Polished steel	0.07	0.09	0.11	0.14

surfaces. For mild steel, rough turned, ε at 15 °C is 0.87; for mild steel, well finished on a lathe, ε at 15 °C is 0.39; and it can be seen from Table 16.4 that when the steel is polished the emissivity, ε, is reduced to 0.07.

16.16 The Stefan–Boltzmann law

It was found experimentally by Stefan, and proved theoretically by Boltzmann, that the emissive power of a black body is directly proportional to the fourth power of its absolute temperature, and this is known as the Stefan–Boltzmann law,

i.e.
$$\dot{E}_B = \sigma T^4 \tag{16.98}$$

the value of σ is 5.67×10^{-8} W/m^2 (K)^4.

The rate of energy emitted by a non-black body is then given by

$$\dot{E} = \varepsilon \sigma T^4 \tag{16.99}$$

where ε is the emissivity of the body.

Consider a body 1 of emissivity ε_1 at a temperature T_1, completely surrounded by black surroundings at a lower temperature T_2. The energy leaving body 1 is completely absorbed by the surroundings, and from equation (16.99)

$$\text{Rate of energy emission} = \varepsilon_1 \sigma T_1^4$$

The rate of energy emitted by the black surroundings is given by equation (16.98)

$$\dot{E}_B = \sigma T_2^4$$

Now the fraction of this energy which is absorbed by body 1 depends on the absorptivity of body 1. For a grey body $\alpha = \varepsilon$ at all temperatures and hence

$$\text{Rate of energy absorption} = \varepsilon \sigma T_2^4 = \varepsilon \alpha T_2^4$$

Then the rate of heat transferred from the body to its surroundings per square metre of the body is

$$\dot{q} = \varepsilon \sigma T_1^4 - \varepsilon \sigma T_2^4$$

i.e.
$$\dot{q} = \varepsilon \sigma (T_1^4 - T_2^4) \tag{16.100}$$

If the emissivity of the body at T_1 is largely different from the emissivity of the body at T_2 then the approximation of the grey body may not be sufficiently accurate. In that case it is a good approximation to take the absorptivity of the body 1 when receiving radiation from a source at T_2 as being equal to the emissivity of body 1 when emitting radiation at T_2.

Then, $\quad \dot{q} = \varepsilon_{T_1} \sigma T_1^4 - \varepsilon_{T_2} \sigma T_2^4$ $\hspace{3cm}$ (16.101)

The absorptivity, while depending mainly on the temperature of the source of radiation, also depends on the temperature of the surface itself. For most metals this factor can be important and it has been shown that the absorptivity of a metal surface at T_1 for radiation from a source at T_2 is approximately equal to the emissivity of the surface when at a temperature, T_3. given by

$$T_3 = \sqrt{(T_1 T_2)} \hspace{3cm} (16.102)$$

Example 16.20

A body at $1100\,°C$ in black surroundings at $550\,°C$ has an emissivity of 0.4 at $1100\,°C$ and an emissivity of 0.7 at $550\,°C$. Calculate the rate of heat loss by radiation per unit surface area:

(i) when the body is assumed to be grey with $\varepsilon = 0.4$;
(ii) when the body is not grey.

Assume that the absorptivity is independent of the surface temperature.

Solution

(i) Using equation (16.100),

$$\dot{q} = \varepsilon\sigma(T_1^4 - T_2^4) = 0.4 \times \frac{5.67}{10^8} \times (1373^4 - 823)^4$$

where $T_1 = 1100 + 273 = 1373$ K, and $T_2 = 550 + 273 = 823$ K

$$\dot{q} = 0.4 \times 5.67 \times (13.73^4 - 8.23^4) = 70\,193 \text{ W/m}^2$$

i.e. $\quad$ Rate of heat loss per square metre by radiation $= 70.19$ kW

(ii) When the body is not grey then the absorptivity when the source is at $550\,°C$ is equal to the emissivity when the body is at $550\,°C$,

i.e. $\quad \alpha = 0.7$

Then $\quad$ Rate of energy emission $= \varepsilon\sigma T_1^4 = 0.4 \times \dfrac{5.67}{10^8} \times 1373^4$

and $\quad$ Rate of energy absorption $= \alpha\sigma T_2^4 = 0.7 \times \dfrac{5.67}{10^8} \times 823^4$

i.e. $\quad \dot{q} = 0.4 \times 5.67 \times 13.73^4 - 0.7 \times 5.67 \times 8.23^4$
$$= 62\,389 \text{ W/m}^2$$

i.e. $\quad$ Rate of heat loss per square metre by radiation $= 62.39$ kW

It can be seen that the grey body assumption of part (i) overestimates by

$$\left(\frac{70.19 - 62.39}{62.39}\right) \times 100 = 12.5\%$$

Example 16.21   Calculate the rate of heat loss by radiation from unit surface area of a body at 1100 °C in black surroundings at 40 °C, when the emissivity at 40 °C is 0.9, and the emissivity at 1100 °C is as in Example 16.20:

(i) when the body is grey with $\varepsilon = 0.4$;
(ii) when the body is not grey.

Assume that the absorptivity is independent of the surface temperature.

Solution   (i) As in Example 16.20,

$$\dot{q} = \varepsilon\sigma(T_1^4 - T_2^4) = 0.4 \times 5.67 \times (13.73^4 - 3.13^4)$$
$$= 80\,380 \text{ W}$$

i.e.   Heat loss per square metre by radiation = 80.38 kW

(ii) As in Example 16.20,

Rate of energy emission = $0.4 \times 5.67 \times 13.73^4 = 80\,598$ W/m^2

and   Rate of energy absorption = $0.9 \times 5.67 \times 3.13^4 = 489.8$ W/m^2

i.e. $\dot{q} = (80\,598 - 490) = 80\,108$ W

i.e.   Rate of heat loss per square metre by radiation = 80.11 kW

Therefore the grey body assumption overestimates by

$$\left(\frac{80.38 - 80.11}{80.11}\right) = 0.337\%$$

It can be seen from Examples 16.20 and 16.21 that the grey body assumption gives a very accurate approximation when one of the temperatures is small compared with the other. The assumption also gives a very accurate approximation when both temperatures are small.

16.17   Lambert's law and the geometric factor

Most surfaces do not emit radiation strongly in all directions; the greater part of the energy emitted is in a direction normal to the surface. Before considering the interchange of energy between two bodies which receive only a part of the radiation emitted by each other, it is necessary to find out how the radiation is distributed in the various directions from the two surfaces. In order to do this the *intensity of radiation, i*, must be defined. The rate of energy emission from unit surface area through unit solid angle, along a normal to the surface, is called the *intensity of normal radiation, i_N*. The intensity of radiation in any other direction at any angle ϕ to the normal is denoted by i_ϕ. (Note: a surface subtends a solid angle at a point distance r from all points on the surface, equal to the surface area divided by r^2. The surface of a sphere is $4\pi r^2$ and hence the solid angle subtended by the surface of the sphere at its centre is 4π.)

The variation in the intensity of radiation is given by *Lambert's cosine law*

$$i_\phi = i_N \cos\phi \tag{16.103}$$

The rate of energy emission from a surface of area dA is then given by

$$\int i_\phi \, \mathrm{d}w \, \mathrm{d}A$$

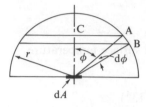

Fig. 16.37 Radiation from a small element to a hemisphere

where dw is a small solid angle.

Consider a small area dA, and consider the radiation from dA which passes through a small element of the surface area of a hemisphere with dA at its centre, as shown in Fig. 16.37. The element subtends an angle ϕ at the centre of the hemisphere and the small increase in angle over the width of the element is then dϕ. The width of the element is the length of the arc, of angle dϕ, and radius r (i.e. AB in Fig. 16.37). Therefore

Width of element, AB = $r \, \mathrm{d}\phi$

The radius of the element is CA = $r \sin \phi$. Hence the surface area of the element is given by

Surface area = (width × circumference) = $r \, \mathrm{d}\phi \times 2\pi r \sin \phi$

i.e.   Solid angle, dw, subtended at d$A = \dfrac{2\pi r^2 \sin \phi \, \mathrm{d}\phi}{r^2}$

i.e. $\mathrm{d}w = 2\pi \sin \phi \, \mathrm{d}\phi$

Hence the rate of total energy emission from dA is given by

$$\dot{E} \, \mathrm{d}A = \int_0^{\pi/2} i_\phi \, \mathrm{d}w \, \mathrm{d}A = \int_0^{\pi/2} \mathrm{d}A \, i_\phi 2\pi \sin \phi \, \mathrm{d}\phi$$

Substituting from equation (16.103), $i_\phi = i_N \cos \phi$, then

$$\dot{E} \, \mathrm{d}A = 2\pi \, \mathrm{d}A \, i_N \int_0^{\pi/2} \cos \phi \sin \phi \, \mathrm{d}\phi$$

or $$\dot{E} \, \mathrm{d}A = 2\pi \, \mathrm{d}A \, i_N \int_0^{\pi/2} \frac{\sin 2\phi}{2} \, \mathrm{d}\phi = \pi i_N \, \mathrm{d}A$$

Now from equation (16.99)

$$\dot{E} = \varepsilon \sigma T^4$$

Therefore

$$\varepsilon \sigma T^4 \, \mathrm{d}A = \pi i_N \, \mathrm{d}A$$

i.e. $$i_N = \frac{\varepsilon \sigma T^4}{\pi} \tag{16.104}$$

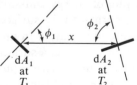

Fig. 16.38 Radiation interchange between two small surfaces in large surroundings

Consider two small black surfaces of area dA_1 and dA_2 at temperatures T_1 and T_2, and distance x apart. The angles of inclination of surfaces are as shown in Fig. 16.38. This is a case where neither body receives all the radiation from the other. Let the surface dA_2 subtend a solid angle dw_1 at the centre of the surface dA_1. Then we have

Rate of energy emission from dA_1 incident on d$A_2 = i_{N_1} \cos \phi_1 \, \mathrm{d}w_1 \, \mathrm{d}A_1$

From equation (16.104), $i_N = \sigma T^4 / \pi$, for a black surface, therefore,

$$\text{Rate of energy incident on } dA_2 = \frac{\cos \phi_1 \, dw_1 \, dA_1 \sigma T_1^4}{\pi}$$

Also, from the definition of solid angle

$$dw_1 = \frac{dA_2 \cos \phi_2}{x^2}$$

Hence   Rate of energy incident on $dA_2 = \dfrac{\cos \phi_1 \cos \phi_2 \, dA_1 \, dA_2 \sigma T_1^4}{\pi x^2}$

Now the rate of total energy emission from dA_1 is $\sigma \, dA_1 \, T_1^4$. The ratio of the energy incident on the second body to the energy emitted by the first is called the *geometric factor*, F_{1-2},

i.e. $F_{1-2} = \dfrac{\cos \phi_1 \cos \phi_2 \, dA_1 \, dA_2 \sigma T_1^4}{\pi x^2 \sigma \, dA_1 \, T_1^4}$

therefore

$$F_{1-2} = \frac{\cos \phi_1 \cos \phi_2 \, dA_2}{\pi x^2} \tag{16.105}$$

In the same way it can be shown that the geometric factor for radiation from surface 2 to surface 1 is given by

$$F_{2-1} = \frac{\cos \phi_1 \cos \phi_2 \, dA_1}{\pi x^2} \tag{16.106}$$

The net rate of energy interchange between the surfaces is given by

$$d\dot{Q}_{1-2} = \frac{\sigma \cos \phi_1 \cos \phi_2 \, dA_1 \, dA_2}{\pi x^2} (T_1^4 - T_2^4)$$

This can be written as

$$d\dot{Q}_{1-2} = F_{1-2} \, dA_1 \sigma (T_1^4 - T_2^4)$$

or $d\dot{Q}_{1-2} = F_{2-1} \, dA_2 \sigma (T_1^4 - T_2^4)$

The geometric factors F_{1-2} and F_{2-1} can be found by a double integration of equations (16.105) and (16.106); this can be done analytically or graphically. For a larger area made up of small areas dA_1 and dA_2, average geometric factors can be defined in the same way as above,

i.e. $\dot{Q}_{1-2} = F_{1-2} A_1 \sigma (T_1^4 - T_2^4)$ $\tag{16.107}$

and $\dot{Q}_{1-2} = F_{2-1} A_2 \sigma (T_1^4 - T_2^4)$ $\tag{16.108}$

From equations (16.107) and (16.108) it can be seen that

$$A_1 F_{1-2} = A_2 F_{2-1} \tag{16.109}$$

This is known as the reciprocal relationship or theorem of reciprocity.

In practice calculating F can be a long and difficult process except for simple shapes; charts are available for some of the more common configurations (see, for example, references 16.1 and 16.10).

When a body, 1, is completely enclosed by other surfaces then

$$F_{1-\text{surfaces}} = 1$$

If the surfaces have separate elements, 2, 3, etc. it follows that

$$F_{1-\text{surfaces}} = F_{1-1} + F_{1-2} + F_{1-3} + \text{etc.} = 1 \qquad (16.110)$$

The term F_{1-1} is necessary in cases where the body 1 can 'see' parts of itself, e.g. a concave body.

In Table 16.5 values of geometric factor, F_{1-2}, are given for some common configurations; for more complex geometries see references 16.1 and 16.10.

Table 16.5 Some geometric factors

Configuration	Geometric factor, F_{1-2}
(a) Body 1 complete enclosed by body 2	1
(b) Parallel circular discs, radii r_1 and r_2, distance x apart on a common-axis	$\dfrac{(x^2 + r_1^2 + r_2^2) - \sqrt{\{(x_1^2 + r_1^2 + r_2^2)^2 - 4r_1^2 r_2^2\}}}{2r_1^2}$
(c) Small disc opposite a parallel circular plate of radius R at a perpendicular distance L	$\dfrac{R^2}{R^2 + L^2}$
(d) Small sphere opposite a circular plate of radius R at a perpendicular distance L	$\dfrac{1}{2}\left\{1 - \dfrac{L}{\sqrt{(L^2 + R^2)}}\right\}$
(e) Small sphere at the centre of the axis of a cylinder of radius R and length $2L$	$\dfrac{L}{\sqrt{(L^2 + R^2)}}$

Example 16.22

A hemispherical cavity of 0.6 m radius is covered by a plate with a hole of 0.2 m diameter drilled in its centre. The inner surface of the plate is maintained at 250 °C by a heater embedded in the surface. The surfaces may be assumed to be black and the hemisphere may be assumed to be well insulated. Calculate:

(i) the temperature of the surface of the hemisphere;
(ii) the power input to the heater.

State any other assumption made.

Solution   (i) Referring to Fig. 16.39, let the inner surface of the plate be 1, the hemisphere surface 2, and the hole projected surface 3, as shown. Then, since surface 1 is completely surrounded, we have

$$F_{1-2} + F_{1-3} = 1$$

or $F_{1-2} = 1$ since surface 1 cannot 'see' surface 3.

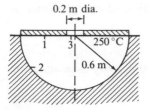

Fig. 16.39
Hemispherical cavity for Example 16.22

From equation (16.109)

$$F_{2-1} = \frac{A_1 F_{1-2}}{A_2} = \frac{\pi(0.6^2 - 0.1^2)}{2\pi \times 0.6^2} = \frac{35}{72}$$

Similarly

$$F_{3-2} = 1 \quad \text{and} \quad A_2 F_{2-3} = A_3 F_{3-2}$$

therefore

$$F_{2-3} = \frac{\pi \times 0.1^2 \times 1}{2\pi \times 0.6^2} = \frac{1}{72}$$

Then   Rate of energy emission from surface $2 = A_2 F_{2-3}\sigma T_2^4 + A_2 F_{2-1}\sigma T_2^4$

$$= A_2 \sigma T_2^4 \left(\frac{1}{72} + \frac{35}{72}\right)$$

$$= A_2 \sigma T_2^4 \times 0.5$$

The rate of energy incident on surface 2 may be taken as the rate of energy emission from surface 1, since the rate of energy entering the hole from outside will be negligible if the surroundings are large and at normal temperature,

i.e.   Rate of energy incident on surface $2 = A_1 F_{1-2}\sigma T_1^4 = A_1 \sigma T_1^4$

Then $A_1 \sigma T_1^4 = A_2 \sigma T_2^4 \times 0.5$ for the steady state

i.e. $$T_2^4 = \frac{T_1^4 \times \pi(0.6^2 - 0.1^2)}{2\pi \times 0.6^2 \times 0.5} = T_1^4 \times \frac{35}{36}$$

$$T_2 = (250 + 273) \times \left(\frac{35}{36}\right)^{1/4} = 519.3 \text{ K} = 246.3 \,°C$$

(ii) Rate of heat input from heater is

$$A_1 F_{1-2}\sigma(T_1^4 - T_2^4) = \pi \times (0.6^2 - 0.1^2) \times \frac{5.67}{10^8} \times 523^4 \left(1 - \frac{35}{36}\right)$$

$$= 129.6 \text{ W}$$

16.18 Radiant interchange between grey bodies

Radiosity, $\dot{J}$, is defined as the total radiant energy leaving a body per unit area per unit time.

Irradiation, $\dot{G}$, is defined as the total radiant energy incident on a body per unit area per unit time. Hence

$$\text{Net heat transfer from body, } \dot{Q} = (\dot{J} - \dot{G})A$$

where A is the area of the body surface. For a black body, from equation (16.98),

$$\dot{J} = \sigma T^4$$

For a grey body, the radiosity must include the fraction of energy which is reflected from the surface,

i.e. $\quad \dot{J} = \varepsilon\sigma T^4 + \rho\dot{G}$

Also, for a grey body, $\varepsilon = \alpha = 1 - \rho$, neglecting transmissivity (see equation (16.95)). Therefore

$$\dot{J} = \varepsilon\sigma T^4 + (1 - \varepsilon)\dot{G}$$

or $\quad \dot{G} = \dfrac{\dot{J} - \varepsilon\sigma T^4}{1 - \varepsilon}$

i.e. $\quad \dfrac{\dot{Q}}{A} = \dot{J} - \dot{G} = \dot{J} - \dfrac{(\dot{J} - \varepsilon\sigma T^4)}{1 - \varepsilon}$

or $\quad \dot{Q} = \dfrac{\varepsilon A}{1 - \varepsilon}(\sigma T^4 - \dot{J})$ $\hfill (16.111)$

For any two bodies 1 and 2, the geometric factor, F_{1-2}, is the fraction of radiation $A_1 \dot{J}_1$ which is intercepted by body 2,

i.e. $\quad \dot{Q}_{1-2} = A_1 F_{1-2}\dot{J}_1 - A_2 F_{2-1}\dot{J}_2$

Using equation (16.109)

$$\dot{Q}_{1-2} = A_1 F_{1-2}(\dot{J}_1 - \dot{J}_2) \qquad (16.112)$$

An electrical analogy can be used based on Ohm's law. For example, from equation (16.111)

$$\text{Resistance due to emissivity of surface} = \frac{1 - \varepsilon}{\varepsilon A} \qquad (16.113)$$

where $\dot{Q}$ is analogous to current and $(\sigma T^4 - J)$ is analogous to potential difference. Similarly, from equation (16.112),

$$\text{Resistance due to geometry} = \frac{1}{A_1 F_{1-2}} \qquad (16.114)$$

Take the simple case of a body 1, completely enclosed by a body 2. Figure 16.40 shows the electrical analogy.

$$\text{Total resistance, } R_T = \frac{1 - \varepsilon_1}{A_1\varepsilon_1} + \frac{1}{A_1 F_{1-2}} + \frac{1 - \varepsilon_2}{A_2\varepsilon_2}$$

Also in this case, $F_{1-2} = 1$, therefore

$$R_T = \frac{1}{A_1}\left(\frac{1}{\varepsilon_1} - 1 + 1 + \frac{A_1}{A_2}\left\{\frac{1}{\varepsilon_2} - 1\right\}\right) = \left(\frac{1}{\varepsilon_1} + \frac{A_1}{A_2}\left\{\frac{1}{\varepsilon_2} - 1\right\}\right)\frac{1}{A_1}$$

$$\dot{Q}_{1-2} = \frac{\sigma(T_1^4 - T_2^4)}{R_T} = \frac{A_1\sigma(T_1^4 - T_2^4)}{(1/\varepsilon_1) + (A_1/A_2)\{(1/\varepsilon_2) - 1\}} \qquad (16.115)$$

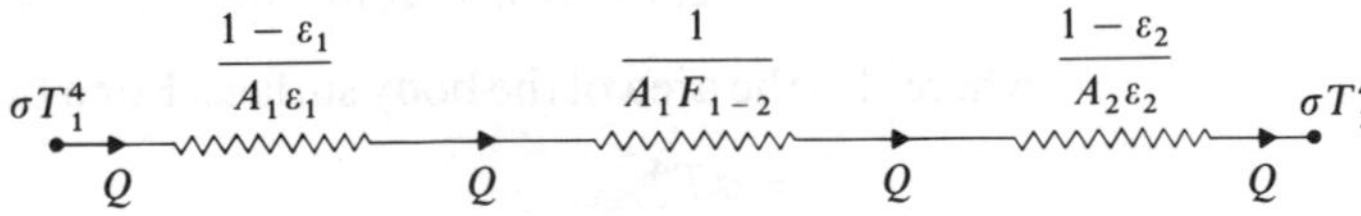

Fig. 16.40 Electrical analogy for radiation from body 1 enclosed by body 2

644

When the bodies are very close together then $A_1 \simeq A_2$,

i.e. $$\dot{Q}_{1-2} = \frac{A_1 \sigma (T_1^4 - T_2^4)}{(1/\varepsilon_1) + (1/\varepsilon_2) - 1}$$

The latter expression for the heat transfer also applies to the case of two large flat parallel surfaces where the size of the surfaces is large compared with their distance apart, i.e. the radiant energy escaping to the surroundings is negligible.

Example 16.23   It is desired to cut down the radiation loss between two parallel surfaces by inserting a sheet of aluminium foil midway between them. The temperatures of the two surfaces are maintained at 40 and 5 °C, and the emissivity of both surfaces is 0.85. The emissivity of aluminium foil is 0.05. Calculate the percentage reduction in heat loss rate by radiation using the aluminium foil, assuming that the surface temperatures are the same in both cases and that all surfaces are grey. Neglect end effects.

Solution   Without the aluminium foil, for two long parallel surfaces, for 1 m² of surface

$$R_T = (1/\varepsilon_1) + (1/\varepsilon_2) - 1 = (2/0.85) - 1$$

i.e. $R_T = 1.353$

Then from equation (16.115)

$$\dot{Q}_{1-2} = \frac{\sigma (T_1^4 - T_2^4)}{R_T} = \frac{5.67 \times (3.13^4 - 2.78^4)}{1.353}$$

where $T_1 = 40 + 273 = 313$ K and $T_2 = 5 + 273 = 278$ K

i.e. $\dot{Q}_{1-2} = 151.92$ W/m²

With the aluminium foil (see Fig. 16.41). Let the temperature of the foil be T K. Now from surface 1 to the foil, from equation (16.115),

$$\dot{Q}_{1-F} = \frac{\sigma (T_1^4 - T^4)}{R_{T_{1-F}}}$$

and from the foil to surface 2

$$\dot{Q}_{F-2} = \frac{\sigma (T^4 - T_2^4)}{R_{T_{F-2}}}$$

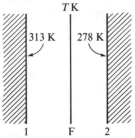

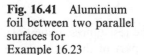

Fig. 16.41 Aluminium foil between two parallel surfaces for Example 16.23

T K

313 K   278 K

1   F   2

Since both sides of the foil act in a similar way, and since $\varepsilon_1 = \varepsilon_2$, then $R_{T_{1-F}} = R_{T_{F-2}}$. Therefore

$$(T_1^4 - T^4) = (T^4 - T_2^4) \quad \text{or} \quad T^4 = \frac{T_1^4 + T_2^4}{2}$$

$$= \frac{(3.13^4 + 2.78^4) \times 10^8}{2}$$

$$T^4 = 77.85 \times 10^8$$

Also $R_{T_{1-F}} = R_{T_{F-2}} = (1/0.85) + (1/0.05) - 1 = 20.176$

Then from equation (16.115)

$$\dot{Q}_{1-F} = \frac{\sigma(T_1^4 - T^4)}{R_T} = \frac{5.67 \times (3.13^4 - 77.85)}{20.176}$$

i.e. $\dot{Q}_{1-F} = 5.09 \, \text{W/m}^2$

Therefore

$$\text{Percentage reduction in heat loss} = \left(\frac{151.92 - 5.09}{151.92}\right) \times 100$$

$$= 96.7\%$$

It can be seen from the above example that a material of low emissivity can act as a very efficient radiation shield. This is used to advantage in many cases in practice (e.g. radiation shields for thermocouples and thermometers).

For the case of body 1 small compared with body 2 then $A_1/A_2 \to 0$, i.e. from equation (16.115)

$$\dot{Q}_{1-2} = \varepsilon_1 A_1 \sigma(T_1^4 - T_2^4)$$

Note that this equation applies even if body 2 is not black, the reason being that a negligible amount of the energy reflected from body 2 is intercepted by body 1 because it is small compared with body 2.

When more than two surfaces exchange heat then an equivalent electric circuit can be drawn using the expressions for resistance given by equations (16.113) and (16.114). For the case shown in Fig. 16.42 a body 1 exchanges heat with body 2, the surroundings 3 being at a different temperature.

Fig. 16.42 Three surfaces exchanging radiation

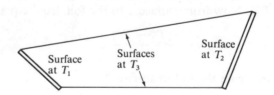

The equivalent circuit is shown in Fig. 16.43 with the resistances, potentials, and currents as shown. Applying Ohm's law to each part of the network we obtain six equations,

i.e. $\sigma T_1^4 - J_1 = I_1 \dfrac{(1 - \varepsilon_1)}{A_1 \varepsilon_1}, \qquad J_2 - \sigma T_2^4 = I_2 \dfrac{(1 - \varepsilon_2)}{A_2 \varepsilon_2}$

$J_3 - \sigma T_3^4 = I_3 \dfrac{(1 - \varepsilon_3)}{A_3 \varepsilon_3}, \qquad J_1 - J_2 = \dfrac{I_5}{A_1 F_{1-2}}$

$J_1 - J_3 = \dfrac{I_4}{A_1 F_{1-3}}, \qquad J_2 - J_3 = \dfrac{I_6}{A_2 F_{2-3}}$

Fig. 16.43 Electrical analogy for the system of Fig. 16.42

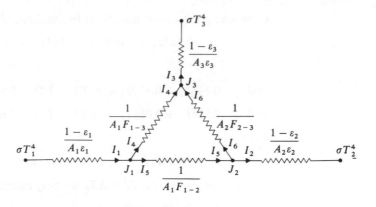

Also, from Kirchhoff's law of electric circuits

$$I_1 = I_4 + I_5, \quad I_2 = I_5 - I_6 \quad \text{and} \quad I_3 = I_4 + I_6$$

Example 16.24

A cylindrical vessel of diameter 2 m contains molten metal whose surface temperature is 1327 °C; the height of the vessel above the level of the liquid metal is 1 m and the vessel is in large surroundings at a mean temperature of 27 °C. The cylindrical sides of the vessel above the liquid metal level are cooled so that the average inside temperature of the surface is 427 °C.

Draw the radiation resistance network for this problem and hence calculate the rate of heat transfer from the liquid metal surface by radiation. Emissivity of liquid metal = 0.3; emissivity of inside surface of vessel = 0.7.

Solution

The vessel is shown in Fig. 16.44; the molten metal is surface 1, the open top of the vessel is surface 2 at the temperature of the surroundings, and the sides of the vessel are surface 3. The equivalent circuit is as shown in Fig. 16.43. From (b) in Table 16.5 (page 642),

$$F_{1-2} = (1 + \tfrac{1}{2}) - \sqrt{\{(1 + \tfrac{1}{2})^2 - 1\}} = 0.382$$

From equation (16.110)

$$F_{1-2} + F_{1-3} = 1 \quad \text{therefore} \quad F_{1-3} = 1 - 0.382 = 0.618$$

Then, from the symmetry of the surfaces, $F_{2-3} = 0.618$. Also

$$\frac{1 - \varepsilon_1}{A_1 \varepsilon_1} = \frac{0.7}{\pi \times 0.3} = 0.743 \qquad \frac{1 - \varepsilon_2}{A_2 \varepsilon_2} = 0 \,(\text{since surroundings are black})$$

$$\frac{1 - \varepsilon_3}{A_3 \varepsilon_3} = \frac{0.3}{\pi \times 2 \times 1 \times 0.7} = 0.068$$

$$\frac{1}{A_1 F_{1-2}} = \frac{1}{\pi \times 0.382} = 0.833 \qquad \frac{1}{A_2 F_{2-3}} = \frac{1}{A_1 F_{1-3}} = \frac{1}{\pi \times 0.618} = 0.515$$

Then
$$\sigma T_1^4 - J_1 = 0.743 I_1, \qquad J_2 - \sigma T_2^4 = 0$$
$$J_3 - \sigma T_3^4 = 0.068 I_3, \qquad J_1 - J_2 = 0.833 I_5$$
$$J_1 - J_3 = 0.515 I_4, \qquad J_2 - J_3 = 0.515 I_6$$

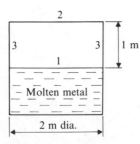

Fig. 16.44 Molten metal in a vessel for Example 16.24

2

3   3   1 m

1

– Molten metal –

2 m dia.

Combining the above equations to eliminate $\dot{J}_1$, $\dot{J}_2$, and $\dot{J}_3$ we have

$$0.743I_1 + 0.068I_3 + 0.515I_4 = \sigma(T_1^4 - T_3^4) = 5.67(16^4 - 7^4)$$

$$= 0.358 \times 10^6$$

and $\quad 0.743I_1 + 0.833I_5 = \sigma(T_1^4 - T_2^4) = 5.67(16^4 - 3^4) = 0.371 \times 10^6$

and $\quad 0.515I_6 + 0.068I_3 = \sigma(T_2^4 - T_3^4) = 5.67(3^4 - 7^4) = -0.0132 \times 10^6$

Also

$$I_1 = I_4 + I_5, \qquad I_2 = I_5 - I_6 \quad \text{and} \quad I_3 = I_4 + I_6$$

Eliminating I_2, I_3, I_4, I_5 and I_6 we may calculate I_1,

i.e. $\qquad I_1 = 0.336 \times 10^6$

Therefore

$$\text{Heat loss by radiation from molten metal} = 0.336 \times 10^6 \text{ W}$$

$$= 336 \text{ kW}$$

If in the above example the sides of the vessel were well insulated rather than cooled, then the equivalent circuit would be as shown in Fig. 16.45. This is a much simpler case; the equivalent resistance of the circuit can be calculated directly since there is no flow of 'current' outwards or inwards at T_3,

i.e. $\qquad$ Equivalent resistance $= 0.743 + \dfrac{1}{\{1/(0.515 \times 2)\} + \{1/0.833\}}$

$$= 0.743 + 0.461 = 1.204$$

Fig. 16.45 Electrical analogy for Example 16.24

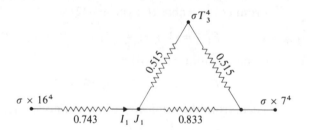

Therefore

$$I_1 = \frac{5.67(16^4 - 3^4)}{1.204} = 0.308 \times 10^6 \text{ W}$$

i.e. $\qquad$ Rate of heat loss by radiation from metal surface $= 308 \text{ kW}$

In this case the temperature of surface 3 will not be 427 °C as before. The 'current' flowing from 1 to 3 is given by

$$308 \times \frac{0.461}{2 \times 0.515} = 137.9 \text{ kW}$$

therefore

$$\sigma(T_3^4 - T_2^4) = 137.9 \times 10^3 \times 0.515 = 0.0710 \times 10^6$$

i.e. $\quad \dfrac{T_3^4}{10^8} = 3^4 + \dfrac{0.0710 \times 10^6}{5.67}$

therefore

$$T_3 = 1059.5 \text{ K} = 786.5\,^\circ\text{C}$$

16.19 Heat transfer coefficient for radiation

A heat transfer coefficient for radiation, α_r, is sometimes defined analogously to the heat transfer coefficient, α, for convection.

From equation (16.107)

$$\dot{Q}_{1-2} = F_{1-2}A_1\sigma(T_1^4 - T_2^4)$$
$$= F_{1-2}A_1\sigma(T_1^2 + T_2^2)(T_1^2 - T_2^2)$$

i.e. $\quad \dot{Q}_{1-2} = F_{1-2}A_1\sigma(T_1^2 + T_2^2)(T_1 + T_2)(T_1 - T_2)$

For convection heat transfer, we have, from equation (16.3),

$$\dot{Q} = \alpha A(t_w - t)$$

Hence, comparing the two equations, we can write

$$\alpha_r = F_{1-2}\sigma(T_1^2 + T_2^2)(T_1 + T_2) \tag{16.116}$$

Therefore

$$\dot{Q} = \alpha_r A_1(t_1 - t_2) \tag{16.117}$$

It should be noted that for convective heat transfer from a surface of surface area A, the total surface area A is used in the calculation, as in equation (16.3). In radiation heat transfer from the same body the area of the surface envelope must be used.

Example 16.25   A ribbed cylinder of outside diameter 0.6 m is at a surface temperature of 260 °C in large surroundings at 20 °C. Calculate the heat transfer coefficient for radiation, and the total heat loss rate due to radiation and convection. The cylinder is 0.9 m long and is made of cast iron of emissivity 0.8. The surface area of the ribbed cylinder is 5 m², and the heat transfer coefficient for convection may be taken as 8.8 W/m² K. Neglect end effects.

Solution   Since the cylinder is small compared with the surroundings, then $F_{1-2} = 1$. Then, from equation (16.116)

$$\alpha_r = \varepsilon_1 F_{1-2}\sigma(T_1^2 + T_2^2)(T_1 + T_2)$$

i.e. $\quad \alpha_r = 0.8 \times \dfrac{5.67}{10^8} \times (533^2 + 293^2)(533 + 293)$

where $T_1 = 260 + 273 = 533$ K, and $T_2 = 20 + 273 = 293$ K,

i.e. $\alpha_r = 13.87$ W/m² K

Then, from equation (16.117)

> Rate of heat loss by radiation $= \alpha_r A_1 (t_1 - t_2)$
> $$= 13.87 \times \pi \times 0.6 \times 0.9 \times (260 - 20)$$
> $$= 5650 \text{ W}$$

i.e.   Rate of heat loss by radiation $= 5.65$ kW

From equation (16.3)

> Rate of heat loss by convection $= \alpha A (t_w - t) = 8.8 \times 5 \times (260 - 20)$

i.e.   Rate of heat loss by convection $= 10\,560$ W $= 10.56$ kW

Therefore,

> Rate of total heat loss $= 5.65 + 10.56 = 16.21$ kW

16.20   Gas radiation

In the problems considered in the previous sections on radiation the effect of the transmission of radiation through the gaseous atmosphere has been neglected; some radiation will be absorbed by the surrounding gases in such cases, but this is normally so small that it can be neglected. Certain types of gases are transparent to thermal radiation; these include inert gases (e.g. argon) and gases with symmetric diatomic molecules (e.g. oxygen and nitrogen). Hence for radiation between surfaces in the normal atmospheric environment the effect of the surrounding gas can be ignored. For gases with certain types of asymmetric molecular structures (e.g. carbon dioxide, carbon monoxide, sulphur dioxide, nitrous oxide, and water vapour), radiation is absorbed from, and emitted to, surrounding surfaces. Due to the rotational and vibrational motions within any gas molecule, the radiation absorbed is dependent on the frequency of the radiation striking the molecule. Absorption and emission of radiation in gases is therefore selective, occurring in only certain bands of wavelengths.

When considering radiation within a furnace, or any enclosure containing combustion gases, it is necessary to allow for the absorption of radiation due to the presence of CO_2, H_2O, and perhaps CO and SO_2. A simplified procedure for the calculation of absorption and emission of radiation in such gases was first suggested by Hottel; this is summarized in the chapter on radiation written by Hottel in reference 16.10. Also included is the radiation from flames made luminous by the thermal decomposition of hydrocarbons.

The greenhouse effect

The so-called greenhouse effect on the earth is caused by the absorption of the sun's rays by gases, mainly carbon dioxide, in the atmosphere. The greenhouse effect is essential for the survival of life on the planet since without it the earth's surface would be rapidly cooled to a temperature of about $-20\,°C$. Gases are selective absorbers of radiation and the major greenhouse gases, such as carbon dioxide and methane, absorb the long-wavelength radiation from the surface of the earth much more readily than the low-wavelength, high-temperature radiation from the sun to the earth's surface. Hence there is a net transfer of heat from the sun to the earth, maintaining the temperatures that sustain life as we know it.

As long as there has been life on earth there has been a build-up of carbon dioxide in the atmosphere; one of our main constituents, carbon, oxidizes to carbon dioxide as we breathe, and whenever wood, vegetable matter, or fossil fuel is burned. Methane is continuously released by sheep and cattle, from any shallow wetland or marsh, and by certain insects such as termites. The reason why there is now international concern is the realization that a dangerously high level of greenhouse gases has built up over the last two centuries, and particularly over the last 50 years or so. This is mainly due to an acceleration in the world-wide combustion of fossil fuel; it is estimated that 19 000 million tonnes per year of carbon dioxide are released into the atmosphere from combustion of coal, oil, and gas. Methane concentration has also increased markedly due to the rapid growth in intensive farming of livestock and of crops such as rice. Other gases that are identified as causing the increase in the greenhouse effect are CFCs (see section 14.11), nitrous oxide from car exhausts and power station chimneys, and ozone from photochemical smog; the Montreal protocol (p. 526) should solve the problem of CFCs by the end of this century, although CFCs are estimated to cause only 15% of the present increase in the greenhouse effect.

The average temperature of the earth's surface is known to be rising although the exact increase due to the increased greenhouse effect is masked by natural cyclic fluctuations due to, for example, changes to the sun itself. The main concern about the relatively small temperature increases occurring in the earth due to the increased greenhouse effect is in the increase in the melting of the polar ice-caps. This will cause increased sea-levels leading to catastrophic flooding; more areas of the world would turn into desert, but other areas would in turn become warmer and able to sustain crops more abundantly. The main solution to this problem is to effect a reduction in the consumption of fossil fuels by first applying the principles of the efficient use of existing energy, and secondly by vigorously introducing and improving renewable energy-conversion methods using solar, wind, tidal, wave, and sea devices (see Ch. 17).

16.21 Further study

It is not possible to cover all aspects of heat transfer in even a cursory manner within the limits of one chapter. Many of the topics considered in this book have been much simplified, and some major topics have been omitted for lack of space. The references at the end of this chapter cover all the topics in the chapter in much greater depth, and also include topics omitted, for example heat transfer with a change of phase, and combined heat and mass transfer. The latter topic applied to cooling towers; humidifying and dehumidifying equipment is covered concisely and analytically in reference 16.11.

Problems

16.1 A furnace wall consists of 250 mm firebrick, 125 mm insulating brick, and 250 mm building brick. The inside wall is at a temperature of 600 °C and the atmospheric temperature is 20 °C. The heat transfer coefficient for the outside surface is 10 W/m² K, and the thermal conductivities of the firebrick, insulating brick, and building brick are 1.4, 0.2, and 0.7 W/m K, respectively. Neglecting radiation, calculate the rate of heat loss per unit wall surface area and the temperature of the outside wall surface of the furnace.

(460 W/m²; 66 °C)

16.2 An electric hot-plate is maintained at a temperature of 350 °C and is used to keep a solution just boiling at 95 °C. The solution is contained in an enamelled cast-iron vessel of wall thickness 25 mm and enamel thickness 0.8 mm. The heat transfer coefficient for the boiling solution is 5.5 kW/m² K, and the thermal conductivities of cast iron and enamel are 50 and 1.05 W/m K respectively. Calculate the resistance to the heat transfer for unit area, and the rate of heat transfer per unit area.

(1.444 m² K/kW; 176.6 kW)

16.3 In Problem 16.2 recalculate the rate of heat transfer per unit area if the base of the cast-iron vessel is not perfectly flat, and the resistance of the resultant air film is 35 m² K/kW.

(7 kW/m²)

16.4 The wall of a house consists of two 125 mm thick brick walls with an inner cavity. The inside wall has a 10 mm coating of plaster, and there is a cement rendering of 5 mm on the outside wall. In one room of the house the external wall is 4 m by 2.5 m, and contains a window of 1.8 m by 1.2 m of 1.5 mm thick glass. The heat transfer coefficients for the inside and outside surfaces of the wall and window are 8.5 and 31 W/m² K respectively. The thermal conductivities of brick, plaster, cement, and glass are 0.43, 0.14, 0.86, and 0.76 W/m K respectively. Assuming that the resistance of the air cavity is 0.16 m² K/W, neglecting all end effects, and neglecting radiation, calculate the proportion of the total heat transfer which is due to the heat loss through the window.

(63.8%)

16.5 Water at 80 °C flows through a 50 mm bore steel pipe of 6 mm thickness, and the atmospheric temperature is 15 °C. The thermal conductivity of steel is 48 W/m K and the inside and outside heat transfer coefficients are 2800 and 17 W/m² K respectively. Neglecting radiation, calculate the rate of heat loss per unit length of pipe.

(0.213 kW/m)

16.6 Calculate the percentage reduction in heat loss for the pipe in Problem 16.5 when a layer of hair felt 12 mm thick, of thermal conductivity 0.03 W/m K, is wrapped round the outside surface. Assume that the heat transfer coefficient for the outside surface remains unchanged.

(85.1%)

16.7 A steam main of 150 mm outside diameter containing wet steam at 28 bar is insulated with an inner layer of diatomaceous earth, 40 mm thick, and an outer layer of 85% magnesia, 25 mm thick. The inside surface of the pipe is at the steam temperature, and the heat transfer coefficient for the outside surface of the lagging is 17 W/m² K. The thermal conductivities of diatomaceous earth and 85% magnesia are 0.09, and 0.06 W/m K respectively. Neglecting radiation, and the thermal resistance of the pipe wall, calculate the rate of heat loss per unit length of the pipe and the temperature of the outside surface of the lagging, when the room temperature is 20 °C.

(0.156 kW/m; 30.5 °C)

16.8 A spherical pressure vessel of 1 m inside diameter is made of 20 mm steel plate. The vessel is lagged with a 25 mm thickness of vermiculite held in position by 10 mm thick asbestos. The heat transfer coefficient for the outside surface is 20 W/m² K, and the thermal conductivities of steel, vermiculite, and asbestos are 48, 0.047, and 0.21 W/m K, respectively. Neglecting radiation, calculate the rate of heat loss from the sphere when the inside surface is at 500 °C, and the room temperature is 20 °C.

(2.744 kW)

16.9 A solid copper conductor of 13 mm diameter carries a current density of 5 A/mm². The conductor is electrically insulated with a thickness of rubber insulation such that the wire temperature is kept to the minimum possible. Assuming that the surrounding air is at 30 °C, calculate:

  (i) the thickness of insulation;

 (ii) the wire temperature at the axis;

(iii) the temperature of the outside surface of the insulation;

(iv) the wire temperature at the axis with the insulation removed and the new steady state reached.

    Give a physical explanation why any larger or smaller thickness of insulation will lead to a higher wire temperature.

Data   Heat transfer coefficient for outside surface of rubber or copper (assumed constant), 20 W/m² K; thermal conductivities of copper and rubber, 380 and 0.2 W/m K; electrical resistivity of copper, 2×10^{-5} Ω mm.

(3.5 mm; 105.6 °C; 82.8 °C; 111.3 °C)

16.10 (a) A gas-cooled nuclear reactor has solid fuel rods of radius, r_F, thermal conductivity, λ_F, sheathed with zirconium of thickness, t, thermal conductivity, λ_Z. The heat transfer coefficient from the sheath to the gas in the surrounding annulus is α.

Assuming that the uniform heat generation rate per unit volume within the fuel is $\dot{q}_g$, show from first principles that the temperature difference between the axis of the fuel rod and the bulk of the coolant at any cross-section is given by

$$\frac{\dot{q}_g r_F^2}{4\lambda_F}\left\{1 + \frac{2\lambda_F}{\lambda_Z}\ln(1 + t/r_F) + \frac{2\lambda_F}{\alpha(r_F + t)}\right\}$$

(b) At a particular cross-section in the reactor the gas mean bulk temperature is 220 °C when the internal heat generation rate in the fuel rod is 30 MW/m³. Using the data below, calculate:

(i) the temperature at the axis of the fuel rod;

(ii) the temperature at the inner and outer surfaces of the sheath.

Data   Thermal conductivities of reactor fuel and zirconium, 33.5 and 18.7 W/m K; radius of fuel rod, 12 mm; thickness of sheath, 3.6 mm; heat transfer coefficient from sheath to cooling gas, 500 W/m² K.

(559.5 °C; 527.3 °C; 496.9 °C)

16.11 The concrete biological shield of a nuclear reactor is 2 m thick and can be considered to be an infinite flat plate of uniform thermal conductivity, $\lambda = 2.0$ W/m K. The rate of heat generation per unit volume due to the incident gamma radiation is given by

$$\dot{q}_g = H e^{-8.5x} \text{ W/m}^3$$

where x is the distance in metres measured from the inside surface, and H is a constant dependent on the gamma radiation.

The maximum temperature difference in the concrete is to be limited to 4 K, and it may be assumed that the outer surface is well insulated.

Calculate the maximum allowable value of the gamma radiation on the inside surface of the concrete in watts per square metre.

$$(68 \text{ W/m}^2)$$

16.12 (a) The temperature–time history of the centre of a large slab of material initially at a constant temperature which is suddenly plunged into a fluid at a different temperature can be shown to be given by

$$\frac{\Delta t_c}{\Delta t_i} = 2 \sum_{n=1}^{\infty} e^{-(p_n L)^2 Fo} \frac{\sin(p_n L)}{(p_n L) + \sin(p_n L)\cos(p_n L)}$$

and $(p_n L)\tan(p_n L) = Bi$.

Where Δt_c is the temperature difference at time τ, between the centre of the slab thickness and the surrounding fluid; Δt_i the temperature difference between slab and fluid initially; Fo the Fourier number, $\kappa \tau / L^2$; κ the thermal diffusivity of slab; L the half-thickness of slab; Bi the Biot number, $\alpha L/\lambda$; α the heat transfer coefficient on slab surfaces; and λ the thermal conductivity of slab material.

Show that for a case where the temperature of the slab surfaces is approximately equal to the fluid temperature (i.e. $\alpha \to \infty$), and for a reasonably long time period

$$\frac{\Delta t_c}{\Delta t_i} = \frac{4}{\pi} e^{-(\pi^2/4)Fo}$$

(b) A large slab of rubber of thickness 40 mm is vulcanized by heating the faces using steam at 330 °C. The required temperature at the centre is 120 °C and the rubber is initially at 20 °C. Calculate the time required for the process:

(i) using the method of (a) above;

(ii) using a numerical method.

Take κ for rubber as $64.5 \times 10^{-9} \text{ m}^2/\text{s}$.

$$(26.4 \text{ min})$$

16.13 A large metal plate of thickness 200 mm is initially at a uniform temperature of 20 °C. One surface of the plate is in contact with ambient air at a constant temperature of 20 °C, while the other surface may be exposed to a constant net radiant heat flux of 100 kW/m^2 from an electric element.

(i) Assuming as an approximation that the plate temperature is uniform throughout at any instant, calculate the time taken for the plate to reach 70 °C from the instant the electric element is switched on;

(ii) estimate the temperature distribution through the plate thickness 9 min after the element is switched on using a numerical method with four space increments and a Fourier number of 0.5.

Data Thermal conductivity of plate, 45 W/m K; density of plate, 7800 kg/m^3; specific heat capacity of plate, 0.5 kJ/kg K; heat transfer coefficient from plate to air, 200 W/m^2 K.

$$(6.85 \text{ min}; 48 \text{°C}; 54 \text{°C}; 91 \text{°C}; 162 \text{°C}; 274 \text{°C})$$

16.14 The wall of a large vessel is 50 mm thick and is initially at 12.5 °C throughout. A hot fluid at a constant temperature of 500 °C is suddenly pumped across the inside surface; the outside surface may be assumed to be perfectly insulated. Using a numerical method, determine the time taken for the junction of wall and insulation to reach 110 °C.

Data   Thermal conductivity of wall, 22 W/m K; thermal diffusivity of wall, 6.22×10^{-6} m^2/s; heat transfer coefficient from fluid to wall, 110 W/m K.

(455 s)

16.15   A thick fin of rectangular cross-section, 1 m × 1 m, projects from a flat surface at 200 °C into a fluid at 20 °C. Using the data below, estimate the temperature distribution in the steady state assuming two-dimensional conduction, and hence calculate the rate of heat loss from the fin surface per unit length.

Data   Thermal conductivity of fin material, 25 W/m K; heat transfer coefficient for all parts of the fin surface, 10 W/m^2 K.

(2.5 kW/m)

16.16   In an oil cooler the oil enters 10 mm diameter tubes at 160 °C and is cooled to 40 °C; the mean velocity of the oil in the tubes is 1.5 m/s. Calculate the heat transfer coefficient. For turbulent flow of a liquid being cooled take:

$$Nu = 0.0265(Re)^{0.8}(Pr)^{0.3}$$

and for laminar flow take $Nu = 3.65$. Take all properties at the mean bulk temperature and use the properties of engine oil given in Table 16.6.

(50 W/m^2 K)

Table 16.6 Information for problem 16.16

$t/(°C)$	$\rho/(kg/m^3)$	$v/(cSt)$	$\lambda/(W/m\ K)$	$c/(kJ/kg\ K)$
40	878	251.0	0.144	1.96
100	839	20.4	0.137	2.22
160	806	5.7	0.131	2.48
1 centistoke (cSt) = 10^{-6} m^2/s				

16.17   In Problem 16.16 the length of each tube is 1.2 m and the tube wall temperature is 20 °C. Allowing for the entry length effect a more accurate expression for the mean Nusselt number over a length, L, for laminar flow is given by

$$Nu = 1.86\left(\frac{\eta}{\eta_w}\right)^{0.14}\{(d/L)(Re\,Pr)\}^{1/3}$$

where properties are at mean bulk temperature with the viscosity of the oil at the tube surface, $\eta_w = 0.8$ kg/m s. Calculate the mean heat transfer coefficient allowing for the entry length.

(177.5 W/m^2 K)

16.18   Air at 15 °C and 1 bar is to be heated to 285 °C while flowing at 34.2 m^3/h through a 25 mm diameter tube which is maintained at 455 °C. Assuming that the simple Reynolds analogy is valid, taking $f = 0.0791/(Re)^{1/4}$, and all properties at the mean bulk temperature, calculate the length of the tube required.

(1.88 m)

16.19   Air flows through a 20 mm diameter tube 2 m long with a mean velocity of 40 m/s. The tube wall temperature is 150 °C and the air temperature increases from 15 to 100 °C. Using the simple Reynolds analogy with all properties at the mean bulk temperature, estimate the pressure loss in millimetres of water in the tube due to friction, and the pumping power required. Take the mean air pressure as 1 atm.

(173 mm water; 21.3 W)

16.20 Air at a temperature of $15\,°C$ is blown across a flat plate with a surface temperature of $550\,°C$ at a mean velocity of $6\,m/s$. Neglecting radiation, calculate the rate of heat transfer per metre width from both sides of the plate over the first $150\,mm$ of the plate. For heat transfer from a flat plate with a large temperature difference between the plate and the fluid, the local Nusselt number is given by

$$Nu = 0.332(Pr)^{1/3}(Re)^{1/2}(T_w/T_s)^{0.117}$$

where all properties are at the mean film temperature, Re is based on the distance from the leading edge of the plate, and T_w and T_s are the absolute temperatures of the plate and the free stream of the air.

$$(4.39\,kW)$$

16.21 A wall $0.6\,m$ high by $3\,m$ wide is maintained at $79\,°C$ in an atmosphere at $15\,°C$. Neglecting end effects and radiation, calculate the rate of heat loss by natural convection. For natural convection from a vertical flat surface take, at any distance, x:

$$Nu_x = 0.509(Pr)^{1/2}(Pr + 0.952)^{-1/4}(Gr_x)^{1/4}$$

where all properties are at the mean film temperature, and $\beta = 1/T$, where T is the absolute temperature of the bulk of the air.

$$(528\,W)$$

16.22 Recalculate Problem 16.21 using the approximations:

$$\alpha = 1.42(\Delta t/l)^{1/4} \quad \text{for } 10^4 < Gr < 10^9$$

or $\quad \alpha = 1.31(\Delta t)^{1/3} \quad \text{for } 10^9 < Gr < 10^{12}$

where α is in $W/m^2\,K$, Δt is in K, and l is in m.

$$(604\,W)$$

16.23 A pipe containing dry saturated steam at $177\,°C$ is $150\,mm$ bore and has a $50\,mm$ thickness of 85% magnesia covering. The steam velocity is $6\,m/s$ and the heat transfer coefficient may be found from

$$Nu = 0.023(Re)^{0.8}(Pr)^{0.4}$$

where all properties are at the mean bulk temperature. The atmospheric temperature is $17\,°C$ and the heat transfer coefficient from a horizontal cylinder is given approximately by

$$\alpha = 1.32(\Delta t/d)^{1/4}$$

where α is in $W/m^2\,K$, Δt is in K, and d is in m.

The pipe wall is $7\,mm$ thick and the thermal conductivity of the pipe metal is $50\,W/m\,K$; the thermal conductivity of the 85% magnesia insulation is $0.06\,W/m\,K$. Neglecting radiation, taking arithmetic mean areas for the pipe wall and lagging, and using a trial-and-error method, calculate:

(i) the temperature of the outside surface of the lagging;
(ii) the rate of heat loss from the pipe per unit length.

$$(46.3\,°C;\ 104\,W/m)$$

16.24 An exhaust pipe of $75\,mm$ outside diameter is cooled by surrounding it by an annular space containing water. The exhaust gas enters the exhaust pipe at $350\,°C$, and the water enters from the mains at $10\,°C$. The heat transfer coefficients for the gases and water may be taken as 0.3 and $1.5\,kW/m^2\,K$, and the pipe thickness may be taken to be negligible. The gases are required to be cooled to $100\,°C$ and the mean specific heat capacity at constant pressure is $1.13\,kJ/kg\,K$. The gas flow rate is $200\,kg/h$ and the water flow rate is 1400 kg/h. Taking the specific heat capacity of water as 4.19 kJ/kg K, calculate:

(i) the required pipe length for parallel-flow;

(ii) the required pipe length for counter-flow.

(1.48 m; 1.44 m)

16.25 In a chemical plant a solution of density 1100 kg/m³ and specific heat capacity 4.6 kJ/kg K is to be heated from 65 °C to 100 °C; the required flow rate of solution is 11.8 kg/s. It is desired to use a tubular heat exchanger, the solution flowing at about 1.2 m/s in 25 mm bore iron tubes, and being heated by wet steam at 115 °C. The length of the tubes must not exceed 3.5 m. Taking the inside and outside heat transfer coefficients as 5 and 10 kW/m² K, and neglecting the thermal resistance of the tube wall, estimate the number of tubes and the number of tube passes required.

(18; 4)

16.26 An oil engine develops 300 kW and the specific fuel consumption is 0.21 kg/kW h. The exhaust from the engine is used in a tubular water heater, flowing through 25 mm diameter tubes, entering with a velocity of 12 m/s, at 340 °C and leaving at 90 °C. The water enters the heater at 10 °C and leaves at 90 °C, flowing in counter-flow to the hot gases. The air–fuel ratio of the engine is 20, and the exhaust pressure is 1.01 bar. The overall heat transfer coefficient of the heat exchanger when designed is found to be 56 W/m² K, but after running for some time a fouling factor of 0.5 m² K/kW must be assumed. Taking the specific heat capacity and the gas constant for the gases as 1.11 kJ/kg K and 0.29 kJ/kg K, and the specific heat capacity for the water as 4.19 kJ/kg K, calculate:

(i) the mass flow rate of water;

(ii) the number of tubes required;

(iii) the required tube length.

(1096 kg/h; 110; 1.457 m)

16.27 In an air cooler the air is blown across a bank of tubes at the rate of 240 kg/h at a velocity of 24 m/s, the air entering at 97 °C and leaving at 27 °C. The cooling water enters the tubes at 10 °C and leaves at 20 °C, at a mean velocity of 0.6 m/s. The tubes are 6 mm diameter and the wall thickness may be neglected. The heat transfer coefficient from the air to the tubes may be calculated from

$$Nu = 0.33(Re)^{0.6}(Pr)^{0.33}$$

with properties at the mean bulk temperature.

The heat transfer coefficient from the water to the tubes is given by

$$St = \frac{f/2}{1 + (Pr)^{-1/6}(Re)^{-1/8}(Pr - 1)}$$

where $f = 0.0791/Re^{1/4}$ and properties are at the mean bulk temperature.

Assuming that the tubes are arranged in six passes, and that the logarithmic mean temperature difference for counter-flow can be assumed, calculate:

(i) the number of tubes required in each pass;

(ii) the necessary tube length.

(7; 0.528 m)

16.28 A two-pass shell-and-tube heat exchanger is used to condense a chemical on the shell side at a rate of 50 kg/s at a saturation temperature of 80 °C. The chemical enters as a dry saturated vapour and is not undercooled during the process. Water at 10 °C and a mass flow rate of 100 kg/s is available as coolant; the velocity of the water is to be approximately 1.5 m/s. Using the data below and taking a nominal tube diameter of 25 mm, neglecting tube wall thickness, calculate:

(i) the number of tubes required;
(ii) the tube length;
(iii) the number of transfer units;
(iv) the effectiveness of the heat exchanger.

Data Specific enthalpy of vaporization of chemical, 417.8 kJ/kg; heat transfer coefficient for shell side, 10 kW/m² K; fouling factor for shell side, 0.1 m² K/kW; fouling factor for tube side, 0.2 m² K/kW.

For turbulent flow in a pipe take

$$Nu = 0.023(Re)^{0.8}(Pr)^{0.4}$$

with properties at the mean bulk temperature.

(274; 13.55 m; 1.253; 71.4%)

16.29 An oil cooler consists of a single-pass, counter-flow shell-and-tube heat exchanger with 300 tubes of internal diameter 7.3 mm and length 8 m. The oil flows in the tube side entering at a mass flow rate of 8 kg/s at a temperature of 70 °C. Cooling water in the shell side enters at a mass flow rate of 12 kg/s at a temperature of 15 °C. Using the data below, calculate:

(i) the number of transfer units;
(ii) the effectiveness of the heat exchanger;
(iii) the outlet temperature of the oil.

Data Shell side heat transfer coefficient, 1000 W/m² K; heat transfer coefficient for the tube side given by $Nu = 0.023(Re)^{0.8}(Pr)^{0.4}$ with properties as follows: specific heat capacity of oil, 3.42 kJ/kg K; density of oil, 900 kg/m³; dynamic viscosity of oil, 1.5×10^{-3} kg/m s; thermal conductivity of oil, 0.15 W/m K.

(1.1; 58.8%; 37.7 °C)

16.30 A double pipe heat exchanger has an effectiveness of 0.5 when the flow is counter-current and the thermal capacity of one fluid is twice that of the other fluid. Calculate the effectiveness of the heat exchanger if the direction of flow of one of the fluids is reversed with the same mass flow rates as before.

(0.469)

16.31 500 kg/h of oil at 120 °C is to be cooled in the annulus of a double pipe counter-flow heat exchanger by water which enters the inside pipe at 10 °C. The inner pipe has an inside diameter of 25 mm and a wall thickness of 2 mm, and the inside diameter of the outer pipe is 50 mm; the effective length is 12 m. Using the data below calculate the exit temperature of the oil.

Data *Oil* take $Nu = 30$, based on an equivalent diameter, d_e, given by

$$d_e = 4 \times \text{(flow area)}/\text{(heat transfer area per unit length)};$$

specific heat capacity, 2.31 kJ/kg K; thermal conductivity, 0.135 W/m K; fouling factor, 0.001 m² K/W.

Water assume the simple Reynolds analogy holds true, taking the velocity as 1 m/s and the friction factor, f, as 0.0002; specific heat capacity, 4.18 kJ/kg K; density, 1000 kg/m³; fouling factor, 0.0002 m² K/W.

Neglect the thermal resistance of the pipe wall.

(98.8 °C)

16.32 A condenser contains four tube passes with tubes 3 m long, 25 mm internal diameter, each pass containing 100 tubes. Cooling water enters the tubes at 20 °C at the rate of 80 kg/s when the shell side vapour is at 50 °C. Before cleaning, the fouling factor on the water side is 0.0005 m² K/W; the outside of the tubes may be taken to be clean.

Neglecting the thermal resistance of the fluid film on the outside of the tubes and the thermal resistance of the tube wall, calculate, using the data below:
(i) the effectiveness of the heat exchanger;
(ii) the condensation rate;
(iii) the fouling factor required on the water side if the effectiveness is to be increased to 0.7 for the same mass flow rate of water.

For heat transfer in the tubes: $Nu = 0.023 Re^{0.8} Pr^{1/3}$.

Data Specific enthalpy of vaporization for shell side fluid, 300 kJ/kg; mean properties of water for the temperature range considered: density, 1000 kg/m³; specific heat capacity, 4.19 kJ/kg K; thermal conductivity, 0.6 W/m K; dynamic viscosity, 0.9×10^{-3} kg/m s.

(0.337; 11.3 kg/s; 0.000 049 m² K/W)

16.33 In a closed-cycle gas turbine plant air from the compressor enters one side of a compact heat exchanger at 150 °C at a mass flow rate of 10 kg/s. The air leaving the turbine enters the heat exchanger at 504 °C and flows in counter-flow to the air. The heat exchanger has a flow area of 0.144 m² and an effective heat transfer area of 115.2 m² per unit length in the direction of flow on both the hot and cold sides of the heat exchanger. Calculate the required length of the heat exchanger to obtain an effectiveness of 0.7.

Assume that the heat exchanger surfaces are clean and neglect the thermal resistance of the separating plates. For flow of air in the heat exchanger passages assume $Nu = 0.023 Re^{0.8} Pr^{0.3}$ based on an equivalent diameter given by 4 × (flow area)/(heated surface area per unit length); take the properties at the mean temperature between the cold air inlet and the hot air inlet.

(1.257 m)

16.34 Circular cross-section studs of radius 10 mm, length 100 m, thermal conductivity 24 W/m K are attached to a flat surface with their axes perpendicular to the surface on a square pitch of 30 mm. The primary surface is at 300 °C.

A fluid at 50 °C is forced across the surface such that the mean heat transfer coefficient is 100 W/m² K. Calculate the rate of heat loss per unit area of studded surface. Assume that the heat transfer coefficient is the same for the primary surface and for the rod surfaces.

(77.66 kW/m²)

16.35 A flat surface at a temperature of 300 °C has rectangular section cooling fins perpendicular to the surface projecting into a fluid at 20 °C. There are 12.5 fins per 100 mm and the fins have a thickness of 3 mm and a length of 30 mm.

The thermal conductivity of the fin material is 26 W/m K and the heat transfer coefficient for all surfaces may be taken as 40 W/m² K.

Neglecting the heat loss at the tip of each fin, calculate:
(i) the fin efficiency;
(ii) the rate of heat loss from unit area of the flat surface;
(iii) the temperature at the tip of each fin.

(77.5%; 72.1 kW/m²; 206.9 °C)

16.36 A cylindrical electrode of radius, r, length, L, is immersed in a liquid which remains at a constant temperature when the current density in the electrode is J. The heat transfer coefficient, α, from the outside surface of the electrode may be assumed to be constant over the entire surface.

659

Assuming steady-state conditions derive the differential equation

$$\frac{d^2(\Delta t)}{dx^2} - \frac{2\alpha}{r\lambda}\Delta t + \frac{J^2 s}{\lambda} = 0$$

where Δt is the temperature difference between the electrode and the fluid at any distance x from the end of the electrode, λ the thermal conductivity of the electrode material, and s the electrical resistivity of the electrode material.

Hence show that for the case when the rate of heat loss through the lead and support at each end is a fraction, y, of the total electrical input, then

$$\Delta t = \frac{J^2 s}{m^2 \lambda}\left\{1 - \frac{mLy \cosh m(x - L/2)}{\sinh mL/2}\right\}$$

where $m = (2\alpha/r\lambda)^{1/2}$.

16.37 A cylindrical storage tank, 1 m diameter by 1.2 m long, has an outside surface temperature of 60 °C, and an emissivity of 0.9. Calculate the rate of heat loss by radiation when the tank is in a large room, the walls of which are at 15 °C. Calculate also the reduction in the rate of heat loss by radiation if the tank is painted with aluminium paint of emissivity 0.4. Assume that the tank is a grey body.

(1474 W; 819 W)

16.38 A copper pipe at 260 °C is in a large room at 15 °C. Calculate the rate of heat loss per unit area of pipe surface by radiation, taking the emissivity of copper as 0.61 at 260 °C, and as 0.56 at 15 °C. Assume that the absorptivity of a surface depends only on the temperature of the source of radiation.

(2571.5 W/m²)

16.39 Calculate the rate of heat transfer per unit surface area by radiation between two brick walls a short distance apart, when the temperatures of the surfaces are 30 °C and 15 °C. The emissivity of brick may be taken as 0.93, and the surfaces may be assumed to be grey.

(76.2 W/m²)

16.40 A thermos flask consists of an inner cylindrical vessel of 60 mm outside diameter and an outer cylindrical vessel of 65 mm inside diameter. Both surfaces are of polished silver, emissivity 0.02. Calculate the rate of heat loss per millimetre length of the flask when it contains boiling water and the temperature of the outside surface is 17 °C. Neglect the thermal resistance of the metal walls of the flask. (NB Polished surfaces reflect specularly and hence in this case the surfaces act as large parallel planes.)

(0.00133 W)

16.41 A gas turbine can-type combustion chamber of 0.3 m diameter reaches a temperature of 500 °C when undergoing a test in large surroundings at 15 °C. The emissivity of the steel surface is 0.79. Calculate the percentage reduction in the rate of radiant heat loss by enclosing the combustion chamber with a cylindrical screen of 0.6 m diameter, the inside and outside surfaces of which are painted with aluminium paint of emissivity 0.4.

(61.3%)

16.42 In a muffle furnace the floor, 4.5 m by 4.5 m, is constructed of refractory material (emissivity = 0.7). Two rows of oxidized steel tubes are placed 3 m above and parallel to the floor, but for the purpose of analysis these can be replaced by a 4.5 m by 4.5 m plane having an effective emissivity of 0.9. The average temperatures for the floor and tubes are 900 and 270 °C respectively.

Taking the geometric factor for radiation from floor to tubes as 0.32, calculate:
(i) the net rate of heat transfer to the tubes;

(ii) the mean temperature of the refractory walls of the furnace, assuming that these are well insulated.

(1009 kW; 687 °C)

16.43 A circular plate of radius 0.1 m is at a temperature of 500 °C in a large room, the walls of which are at 10 °C. The air in the room is at a mean temperature of 15 °C. A small, spherical thermocouple junction is placed at a distance of 0.1 m from the centre of the plate. Show that the temperature recorded by the thermocouple is approximately 100 °C. The heat transfer coefficient from the thermocouple to the air is 25.6 W/m² K, and the plate surface may be assumed to be black for thermal radiation. Neglect conduction through the thermocouple leads. The geometric factor is given in Table 16.3.

16.44 An electric heater 25 mm diameter and 0.3 m long is used to heat a room. Calculate the electrical input to the heater when the bulk of the air in the room is at 20 °C, the walls are at 15 °C, and the surface of the heater is at 540 °C. For convective heat transfer from the heater, assume that

$$Nu = 0.4(Gr)^{1/4}$$

where all properties are at mean film temperature and $\beta = 1/T$, where T K is the bulk temperature of the air.

Take the emissivity of the heater surface as 0.55 and assume that the surroundings are black.

(481 W)

16.45 Calculate the radiation heat transfer coefficient for the flat plate in Problem 16.20, assuming that the surroundings are large and are at the air temperature, and compare this with the heat transfer coefficient for convection. Take the emissivity of the plate surface as 0.6.

(28.75 W/m² K; $\alpha_r/\alpha = 1.05$)

16.46 Calculate the radiation heat transfer coefficient for the vertical wall in Problem 16.21, assuming that the wall radiates into black surroundings at 15 °C, and the emissivity of the wall surface is 0.93. Compare this value with the heat transfer coefficient for convection.

(6.98 W/m² K; $\alpha_r/\alpha = 1.52$)

16.47 A hot-water heater 150 mm wide by 1.2 m long by 1 m high is at a surface temperature of 50 °C in surroundings at 20 °C. The walls of the room are at 13 °C. The surface area of the heater is 7 m² and the heat transfer coefficient for convection is given by

$$\alpha = 1.31(\Delta t)^{1/3}$$

where α is in W/m² K, and Δt is in K.

Calculate the rate of heat transfer from the heater. Take the emissivity of the heater as 0.95 and assume that it is completely surrounded by black surroundings.

(1.545 kW)

References

16.1 WELTY J R 1984 *Fundamentals of Momentum, Heat and Mass Transfer* 3rd edn John Wiley

16.2 CROFT D R and LILLEY D G 1986 *Heat Transfer Calculations Using Finite Difference Equations* Pavic Publications

16.3 INCROPERA F P and DE WITT D P 1990 *Fundamentals of Heat and Mass Transfer* 3rd edn John Wiley

16.4 ROGERS G F C and MAYHEW Y R 1987 *Thermodynamic and Transport Properties of Fluids* 4th edn Basil Blackwell

16.5 ECKERT E R and DRAKE R M 1971 *Analysis of Heat and Mass Transfer* Taylor and Francis

16.6 KERN D Q 1950 *Process Heat Transfer* McGraw-Hill

16.7 WALKER G 1990 *Industrial Heat Exchangers* 2nd edn McGraw-Hill

16.8 KAYS W M and LONDON A L 1984 *Compact Heat Exchangers* 3rd edn McGraw-Hill

16.9 EASTOP T D and CROFT D R 1990 *Energy Efficiency* Longman

16.10 McADAMS W H 1954 *Heat Transmission* 3rd edn McGraw-Hill

16.11 EASTOP T D and WATSON W E 1992 *Mechanical Services for Buildings* Longman

17

The Sources, Use, and Management of Energy

The most common source of heat energy is the chemical energy of substances called fuels which is released upon the combustion of the fuel in air as described in Chapter 7. Fuels such as coal, oil, and natural gas are created by natural processes deep down in the earth after many thousands of years and as such are described as non-renewable fossil fuels. However large the world's resources of fossil fuels may be, they are being consumed at a high rate and one day the fuel resources will become so depleted that the normal existence of energy-dependent countries will be seriously disrupted unless other energy sources have become available on the scale necessary to meet world demand. The fossil fuels are also sources of chemical substances to be used, other than as fuels, in the manufacture of goods such as plastics or chemicals for agriculture and animal foodstuffs. A comprehensive understanding of the supply problem includes all natural products including fuels. Many warnings have been given over the years about the world rate of fuel consumption and the prospects of a fuel shortage, but each time the crisis has passed and the world has continued on its way, hopefully more aware of the importance of its energy sources, the need for the efficient use of fuels and materials and for long-term planning for the future.

In the early 1970s the oil industry warned users that the exponential growth in oil consumption, which was doubling every seven years, could not be sustained. In 1979 the oil suppliers doubled the price of oil and shocked those countries, including the UK and the USA, which had based their industrial economy on the cheap oil available from the Middle Eastern countries. In equivalent costs in dollars per barrel of oil the relative costs of fuels in 1978 was: Middle East oil 1.3; North Sea oil 5.3; imported coal (north-west Europe) 8.3; indigenous coal (north-west Europe) 11.3; nuclear 7.6; imported natural gas 15; natural gas from indigenous coal 24.3; biomass (crops grown for fuel) 43; solar hot water 40.

The increase in the cost of oil in the late 1970s resulted in a fall in demand and a world-wide recession in the manufacturing industries leading to a considerable conservation not only of fuels but of other basic manufacturing materials such as iron, tin, zinc, aluminium, silver, gold, lead, and copper which are also in finite supply.

It was estimated in 1973 that by the mid-1980s the world oil consumption would be three times as much as it actually became and the electricity consumption would be twice as much. Instead there was a surplus of oil-tankers and bulk carriers, the demands on oil refineries and power stations decreased and so did the demand for steel, cement, and building materials, leading to a further reduction in the demand for energy. During this period the UK became an oil- and natural-gas-producing nation due to the exploitation of finds under the North Sea.

A greater awareness of the value of natural resources grew in the mid-1970s, and probably an even greater one for the influence of the cost of fuels on manufactured goods, for it was realized that an efficient use of fuels could affect industry and society substantially in relation to basic costs.

A cheap supply of energy does not encourage an economical attitude to energy use and methods of energy conservation, but the balance between demand, supply and cost is one which can change rapidly and must be kept under constant review by nations and their industries. If the balance of supply from various sources is to be changed, the technology required to make the change must be available and a new technology can take a long time and need high investments. Since the 1970s a greater interest in alternative sources of energy has grown, many developments have been made and considerable research has taken place, but by the mid-1980s no alternative solutions to existing sources had appeared that made any real impact on the energy scene – other than those for nuclear energy which began in the UK in the 1950s. The development work done since the 1970s on alternative forms of energy may show benefit in the future.

The following sections will cover some aspects of energy supply, demand, use, conservation and management, etc. but such a wide-ranging subject cannot be given a comprehensive treatment in this book and readers are advised to pursue the subject by reading some of the many specialist texts (references 17.1–17.4), and the current reports of government departments and research institutions, such as the Research Councils.

17.1 Sources of energy supply, and energy demands

The primary source of energy is the sun from which all of the earth's energy requirement is finally obtained. Only a negligible fraction of the daily need is obtained directly, even including the secondary supplies from power generation by winds, waves, and rivers. The vast majority is obtained from the combustion of fossil fuels including coal, oil, natural gas, peat, wood, natural waste, etc. and a lesser amount from nuclear reactions. Additional known sources of energy are the tides and the natural geothermal gradient of the earth. It has been estimated that the amount of solar energy falling on the earth in three days is equal to the known fossil-fuel reserves of the world. The sun is an ample provider of energy for the earth; the problem is how to collect and store solar energy so that it can be released in the right form to meet the world demand for heat and power.

Energy resources and energy consumption

It is highly desirable at any time to be able to say what the world's fuel resources are, and the corresponding demand and the rate at which it is changing. The determination of this information is a complex exercise, but it is done from time to time and the most authoritative source available should be consulted. Any estimate is likely to be out of date when it is published. New finds of coal, gas, and oil are being made and old sources abandoned; the extent of deposits in known sources is unknown in many places. Not many years ago there was no indication of gas and oil under the North Sea; now it is possible that known reserves are only a fraction of the total. It is believed that the experience of the North Sea will be repeated elsewhere in the world. On the mainland of the UK large deposits of coal have been found and there is an increasing possibility of oil and gas being found in commercially viable quantities.

The difficulty of forecasting the world's energy resources is shown by the experiences following the major oil crisis in 1974. At that time, estimates of the earth's store of fossil fuels ranged from about 50×10^{15} MJ to 170×10^{15} MJ, and it was calculated that the life of these resources, based on an annual growth rate of 5%, was between 30 and 65 years. Many new finds of oil and natural gas since then have transformed the picture; the energy resources crisis has been replaced by the environmental crisis caused by our use of fossil fuels. Fortunately the solution for the latter, which is to use less energy particularly from carbon-based fuels, will also help to maintain reserves of fossil fuels over a longer time period.

The final annual energy consumption for the UK, the USA, and the EC for the year 1989, taken from reference 17.5, is given in Table 17.1, the figures have been put into the same units of terajoules (1 TJ $= 10^{12}$ J), for ease of comparison. The OECD figures are given in megatonnes for coal and oil, in terajoules for natural gas, and in gigawatt hours for electricity; to convert the coal and oil figures the given OECD average calorific values of coal and oil of 25 427 megajoules per megatonnes and 42 835 megajoules per megatonnes have been used. 'Final energy consumption' is defined by the OECD as the energy

Table 17.1 Final annual energy consumption of the UK, USA, and the EC for 1989

	Energy consumed/(TJ)					
	UK		USA		EC	
Energy type	Total	Per head	Total	Per head	Total	Per head
Coal	0.51×10^6	0.009	2.41×10^6	0.010	2.20×10^6	0.006
Petroleum	2.83×10^6	0.050	28.99×10^6	0.117	17.46×10^6	0.051
Natural gas	2.00×10^6	0.035	15.15×10^6	0.061	8.11×10^6	0.024
Electricity	0.97×10^6	0.017	9.23×10^6	0.037	5.35×10^6	0.016
Total	6.31×10^6	0.110	55.78×10^6	0.224	33.12×10^6	0.097

665

actually delivered to the consumers of energy; the figures for electricity consumed therefore include a further consumption of fossil fuels. Table 17.2, also taken from reference 17.5, shows the electricity produced by the various methods expressed as a percentage; from the figures it is seen that for example for the UK about 75% of the electricity produced is from fossil fuels. Since the efficiency of production of electricity is only about 30%, then the electricity consumption figures in Table 17.1, for the UK for example, represent a use of fossil fuel per year of about

$$(0.75/0.3) \times 0.97 \times 10^6 = 2.425 \times 10^6 \text{ TJ}$$

Table 17.2 Methods of electricity production for the UK, USA, and the EC in percentage terms

Method of electricity production	UK	USA	EC
Hydro	2.10%	9.06%	8.22%
Nuclear	22.94%	18.99%	35.66%
Wind/solar/wave/tide geothermal	Negligible	0.34%	0.24%
Conventional thermal	74.96%	71.61%	55.88%
	100.00%	100.00%	100.00%

The populations of the UK, the USA, and the EC in 1989 are given by the OECD as 57.2×10^6, 248.8×10^6, and 342.6×10^6 respectively. Table 17.1 includes a crude measure of energy consumption per head obtained by dividing the energy consumed by the population. It can be seen that the UK is above the EC average, but the figure for the USA is more than twice as high as that of the UK; this is as expected since energy consumption is proportional to the gross domestic product of a country. The figures in Table 17.1 for annual energy consumption per head when converted into rates of energy used give the following:

Country	UK	USA	EC
Rate of energy used per head/(kW)	3.50	7.11	3.07

These figures compare with a figure of 0.05 kW for a typical Third World country.

Table 17.2 shows the comparatively high dependence of the EC on nuclear power; this is due mainly to the policies of France and, to a lesser extent, Germany. Public opinion in most countries is now moving against the increased use of nuclear power, for economic as well as for safety reasons, and before the end of the century the figure for renewable energy is forecast to increase as the nuclear and fossil fuel percentages decrease. Renewable energy sources and nuclear power are considered further in sections 17.6 and 17.7.

The dependence on energy from one particular source can be varied, but oil is required for other uses than just power production. Currently about 55% of the energy consumption of the EC is provided by oil; 22% of this oil consumption is for transport, lubrication, bitumen, and chemical feedstocks, none of which can be replaced by other energy sources. The breakdown of the UK energy statistics gives: industry, 42%; transport, 24%; domestic and business, 33%. Industry covers all products manufactured for home use and for export, electrical power, agriculture, iron and steel, materials manufacture, food, general engineering, chemicals, textiles, paper, bricks, etc. Transport covers road, rail, water, and air. Domestic and business covers heating, lighting, cooking, private vehicles, and entertainment.

Forecasts of energy resources and energy usage on a world-wide basis are fraught with difficulty, but the essential fact remains that all fossil fuels will eventually be consumed. Other energy sources must be found and energy should be used more efficiently. There is a widespread acceptance of these facts and many countries are implementing energy policies and encouraging research and development work to contribute to an energy-efficient future. The following list of factors shows the complexity of the problem:

1. The world-wide distribution of fuel resources is very different from the pattern of demand for energy.
2. The accessibility of fuels and the cost of exploration.
3. Different countries have different energy situations, e.g. Switzerland is low in natural fuel supplies while the UK is sitting upon large reserves of coal, oil, and gas. One country may therefore have to develop new technologies as an urgency, but another may rely on cheap fuel supplies and neglect its technology.
4. The effect of the demands for energy by the developing countries.
5. Supplies from different parts of the world are subject to political attitudes and the political stability of supplying nations.
6. Distant supplies are vulnerable to energy action and sabotage.
7. Storage capacity is limited.
8. Demand is not really controlled and is subject to variations over the day and over the year.
9. Government decisions can affect the demand for and provision of basic fuels.
10. Changes in technology are not easily regulated to need.
11. The growing unpopularity of fuels that may pollute the environment.
12. The price of fuel is an important feature in its selection, but this is subject very much to the world market forces.
13. Although 'alternative methods' of power generation and heat supply are attractive in principle, and it is essential to know what they can offer as part of energy policy development, there are good reasons for supporting existing methods of energy supply based on fossil and nuclear fuels, such as the following:

    (a) the resources are known to a reliable extent;
    (b) the technology is established and the economics are understood, i.e. the cost–demand balance and labour needs are known;
    (c) there is a reasonable variety in form (gas, solid, and liquid), use and main features;
    (d) fundamental to all considerations is the high potential energy per unit mass of the fossil and nuclear fuels.

667

Sankey diagrams

A high proportion of fuel is used in generating electricity (33% in the UK) and it is shown in earlier chapters that only about 30% of the fuel energy is converted into electricity, i.e. 70% goes to waste! If the overall losses could be reduced the useful output would be increased. The invitation to improve known methods of power generation is evident, but there are fundamental limitations to gains to be obtained from conventional cycles as described in Chapter 8. Figures 8.8 and 8.10 (pp. 242 and 244), show the variations in efficiency with boiler pressure and steam temperature respectively. It is seen that increases in efficiency by raising steam pressure and temperature are likely to be very expensive in terms of capital investment, and the amount of improvement in efficiency small.

A steam power plant working continuously on full load may have an overall efficiency of about 30%. This is low due to the large percentage of energy which is rejected to the condenser. The power plant energy distribution can be shown diagrammatically by a Sankey diagram as shown in Fig. 17.1; the system shown has feed heating, an air pre-heater, and an economizer.

Fig. 17.1 Sankey diagram for condensing power plant

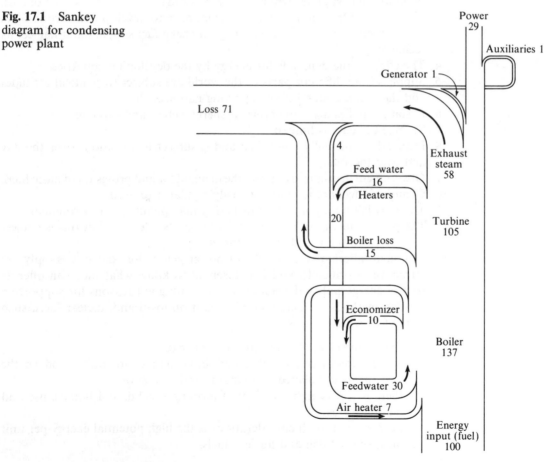

For a factory or large process plant generating its own power and with a requirement for process steam the overall efficiency of energy use is much greater, as shown by the Sankey diagram in Fig. 17.2. The process energy used plus the power output is 81% of the energy input; the heat to power ratio is just over 10. Combined power and heating is considered in more detail in section 17.3.

Fig. 17.2 Sankey diagram for combined power and process steam plant

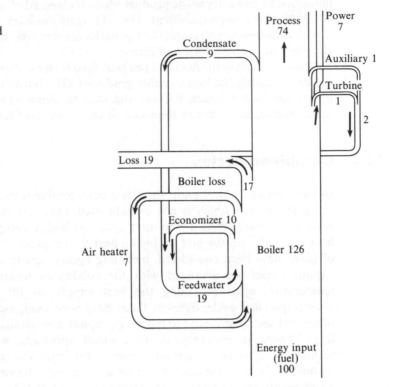

Fluidized boiler

Prospects for energy saving and the efficient use of fuels are not necessarily to be found only in areas of completely new technology. A great deal of research work has been done over the years to re-establish coal as a primary source of fuel, particularly if it is an alternative to the more expensive fuel oil. One of the developments has been the firing of coal in a fluid bed as an alternative to the normal combustion in a steam boiler which creates soot, ash, and noxious gases like oxides of sulphur (SO_x) and oxides of nitrogen (NO_x).

The *fluidized bed boiler* is an economical venture in the 10–20 MW range which is particularly useful for district heating systems and industrial energy plants.

In fluid-bed firing the coal is mixed with fine particles of sand, limestone, and ash and is burned in a suspended state in one or more fluid compartments.

669

The combustion air is supplied through a nozzle causing the air to swirl and hence the coal mixture also. Water tubes are immersed in the fluid bed which absorb heat directly and control the temperature of the bed. The heat transfer rate is good due to the swirling movement of the charge. The fluid bed is convective in action also as in conventional combustion.

At the combustion temperature, 800–900 °C, the sulphur combines with the limestone to give a dry waste product which is disposed of with the ash removing 80–90% of the sulphur content. The NO_x emission level is low because of the low combustion temperature. Dust particles are removed by cyclone separators, electrostatic precipitators, and filters.

The attraction of the fluidized bed coal-fired boiler is the relatively low price of coal, particularly the lower quality grades which it burns very well. The capital outlay for such a boiler, the ash and coal handling equipment, control and maintenance, etc. is higher than that of an oil- or gas-fired boiler.

17.2 Combined cycles

In conventional power plant there is a considerable wastage of energy due to the heat rejected which is not usefully used. One way of making the power production more efficient is to use a cycle combining two power units with the heat rejected from the first supplying heat to the second. Many different types of plant have been considered, including binary vapour cycles in which two vapour cycles are arranged with the condensation process for the high-temperature cycle providing the heat supply to the evaporator of the low-temperature cycle; different fluids have been used, usually steam and one other, but such plants tend to have high capital costs compared with the overall improvement in efficiency. A more direct approach, with existing proven technology, is to use the gas and steam turbine cycles in combination. The gas turbine is the higher-temperature unit and the gases leaving the turbine are at a sufficiently high temperature to be used as a source of heat for the production of steam at a suitable pressure and temperature. Figure 17.3 shows such a combination using a closed-cycle gas turbine unit, and Fig. 17.4 shows a more common arrangement with an open-cycle gas turbine unit in which the exhaust gases pass directly to the steam generator. The open cycle unit also allows the burning of further fuel in the exhaust gas stream in the generator since

Fig. 17.3 Closed-cycle gas turbine plant combined with a steam plant

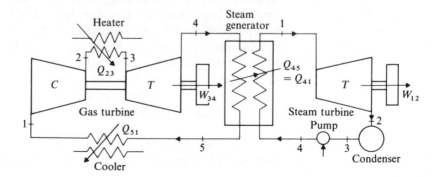

670

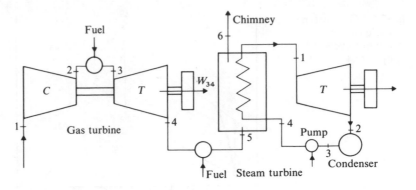

Fig. 17.4 Open-cycle gas turbine plant combined with a steam plant

the air–fuel ratio of a gas turbine is high and there is sufficient oxygen in the turbine exhaust to sustain further combustion.

The overall thermal efficiency is increased and since the installed cost of a gas turbine per unit power output is less than that of a steam turbine, there are obvious economic advantages. Disadvantages of a combined cycle are the greater complexity leading to loss of flexibility and reliability; the gas turbine in the past has also used a more expensive fuel although more and more turbines are now operating on natural gas which improves the economics of the overall plant.

Example 17.1   A combined power plant consists of a gas turbine unit and a steam turbine unit, the exhaust from the gas turbine being supplied to the steam generator. Using the data below, neglecting the mass flow rate of fuel, feed pump work, and all pressure losses, calculate:

(i) the cycle efficiency for the gas turbine cycle;
(ii) the cycle efficiency for the steam cycle if the heat supplied in the generator were supplied by an external fuel supply;
(iii) the mass flow rates of air to the gas turbine and steam to the steam turbine;
(iv) the overall efficiency of the combined cycle.

Data   Pressure ratio for the gas turbine cycle, 8; inlet air temperature to compressor, 15 °C; maximum cycle temperature for gas turbine cycle, 800 °C; temperature of gases leaving steam generator, 160 °C; steam conditions at entry to turbine, 20 bar and 400 °C; condenser pressure, 0.05 bar; total power output of the plant, 50 MW; isentropic efficiencies of air compressor, gas turbine, and steam turbine, 80%, 82%, and 80% respectively; c_p and γ for the combustion gases, 1.11 kJ/kg K and 1.333.

Solution   (i) The gas turbine cycle is shown in Fig. 17.5(a).

$$T_{2s} = T_1(p_2/p_1)^{(\gamma-1)/\gamma}$$

where $T_1 = 15 + 273 = 288$ K.

$$T_{2s} = 288(8)^{0.4/1.4} = 521.7 \text{ K}$$

671

Fig. 17.5 Gas turbine unit cycle (a) and steam plant cycle (b) for Example 17.1

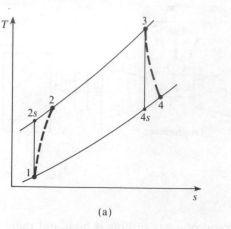

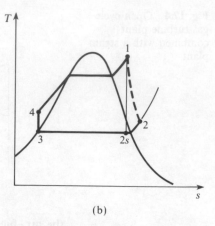

(a)

(b)

$$T_2 = 288 + \frac{(521.7 - 288)}{0.8} = 580.1 \text{ K}$$

$$T_{4s} = \frac{T_3}{(p_3/p_4)^{(\gamma - 1)/\gamma}}$$

where $T_3 = 800 + 273 = 1073 \text{ K}$

$T_{4s} = 1073/(8)^{0.333/1.333} = 638.3 \text{ K}$

therefore

$$T_4 = 1073 - 0.82(1073 - 638.3) = 716.5 \text{ K } (443.5\,°C)$$

Then Work output, $-W = c_{pg}(T_3 - T_4) - c_{pa}(T_2 - T_1)$

$$= 1.11(1073 - 716.5) - 1.005(580.1 - 288)$$

$$= 102.2 \text{ kJ/kg}$$

and Heat supplied $= c_{pg}(T_3 - T_2) = 1.11(1073 - 580.1) = 547.1 \text{ kJ/kg}$

Gas turbine cycle efficiency $= 102.2/547.1 = 18.7\%$

(ii) The steam cycle is shown in Fig. 17.5(b). From tables or h–s chart:

$h_1 = 3248 \text{ kJ/kg}, \qquad h_3 = h_4 = 138 \text{ kJ/kg}$

$h_{2s} = 2173 \text{ kJ/kg}$

$h_1 - h_2 = 0.8(3248 - 2173) = 860 \text{ kJ/kg}$

i.e. Work output, $-W = 860 \text{ kJ/kg}$

The heat to be supplied in the steam generator is given by; $(h_1 - h_3) = 3248 - 138 = 3110 \text{ kJ/kg}$, therefore, if the heat is supplied from an external fuel supply

Steam cycle efficiency $= 860/3110 = 27.7\%$

(iii) Let the mass flow rates of air and steam be $\dot{m}_a$ and $\dot{m}_s$. For an energy balance on the steam generator

$$\dot{m}_a \times 1.11 \times (443.5 - 160) = \dot{m}_s(3110)$$

i.e. $\dot{m}_a/\dot{m}_s = 9.883$

The total power output is 50 MW, therefore

$$\dot{m}_a(102.2) + \dot{m}_s(860) = 50\,000$$

i.e. $\dot{m}_s\{(9.883 \times 102.2) + 860\} = 50\,000$

therefore

$$\dot{m}_s = 26.74 \, \text{kg/s}$$

and $\dot{m}_a = 9.883 \times 26.74 = 264.3 \, \text{kg/s}$

(iv) Combined cycle efficiency $= \dfrac{50\,000}{264.3 \times 547.1} = 34.6\%$

17.3 Combined heat and power (co-generation)

The most efficient modern heat engine converts about 40% of the fuel energy input into useful power and rejects 60% as waste heat. If there is a requirement for heat as well as power then it is more efficient to design one plant to provide both rather than provide each separately. Such a system is known as combined heat and power, CHP, or co-generation (in North America).

The optimization of use of the energy supplied to a system is called the 'total energy' approach (see reference 17.4), and has the objective of using all of the thermal energy in a power system at the different temperatures at which it becomes available, to produce work, or steam, or the heating of air or water, thereby rejecting the minimum of waste energy. For many years until the late 1970s energy-intensive industries such as iron and steel, oil refining, paper production, and chemical processes, used CHP plant with heat to power ratios in the range from about 5 to 10. In the 1980s, due to a growing emphasis on energy efficiency and a move away from energy-intensive systems, such high heat to power ratios were not required and there was a decline in industrial use of CHP. During the same period there was a rapid increase in the use of plant combining gas and steam turbines, with heat production for space heating and hot-water supply; the heat to power ratio for such plant is about half that of the previous industrial schemes using steam turbines with process steam. Such large CHP schemes have been widely introduced for providing power, heating, and hot water for districts, towns and even cities (see later).

Micro-CHP

Small scale CHP systems, known as micro-CHP, are packaged units with comparatively small power outputs (e.g. up to about 200 kW), with heat to power ratios in the range 2–3. These systems are suitable for complete buildings such as hotels, hospitals, colleges, swimming-pools, and other public buildings. The power generator can be linked to the electric grid so that power may be purchased if demand exceeds the capacity of the unit, or sold to the grid if demand falls. In the UK the 1983 Energy Act made possible the sale of electricity to the national grid and led to the rapid development of such schemes.

The engines used are industrial oil engines or automotive engines adapted to run on natural gas. Heat is transferred to water from the engine cooling jacket, from the engine oil cooler, and from the engine exhaust gases; a typical system is shown in Fig. 17.6.

Fig. 17.6 Typical Micro-CHP system

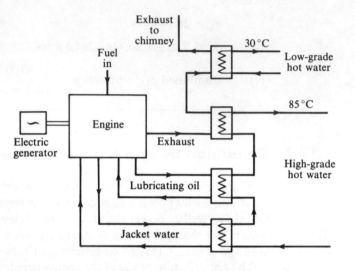

Example 17.2　A large hotel has a maximum demand of 90 kW of power and 240 kW of heat for space heating and hot water. It is decided to use six micro-CHP units running on natural gas, each developing 15 kW of power and each with a heat output of 40 kW. The jacket and oil cooling systems provide thermal energy equivalent to 10% of the fuel energy input, and the exhaust cooler provides thermal energy equivalent to 55% of the fuel energy input. Water enters each engine at 30 °C and leaves at 85 °C. Taking the specific heat capacity of water as 4.18 kJ/kg K, and the calorific value of natural gas as 38.5 MJ/m^3, calculate:

(i) the rate of volume flow of natural gas used;
(ii) the overall efficiency of the system;
(iii) the mass flow rate of water available at 85 °C.

Solution　(i) Each engine provides 40 kW of thermal energy which is made up of 10% from the jacket and oil coolers, and 55% from the exhaust heat exchanger.
Hence, for each engine

$$\text{Fuel energy input} = \frac{40}{(0.1 + 0.55)} = 61.54 \text{ kW}$$

Then   Rate of natural gas used $= \dfrac{6 \times 61.54}{38.5 \times 10^3} = 0.0096 \text{ m}^3/\text{s}$

In the UK gas consumption is measured in therms where 1 therm is equivalent to 29.307 kW h. Therefore in this case

$$\text{Rate of gas consumption} = \frac{6 \times 61.54}{29.307} = 12.6 \text{ therm/h}$$

(ii) The overall efficiency of the system is the total useful energy divided by the energy input from the gas,

i.e.   Overall efficiency $= \dfrac{(15 + 40)}{61.54} = 89.4\%$

(iii) The energy transferred to the water in each engine is 40 kW, hence

$$6 \times 40 = \dot{m}_\text{w} \times 4.18 \times (85 - 30)$$

i.e. $\dot{m}_\text{w} = 1.044 \text{ kg/s}$

Large-scale CHP and district heating

CHP combined with district heating has been slow to develop on a large scale in the UK. Some of the reasons why this is so are as follows:

(a) The generating system set up in the 1950s is highly efficient with a high load factor, easily adjustable to adapt to load changes, and with most power stations remote from centres of population; retrofitting such stations to provide district heating would be difficult and not cost effective for these reasons.
(b) The UK does not have extreme temperatures in winter on a continuous basis, unlike countries such as those of Central and Eastern Europe where district heating is commonplace.
(c) As long as the existing power stations can satisfy the current demand for electricity, and heating can be provided comparatively cheaply by gas or oil, the relative returns on an investment in a district heating scheme would not be high.

Government reports in the late 1970s and early 1980s, the Marshall Report and the Atkins Report (see references 17.6 and 17.7), concluded that complete city schemes were viable and cost effective. A limited amount of funding was provided centrally to enable three pilot schemes to be started; in one case this was supplemented by a further EC grant. Progress since then has been slow. Some of the factors which may have contributed to the slow progress are listed below:

1. Load factors are low because of variations in demand, particularly for heating; standby heating may be necessary to top up when weather conditions are severe.

2. When heat is provided centrally it must be metered accurately; systems in which flats are metered as a whole and each flat holder pays a proportion cause problems; a system of accurate metering of each household is expensive.

3. For heating, private consumers have become used to having a choice of fuel and when and where to heat; there is an in-built resistance to heating systems which are imposed from the local administration.

4. The capital cost, and necessary upheaval, of laying water mains for heating throughout an existing city is a major factor; previous experience of corrosion in mains has also led to fears of expensive maintenance.

Most European countries have developed district heating on a step-by-step basis: initially individual boiler houses provide heating only for small districts; power stations are then built and the boiler stations are gradually shut down leaving some for stand-by heating; finally heat exchangers are added in the power stations to supply the whole network. Once a complete system is in operation new customers can be added by extending the pipework.

A typical example of a large-scale CHP scheme with district heating is that of The Hague in The Netherlands. A diagrammatic arrangement of the plant is shown in Fig. 17.7. Power is provided by two gas turbine units, each developing 25 MW, and a steam turbine unit also developing 25 MW; the three electric generators are identical for ease of maintenance and control. The exhaust from each gas turbine passes through a steam generator and water heater; additional heating of the water is provided by bleed steam from the exit of the HP turbine. The following example is based on the data from The Hague scheme.

Fig. 17.7
Diagrammatic
representation of a CHP
plant at The Hague,
Netherlands

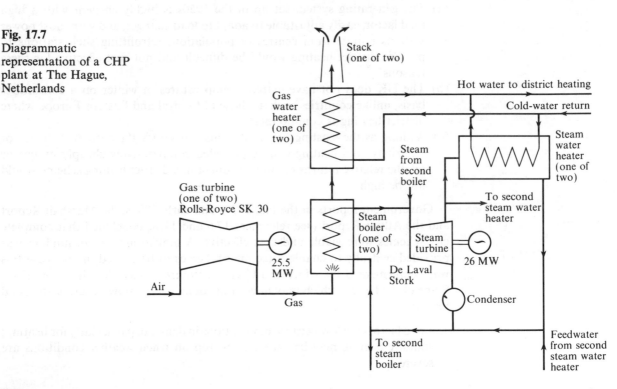

Example 17.3   The data below refers to a plant similar to the one shown in Fig. 17.7. It may be assumed that the water leaving the heat exchanger is saturated at 2 bar and is mixed with the condensate from the condenser before being pumped back to the steam generator. Neglecting the mass flow rate of fuel to the gas turbines, feed-pump work in the steam system, and all heat and pressure losses, calculate:

(i) the airflow induced into each gas turbine compressor;
(ii) the total steam flow required;
(iii) the total heat supply available for the district heating scheme;
(iv) the mass flow rate of water to the district heating scheme;
(v) the overall efficiency of the entire system.

Data   *Gas turbine:* electric power output of each turbine, 25 MW; air temperature at entry, 15 °C; pressure ratio, 12; compressor and turbine isentropic efficiencies, 83% and 85%; maximum cycle temperature, 1013 °C; exhaust gas temperature entering chimney, 83 °C; specific heat and isentropic index for gases, 1.11 kJ/kg K and 4/3; combustion efficiency, 98%.

Steam plant: electric power output of turbine, 25 MW; steam supply conditions, 30 bar and 450 °C; pressure of bleed steam to heat exchanger, 2 bar; mass flow rate of bleed steam to heat exchanger, 17 kg/s; overall isentropic efficiency of steam expansion process, 80%; condenser saturation temperature, 25 °C; no undercooling in condenser; take the expansion process in the turbines as a straight line on the *h–s* chart.

Electric generators: combined mechanical and electrical efficiency, 95%.

District heating: flow temperature, 115 °C; return temperature, 75 °C; mean specific heat capacity of water, 4.2 kJ/kg K.

Solution   (i) The cycle for each gas turbine unit is shown on a *T–s* diagram in Fig. 17.8.

$$T_{2s} = (15 + 273)(12)^{0.4/1.4} = 585.77 \text{ K}$$

Fig. 17.8 Gas turbine cycle on a *T–s* diagram for Example 17.3

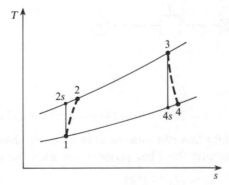

therefore

$$T_2 = 288 + \frac{(585.77 - 288)}{0.83} = 646.76 \text{ K}$$

$$T_{4s} = (1013 + 273)/(12)^{0.25} = 690.95 \text{ K}$$

therefore

$$T_4 = 1286 - 0.85(1286 - 690.95) = 780.21 \text{ K } (507.21\,^\circ\text{C})$$

Then   Net gas power output from each turbine

$$= \frac{25\,000}{0.95} \text{ kW} = \dot{m}_\text{a}\{1.11(1286 - 780.21) - 1.005(646.76 - 288)\}$$

therefore

$$\dot{m}_\text{a} = 131.0 \text{ kg/s}$$

i.e.   Airflow rate into each compressor $= 131$ kg/s

Fig. 17.9 Plant diagram (a) and steam turbine process on an h–s chart (b) for Example 17.3

(ii) For the steam cycle (shown diagrammatically in Fig. 17.9(a) and on a sketch of the h–s chart in Fig. 17.9(b)), we have, from the h–s chart:

$$h_1 = 3343 \text{ kJ/kg} \qquad h_{3s} = 2107 \text{ kJ/kg}$$

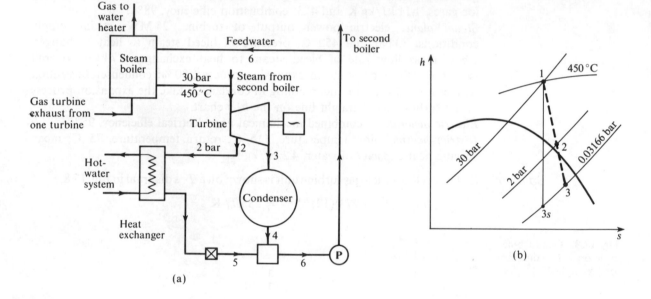

(a)

(b)

therefore

$$h_3 = 3343 - 0.8(3343 - 2107) = 2354 \text{ kJ/kg}$$

A straight line can now be drawn on the chart joining points 1 and 3. Where this line cuts the 2 bar pressure line fixes the point 2,

i.e. $h_2 = 2865$ kJ/kg

Let the total steam flow required be $\dot{m}_\text{s}$ kg/s, then

Power output from steam

$$= \frac{25\,000}{0.95} = \dot{m}_\text{s}(3343 - 2865) + (\dot{m}_\text{s} - 17)(2865 - 2354)$$

$$\dot{m}_s = 35.39 \text{ kg/s}$$

i.e.   Mass flow rate of steam required $= 35.39$ kg/s

(iii) To find the heat supplied to the water of the district heating system from the gas turbine exhaust gases it is necessary first to calculate the heat received by the steam in the generators; to do this we require the temperature of the feed water entering the boiler at state 6.

Applying a heat balance to the mixing process after the condenser we have

$$\{(35.39 - 17) \times 104.8\} + (17 \times 505) = 35.39 \times h_6$$

where $h_4 = h_f$ at $25\,^{\circ}\text{C} = 104.8$ kJ/kg, and $h_5 = h_f$ at 2 bar $= 505$ kJ/kg, therefore

$$h_6 = 297 \text{ kJ/kg}$$

Then   Total heat supplied to steam in both generators

$$= \dot{m}_s(h_1 - h_6) = 35.39(3343 - 297)$$

$$= 107\,798 \text{ kW}$$

The gas turbine exhaust gas enters the chimney at $83\,^{\circ}\text{C}$, therefore we have

Total heat supplied by exhaust gases

$$= 2 \times \dot{m}_a \times 1.11 \times (t_4 - 83)$$

$$= 2 \times 131 \times 1.11 \times (507.21 - 83)$$

$$= 123\,369 \text{ kW}$$

Hence by difference

Heat supplied to district heating water by gases

$$= 123\,369 - 107\,798 = 15.57 \text{ MW}$$

The heat supplied to the district heating water in the steam heat exchanger is given by

$$\dot{m}_s(h_2 - h_5) = 17(2865 - 505) = 40.12 \text{ MW}$$

i.e.   Total heat supplied for district heating

$$= 15.57 + 40.12 = 55.69 \text{ MW}$$

(iv)   Mass flow of water for district heating

$$= 55\,690/4.2(115 - 75) = 331.5 \text{ kg/s}.$$

(v) The overall efficiency is given by the total power plus heat supply divided by the thermal energy supplied by the fuel.

Heat supplied by fuel in the gas turbine cycles

$$= \frac{2 \times \dot{m}_a \times c_{pg}(T_3 - T_2)}{0.98}$$

$$= \frac{2 \times 131 \times 1.11(1286 - 646.76)}{0.98}$$

679

$$= 189.7 \text{ MW}$$

Therefore

$$\text{Overall efficiency} = \frac{(3 \times 25) + 55.69}{189.7}$$

$$= 68.9\%$$

Energy from waste

One of the outcomes of a developing energy programme over recent years has been the renewed interest in using municipal, industrial, hospital, and agricultural waste as a source of heat energy. While exploitation of this resource in the UK has been slow there are some notable examples of successful refuse incineration heat-recovery plants at Edmonton (see ref. 17.4), Sheffield, and Coventry. Small-scale incineration plants with heat recovery are also becoming more common particularly in hospitals. Solid waste has to be disposed of and collection and disposal is expensive, so the financial return possible from burning the waste to give a useful supply of energy is worth considering. The nature of the waste material brings its own problems when used as a fuel since waste is highly heterogeneous and on combustion produces corrosive gases. This introduces restrictive design criteria not present in the design of boilers and ancillary equipment for plant using conventional fuels.

Another source of energy from waste is the methane generated when waste decomposes in a landfill site. For further information on this and other energy-saving waste disposal schemes consult reference 17.4.

17.4 Energy management and energy audits

Energy-consuming systems must be effectively managed. Supplies should be abundant, of low cost and safe to handle. Each energy-consuming system should be designed in the beginning with efficient energy objectives for the most economic manufacture of products or heating service. Governments should have national policies involving those of cities, towns, etc. and, in turn, each company, public or private building, should have complementary energy policies. A well-designed system will include a means of monitoring the energy demand and the supply and distribution so that it can be adjusted to operate to selected values of the main parameters which will usually be the temperatures at the main points in the system. This requires the services of technologists in the design and operation of energy systems who will require a knowledge of thermodynamics, heat and mass transfer, fuels, fluid flow, systems analysis, and control. Some of these topics are dealt with in other chapters in this book and the remainder will be touched upon in this and later sections.

Most energy-using systems and buildings could be made more efficient immediately by an application of common-sense methods of preventing heat loss to the surroundings and lowering temperature levels where possible since heat loss is proportionate to temperature difference (Ch. 16).

A basis of organization of an energy policy is the energy audit which is a formal account of the energy consumption and costs of a building or company over a period of a year, or shorter periods if necessary. The account or audit is broken down into different sections. Table 17.3 shows the form of an energy consumption and cost distribution account.

Table 17.3 Energy consumption and cost

Energy	Quantity	Heating value	Price/ unit	Cost (£)	Energy (MJ)	Cost (£/MJ)
Coal	A tonne	27 500 MJ/tonne	a £/tonne	A × a	27 500 × A	a/27 500
Heavy oil	B tonne	43 200 MJ/tonne	b £/tonne	B × b	43 200 × B	b/43 200
Medium oil	C tonne	43 600 MJ/tonne	c £/tonne	C × c	43 600 × C	c/43 600
Gas oil	D tonne	45 480 MJ/tonne	d £/tonne	D × d	45 480 × D	d/45 480
Gas	E m^3	38.5 MJ/m^3	e £/m^3	E × e	38.5 × E	e/38.5
Electricity	F kW h	3.6 MJ/kW h	f £/kW h	F × f	3.6 F	f/3.6
			Totals			

Numerical values are substituted for A to F for the actual quantities used and the current prices/unit for *a* to *f*. The totals evaluated on this basis are readily computed, but perhaps what is important is the cost for each form of energy used as this suggests where economies may be made; changing to a different, cheaper form of energy may also be a possibility.

The next part of the audit is a breakdown of the different forms of energy and where it is consumed. An example for a manufacturing company is shown in Table 17.4. and is a simplified example of a way in which the energy distribution can be examined; an actual audit would be more detailed.

Table 17.4 Energy distribution through Company A

Energy	Workshop or department				
	Workshop 1	Workshop 2	Boiler- house	Stores	Office complex
Electricity (kWh)					
Machines					
Lighting					
Compressed air					
Gas (MJ)					
Oil (MJ)					
Coal (MJ)					

Table 17.5 Fuels and lubricating oils used for the transport for Company A including garaging

Energy (quantity)	Internal	External delivery, staff	Cost/(£)
Petrol/(litre)			
Diesel/(litre)			
Lubricant/(litre)			
Electricity/(kW h)			
Vehicle miles			
Load/(tonne)			

Another audit would be made for the transport used by the company as shown in Table 17.5.

Tables 17.4 and 17.5 should be completed monthly and comparisons can then be made between the months and across the years. Changes should be explained and any corrective action taken to reduce fuel consumption. Such action may involve improving the maintenance standards, checking control equipment and level settings, modifying plant or operation control, installing new plant, changing process methods, installing more meters, etc.

Table 17.6 gives some energy values for different fuels and useful conversion factors for the kinds of calculations covered in this chapter. Energy values vary with the quality of the particular fuel and the values quoted are representative examples only. The values quoted here are taken from reference 17.8.

Table 17.6 Energy values and conversion factors

Energy source	Energy content
Coal	
Anthracite	32 000 MJ/tonne
Good bituminous	30 000 MJ/tonne
Average industrial	28 000 MJ/tonne
Poor industrial	21 000 MJ/tonne
Oil (relative density)	
Gas (0.835)	45 600 MJ/tonne
Light (0.935)	43 500 MJ/tonne
Medium (0.95)	43 000 MJ/tonne
Heavy (0.97)	42 600 MJ/tonne
Gas	
North Sea	38.5 MJ/m^3

Conversion factors

1 Btu = 1.055 kJ	1 therm = 100 000 Btu
1 Btu/ft^3 = 37.259 kJ/m^3	1 therm = 105.506 MJ
1 kW h = 3.6 MJ	1 tonne = 1000 kg
1 lb = 0.4536 kg	1 ton = 1.016 tonne
1 imperial gallon = 4.546 l	1 therm/ton = 103.839 kJ/kg

Degree days

A comparison of energy audits for different periods may have limited significance and may even be misleading as the conditions under which they were obtained may have been quite different.

In the UK it is assumed that heating in buildings is necessary only when the outside temperature falls below 15.5 °C (60 °F). The normal inside temperature is taken as 18.3 °C (65 °F), and the heat from internal sources such as occupants, lighting, machinery, plant, etc. is assumed to be sufficient to maintain the inside at 18.3 °C when the outside temperature is at 15.5 °C.

When the outside temperature falls below 15.5 °C heating is necessary and the fuel used is proportional to the temperature difference between 15.5 °C and the actual outside temperature. This is so since the building heat losses are proportional to the temperature difference between the inside and outside temperatures. If the outside temperature averaged over 24 h was 14.5 °C, say, then this would represent 1 degree day. If maintained for a week the cumulative heat loss would be proportional to 7 degree days and so on. If degree days and the fuel consumed are measured over the same period the quantity of fuel per degree day can be obtained and is a criterion of the energy consumption corrected for climatic conditions. Outside temperatures do not remain constant for 24 h periods and averages are taken by one of three formulae recommended by the Meteorological Office, see references 17.4 and 17.8.

Degree-day measurements are taken at 17 meteorological stations spread over England, Wales, Scotland, and Northern Ireland and the reports for the regions are available from the Meteorological Office. Table 17.7 shows the average values of degree days taken over a 20-year period for a selection of the 17 stations.

Table 17.7 Degree days averaged over a 20-year period

Region	Jan	Feb	Mar	Apr	May	June	July	Aug	Sept	Oct	Nov	Dec	Total
1 Thames Valley	338	306	273	203	115	51	22	24	54	133	243	301	2163
4 South Western	288	270	260	205	131	62	27	29	56	118	204	250	2000
6 Midlands	369	340	306	242	158	83	43	49	89	178	277	333	2467
14 East Scotland	380	342	322	259	190	108	61	66	110	196	297	349	2680
16 Wales	325	307	297	239	165	92	49	46	78	150	233	287	2268
17 Northern Ireland	359	320	309	240	167	91	52	59	100	178	279	324	2478

The reference temperature of 15.5 °C is usually used, but for some buildings higher inside temperatures than 18.3 °C are necessary, e.g. in industrial buildings or hospitals. The degree days tabulated may be referred to another base temperature as described in reference 17.8.

The fuel used per degree day is a useful guide to the fuel consumption figures for buildings with similar use and size. Degree days should not be used for short period tests and at least monthly figures are necessary. If combined heating and hot-water systems are installed the two parts of the heating load as expressed in degree days cannot be separated due to the fluctuations in demand. If the incidental heating is not equivalent to 2.8 K the degree-day method will be less reliable. Heat losses and hence fuel consumptions are affected by other influences such as prevailing winds, humidity, solar radiation, cloud, fluctuating demands, thermal capacity, etc. By their nature degree days need to be used with caution in calculating fuel consumptions and cannot be used to forecast weather conditions.

Example 17.4 A central heating boiler-house provides steam for both the process and space-heating requirements of a factory complex. The boilers are fired with gas oil costing £3/GJ and run at an average efficiency of 71%. The monthly fuel consumptions and corresponding degree days (D days) referred to 15.5 °C for a previous period are as given in Table 17.8.

Table 17.8 Data for Example 17.4

Month	Consumption/(GJ)	D days	20-year average D days
S	12 040	102	94
O	13 100	156	171
N	NA*	300	286
D	17 460	370	360
J	NA	370	379
F	17 000	350	343
M	NA	312	320
A	NA	215	238
M	12 600	132	156
J	NA	60	79
J	NA	28	48
A	10 600	30	53
			2527

NA = not available.

Recent modifications at the factory have included the installation of a new process consuming 1000 GJ steam/month and the installation of an economizer to increase the boiler efficiency to 75%.

Using the degree days as being applicable to all space heating loads, state any assumptions made and determine:

(i) the expected annual fuel consumption of the modified factory;
(ii) the maximum monthly fuel consumption of the modified factory;
(iii) the payback period in years for the economizer if the capital cost is £40 000.

Solution A graph of the fuel consumption in gigajoules against the recorded degree days is drawn as shown in Fig. 17.10 and gives a base load at zero degree days of 10 000 GJ/month which is a measure of the process energy required. The slope of the graph gives a space heating factor of 20 GJ/D day. The actual energy

Fig. 17.10 Oil consumption against Degree days for Example 17.4

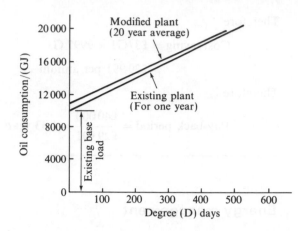

requirements taking into account the boiler efficiency of 71% are as follows:

Space heating factor $= 0.71 \times 20 = 14.2$ GJ/D day

Base load $= 0.71 \times 10\,000 = 7100$ GJ/month

Base load for the new process $= 7100 + 1000 = 8100$ GJ/month

and for a new boiler efficiency of 75% the new fuel consumption will be

For the base load $= \dfrac{8100}{0.75} = 10\,800$ GJ/month

For space heating $= \dfrac{14.2}{0.75} = 18.933$ GJ/D day

The projected annual figures are based on the 20-year average degree-day figure quoted:

(i) The annual consumption $= 12$ months $\times\ 10\,800$ GJ/month
(base load + space $\qquad + 2527$ D days $\times\ 18.933$ GJ/D days
heating) $\qquad\qquad\qquad = 129\,600 + 47\,844$
$\qquad\qquad\qquad = 177\,444$ GJ/annum

(ii) The fuel consumption is at a maximum in January. Therefore,

Maximum monthly fuel consumption
$= 10\,800 + (379 \times 18.933)$
$= 17\,976$ GJ/month

(iii) The savings to be expected from fitting the economizer at £40\,000 are based on the existing cost of fuel at £3/GJ. The energy consumption required for the new base load at the original plant efficiency of 71% would be $12 \times 8100/0.71 = 136\,901$ GJ and for the space heating, based on the original space heating factor of 20 GJ/D is $2527 \times 20 = 50\,540$ GJ. Therefore

Energy saving $= 136\,901 + 50\,540 - 177\,444$
$= 9997$ GJ/annum

Therefore

$$\text{Cost saving} = £3/GJ \times 9997\,GJ$$

$$= £29\,991 \text{ per annum}$$

Therefore

$$\text{Pay-back period} = \frac{£40\,000}{£29\,991} = 1.33 \text{ years}$$

Energy management

Efficient energy management must cover the whole industrial, domestic or service system used and a consideration here of the requirements will be based on an industrial unit as a comprehensive example. Energy considerations should be taken into account at the start of a new industrial project and may even be influential in deciding on the geographical location in the first place. The problem of improving the energy audit of an existing plant is quite different and has inherent constraints. Even for a new enterprise energy conservation may not be able to be taken to its limit because of the investment capital involved even though the long-term saving may be attractive, so from the outset a compromise solution may have to be accepted. The energy analysis includes the energy required for the manufacture of the product, the energy used in the manufacturing process, the energy needed to support the manufacturing environment, and the effect of the industrial unit on its external surroundings including the economic disposal of waste. With existing plant the opportunities for improving the energy consumption may be more restrictive and relatively more expensive if constructional changes, the replacement of plant, and the changing of fixed attitudes and practices are concerned. Sankey diagrams, as shown in section 17.1, illustrate energy use.

The industrial recession in Western countries in the mid-1970s created an active interest in the relative industrial performances of the Western nations and Japan, which emerged as a leading industrial nation producing an increasing range of high-quality manufactured goods at low prices. The manufacture of motor vehicles can be taken as an example of particular significance to the UK and the USA and this example has been subject to some analysis. One of the features of the study was the energy consumption, although it is not the main factor in relation to others like the productivity of more modern plants, labour costs, and employee attitudes and practices.

Japan had to be energy conscious as 90% of its energy was imported – about 75% being oil – and so its industry implemented conservation techniques more rigorously than did industry in the West. Waste-heat recovery systems, the burning of waste paint solvents, and the purification and recycling of heated air all contributed to making the energy costs of Japanese products 20% less than those of many of their competitors. Between 1973 and 1978 the energy consumption per vehicle dropped from 13.6 to 10.8 GJ. For a similar vehicle built in the USA in 1978 the energy consumption was 31.44 GJ/vehicle, an improvement on 40.32 GJ/vehicle for 1972, and even taking into account the mass ratio of 1.8 for the US to Japanese vehicles the energy consumption for the Japanese is better by 1.9 to 1. For one UK manufacturer the figure in the same year was 22 GJ/car and for another, working at less than full production capacity, the energy consumed was 58 GJ/car. The relative cost advantages to the Japanese car were estimated at £45–£85 depending on the capacity working. This difference is significant but is swamped by the much lower labour cost savings of £1200–£1600 of a high production unit employing efficient manufacturing methods.

Some of the factors contributing to better energy use by the Japanese manufacturers are instructive.

(a) The holding of buffer stocks of parts and materials is held to a very low period of 3–4 h, where UK manufacturers hold several weeks' supply of the main items. Cash flow is reduced and savings are made in buildings, heating, lighting, and employees.
(b) The factories are compact and do not have the long lines connecting some UK companies where car bodies are made in one place and then transported long distances for assembly.
(c) The factories are sited in a temperate climate giving a 20% saving on similar size plants in the USA and 15% on the UK plants. They have better low-grade heat recovery, better insulation, and more use of recirculated air.
(d) The working periods are continuous – two shifts instead of one – therefore there are shorter shut-down periods, giving about 30% reduction in the energy for the working environment.
(e) The energy cost of the basic materials such as steel is less for the Japanese industry due to a high investment in modern energy-saving plant which give higher yields than their Western competitors of the high-quality steels used in vehicle manufacture. The relative energy costs for a tonne of sheet steel was UK 40 GJ, US 38 GJ, and Japan 28.5 GJ.

It is hoped that the case just described suggests to the reader the economic importance of energy to a manufacturing nation which has to be competitive in world markets. It is not possible to give a complete treatment of the subject, but the following points may be made:

1. Energy production itself is the greatest user of fuel energy.
2. The production of materials is the second largest consumer of energy.
3. Every product has an energy cost, e.g. 8 MJ for a milk bottle, 25 000 MJ for a colour TV. The energy savings of double glazing are enthusiastically quoted by salesmen, but an overall analysis would take into account that it costs 6000 MJ/m² to manufacture double glazing units.
4. The energy cost of manufactured items should be constantly reviewed as the opportunites for waste and hence saving are many, complex, and interrelated. Manufactured waste should be a minimum and different methods of forming should be investigated, e.g. casting and machining, extrusion, drawing or continuous casting.

5. The reduction of losses by insulation and using 'waste' heat for useful applications.
6. Space heating takes about 30% of the total energy used and an estimate is that 50% of this could be saved by using lower room temperatures, preventing draughts, insulation, controlled ventilation rates, efficient air and water distribution, and planned systematic maintenance (see reference 17.4).
7. The need for an energy-conscious transport system – a massive study in itself.
8. The possibilities for re-use, recycling, and reclamation of discarded commodities.
9. The pollution of the environment by thermal output, solid waste disposal, oil, smoke and gases such as sulphur dioxide, nitrogen oxide and carbon monoxide. Radioactive wastes are a particular problem, being gaseous, liquid and solid.

17.5 The technology of energy saving

A system producing work and useful heat from the combustion of fuel should be designed to reduce energy wastage to a minimum. Many improvements in overall efficiency can be effected by simple measures such as improved insulation, operating with correct air–fuel ratios in the burners, and a good control system. In a more complex plant involving several processes each producing waste heat (e.g. a brewery), the design of the system should be based on pinch technology (see later).

In most existing plant, energy is wasted in the form of hot fluids: cooling water is passed to waste or to cooling towers; exhaust gases are allowed to pass to the chimney at a high temperature; many systems produce hot fluids as a by-product and these are usually allowed to go to waste (e.g. hot waste water in a laundry or hot air from a swimming-pool hall). As the emphasis on energy efficiency has increased, the technology of energy recovery has become more important. Boilers with economizers and air pre-heaters to recover some of the energy of the flue gases have been in use for many years, and heat recovery from waste fluids has been commonplace in large process plants, but only comparatively recently has it been realized that energy savings and hence financial savings can be made in many other cases. In all cases the capital cost of adding energy recovery equipment to new or existing plant must be recouped in a reasonably short pay-back period (say 3–5 years). A simple treatment of methods for the calculation of the economics of energy-saving schemes is given in reference 17.4.

Energy recovery

Energy-recovery equipment is covered in more detail in other books (references 17.3, 17.4, 17.14) so only a brief summary is given in this section. The use of heat exchangers is covered in Chapter 16 and the various types are described; compact heat exchangers such as the plate type and the plate-fin type (see Figs 16.23 and 16.24, p. 621) are commonly used for energy recovery where the hot and cold fluid streams are adjacent to each other and the temperature of the hot fluid is high enough to give the required temperature difference for heat transfer. Shell-and-tube boilers and shell-and-tube water heaters are commonly used to recover energy from hot gases, and special waste-heat boilers have been designed to recover energy from hot waste gases. A special type of heat exchanger for use when two fluid streams are some distance apart is the *run-around coil* shown diagrammatically in Fig. 17.11. A secondary fluid with good heat transfer properties and a low freezing-point is chosen; the heat exchangers may be standard items which reduces the cost of the system. Some typical examples of the use of run-around coils are air-to-air heat recovery in buildings, recovery of heat from drying chambers, ovens, or other process heaters, and the transfer of heat from corrosive gas streams to water.

Fig. 17.11
Run-around coil system
of heat recovery

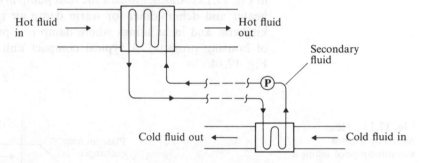

Regenerative heat exchangers are also used in energy recovery. This type of heat exchanger is described in section 16.11 and examples are given for use in industry. The rotary regenerator, frequently known as the *thermal wheel*, is now widely used for air-to-air heat recovery in buildings; the principle is the same as for the Ljungström air pre-heater described in section 16.11. The thermal wheel is also capable of transferring moisture between the two streams when the surfaces are coated with an absorbent such as lithium chloride; for a simple illustration of this effect see reference 17.9. A typical example of the use of a static regenerator for air-to-air heat recovery in a building is given in Fig. 17.12. Referring to Fig. 17.12(a), hot air from the building is blown across matrix *A*

Fig. 17.12 Double-
accumulator
regenerative heat
exchanger

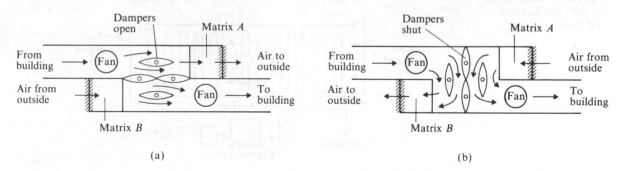

(a)   (b)

thus heating it, while cold air from outside is heated by flowing across matrix B before entering the building. After a period of time the damper positions are altered to that shown in Fig. 17.12(b); air from the building is now driven across matrix B thus heating it up again, while cold air from outside passes across matrix A cooling it down again and being heated in the process. The effect is to heat the outside air continuously, and the system is known as a double-accumulator regenerative heat exchanger. The advantages are in the large surface area to volume ratio for the heat transfer, and the self-cleaning property due to the continual flow reversal. A non-metallic matrix with a desiccant coating can be used to reclaim the enthalpy of vaporization of an air stream with a high relative humidity.

The heat pump (see section 14.1) is particularly suited to energy recovery, frequently in combination with a run-around coil or other form of heat exchanger (see for example the swimming-pool hall heating system shown diagrammatically in Fig. 17.13). Another use for the heat pump in energy recovery is as a combined heater and dehumidifier for warm damp air in rooms such as laundries and kitchens, and in buildings where damp is a problem (e.g. certain older types of housing projects). A typical compact unit is shown diagrammatically in Fig. 17.14.

Fig. 17.13 Heat recovery from a swimming-pool using a heat pump and a plate-fin heat exchanger

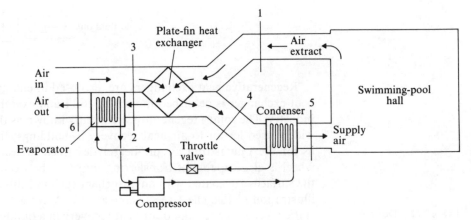

Fig. 17.14 Heat pump for dehumidification and heating

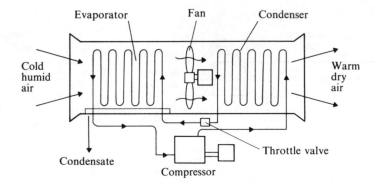

Example 17.5

A particular process requires air at 40 °C at a flow rate of 0.5 kg/s, the air leaving the process at 35 °C. The supply of air is from the atmosphere at 15 °C. Currently the air is heated electrically, but it is proposed to replace the electric heater with a heat pump system linking the inlet and exit air streams. The refrigerant to be used is R134a operating with an evaporator temperature of 25 °C and a condenser saturation temperature of 45 °C. The vapour is saturated at entry to the compressor and there is no undercooling in the condenser; isentropic compression may be assumed for the compression process. The combined mechanical–electrical efficiency of the compressor electric motor is 95%. Taking the specific heat capacity of air as 1.005 kJ/kg K, the properties of R134a from Table 14.1 (p. 498), and neglecting all heat losses, calculate:

   (i) the mass flow rate of refrigerant required;
 (ii) the electrical power input required;
(iii) the temperature of the air leaving the evaporator coils;
(iv) the percentage saving in running costs using the heat pump instead of electric heating.

Solution

(i) The system is shown diagrammatically in Fig. 17.15(a) with the $T–s$ diagram for the heat pump cycle in Fig. 17.15(b). From Table 14.1:

$$h_1 = h_g \text{ at } 25 \,°C = 309.03 \text{ kJ/kg}$$

$$s_1 = s_g \text{ at } 25 \,°C = 1.7051 \text{ kJ/kg K} = s_2$$

Fig. 17.15 System (a) and $T–s$ diagram (b) for Example 17.5

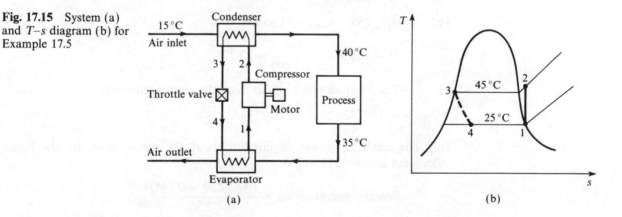

Therefore interpolating for superheated vapour at 11.5447 bar,

$$h_2 = 319.54 + \frac{(1.7051 - 1.7028)}{(1.7440 - 1.7028)}(332.87 - 319.54)$$

$$= 320.28 \text{ kJ/kg}$$

Also $h_3 = h_f \text{ at } 45 \,°C = h_4 = 162.93 \text{ kJ/kg}$

691

The heat required to heat the air is given by

$$\dot{m}_a c_{pa} t_a = 0.5 \times 1.005(40 - 15)$$

$$= 12.56 \text{ kW}$$

i.e. $\dot{m}_{ref}(h_2 - h_3) = 12.56 \text{ kW}$

therefore

$$\dot{m}_{ref} = \frac{12.56}{320.28 - 162.93} = 0.0798 \text{ kg/s}$$

i.e.   Mass flow rate of refrigerant = 0.0798 kg/s

(ii)   Power input required $= \dot{m}_{ref}(h_2 - h_1)$

$$= 0.0798(320.28 - 309.03)$$

$$= 0.898 \text{ kW}$$

i.e.   Electrical power input required

$$= 0.898/0.95 = 0.945 \text{ kW}$$

(iii) Heat required to evaporate the refrigerant

$$= \dot{m}_{ref}(h_1 - h_4)$$

$$= 0.0798(309.03 - 162.93)$$

$$= 11.66 \text{ kW}$$

i.e. $\dot{m}_a c_{pa}(35 - t_{exit}) = 11.66 \text{ kW}$

$$35 - t_{exit} = \frac{11.66}{0.5 \times 1.005} = 22.2 \text{ K}$$

i.e.   Temperature of air at exit from the evaporator coils

$$= 12.8 \,^\circ\text{C}$$

(iv) The electrical power required to heat the air is given by the figure calculated above of 12.56 kW

$$\text{Percentage cost saving} = \frac{100(12.56 - 0.945)}{12.56}$$

$$= 92.5\%$$

The annual cost saving must be set against the capital cost of the heat pump using one of the discounted cash flow methods such as net present value (see reference 17.4), before deciding to replace the existing heater with the heat pump.

The *heat pipe* is now being used for air-to-air heat recovery. Referring to Fig. 17.16, liquid in the heat pipe is evaporated by the hot air stream; the vapour flows through the pipe to the cold end where it condenses and then

Fig. 17.16 Heat pipe

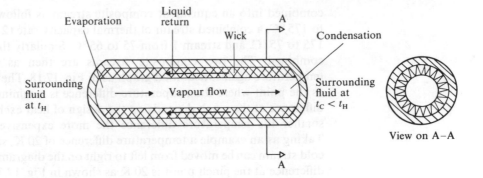

View on A–A

returns as a liquid via an annular wick. A typical air-to-air heat-recovery system using a bank of heat pipes is shown in Fig. 17.17. For a complete treatment see reference 17.10.

Fig. 17.17 Bank of finned heat pipes for air-to-air heat recovery

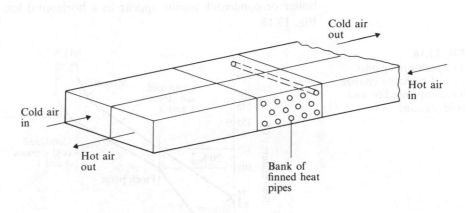

Pinch technology

Pinch technology, or *process integration*, is the name given to a technique developed by Professor Linnhof and co-workers (reference 17.11) to optimize heat recovery in large complex plants with several hot and cold streams of fluids. To illustrate the basic principle take a case of a plant with two hot streams and two cold streams, as shown in Table 17.9. The hot streams can be

Table 17.9 Data for four fluid streams

Stream number	Initial temp. (°C)	Final temp. (°C)	Mass flow rate (kg/s)	Specific heat cap. (kJ/kg K)	Heat cap. rate (kW/K)	Rate of enthalpy increase (kW)
1	205	65	2.00	1.00	2.0	−280
2	175	75	3.20	1.25	4.0	−400
3	45	180	3.75	0.80	3.0	+405
4	105	155	3.00	1.50	4.5	+225
						−50

combined into an equivalent composite stream as follows: stream 1 from 205 to 175 °C, a combined stream of thermal capacity rate $(2 + 4) = 6$ kW/K from 175 to 75 °C, and stream 1 from 75 to 65 °C. Similarly the cold streams can be combined. The two composite streams are then as shown plotted on a temperature against heat load graph as in Fig. 17.18. The pinch point is defined as the point where the temperature difference is a minimum. The temperature difference at the pinch depends on the design of heat exchanger; in general, the smaller the temperature difference the more expensive the heat exchanger. Taking as an example a temperature difference of 20 K, say, then the combined cold stream can be moved from left to right on the diagram until the temperature difference at the pinch point is 20 K as shown in Fig. 17.18. It can then be seen that external heating of 90 kW and external cooling of 140 kW are required; all other energy changes can be achieved by heat exchangers between the various streams; the difference between the external heating and the external cooling is $(90-140) = -50$ kW, the same as in Table 17.9. Note that the process in a boiler or condenser would appear as a horizontal line on a diagram such as Fig. 17.18.

Fig. 17.18
Temperature against rate of enthalpy change for composite hot and cold streams

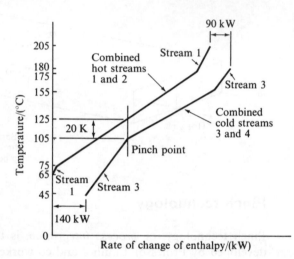

The required heating and cooling of each stream above and below the pinch point is shown in Table 17.10. It can be seen that the cooling of stream 1 above the pinch can be supplied by heat exchange with stream 3; the cooling of

Table 17.10 Cooling and heating above and below the pinch

	Above the pinch	Below the pinch
Stream 1	Cooling: $2(205 - 125) = 160$ kW	Cooling: $2(125 - 65) = 120$ kW
Stream 2	Cooling: $4(175 - 125) = 200$ kW	Cooling: $4(125 - 75) = 200$ kW
Stream 3	Heating: $3(180 - 105) = 225$ kW	Heating: $3(105 - 45) = 180$ kW
Stream 4	Heating: $4.5(155 - 105) = 225$ kW	zero

stream 2 above the pinch can be supplied by stream 4. Similarly, the cooling of stream 2 below the pinch can be supplied by stream 3. External cooling is then required for stream 1 below the pinch (i.e. 120 kW), and for part of the cooling of stream 2 below the pinch, (i.e. 200 − 180 = 20 kW), giving a total external cooling requirement of 140 kW as shown in Fig. 17.18. External heating is required for part of stream 3 (i.e. 225 − 160 = 65 kW), and for part of stream 4 (i.e. 225 − 200 = 25 kW), giving a total external heating requirement of 90 kW as shown in Fig. 17.18. The final arrangement of heat exchangers and external cooling and heating devices is shown diagrammatically in Fig. 17.19.

Fig. 17.19 Possible plant to heat and cool four fluid streams for a minimum 20 K temperature difference

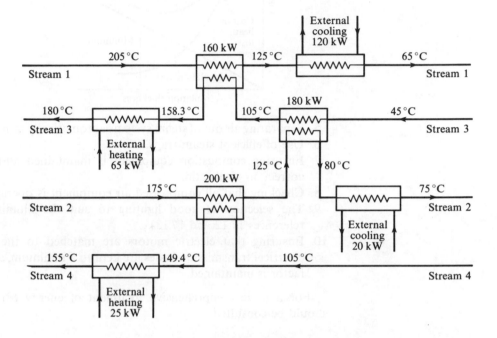

It is not possible in the limited space of this book to give a comprehensive treatment of pinch technology; reference 17.4 gives a concise analytical treatment. The following simple rules can be shown to apply:

(a) do not transfer heat from one fluid to another across the pinch point;
(b) no external heating below the pinch;
(c) no external cooling above the pinch;
(d) a heat engine should operate on one side of the pinch, either taking a heat supply from below the pinch, or rejecting heat to a fluid above the pinch;
(e) a heat pump should operate across the pinch from a cold stream below the pinch to a hot stream above the pinch.

Other energy-saving points

1. The use of good control systems to keep temperatures and pressures at the design values.

2. Maximizing the energy recovery of flash steam, vapour, and condensate.
3. Preventing energy loss due to steam leakage.
4. Ensuring that lagging is of the economic thickness throughout, see Fig. 17.20.

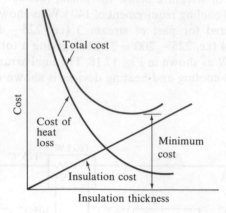

Fig. 17.20 Economic thickness of insulation

5. Separating air out of steam systems automatically to maximize heat transfer.
6. Use of efficient steam traps.
7. Ensuring combustion equipment is maintained and is operating at the correct air–fuel ratio.
8. Checking that all compressed air equipment is operating without leakage.
9. The selection of good lighting to suit the illumination required (see references 17.12 and 17.13).
10. Ensuring that electric motors are matched to the duty required, that electrical transmission losses are kept to a minimum, and that a high power factor is maintained.

For a more comprehensive treatment of energy efficiency reference 17.4 should be consulted.

17.6 Alternative energy sources

Fossil fuels supply a very high percentage of the world's energy needs at present, but there are two main reasons why alternative sources of energy are required for the future: first, the production of gases such as carbon dioxide is causing a worsening of the greenhouse effect, and there is a general risk to the ecosystem from the other gaseous products of combustion; secondly, the earth's supply of fossil fuel is finite.

The present system of power production from fossil or nuclear fuels is highly developed and there is a high potential energy from a given mass of fuel. In a hydroelectric scheme for example, the equivalent mass of water required for the same energy output requires a very large-scale structure to store and handle the water. Collection, storage, and distribution are other factors to be taken into account (see references 17.14 and 17.15).

Some of the possible alternative energy sources are briefly discussed in the following sections.

Solar energy

The earth receives energy directly from the sun. It is silent, inexhaustible, and non-polluting. The means of collecting and distributing solar energy are known, but the cost is about twice that of conventional electricity generation. The sun's rays fall on 'collectors' which are mirrors and reflect the rays to a central receiver. It is necessary for the collectors to be able to 'track' the sun to ensure a continuous maximum reception. Solar energy can be used on small-scale units, and even in the UK solar heating panels are being fitted into the roofs of houses and other buildings to contribute about 50% of the water-heating load. An increasing number of buildings are being designed using passive solar heating, i.e. a positive use of glazing, conservatories, etc. in south-facing walls with fans distributing hot air into the interior of the building. In other, sunnier, countries the use of solar heating is more widespread. The application to large power plants is likely to be slow. An ambitious study has been made of using a solar satellite which is continuously in direct sunlight to collect the energy, convert it to electricity and direct a microwave beam to a receiver on earth where it would be reconverted to electricity. The cost of such a scheme is likely to prohibit its realization.

Three experimental solar power stations have been erected in Almeria, Spain, one of 1000 kW and two of 500 kW rating. The irradiation density is about 1 kW/m² and the locality has 3000 h of sunshine per annum. The 500 kW station has 93 collectors or heliostats which are computer controlled, each of them having 12 mirrors directed to reflect sunlight over the year into the receiver which is mounted on a tower 43 m high. The collector consists of a bank of tubes behind which is a ceramic wall that absorbs the radiation passing the tubes, amounting to about 5%, and serves as a heat store which is insulated to prevent loss. Liquid sodium is circulated through the receiver tubes where the temperature is raised to 530 °C and then through the steam generator to a cold storage tank before returning to the receiver at 270 °C for heating. The steam drives a steam turbine which drives the power generator. At the receiver the thermal output is 2.7 MW, the mass flow rate of sodium is 7.34 kg/s. Sodium is used because of its good heat transfer properties and its ability to be stored. Thus the plant handling the sodium is compact.

It is anticipated that in some areas the solar power station should compete with oil, but perhaps not with large coal or nuclear stations. However, there are areas in the world, as in the developing countries, where conventional power stations are less likely to be chosen and the solar station could be very attractive. In the UK sunshine hours are about 1500 per annum in comparison with 3000 h/annum elsewhere (see references 17.15 and 17.16).

In some parts of the world the sea temperatures are such that at the surface readings of 24–32 °C are measured and at depths of 300–400 m temperatures of 4–7 °C exist. This temperature gradient is a source of power. A boiler plant using propane or ammonia under high pressure is required, the hot water heating the boiler and the condenser situated in the cold water. The potential power by this means is very high, about 300 times the world's present power consumption, but the associated engineering problems have to be overcome which includes transmitting the power developed to the land.

Wind power

Winds possess high kinetic energy and windmills have been used for many years to drive mill mechanisms. The search for alternative power sources has led to the rediscovery of wind power and many wind-driven power stations, large and small, have been built, and are generating power. The modern windmills are much more technically sound than their historical counterparts and have benefited from established knowledge of aerodynamic blade design. They include automatic control of the rotor position to suit chainging wind directions and for adjustment to the blade pitch. The behaviour of the unit can be controlled and monitored by computers.

A joint project by German companies built Growian near Marne in 1982 on the North Sea coast which has a 3 MW output. The tower is 100 m high with two blades 50 m long, 5.2 m at the root, and 1.3 m at the top. It generates 6.3 kV at 50 Hz for a speed of 1500 rev/min with wind speed from 6.3 to 24 m/s (nominal speed 12 m/s). In Denmark two 630 kW wind-power plants have been constructed near Nibe and were commissioned in 1979 and 1980. The Netherlands, which has smaller natural reserves of fossil fuel than most other countries, has put in hand an ambitious programme for wind farms. The reasons for wind power developments vary between different countries and localities within the countries and quite different generators have been evolved to meet the demands. There is a particular market for small systems for use in remote areas where diesel generators and thermo-electric generators have been used traditionally, but both need fuel to be transported to them. The wind generator is taking its place with the other two in an integrated system. For this application the relative costs with the diesel generator as comparison are: diesel generators, £1.00; solar generators, £1.30; wind generators, £0.35; thermoelectric generators, £1.25.

In the UK a Wind Energy Group of Taylor Woodrow Construction Ltd, Greenford, have designed and built a 20 m diameter wind turbine generator (250 kW) on Orkney at Burgar Hill and a 3 MW unit with a 60 m diameter blade. Both units are extensively monitored by a microprocessor-based system to give blade loads, power levels, gearbox, nacelle and tower accelerations, blade pitch angles, etc. (see reference 17.17 for complete details). Machines of 20 and 60 kW are being built by other companies. In 1992, there are 49 projects under way on wind-power generation with a total of 82 MW. This is an encouraging trend but should be compared with a much faster growth in other countries; for example, California alone has wind farms running with a capacity of 1400 MW.

Water power

The movement of large quantities of water naturally or by design affords the opportunity to generate electrical power. Water flowing from reservoirs or in rivers passes through hydroelectric generators to produce electricity. In the UK most potential sites have been developed but the hydroelectric power is a small proportion of the total used.

In the UK pumped-storage systems have been constructed, e.g. at Dinorwic, Snowdonia, and on Loch Ness in Scotland. They are not net producers of power but are used to smooth out the load requirements on conventional plant as they pump water to the high storage reservoir when the external load is low and return it through the hydro-generators when the external load is high.

Tidal power has been traditionally harnessed, along with windmills and watermills, for small power units and several are still in use in England and Wales. The fluctuations in tidal behaviour are not compatible with the continuous operation mode of the mill. Modern tidal power systems are large-scale constructions and there are two, one in the Rance estuary in France (544×10^6 kW h/annum from twenty four 10 MW units) and at Kislaya near Murmansk, Russia (400 kW). There are a number of countries which have suitable sites for barrage schemes including Morecambe Bay, the Solway Firth, and the Wash in the UK but these could also be freshwater storage sites.

The site requires a natural coastal basin formed by a short dam to separate it from the sea which should have a mean tidal range >7 m. The water-driven generators are sited in the dam and are driven by the water as it passes in each direction between high and low tide. The potential power is very high, but so is the capital investment.

Wave generators received a great deal of attention in the UK in the 1970s but were sadly neglected in the 1980s. Devices to extract power from wave movement have been proposed for over a century. In some parts of the world, e.g. the Orkneys (77 kW/m), the estimates of possible power outputs are high, particularly in winter when the energy demands are greatest, but the variation in output would mean that other systems would be required as well. The purpose of the wave generators is to accept the movement of the waves in some way and use this to generate electricity by a mechanical means or by displacing air from a bag through an air turbine into another bag. There are many different designs built on different principles such as Cockerell's rafts, Masuda's ring buoy, and the Salter duck. The Salter duck is mechanically sophisticated with hydraulic pumps, motors, and generators and is used in banks of units about 500 m long with power take-off links to reach the shore. Reference 17.18 gives information on other devices being developed.

In 1991 an advanced wave-energy scheme was opened off the Scottish island of Islay; this uses an oscillating water column to compress air which then drives a turbine. Similar schemes operate in Norway and Japan.

17.7   Nuclear power plant

The first nuclear power plants came into operation in the UK in the late 1950s with the Magnox gas-cooled reactors. Public opinion has moved against nuclear power following several serious incidents; the accident at Three Mile Island in the USA was followed by a disastrous explosion at Chernobyl in the former Soviet Union. In the UK in the late 1980s the nuclear industry suffered a further blow after the electricity industry was privatized when it became clear that the costs of nuclear power compared unfavourably with other means of producing power. The higher and higher costs of more and more stringent safety requirements plus the cost of decommissioning the existing ageing Magnox reactors cast serious doubts on the economic viability of nuclear power. Although there are increasing environmental worries over the use of fossil fuels (e.g. the greenhouse effect, acid rain, etc.), the greater worries over nuclear radiation leaks and the possibility of a catastrophe similar to Chernobyl have swung public opinion firmly against the proliferation of nuclear power stations.

In the early 1990s electricity generating sources in the UK are as follows: 76.4% coal; 17.3% nuclear fuel; 6.2% oil; 0.1% hydro, but the privatization of electricity has led to a rapid increase in the use of natural gas in combined gas turbine and steam turbine power stations; generation by wind is less than 1 GW h but is also expected to grow rapidly.

In a nuclear power plant the reactor and heat exchanger take the place of the conventional boiler, and the steam generated is then expanded through a conventional turbine. In the Magnox reactors, which are now reaching the end of their safe working life and are being shut down, the fuel used is uranium clad in a magnesium alloy (Magnox), with a low neutron absorption. The core is made of graphite and the coolant used is carbon dioxide. The second generation of reactors in the UK is the advanced gas-cooled reactor (AGR), using fuel of uranium oxide clad in stainless steel.

A simplified sketch of a gas-cooled reactor is given in Fig. 17.21. The core of the reactor consists of a moderating material (e.g. carbon), which slows down the neutrons to speeds which will give controlled fission. The holes in the core carry the fuel elements in suitable casings. Through some of these holes are passed control rods which are made of a material which will absorb neutrons and so control the rate at which the reaction takes place. The function of the reflector is to bounce back escaping neutrons into the core, and the shielding is to prevent transmission of harmful particle radiation, such as α and

Fig. 17.21 Diagram of a gas-cooled reactor core

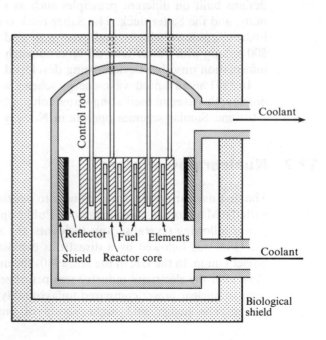

β particles, and γ radiation. The whole of the reactor vessel is surrounded by shielding of steel and concrete. The coolant to the reactor, carbon dioxide, removes heat continuously from the core and then passes through the heat exchanger where steam is generated. In the AGR the heat exchangers and the carbon dioxide circulators are situated inside the pre-stressed concrete pressure vessel so that the carbon dioxide is completely contained within the pressure vessel.

In the 1980s the decision was taken in the UK to move to pressurized water reactors (PWR), but so far only one is under construction at Sizewell in Suffolk. The PWR is fuelled with uranium dioxide clad in zircaloy, an alloy of zirconium, and the moderator is water at a pressure of 154 bar. A simplified diagram of a PWR is shown in Fig. 17.22.

For a full treatment of nuclear power consult reference 17.19.

Fig. 17.22 Diagram of a pressurized water reactor (PWR)

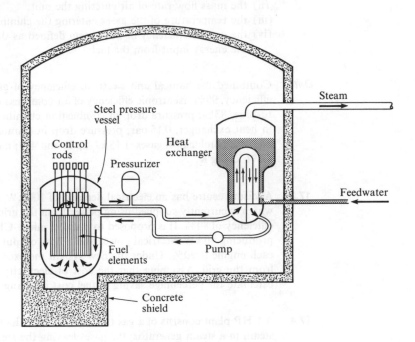

Problems

17.1 A combined power plant consists of a gas turbine unit and a steam turbine unit. The exhaust gas from the open-cycle gas turbine is the supply gas to the steam generator of the steam turbine cycle at which additional fuel is burned in the gas. The pressure ratio for the turbine is 7.5, the air inlet temperature is 15 °C, and the maximum cycle temperature is 750 °C. Combustion in the steam generator raises the gas temperature to 750 °C and the gas leaves the generator to the chimney at 100 °C. Steam is supplied to the steam turbine at 50 bar, 600 °C, and the condenser pressure is 0.1 bar.

The isentropic efficiencies of the air compressor, gas turbine, and steam turbine are 83%, 86%, and 85% respectively.

Taking $c_p = 1.11$ kJ/kg K and $\gamma = 1.33$ for the combustion gases, and neglecting the effect of the mass flow rate of fuel, feed-pump work, and all pressure losses, calculate:

(i) the required flow rates of air and steam for a total power output of 200 MW;
(ii) the power output of each unit;
(iii) the overall efficiency of the plant.

(568 kg/s, 117 kg/s; 64.1 MW, 135.9 MW; 39.5%)

17.2 A gas turbine plant consists of a compressor with a pressure ratio of 10, a combustion chamber, and a turbine mounted on the same shaft as the compressor; the net electrical power of the unit is 20 MW. The inlet air conditions are 1.013 bar and 15 °C and the maximum cycle temperature is 1100 K. The exhaust gases from the turbine are passed through a heat exchanger to heat water for space heating before passing to the chimney; by this means water at 60 °C flowing at a rate of 2×10^6 kg/h is heated to 80 °C. Using the further data below and neglecting the mass flow rate of fuel, calculate:

(i) the temperature of the gases leaving the turbine;
(ii) the mass flow rate of air entering the unit;
(iii) the temperature of the gases entering the chimney;
(iv) the overall efficiency of the system defined as the useful energy output divided by the energy input from the fuel.

Data Combined mechanical and electrical efficiency of gas turbine unit, 90%; combustion efficiency, 99%; isentropic efficiency of air compressor, 80%; isentropic efficiency of gas turbine, 83%; pressure drop in combustion chamber, 0.20 bar; pressure drop of gases in heat exchanger, 0.15 bar; pressure drop in chimney, 0.05 bar; specific heat capacity and γ of combustion gases, 1.15 kJ/kg K and 4/3; mean specific heat capacity of water, 4.191 kJ/kg K.

(453.8 °C; 240 kg/s; 285 °C; 50%)

17.3 A leisure centre has an electrical demand of 100 kW and a heat requirement of 190 kW which is currently supplied using power from the grid and a boiler burning gas with an efficiency of 80%. It is proposed to install a micro-CHP system of two gas engines each producing 40 kW electrical power and a heat output of 95 kW; the overall efficiency of each engine is 90%. Under the new system the excess power requirements will be met from the grid. Assuming that the cost of electricity is four times that of natural gas, calculate the percentage saving in fuel cost in changing to the new system.

(40.4%)

17.4 A CHP plant consists of a gas turbine unit the exhaust gases of which are used to raise steam in a steam generator; the gases leaving the steam generator then pass through a heat exchanger to heat water for space heating before entering the chimney. In the steam cycle steam is bled off at an intermediate pressure and passed to a de-aerator/open-feed heater; the water leaving the heater is saturated at the bleed pressure. Using the data below and neglecting the mass flow rate of fuel, feed-pump work, and all thermal and pressure losses not listed below, calculate:

(i) the required mass flow rate of steam;
(ii) the total power output;
(iii) the overall efficiency defined as the total useful energy output divided by the fuel energy input.

Data *Gas turbine cycle*: pressure ratio for air compressor, 12; air inlet conditions, 1.013 bar and 15 °C; maximum cycle temperature, 1000 °C; isentropic efficiency of compressor, 85%; isentropic efficiency of turbine, 88%; power output, 22 MW; pressure drop in combustion chamber, 0.22 bar; pressure drop in generator and heat exchanger combined,

0.30 bar; temperature of gases entering chimney, 130 °C; specific heat and γ for gases, 1.15 kJ/kg K and 4/3.

Steam turbine cycle: steam conditions at entry to turbine, 40 bar, 450 °C; condenser pressure, 0.03 bar; isentropic efficiency of steam turbine, 85%; bleed pressure to feed heater, 2 bar; take the turbine expansion line as straight on h–s chart; there is no undercooling in the condenser.

Space heating: flow temperature, 85 °C; return temperature, 60 °C; mass flow rate of water, 250 kg/s.

(8.9 kg/s; 31 MW; 69.2%)

17.5 Figure 17.23 shows an arrangement whereby a supply of process water at 55 °C is obtained from river water initially at 10 °C. The steam turbine drives a heat pump which uses Refrigerant 134a.

Fig. 17.23 Plant diagram for Problem 17.5

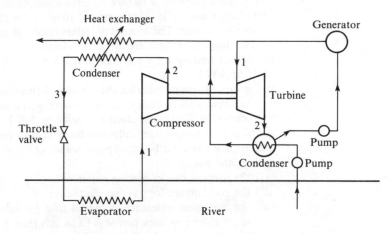

The heat pump uses the river as the source of heat and operates between evaporator and condenser *saturation* temperatures of 0 and 65 °C. The refrigerant is dry saturated at the compressor intake and the liquid is just saturated at inlet to the expansion valve. Take the actual COP_{hp} as 0.7 of the ideal value.

The steam supply is at 60 bar and 450 °C, the condenser pressure is 0.07 bar and the condensate is saturated liquid at inlet to the feed pump, the power input to which can be neglected. The turbine isentropic efficiency is 85%. The process water, drawn from the river, is pumped through the steam condenser as cooling water before passing to the heat pump as the condensing medium. Neglecting the pump work required for the process water, and assuming isentropic compression in the refrigeration compressor, compare the performance of the plant with direct heating of the water assuming that the efficiency of the direct water heater is the same as that of the steam generator.

(1.51)

17.6 A sports centre includes an ice-rink and swimming-pools and it is proposed to use a mechanical heat pump to provide the combined effect of maintaining the ice in the rink and some of the heating requirement for the water of the swimming-pools and for the buildings.

The power input to the refrigerant vapour is 150 kW and the mechanical efficiency of the compressor is 96%. The refrigerant is ammonia, the evaporation temperature for which is −7 °C and the condensation temperature is 48 °C. The ammonia can be assumed to be dry saturated at entry to the compressor and the liquid ammonia is cooled to

703

46 °C on leaving the condenser. The refrigeration rate in the ice-rink for this power input is 460 kW. Calculate:

(i) the mass flow rate of refrigerant;

(ii) the isentropic efficiency of the compression process;

(iii) the heat available for heating purposes.

(iv) compare this method of providing the heating by the direct heating of water in a boiler for 1 kW of primary fuel energy input. Electrical power is generated at a power station overall efficiency of 40%, the transmission efficiency to the sports centre is 83% and the water boiler has an overall efficiency of 90%.

What other factors would be taken into account before reaching a decision on the proposed method?

$$(0.445 \text{ kg/s}; 79.4\%; 610 \text{ kW}; 1.3 \text{ kW by heat pump}, 0.9 \text{ kW for direct heating})$$

17.7 For process purposes a factory requires a supply of air at an atmospheric pressure of 1.013 25 bar and at 35 °C at a rate of 10 m³/s for 16 h per day, 5 days per week, and 48 weeks per year. The average air temperature at entry is 11 °C.

Two heating systems are to be considered:

(a) Direct heating by a gas-fired heater which is 80% efficient and uses gas costing 1.5p/kW h.

(b) A vapour-compression heat pump using Refrigerant 134a driven by an electric motor. The heat is taken from an outside source which has an average temperature over the year of 10 °C. Assume electricity costs 5p/kW h.

Allow a 10 K temperature difference for heat transfer at the evaporator and condenser and assume a real coefficient of performance of 0.6 of that based on a simple heat pump cycle. Calculate:

(i) the average rate of heating required;

(ii) the cost/annum for gas and electricity;

(iii) the maximum difference in capital cost per kilowatt of heating load that can be spent if the pay-back period is to be less than 5 years.

What other factors would you take into account before making a recommendation on the choice of installation?

$$(276.5 \text{ kW}; £19\,900, £15\,540; £78.8 \text{ kW})$$

17.8 A manufacturer quoted the following energy balance for a diesel engine at full power:

Fuel energy supplied (kW)	Power (kW)	Coolant (kW)	Oil (kW)	Exhaust (kW)	Radiation (kW)
190.6	64.7	30.6	5.1	77.3	12.9

The atmospheric pressure during the test was 1013 mbar and the temperature was 20 °C. The specific fuel consumption was 0.236 kg/kW h for gas oil of relative density 0.854. The airflow to the engine was 4.6 m³/min and the exhaust gas temperature was 640 °C. c_p for the exhaust gas can be taken as 1.175 kJ/kg K.

A scheme to supply hot water to a heating system is to be considered using the engine to drive the compressor of a heat pump which takes heat from a large pond at 10 °C. The heat pump uses Refrigerant 134a operating between saturation temperatures of −5 and 45 °C. Assume that the actual coefficient of performance is 0.7 of that for the idealized vapour compression cycle with dry vapour entering the compressor and saturated liquid leaving the condenser. The scheme is shown in Fig. 17.24 with water at 10 °C entering the condenser as a coolant and then passing through two heat exchangers where

Fig. 17.24 Plant diagram for Problem 17.8

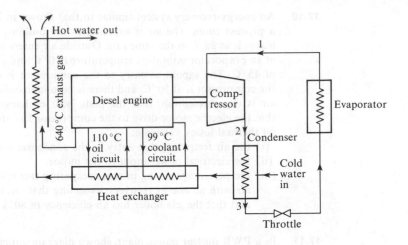

it receives additional heating from the engine coolant and lubrication circuits before receiving heat from the engine exhaust gas. If the exhaust gas is to be cooled to 150 °C and the water is required at 90 °C calculate:

  (i) the heating available for the water;
 (ii) the rate of flow of hot water;
(iii) the ratio of the heat available to the fuel energy supplied;
(iv) the cost advantage to the engine/heat pump proposal over direct heating in an oil-fired water boiler of 80% efficiency.

(319.6 kW; 0.954 kg/s; 1.68; 2.10)

17.9   An office building is space heated between October and April (inclusive). The degree days and the corresponding fuel consumptions in units of 1000 litres for two consecutive years are shown in Table 17.11. In between the two seasons the heating plant was modernized and insulation work was carried out which cost £27 000. The cost of fuel was 22p/litre when the costing was done. Plot the fuel consumption/degree-day characteristics for the two seasons and calculate:

  (i) the average fuel consumptions for the two seasons in litres per degree day;
 (ii) the saving to be expected per annum based on the average degree-day figures for the district tabulated, for the same months, over a 20-year period;
(iii) the period of repayment for the cost of the modifications.

(83 litres/D day, 50 litres/D day; £13 046; 2.07 years)

Table 17.11 Data for Problem 17.9

	Oct	Nov	Dec	Jan	Feb	Mar	Apr	Total
First year								
Degree days	175	290	320	280	275	300	200	1840
Fuel/(10^3 litre)	14.5	24.0	26.5	23.3	23.0	25.0	16.5	152.8
Second year								
Degree days	240	240	220	255	290	315	215	1775
Fuel/(10^3 litre)	12.0	12.0	11.0	12.8	14.5	15.8	10.8	88.9
20-year average								
Degree days	133	243	301	338	306	273	203	1797

17.10 An energy-recovery system similar to that shown in Fig. 17.13 (p. 690), is used to heat a process room. The air is supplied to the building at 31 °C at a rate of 5 kg/s, and leaves it at 28 °C at the same rate. Outside air enters at 4 °C. The heat pump uses R134a at an evaporator saturation temperature of 0 °C and a condenser saturation temperature of 45 °C. The vapour at entry to the compressor is dry saturated, the vapour leaving the compressor is at 50 °C, and there is no undercooling in the condenser. Assuming the air is dry throughout, that the plate heat exchanger has an effectiveness of 57.5%, that the electric motor drive to the compressor has an efficiency of 95%, and neglecting all thermal losses, calculate:

  (i) the air temperature at entry to the condenser coils;

  (ii) the electrical power input to the motor;

  (iii) the percentage saving in cost comparing this system with direct gas heating of the air with no energy recovery, assuming that electricity costs four times natural gas and that the gas boiler has an efficiency of 80%.

(17.8 °C; 13.48 kW; 68.2%)

17.11 In a PWR nuclear power plant, shown diagrammatically in Fig. 17.25, water leaves the reactor core at 160 bar and a temperature of 314.6 °C, and enters the reactor core at 275.6 °C. The steam leaving the heat exchanger at 50 bar is dry saturated and the condensate is a saturated liquid at 0.05 bar. The feedwater is pre-heated to the saturation temperature of the bleed steam at 5 bar in the open feed heater. Taking an isentropic efficiency for the turbine of 80% with the process on the h–s chart as a straight line, and neglecting all pressure losses and pump work, calculate:

  (i) the ratio of the mass flow rates of the working fluids in the two circuits;

  (ii) the fraction of the steam supply which is bled from the turbine to the feed heater;

  (iii) the cycle efficiency.

(10; 0.213; 31.4%)

Fig. 17.25 Plant diagram for Problem 17.11

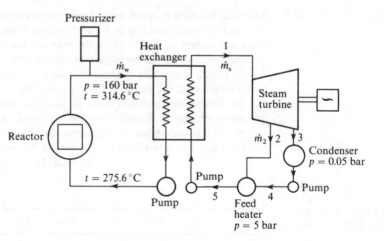

References

17.1 *Energy 2000 – A Global Strategy for Sustainable Development* 1987 Report for the World Commission on Environment and Development ZED Books

17.2 DUNN P D 1986 *Renewable Energies: Sources, Conversion and Applications* Peter Peregrines.

17.3   CULP (jr) A R 1980 *Principles of Energy Conversion* McGraw-Hill

17.4   EASTOP T D and CROFT D R 1990 *Energy Efficiency* Longman

17.5   *Energy Statistics of OECD Countries, 1991* International Energy Agency

17.6   MARSHALL W 1979 *Combined Heat and Power Generation in the UK* Energy Paper 35 HMSO

17.7   ATKINS G 1986 The advantages of CHP systems *Proc. I Mech E Symposium on CHP* Sheffield

17.8   Department of Energy *Fuel Efficiency Booklets*

17.9   EASTOP T D and WATSON W E 1992 *Mechanical Services for Buildings* Longman

17.10 DUNN P D and REAY D A 1982 *Heat Pipes* 3rd edn Pergamon Press

17.11 LINNHOFF B, TOWNSEND D W, BOLAND D, HEWITT G F, THOMAS B E A, GUY A R and MARSLAND R H 1982 *User Guide on Process Integration for the Efficient Use of Energy* Inst. Chem. Engrs.

17.12 CIBSE 1984 *Code for Interior Lighting*

17.13 CIBSE 1989 *Lighting Guide: the Industrial Environment*

17.14 O'CALLAGHAN P W 1981 *Design and Management for Energy Conservation* Pergamon Press

17.15 MESSEL H 1979 *Energy for Survival* Pergamon Press

17.16 STAMBOLIS C 1981 *Solar Energy in the 80s* Pergamon Press

17.17 LINDLEY D 1984 *The 250 kW* and *3 MW Wind Turbines on Burgar Hill, Orkney* I Mech E

17.18 DUCKERS L J, LOCKETT F P, LOUGHRIDGE B W, PEATFIELD A M, WEST M J and WHITE P R S *Novel Wave and Hydro Devices* World Renewable Energy Congress, Kobe, Japan 1989

17.19 BENNET D J and THOMSON J R 1989 *Elements of Nuclear Power* 3rd edn Longman

Index

Absolute pressure, 3
Absolute temperature scale, 8, 127
Absolute zero, 7–8
Absorption refrigerator, 511–17
Absorptivity, 633
  monochromatic, 634
Accelerating well (pump), 456
Ackroyd-Stuart, 138
Additives, 445
Adiabatic process, 59
  free expansion, 74
  irreversible flow, 75
  mixing, 78, 162–6
  reversible, non-flow (isentropic),
    59
    for a perfect gas, 60–3, 103–4
    for a vapour, 64–6, 103–4
  throttling, 75
Admission
  full, 346
  partial, 343
Advanced gas-cooled reactor
  (AGR), 721–2
After burning, 321–2
After-cooler, 397
Air
  conditioning, 542–53
  composition of, 148–9
  compressors, 381–412
  molar mass of, 148
  motors
    reciprocating, 412–15
    rotary, 415–16
  pollution, 470–5, 524–8, 650–1
  preheater, 254, 623
  saturated, 166–7, 533–6
Air/fuel ratio
  for combustion, 180–3
  in C.I. engines, 441–2
  in gas turbines, 282
  in S.I. engines, 439–41
  measurement of, 460–3
  stoichiometric (chemically
    correct), 182–3, 439—40

Air-fuel vapour mixtures, 228–30
Air-standard cycles, 133–45
  comparison of real cycles, 447—50
  diesel, 136–8
  dual or mixed, 138–41
  Ericsson, 145
  Joule, 130–3
  Otto, 135–6
  Stirling, 143–5
Air-vapour mixtures, 533–42
Alcohol-petrol mixtures, 450–1
Alternative energy, 696–9
Amagat's law, 151
Amount of substance, 40
Analogy
  electrical, of conduction,
    568–71
  Reynolds, 603–8
Analysis
  by mass (gravimetric), 148,
    183–92
  dimensional, 599–602
  of air, 148–9
  of exhaust and flue gases,
    183–92
  of fuels, 179
  practical, of combustion products,
    192–200
  proximate, 178–9
  ultimate, 178–9
  volumetric, 150, 183–92
Apparatus dew point, 545
Applied thermodynamics
  (definition), 1
Arithmetic mean radius, 574
Atmospheric nitrogen, 180
Atoms, 177
Audits (energy), 680–8
Availability, 115
Avogadro's hypothesis, 40, 151
Axial-flow
  compressor, 360–3
  turbine, 346–58
Axial thrust, 346

Back pressure turbine, 255
Backward difference, 584–5, 593
Balance point (heat pump),
  489–90
Barometer, 4
Bell-Coleman refrigerator, 518
Benzole, 451
Black body radiation, 633–4
Blade height (turbine), 343–6
Blade profiles, 332, 347, 354,
  362
Blade-speed ratio, 335
  optimum, impulse stage, 337–8
  optimum, reaction stage, 351–2
  optimum, velocity-compounded
    stage, 340–3
Blade velocity coefficient, 333
Blade velocity diagrams, 332–50
  optimum operating conditions,
    337, 340, 351
Blades, stator and rotor, 351
Bown-down (air motor), 412
Blower, Roots, 407–8
Boiler
  capacity, 255
  economizer, 254
  efficiency, 254
  equivalent evaporation, 255
  feed pump, 235–7
  fluidized bed, 669–90
  preheater, 254
  waste, 680
Bomb calorimeter, 223–6
Bottom dead centre, 420
Boundary of system, 2–3
Bourdon gauge, 3
Boys' calorimeter, 227–8
Brake
  mean effective pressure, 430–3
  power, 429
  thermal efficiency, 430–3
Brayton or Joule cycle, 130–3
Bulk temperature, mean, 601
By-pass engine, 325

Calorific value of fuels, 221
  gross and net, 221, 254, 450
Calorifier, 255
Calorimeters
  bomb, 223–6
  Boys', 227–8
  gas, 227–8
  separating, 76–8
  throttling, 76–8
Capacity, boiler, 255
Carbon dioxide recorder, 194
Carburation, 452–6
Carburettors
  accelerating pump (well), 456
  constant vacuum, 454
  fixed choke, 454
  S.U., 454
Carnot cycle, 125–30
  perfect gas, 128–30
  wet vapour, 126
  work ratio, for, 128
  efficiency, 126
  reversed, 487
Cathode ray oscilloscope, 428
Celsius or Centigrade scale, 7
Central difference, 584–5
Centrifugal compressor, 372–5
Cetane number, 447
CFCs, 524
Change of phase, 27
Characteristic equation of state,
  39–40
Charge stratification, 459–60
Chemical
  energy, 2, 176, 431
  equations of combustion, 180–2
  reaction, 176, 180–2
Chemiluminescent analyser, 196
Choked flow, 295
CHP (cogeneration), 673–80
Clausius statement of Second Law,
  89
Clearance ratio, 388–9
Clearance volume, 135, 388–9
Closed cycle, 13
Closed cycle, gas turbine, 130–1
Closed feed water heater, 251–3
Closed system, 3, 16
Coefficient
  blade velocity, 333
  of discharge, 301
  of heat transfer, 565
  of performance, heat pump, 487
  of performance, refrigerator, 486
  of velocity, 301
Cogeneration (CHP), 673–80
Coil by-pass factor, 545
Combined cycles, 670–3
Combined heat and power, 673–80
Combustion, 176–230
  basic chemistry, 177–8
  by self ignition, 444
  efficiency, 282
  equations of, 180–2

fundamental loss, 283
  in C.I. engines, 446–7
  in gas turbines, 281–3
  in S.I. engines, 444–6
  intensity, 282
  mixture strength, 183
  products of, 192–200
Composite streams, 694
Composite wall, 568–71
Compounded impulse turbine,
  338–43
Compound engines, 468
Compression-ignition engine, 419
Compression ratio, 135
  influence on performance, 442–3
Compressor stage, 396–7
Compressors, positive displacement,
  382–411
  reciprocating machines, 382–406
    actual indicator diagram, 392
    condition for minimum work,
      402
    effect of clearance, 388–96
    energy balance, two stage
      machine, 403–5
    for refrigeration, 503–5
    free air delivery, 393
    ideal indicator diagram for, 389,
      397
    ideal intermediate pressure,
      401–2
    isothermal efficiency, 387–8
    multi-stage compression,
      396–403
    volumetric efficiency, 392–6
  rotary machines, 406–12, 415–16
    Roots blower, 407–8
    Roots efficiency, 408
    vacuum pump, 411–12
    vane type, 408–11
Compressors, rotary
  axial flow, 360–3
  centrifugal flow, 372–5
  positive displacement, 382–406
Condenser (steam), 170
Conduction of heat, 561, 562–99
  Fourier's law, 562–3
  general equation, 577–81
  numerical methods, 584–99
  radial (cylinder and sphere), 572–7
  rod analogy, 587–8
  through composite wall, 568–71
  transient, 593–9
  two-dimensional steady, 587–93
Conductivity, thermal, 563
  values of, 564
Conservation of energy, 15–17
Constant pressure (Joule) cycle,
  130–3
Constant pressure process, 52–4
Constant temperature process, 55–9
Constant volume combustion,
  208–10, 221, 223–6
Constant volume process, 51–2

Consumption loop, 439–40, 442
Contact factor, 545
Continuity of mass equation, 21
Continuous injection, 458–9
Control of refrigerating capacity,
  528–9
Control volume and surface, 19
Convection
  forced, 562, 599–610
  natural, 562, 610–13
Convergent-divergent nozzle,
  288–94, 299–300
Convergent nozzle, 295–8
Cooling correction (combustion),
  225–6
Cooling, Newton's law of, 565–7
Cooling towers, 553–6
Corona discharge, 463
Counter-flow heat exchanger, 614–18
Criteria of performance, 427–34
Critical
  point, 28
  pressure, 28
  pressure ratio, 289–94
  temperature, 29
  temperature ratio, 291
  velocity, 293
Cross-flow recuperator, 618–19
Cross-flow scavenge, 425
Curtis turbine, 339–43
Cushion air, 412
Cut-off ratio
  in air motor, 413
  in diesel cycle, 137
Cycle
  air-standard, 133–45
  Carnot, 125–30
  closed, 13, 130–1
  constant pressure, 130–3
  definition of, 13
  diesel, 136–8
  dual (mixed), 138–41
  dual (refrigeration), 507–11
  efficiency, 89
  gas refrigeration, 517–20
  gas turbine, 130–3
  ideal (Carnot), 125–30
  Joule (Brayton), 130–3
  open, 132–3
  Otto, 135–6
  Rankine, 235–45
  refrigeration, vapour absorption,
    511–17
  refrigeration, vapour compression,
    491–9
  regenerative (steam), 248–53
  reheat, 246–8
  reversed heat engine, 89
  reversible, 13
  steam, 16–17, 234–55
  thermodynamic, 13

Dalton's law, 147
Dead centres, definitions, 420

Degradation of energy, 115
Degree days, 683–6
Degree of reaction, 347
  half degree of, 347
Degree of super-cooling, 307
Degree of superheat, 29
Degree of supersaturation, 307
Dehumidification, 690
Delay-period, 444, 446
De Laval turbine, 330
Density, 4
Detonation, 444
Dew point, 534
Diagram efficiency, 334–5
Diagram of properties
  enthalpy-concentration
    (LiBr-H₂O), 514
  enthalpy-entropy (steam), 246
  pressure-enthalpy (R134a), 502
  psychrometric chart, 539
Diesel cycle, 136–8
Diesel fuel, 451
Diesel knock, 447
Diffuser, 287
Dimensional analysis, 599–602
Discharge coefficient, 301
Dissociation, 200–7, 217–19
District heating, 675–80
Dry bulb thermometer, 537
Dryness fraction, 29–30
Dual combustion cycle, 138–41
Dual fuel engine, 426
Ducted fan engine, 325

Economic thickness of insulation,
  696
Economizer, 254
Effectiveness
  of a heat exchanger, 274–5, 623–7
  of a process, 117–21
Efficiency
  air-standard, 133–75
  boiler, 254
  brake thermal, 430–3
  Carnot, 126
  combustion, 282
  cycle, 89
  diagram, 334–5
  engine, 434–7
  fin, 632
  indicated thermal, 431
  intake, 314
  isentropic, 238, 262–3
  isothermal, 387–8
  jet pipe, 315
  mechanical, 385, 429–30
  nozzle, 301
  overall, 363–4
  plant, 253–5
  propulsive, 312
  Rankine, 235–8
  ratio, 238
  Roots, 407–8
  stage, 364–6

thermal, 221–3
  brake, 430–3
  indicated, 431–3
  volumetric, 392–6, 433–4, 504
Electrical analogy (conduction of
  heat), 568–71
Electrolux refrigerator, 512–3
Emissions, 470–5
Emissive power, 633
Emissivity, 634
  monochromatic, 634
Emitter, selective, 636
Energy
  alternative sources, 696–9
  audits, 680–8
  balance (IC engine), 437–8
  conservation of, 15–17
  conversion, 2
  degradation of, 115
  demands, 664–7
  equation
    non-flow, 17–19
    steady flow, 19–23
  from waste, 680
  in transition, 4
  internal of reaction, 208
  intrinsic, 4–5
  kinetic, 19
  management, 680–8
  potential, 19
  recovery, 688–93
  reservoir, 88–9
  saving, 688–96
  solar, 697
  sources, 663–7
  total, 673
  values, 682
Engine
  by-pass, 325
  consumption loop, 439–40, 442
  criteria of performance, 427–34
  dual fuel, 426
  efficiency, 434–7
  emissions, 470–5
  energy balance, 437–8
  factors influencing performance,
    442–7
  four- and two-stroke cycle,
    421–4, 424–6
  free piston, 427, 477–8
  jet, 311–25
  legal requirements in I.C. engines,
    470–5
  modern developments, 479–81
  multi-fuel, 426
  output and efficiencies, 434–7
  performance characteristics,
    437–42
    compression ignition (C.I.),
      441–2
    spark ignition (S.I.), 439–41
  rotary, 475–7
  sleeve valve, 426
  supercharging, 436–70

turbo-prop, 322–5
two-stroke, 424–6
vehicle, 435–6
volumetric efficiency, 433–4
Wankel, 475–7
Enthalpy, 20
  and change of phase, 27, 31–2
  datum of, 32
  -entropy (h-s) chart, 245–6
  of formation, 219–20
  of mixtures, 150, 158
  of moist air, 540–1
  of perfect gas, 44–5
  of reaction, 208–19
  of vaporization, 27–9
  of wet vapour, 31
  specific, 20
Entropy, 90–3
  and irreversibility, 109–121
  as criterion of reversibility, 113–15
  datum of, 93
  of mixtures, 150, 159–60
  of perfect gas, 96–9
  of wet vapour, 93–6
  specific, 92
Entry length, 609–10
Equation, flow, 19–23
Equation, non-flow, 17–19
Equation of continuity, 21
Equation of state, of perfect gas,
  39–40
Equilibrium
  constant, 203
  metastable, 306
  running, 325
  thermal, 2
Equivalent evaporation, 255
Ericsson cycle, 145
Ethyl alcohol, 220, 229, 450–1
Excess air (definition), 182
Exergy, 115–21
Exhaust and flue gas analysis,
  183–92
Expansion
  free or unresisted, 15, 74–5
  supersaturated, 305–9
  throttling, 75–8
Extended surface recuperator, 621–2
Extended surfaces, 627–33
External irreversibility, 10–11

Feed pump (boiler), 235–7
Feed water heater, open, 248–51
Feed water heater, closed, 251–3
Film temperatures, 601
Fin efficiency, 632
Finite difference (conduction),
  584–99
First Law of Thermodynamics, 16
Fixed choke carburettor, 454
Flame ionization detector (FID), 196
Flash chamber, 507–11
Flow equation, 19–23
Flow meters (IC engines), 460–3

Flow processes reversible, 72–3
Flue gas analysis, 183–92
Fluidized bed boiler, 669–70
Forced convection, 599–610
Forward difference, 584–5
Fouling resistance, 620–1
Fourier's law of conduction, 562–3
Four-stroke cycle, 421–4
Fraction
  dryness, 29–30
  mass, 152
  volume, 153
  wetness, 29–30
Free air delivery, 393
Free convection, 610–13
Free expansion, 15, 74–5
Free piston engine, 427, 477–8
Free vortex blading, 355
Friction
  power, 385, 429–30
  losses in turbines, 358–9
Fuel consumption, specific, 431
Fuel injection, 456–9
Fuel systems (IC engines), 452–60
Fuels, 178–9
  analysis of, 179
  self-ignition, 444
  properties of, 179, 450–1
  volatility, 451
Full admission, 346
Fundamental loss (combustion), 283

Gas,
  calorimeter (Boys'), 227–8
  constant, molar, 40
  constant, specific, 40–1
  power, 478
  liquefaction of, 520–1
  perfect, 39–47
    specific heat capacities of, 42–3
  radiation, 650–1
  thermometer, 7–8
Gas analysis
  by flame ionization (FID), 196
  by infra-red spectra (NDIR),
    192–4
  chemiluminescent, 196–7
    by magnetic method, (O_2), 195
    by thermal conductivity method,
      (CO_2), 194
    by zircon cell, 196
Gas mixtures, 147–66
  adiabatic mixing of perfect, 162–6
  molar heat capacities, 160
  molar mass, 151–7
  partial pressure in, 147–8
  partial volumes of, 151
  specific gas constant, 151–7
  specific heat capacities of, 157–62
  volumetric analysis of, 150–1
  with vapour, 166–73
Gas oil, 451, 682
Gas radiation, 650–1
Gas refrigeration cycle, 517–20

Gas turbine cycle, 260–83
  closed, 130–1
  modifications to basic cycle,
    269–80
  open, 132–3
  parallel flow, 269
  practical, 260–9
  pressure ratio, 131, 269–70
  simple, 130–1
  with heat exchanger, 273–9
  with intercooling, 271–2
  with reheating, 272–3
  work ratio, 132
Gasifier, 477–8
Gauge, pressure, 3
Geometric factor, 641
Geometric mean radius, 576
Gibbs-Dalton law, 147
Governing
  quality, 441–2
  quantity, 439
Grashof number, 611
Gravimetric analysis, 148, 183–92
Greenhouse effect, 650–1
Grey body, 634–7
Gross Calorific Value (GCV), 221

Half degree reaction, 347
HCFCs, 524
Heat,
  definition, 2
  flow, resistance to, 568
  recovery, 688–93
  sign convention, 5–6
  transfer of, 561–651
Heat engine, 1,
  cycle, 88, 125–45
  efficiency, 89
  reversed, 88–90
Heat exchanger, 613–27
  effectiveness, 623–7
  fouling resistance, 620–1
  recuperator, 614–18
    compact, 621–2
    cross flow, 618–19
    extended surface, 621–2
    mixed flow, 619–20
    multi-pass, 619–20
    parallel and counter flow, 614–18
    plate, 622
    plate-fin, 621, 690
  number of transfer units, 623
  regenerators, 621–3
  thermal capacity, ratio, 624
  thermal ratio, 275
Heat pipe, 692
Heat pump, 90, 486–91, 690–2
Heat transfer
  by conduction, 561, 562–99
  by convection, 562, 599–613
  by radiation, 562, 633–51
  coefficient for convection, 565
  coefficient for radiation, 649–50
  extended surfaces, 627–33

general conductor equation,
  577–81
through cylinder and sphere,
  572–7
transient conduction, 593–9
two-dimensional steady
  conduction, 587–93
Hero of Alexandria, 330
Hot-well, 243
Humidity
  relative, 535
  specific, 534
  measurement of, 537–8
Hygrometer, 537–8
Hygrometry, 533–42

Ideal cycle (Carnot), 125–7, 128–30
Ignitability, 447
Ignition
  advance, 444, 448–9
  CI engines, 446
  delay, 444, 446
  SI engines, 444, 448–9
Impeller (centrifugal compressor),
  372–3
Impulse steam turbine, 332–8
Index of expansion or compression
  isentropic, 61, 104
  polytropic, 66
Indicated
  power, 427–8
  mean effective pressure, 428
  thermal efficiency, 431
Indicator diagram, 427–8
Injection, fuel, 456–9
Inner dead centre, 420
Intake efficiency, 314
Intensity of radiation, 639
Intercooling, 271–2, 397–400
Internal
  reversibility, 10–11, 73–4
  turbine losses, 358–60
Internal-combustion engines,
  419–81
  brake mean effective pressure,
    430–1
  brake power, 429
  British Standards for, 434–6
  comparison with air standard
    cycles, 447–50
  compression-ignition, 419,
    423–4, 426
  criteria of performance, 427–34
  consumption loops, 439–40, 442
  cross-flow scavenge, 425
  DIN, 434–6
  detonation, 444
  developments, 479–81
  dual fuel, 426
  energy balance, 437–8
  factors influencing performance,
    442–7
  four-stroke cycle, 421–4
  free piston, 427, 477–8

Internal-combustion engines, *continued*
  friction power, 429
  fuel systems, 452–60
  indicated mean effective pressure, 428
  indicated power, 427–8
  loop-scavenge, 426
  mechanical efficiency, 429
  Morse test, 430
  multi-fuel engine, 426
  output and efficiencies, 434–7
  performance characteristics, 437–42
  piston speeds, 436, 443
  properties of fuels for, 450–1
  ram jet, 312–3
  reciprocating, 419–81
  self-ignition, 444
  sleeve valve, 426
  spark ignition, 419
  specific fuel consumption, 431
  stratification, 459–60
  supercharging, 463–70
  thermal efficiency, 430–3
  timing diagrams, 422–6
  torque, 429, 431, 435–6
  two-stroke cycle, 424–6
  uni-flow scavenge, 426
  volumetric efficiency, 433–4
  Willan's line, 430
Internal energy, 17
  datum for, 32
  of a wet vapour, 33
  of mixtures, 150, 160
  of perfect gas, 43–4
  of reaction, 208–19
International scale of temperature, 8, 127–8
Intrinsic energy, 4–5
Inversion temperature, 521
Inward flow turbine, 375–6
Irradiation, 643
Irreversibility, 10, 73–8
  and work, 15
  external and internal, 11, 73–8
  in mixing processes, 78
Irreversible processes, 10, 73–8
Isentropic
  efficiency, 238, 262–3
  flow process, 72–3
  index of expansion or compression, 61
  non-flow process, 59–66
Isothermal
  efficiency, 387–8
  non-flow process, 55–9, 99–103

Jet engine, 311–25
Jet pipe efficiency, 315
Jet propulsion, 311–22
Jet refrigeration, 522
Joule cycle, 130–3
Joule's law, 43

Kadenacy effect, 426
Kelvin temperature scale, 8
Kerosene, 229, 282, 451
Kinetic energy, 19
Kirchhoff's law, 634
Knock in IC engines, 444, 447

Labyrinth gland, 359–60
Lambert's cosine law, 639–43
Laminar flow, 602, 613
Laminar sub-layer, 604
Large-scale CHP, 675–80
Law of conservation of energy, 15–17
Law of partial volumes, 151
Laws of thermodynamics,
  First Law, 16
  Second Law, 89
Lead in petrol, 451, 470–1
Leduc's law, 151
Linde process, 521
Liquefaction of gases, 520–1
Liquid metals, 608
Lithium bromide-water chart, 514
Ljungström
  air preheater, 622–3
  turbine, 376
Logarithmic mean
  area, 573
  radius, 573
  temperature difference, 616
Loop scavenge, 426
Losses in turbines, 359–60

Mach number, 310
Machine cycle, 314
Management of energy, 680–8
Manifold injection, 457–9
Manometer, 4
Mass
  continuity of, equation, 21
  fraction, 152
Mean bulk temperature, 601
Mean film temperature, 601
Mean effective pressure, 141–2
  brake, 430–3
  indicated, 428
Mechanical efficiency, 385, 429–30
Metallurgical limit, 134, 270
Metastable state, 306
Micro-CHP, 673–5
Mixed-flow recuperators, 619–20
Mixing of gases, 78, 162–6
Mixture strength, 183
Mixtures, 147–73
  enthalpy of, 150, 158
  entropy of, 150, 159–60
  gas constant of, 151–7
  gravimetric analysis of, 148, 183–92
  internal energy of, 150, 160
  molar heat capacities of, 160
  molar mass of, 151–7
  of air and water vapour, 166–73
  of gases, 162–6

  partial pressure of, 145–51
  partial volume of, 150–1
  psychrometric, 533–42
  saturated, 166–7
  specific heat capacities of, 157–62
  stoichiometric, 182–3
  volumetric analysis of, 150–1
Moisture content, 534
Molar gas constant, 40
Molar heat capacities, 160
Molar mass, 40, 151–7
  of air, 148
  of a gas mixture, 151–7
Molecules, 177
Momentum
  rate of change of, 329–30
  thrust, 312
  transfer, 603–6
Monochromatic
  absorptivity, 634
  emissivity, 634
Montreal protocol, 526
Morse test, 430
Multi-fuel engines, 426
Multi-pass recuperators, 619–20
Multi-stage compression, 396–403

Natural convection, 610–3
Net Calorific Value (NCV), 221
Newtonian heating and cooling, 582–4
Newton's law of cooling, 565–7
Non-dispersive infra-red (NDIR), 192
Non-flow energy equation, 17–19
Non-flow exergy, 117
Non-flow process, 51–72, 99–109
  irreversible, 10, 73–8, 109–15
  isentropic, 59–66, 103–4
  on the T-s diagram, 93–109
  polytropic, 66–72, 104–9
  reversible adiabatic, 59–66, 103–4
  reversible constant pressure, 52–4
  reversible constant volume, 51–2
  reversible isothermal, 55–9
Nonsteady-flow process, 78–84
NO_x, 196–7, 204, 470–5
Nozzles, 287–311
  choked flow in, 295–8
  coefficient of discharge, 301
  coefficient of velocity, 301
  convergent, 295–8
  convergent-divergent, 288–94, 299–300
  critical pressure ratio, 289–94
  critical temperature ratio, 291
  critical velocity, 293
  efficiency of, 300–4
  maximum mass flow in, 295–8
  off the design point, 298–300
  pressure and velocity variations, 288
  shape, 287–9
  shock waves in, 298–300

Nozzles, *continued*
  stagnation conditions, 309–11
  steam, 304–9
  supersaturation in, 305–9
Nuclear power plant, 699–701
Nuclear reactors, 699–701
Number of transfer units, 623
Numerical methods for conduction, 584–99
  boundary conditions, 588–9, 594–6
  choice of, 590
  Crank-Nicholson, 597
  errors, 586
  Euler solution, 593
  explicit solution, 593
  Gaussian elimination, 585, 598
  Guass–Siedel iteration, 586, 590
  implicit solution, 596–9
  matrix inversion, 585
  notation, 586
Nusselt number, 601

Octane number, 445
ODP, 526
One-dimensional flow, 287
Open
  cycle, 132–3, 260–1
  feed heater, 248–51
  system, 3, 19
Otto cycle, 135–6
Outer dead centre, 420
Outward-flow radial turbine, 375
Overexpansion, 298
Overall efficiency, 363–8
Overall heat transfer coefficient, 567
Oxides of nitrogen, 470–6
Oxygen
  in air, 148–9
  in combustion products, 183–92
  in fuels, 179
  recorders, 195
Ozone depletion potential (ODP), 526
Ozone layer, 525

Paddle-wheel work, 5
Parallel-flow recuperators, 614–8
Paramagnetic gas analysis, 195
Parson's turbine, 346–7
Partial admission, 343
Partial pressure, 147
Partial volumes, 150–1
Pass-out turbine, 255
Path of a process, 10
Percentage saturation, 536
Perfect gas, 39–47
  characteristic equation of state, 39–40
  enthalpy, 44–5
  entropy, 96–7
  gas constant, 40–1
  internal energy, 43–4
  Joule's law, 43–4

mixtures, 147–66
molar gas constant, 40
molar heat, capacities, 160
process,
  non-flow, 51–72
  steady-flow, 72–3
  specific heat capacities, 42–3
Performance
  characteristics (IC engines), 437–42
  criteria (IC engines), 427–34
  number, 445
  ratio, 487
Petrol injection, 457–9
Pinch technology, 693
Piston speed, 436, 443
Plate heat exchanger, 622
Plate-fin heat exchanger, 621, 690
Pollution, 470–5, 524–8, 650–1
Polytropic efficiency, 368–72
Polytropic process, 66–72
Positive displacement machines, 381–416
  motors, 412–6
  reciprocating compressors, 382–406
  rotary compressors, 406–11
  vacuum pump, 411–12
Potential energy, 19
Power
  brake, 429
  friction, 385, 429–30
  indicated, 427–8
  nuclear, 699–701
  shaft, 385
  solar, 697
  water, 698–9
  wind, 698
Prandtl number, 601
Pre-ignition, 444
Pre-whirl, 374
Pressure
  absolute, 3
  barometric, 4
  compounding, 338–9
  gauge, 3
  mean effective, 141–2
    brake, 430–3
    indicated, 428
  partial, 147
  ratio
    gas turbine, 131, 269–70
    critical, 289–94
  saturation, 29
  stagnation, 309–11
  thrust, 315–6
  vacuum, 4
Pressure-enthalpy diagram, 502
Principle of the conservation of energy, 15–17
Process
  reversible and irreversible, 10–11, 51–73
  non-flow, 51–72

nonsteady-flow, 78–84
steady-flow, 72–3
Process integration, 693
Process steam, 255–7
Products of combustion, 192–200
Properties, 9
  ethyl alcohol, 220, 229
  liquid and vapour, 30–1
  of fuels for I.C. engines, 179, 450–1
  of refrigerants, 498, 502, 505, 514, 522–3
Propulsion, jet, 311–22
Propulsive efficiency, 312
Proximate analysis, 178–9
Psychrometer, 537
Psychrometric chart, 538–40
Psychrometry, 533–42
Pulsed injection, 458–9
Pump work, 236–7
Pumping loop, 427–8
Pumping power, 606

Quality governing, 441–2
Quantity governing, 439

Radial-flow turbine, 375–6
Radiation, 562, 633–51
  absorptivity, 633
  black body, 633–4
  emissive power, 633
  emissivity, 634
  from gases, 650–1
  geometric factor, 639–43
  grey body, 634–7
  heat transfer coefficient, 649–50
  intensity, 639
  irradiation, 643
  Kirchhoff's law, 634
  Lambert's cosine law, 639
  radiosity, 643
  reflectivity, 633
  selective emitter, 636
  Stefan-Boltzmann law, 637–9
  transmissitivity, 633
  wavelength, 635
  Wien's law, 635
Radiosity, 643
Ram effect, 313
Ram-jet engine, 312–13
Rankine cycle, 235–45
  with superheat, 243–5
Rateau turbine, 338–9
Ratio
  air/fuel, 180–3, 282, 439–42, 460–3
  blade-speed, 335
  clearance, 388–9
  compression, 135
  critical pressure, 289–94
  critical temperature, 291
  cut-off, 137, 413
  efficiency, 238
  of specific heats, 45

Ratio, *continued*
  pressure, 131, 269–70
  work, 128
Reaction
  chemical, 176, 180–2
  degree of, 347
  turbine, 330–1, 346–58
Reactors
  advanced gas cooled, 700–1
  Magnox, 699–700
  pressurized water, 701
Reciprocating
  air motor, 412–14
  compressor, 382–406
  I.C. engine, 419–81
Recuperator, 614–21
Reflectivity, 633
Refrigerants, 522–8
Refrigeration and heat pumps,
    485–529
  coefficient of performance, 486–7
  compressor type, 503–7
  control, 528–9
  dual cycles, 507–11
  Electrolux, 512–13
  flash chamber, 507–11
  gas cycles, 517–20
  load, 499–500
  pressure-enthalpy diagram, 501–3
  steam jet, 522
  throttle valve, 491
  undercooling, 492
  vapour absorption, 511–17
  vapour compression, 491–9
Refrigerating effect, 485
Regenerative cycle, 248–53
Regenerator
  rotating matrix, 622–3
  stationary matrix, 621–2
Regnault and Pfaundler's
    correction, 225–6
Reheat
  factor, 365–8
  gas turbine cycle, 272–3
  steam cycle, 246–8
Relative atomic mass, 40
Relative humdity, 535
Relative molecular mass, 40
Reservoir of energy, 88–9
Resistance to heat flow, 568–9
Reversibility, 10–11
  and chemical reaction, 180
  and heat, 11, 16
  and mixing, 78, 110–11, 162–3
  and work, 11–15
  criteria of, 10–11
  internal and external, 11, 73–4,
    113–15
Reversible heat engine, 89, 486–7
Reversible processes,
  non-flow, 51–72
  nonsteady-flow, 78–84
  steady-flow, 72–3
Reynolds analogy, 603–8

Reynolds number, 601
Rich mixture, 182
Room ratio line, 544
Roots blower, 407–8
Rotadynamic machinery, 328–76
Rotary
  air compressor, 406–11
  air motor, 415–16
  engine, 475–7
Rotating matrix, 622–3
Rumble or pounding, 446
Run-around coil, 689

Sankey diagram, 668–9
Saturated
  air, 166–7, 533–6
  liquid and vapour, 28–9
  mixture, 166–7
Saturation pressure and
    temperature, 28–9
Scavenging, 425–6
Second Law of Thermodynamics,
    88–121
  and chemical reaction, 202
  statements, 89
Secondary injection, 457
Selective emission, 636
Self-ignition, 444
Shear stress in fluid, 604
Shock-wave, 298
Sleeve valve, 426
Sling psychrometer, 538
Slip factor, 374
Small-scale CHP, 673–5
Smog, 470
Smoke limit, 442
Solar energy, 697
Solid angle, 640
Sonic velocity, 292
Sources of energy, 663–7
Spark-ignition engine, 419
Specific enthalpy of moist air,
    540–1
Specific enthalpy of vaporization,
    27–9
Specific fuel consumption, 431
Specific heat capacities
  of gas mixtures, 157–62
  of gases, 42–7
  of moist air, 541–2
  ratio of, 45
Specific humidity, 534
Specific steam consumption, 238–9
Specific volume, 4
  of moist air, 542
  of wet vapour, 32
Stage
  compressor, 396–7
  efficiency, 364–8
  impulse, 338
  reaction, 346–7
  turbine, 338
Stagnation conditions, 309–11
Stanton number, 605

State, 9
  equation of, 39–40
Steady conduction of heat, 562–76
Steady flow, 72–3
  energy equation, 19–23
  exergy, 117
  mixing, 78
  processes, 72–3
Steam
  condenser, 170–3
  consumption, 238–9
  cycle, 16–7, 234–55
  diagram of properties, 246
  for heating and process work, 255–7
  jet refrigeration, 522
  nozzle, 304–9
  tables of properties, 30–9
  turbines, 332–54
Stefan-Boltzmann law, 637–9
Stirling cycle, 143–5
Stoichiometric air/fuel ratio, 182–3
Stratification, 459–60
S.U. carburettor, 454
Sub layer, laminar, 604
Summer air conditioning, 543–6,
    550–2
Supercharging, 463–70
Supercooling, degree of, 307
Superheat
  degree of, 29
  tables, 30–9
Supersaturation, 305–9
Surroundings, 4, 10–11
Swept volume, 142, 389–90, 433
System, 2
  closed, 3
  open, 3

Tables of properties,
  air, 148–9
  refrigerants, 498, 505
  steam, 30–9
Temperature
  absolute, 8, 127
  bulk mean, 601
  Celsius (centigrade), 7
  critical, 29
  dew point, 534
  equivalent of velocity, 309
  film, 601
  international scale, 8, 127–8
  Kelvin, 8
  measurement, 7–8
  of inversion, 521
  saturation, 28–9
  thermodynamic, 127
  wet- and dry-bulb, 537
  zero, absolute, 8
Temperature-entropy diagram, 93–9
Thermal conductivity, 563
Thermal efficiency, 221–3
  brake, 430–3
  indicated, 431
  power plant, 221